Birkhäuser

ISNM
International Series of Numerical Mathematics

Volume 160

Managing Editors:
K.-H. Hoffmann, München
G. Leugering, Erlangen-Nürnberg

Associate Editors:
Z. Chen, Beijing
R. H. W. Hoppe, Augsburg / Houston
N. Kenmochi, Chiba
V. Starovoitov, Novosibirsk

Honorary Editor:
J. Todd, Pasadena †

Constrained Optimization and Optimal Control for Partial Differential Equations

Günter Leugering
Sebastian Engell
Andreas Griewank
Michael Hinze
Rolf Rannacher
Volker Schulz
Michael Ulbrich
Stefan Ulbrich
Editors

Birkhäuser

Editors
Günter Leugering
Universität Erlangen-Nürnberg
Department Mathematik
Lehrstuhl für Angewandte Mathematik 2
Martensstr. 3
91058 Erlangen, Germany

Sebastian Engell
Technische Universität Dortmund
Fakultät Bio- und Chemieingenieurwesen
Lehrstuhl für Systemdynamik und Prozessführung
Geschossbau 2
44221 Dortmund, Germany

Andreas Griewank
Humboldt-Universität zu Berlin
Institut für Mathematik
Rudower Chaussee 25
12489 Berlin, Germany

Michael Hinze
Universität Hamburg
Fachbereich Mathematik
Optimierung und Approximation
Bundesstraße 55
20146 Hamburg, Germany

Rolf Rannacher
Ruprecht-Karls-Universität Heidelberg
Institut für Angewandte Mathematik
Im Neuenheimer Feld 293
69120 Heidelberg, Germany

Volker Schulz
Universität Trier
FB 4 - Department Mathematik
54286 Trier, Germany

Michael Ulbrich
Technische Universität München
Lehrstuhl für Mathematische Optimierung
Fakultät für Mathematik, M1
Boltzmannstr. 3
85748 Garching b. München, Germany

Stefan Ulbrich
Technische Universität Darmstadt
Fachbereich Mathematik
Dolivostr. 15
64293 Darmstadt, Germany

ISBN 978-3-0348-0132-4 e-ISBN 978-3-0348-0133-1
DOI 10.1007/978-3-0348-0133-1
Springer Basel Dordrecht Heidelberg London New York

Library of Congress Control Number: 2011941447

Mathematics Subject Classification (2010): 35K55, 35Q93, 49J20, 49K20, 49M25, 65K10, 65M60, 65N15, 65N30, 76D55, 90C30

Printed on acid-free paper

Springer Basel AG is part of Springer Science+Business Media
(www.birkhauser-science.com)

Contents

Introduction

Solving optimization problems subject to constraints involving distributed parameter systems is one of the most challenging problems in the context of industrial, medical and economical applications. In particular, in the design of aircraft, 'moving bed' processes in chemical engineering, crystal growth etc., the transition from model-based numerical simulations to model-based design and optimal control is crucial. All of these applications involve modeling based on partial differential equations (PDEs), and can be classified as form and topology optimization problems or optimal control problems in an infinite-dimensional setting.

For the treatment of such problems the interaction of optimization techniques and numerical simulation is crucial. After proper discretization, the number of optimization variables varies between 10^3 and 10^{10}. It is only very recently that the enormous advances in computing power have made it possible to attack problems of this size. However, in order to accomplish this task it is crucial to utilize and further explore the specific mathematical structure of prototype applications and to develop new mathematical approaches concerning structure-exploiting algorithms, adaptivity in optimization, and handling of control and state constraints. It is necessary to combine numerical analysis of PDEs and optimization governed by prototype applications so that the most recent activities in the fields are merged and further explored.

On the one hand, new models, and along with them new mathematical challenges, also require basic theoretical research. The synergy between theoretical developments, numerical analysis and its realizations make the group of contributors involved in the program, and hence these research notes, a unique collection of ideas and discussions. The organization of this volume is as follows: there are five topics, namely,

- Constrained Optimization, Identification and Control
- Shape and Topology Optimization
- Model Reduction
- Discretization: Concepts and Analysis
- Applications

which are covered by overview articles complemented by short communications. This gives the opportunity to provide an introduction to each topical area along with a specialization. Below we provide an executive summary for each topical area and then give a more detailed description of the actual content.

The research documented in this volume has been and continues to be funded by the "Deutsche Forschungsgemeinschaft" (DFG) within the priority program "Optimization with Partial Differential Equations" (DFG-PP 1253). The editors and the authors express their gratitude to the DFG for their financial support and, particularly, to all involved referees. The book benefited substantially from their valuable comments, evaluations and suggestions. They also thank Dipl.-Technomath. Michael Gröschel for his continuous work in organizing this special volume.

G. Leugering, Spokesman

Constrained Optimization, Identification and Control

Several projects of the priority program consider the development of novel algorithmic concepts for the numerical solution of PDE constrained optimization, identification, and optimal control problems. Semi-smooth Newton techniques are developed and analyzed for optimization problems with PDE- and control constraints as well as for time-dependent variational equations. In particular, optimization techniques that use multigrid methods, adaptivity, model reduction, and structure-exploiting algorithmic differentiation are proposed for complex time-dependent optimal control problems with applications to fluid dynamics and PDAEs. Moreover, the development of one-shot techniques based on iterative solvers for the governing PDE-solver is considered. Furthermore, analytical and algorithmic advances in challenging applications such as the optimal control of networks of hyperbolic systems, optimum experimental design for PDE-constrained problems, and the feedback stabilization of incompressible flows are presented.

M. Ulbrich, S. Ulbrich and A. Griewank

Shape and Topology Optimization

Here, novel algorithmic approaches towards the practically highly important class of shape and topology optimization are presented and also generalized to a stochastic setting. Allen-Cahn and Cahn-Hilliard-formulations of topology optimization problems based on the homogenization approach are investigated. Free boundary value problems based on the Poisson equation are treated by the use of shape calculus of first and second order. Furthermore, new algorithmic schemes are developed and analyzed for parametric and non-parametric shape optimization in the context of flow problems.

V. Schulz

Model Reduction

Within this special volume, recent developments are presented in the adaptive discretization of elliptic or parabolic PDE-constrained optimization problems. The possibility of additional pointwise control or state constraints is included. Particular emphasis is put on automatic model reduction by adaptive finite element discretization for the accurate computation of a target quantity ("goal-oriented adaptivity").

R. Rannacher

Discretization: Concepts and Analysis

Within the priority program, recent trends and future research directions in the field of tailored discrete concepts for PDE constrained optimization with elliptic and parabolic PDEs in the presence of pointwise constraints are addressed. This includes the treatment of mesh grading and relaxation techniques, of adaptive a posteriori finite element concepts, and of a modern treatment of optimal control problems with parabolic PDEs. Tailoring discrete concepts is indispensable for a proper and effective treatment of complex model problems and applications.

M. Hinze

Applications

As explained above, the entire project, while mathematically oriented, ultimately aims at and is justified by challenging applications. The participating research groups are convinced that the relation between modern mathematics and applications is a two-way interaction where both sides profit from each other: new mathematics is created by pushing the boundary of applications that can be tackled. In this sense, several applications in the domain of optimization and control with partial differential equations are provided here. They range from the nano-scale to the macro-scale. The areas of application are chosen from electrical, mechanical and chemical engineering as well as from medical and biological sciences.

S. Engell and G. Leugering

Part I

Constrained Optimization, Identification and Control

Introduction to Part I
Constrained Optimization, Identification and Control

This part presents novel algorithmic developments within the priority program for the efficient solution of PDE constrained optimization, identification and control problems.

Eberhard Bänsch and Peter Benner consider, in *Stabilization of Incompressible Flow Problems by Riccati-based Feedback*, optimal control-based boundary feedback stabilization of flow problems for incompressible fluids and present algorithmic advances in solving the associated algebraic Riccati equations.

Luise Blank, Martin Butz, Harald Garcke, Lavinia Sarbu and Vanessa Styles formulate, in *Allen-Cahn and Cahn-Hilliard Variational Inequalities Solved with Optimization Techniques*, time stepping schemes for Allen-Cahn and Cahn-Hilliard variational inequalities as optimal control problems with PDE and inequality constraints and apply a primal dual active set strategy for their solution.

Hans Georg Bock, Andreas Potschka, Sebastian Sager and Johannes Schlöder study, in *On the Connection between Forward and Optimization Problem in One-shot One-step Methods*, the relation between the contraction of the forward problem solver and simultaneous one-step optimization methods. They analyze strategies to ensure convergence under appropriate assumptions and show that in general the forward problem solver has to be used with adaptive accuracy controlled by the optimization method.

Debora Clever, Jens Lang, Stefan Ulbrich and Carsten Ziems present, in *Generalized Multilevel SQP-methods for PDAE-constrained Optimization Based on Space-Time Adaptive PDAE Solvers*, an adaptive multilevel optimization approach for complex optimal control problems with time-dependent nonlinear PDAEs and couple it with the PDAE-solver KARDOS. The algorithm is applied to a glass cooling problem with radiation.

Thomas Franke, Ronald H. W. Hoppe, Christopher Linsenmann and Achim Wixforth combine, in *Projection Based Model Reduction for Optimal Design of the Time-dependent Stokes System*, concepts of domain decomposition and model reduction by balanced truncation for the efficient solution of shape optimization problems where the design is restricted to a relatively small portion of the computational domain. The approach is applied to the optimal design of capillary barriers as part of a network of microchannels and reservoirs on microfluidic biochips.

Nicolas Gauger, Andreas Griewank, Adel Hamdi, Claudia Kratzenstein, Emre Özkaya and Thomas Slawig consider, in *Automated Extension of Fixed Point PDE Solvers for Optimal Design with Bounded Retardation*, PDE-constrained optimization problems where the state equation is solved by a pseudo-time stepping or fixed point iteration and develop a coupled iteration for the optimality system with the goal to achieve bounded retardation compared to the state equation solver.

Martin Gugat, Michael Herty, Axel Klar, Günter Leugering and Veronika Schleper present, in *Well-posedness of Networked Hyperbolic Systems of Balance*

Laws, an overview on recent existence, uniqueness and stability results for hyperbolic systems on networks based on wave-front tracking.

Michael Hinze, Michael Köster and Stefan Turek combine, in *A Space-Time Multigrid Method for Optimal Flow Control*, a Newton solver for the treatment of the nonlinearity with a space-time multigrid solver for linear subproblems to obtain a robust solver for instationary flow control problems. The algorithm is applied to several test cases.

Michael Hinze and Morton Vierling address, in *A Globalized Semi-smooth Newton Method for Variational Discretization of Control Constrained Elliptic Optimal Control Problems*, the implementation, convergence and globalization of semismooth Newton methods for elliptic PDE-constrained optimization with control constraints.

Denise Holfeld, Philipp Stumm and Andrea Walther summarize, in *Structure Exploiting Adjoints for Finite Element Discretizations*, details for the development, analysis, and implementation of efficient numerical optimization algorithms using algorithmic differentiation (AD) in the context of PDE constrained optimization. In particular, multistage and online checkpointing approaches are considered.

Ekaterina Kostina and Olga Kostyukova propose, in *Computing Covariance Matrices for Constrained Nonlinear Large Scale Parameter Estimation Problems Using Krylov Subspace Methods*, an algorithm for the fast computation of the covariance matrix of parameter estimates, which is crucial for efficient methods of optimum experimental design.

M. Ulbrich, S. Ulbrich and A. Griewank

International Series of Numerical Mathematics, Vol. 160, 5–20

Stabilization of Incompressible Flow Problems by Riccati-based Feedback

Eberhard Bänsch and Peter Benner

Abstract. We consider optimal control-based boundary feedback stabilization of flow problems for incompressible fluids. We follow an analytical approach laid out during the last years in a series of papers by Barbu, Lasiecka, Triggiani, Raymond, and others. They have shown that it is possible to stabilize perturbed flows described by Navier-Stokes equations by designing a stabilizing controller based on a corresponding linear-quadratic optimal control problem. For this purpose, algorithmic advances in solving the associated algebraic Riccati equations are needed and investigated here. The computational complexity of the new algorithms is essentially proportional to the simulation of the forward problem.

Mathematics Subject Classification (2000). 76D55,93D15,93C20,15A124.

Keywords. Flow control, feedback, Navier-Stokes equations, Riccati equation.

1. Introduction

The aim of this work is to develop numerical methods for the stabilization of solutions to flow problems. This is to be achieved by action of boundary control using feedback mechanisms. In recent work by Raymond [1, 2, 3] and earlier attempts by Barbu [4], Barbu and Triggiani [5] and Barbu, Lasiecka, and Triggiani [6], it is shown analytically that it is possible to construct a linear-quadratic optimal control problem associated to the linearized Navier-Stokes equation so that the resulting feedback law, applied to the instationary Navier-Stokes equation, is able to exponentially stabilize solution trajectories assuming smallness of initial values.

The work described in this paper was supported by *Deutsche Forschungsgemeinschaft*, Priority Programme 1253, project BA1727/4-1 and BE2174/8-1 "Optimal Control-Based Feedback Stabilization in Multi-Field Flow Problems".

To be more precise, consider the following situation. The flow velocity field v and pressure χ fulfill the incompressible Navier-Stokes equations

$$\partial_t v + v \cdot \nabla v - \frac{1}{\mathrm{Re}} \Delta v + \nabla \chi = f, \tag{1a}$$

$$\operatorname{div} v = 0, \tag{1b}$$

on $Q_\infty := \Omega \times (0, \infty)$ with a bounded and connected domain $\Omega \subseteq \mathbb{R}^d$, $d = 2, 3$, with boundary $\Gamma := \partial\Omega$ of class C^4, a Dirichlet boundary condition $v = g$ on $\Sigma_\infty := \Gamma \times (0, \infty)$, and appropriate initial conditions (the latter are discussed, e.g., in [7]).

Now assume we are given a regular solution w of the stationary Navier-Stokes equations

$$w \cdot \nabla w - \frac{1}{\mathrm{Re}} \Delta w + \nabla \chi_s = f, \tag{2a}$$

$$\operatorname{div} w = 0, \tag{2b}$$

with Dirichlet boundary condition $w = g$ on Γ. Moreover, the given flow field w is assumed to be a (possibly unstable) solution of (1).

If one can determine a Dirichlet boundary control u so that the corresponding controlled system

$$\partial_t z + (z \cdot \nabla) w + (w \cdot \nabla) z + (z \cdot \nabla) z - \frac{1}{\mathrm{Re}} \Delta z + \nabla p = 0 \quad \text{in } Q_\infty, \tag{3a}$$

$$\operatorname{div} z = 0 \quad \text{in } Q_\infty, \tag{3b}$$

$$z = bu \quad \text{in } \Sigma_\infty, \tag{3c}$$

$$z(0) = z_0 \quad \text{in } \Omega, \tag{3d}$$

is stable for initial values z_0 sufficiently small in an appropriate subspace $X(\Omega)$ of the space of divergence-free L_2 functions with $z \cdot n = 0$, called here $V_n^0(\Omega)$, then it can be shown for several situations (see below) that z decreases exponentially in the norm of $X(\Omega)$ and thus the solution to the instationary Navier-Stokes equations (2) with flow field $v = w+z$, pressure $\chi = \chi_s+p$, and initial condition $v(0) = w+z_0$ in Ω is controlled to the stationary solution w. The initial value z_0 can be seen as a small perturbation of the steady-state flow w and the considered problem can be interpreted as designing a feedback control in order to stabilize the perturbed flow back to the steady-state solution. Note that the operator b in the boundary control formulation (3c) is the identity operator if the control acts on the whole boundary and can be considered as a restriction operator if the control is localized in part of the boundary.

The following analytical solutions to the control problem described above are discussed in the literature. For $w \in L_\infty(\Omega)$ and $z_0 \in V_n^0(\Omega) \cap L_4(\Omega) =: X(\Omega)$, the existence of a stabilizing boundary control is proved in [8], but no constructive way to derive a boundary control in feedback form is derived. In the 3D case, the existence of an exponentially stabilizing feedback control for an appropriately defined subspace $X(\Omega)$ is proved in [9].

Stabilization of the Navier-Stokes system with feedback control based on an associated linear-quadratic optimal control problem has been recently discussed by several authors. The situation as described above with a distributed control localized in an open subset $\Omega_u \subset \Omega$ of positive measure instead of a boundary control is discussed in [4, 10, 5]. The only numerical study of such an approach known to the authors is [11] and treats the special case of a rectangular driven cavity and a control with zero normal component. Note that also numerous other strategies for stabilization of flow problems using feedback control have been discussed in the literature, including backstepping, instantaneous and model predictive control, Lyapunov methods, etc., see, e.g., [12, 13, 14, 15, 16]. The problem considered above with Dirichlet boundary control, b being the identity operator, and a control with zero normal component is studied in [6]. The first treatment of the problem considered above in the case when the control has nonzero normal components (which is often the relevant case in engineering applications) and is localized on parts of the boundary is given in [1, 2, 3, 7]. The stabilization results described in [1, 2, 3, 7] are constructive in that they allow the computation of a Dirichlet feedback law which stabilizes the Navier-Stokes system (1) in the sense that its solution (v, χ) is controlled to the steady-state solution (w, χ_s) associated to (2). We will briefly outline this construction for the 2D case as described in [1, 2] – the 3D case is treated similarly [3], but the derivation is quite more involved. The stabilizing feedback law is derived from a perturbed linearization of the Navier-Stokes system (3), given by

$$\partial_t z + (z \cdot \nabla) w + (w \cdot \nabla) z - \frac{1}{Re} \Delta z - \alpha z + \nabla p = 0 \quad \text{in } Q_\infty, \tag{4a}$$

$$\operatorname{div} z = 0 \quad \text{in } Q_\infty, \tag{4b}$$

$$z = bu \quad \text{in } \Sigma_\infty, \tag{4c}$$

$$z(0) = z_0 \quad \text{in } \Omega, \tag{4d}$$

where $\alpha > 0$ is a positive constant. The perturbation term $-\alpha z$ (note the sign!) is required to achieve exponential stabilization of the feedback law. Together with the cost functional

$$J(z, u) = \frac{1}{2} \int_0^\infty \langle z, z \rangle_{L_2(\Omega)} + \langle u, u \rangle_{L_2(\Gamma)} \, dt, \tag{5}$$

the linear-quadratic optimal control problem associated to (3) becomes

$$\inf \left\{ J(z, u) \mid (z, u) \text{ satisfies (4)}, \; u \in L_2((0, \infty); V^0(\Gamma)) \right\}, \tag{6}$$

where $V^0(\Gamma) = \{g \in L_2(\Gamma) \mid \langle g \cdot n, 1 \rangle_{H^{-\frac{1}{2}}(\Gamma), H^{\frac{1}{2}}(\Gamma)} = 0\}$. Then it is shown in [1, 2] that the feedback law

$$u = -R_A^{-1} b B^* \Pi P z \tag{7}$$

is exponentially stabilizing for small enough initial values z_0.

The operators defining the feedback law are

- the linearized Navier-Stokes operator A;
- the Helmholtz projector $P : L_2(\Omega) \mapsto V_n^0(\Omega)$;
- the control operator $B := (\lambda_0 I - A)D_A$, where D_A is the Dirichlet operator associated with $\lambda_0 I - A$ and $\lambda_0 > 0$ is a constant;
- $R_A := bD_A^*(I - P)D_A b + I$;
- the Riccati operator $\Pi = \Pi^* \in \mathcal{L}(V_n^0(\Omega))$ which is the unique nonnegative semidefinite weak solution of the operator Riccati equation

$$0 = I + (A + \alpha I)^*\Pi + \Pi(A + \alpha I) - \Pi(B_\tau B_\tau^* + B_n R_A^{-1} B_n^*)\Pi, \tag{8}$$

  where B_τ and B_n correspond to the projection of the control action in the tangential and normal directions derived from the control operator B.

Note that a simpler version of the Riccati equation and feedback control law with $R_A = I$ is discussed in [7].

Thus, the numerical solution of the stabilization problem for the Navier-Stokes equation (1) requires the numerical solution of the operator Riccati equation (8). To achieve this, we propose to proceed as in classical approaches to linear-quadratic optimal control problems for parabolic systems described, e.g., in [17, 18, 19, 20]. That is, first we discretize (4) in space using a finite element (FEM) Galerkin scheme, then solve the associated finite-dimensional algebraic Riccati equation (ARE) from which we obtain a finite-dimensional controller in feedback form as in (7). The finite-dimensional controller is then applied to (1). The stabilizing properties of such a finite-dimensional controller in case of internal control are discussed in [5] and for linear parabolic systems in [21, 22]. For the boundary control problem considered here, the approximation and convergence theory is an open problem. As far as the numerical solution of the described stabilization problems is concerned, we are only aware of the above-mentioned paper [11] and preliminary results for stabilization of Kármán vortex streets presented by Raymond in [23] which, in both cases, lead to dimensions of the FEM ansatz space so that the associated ARE can still be solved with classical methods based on associated eigenvalue problems (see [24, 25, 26] for overviews of available methods).

We also would like to point out that there exists a variety of different approaches to the stabilization of flow problems based, e.g., on gradient or adjoint methods, see the recent books [12, 13] and the references therein. Another recent contribution treats the 3D Navier-Stokes system in the exterior of a bounded domain [27]. Here, a control is derived based on an optimality system involving the adjoint Oseen equations. In contrast to all these approaches, the ansatz followed here does not require the sometimes infeasible computation of gradient systems or the solution of adjoint systems. We pursue the feedback stabilization approach as it allows to incorporate current information on the perturbed flow, which is not possible when using a pre-computed open-loop control as often obtained from optimization-based methods. That is, our approach allows to also deal with perturbations "on the fly" in contrast to open-loop control schemes. Of course, in practical situations, our approach will be best utilized in order to compensate for perturbations of an optimized flow profile that is obtained through open-loop control, i.e., the feedback control loop is added to optimal control obtained by other methods. If the deviation from the optimized flow profile is sensed fast enough, the smallness assumption of the initial values is well justified and our scheme will steer the flow back to the desired flow exponentially fast.

In the following section we describe a strategy for the numerical computation of the stabilizing feedback boundary control law based on solving a discretized version of (8). We will see that here, several problems occur as compared to the standard case of discretized operator Riccati equations associated to linear parabolic linear-quadratic optimal control problems. Remedies for those problems are described in Section 3. We provide an outlook on future work in Section 4.

2. Numerical solution of the operator Riccati equation

The general problem encountered with the Riccati approach in the context of feedback control for PDEs is that the expected dimensions of the FEM ansatz spaces, called n, required to achieve a reasonable approximation error will be far beyond the abilities of ARE solvers known from systems and control theory as the number of unknowns in the Riccati solution is essentially n^2. In the following, we will therefore discuss the computational challenges encountered after the spatial FEM discretization is performed.

First note that a Galerkin FEM discretization of (4) (employing the Helmholtz decomposition so that Pz becomes the new state variable) yields a finite-dimensional linear-quadratic optimal control problem for an evolution equation of the form

$$M_h \dot{z}_h(t) = -K_h z_h(t) + B_h u_h(t), \quad z_h(0) = Q_h z_0, \tag{9}$$

where M_h, K_h are the mass and stiffness matrices, B_h is the discretized boundary control operator, and Q_h is the canonical projection onto the FEM ansatz space. The stabilizing feedback law for the finite-dimensional approximation to the optimal control problem (6) then becomes

$$u_h(t) = -B_h^T \Pi_h M_h z_h(t) =: -F_h z_h(t), \tag{10}$$

where $\Pi_h = \Pi_h^T$ is the unique nonnegative semidefinite solution to the ARE

$$0 = \mathcal{R}(\Pi_h) = M_h - K_h^T \Pi_h M_h - M_h \Pi_h K_h - M_h \Pi_h B_h B_h^T \Pi_h M_h. \tag{11}$$

This matrix-valued nonlinear system of equations can be considered as a finite-dimensional approximation to the operator Riccati equation (8). Under mild assumptions, it can be proved that for linear parabolic control systems, Π_h converges uniformly to Π in the strong operator topology, see, e.g., [18, 28, 21, 22, 19, 20]. Convergence rates and a priori error estimates are derived in [29] under mild assumptions for linear parabolic control systems. Note that similar results are not yet available for the situation considered above despite first results in the stationary case [30]. The major computational challenge in this approach is that in order to solve the finite-dimensional problem for computing a feedback law u_h as in (10) for approximating u from (7), we need to solve (11) numerically. As the solution is a symmetric matrix, we are faced with $n(n+1)/2$ unknowns, where n is the dimension of the FEM ansatz space. This is infeasible for 3D problems and even most 2D problems. Even if there were algorithms to handle this complexity, it would in general not be possible to store the solution matrix Π_h in main memory or even on a contemporary hard disk. In recent years, several approaches to circumvent this problem have been proposed.

One major ingredient necessary to solve AREs for large-scale systems arising from PDE-constrained optimal control problems is based on the observation that often the eigenvalues of Π_h decay to zero rapidly. Thus, Π_h can be approximated by a low-rank matrix $Z_h Z_h^T$, where $Z_h \in \mathbb{R}^{n \times r}$, $r \ll n$. Current research focuses on numerical algorithms to compute Z_h directly without ever forming Π_h. A promising possibility in this direction is to employ Newton's method for AREs. Such methods have been developed in recent years, see [17, 31, 32, 33]. The basis for these methods is the efficient solution of the Lyapunov equations to be solved in each Newton step,

$$(K_h + B_h B_h^T \Pi_{h,j} M_h)^T N_j M_h + M_h N_j (K_h + B_h B_h^T \Pi_{h,j} M_h) - \mathcal{R}(\Pi_{h,j}) = 0, \quad (12)$$

where $\Pi_{h,j}$ are the Newton iterates and $\Pi_{h,j+1} = \Pi_{h,j} + N_j$. It is mandatory to employ the structure in the coefficient matrices: sparse (stiffness, mass) matrix plus low-rank update. This can be achieved utilizing methods based on the ADI iteration [31, 34, 35], so that the computational complexity of the Newton step can be decreased from an infeasible $\mathcal{O}(n^3)$ to merely the complexity of solving the corresponding stationary PDE. Note that the method used in [17] is similar to the one discussed in [31, 32, 33], but the latter one exhibits a better numerical robustness and enhanced efficiency. In the situation considered here, the cost for one Newton step is expected to become proportional to solving the stationary linearized Navier-Stokes problem. This requires some research into the direction of appropriate linear solvers for nonsymmetric saddle point problems, including preconditioners as discussed in [36]. Another issue is that using standard finite-element methods (like for instance the Taylor-Hood elements) for Navier-Stokes equations, the mass matrices in (11) and (12) are singular due to the incompressibility condition. None of the existing solvers for large-scale Lyapunov equations and AREs is able to deal with these issues. Therefore, we first intended to use divergence-free elements computed using explicit Helmholtz projection. A recent new algorithmic idea [37] in this direction inspired a new approach (see Subsection 3.1 below).

Moreover, the efficiency of the discussed approach is much enhanced, when instead of the identity operator I, a nonnegative semidefinite operator W or W_h is employed in (8) or (11). This is the case, e.g., if an observation equation is employed: $y = \mathcal{C}z$, where $\mathcal{C}$ is a restriction operator, so that only information localized in some subdomain of Ω or Γ is used in the cost functional. Note that it is quite natural in technical applications that not the complete state is available for optimization and control, but only selected measurements can be employed. Thus, it is reasonable to consider such an observation equation.

Other approaches for large-scale AREs arising in a PDE control context are multigrid approaches [38, 39] and a sign function implementation based on formatted arithmetics for hierarchical matrices proposed in [40]. Both approaches seem not feasible for the matrix structures that will result from the FEM discretization of the linearized Navier-Stokes problems considered here. Also, approaches based on Krylov subspace methods have been investigated for solving large-scale Riccati and Lyapunov equations [41, 42, 43]. These methods are not guaranteed to compute the requested nonnegative semidefinite solution of the ARE and in general are not competitive to the Newton-ADI approach. For these approaches, too, the singularity of the mass matrices is an obstacle and would require deeper investigations itself. Therefore, we will concentrate on the modification of the Newton-ADI approach using the ideas from [37] as this appears to be most promising.

Even though the computational complexity of new methods for large-scale AREs has decreased considerably, the number of columns in Z_h may still be too large in case of 3D problems or if convection is the dominating effect in the flow dynamics. Nonetheless, there are some approaches that allow a direct iteration on the approximate feedback operator F_h [17, 31, 33]. Thus, Z_h need not be computed, and the storage requirements are limited by the number of rows in F_h which equals the number of input variables. It is therefore of vital importance to model the problem such that the number of resulting control variables is as small as possible. If this can be achieved, then the method discussed in [31, 33] is applicable also for 2D and 3D problems while in [17, 29, 38], only 1D problems are considered. Further investigation of the case of a positive definite W and W_h as in the case of $W = I$ is required here. The feedback iteration, too, needs to be adapted in order to be able to deal with the singular mass matrix.

In order to apply our control strategy in practice, we need to find the stabilizing solution of the operator Riccati equation (8). If for the moment we assume that we discretize this equation after the Helmholtz decomposition is performed so that we work with the state variable Px in the set of divergence-free functions, in the end we have to solve numerically the ARE (11). In order to simplify notation, in the following we will write this *Helmholtz-projected ARE* in the form

$$0 = M + (A + \alpha M)^T XM + MX(A + \alpha M) - MXBB^T XM. \tag{13}$$

Before solving this equation with the now well-established low-rank Newton-ADI method, several problems have to be solved:

1. In order to arrive at the matrix representation (13) of (8), the discretization of the Helmholtz-projected Oseen equations (4) would require divergence-free finite elements. As our approach should work with a standard Navier-Stokes solver like NAVIER [44], where Taylor-Hood elements are available, we have to deal in some way with the Helmholtz projector P.

2. Each step of the Newton-ADI iteration with $A_0 := A + \alpha M$ requires the solution of

$$A_j^T Z_{j+1} Z_{j+1}^T M + M Z_{j+1} Z_{j+1}^T A_j = -W_j W_j^T$$
$$= -M - M(Z_j Z_j^T B)(Z_j Z_j^T B)^T M,$$

where $n_v :=$ rank $M =$ dim of ansatz space for velocities. This leads to the need to solve $n_v + m$ linear systems of equations in each step of the Newton-ADI iteration, making the approach less efficient.

3. The linearized system associated with $A + \alpha M$ is unstable in general. But to start the Newton iteration, a stabilizing initial guess is needed.

In order to deal with the Helmholtz projector, the first strategy employed was to explicitly project the ansatz functions. This requires the solution of one saddle-point problem per ansatz function. Thus, this approach becomes fairly expensive already for coarse-grain discretizations. Nevertheless, we could verify our stabilization strategy using these ideas, first presented in [45]. In particular, with this strategy, the third issue from above is not problematic as several stabilization procedures available for standard linear dynamical systems can be employed. Here, we chose an approach based on the homogeneous counterpart of (13), i.e., the *algebraic Bernoulli equation*

$$0 = (A + \alpha M)^T X M + M X (A + \alpha M) - M X B B^T X M, \tag{14}$$

which can be solved for the stabilizing solution X efficiently, e.g., with methods described in [46, 47] if the dimension of the problem is not too large.

In order to evaluate the solution strategies, we decided to test the developed numerical algorithms first for the Kármán vortex street. An uncontrolled flow profile is shown in Figure 1. Clearly, the obstacle produces the expected vortex shedding behind the obstacle.

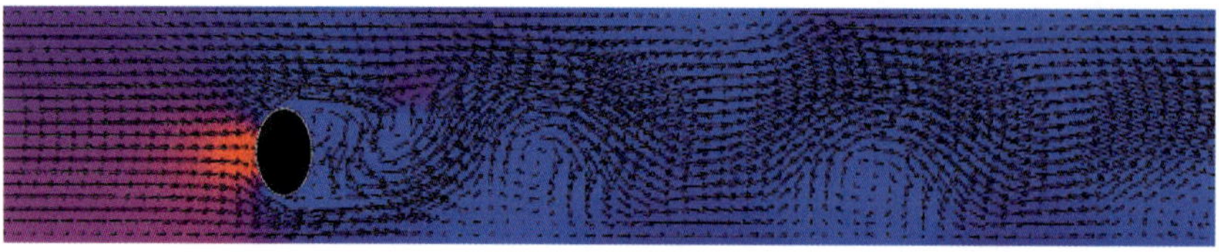

FIGURE 1. Kármán vortex street, uncontrolled.

For testing our feedback stabilization strategy, we use a slightly different setting from the one shown in Figure 1. Here, the obstacle is tilted to the right. We try to stabilize the flow behind the elliptic obstacle at Re = 500. We use a parabolic inflow profile, a "do-nothing" outflow condition, and chose w as solution of the stationary Navier-Stokes equations for Re = 1.

As a coarse mesh was employed (approx. 5.000 degrees of freedom for the velocities), the explicit projection of all ansatz functions onto the set of divergence-free functions is possible, so that we arrive at a standard ARE as in (11) that can

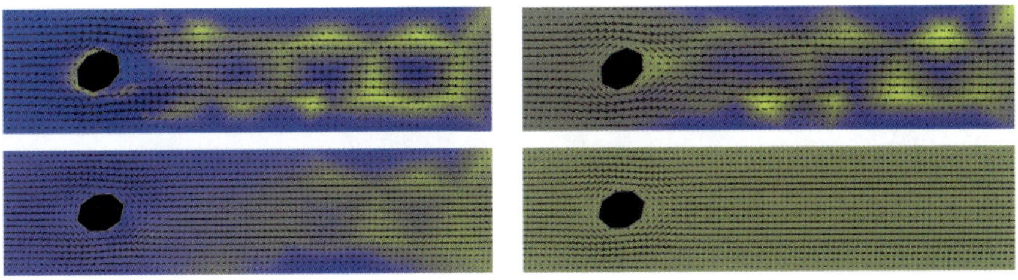

FIGURE 2. Kármán vortex street, controlled, at $t = 1$ (top left), $t = 5$ (top right), $t = 8$ (bottom left), and $t = 10$ (bottom right).

be solved, e.g., by the method described in [32, 33, 31]. The controlled system using the Dirichlet boundary feedback control approach based on the solution of this ARE is displayed in Figure 2. The figures show the velocity field v of the stabilized flow at various time instances, where the control acts by blowing or sucking at two positions: one at the top of the obstacle, and one at its bottom. Clearly, the desired stabilization of the flow behind the obstacle can be observed.

In the following section, we will describe the developed solution strategies for the encountered problems as described above.

3. Remedies of problems encountered by the ARE solver

3.1. Avoiding divergence-free FE

A standard FE discretization of the linearized Navier-Stokes system (4) using Taylor-Hood elements yields the following system of differential-algebraic equations (DAEs):

$$\begin{aligned} E_{11}\dot{z}_h(t) &= A_{11}z_h(t) + A_{12}p_h(t) + B_1u(t) \\ 0 &= A_{12}^T z_h(t) + B_2u(t) \\ z_h(0) &= z_{h,0}, \end{aligned} \tag{15}$$

where $E_{11} \in \mathbb{R}^{n_v \times n_v}$ is symmetric positive definite, $A_{12}^T \in \mathbb{R}^{n_p \times n_v}$ is of rank n_p. $z_h \in \mathbb{R}^{n_v}$ and $p_h \in \mathbb{R}^{n_p}$ are the states related to the FEM discretization of velocities and pressure, and $u \in \mathbb{R}^{n_g}$ the system input derived from the Dirichlet boundary control.

Unfortunately, the ARE corresponding to the DAE (15) in the form (13) (with $M = \operatorname{diag}(E_{11}, 0)$, $A + \alpha M = \begin{bmatrix} A_{11} & A_{12} \\ A_{12}^T & 0 \end{bmatrix}$, etc., does not yield the information required about the stabilizing feedback. It is well known (see, e.g., [48]) that if M is singular, (13) may or may not have solutions, no matter whether the corresponding DAE is stabilizable. Moreover, the positive semidefinite solution of (13), if it exists, may not be stabilizing. In [49] we suggest a projected ARE using spectral projectors onto the deflating subspaces of the matrix pair (A, M) corresponding to the finite eigenvalues. It can then be proved that under standard assumptions on stabilizability of descriptor systems (linear systems in DAE form), the so-obtained ARE has a unique stabilizing positive semidefinite solution as in the regular case of invertible M. As a corollary, we can show that the projected version of (13) has a unique stabilizing positive semidefinite solution. Unfortunately, the theory and methods derived in [49] require the explicit availability of the spectral projectors.

Inspired by a similar approach in [37], it turns out that one can avoid the computation of these spectral projectors. First observe that forming these projectors in the form

$$P_h := I_{n_v} - A_{12}(A_{12}^T E_{11}^{-1} A_{12})^{-1} A_{12}^T E_{11}^{-1},$$

and applying the Newton-ADI iteration to the projected version of the ARE (13), it turns out that in each step of the iteration, we have to solve the Lyapunov equation

$$A_j^T Z_{j+1} Z_{j+1}^T P_h E_{11} P_h^T + P_h E_{11} P_h^T Z_{j+1} Z_{j+1}^T A_j = -W_j W_j^T,$$

where

$$\begin{aligned} A_j &:= P_h(A_{11} - B_1 B_1^T P_h Z_j Z_j^T P_h E_{11}) P_h, \\ W_j &:= \left[\begin{array}{cc} P_h C^T & P_h E_{11} P_h Z_j Z_j^T P_h B_1 \end{array}\right]. \end{aligned}$$

Thus, a low-rank factor so that $X_{j+1} \approx Z_{j+1} Z_{j+1}^T$ can be computed as

$$Z_{j+1} = \sqrt{\mu}\left[B_{j,\mu},\, A_{j,\mu} B_{j,\mu},\, A_{j,\mu}^2 B_{j,\mu},\, \ldots,\, A_{j,\mu}^j B_{j,\mu}\right],$$

where $B_{j,\mu}$ solves the saddle point problem

$$\left[\begin{array}{cc} E_{11} + \mu(A_{11} - B_1 B_1^T Z_j Z_j^T E_{11}) & A_{12} \\ A_{12}^T & 0 \end{array}\right] \left[\begin{array}{c} B_{j,\mu} \\ * \end{array}\right] = \left[\begin{array}{cc} C^T & E_{11} Z_j Z_j^T B_1 \\ 0 & 0 \end{array}\right],$$

The multiplication by $A_{j,\mu}$ is realized by the solution of a saddle-point problem with the same coefficient matrix. Hence, the projector P_h needs not be formed explicitly, and we can advance the Newton iteration by solving saddle point problems associated to stationary Oseen-like problems.

3.2. Removing M_h from the right-hand side of Lyapunov equations

The solution strategy employed here can be based on a strategy already suggested for standard AREs

$$0 = AX + XA^T - XBB^T X + W. \qquad (16)$$

Thus, for simplicity, consider the following Lyapunov equation arising in the Newton step when applying Newton-ADI to (16):

$$A_j^T \underbrace{(X_j + N_j)}_{=X_{j+1}} + X_{j+1} A_j = -W - X_j BB^T X_j \qquad \text{for } j = 1, 2, \ldots$$

By subtracting the two consecutive Lyapunov equations for $j-1, j$ from each other, we obtain

$$A_j^T N_j + N_j A_j = N_{j-1} BB^T N_{j-1} \qquad \text{for } j = 1, 2, \ldots \qquad (17)$$

See [17, 50, 51] for details and applications of this variant. By the subtraction, the constant term W vanishes from the right-hand side of the Lyapunov equation to be solved in the Newton-ADI step. Thus, we are now facing a Lyapunov equation with low-rank right-hand side as desirable for an efficient application of the low-rank ADI method. In particular, this rank equals the number of inputs used for stabilization. In general, this will be fairly low: mostly, $1 \leq m \leq 10$. This strategy can be applied without changes to our situation. Also note that (17) can be written in factored form as required by the low-rank ADI method for Lyapunov equations.

One remaining problem is that in order to start the Newton-ADI iteration based on (17), we need a guess for $N_0 = X_1 - X_0$, i.e., besides a stabilizing initial guess $X_0 = Z_0 Z_0^T$, we also need $X_1 = Z_1 Z_1^T$. Finding X_0 is a task on its own (this is actually Problem 3 in the above list and will be discussed in the following subsection).

One possibility here is to compute X_0 and X_1 from the full, dense Lyapunov equations obtained from a coarse grid discretization of (4), and prolongate these to the fine grid, leading naturally to low-rank factorized forms as required. It remains to investigate the accuracy required for the approximation of X_1. Some results in this direction can probably be derived using an interpretation of this approach as inexact Newton method. Recent work on inexact Newton-Kleinman for standard AREs [52] sheds some light on this situation.

A possible refinement of the proposed strategy could involve coarse grid corrections using Richardson iteration or a nested iteration for AREs as in [39].

3.3. Computing a stabilizing initial feedback

If we use the same notation as in (13), this task can be described as finding a matrix F_0 such that all finite eigenvalues of the matrix pair $(A - BB^T F_0, M)$ are contained in the open left half of the complex plane.

There are basically three approaches discussed in the literature to compute such an F_0:

- pole placement (see [25] for an overview of algorithms in the standard case that M is invertible, for descriptor systems see [53]);
- Bass-type algorithms based on solving a Lyapunov equation, described for standard systems in [54] while for descriptor systems, see [53, 55, 49];
- $F_0 = B^T X_0$, where X_0 is the stabilizing solution of the algebraic Bernoulli equation (14). For standard systems, see [46, 47] for algorithms to compute X_0, while descriptor systems are treated in [55].

The Bernoulli stabilization algorithm from [55] was tested so far for a descriptor system obtained from discretizing a Stokes-like equation that can be derived from the linearized Navier-Stokes equations (4) by neglecting the convective terms in (4a). Figure 3 shows that stabilization is achieved even in a case where an artificially strong de-stabilization parameter $\alpha = 1000$ is used. This algorithm will be our first choice as stabilization procedure as it turns out to be a lot more robust to the effects of the ill-conditioning of the stabilization problem than all pole

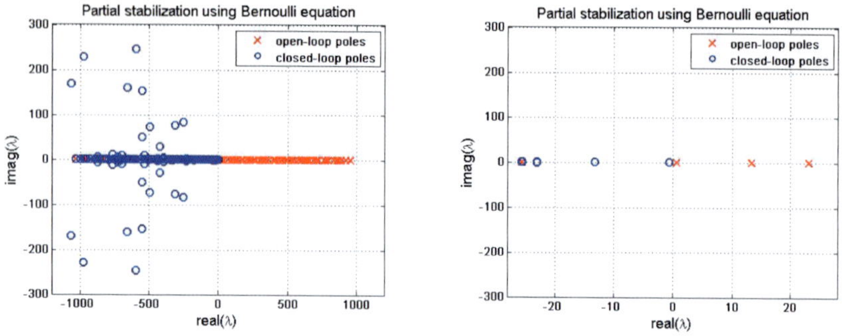

FIGURE 3. Bernoulli stabilization of a Stokes-/Oseen-type descriptor system: shown are the finite eigenvalues of (A, M) (open-loop poles) and of $(A - BB^T F_0, M)$ (closed-loop poles) (left) with a close-up around the origin (right).

placement and Bass-type algorithms; for details see the numerical experiments in [55].

It remains to apply the described strategies in the situation discussed here, i.e., to perturbed flow problems described by the Navier-Stokes system (1). At this writing, the implementation of the ideas described in this section is underway and numerical results obtained with this will be reported elsewhere.

4. Conclusions and outlook

We have described a method to stabilize perturbed flow problems, described by the incompressible Navier-Stokes equations, by Dirichlet boundary control. The control may act tangential as well as normal to the boundary. It allows to compensate for disturbances of optimized flow profiles, where an open-loop control may have been obtained by optimization methods. The stabilizing control is of feedback-type so that a closed-loop control system is obtained. The feedback is computed using an associated linear-quadratic optimal control problem for the linearized Navier-Stokes system, projected onto the space of divergence-free functions, arising from linearizing the actual system about a desired stationary solution. The numerical solution of this linear-quadratic optimal control problem is obtained by solving a large-scale algebraic Riccati equation. Several modifications of existing algorithms for this task are necessary in order to achieve an efficient and applicable solution strategy. We have described these modifications here. Their implementation and application to realistic flow control problems is under current investigation.

In the future, we plan to extend the techniques described in this paper to flow problems governed by the incompressible Navier-Stokes equations coupled with field equations describing reactive, diffusive, and convective processes.

Further improvements in the efficiency of the Lyapunov and Riccati solvers using the ADI-Galerkin hybrid method suggested in [56, 57] which often significantly accelerate the ADI and Newton iteration for the Lyapunov and Riccati equations will also be investigated.

Acknowledgment

We would like to thank Heiko Weichelt for careful reading of the manuscript and detecting several mistakes. Moreover, we gratefully acknowledge the helpful remarks of an anonymous reviewer whose suggestions improved the quality of this paper.

References

[1] Raymond, J.P.: Local boundary feedback stabilization of the Navier-Stokes equations. In: Control Systems: Theory, Numerics and Applications, Rome, 30 March–1 April 2005, Proceedings of Science. SISSA (2005). URL http://pos.sissa.it

[2] Raymond, J.P.: Feedback boundary stabilization of the two-dimensional Navier-Stokes equations. SIAM J. Control Optim. **45**(3), 790–828 (2006)

[3] Raymond, J.P.: Feedback boundary stabilization of the three-dimensional incompressible Navier-Stokes equations. J. Math. Pures Appl. (9) **87**(6), 627–669 (2007)

[4] Barbu, V.: Feedback stabilization of the Navier-Stokes equations. ESAIM: Control Optim. Calc. Var. **9**, 197–206 (2003)

[5] Barbu, V., Triggiani, R.: Internal stabilization of Navier-Stokes equations with finite-dimensional controllers. Indiana Univ. Math. J. **53**(5), 1443–1494 (2004)

[6] Barbu, V., Lasiecka, I., Triggiani, R.: Tangential boundary stabilization of Navier-Stokes equations. Mem. Amer. Math. Soc. **181**(852), x+128 (2006)

[7] Badra, M.: Lyapunov function and local feedback boundary stabilization of the Navier-Stokes equations. SIAM J. Control Optim. **48**(3), 1797–1830 (2009)

[8] Fernández-Cara, Y., Guerrero, S., Imanuvilov, O., Puel, J.P.: Local exact controllability of the Navier-Stokes system. J. Math. Pures Appl., IX. Sér. **83**(12), 1501–1542 (2004)

[9] Fursikov, A.: Stabilization for the 3D Navier-Stokes system by feedback boundary control. Discrete Contin. Dyn. Syst. **10**(1–2), 289–314 (2004)

[10] Barbu, V., Sritharan, S.S.: H^∞-control theory of fluid dynamics. Proc. R. Soc. Lond. A **454**, 3009–3033 (1998)

[11] King, B., Ou, Y.R.: Nonlinear dynamic compensator design for flow control in a driven cavity. In: Proc. 34th IEEE Conf. Decision Contr., pp. 3741–3746 (1995)

[12] Aamo, O.M., Krstic, M.: Flow Control by Feedback. Stabilization and Mixing. Springer-Verlag, London (2003)

[13] Gunzburger, M.: Perspectives in Flow Control and Optimization. SIAM Publications, Philadelphia, PA (2003)

[14] Hinze, M.: Instantaneous closed loop control of the Navier-Stokes system. SIAM J. Control Optim. **44**(2), 564–583 (2005)

[15] Hinze, M., Matthes, U.: Optimal and model predictive control of the Boussinesq approximation. In: Control of Coupled Partial Differential Equations, *Internat. Ser. Numer. Math.*, vol. 155, pp. 149–174. Birkhäuser, Basel (2007)

[16] Krstic, M., Smyshlyaev, A.: Boundary Control of PDEs. A course on backstepping designs, *Advances in Design and Control*, vol. 16. Society for Industrial and Applied Mathematics (SIAM), Philadelphia, PA (2008)

[17] Banks, H., Ito, K.: A numerical algorithm for optimal feedback gains in high-dimensional linear quadratic regulator problems. SIAM J. Cont. Optim. **29**(3), 499–515 (1991)

[18] Banks, H., Kunisch, K.: The linear regulator problem for parabolic systems. SIAM J. Cont. Optim. **22**, 684–698 (1984)

[19] Lasiecka, I., Triggiani, R.: Differential and Algebraic Riccati Equations with Application to Boundary/Point Control Problems: Continuous Theory and Approximation Theory. No. 164 in Lecture Notes in Control and Information Sciences. Springer-Verlag, Berlin (1991)

[20] Lasiecka, I., Triggiani, R.: Control Theory for Partial Differential Equations: Continuous and Approximation Theories I. Abstract Parabolic Systems. Cambridge University Press, Cambridge, UK (2000)

[21] Ito, K.: Finite-dimensional compensators for infinite-dimensional systems via Galerkin-type approximation. SIAM J. Cont. Optim. **28**, 1251–1269 (1990)

[22] Morris, K.: Design of finite-dimensional controllers for infinite-dimensional systems by approximation. J. Math. Syst., Estim., and Control **4**, 1–30 (1994)

[23] Raymond, J.P.: Boundary feedback stabilization of the incompressible Navier-Stokes equations. Talk at TC7 IFIP Conference, Torino, July 18-22, 2005

[24] Benner, P.: Computational methods for linear-quadratic optimization. Supplemento ai Rendiconti del Circolo Matematico di Palermo, Serie II **No. 58**, 21–56 (1999)

[25] Datta, B.: Numerical Methods for Linear Control Systems. Elsevier Academic Press (2004)

[26] Sima, V.: Algorithms for Linear-Quadratic Optimization, *Pure and Applied Mathematics*, vol. 200. Marcel Dekker, Inc., New York, NY (1996)

[27] Fursikov, A., Gunzburger, M., Hou, L.: Optimal boundary control for the evolutionary Navier-Stokes system: The three-dimensional case. SIAM J. Cont. Optim. **43**(6), 2191–2232 (2005)

[28] Benner, P., Saak, J.: Linear-quadratic regulator design for optimal cooling of steel profiles. Tech. Rep. SFB393/05-05, Sonderforschungsbereich 393 *Parallele Numerische Simulation für Physik und Kontinuumsmechanik*, TU Chemnitz, 09107 Chemnitz, FRG (2005). URL http://www.tu-chemnitz.de/sfb393

[29] Kroller, M., Kunisch, K.: Convergence rates for the feedback operators arising in the linear quadratic regulator problem governed by parabolic equations. SIAM J. Numer. Anal. **28**(5), 1350–1385 (1991)

[30] Casas, E., Mateos, M., Raymond, J.P.: Error estimates for the numerical approximation of a distributed control problem for the steady-state Navier-Stokes equations. SIAM J. Control Optim. **46**(3), 952–982 (2007)

[31] Benner, P., Li, J.R., Penzl, T.: Numerical solution of large Lyapunov equations, Riccati equations, and linear-quadratic control problems. Numer. Lin. Alg. Appl. **15**, 755–777 (2008)

[32] Benner, P.: Efficient algorithms for large-scale quadratic matrix equations. Proc. Appl. Math. Mech. **1**(1), 492–495 (2002)

[33] Benner, P.: Solving large-scale control problems. IEEE Control Systems Magazine **14**(1), 44–59 (2004)

[34] Li, J.R., White, J.: Low rank solution of Lyapunov equations. SIAM J. Matrix Anal. Appl. **24**(1), 260–280 (2002)

[35] Wachspress, E.: Iterative solution of the Lyapunov matrix equation. Appl. Math. Letters **107**, 87–90 (1988)

[36] Elman, H., Silvester, D., Wathen, A.: Finite Elements and Fast Iterative Solvers: with applications in incompressible fluid dynamics. Oxford University Press, Oxford (2005)

[37] Heinkenschloss, M., Sorensen, D., Sun, K.: Balanced truncation model reduction for a class of descriptor systems with application to the Oseen equations. SIAM J. Sci. Comput. **30**(2), 1038–1063 (2008)

[38] Rosen, I., Wang, C.: A multi-level technique for the approximate solution of operator Lyapunov and algebraic Riccati equations. SIAM J. Numer. Anal. **32**(2), 514–541 (1995)

[39] Grasedyck, L.: Nonlinear multigrid for the solution of large-scale Riccati equations in low-rank and $\mathcal{H}$-matrix format. Numer. Lin. Alg. Appl. **15**, 779–807 (2008)

[40] Grasedyck, L., Hackbusch, W., Khoromskij, B.: Solution of large scale algebraic matrix Riccati equations by use of hierarchical matrices. Computing **70**, 121–165 (2003)

[41] Benner, P., Faßbender, H.: An implicitly restarted symplectic Lanczos method for the Hamiltonian eigenvalue problem. Linear Algebra Appl. **263**, 75–111 (1997)

[42] Jaimoukha, I., Kasenally, E.: Krylov subspace methods for solving large Lyapunov equations. SIAM J. Numer. Anal. **31**, 227–251 (1994)

[43] Jbilou, K., Riquet, A.: Projection methods for large Lyapunov matrix equations. Linear Algebra Appl. **415**(2-3), 344–358 (2006)

[44] Bänsch, E.: Simulation of instationary, incompressible flows. Acta Math. Univ. Comenianae **LXVII**(1), 101–114 (1998)

[45] Bänsch, E., Benner, P., Heubner, A.: Riccati-based feedback stabilization of flow problems. In: 23rd IFIP TC 7 Conference, July 23–27, 2007, Cracow, Poland (2007)

[46] Barrachina, S., Benner, P., Quintana-Ortí, E.: Efficient algorithms for generalized algebraic Bernoulli equations based on the matrix sign function. Numer. Algorithms **46**(4), 351–368 (2007)

[47] Baur, U., Benner, P.: Efficient solution of algebraic Bernoulli equations using H-matrix arithmetic. In: K. Kunisch, G. Of, O. Steinbach (eds.) Numerical Mathematics and Advanced Applications. Proceedings of ENUMATH 2007, the 7th European Conference on Numerical Mathematics and Advanced Applications, Graz, Austria, September 2007, pp. 127–134. Springer-Verlag, Heidelberg (2008)

[48] Mehrmann, V.: The Autonomous Linear Quadratic Control Problem, Theory and Numerical Solution. No. 163 in Lecture Notes in Control and Information Sciences. Springer-Verlag, Heidelberg (1991)

[49] Benner, P., Stykel, T.: Numerical algorithms for projected generalized Riccati equations. In preparation

[50] Benner, P., Hernández, V., Pastor, A.: On the Kleinman iteration for nonstabilizable systems. Math. Control, Signals, Sys. **16**, 76–93 (2003)

[51] Morris, K., Navasca, C.: Solution of algebraic Riccati equations arising in control of partial differential equations. In: Control and Boundary Analysis, *Lect. Notes Pure Appl. Math.*, vol. 240, pp. 257–280. Chapman & Hall/CRC, Boca Raton, FL (2005)

[52] Feitzinger, F., Hylla, T., Sachs, E.: Inexact Kleinman-Newton method for Riccati equations. SIAM J. Matrix Anal. Appl. **31**, 272–288 (2009)

[53] Varga, A.: On stabilization methods of descriptor systems. Sys. Control Lett. **24**, 133–138 (1995)

[54] Armstrong, E.: An extension of Bass' algorithm for stabilizing linear continuous constant systems. IEEE Trans. Automat. Control **AC–20**, 153–154 (1975)

[55] Benner, P.: Partial stabilization of descriptor systems using spectral projectors. In: S. Bhattacharyya, R. Chan, P. Van Dooren, V. Olshevsky, A. Routray (eds.) Numerical Linear Algebra in Signals, Systems and Control, Lecture Notes in Electrical Engineering, vol. 80, pp. 55–76. Springer-Verlag, Dordrecht (2011)

[56] Benner, P., Li, R.C., Truhar, N.: On the ADI method for Sylvester equations. J. Comput. Appl. Math. **233**(4), 1035–1045 (2009)

[57] Benner, P., Saak, J.: A Galerkin-Newton-ADI Method for Solving Large-Scale Algebraic Riccati Equations. Preprint SPP1253–090, DFG Priority Programme *Optimization with Partial Differential Equations* (2010). URL http://www.am.uni-erlangen.de/home/spp1253/wiki/index.php/Preprints

Eberhard Bänsch
FAU Erlangen
Lehrstuhl Angewandte Mathematik III
Haberstr. 2
D-91058 Erlangen, Germany
e-mail: `baensch@am.uni-erlangen.de`

Peter Benner
Max Planck Institute for Dynamics of Complex Technical Systems
Sandtorstr. 1
D-39106 Magdeburg, Germany
e-mail: `benner@mpi-magdeburg.mpg.de`

International Series of Numerical Mathematics, Vol. 160, 21–35

Allen-Cahn and Cahn-Hilliard Variational Inequalities Solved with Optimization Techniques

Luise Blank, Martin Butz, Harald Garcke,
Lavinia Sarbu and Vanessa Styles

Abstract. Parabolic variational inequalities of Allen-Cahn and Cahn-Hilliard type are solved using methods involving constrained optimization. Time discrete variants are formulated with the help of Lagrange multipliers for local and non-local equality and inequality constraints. Fully discrete problems resulting from finite element discretizations in space are solved with the help of a primal-dual active set approach. We show several numerical computations also involving systems of parabolic variational inequalities.

Mathematics Subject Classification (2000). 35K55, 35S85, 65K10, 90C33, 90C53, 49N90, 65M60.

Keywords. Optimization, Phase-field method, Allen-Cahn model, Cahn-Hilliard model, variational inequalities, active set methods.

1. Introduction

Interface evolution can be described with the help of phase field approaches, see, e.g., [12]. An interface, in which a phase field or order parameter rapidly changes its value, is modelled to have a thickness of order ε, where $\varepsilon > 0$ is a small parameter. The model is based on the non-convex Ginzburg-Landau energy E which has the form

$$E(u) := \int_\Omega \left(\frac{\gamma\varepsilon}{2} |\nabla u|^2 + \frac{\gamma}{\varepsilon} \psi(u) \right) dx \,, \tag{1.1}$$

where $\Omega \subset \mathbb{R}^d$ is a bounded domain, $\gamma > 0$ is a parameter related to the interfacial energy and $u : \Omega \to \mathbb{R}$ is the phase field variable, also called order parameter. The

This work was supported by the DFG within SPP 1253 "Optimization with partial differential equations" under BL433/2-1 and by the Vielberth foundation. The fifth author was also supported by the EPSRC grant EP/D078334/1.

different phases correspond to the values $u = \pm 1$. In interfacial regions solutions rapidly change from values close to 1 to values close to -1 and the thickness of this interfacial region is proportional to the parameter ε. The potential function ψ can be a smooth double well potential, e.g., $\psi(u) = (1-u^2)^2$ or an obstacle potential, e.g.,

$$\psi(u) = \psi_0(u) + I_{[-1,1]}(u)\,, \tag{1.2}$$

where $\psi_0 = \frac{1}{2}(1-u^2)$ or another smooth, non-convex function and $I_{[-1,1]}$ is the indicator function, for the interval $[-1,1]$. The interface evolution is then given by the gradient flow equation, i.e., the phase field tries to minimize the energy in time with respect to an inner product corresponding to a vector space $\mathbf{Z}$. More specifically we obtain

$$\partial_t u(t) = -\mathrm{grad}_{\mathbf{Z}} E(u(t))\,. \tag{1.3}$$

Considering a scaled L^2-inner product and the obstacle potential we obtain the *Allen-Cahn variational inequality*

$$(\varepsilon\partial_t u, \chi - u) + \gamma\varepsilon(\nabla u, \nabla(\chi - u)) + \tfrac{\gamma}{\varepsilon}(\psi_0'(u), \chi - u) \geq 0\,, \tag{1.4}$$

which has to hold for almost all t and all $\chi \in H^1(\Omega)$ with $|\chi| \leq 1$. Here and in the following $(.,.)$ denotes the L^2-inner product. The mass-conserving H^{-1}-inner product yields in the obstacle case the fourth-order *Cahn-Hilliard variational inequality*:

$$\partial_t u = \Delta w \quad \text{a.e.}, \tag{1.5}$$

$$(w - \tfrac{\gamma}{\varepsilon}\psi_0'(u), \xi - u) \leq \gamma\varepsilon(\nabla u, \nabla(\xi - u)) \quad \forall\, \xi \in H^1(\Omega), |\xi| \leq 1\,, \tag{1.6}$$

together with $|u| \leq 1$ a.e. For these formulations it can be shown that under Neumann boundary conditions for w and initial conditions for u a unique solution (u, w) of (1.5)-(1.6) exists where u is H^2-regular in space, see [8, 4].

Using the H^2-regularity the formulation (1.4) and (1.5), (1.6) can be restated in the *complementary form*, where the equalities and inequalities have to hold almost everywhere:

$$\varepsilon\partial_t u = \gamma\varepsilon\Delta u - \tfrac{1}{\varepsilon}(\gamma\psi_0'(u) + \mu)\,, \tag{1.7}$$

respectively

$$\partial_t u = \Delta w\,, \quad w = -\gamma\varepsilon\Delta u + \tfrac{1}{\varepsilon}(\gamma\psi_0'(u) + \mu)\,, \tag{1.8}$$

$$\mu = \mu_+ - \mu_-,\ \mu_+ \geq 0,\ \mu_- \geq 0, \quad |u| \leq 1\,, \tag{1.9}$$

$$\mu_+(u-1) = 0,\ \mu_-(u+1) = 0\,, \tag{1.10}$$

and homogeneous Neumann boundary conditions for u and w together with an initial phase distribution $u(0) = u_0$. Here, $\frac{1}{\varepsilon}\mu$ can be interpreted as a scaled Lagrange multiplier for the pointwise box-constraints. The scaling of the Lagrange multiplier μ with respect to ε is motivated by formal asymptotic expansions for obstacle potentials, see Blowey and Elliott [10] and Barrett, Garcke and Nürnberg [2]. Our scaling guarantees that μ is of order one. If we replace $\frac{1}{\varepsilon}\mu$ by μ in (1.7) we would observe a severe ε-dependence of μ. In practice this often results in oscillations in the primal-dual active set method for the discretized problem. In fact iterates would oscillate between the bounds ± 1 and no convergence takes place.

For an arbitrary constant $c > 0$ we introduce the *primal-dual active sets* employing the primal variable u and the dual variable μ

$$A_+(t) = \{x \in \Omega \mid c(u-1) + \mu > 0\}\,, \qquad A_-(t) = \{x \in \Omega \mid c(u+1) + \mu < 0\}$$

and the inactive set $I := \Omega \setminus (A_+ \cup A_-)$. The restrictions (1.9)–(1.10) can now be reformulated as

$$u(x) = \pm 1 \text{ for almost all } x \in A_\pm\,, \quad \mu(x) = 0 \text{ for almost all } x \in I\,. \tag{1.11}$$

A discrete version of (1.11) will lead later on to the primal-dual active set algorithm (PDAS).

Another reformulation of (1.9)–(1.10) is given with the help of a *non-smooth equation* as follows

$$H(u,\mu) := \mu - (\max(0, \mu + c(u-1)) + \min(0, \mu + c(u+1))) = 0\,, \tag{1.12}$$

which allows us to interpret the following PDAS-method for the discretized problem as a semi-smooth Newton method and provides then local convergence, see [17], for a different context.

Given discrete times $t_n = n\tau, n \in \mathbb{N}_0$, where $\tau > 0$ is a given time step, and denoting by u^n an approximation of $u(t_n, \cdot)$, the backward Euler discretization of the gradient flow equation (1.3) is given as

$$\tfrac{1}{\tau}(u^n - u^{n-1}) = -\mathrm{grad}_{\mathbf{Z}} E(u(t))\,. \tag{1.13}$$

This time discretization has a natural variational structure. In fact one can compute a solution u^n as the solution of the minimization problem

$$\min_{|u|\le 1} \{E(u) + \tfrac{1}{2\tau}\|u - u^{n-1}\|^2_{\mathbf{Z}}\}\,. \tag{1.14}$$

One hence tries to decrease the energy E whilst taking into account the fact that deviations from the solution at the old time step costs, where the cost depends on the norm on **Z**.

In particular for the Cahn-Hilliard formulation we obtain a non-standard PDE-constraint optimization problem as follows

$$\min \left\{ \tfrac{\gamma\varepsilon}{2}\int_\Omega |\nabla u|^2 + \tfrac{\gamma}{\varepsilon}\int_\Omega \psi_0(u) + \tfrac{\tau}{2}\int_\Omega |\nabla v|^2 \right\} \tag{1.15}$$

such that

$$\begin{aligned} \Delta v &= \tfrac{1}{\tau}(u - u^{n-1}) \text{ a.e.}\,, \\ |u| &\le 1 \text{ a.e.}\,,\ \textstyle\fint_\Omega u = m\,, \end{aligned} \tag{1.16}$$

with

$$\tfrac{\partial v}{\partial n} = 0 \ \text{ a.e. on } \partial\Omega \ \text{ and } \ \textstyle\fint_\Omega v = 0.$$

This formulation has the form of an optimal control problem where u is the control and v is the state. In particular one has the difficulty that L^2-norms of gradients enter the cost functional. We also remark that the non-local mean value constraints appear since the Cahn-Hilliard evolution variational inequality conserves mass. It turns out that the Lagrange multiplier w for the PDE equality constraint (1.16) equals v up to a constant, see also [16]. We hence obtain the reduced KKT-system (1.17)–(1.19), see [4] for details:

$$\frac{1}{\tau}(u - u^{n-1}) \;=\; \Delta w \text{ in } \Omega,\ \tfrac{\partial w}{\partial n} = 0 \text{ on } \partial\Omega\,, \tag{1.17}$$

$$w + \gamma\varepsilon\Delta u - \frac{\gamma}{\varepsilon}\psi_0'(u) - \frac{1}{\varepsilon}\mu \;=\; 0 \quad \text{in } \Omega,\ \tfrac{\partial u}{\partial n} = 0 \text{ on } \partial\Omega\,, \tag{1.18}$$

$$u(x) = \pm 1 \quad \text{if } x \in A_{\pm}\,, \qquad \mu(x) = 0 \quad \text{if } x \in I\,, \tag{1.19}$$

where all equalities and inequalities have to hold almost everywhere. This system is a time discretization of the Cahn-Hilliard model (1.8), (1.11). We obtain a corresponding system for the time discretized Allen-Cahn variational inequality.

2. Primal-dual active set method (PDAS method)

The idea is now to apply the PDAS-algorithm (see below) to (1.14). However as is known for control problems with state constraint and for obstacle problems this strategy is not applicable in functions space as the iterates for the Lagrange multiplier μ are in general only measures. Therefore we apply the method to the fully discretized problems. Since we consider here evolution processes, where good preinitialization is available from the previous time steps, we avoid additional regularization or penalization techniques (see [19, 22, 23]) and still numerically obtain mesh independence.

We now use a finite element approximation in space with piecewise linear, continuous finite elements S_h with nodes $p_1, \ldots, p_{J_h}$ and nodal basis function $\chi_j \in S_h, j \in \mathcal{J}_h := \{1, \ldots, J_h\}$, and introduce a mass lumped inner product $(.,.)_h$. We can then formulate a discrete primal-dual active set method for iterates $(u^{(k)}, \mu^{(k)}) \in S_h \times S_h$ based on active nodes with indices $\mathcal{A}_{\pm}^{(k)}$ and inactive nodes with indices $\mathcal{I}^{(k)}$ as follows.

Primal-Dual Active Set Algorithm (PDAS):

1. Set $k = 0$, initialize $\mathcal{A}_{\pm}^{(0)}$ and define $\mathcal{I}^{(0)} = \mathcal{J}_h \setminus (\mathcal{A}_{+}^{(0)} \cup \mathcal{A}_{-}^{(0)})$.
2. Set $u^{(k)}(p_j) = \pm 1$ for $j \in \mathcal{A}_{\pm}^{(k)}$ and $\mu^{(k)}(p_j) = 0$ for $j \in \mathcal{I}^{(k)}$.
3. Solve the fully discretized version of the coupled system of PDEs (1.17)–(1.18), respectively of the system (1.7) to obtain $u^{(k)}(p_j)$ for $j \in \mathcal{I}^{(k)}$, $\mu^{(k)}(p_j)$ for $j \in \mathcal{A}_{+}^{(k)} \cup \mathcal{A}_{-}^{(k)}$ and $w^{(k)} \in S_h$.
4. Set $\mathcal{A}_{+}^{(k+1)} := \{j \in \mathcal{J}_h \mid c(u^{(k)}(p_j) - 1) + \mu^{(k)}(p_j) > 0\}$,
   $\mathcal{A}_{-}^{(k+1)} := \{j \in \mathcal{J}_h \mid c(u^{(k)}(p_j) + 1) + \mu^{(k)}(p_j) < 0\}$ and
   $\mathcal{I}^{(k+1)} := \mathcal{J}_h \setminus (\mathcal{A}_{+}^{(k+1)} \cup \mathcal{A}_{-}^{(k+1)})$.
5. If $\mathcal{A}_{\pm}^{(k+1)} = \mathcal{A}_{\pm}^{(k)}$ stop, otherwise set $k = k + 1$ and goto 2.

In the above algorithm Step 3 can be split into two steps. The first is to solve for u and w and the second is to determine μ. We give the details only for the

Cahn-Hilliard problem. For the Allen-Cahn formulation there is a corresponding system.

3a. Solve for $w^{(k)} \in S_h$ and $u^{(k)}(p_j)$ with $j \in \mathcal{I}^{(k)}$, the system

$$\begin{aligned} \tfrac{1}{\tau}(u^{(k)} - u_h^{n-1}, \chi)_h + (\nabla w^{(k)}, \nabla \chi) &= 0 \qquad \forall\, \chi \in S_h\,, &(2.1)\\ (w^{(k)}, \tilde{\chi})_h - \gamma\varepsilon(\nabla u^{(k)}, \nabla \tilde{\chi}) - \tfrac{\gamma}{\varepsilon}(\psi_0'(u_h^*), \tilde{\chi})_h &= 0 \qquad \forall\, \tilde{\chi} \in \tilde{S}^{(k)} &(2.2)\\ \text{with } \tilde{S}^{(k)} := \operatorname{span}\{\chi_i \mid i \in \mathcal{I}^{(k)}\}\,. \end{aligned}$$

3b. Define $\mu^{(k)}$ on the active sets such that for all $j \in \mathcal{A}^{(k)}$

$$\mu^{(k)}(p_j)\,(1, \chi_j)_h = (\varepsilon w^{(k)} - \gamma\psi_0'(u_h^*), \chi_j)_h - \gamma\varepsilon^2(\nabla u^{(k)}, \nabla \chi_j). \qquad (2.3)$$

In the above we either consider an implicit or an explicit discretization of the term $\psi_0'(u)$, i.e., we choose $\psi_0'(u_h^*)$ where $* \in \{n-1, n\}$. Figure 1 shows the structure of the system. The discretized elliptic equation (2.1) for $w^{(k)}$ is defined on the whole of $\mathcal{J}_h$ whereas the discrete elliptic equation (2.2) is defined only on the inactive set corresponding to $\mathcal{I}^{(k)}$ which is an approximation of the interface. The two equations are coupled in a way which leads to an overall symmetric system which will be used later when we propose numerical algorithms. For the Allen-Cahn system we have to solve (1.7) only on the approximated interface.

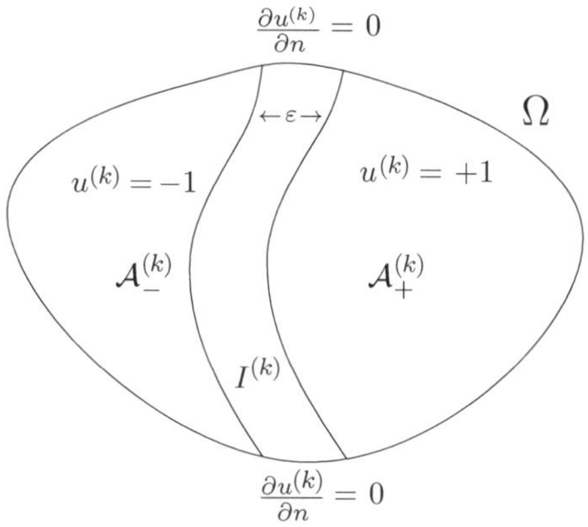

FIGURE 1. Structure of active and inactive sets.

3. Results for the Cahn-Hilliard variational inequality

It can be shown, see [4] for details, that the following results hold.

Lemma 3.1. *For all $u_h^{n-1} \in S_h$ and $\mathcal{A}_\pm^{(k)}$ there exists a unique solution $(u^{(k)}, w^{(k)}) \in S^h \times S^h$ of (2.1)-(2.2) with $* = (n-1)$ provided that $\mathcal{I}^{(k)} = \mathcal{J}_h \setminus (\mathcal{A}_+^{(k)} \cup \mathcal{A}_-^{(k)}) \neq \emptyset$.*

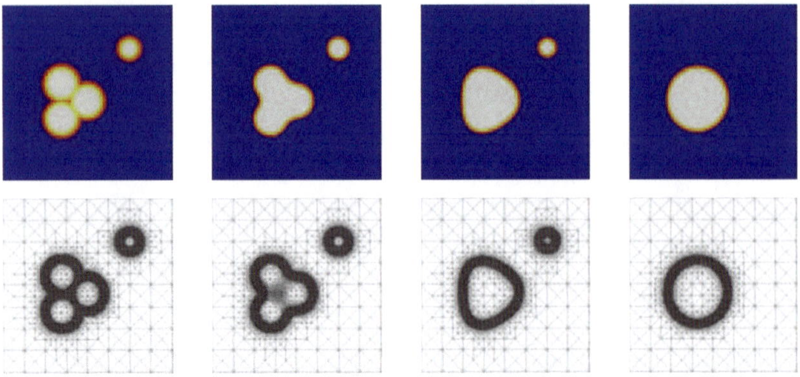

FIGURE 2. The upper row shows the evolution of the concentration and the lower row shows the corresponding meshes.

The assumption $\mathcal{I}^{(k)} \neq \emptyset$ guarantees that the condition $\fint_\Omega u^{(k)} = m$ can be fulfilled. Otherwise (2.1) may not be solvable. Furthermore we have shown in [4] using the equivalence of the PDAS-algorithm to a semi-smooth Newton method:

Theorem 3.2. *The primal-dual active set algorithm (PDAS) converges locally.*

Global convergence is not of large interest here, as we study a discrete time evolution and hence we always have good starting values from the previous time-step. However, the appropriate scaling of the Lagrange multiplier μ by $\frac{1}{\varepsilon}$, or respectively the choice of the parameter c is essential to avoid oscillatory behaviour due to bilateral constraints (see [4, 5]).

As far as we can compare the results with other methods the PDAS-method outperformed previous approaches, see [4]. One of the bottle necks for a speed-up is the linear algebra solver. The linear system to be solved is symmetric and has a saddle point structure. Efficient preconditioning of the system is difficult and has to be addressed in the future.

Finally let us mention further methods to solve Cahn-Hilliard variational inequalities. It is also possible to use a projected block Gauss-Seidel type scheme to solve the variational inequality directly and hence at each node a variational inequality for a vector with two components has to be solved. Another approach is a splitting method due to Lions and Mercier [21] (see [9], [4]) and Gräser and Kornhuber [16] use preconditioned Uzawa-type iterations coupled to a monotone multigrid method. The latter approach is similar to our approach as it is also an active set approach. Although, unlike our approach, Gräser and Kornhuber [16] have to solve a second-order variational inequality in each step in order to update the active set. Finally, we also mention a recent multigrid method of Banas and Nürnberg [1] and an approach based on regularization by Hintermüller, Hinze and Tber [18].

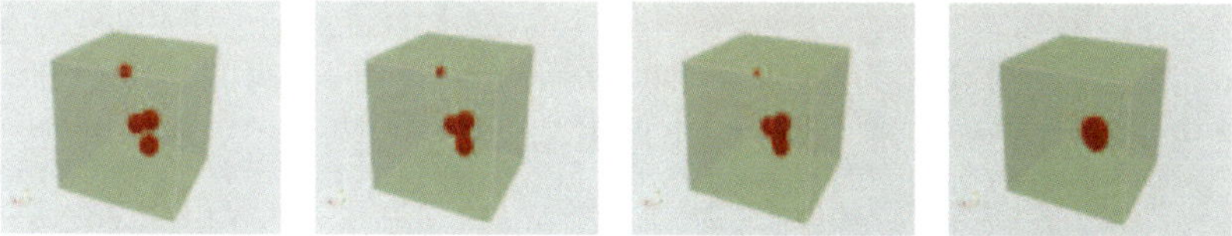

(a) $n = 0$ (b) $n = 50$ (c) $n = 100$ (d) $n = 300$

FIGURE 3. 3d simulation with 4 spheres as initial data on an adaptive mesh.

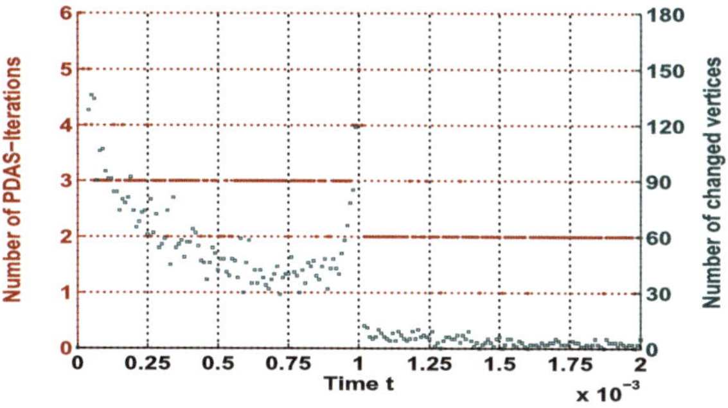

FIGURE 4. PDAS-iterations count per time step for the 2d simulation in Figure 2.

4. Results for the Allen-Cahn variational inequality

4.1. Scalar Allen-Cahn problem without local constraints

If we consider interface evolution given by the Allen-Cahn variational inequality (1.4) we obtain corresponding results to the Cahn-Hilliard problem. However, the L^2-inner product does not conserve the mass, and hence, e.g., given circles or spheres as initial data they will vanish in finite time. For the example of a shrinking circle we discuss the issue of mesh independence of the PDAS-method applied to the fully discretized problem. The number of PDAS iterations might depend on the mesh size h. There is still a lack of analytical results. However, we can investigate this numerically comparing various uniform meshes of a maximal diameter h. We choose the radius 0.45 of the circle at $t = 0$. The time where the circle disappears is 0.10125. Table 1 shows the average number of PDAS iterations up to $t = 0.03$ for fixed $\varepsilon = \frac{1}{16\pi}$. In the third column we list the results fixing also $\tau = 5 \cdot 10^{-4}$. Although the number of PDAS iterations increases for smaller mesh size, this increase is only by a factor of approximately 1.3. However, in our applications the time step τ and the space discretization are in general coupled. Hence it is more appropriate to look at the number of Newton iterations when both τ and h are driven to zero. We see in the last column that the iteration number is almost constant. This is due to the time evolution, since good initial data on the current time step are given from the solution of the previous time step. Hence our numerical investigations indicate that the proposed method is mesh independent for this setting. We remark that Table 1 shows the average number of Newton iterations. There might be situations, in particular if one starts with irregular initial data, where a slight dependence of the number of Newton iterations on the time step size is possible for certain time steps. In all our simulations we hardly observed a mesh dependence of the number of Newton iterations even for irregular initial data if time step and space step size were reduced simultaneously. An analytical result showing mesh independence for arbitrary initializations is however not available.

h	DOFs	PDAS iter. ø for $\tau = 5 \cdot 10^{-4}$	τ	PDAS iter. ø for varying τ
1/128	66049	2.57	$1 \cdot 10^{-3}$	3.20
1/256	263169	3.10	$5 \cdot 10^{-4}$	3.10
1/512	1050625	4.02	$2.5 \cdot 10^{-4}$	3.30
1/1024	4198401	5.18	$1.25 \cdot 10^{-4}$	3.37

TABLE 1. Average number of PDAS iterations.

4.2. Scalar Allen-Cahn problem with mass constraint

While the Cahn-Hilliard approach incorporates mass conservation by using the H^{-1}-norm, we can also use the L^2-norm and enforce in addition the mass conservation as a non-local constraint during the gradient flow. This leads to the Allen-Cahn variational inequality (1.4) with the additional constraint $\fint_\Omega u dx := \frac{1}{|\Omega|}\int_\Omega u dx = m$ where $m \in (-1, 1)$ is the mean value of the initial data u_0. We can introduce for this constraint a Lagrange multiplier λ and we can restate the problem as

$$\varepsilon \partial_t u = \gamma \varepsilon \Delta u - \tfrac{1}{\varepsilon}(\gamma \psi_0'(u) + \mu - \lambda) \text{ a.e.}, \tag{4.1}$$

$$\fint_\Omega u dx = m \qquad \text{for almost all } t \in [0, T], \tag{4.2}$$

where also the complementarity conditions (1.9)–(1.10) hold.

Using a penalty approach for the inequality constraints and projecting the mass constraint we have shown in [5] the existence, uniqueness and regularity of the solution of the KKT-system which is non-standard due to the coupling of non-local equality and local inequality constraints.

Theorem 4.1. *Let $T > 0$ and Ω be a domain which is bounded and either convex or has a $C^{1,1}$-boundary. Furthermore the initial data $u_0 \in H^1(\Omega)$ fulfill $|u_0| \leq 1$ a.e. and $\fint_\Omega u_0 = m$ for a given $m \in (-1, 1)$. Then there exists a unique solution (u, μ, λ) of the KKT-system (4.1), (4.2), (1.9), (1.10) with $\mu \in L^2(\Omega_T)$, $\lambda \in L^2(0, T)$ and $u \in L^2(0, T; H^2(\Omega)) \cap L^\infty(0, T; H^1(\Omega)) \cap H^1(\Omega_T)$.*

Using the presented implicit time discretization (1.13) and the given finite element approximation with mass lumping we apply, similar as above, a PDAS algorithm. We define $m_j := (1, \chi_j)$ and $a_{ij} = (\nabla \chi_j, \nabla \chi_i)$ and denote by $u_j^{(k)}$ the coefficients of $u^{(k)} = \sum_{j \in \mathcal{J}_h} u_j^{(k)} \chi_j$. Then we obtain as Step 3:

3a. Solve for $u_j^{(k)}$ for $j \in \mathcal{I}^{(k)}$ and $\lambda^{(k)}$:

$$(\tfrac{\varepsilon}{\tau} - \tfrac{\gamma}{\varepsilon}) m_j u_j^{(k)} + \gamma\varepsilon \sum_{i \in \mathcal{I}^{(k)}} a_{ij} u_i^{(k)} - \tfrac{1}{\varepsilon} m_j \lambda^{(k)} \\ = \tfrac{\varepsilon}{\tau} m_j u_j^{n-1} + \gamma\varepsilon \Big(\sum_{i \in \mathcal{A}_-^{(k)}} a_{ij} - \sum_{i \in \mathcal{A}_+^{(k)}} a_{ij} \Big) \quad \forall j \in \mathcal{I}^{(k)}, \tag{4.3}$$

$$\sum_{i \in \mathcal{I}^{(k)}} m_i u_i^{(k)} = m \sum_{i \in \mathcal{J}_h} m_i - \sum_{i \in \mathcal{A}_+^{(k)}} m_i + \sum_{i \in \mathcal{A}_-^{(k)}} m_i . \tag{4.4}$$

3b. Define $\mu_j^{(k)}$ for $j \in \mathcal{A}_\pm^{(k)}$ using:

$$\mu_j^{(k)} = -(\tfrac{\varepsilon^2}{\tau} - \gamma) u_j^{(k)} - \gamma\varepsilon^2 \tfrac{1}{m_j} \sum_{i \in \mathcal{J}_h} a_{ij} u_i^{(k)} + \lambda^{(k)} + \tfrac{\varepsilon^2}{\tau} u_j^{n-1} . \tag{4.5}$$

Hence the main effort when applying the PDAS algorithm is solving the system (4.3)–(4.4) where the size $|\mathcal{I}^{(k)}| + 1$ is given by the size of the approximated interface.

Like in the Cahn-Hilliard case we can show local convergence of the method by interpreting the algorithm as a Newton method. However using the presented implicit time discretization we obtain analytically the following restriction on the time step (see [5] Theorem 4.2):

$$\tau \Big(1 - \tfrac{\varepsilon^2}{c_h^p} \Big) < \tfrac{\varepsilon^2}{\gamma} \tag{4.6}$$

where $c_h^p > 0$ is the Poincaré constant such that $(v,v)_h \le c_h^p (\nabla v, \nabla v)$ for all $v \in S^h$ with $\int_\Omega v = 0$ and $v(p_j) = 0$ for active nodes p_j. In [5] the size of c_h^p is discussed in more detail. For example in one dimension given a good numerical approximation of $\mathcal{I}$ no restriction at all has to be enforced for the time step τ. We can also use a semi-implicit discretization with a primal-dual active set algorithm. In this case no time restrictions have to be enforced for the algorithm. However it turns out that the fully implicit time discretization is much more accurate [5].

We give two numerical simulations. In Figure 5 we show interface evolution in two dimensions where the initial phase distribution is random and no pure phases are present. Already at time $t = 0.002$ grains start to form and grow and at $t = 0.003$ we have two phases separated by a diffuse interface. Now the interface moves according to motion by mean curvature but preserving the volume of both phases. That means that closed curves turn into circles and shapes with less volume shrink and disappear while at the same time shapes with the highest volume will grow. At the end (i.e., when the problem becomes stationary) there are three different shapes we can obtain: a circle, a quarter of a circle in one of the corners (see Figure 5) or a straight vertical or horizontal line dividing the two phases. The computation in Figure 6 presents the evolution for a dumbbell. Without the volume conservation the dumbbell would dissect and the two spheres

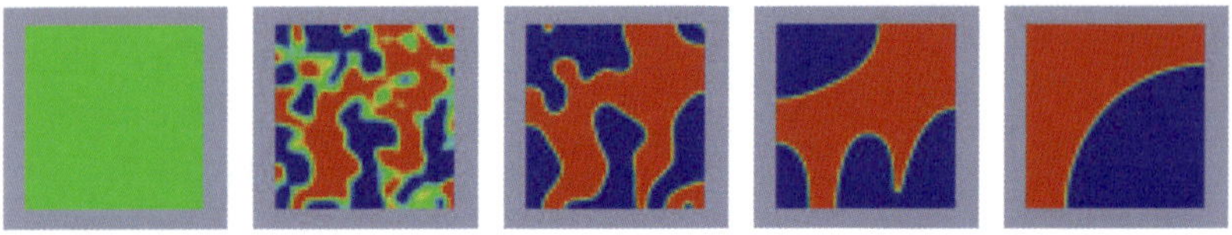

$t = 0.000$ $t = 0.003$ $t = 0{,}010$ $t = 0.090$ $t = 3.000$

FIGURE 5. Volume controlled Allen-Cahn equation (2d) with random initial data (varying between -0.1 and 0.1).

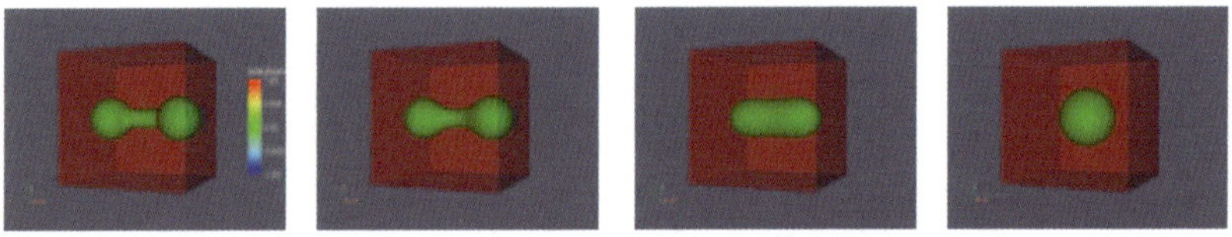

$t = 0.001$ $t = 0.010$ $t = 0.100$ $t = 0.500$

FIGURE 6. Volume controlled Allen-Cahn equation with a dumbbell as initial data.

would shrink and disappear. The volume conservation forces the dumbbell to turn into an ellipsoid before turning into a sphere and finally becoming stationary.

Finally, we briefly would like to mention that our approach can be used to solve problems in structural topology optimization. There the mean compliance penalized with the Ginzburg-Landau energy E (1.1) has to be minimized. The gradient approach can be seen as a pseudo time stepping approach and results in a time discretized Allen-Cahn variational inequality coupled with elasticity and mass constraints, which can be solved with the above method (see [5, 6, 3]).

4.3. Systems of Allen-Cahn variational inequalities

In many applications more than two phases or materials appear, see [11, 13] and the references therein. For numerical approaches to systems of Allen-Cahn variational inequalities we refer to [14, 15, 13] where explicit in time discretizations have been used, and to the work of Kornhuber and Krause [20] who discuss Gauss-Seidel and multigrid methods. In what follows we want to introduce a primal-dual active set method for systems of Allen-Cahn variational inequalities which in contrast to earlier approaches does not need an explicit handling of the geometry of the Gibbs simplex.

Therefore we introduce a concentration vector $\mathbf{u} = (u_1, \ldots, u_N)^T : \Omega \to \mathbb{R}^N$ with the property $u_i \geq 0$, $\sum_{i=1}^N u_i = 1$, i.e., $\mathbf{u}(x,t)$ lies on the Gibbs simplex

$$G := \{\boldsymbol{\xi} \in \mathbb{R}^N : \boldsymbol{\xi} \geq \mathbf{0}, \boldsymbol{\xi} \cdot \mathbf{1} = 1\}.$$

For the bulk potential $\psi : \mathbb{R}^N \to \mathbb{R}_0^+ \cup \{\infty\}$ we consider the multi-obstacle potential

$$\psi(\boldsymbol{\xi}) = \begin{cases} \psi_0(\boldsymbol{\xi}) := -\frac{1}{2}\boldsymbol{\xi} \cdot \mathbf{A}\boldsymbol{\xi} & \text{for } \boldsymbol{\xi} \geq \mathbf{0}, \boldsymbol{\xi} \cdot \mathbf{1} = 1, \\ \infty & \text{otherwise,} \end{cases} \tag{4.7}$$

with $\mathbf{A}$ being a symmetric constant $N \times N$ matrix. We remark that different phases which correspond to minima of ψ only occur if $\mathbf{A}$ has at least one positive eigenvalue. The total underlying non-convex energy is given similar to (1.1) by

$$E(\mathbf{u}) := \int_\Omega \left(\frac{\gamma\varepsilon}{2} |\nabla \mathbf{u}|^2 + \frac{\gamma}{\varepsilon} \psi(\mathbf{u}) \right) dx \,.$$

We also consider systems in which the total spatial amount of the phases are conserved. In this case one studies the steepest descent of E under the constraint $\fint_\Omega \mathbf{u}\, dx = \mathbf{m} = (m^1, \ldots, m^N)^T$ where $m^i \in (0,1)$ for $i \in \{1, \ldots, N\}$ is a fixed number. We now define

$$\mathcal{G}^{\mathbf{m}} := \{\mathbf{v} \in H^1(\Omega, \mathbb{R}^N) : \fint_\Omega \mathbf{v} = \mathbf{m}, \sum_{i=1}^N v_i = 1, \mathbf{v} \geq \mathbf{0}\}$$

and note that for $\mathbf{u} \in \mathcal{G}^{\mathbf{m}}$ it follows $\fint_\Omega \mathbf{u} - \mathbf{m} \in S := \{\mathbf{v} \in \mathbb{R}^N : \sum_{i=1}^N v_i = 0\}$. Then the interface evolution with mass conservation can be formulated as the following variational inequality: For given $\mathbf{u}_0 \in \mathcal{G}^{\mathbf{m}}$ find $\mathbf{u} \in L^2(0,T; \mathcal{G}^{\mathbf{m}}) \cap H^1(0,T; \mathbf{L}^2(\Omega))$ such that $\mathbf{u}(.,0) = \mathbf{u}_0$ and such that for almost all $t \in (0,T)$ it holds

$$\varepsilon(\tfrac{\partial \mathbf{u}}{\partial t}, \boldsymbol{\chi} - \mathbf{u}) + \gamma\varepsilon(\nabla \mathbf{u}, \nabla(\boldsymbol{\chi} - \mathbf{u})) - \tfrac{\gamma}{\varepsilon}(\mathbf{A}\mathbf{u}, \boldsymbol{\chi} - \mathbf{u}) \geq 0 \quad \forall \boldsymbol{\chi} \in \mathcal{G}^{\mathbf{m}} \,. \tag{4.8}$$

Our numerical approach again depends on a reformulation of (4.8) with the help of Lagrange multipliers. We introduce Lagrange multipliers $\boldsymbol{\mu}$ and Λ corresponding to the constraints $\mathbf{u} \geq \mathbf{0}$ and $\sum_{i=1}^N u_i = 1$ respectively. Taking into account the condition $\sum_{i=1}^N u_i = 1$ we use for the mass constraints the projected version $\mathbf{P}_S(\fint_\Omega \mathbf{u} - \mathbf{m}) = \mathbf{0}$, where $\mathbf{P}_S$ is a projection onto S and introduce for this condition a Lagrange multiplier $\boldsymbol{\lambda} \in S$. In [6] we prove the following theorem in which $\mathbf{L}^2(\Omega)$, $\mathbf{H}^1(\Omega)$, etc. denote spaces of vector-valued functions.

Theorem 4.2. *Let $\Omega \subset \mathbb{R}^d$ be a bounded domain and assume that either Ω is convex or $\partial\Omega \in C^{1,1}$. Let $\mathbf{u}_0 \in \mathcal{G}^{\mathbf{m}}$ such that $\int_\Omega \mathbf{u}_0 > \mathbf{0}$. Then there exists a unique solution $(\mathbf{u}, \boldsymbol{\mu}, \boldsymbol{\lambda}, \Lambda)$ with*

$$\mathbf{u} \in L^\infty(0,T; \mathbf{H}^1(\Omega)) \cap H^1(0,T; \mathbf{L}^2(\Omega)) \cap L^2(0,T; \mathbf{H}^2(\Omega)), \tag{4.9}$$

$$\boldsymbol{\mu} \in L^2(0,T; \mathbf{L}^2(\Omega)), \tag{4.10}$$

$$\boldsymbol{\lambda} \in \mathbf{L}^2(0,T) \text{ and } \sum_{i=1}^N \lambda_i = 0 \text{ for almost all } t \in (0,T), \tag{4.11}$$

$$\Lambda \in L^2(0,T; L^2(\Omega)) \tag{4.12}$$

such that on $\Omega_T := \Omega \times (0,T)$ *we have*

$$\varepsilon \tfrac{\partial \mathbf{u}}{\partial t} - \gamma\varepsilon\Delta\mathbf{u} - \tfrac{\gamma}{\varepsilon}\mathbf{A}\mathbf{u} - \tfrac{1}{\varepsilon}\boldsymbol{\mu} - \tfrac{1}{\varepsilon}\Lambda\mathbf{1} - \tfrac{1}{\varepsilon}\boldsymbol{\lambda} = \mathbf{0} \quad a.e.\ in\ \Omega_T, \tag{4.13}$$

$$\mathbf{u}(0) = \mathbf{u}_0,\ \tfrac{\partial \mathbf{u}}{\partial \nu} = \mathbf{0} \qquad a.e.\ on\ \partial\Omega \times (0,T), \tag{4.14}$$

$$\sum_{i=1}^{N} u_i = 1,\ \mathbf{u} \geq \mathbf{0},\ \boldsymbol{\mu} \geq \mathbf{0} \quad a.e.\ in\ \Omega_T, \tag{4.15}$$

$$\mathbf{P}_S(\textstyle\fint_\Omega \mathbf{u} - \mathbf{m}) = \mathbf{0},\ (\boldsymbol{\mu}, \mathbf{u}) = 0 \quad for\ almost\ all\ t. \tag{4.16}$$

The proof is based on a penalty approach where the main difficulty is to show that approximations of the Lagrange multipliers $\boldsymbol{\mu}, \boldsymbol{\lambda}$ and Λ can be bounded. This issue is related to the question whether the constraints are in a suitable sense independent of each other. In order to show that Lagrange multipliers are unique one has to use some graph theory in order to show that the undirected graph with vertices $\{1,\ldots,N\}$ and edges $\{\{i,j\} \mid$ there is an interface between i and $j\}$ is connected.

Similar to the previous sections we now discretize (4.13)–(4.16) in time and space and we use a PDAS algorithm. However, for each component u_i we have to consider its own active and inactive sets $\mathcal{A}_i := \{j \in \mathcal{J}_h \mid c(u_i)_j + (\mu_i)_j < 0\}$ and $\mathcal{I}_i := \mathcal{J}_h \setminus \mathcal{A}_i$. In the following we use the notation

$$\mathbf{u}^{(k)} = \sum_{i=1}^{N} \sum_{j \in \mathcal{J}_h} (u_i^{(k)})_j \chi_j \mathbf{e}_i$$

for the kth iterate $\mathbf{u}^{(k)} \in (S_h)^N$ in the vector-valued PDAS algorithm.

Primal-Dual Active Set Algorithm (PDAS-Vector):

1. Set $k = 0$, initialize $\mathcal{A}_i^{(0)}$ and define $\mathcal{I}_i^{(0)} = \mathcal{J}_h \setminus \mathcal{A}_i^{(0)}$ for all $i \in \{1,\ldots,N\}$.
2. Set $(u_i^{(k)})_j = 0$ for $j \in \mathcal{A}_i^{(k)}$ and $(\mu_i^{(k)})_j = 0$ for $j \in \mathcal{I}_i^{(k)}$ for all $i \in \{1,\ldots,N\}$.

3a. To obtain $(\Lambda^{(k)})_j$ for all $j \in \mathcal{J}_h$, $\lambda_i^{(k)}$ for all $i \in \{1,\ldots,N-1\}$ and $(u_i^{(k)})_j$ for all $i = 1,\ldots,N$ and $j \in \mathcal{I}_i^{(k)}$ we solve

$$\tfrac{\varepsilon^2}{\tau}(u_i^{(k)})_j - \gamma \sum_{m=1}^{N} A_{im}(u_m^{(k)})_j + \tfrac{\gamma\varepsilon}{m_j} \sum_{l \in \mathcal{I}_i^{(k)}} a_{lj}(u_i^{(k)})_l - [\lambda_i^{(k)} + (\Lambda^{(k)})_j]$$

$$= \tfrac{\varepsilon^2}{\tau}(u_i^{n-1})_j, \quad i = 1,\ldots,N, j \in \mathcal{I}_i^{(k)}, \tag{4.17}$$

$$\sum_{j \in \mathcal{J}_h} m_j((u_i^{(k)})_j - (u_N^{(k)})_j) = \sum_{j \in \mathcal{J}_h} m_j(m^i - m^N),\ i = 1,\ldots,N-1,$$

$$\sum_{i=1}^{N} (u_i^{(k)})_j = 1, \quad j \in \mathcal{J}_h, \tag{4.18}$$

where we replace $\lambda_N^{(k)}$ by $\lambda_N^{(k)} = -\lambda_1^{(k)} - \cdots - \lambda_{N-1}^{(k)}$.

$t = 0.001$ $t = 0.010$ $t = 0.040$ $t = 0.070$ $t = 0.080$

FIGURE 7. Vector-valued Allen-Cahn variational inequality with a Voronoi partitioning as initial data (30 order parameters).

3b. Set $\lambda_N^{(k)} = -\lambda_1^{(k)} - \cdots - \lambda_{N-1}^{(k)}$ and determine the values

$$(\mu_i^{(k)})_j = \tfrac{\varepsilon^2}{\tau}(u_i^{(k)})_j - \gamma \sum_{m=1}^{N} A_{im}(u_m^{(k)})_j + \tfrac{\gamma\varepsilon^2}{m_j} \sum_{l \in \mathcal{J}_h} a_{lj}(u_i^{(k)})_l$$
$$- \lambda_i^{(k)} - \Lambda_j^{(k)} - \tfrac{\varepsilon^2}{\tau}(u_i^{n-1})_j$$

for $i = 1, \ldots, N$ and $j \in \mathcal{A}_i^{(k)}$.

4. Set $\mathcal{A}_i^{(k+1)} := \{j \in \mathcal{J}_h : c(u_i^{(k)})_j - (\mu_i^{(k)})_j < 0\}$ and $\mathcal{I}_i^{(k+1)} := \mathcal{J}_h \setminus \mathcal{A}_i^{(k+1)}$.
5. If $\mathcal{A}_i^{(k+1)} := \mathcal{A}_i^{(k)}$ for all $i \in \{1, \ldots, N\}$ stop, otherwise set $k = k + 1$ and goto 2.

Remark 4.1. i) In each node p_j for $j \in \mathcal{J}_h$ some components are active and the others are inactive. The number of components which are active can vary from point to point. Only for each individual component can we split the set of nodes into nodes which are active (for this component) and its complement. The resulting linear system is hence quite complex but can be solved efficiently with the help of MINRES, see [7].

ii) There is a straightforward variant of (PDAS-Vector) without mass constraints. In this case we omit the first conditions in (4.18) and the Lagrange multipliers $\boldsymbol{\lambda} \in S$.

In Figure 7 we use a Voronoi partitioning algorithm to randomly define initial data in a 2d computational domain. We use 30 order parameters for this computation and show the time evolution in Figure 7. In Figure 8 we begin the computation with a sphere that is divided into three equal spherical wedges. Each of these wedges is represented by a different phase, i.e., we have three phases in the sphere and one phase outside. The evolution converges for large times to a triple bubble which is the least area way to separate three given volumes.

$t = 0.001$ $t = 0.020$ $t = 0.040$ $t = 0.500$

FIGURE 8. Triple bubble; vector-valued Allen-Cahn with volume constraints, 4 order parameter.

References

[1] L. Banas and R. Nürnberg, *A multigrid method for the Cahn-Hilliard equation with obstacle potential.* Appl. Math. Comput. **213** (2009), 290–303.

[2] J.W. Barrett, H. Garcke, and R. Nürnberg, *On sharp interface limits of Allen-Cahn/Cahn-Hilliard variational inequalities.* Discrete Contin. Dyn. Syst. S **1** (1) (2008), 1–14.

[3] L. Blank, H. Garcke, L. Sarbu, T. Srisupattarawanit, V. Styles, and A. Voigt, *Phase-field approaches to structural topology optimization*, a contribution in this book.

[4] L. Blank, M. Butz, and H. Garcke, *Solving the Cahn-Hilliard variational inequality with a semi-smooth Newton method*, ESAIM: Control, Optimization and Calculus of Variations, DOI: 10.1051/COCV/2010032.

[5] L. Blank, H. Garcke, L. Sarbu, and V. Styles, *Primal-dual active set methods for Allen-Cahn variational inequalities with non-local constraints.* DFG priority program 1253 "Optimization with PDEs", Preprint No. 1253-09-01.

[6] L. Blank, H. Garcke, L. Sarbu, and V. Styles, *Non-local Allen-Cahn systems: Analysis and a primal dual active set method.* Preprint No. 02/2011, Universität Regensburg.

[7] L. Blank, L. Sarbu, and M. Stoll, *Preconditioning for Allen-Cahn variational inequalities with non-local constraints.* Preprint Nr.11/2010, Universität Regensburg, Mathematik.

[8] J.F. Blowey and C.M. Elliott, *The Cahn-Hilliard gradient theory for phase separation with nonsmooth free energy I. Mathematical analysis.* European J. Appl. Math. **2**, no. 3 (1991), 233–280.

[9] J.F. Blowey and C.M. Elliott, *The Cahn-Hilliard gradient theory for phase separation with nonsmooth free energy II. Mathematical analysis.* European J. Appl. Math. **3**, no. 2 (1991), 147–179.

[10] J.F. Blowey and C.M. Elliott, *A phase-field model with a double obstacle potential.* Motion by mean curvature and related topis (Trento, 1992), 1–22, de Gruyter, Berlin 1994.

[11] L. Bronsard and F. Reitich, *On three-phase boundary motion and the singular limit of a vector-valued Ginzburg-Landau equation.* Arch. Rat. Mech. Anal. **124** (1993), 355–379.

[12] L.Q. Chen, *Phase-field models for microstructure evolution.* Annu. Rev. Mater. Res. 32 (2001), 113–140.

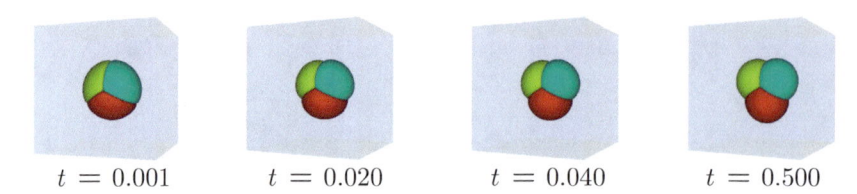

[13] H. Garcke, B. Nestler, B. Stinner, and F. Wendler, *Allen-Cahn systems with volume constraints.* Mathematical Models and Methods in Applied Sciences **18**, no. 8 (2008), 1347–1381.

[14] H. Garcke, B. Nestler, and B. Stoth. *Anisotropy in multi-phase systems: a phase field approach.* Interfaces Free Bound. **1** (1999), 175–198.

[15] H. Garcke and V. Styles. *Bi-directional diffusion induced grain boundary motion with triple junctions.* Interfaces Free Bound. **6**, no. 3 (2004), 271–294.

[16] C. Gräser and R. Kornhuber, *On preconditioned Uzawa-type iterations for a saddle point problem with inequality constraints.* Domain decomposition methods in science and engineering XVI, 91–102, Lect. Notes Comput. Sci. Engl. **55**, Springer, Berlin 2007.

[17] M. Hintermüller, K. Ito, and K. Kunisch, *The primal-dual active set strategy as a semismooth Newton method*, SIAM J. Optim. **13** (2002), no. 3 (2003), 865–888 (electronic).

[18] M. Hintermüller, M. Hinze, and M.H. Tber, *An adaptive finite element Moreau-Yosida-based solver for a non-smooth Cahn-Hilliard problem.* Matheon preprint Nr. 670, Berlin (2009), to appear in Optimization Methods and Software.

[19] K. Ito and K. Kunisch, *Semi-smooth Newton methods for variational inequalities of the first kind.* M2AN Math. Model. Numer. Anal. **37**, no. 1 (2003), 41–62.

[20] R. Kornhuber and R. Krause. *On multigrid methods for vector-valued Allen-Cahn equations with obstacle potential.* Domain decomposition methods in science and engineering, 307–314 (electronic), Natl. Auton. Univ. Mex., México, 2003.

[21] P.-L. Lions and B. Mercier, *Splitting algorithms for the sum of two non-linear operators.* SIAM J. Numer. Anal. **16** (1979), 964–979.

[22] C. Meyer, A. Rösch, and F. Tröltzsch, *Optimal control problems of PDEs with regularized pointwise state constraints.* Comput. Optim. Appl. **33**, (2006), 209–228.

[23] I. Neitzel and F. Tröltzsch, *On regularization methods for the numerical solution of parabolic control problems with pointwise state constraints.* ESAIM: COCV **15** (2) (2009), 426–453.

[24] A. Schmidt and K.G. Siebert, *Design of adaptive finite element software. The finite element toolbar.* ALBERTA, Lecture Notes in Computational Science and Engineering **42**, Springer-Verlag, Berlin 2005.

Luise Blank, Martin Butz and Harald Garcke
Fakultät für Mathematik, Universität Regensburg
D-93040 Regensburg, Germany
e-mail: `luise.blank@mathematik.uni-r.de`
`martin.butz@mathematik.uni-r.de`
`harald.garcke@mathematik.uni-r.de`

Lavinia Sarbu and Vanessa Styles
Department of Mathematics
University of Sussex
Brighton BN1 9RF, UK
e-mail: `ls99@sussex.ac.uk`
`v.styles@sussex.ac.uk`

International Series of Numerical Mathematics, Vol. 160, 37–49

On the Connection Between Forward and Optimization Problem in One-shot One-step Methods

Hans Georg Bock, Andreas Potschka, Sebastian Sager and Johannes P. Schlöder

Abstract. In certain applications of PDE constrained optimization one would like to base an optimization method on an already existing contractive method (solver) for the forward problem. The forward problem consists of finding a feasible point with some parts of the variables (e.g., design variables) held fixed. This approach often leads to so-called simultaneous, all-at-once, or one-shot optimization methods. If only one iteration of the forward method per optimization iteration is necessary, a simultaneous method is called one-step. We present three illustrative linear examples in four dimensions with two constraints which highlight that in general there is only little connection between contraction of forward problem method and simultaneous one-step optimization method. We analyze the asymptotics of three prototypical regularization strategies to possibly recover convergence and compare them with Griewank's One-Step One-Shot projected Hessian preconditioners. We present de facto loss of convergence for all of these methods, which leads to the conclusion that, at least for fast contracting forward methods, the forward problem solver must be used with adaptive accuracy controlled by the optimization method.

Mathematics Subject Classification (2000). 65K05.

Keywords. Inexact SQP, simultaneous, one-shot, all-at-once.

1. Introduction

Many nonlinear problems

$$g(x, y) = 0, \quad x \in \mathbb{R}^n, y \in \mathbb{R}^m, g \in \mathcal{C}^1(\mathbb{R}^{n+m}, \mathbb{R}^n), \tag{1}$$

This work was supported by the German Research Foundation (DFG) within the priority program SPP1253 under grant BO864/12-1, by the Heidelberg Graduate School for Mathematical and Computational Methods in the Sciences, and by the German Federal Ministry of Education and Research (BMBF) under grant 03BONCHD. The authors would like to thank M. Diehl for discussions which started this work and for valuable suggestions to improve this paper.

with fixed y can be successfully solved with Newton-type methods

$$\text{given } x_0, \quad x_{k+1} = x_k - J_k^{-1} g(x_k, y). \tag{2}$$

In most cases, a cheap approximation $J_k \approx \frac{\partial g}{\partial x}(x_k, y)$ with linear contraction rate of, say, $\kappa = 0.8$ is already good enough to produce an efficient numerical method. In general, cheaper computation of the action of J_k^{-1} on the residual compared to the action of $(\frac{\partial g}{\partial x})^{-1}$ must compensate for the loss of locally quadratic convergence of a Newton method to obtain an overall performance gain within the desired accuracy. It is a tempting idea to use the same Jacobian approximations J_k from the Newton-type method in an inexact SQP method for the optimization problem with the same constraint

$$\min_{z=(x,y)\in\mathbb{R}^{n+m}} f(z) \quad \text{s.t. } g(z) = 0. \tag{3}$$

From this point of view we call problem (1) the *forward problem* of optimization problem (3) and we will refer to the variables y as *control* or *design* variables and to x as *state* variables.

Reusing J_k in an inexact SQP method for (3) is usually called *simultaneous*, *all-at-once*, or *one-shot* approach and has proven to be successful for some applications in PDE constrained optimization, e.g., in aerodynamic shape optimization [6, 10], chemical engineering [13], or for a model problem [14].

Any inexact SQP method for equality constrained problems of the form (3) is equivalent to a Newton-type method for the necessary optimality conditions

$$\nabla_z L(z, \lambda) = 0, \quad g(z) = 0,$$

where the Lagrange multiplier λ (the so-called *dual* variable) is an n-vector and the Lagrangian functional is defined as

$$L(z, \lambda) = f(z) + \lambda^{\mathrm{T}} g(z).$$

This leads to a Newton-type iteration for the primal-dual variables $w = (z, \lambda)$

$$w^{k+1} = w^k - \begin{pmatrix} H_k & A_k^{\mathrm{T}} \\ A_k & 0 \end{pmatrix}^{-1} \begin{pmatrix} \nabla_z L(w^k) \\ g(z^k) \end{pmatrix}, \tag{4}$$

where H_k is an approximation of the Hessian of the Lagrangian L and A_k is an approximation of the constraint Jacobian

$$\frac{\mathrm{d}g}{\mathrm{d}z} \approx \begin{pmatrix} A_{1k} & A_{2k} \end{pmatrix}.$$

If $A_{1k} = J_k$ holds, the method is called *one-step* because exactly one step of the solver for the forward and the adjoint problem is performed per optimization iteration.

In this paper, we illustrate with examples that in general only little connection exists between the convergence of Newton-type methods (2) for the forward problem and the convergence of simultaneous one-step inexact SQP methods (4) for the optimization problem, because the coupling of control, state, and dual variables gives rise to an intricate feedback between each other within the optimization.

Griewank [9] discusses that in order to guarantee convergence of the simultaneous optimization method this feedback must be broken up, e.g., by keeping the design y fixed for several optimization steps, or by at least damping the feedback in the update of the design y by the use of "preconditioners" for which he derives a necessary condition for convergence based on an eigenvalue analysis.

Other strategies for enforcing convergence of inexact SQP methods are based on globalization techniques like trust region [12, 11, 15, 8] or line-search [4, 7]. These approaches assume that the approximation of the Jacobian can be made arbitrarily accurate in certain directions, thus violating the one-step principle. The explicit algorithmic control of Jacobian approximations is usually enforced via an adaptively chosen termination criterion for an inner preconditioned Krylov solver for the solution of the linear system in equation (4). In some applications, efficient preconditioners are available which cluster the eigenvalues of the preconditioned system and thus effectively reduce the number of inner Krylov iterations necessary to solve the linear system exactly [1, 2, 3].

In this paper, we are interested in the different but important case where there exists a contractive method for the forward problem (e.g., [6, 10, 13, 14]). If applied to the linearized forward problem, we obtain preconditioners which are contractive, i.e., the eigenvalues of the preconditioned system lie in a ball around 1 with radius less than 1. However, contractive preconditioners do not necessarily lead to clustering of eigenvalues other than at 1. One would like to believe that the contraction property suggests the use of a simultaneous one-step approach. However, we will show that contraction for the forward problem is neither sufficient nor necessary for convergence of the simultaneous one-step method.

The structure of this paper is the following: We shortly review local convergence theory for Newton-type methods in Section 2. In Section 3 we present illustrative, counter-intuitive examples of convergence and divergence for the forward and optimization problem which form the basis for the later investigations on recovery of convergence. We continue with presenting a third example and three prototypical subproblem regularization strategies in Section 4 and perform an asymptotic analysis for large regularization parameters in Section 5. We also show de facto loss of convergence for one of the examples and compare the regularization approaches to Griewank's One-Step One-Shot preconditioner.

2. Convergence of Newton-type methods

In the following, let $F : D \subset \mathbb{R}^n \to \mathbb{R}^n$ be continuously differentiable with Jacobian $J(x)$. Let $M(x)$ be a matrix, usually an approximation of the inverse $J^{-1}(x)$ in the case that $J(x)$ is invertible. In order to investigate the convergence of Newton-type methods of the form

$$\text{given } x^0, \quad x^{k+1} = x^k + \Delta x^k, \quad \text{where } \Delta x^k = -M(x^k)F(x^k),$$

we use a variant of the Local Contraction Theorem (see [5]). Let the set of Newton pairs be defined according to

$$N = \{(x, y) \in D \times D \,|\, y = x - M(x)F(x)\}.$$

We need two conditions on J and M:

Definition 1 (Lipschitz condition: ω-condition). The Jacobian J together with the approximation M satisfy the ω-condition in D if there exists $\omega < \infty$ such that

$$\|M(y)\left(J(x + t(y - x)) - J(x)\right)(x - y)\| \le \omega t \|x - y\|^2, \quad \forall t \in [0, 1], (x, y) \in N.$$

Definition 2 (Compatibility condition: κ-condition). The approximation M satisfies the κ-condition in D if there exists $\kappa < 1$ such that

$$\|M(y)(\mathbb{I} - J(x)M(x))F(x)\| \le \kappa \|x - y\|, \quad \forall (x, y) \in N.$$

Remark 3. If $M(x)$ is invertible, then the κ-condition can also be written as

$$\|M(y)(M^{-1}(x) - J(x))(x - y)\| \le \kappa \|x - y\|, \quad \forall (x, y) \in N.$$

With the constants from the previous two definitions, we define

$$\delta_k = \kappa + (\omega/2)\|\Delta x^k\|$$

and the closed ball

$$D_0 = \overline{\mathrm{B}}(x^0; \|\Delta x^0\|/(1 - \delta_0)).$$

Theorem 4 (Local Contraction Theorem). *Let J and M satisfy the ω-κ-conditions in D and let $x^0 \in D$. If $\delta_0 < 1$ and $D_0 \subset D$, then $x^k \in D_0$ and the sequence $(x^k)_k$ converges to some $x^* \in D_0$ with convergence rate*

$$\|\Delta x^{k+1}\| \le \delta_k \|\Delta x^k\| = \kappa \|\Delta x^k\| + (\omega/2)\|\Delta x^k\|^2.$$

Furthermore, the a priori estimate

$$\|x^{k+j} - x^*\| \le \frac{\delta_k^j}{1 - \delta_k}\|\Delta x^k\| \le \frac{\delta_0^{k+j}}{1 - \delta_0}\|\Delta x^0\|$$

holds. If additionally $M(x)$ is continuous and non-singular in x^*, then

$$F(x^*) = 0.$$

Remark 5. If F is linear we obtain $\omega = 0$ and if furthermore $M(x)$ is constant we can choose a norm such that κ is arbitrarily close to the spectral radius $\sigma_{\mathrm{r}}(\mathbb{I} - MJ)$. Under these assumptions, $\kappa > 1$ leads to divergence. If $\kappa < 1$ and additionally MJ is diagonalizable then the contraction is monotonic without transient behaviour in the first iterations.

3. Illustrative, counter-intuitive examples in low dimensions

Consider the following linear-quadratic optimization problem

$$\min_{z=(x,y)\in\mathbb{R}^{n+m}} \frac{1}{2}z^{\mathrm{T}}Hz, \quad \text{s.t. } A_1x + A_2y = 0 \tag{5}$$

with symmetric positive-definite Hessian H and invertible A_1. The unique solution is $z = 0$. As in Section 1 we approximate A_1 with $\widetilde{A}_1$ such that we obtain a contracting method for the forward problem. Without loss of generality, let $\widetilde{A}_1 = \mathbb{I}$ (otherwise multiply the constraint in (5) with $\widetilde{A}_1^{-1}$ from the left). We shall now have a look at instances of problem (5) with $n = m = 2$. They were constructed by the use of optimization methods, an interesting topic in itself which we want to elaborate in a future paper. We stress that there is nothing obviously pathologic about the following examples. The exact and approximated constraint Jacobians have full rank, the Hessians are symmetric positive-definite, and A_1 is always diagonalizable or even symmetric. We use the denotations

$$A = \begin{pmatrix} A_1 & A_2 \end{pmatrix}, \quad \widetilde{A} = \begin{pmatrix} \widetilde{A}_1 & A_2 \end{pmatrix}, \quad K = \begin{pmatrix} H & A^{\mathrm{T}} \\ A & 0 \end{pmatrix}, \quad \widetilde{K} = \begin{pmatrix} H & \widetilde{A}^{\mathrm{T}} \\ \widetilde{A} & 0 \end{pmatrix}.$$

In all examples, the condition numbers of K and $\widetilde{K}$ are below 600.

3.1. Fast forward convergence, optimization divergence

As a first instance, we investigate problem (5) for the special choice of

$$\begin{pmatrix} H & A^{\mathrm{T}} \end{pmatrix} = \left(\begin{array}{cccc|cc} 0.67 & 0.69 & -0.86 & -0.13 & 1 & -0.072 \\ 0.69 & 19 & 2.1 & -1.6 & -0.072 & 0.99 \\ \hline -0.86 & 2.1 & 1.8 & -0.33 & -0.95 & 0.26 \\ -0.13 & -1.6 & -0.33 & 0.78 & -1.1 & -0.19 \end{array}\right), \tag{Ex1}$$

According to Remark 5 the choice of $\widetilde{A}_1 = \mathbb{I}$ leads to a fast linear contraction rate for the forward problem of

$$\kappa_{\mathrm{F}} = \sigma_{\mathrm{r}}(\mathbb{I} - \widetilde{A}_1^{-1}A_1) = \sigma_{\mathrm{r}}(\mathbb{I} - A_1) \approx 0.077 < 1.$$

However, for the contraction rate of the inexact SQP method with exact Hessian and exact constraint derivative with respect to y, we get

$$\kappa_{\mathrm{O}} = \sigma_{\mathrm{r}}(\mathbb{I} - \widetilde{K}^{-1}K) \approx 1.07 > 1.$$

Thus the full-step inexact SQP method does not have the property of linear local convergence. In fact, it diverges if the starting point z^0 has a non-vanishing component in the direction of any generalized eigenvector of $\mathbb{I} - \widetilde{K}^{-1}K$ corresponding to a Jordan block with diagonal entries greater than 1.

3.2. Forward divergence, fast optimization convergence

Our second example is

$$(H \quad A^{\mathrm{T}}) = \left(\begin{array}{cccc|cc} 17 & 13 & 1.5 & -0.59 & 0.27 & -0.6 \\ 13 & 63 & 7.3 & -4.9 & -0.6 & 0.56 \\ \hline 1.5 & 7.3 & 1.2 & -0.74 & -0.73 & -3.5 \\ -0.59 & -4.9 & -0.74 & 0.5 & -1.4 & -0.0032 \end{array}\right). \tag{Ex2}$$

We obtain

$$\kappa_{\mathrm{F}} = \sigma_{\mathrm{r}}(\mathbb{I} - \widetilde{A}_1^{-1} A_1) \approx 1.20 > 1, \qquad \kappa_{\mathrm{O}} = \sigma_{\mathrm{r}}(\mathbb{I} - \widetilde{K}^{-1} K) \approx 0.014 < 1,$$

i.e., fast convergence of the method for the optimization problem but divergence of the method for the forward problem. From these two examples we see that in general only little can be said about the connection between contraction for the forward and the optimization problem.

4. Subproblem regularization without changing the Jacobian approximation

We consider another example which exhibits de facto loss of convergence for Griewank's One-Step One-Shot method and for certain subproblem regularizations. By de facto loss of convergence we mean that although κ_{F} is well below 1 (e.g., below 0.5), κ_{O} is greater than 0.99. With

$$(H \quad A^{\mathrm{T}}) = \left(\begin{array}{cccc|cc} 0.83 & 0.083 & 0.34 & -0.21 & 1.1 & 0 \\ 0.083 & 0.4 & -0.34 & -0.4 & 1.7 & 0.52 \\ \hline 0.34 & -0.34 & 0.65 & 0.48 & -0.55 & -1.4 \\ -0.21 & -0.4 & 0.48 & 0.75 & -0.99 & -1.8 \end{array}\right) \tag{Ex3}$$

we obtain

$$\kappa_{\mathrm{F}} = \sigma_{\mathrm{r}}(\mathbb{I} - \widetilde{A}_1^{-1} A_1) \approx 0.48 < 1, \qquad \kappa_{\mathrm{O}} = \sigma_{\mathrm{r}}(\mathbb{I} - \widetilde{K}^{-1} K) \approx 1.54 > 1.$$

The quantities N_{xx}, G_y, G_u in the denotation of [9] are

$$N_{xx} = \mu H, \quad G_y = \mathbb{I} - \widetilde{A}_1^{-1} A_1, \quad G_u = -A_2,$$

where $\mu > 0$ is some chosen weighting factor for relative scaling of primal and dual variables. Based on

$$Z(\lambda) = (\lambda\mathbb{I} - G_y)^{-1} G_u, \quad H(\lambda) = (Z(\lambda)^{\mathrm{T}}, \mathbb{I}) N_{xx} (Z(\lambda)^{\mathrm{T}}, \mathbb{I})^{\mathrm{T}},$$

we can numerically verify that the projected Hessian preconditioners $H(\lambda), \lambda \in [-1, 1]$, do not restore contraction. The lowest spectral radius of the iteration matrix is 1.17 for $\lambda = -0.57$ and larger for all other values (compare Figure 1).

We now investigate three different modifications of the subproblems which do not alter the Jacobian blocks of the KKT systems. These modifications are based on

$$\kappa_{\mathrm{O}} = \sigma_{\mathrm{r}}(\mathbb{I} - \widetilde{K}^{-1} K) = \sigma_{\mathrm{r}}(\widetilde{K}^{-1}(\widetilde{K} - K)),$$

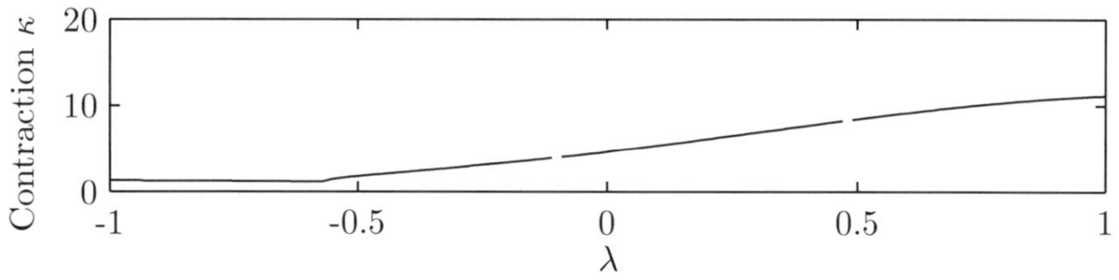

FIGURE 1. Contraction rates for One-Step One-Shot preconditioning with the projected Hessians $H(\lambda)$. The two gaps are due to the two eigenvalues of G_y rendering $\lambda\mathbb{I} - G_y$ singular.

which suggests that small eigenvalues of $\widetilde{K}$ might lead to large κ_{O}. Thus we regularize $\widetilde{K}$ such that the inverse $\widetilde{K}^{-1}$ does not have large eigenvalues in the directions of inexactness of $\Delta K = \widetilde{K} - K$.

We consider three prototypical regularization methods here which all add a positive multiple α of a matrix Λ to $\widetilde{K}$. The regularizing matrices are

$$\Lambda_{\mathrm{p}} = \begin{pmatrix} \mathbb{I} & 0 & 0 \\ 0 & \mathbb{I} & 0 \\ 0 & 0 & 0 \end{pmatrix}, \qquad \Lambda_{\mathrm{pd}} = \begin{pmatrix} \mathbb{I} & 0 & 0 \\ 0 & \mathbb{I} & 0 \\ 0 & 0 & -\mathbb{I} \end{pmatrix}, \qquad \Lambda_{\mathrm{hp}} = \begin{pmatrix} 0 & 0 & 0 \\ 0 & \mathbb{I} & 0 \\ 0 & 0 & 0 \end{pmatrix},$$

where the subscripts stand for primal, primal-dual, and hemi-primal (i.e., only in the space of design variables), respectively.

5. Analysis of the regularized subproblems

We investigate the asymptotic behavior of the subproblem solution for $\alpha \to \infty$ for the primal, primal-dual, and hemi-primal regularization. We assume invertibility of the approximation $\widetilde{A}_{1k}$ and drop the iteration index k. We generally assume that H is positive-definite on the nullspace of the approximation $\widetilde{A}$.

Consider the α-dependent linear system for the step determination of the inexact SQP method

$$\left(\widetilde{K} + \alpha\Lambda\right) \begin{pmatrix} \Delta z(\alpha) \\ \Delta\lambda(\alpha) \end{pmatrix} = \begin{pmatrix} -\ell \\ -r \end{pmatrix}, \tag{6}$$

where ℓ is the current Lagrange gradient and r is the current residual of the equality constraint. We use a nullspace method to solve the α-dependent system (6). Let matrices $Y \in \mathbb{R}^{(n+m)\times n}$ and $Z \in \mathbb{R}^{(n+m)\times m}$ have the properties

$$\widetilde{A}Z = 0, \quad (Z\ Y)^{\mathrm{T}}(Z\ Y) = \begin{pmatrix} Z^{\mathrm{T}}Z & 0 \\ 0 & Y^{\mathrm{T}}Y \end{pmatrix}, \quad \det(Y\ Z) \neq 0.$$

In other words, the columns of Z span the nullspace of $\widetilde{A}$. These are completed to form a basis of $\mathbb{R}^{n+m}$ by the columns of Y which are orthogonal to the columns of Z. In the new basis, we have $\Delta z = Yp + Zq$, with $(p, q) \in \mathbb{R}^{n+m}$.

5.1. Primal regularization

The motivation for the primal regularization stems from an analogy to the Levenberg-Marquardt method which, in the case of unconstrained minimization, is equivalent to a trust-region modification of the subproblem. It will turn out that the regularization with Λ_{p} will bend the primal subproblem solutions towards the step of smallest norm onto the linearized feasible set. However, it leads to a blow-up in the dual solution. From the following Lemma we observe that the primal step for large α is close to the step obtained by the Moore-Penrose-Pseudoinverse $\widetilde{A}^{+} = \widetilde{A}^{\mathrm{T}}(\widetilde{A}\widetilde{A}^{\mathrm{T}})^{-1}$ for the underdetermined system (1) and that the step in the Lagrange multiplier blows up for $r \neq 0$, and thus, convergence cannot be expected.

Lemma 6. *Under the general assumptions of Section* 5*, the solution of equation* (6) *for the primal regularization for large* α *is asymptotically given by*

$$\Delta z(\alpha) = -\widetilde{A}^{+}r + (1/\alpha)ZZ^{+}\big(H\widetilde{A}^{+}r - \ell\big) + o(1/\alpha),$$
$$\Delta\lambda(\alpha) = \alpha(\widetilde{A}\widetilde{A}^{\mathrm{T}})^{-1}r + (\widetilde{A}^{+})^{\mathrm{T}}(H\widetilde{A}^{+}r - \ell) + o(1).$$

Proof. From the second block-row of equation (6) and the fact that $\widetilde{A}Y$ is invertible due to $\widetilde{A}$ having full rank we obtain $p = -(\widetilde{A}Y)^{-1}r$. Premultiplying the first block-row of equation (6) with Z^{T} from the left yields the α-dependent equation

$$Z^{\mathrm{T}}HYp + Z^{\mathrm{T}}HZq + \alpha Z^{\mathrm{T}}Zq + Z^{\mathrm{T}}\ell = 0. \tag{7}$$

Let $\alpha > 0$ and $\beta = 1/\alpha$. Solutions of equation (7) satisfy

$$\Phi(q,\beta) := \big(\beta Z^{\mathrm{T}}HZ + Z^{\mathrm{T}}Z\big)\,q + \beta Z^{\mathrm{T}}\big(\ell - HY(\widetilde{A}Y)^{-1}r\big) = 0.$$

It holds that $\Phi(0,0) = 0$ and $\frac{\partial\Phi}{\partial q}(0,0) = Z^{\mathrm{T}}Z$ is invertible, as Z has full rank. Therefore, the Implicit Function Theorem yields the existence of a neighborhood $U \subset \mathbb{R}$ of 0 and a continuously differentiable function $\bar{q} : U \to \mathbb{R}^{m}$ such that $\bar{q}(0) = 0$ and

$$\Psi(\beta) := \Phi(\bar{q}(\beta),\beta) = 0 \quad \forall \beta \in U.$$

Using $0 = \frac{\mathrm{d}\Psi}{\mathrm{d}\beta} = \frac{\partial\Phi}{\partial q}\frac{\mathrm{d}\bar{q}}{\mathrm{d}\beta} + \frac{\partial\Phi}{\partial\beta}$ and Taylor's Theorem we have

$$\bar{q}(\beta) = \bar{q}(0) + \frac{\mathrm{d}\bar{q}}{\mathrm{d}\beta}(0)\beta + o(\beta) = \beta(Z^{\mathrm{T}}Z)^{-1}Z^{\mathrm{T}}\big(HY(\widetilde{A}Y)^{-1}r - \ell\big) + o(\beta),$$

which lends itself to the asymptotic

$$\Delta z(\alpha) = -Y(\widetilde{A}Y)^{-1}r + (1/\alpha)Z(Z^{\mathrm{T}}Z)^{-1}Z^{\mathrm{T}}\big(HY(\widetilde{A}Y)^{-1}r - \ell\big) + o(1/\alpha) \tag{8}$$

of the primal solution of equation (6) for large regularization parameters α.

Consider a special choice for the matrices Y and Z based on the QR decomposition $\widetilde{A} = Q\begin{pmatrix}R & B\end{pmatrix}$ with unitary Q and invertible R. We define

$$Z = \begin{pmatrix}-R^{-1}B\\ \mathbb{I}\end{pmatrix}, \qquad Y = \begin{pmatrix}R^{\mathrm{T}}\\ B^{\mathrm{T}}\end{pmatrix} = \widetilde{A}^{\mathrm{T}}Q$$

and obtain $Y(\widetilde{A}Y)^{-1} = \widetilde{A}^{\mathrm{T}}QQ^{-1}(\widetilde{A}\widetilde{A}^{\mathrm{T}})^{-1} = \widetilde{A}^{+}$, which yields the first assertion of the lemma.

For the corresponding dual solution we multiply the first block-row of equation (6) with Y^{T} from the left to obtain

$$Y^{\mathrm{T}}(H+\alpha\mathbb{I})\Delta z(\alpha)+(\widetilde{A}Y)^{\mathrm{T}}\Delta\lambda(\alpha)+Y^{\mathrm{T}}\ell=0.$$

After some rearrangements and with the help of the identity

$$(\widetilde{A}^{+})^{\mathrm{T}}(\mathbb{I}-ZZ^{+})=(\widetilde{A}Y)^{-\mathrm{T}}Y^{T}(\mathbb{I}-Z(Z^{\mathrm{T}}Z)^{-1}Z^{\mathrm{T}})=(\widetilde{A}^{+})^{\mathrm{T}}$$

we obtain the second assertion of the lemma. □

5.2. Primal-dual regularization

The primal-dual regularization is motivated by moving all eigenvalues of the regularized KKT matrix away from zero. It is well known that under our assumptions the matrix $\widetilde{K}$ has $n+m$ positive and n negative eigenvalues. The primal regularization method only moves the positive eigenvalues away from zero. By adding the $-\mathbb{I}$ term to the lower right block, also the negative eigenvalues can be moved away from zero while conserving the inertia of $\widetilde{K}$.

Lemma 7. *Under the general assumptions of Section* 5, *the solution of equation* (6) *for the primal-dual regularization is for large α asymptotically given by*

$$\begin{pmatrix}\Delta z(\alpha)\\ \Delta\lambda(\alpha)\end{pmatrix}=-\frac{1}{\alpha}\Lambda_{\mathrm{pd}}\begin{pmatrix}\ell\\ r\end{pmatrix}+o(1/\alpha)=\frac{1}{\alpha}\begin{pmatrix}-\ell\\ r\end{pmatrix}+o(1/\alpha).$$

Proof. Define again $\beta=1/\alpha$, $w=(\Delta z,\Delta\lambda)$, and

$$\Phi(w,\beta)=(\beta\widetilde{K}+\Lambda_{\mathrm{pd}})w+\beta\begin{pmatrix}\ell\\ r\end{pmatrix}.$$

It holds that

$$\Phi(0,0)=0,\qquad \frac{\partial\Phi}{\partial w}=\beta\widetilde{K}+\Lambda_{\mathrm{pd}},\qquad \frac{\partial\Phi}{\partial w}(0,0)=\Lambda_{\mathrm{pd}}.$$

The Implicit Function Theorem and Taylor's Theorem yield the assertion. □

Consider the limit case

$$\begin{pmatrix}\Delta z(\alpha)\\ \Delta\lambda(\alpha)\end{pmatrix}=-\frac{1}{\alpha}\Lambda_{\mathrm{pd}}\begin{pmatrix}\ell\\ r\end{pmatrix}$$

and the corresponding local contraction rate

$$\kappa_{\mathrm{pd}}=\sigma_{\mathrm{r}}(\mathbb{I}-(1/\alpha)\Lambda_{\mathrm{pd}}\widetilde{K}).$$

If all the real parts of the (potentially complex) eigenvalues of the matrix $\Lambda_{\mathrm{pd}}\widetilde{K}$ are larger than 0, contraction for large α can be recovered although contraction may be extremely slow, leading to de facto loss of convergence.

5.3. Hemi-primal regularization

In this section we are interested in a regularization of $\widetilde{K}$ only on the design variables y with Λ_{hp}. From the following lemma we observe that for large α, the primal solution of the hemi-primal regularized subproblem tends toward the step obtained from equation (2) for the underdetermined system (1) and that the dual variables do not blow up for large α in the hemi-primal regularization.

Lemma 8. *Under the general assumptions of Section 5, the solution of equation (6) for the hemi-primal regularization is for large α asymptotically given by*

$$\Delta z(\alpha) = \begin{pmatrix} -\widetilde{A}_1^{-1} r \\ 0 \end{pmatrix} + (1/\alpha) Z Z^{\mathrm{T}} \left(H \begin{pmatrix} \widetilde{A}_1^{-1} r \\ 0 \end{pmatrix} - \ell \right) + o(1/\alpha), \tag{9a}$$

$$\Delta\lambda(\alpha) = \begin{pmatrix} \widetilde{A}_1^{-1} \\ 0 \end{pmatrix}^{\mathrm{T}} \left(H \begin{pmatrix} \widetilde{A}_1^{-1} r \\ 0 \end{pmatrix} - \ell \right) + o(1), \tag{9b}$$

with the choice $Z = \left((-\widetilde{A}_1^{-1} A_2)^{\mathrm{T}} \ \mathbb{I}\right)^{\mathrm{T}}$ *and* $Y = (\widetilde{A}_1 \ A_2)^{\mathrm{T}} = \widetilde{A}^{\mathrm{T}}$.

Proof. By our general assumption, $\widetilde{A}_1$ is invertible and the previous assumptions on Y and Z are satisfied. Again, it holds that $Y(\widetilde{A}Y)^{-1} = \widetilde{A}^{+}$. We recover p as before. Let $\beta = 1/\alpha$. We can define an implicit function to determine $q(\beta)$ asymptotically via

$$\Phi(q, \beta) = (\beta Z^{\mathrm{T}} H Z + \mathbb{I}) q + Y_2 p + \beta Z^{\mathrm{T}} (H Y p + \ell),$$

where we used that $Z_2^{\mathrm{T}} Z_2 = \mathbb{I}$. It holds that $\Phi(-A_2^{\mathrm{T}} p, 0) = 0$ and $\frac{\partial \Phi}{\partial q}(-A_2^{\mathrm{T}} p, 0) = \mathbb{I}$. Thus the Implicit Function Theorem together with Taylor's Theorem yields

$$q(\beta) = -A_2^{\mathrm{T}} p - \beta Z^{\mathrm{T}} \left(H Y p + \ell - H Z A_2^{\mathrm{T}} p\right) + o(\beta).$$

By resubstitution of p and $q(1/\alpha)$ by the use of the identity $(Y - Z A_2^{\mathrm{T}})(\widetilde{A}Y)^{-1} = (\widetilde{A}_1^{-\mathrm{T}} \ 0)^{\mathrm{T}}$ we recover the first assertion of the lemma.

For the dual solution, we again multiply the first block-row of equation (6) with Y^{T} from the left to obtain

$$Y^{\mathrm{T}} \left(H + \alpha \begin{pmatrix} 0 & 0 \\ 0 & \mathbb{I} \end{pmatrix} \right) \Delta z(\alpha) + (\widetilde{A}Y)^{\mathrm{T}} \Delta\lambda(\alpha) + Y^{\mathrm{T}} \ell = 0,$$

which after some rearrangements yields the second assertion. □

Consider the limit case $\alpha \to \infty$. We recover from equation (9a) that

$$\Delta z_k = \begin{pmatrix} -\widetilde{A}_1^{-1} r_k \\ 0 \end{pmatrix}.$$

Hence $y_k = y^*$ stays constant and x_k converges to a feasible point x^* with the contraction rate κ_{F} of the Newton-type method for problem (1). For the asymptotic step in the dual variables we then obtain

$$\Delta\lambda_k = -\widetilde{A}_1^{-\mathrm{T}} \left(\nabla_x f_k + \nabla_x g_k \lambda_k\right) + \left(\widetilde{A}_1^{-\mathrm{T}} \ 0\right) H \left(\widetilde{A}_1^{-\mathrm{T}} \ 0\right)^{\mathrm{T}} r_k.$$

For the convergence of the coupled system with x_k and λ_k let us consider the Jacobian of the iteration $(x_{k+1}, \lambda_{k+1}) = T(x_k, \lambda_k)$ (with suitably defined T)

$$\frac{\mathrm{d}T}{\mathrm{d}(x, \lambda)} = \begin{pmatrix} \mathbb{I} - \widetilde{A}_1^{-1} \nabla_x g_k^{\mathrm{T}} & 0 \\ * & \mathbb{I} - \widetilde{A}_1^{-\mathrm{T}} \nabla_x g_k \end{pmatrix}.$$

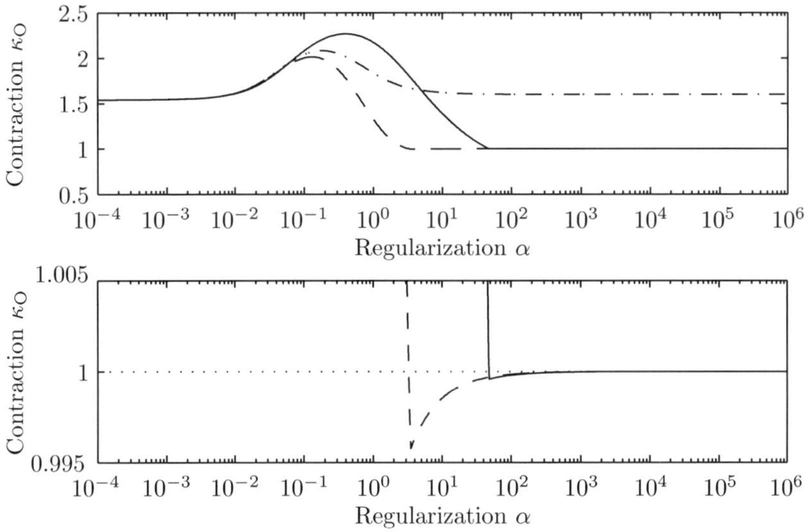

FIGURE 2. Divergence of the primal regularization (–·) and de facto loss of convergence for primal-dual (– –) and hemi-primal (–) regularization for example Ex3 depending on the regularization magnitude α. The lower diagram is a vertical close-up around $\kappa_O = 1$ of the upper diagram.

Hence (x_k, λ_k) converges with linear convergence rate κ_F, and λ_k converges to

$$\lambda^* = -\widetilde{A}_1^{-T} \nabla_x f(x^*, y^*).$$

Thus the primal-dual iterates converge to a point which is feasible and stationary with respect to x but not necessarily to y. Taking α large but finite we see that the hemi-primal regularization damps design updates while correcting state and dual variables with a contraction of almost κ_F.

5.4. Divergence and de facto loss of convergence for subproblem regularizations

Figure 2 depicts the dependence of κ_O of the optimization method on the regularization parameter α and the choice of regularization (primal, primal-dual, and hemi-primal) on example Ex3. The example was specifically constructed to show de facto loss of convergence for all three regularizations. Obviously the primal regularization does not even reach $\kappa_O = 1$.

We want to remark that the above discussion is not a proof of convergence for the primal-dual or hemi-primal regularization approach. Nonetheless, we have given a counter-example which shows failure of the primal regularization approach. With the de facto loss of convergence in mind, we believe that a possible proof of convergence for the other regularization strategies is of only limited practical importance.

6. Conclusion

We have presented three small, illustrative, linear examples which show that in general in a simultaneous one-step optimization approach the forward problem method might converge fast while the optimization method diverges and vice versa. This leads to the conclusion that in general the connections between contraction of the forward problem method and the simultaneous one-step optimization method are very limited. On the other hand, in certain applications, simultaneous one-step optimization methods can be proven to converge as fast as the forward problem method [14], or simply work well in practice [6, 10, 13]. We have investigated three prototypical subproblem regularization approaches: the primal, primal-dual, and hemi-primal approach. They all lead to de facto loss of convergence on example Ex3. The primal regularization even leads to divergence. Moreover, One-Step One-Shot projected Hessian preconditioners as proposed in [9] cannot restore convergence for Ex3. We conclude that, at least for fast contracting forward methods, the accuracy of the forward problem solver must be adaptively controlled by the optimization method like in [1, 2, 3, 4, 11], even though this may necessitate many iterations in the forward method.

References

[1] A. Battermann and M. Heinkenschloss. Preconditioners for Karush-Kuhn-Tucker matrices arising in the optimal control of distributed systems. In *Control and estimation of distributed parameter systems (Vorau,* 1996), volume 126 of *Internat. Ser. Numer. Math.*, pages 15–32. Birkhäuser, Basel, 1998.

[2] A. Battermann and E.W. Sachs. Block preconditioners for KKT systems in PDE-governed optimal control problems. In *Fast solution of discretized optimization problems (Berlin,* 2000), volume 138 of *Internat. Ser. Numer. Math.*, pages 1–18. Birkhäuser, Basel, 2001.

[3] G. Biros and O. Ghattas. Parallel Lagrange-Newton-Krylov-Schur methods for PDE-constrained optimization. Part I: The Krylov-Schur solver. *SIAM Journal on Scientific Computing*, 27(2):687–713, 2005.

[4] G. Biros and O. Ghattas. Parallel Lagrange-Newton-Krylov-Schur methods for PDE-constrained optimization. Part II: The Lagrange-Newton solver and its application to optimal control of steady viscous flows. *SIAM Journal on Scientific Computing*, 27(2):714–739, 2005.

[5] H.G. Bock. *Randwertproblemmethoden zur Parameteridentifizierung in Systemen nichtlinearer Differentialgleichungen*, volume 183 of *Bonner Mathematische Schriften.* Universität Bonn, Bonn, 1987.

[6] H.G. Bock, W. Egartner, W. Kappis, and V. Schulz. Practical shape optimization for turbine and compressor blades by the use of PRSQP methods. *Optimization and Engineering*, 3(4):395–414, 2002.

[7] R.H. Byrd, F.E. Curtis, and J. Nocedal. An inexact SQP method for equality constrained optimization. *SIAM Journal on Optimization*, 19(1):351–369, 2008.

[8] N.I.M. Gould and Ph.L. Toint. Nonlinear programming without a penalty function or a filter. *Mathematical Programming, Series A*, 122:155–196, 2010.

[9] A. Griewank. Projected Hessians for preconditioning in One-Step One-Shot design optimization. In *Large-Scale Nonlinear Optimization*, volume 83 of *Nonconvex Optimization and Its Applications*, pages 151–171. Springer US, 2006.

[10] S.B. Hazra, V. Schulz, J. Brezillon, and N.R. Gauger. Aerodynamic shape optimization using simultaneous pseudo-timestepping. *Journal of Computational Physics*, 204(1):46–64, 2005.

[11] M. Heinkenschloss and D. Ridzal. An Inexact Trust-Region SQP method with applications to PDE-constrained optimization. In K. Kunisch, G. Of, and O. Steinbach, editors, *Proceedings of ENUMATH 2007, the 7th European Conference on Numerical Mathematics and Advanced Applications, Graz, Austria, September 2007*. Springer Berlin Heidelberg, 2008.

[12] M. Heinkenschloss and L.N. Vicente. Analysis of Inexact Trust-Region SQP algorithms. *SIAM Journal on Optimization*, 12(2):283–302, 2002.

[13] A. Potschka, A. Küpper, J.P. Schlöder, H.G. Bock, and S. Engell. Optimal control of periodic adsorption processes: The Newton-Picard inexact SQP method. In *Recent Advances in Optimization and its Applications in Engineering*, pages 361–378. Springer Verlag Berlin Heidelberg, 2010.

[14] A. Potschka, M.S. Mommer, J.P. Schlöder, and H.G. Bock. A Newton-Picard approach for efficient numerical solution of time-periodic parabolic PDE constrained optimization problems. Technical Report 2010-03-2570, Interdisciplinary Center for Scientific Computing (IWR), Heidelberg University, 2010. Preprint, `http://www.optimization-online.org/DB_HTML/2010/03/2570.html`.

[15] A. Walther. A first-order convergence analysis of Trust-Region methods with inexact Jacobians. *SIAM Journal on Optimization*, 19(1):307–325, 2008.

Hans Georg Bock, Andreas Potschka, Sebastian Sager and Johannes P. Schlöder
Interdisciplinary Center for Scientific Computing (IWR)
Heidelberg University
Im Neuenheimer Feld 368
D-69120 Heidelberg, Germany
e-mail: `potschka@iwr.uni-heidelberg.de`

International Series of Numerical Mathematics, Vol. 160, 51–74

Generalized Multilevel SQP-methods for PDAE-constrained Optimization Based on Space-Time Adaptive PDAE Solvers

Debora Clever, Jens Lang, Stefan Ulbrich and Carsten Ziems

Abstract. In this work, we present an all-in-one optimization approach suitable to solve complex optimal control problems with time-dependent nonlinear partial differential algebraic equations and point-wise control constraints. A newly developed generalized SQP-method is combined with an error based multilevel strategy and the state-of-the-art software package KARDOS to allow the efficient resolution of different space and time scales in an adaptive manner. The numerical performance of the method is demonstrated and analyzed for a real-life two-dimensional radiative heat transfer problem modelling the optimal boundary control for a cooling process in glass manufacturing.

Mathematics Subject Classification (2000). 49J20, 49M25, 65C20, 65K10, 65M60, 90C46, 90C55.

Keywords. Adaptive multilevel finite elements, Rosenbrock method, glass cooling, radiation, multilevel optimization, generalized SQP method, control constraints, trust region methods, PDAE constrained optimization.

1. Introduction

To explore the fundamental scientific issues of high-dimensional complex engineering applications such as optimal control problems with time-dependent partial differential algebraic equations (PDAEs) scalable numerical algorithms are requested. This means that the work necessary to solve increasingly larger problems should grow all but linearly – the optimal rate. Therefore, we combine modern solution strategies to solve time-dependent systems of partial differential algebraic equations such as adaptive multilevel finite elements methods and error-controlled linearly implicit time integrators of higher order with novel generalized adaptive multilevel inexact SQP methods, which provide an efficient handling of control constraints by a projected inexact Newton strategy. Reduced gradient and reduced

Hessian are evaluated following the adjoint approach and a first-optimize-then-discretize strategy. From a numerical point of view this strategy is of great advantage, since now the discretization schemes can be chosen in such a way that they perfectly exploit the structure of the underlying PDAE which can vary significantly between state, linearized state and adjoint equations. This always guarantees consistency with the differential and algebraic equations [29]. Furthermore, allowing for different spatial meshes at different points of time and for different systems of PDAEs leads to substantial savings in computation time and memory.

From an optimization point of view, the first-optimize-then-discretize strategy introduces inconsistencies between discrete reduced gradient and optimization problem itself. Therefore, we propose a generalized multilevel trust-region SQP method that controls the inaccuracy of state, adjoint and criticality measure in an appropriate way by adapting the accuracy of the PDAE-solver in a discretization based multilevel strategy. The accuracy criteria of the optimization method can be implemented by using a posteriori error estimators within the PDAE solver. In contrast to a classical trust-region SQP method where the state equation is linearized to approach feasibility, in the generalized multilevel SQP algorithm presented in this paper a new state iterate is determined directly as the solution of the discretized nonlinear state equation. In the context of linearly implicit time integrators, which the considered software package KARDOS is based on, this modification does not introduce any additional costs, since the idea of linearization is already handled by the time integrator itself. The resulting generalized SQP method is globally convergent for general nonlinear problems and approaches the solution of the PDAE constrained optimization problem in an efficient way by generating a hierarchy of adaptively refined discretizations. A detailed description of the algorithmic structure and a convergence statement are given in Section 2.

For the determination of an optimal control to a PDAE-constrained optimization problem, the involved systems of partial differential algebraic equations are solved by applying the state-of-the-art solver KARDOS [13]. This software package uses adaptive Rosenbrock methods for the temporal and adaptive multilevel linear finite elements for the spatial integration. Originally, KARDOS is implemented as a pure PDAE-solver for a predefined problem. For optimization purposes, it is necessary to augment the application in such a way that related problems are generated automatically, based on results from up to three previous PDAE-solves. Since for every solve we allow independent adaption of space and time grids due to local error estimates, data is exploited on various grids. Hence, a major task of the coordination between PDAE-solver and optimization algorithm is the data and grid management. Fast access is ensured using quadtrees. To optimize time and space grids in order to reach a specified error tolerance prescribed by the optimizer, local tolerances are adapted automatically, relying on tolerance proportionality as described in Section 3. Therefore, we use global estimates from a previous run to automatically modify current local tolerances in such a way that global errors drop below a specified error tolerance, generally, in no more than one additional run. Global error based accuracy control has been derived in [8, 24]. For efficiency reasons, all function evaluations like objective, gradient or Hessian values are computed directly within the PDAE solver and handed to the optimization tool.

To test our newly developed all-in-one optimization approach, an optimal boundary control problem of the cooling down process of glass modelled by radiative heat transfer serves as showcase engineering applications. The model is given by a system of time-dependent partial differential algebraic equations with pointwise restrictions on the control. Results are discussed in Section 4.

2. Adaptive multilevel generalized SQP-method

In this section we present a class of adaptive multilevel generalized trust-region sequential quadratic programming (SQP) methods for the efficient solution of optimization problems governed by nonlinear time-dependent partial differential algebraic equations (PDAEs) with control constraints. A generalized SQP method is signified here as an SQP method where instead of the quasi-normal step of a classical trust-region SQP method ([4, 28, 9, 7]) which results in solving the linearized PDAE sufficiently well, a nonlinear solver is applied to the current discretization of the PDAE. The algorithm is based on the ideas of the adaptive multilevel inexact SQP-method from [39] and on [17, 32]. It is inspired by and designed for optimization problems governed by parabolic partial differential algebraic equations since solving a linearized parabolic PDAE or the nonlinear parabolic PDAE itself numerically with linear implicit methods in time on a given spatial discretization has about the same computational costs. This generalized SQP method is designed to combine an elaborated optimization technique with efficient PDAE solvers in a stable framework. Therefore, the possibility to use different solvers for the PDAE and the adjoint PDAE is given. State of the art techniques in solving PDAEs can be applied. The occurring inexactness in the reduced gradient on a fixed discretization level is controlled. Modern adaptive discretization techniques for partial differential equations based on a posteriori error estimators [8, 35, 27, 1, 36, 37] are integrated in the algorithm. This offers the possibility to perform most of the optimization iterations and PDAE solves on coarse meshes. Moreover, the optimization problem is always well represented and the infinite-dimensional problem is approached during the optimization in an efficient way.

In recent years, multilevel techniques in optimization have received considerable attention [2, 3, 12, 15, 14, 26, 33]. These approaches use a fixed predefined hierarchy of discretizations to solve an optimization problem on the finest grid. The combination with adaptive error control techniques is not considered. On the other hand, a posteriori error estimators in the context of parabolic PDAEs are an active research area [8, 35, 27, 1, 36, 37]. However, the rigorous embedding of such error estimators in globally convergent multilevel optimization methods was to the best of our knowledge not considered so far. The algorithmic approach presented here differs to the best of our knowledge significantly from existing work in this area since different discretizations for the PDAE and the adjoint PDAE are allowed, while adaptive refinements based on error estimators are used. Moreover, global convergence can be shown.

In the following we consider PDAE-constrained optimal control problems of the form

$$\min_{y\in Y, u\in U} J(y,u) \quad \text{subject to} \quad e(y,u)=0,\ u\in U_{\text{ad}}, \tag{2.1}$$

where U is the control space, $U_{\text{ad}} \subset U$ a closed and convex subset representing the set of admissible controls, Y is the state space, $J : Y\times U \to \mathbb{R}$ is the objective function. The state equation $e : Y\times U \to V^*, e(y,u)=0$ comprises a system of partial differential equations supplemented with appropriate initial and boundary conditions in a variational formulation with V as the set of test functions. V^* denotes the dual space of V. Y and U are assumed to be Hilbert spaces and V a reflexive Banach space. For abbreviation let $X = Y\times U$. Furthermore, we assume that $J(y,u)$ and $e(y,u)$ are twice continuously Fréchet differentiable, the Fréchet derivative of the constraint e with respect to the state y, $e_y(y,u)\in\mathcal{L}(Y,V^*)$, has a bounded inverse and that the PDAE admits a unique and Fréchet differentiable solution operator $S : u\in U \mapsto S(u)\in Y$. Let

$$l : Y\times U\times V\to\mathbb{R}, \quad l(y,u,\lambda) = J(y,u) + \langle\lambda, e(y,u)\rangle_{V,V^*} \tag{2.2}$$

denote the Lagrangian function. Let $(\bar y,\bar u)\in Y\times U_{\text{ad}}$ be an optimal solution of problem (2.1). Then the following first-order necessary optimality conditions hold ([18, 31]): There exists an adjoint state $\bar\lambda\in V$ such that

$$\begin{aligned} e(\bar y,\bar u) &= 0 && \text{(state equation)},\\ l_y(\bar y,\bar u,\bar\lambda) &= 0 && \text{(adjoint equation)},\\ P_{U_{\text{ad}}-\bar u}(-l_u(\bar y,\bar u,\bar\lambda)) &= 0 && \text{(stationarity)}, \end{aligned} \tag{2.3}$$

where $P_{U_{\text{ad}}} : U\to U_{\text{ad}}$ denotes the projection onto the closed and convex set U_{ad}, the set $U_{\text{ad}}-\bar u$ is given as $U_{\text{ad}}-\bar u = \{u\in U : u+\bar u\in U_{\text{ad}}\}$ and $l_u(\bar y,\bar u,\bar\lambda)$ denotes the Riesz representation of the control gradient of the Lagrangian function in U. The adjoint state $\bar\lambda$ is uniquely determined by $\bar\lambda = -e_y^{-*}(\bar y,\bar u)J_y(\bar y,\bar u)$, since $e_y(\bar y,\bar u)$ has a bounded inverse. We refer to $\|P_{U_{\text{ad}}-u}(-l_u(y,u,\lambda))\|_U$ as a criticality measure. If there are no control constraints, i.e., $U_{\text{ad}} = U$, then the criticality measure is simply the norm of the control gradient of the Lagrangian function.

The proposed multilevel SQP-algorithm for (2.1) generates a hierarchy of finite-dimensional approximations of problem (2.1) on adaptively refined meshes. Let e^h comprise the discretized formulation of the state equation depending on the solving and discretization technique such that $e^h(y_k,u_k)=0$ represents solving the PDAE on the given discretization. Using suitable interpolation, a conformal finite-dimensional subspace $Y_h\subset Y$ can be defined, cf. [38]. Similarly, the finite-dimensional space $V_h\subset V$ can be defined as solution space for the discretized adjoint PDAE. Moreover, finite-dimensional subspaces $U_h\subset U$ of the control space and $U_{\text{ad}}^h\subset U_{\text{ad}}$ are chosen. First, a generalized reduction step towards feasibility is computed by applying a (nonlinear) solver to the discretized state equation $e^h(y_k, u_k) = 0$ for the current control iterate $u_k \in U^h_{\text{ad}}$. The optimization problem is then approximated in the full iterate (y_k, u_k) locally by a quadratic subproblem with trust-region constraint. The quadratic model is inexactly reduced to the control component by computing λ_k as solution of an independent discretization of the adjoint PDAE $l_y(y_k, u_k, \lambda_k) = 0$ and using an approximation $\hat{H}_k$ of a discretization of the reduced Hessian of the Lagrangian function $W_k^* l_{xx}(y_k, u_k, \lambda_k) W_k$ with $W_k = (-e_y(y_k, u_k)^{-1} e_u(y_k, u_k), I) \in \mathcal{L}(U, Y \times U)$. We require $\langle s_u, \hat{H}_k s_u \rangle_{U,U^*} \le \xi \|s_u\|_U^2$ for all control steps s_u and some fixed $\xi > 0$. The local (generalized) SQP problem at (y_k, u_k, λ_k) then is

$$\begin{aligned} &\min_{s_u \in U_h} \; q_k(s_u) := J(y_k, u_k) + \langle g_k, s_u \rangle_{U^*,U} + \tfrac{1}{2} \langle s_u, \hat{H}_k s_u \rangle_{U,U^*} \\ &\text{subject to} \quad \|s_u\|_U \le \Delta_k, \; u_k + s_u \in U^h_{\text{ad}}, \end{aligned} \tag{2.4}$$

where

$$g_k := l_u(y_k, u_k, \lambda_k) \tag{2.5}$$

denotes the inexact reduced gradient and Δ_k is the trust-region radius. The approximate solution $s_{u,k}$ of (2.4) must satisfy the generalized Cauchy decrease condition in the approximate model q_k, i.e.,

$$q_k(0) - q_k(s_{u,k}) \ge \kappa_1 \|P_{U^h_{\text{ad}} - u_k}(-g_k)\|_U \min \left\{ \kappa_2 \|P_{U^h_{\text{ad}} - u_k}(-g_k)\|_U, \kappa_3 \Delta_k \right\} \; \forall k \in \mathbb{N}, \tag{2.6}$$

where $\kappa_1, \kappa_2, \kappa_3$ are positive constants independent of k and the grid, Δ_k is the trust-region radius and $\|P_{U^h_{\text{ad}} - u_k}(-g_k)\|_U$ the discrete criticality measure.

If there are no control constraints present, i.e., $U_{\text{ad}} = U$, then (2.6) is just the Cauchy decrease condition, that originates from the approximate model q_k providing a fraction of the decrease that is possible along the direction of steepest descent inside the trust-region. If control constraints are considered this condition is generalized to the projected negative gradient path direction. Condition (2.6) can often easily be satisfied. Indeed, without control constraints, steps can be computed in a Steihaug-type manner, cf., e.g., [38, 39, 30, 17, 7], the simplest steps are gradient steps with optimal steplength. In the presence of control constraints projected Newton steps (cf. [19]) with Armijo or Goldstein-type linesearch (cf. [7]) can be used if (2.6) is checked, compare [38]. Projected negative gradient steps with Armijo or Goldstein-type linesearch satisfy (2.6) as long as the reduced Hessians $\hat{H}_k$ are bounded (cf. [38]).

A new control $u_{k+1} := u_k + s_{u,k}$ is accepted, if the actual reduction

$$\text{ared}_k = J(y_k, u_k) - J(y_{k+1}, u_{k+1})$$

is at least a fraction of the model based predicted reduction

$$\text{pred}_k = q_k(0) - q_k(s_{u,k}).$$

The ratio of actual and predicted reduction is used in a standard fashion to adjust the trust region radius, cf., e.g., [4, 28, 9, 7, 17, 39, 38].

Allowing independent discretizations for state and adjoint equations introduces inconsistencies between reduced derivatives and minimization problem itself. To guarantee convergence nevertheless, we have developed a multilevel strategy which automatically refines computational grids if the ratio between the discrete criticality measure and global error estimates indicates severe inconsistencies of discrete and continuous problem. Details and a convergence proof of the presented algorithm with these refinement criteria can be found in [38].

The main idea for refinement is to control the infinite-dimensional norms of the residuals in the infinite-dimensional optimality system (2.3) with the discrete criticality measure $\|P_{U^h_{\mathrm{ad}}-u_k}(-g_k)\|_U$. In case of no control constraints the criticality measure is just the norm of the reduced gradient. Thus, if the norm of the criticality measure is large enough compared to the infinite-dimensional residuals, the current discretization will be good enough to compute sufficient descent. On the other hand, if the norm of the discrete criticality measure on the current grid is small compared to the continuous norms, one has to ensure by refinement of the discretizations that the infinite-dimensional problem and, in particular, the infinite-dimensional reduced gradient and its projection are well represented in the current discretization such that reasonable steps can be computed. Observe that based on the optimality system (2.3) the inexact reduced gradient g_k depends on the (inexact) adjoint λ_k and, thus, also on the (inexact) state y_k. Therefore, the residual norms of the infinite-dimensional state- and adjoint equation must be controlled. Since these residual norms generally cannot be computed directly, we will use reliable error estimators instead.

Hence, we would like to ensure the following inequalities

$$\begin{array}{lcl} \|e(y_k,u_k)\|_{V^*} & \leq & K_y\|P_{U^h_{\mathrm{ad}}-u_k}(-g_k)\|_U \\ \|l_y(y_k,u_k,\lambda_k)\|_{Y^*} & \leq & K_\lambda\|P_{U^h_{\mathrm{ad}}-u_k}(-g_k)\|_U \\ \|P_{U_{\mathrm{ad}}-u_k}(-g_k)-P_{U^h_{\mathrm{ad}}-u_k}(-g_k)\|_U & \leq & K_u\|P_{U^h_{\mathrm{ad}}-u_k}(-g_k)\|_U \end{array} \tag{2.7}$$

with fixed (unknown) constants $K_y, K_\lambda, K_u > 0$, where y_k denotes the solution of the discretized PDAE and λ_k the solution of the discretized adjoint PDAE. In the third equation, the difference between the finite and the infinite-dimensional projections of the inexact reduced gradient is controlled. Note that using the triangle inequality, the third equation of (2.7) implies

$$\|P_{U_{\mathrm{ad}}-u_k}(-g_k)\|_U \leq (K_u+1)\|P_{U^h_{\mathrm{ad}}-u_k}(-g_k)\|_U. \tag{2.8}$$

If there are no control constraints, then the discrete criticality measure simply becomes the norm of the Riesz representation in U_h of the inexact reduced gradient g_k. Let $y_k^{\mathrm{ex}} = y_k^{\mathrm{ex}}(u_k)$ and $\lambda_k^{\mathrm{ex}} = \lambda_k^{\mathrm{ex}}(y_k,u_k)$ denote the exact solutions of the state equation and adjoint equation for the given state and control iterate (y_k,u_k), respectively. Since it is sometimes preferable to compute or estimate the norm of the difference of y_k and the exact solution y_k^{ex} of the PDAE and the difference of λ_k and λ_k^{ex} instead of controlling the infinite-dimensional norms of the residuals, we use continuity arguments to provide a sufficient condition with such quantities. We assume that $e(y,u)$ is uniformly Lipschitz continuous in y with Lipschitz constant $L_y > 0$ independent of $u \in D_U$, D_U an open convex subset of U that contains all control iterates. Let S^h denote the discrete solution operator. Then also $l_y(S^h(u^h), u^h, \lambda)$ is assumed to be uniformly Lipschitz continuous with Lipschitz constant $L_\lambda > 0$ independent of $(S^h(u^h), u^h) \in D = D_Y \times D_U$, where D_Y is an open convex subset of Y that contains all state iterates, cf. assumption A.4. Since $e(y_k^{\text{ex}}, u_k) = 0$ in V^* and $l_y(y_k, u_k, \lambda_k^{\text{ex}}) = 0$ in Y^*, the assumptions on the Lipschitz continuity yield

$$\begin{aligned} \|e(y_k, u_k)\|_{V^*} &= \|e(y_k, u_k) - e(y_k^{\text{ex}}, u_k)\|_{V^*} \leq L_y \|y_k - y_k^{\text{ex}}\|_Y \\ \|l_y(y_k, u_k, \lambda_k)\|_{Y^*} &= \|l_y(y_k, u_k, \lambda_k) - l_y(y_k, u_k, \lambda_k^{\text{ex}})\|_{Y^*} \leq L_\lambda \|\lambda_k - \lambda_k^{\text{ex}}\|_V. \end{aligned} \tag{2.9}$$

Using (2.9), the refinement conditions are then given by

$$\begin{aligned} \|y_k - y_k^{\text{ex}}\|_Y &\leq \tilde{K}_y \|P_{U_{\text{ad}}^h - u_k}(-g_k)\|_U \\ \|\lambda_k - \lambda_k^{\text{ex}}\|_V &\leq \tilde{K}_\lambda \|P_{U_{\text{ad}}^h - u_k}(-g_k)\|_U \\ \|P_{U_{\text{ad}} - u_k}(-g_k) - P_{U_{\text{ad}}^h - u_k}(-g_k)\|_U &\leq \tilde{K}_u \|P_{U_{\text{ad}}^h - u_k}(-g_k)\|_U \end{aligned} \tag{2.10}$$

with fixed (unknown) constants $\tilde{K}_y, \tilde{K}_\lambda, \tilde{K}_u > 0$. Since the infinite-dimensional solutions of the state and adjoint PDAE are not known we assume that there are global error estimators η_y, η_λ and a computation or estimation technique η_u available with

$$\|y_k - y_k^{\text{ex}}\|_Y \leq c_y \, \eta_y(y_k) \tag{2.11a}$$

$$\|\lambda_k - \lambda_k^{\text{ex}}\|_V \leq c_\lambda \, \eta_\lambda(\lambda_k) \tag{2.11b}$$

$$\|P_{U_{\text{ad}} - u_k}(-g_k) - P_{U_{\text{ad}}^h - u_k}(-g_k)\|_U \leq c_u \, \eta_u(g_k) \tag{2.11c}$$

with unknown, bounded constants $c_y, c_\lambda, c_u > 0$, where y_k^{ex} and λ_k^{ex} are the exact solutions of state equation and adjoint equation for the given control and state iterate (y_k, u_k), respectively. Inserting the inequalities (2.11) in (2.10), we require the following implementable and sufficient refinement conditions

$$\eta_y(y_k) \leq \tilde{c}_y \, \|P_{U_{\text{ad}}^h - u_k}(-g_k)\|_U \tag{2.12a}$$

$$\eta_\lambda(\lambda_k) \leq \tilde{c}_\lambda \, \|P_{U_{\text{ad}}^h - u_k}(-g_k)\|_U \tag{2.12b}$$

$$\eta_u(g_k) \leq \tilde{c}_u \, \|P_{U_{\text{ad}}^h - u_k}(-g_k)\|_U \tag{2.12c}$$

with fixed constants $\tilde{c}_y, \tilde{c}_\lambda, \tilde{c}_u > 0$. If the refinement conditions (2.12) are not satisfied, then the discretizations of Y_h, V_h, U_h are refined adaptively, i.e., the corresponding space-time grid is refined adaptively, and the data is recomputed until the conditions hold. Thus, if the norm of the discrete criticality measure is small compared to the residuals in the state and the adjoint computation, then the discretizations need to be refined such that the infinite-dimensional problem and, in particular, the reduced gradient are well represented. Whereas sufficient descent can be computed as long as the norm of the discrete criticality measure is large compared to these residuals.

We assume that all functions and values of the discretized problem converge to the infinite-dimensional counterparts as the overall fineness of the discretization tends to zero, i.e., $h \searrow 0$, where h denotes the maximal meshsize, in particular the solutions of the discretized equations converge to the infinite-dimensional solutions. Moreover, we require that the error estimators in (2.11) tend to zero as $h \searrow 0$. Thus, the refinement conditions (2.12) can always be satisfied by sufficient refinement.

Remark 2.1.

1. Note that the conditions (2.7) and (2.8) that lead to the optimality conditions (2.3) are implied by our refinement conditions (2.12). In particular, convergence of the discrete state and discrete adjoint state towards the continuous solutions in Y and V, respectively, implies convergence of the discrete inexact reduced gradient towards the continuous reduced gradient of the infinite-dimensional problem.
2. Observe that the assumptions on the convergence of the discretization guarantee that (2.12a) can always be satisfied by sufficient refinement. Moreover, the locally uniform convergence of the solution operator of the adjoint PDAE together with its continuity in the state variable ensure the satisfiability of (2.12b) by sufficient refinement. Condition (2.12c) can be satisfied by sufficient refinement since the finite-dimensional projection operator approximates the infinite-dimensional projection operator as the meshsize of the discretization tends to zero.

Due to the independent discretizations of state and adjoint PDAE, g_k is not an exact reduced gradient on the discrete level. Indeed, since we consider the constraint PDAE in our method in the variational formulation $e(y_k, u_k) \in V^*$ and in dual pairings of the form $\langle v^h, e(y_k, u_k)\rangle_{V,V^*}$ with $v^h \in V_h$ and V_h a finite-dimensional subspace of V, the discrete adjoint equation would be given by

$$\langle J_y(y_k, u_k) + e_y^*(y_k, u_k)\tilde{\lambda}_k, w^h\rangle_{Y^*,Y} = 0 \quad \text{for all } w^h \in Y_h. \tag{2.13}$$

Accordingly, the discrete tangent equation would be given by

$$\langle e_y(y_k, u_k)\hat{s}_y + e_u(y_k, u_k)s_u, v^h\rangle_{V^*,V} = 0 \quad \text{for all } v^h \in V_h. \tag{2.14}$$

for given $(y_k, u_k) \in Y_h \times U_h$ and $s_u \in U_h$. Observe that the derivative of the cost functional J in direction $(\hat{s}_y, s_u)$ can be evaluated as

$$\langle J_x(y_k, u_k), (\hat{s}_y, s_u)\rangle_{X^*,X} = \langle l_u(y_k, u_k, \tilde{\lambda}_k), s_u\rangle_{U^*,U},$$

since $\tilde{\lambda}_k$ solves the discrete adjoint equation. Then, in particular, this yields

$$|\langle J_x(y_k, u_k), (\hat{s}_h, s_u)\rangle_{X^*,X} - \langle l_u(y_k, u_k, \tilde{\lambda}_k), s_u\rangle_{U^*,U}| = 0. \tag{2.15}$$

In this generalized SQP method we solve a discretization of the state PDAE described by $e^h(S^h(u_k), u_k) = 0$, where S^h denotes the discrete solution operator given by the discretization scheme. Thus, we obtain as discrete tangential equation in $x_k = (y_k, u_k)$ with $y_k = S^h(u_k)$

$$e_y^h(x_k)S^h(u_k)'s_u + e_u^h(x_k)s_u = 0, \quad s_u \in U_h. \tag{2.16}$$

Moreover, in our method, when λ_k solves an independent discretization of the adjoint PDAE, the equality

$$\langle J_y(y_k, u_k) + e_y^*(y_k, u_k)\lambda_k, w^h\rangle_{Y^*,Y} = 0 \quad \text{for all } w^h \in Y_h$$

is generally not true. Consequently, we require the following condition based on (2.15). If a step is not accepted, one has to verify that the following gradient condition in direction of the tangential step $\hat{s}_k = (\hat{s}_{y,k}, s_{u,k}) \in X_h = Y_h \times U_h$ to the computed current step $s_{u,k}$

$$|\langle J_x(y_k, u_k), \hat{s}_k\rangle_{X_h^*,X_h} - \langle g_k, s_{u,k}\rangle_{U_h^*,U_h}| \leq \zeta \min\{\|P_{U_{\text{ad}}^h - u_k}(-g_k)\|_U, \Delta_k\}\|s_{u,k}\|_U \tag{2.17}$$

is satisfied for some $\zeta > 0$. Here, $\hat{s}_{y,k} = S^h(u_k)' s_{u,k}$ is the solution of the discrete tangential equation (2.16) for a given control step $s_{u,k}$. If (2.17) does not hold, then the discretizations of Y for the state and of V for the adjoint are refined and the iteration is recomputed until either the stopping criterion of the algorithm is satisfied or the trial step is accepted or the gradient condition (2.17) is satisfied. The latter can be achieved by sufficient refinement, cf. [38].

Furthermore, the refinements of the grids need to be such that the descent on the discrete level implies some fraction of this descent in the infinite-dimensional setting. Therefore, after the computation of a successful step, one needs to verify

$$\begin{aligned} \text{ared}_k \geq (1+\delta)\big(\big(f(S(u_k + s_{u,k}), u_k + s_{u,k}) - f(y_{k+1}, u_k + s_{u,k})\big) \\ -\big(f(S(u_k), u_k) - f(y_k, u_k)\big)\big) \end{aligned} \tag{2.18}$$

with $0 < \delta \ll 1$, where S denotes the solution operator of the given PDAE. If criterion (2.18) is not satisfied the Y-grid needs to be refined properly such that the next iterate can be represented well (at least) with respect to the cost functional. Thus, we check after a successful step if the current discretization was suitable to compute sufficient descent in the cost functional. And, hence, this criterion guarantees suitable (adaptive) refinements. Generally, if one refines reasonably, criterion (2.18) is always satisfied and, therefore, does not need to be implemented. However, an implementable version can be found in [38].

Algorithm 2.2 (Adaptive multilevel generalized trust-region SQP algorithm).

S.0 Initialization: Choose $\varepsilon_{\text{tol}} > 0$, $\tilde{c}_1, \tilde{c}_2, \tilde{c}_3 > 0$, an initial discretization, $u_0 \in U_{\text{ad}}^h$ and $\Delta_0 \in [\Delta_{\min}, \Delta_{\max}]$. Set $k := 0$.

For $k = 0, 1, 2, \ldots$

S.1 Compute a generalized reduction step y_k as solution of the discretized state equation $e^h(y_k, u_k) = 0$ and $\eta_y(y_k)$ (if not already done).

S.2 Compute an adjoint state λ_k as solution of the discretized adjoint equation and $\eta_\lambda(\lambda_k)$. Determine the inexact reduced gradient g_k by (2.5) and the criticality measure $\|P_{U_{\text{ad}}^h - u_k}(-g_k)\|_U$.

S.3 If the refinement condition (2.12a) holds, then goto S.4. Otherwise refine the Y-grid adaptively and goto S.1.

S.4 If the refinement condition (2.12b) and (2.12c) hold, then goto S.5. Otherwise refine the V- and U-grid adaptively and goto S.2.
S.5 If $\|P_{U^h_{\mathrm{ad}}-u_k}(-g_k)\|_U \leq \varepsilon_{\mathrm{tol}}$ then stop and return (y_k, u_k) as approximate solution of problem (2.1).
S.6 Compute $s_{u,k}$ as inexact solution of (2.4) satisfying (2.6).
S.7 Compute a discrete state y_{k+1} to $u_k + s_{u,k}$ and ared_k.
S.8 Compute pred_k. If $s_{u,k}$ is accepted by the reduction ratio of ared_k and pred_k then goto S.9. Otherwise, if the gradient condition (2.17) is satisfied, then reject the step $s_{u,k}$, reduce the trust-region and go back to S.6 with (y_k, u_k), but if the condition (2.17) is not satisfied, then refine the Y- and V-grid properly and go back to S.1 with u_k.
S.9 If (2.18) is satisfied, then accept $s_{u,k}$ and go to S.2 with $(y_{k+1}, u_{k+1}) = (y_{k+1}, u_k + s_{u,k})$. Otherwise reject $s_{u,k}$, refine the Y-grid and go back to S.1 with u_k.

Remark 2.3. If the refinement of the V- and U-grid in S.4 of Algorithm 2.2 implies a refinement of the Y-grid, e.g., due to implementation issues, then, after a refinement in S.4, one needs to go to S.1.

Convergence can be shown under the following set of standard assumptions enhanced by a few basic assumptions on the convergence of the discretization.

Assumptions on the problem and the functionals. For all iterations k we assume that $u_k, u_k + s_{u,k} \in D_U$, where D_U is an open, convex subset of U. Moreover, we assume that $y_k \in D_Y$ where D_Y is an open convex subset of Y. We set $D = D_Y \times D_U$.

A.1. The functionals J, e are twice continuously Fréchet differentiable in D.
A.2. The partial Jacobian $e_y(y, u)$ has an inverse for all $(y, u) \in D$.
A.3. The functionals and operators J, J_x, J_{xx}, e, e_u are bounded in D.
A.4. $e(y, u)$ is uniformly Lipschitz continuous in y with Lipschitz constant $L_y > 0$ independent of $u \in D_U$. $l_y(y^h, u, \lambda)$ is uniformly Lipschitz continuous with Lipschitz constant $L_\lambda > 0$ independent of $(y^h, u) \in D$ with $e^h(y^h, u) = 0$.
A.5. The sequence $\{\hat{H}_k\}$ is bounded.

Assumptions on the discretization, refinement and solution operators. Let S denote the solution operator for the given state PDAE $e(y, u) = 0$, $S_{e_x}(y, u)$ the solution operator for the tangent PDAE in (y, u), $e_y(y, u)s_y + e_u(y, u)s_u = 0$, and S_{l_y} the solution operator for the adjoint PDAE $l_y(y, u, \lambda) = 0$. Let S^h, $S^h_{e^h_x}$, $S^h_{l_y}$ denote the discrete solution operators of the corresponding discretizations of the PDAEs. Moreover, let $h \searrow 0$ denote that the overall maximal meshsize of the space-time discretization tends to zero. For the discrete solution operators we then require the following continuity properties and convergence that is locally uniform in the state component $y^h = S^h(u^H)$:

Let $u^H, s_u \in U_H$, $H > 0$, $U_H \subset U_h$ for $h < H$.

D.1. $S(u^h)$, $S_{e_x}(y^h, u^h)$, $S_{l_y}(y^h, u^h)$, $S^h(u^h)$, $S^h_{e^h_x}(y^h, u^h)$, $S^h_{l_y}(y^h, u^h)$ are continuous in the control $u^h \in U_h$ and the state $y^h \in Y_h$.

D.2. The solution operator S^h is twice continuously Fréchet differentiable in D_U.

D.3. The operators $(S^h(u^h))'$ and $(S^h(u^h))''$ are bounded for all $u^h \in D_U$.

D.4. $S^h(u^H) \to S(u^H)$ as $h \searrow 0$.
$S^h_{e^h_x}(S^h(u^H), u^H)s_u \to S_{e_x}(S(u^H), u^H)s_u$ as $h \searrow 0$.
$S^h_{l_y}(S^h(u^H), u^H) \to S_{l_y}(S(u^H), u^H)$ as $h \searrow 0$.

Assumptions on the convergence of the discretization. We assume that all functions and values of the discretized problem converge to the infinite-dimensional counterparts in the corresponding infinite-dimensional spaces as the overall fineness of the discretization tends to zero, i.e., $h \searrow 0$, in particular the solutions of the discretized state and adjoint equations converge to the infinite-dimensional solutions in Y and V, respectively.

The following theorem states that there exists a subsequence of the iterates that satisfies the first-order necessary optimality conditions (2.3) of the given problem (2.1) in the limit if $\varepsilon_{\text{tol}} = 0$.

Theorem 2.4. *Let the assumptions on the problem and the discretization hold. Then for $\varepsilon_{\text{tol}} > 0$ Algorithm 2.2 terminates finitely and for $\varepsilon_{\text{tol}} = 0$ the algorithm terminates finitely with a stationary point of problem (2.1) or the sequence of iterates (y_k, u_k, λ_k) generated by Algorithm 2.2 satisfies*

$$\liminf_{k\to\infty} \left(\left\|e(y_k, u_k)\right\|_{V^*} + \left\|l_y(y_k, u_k, \lambda_k)\right\|_{Y^*} + \left\|P_{U_{\text{ad}}-u_k}(-l_u(y_k, u_k, \lambda_k))\right\|_U \right) = 0.$$

A proof can be found in [38].

In the following, the presented Algorithm 2.2 will be coupled with the PDAE solver KARDOS.

3. Coupling of the SQP method with a PDAE solver

Even though the use of the generalized SQP method presented in Section 2 reduces the number of PDAE solves significantly, it is essential to solve the involved PDAEs as efficient as possible while approaching the required accuracy. We apply a semi-discretization following Rothe's method. The time interval $[0, t_e]$ is discretized using variable time step Rosenbrock schemes of high order. The discretization of the spatial domain $\Omega \subset \mathbb{R}^d$, $d = 2, 3$ is based on adaptive multilevel finite elements of appropriate order.

For the development of an all-in-one optimization tool, based on a state-of-the-art PDAE solver we build on the software package KARDOS. Especially for boundary or spatially constant control problems, the control component is of lower dimension than the state. Therefore, we took care that all computations and evaluations concerning states and adjoint states are carried out within the

KARDOS environment itself. Furthermore, the adaptive grid refinement based on local tolerances $\mathrm{tol}_{x,y}$, $\mathrm{tol}_{x,\lambda}$ and tol_t is autonomously managed within the solver. Making use of tolerance proportionality the estimated local errors can be used to derive global error estimators. Now, the local tolerances can be chosen in such a way that the resulting adaptive grids lead to global error estimates which meet formula (2.11) and (2.12).

3.1. Adaptive discretization based on local error estimates

Considering problems of parabolic partial differential algebraic type, all systems, state as well as adjoint systems can be written in the abstract form

$$H\partial_t v = R(v) \quad \text{for } (x,t) \in \Omega \times (0, t_e), \tag{3.1}$$

$$v(t=0) = v^{(0)} \quad \text{for } x \in \Omega, \tag{3.2}$$

with singular matrix H and $v := y$ for the state, $v := s_y$ for the linearized state, $v := \lambda$ for the adjoint and $v := w$ for the second adjoint equation, which is used to evaluate an application of the reduced Hessian (see Section 3.3). The source vector $R(v)$ includes all differential operators supplemented with their boundary conditions.

Rosenbrock methods which are suitable to solve partial differential algebraic equations are described by the recursive linear implicit one-step scheme

$$v_{n+1} = v_n + \sum_{i=1}^{s} b_i V_i^n, \tag{3.3}$$

$$\left(\frac{H}{\gamma\tau_n} - \partial_v R(v_n)\right) V_i^n = \sum_{j=1}^{i-1} \frac{c_{ij}}{\tau_n} H V_j^n + R\big(v_n + \sum_{j=1}^{i-1} a_{ij} V_j^n\big),\ i = 1, \dots, s, \tag{3.4}$$

where τ_n denotes the step size, v_n the approximation to $v(t_n)$ at $t_n = \sum_{i=0,\dots,n-1} \tau_i$, and s is the number of stages. Rosenbrock methods are designed by inserting the exact Jacobian directly into the formula. This means that for the calculation of the intermediate values V_i^n, $i = 1, \dots, s$, only a sequence of linear systems with one and the same operator has to be solved. Furthermore, the one-step character of the scheme gives us the opportunity to quickly change the step size in every step. This step size is adapted automatically according to the local temporal discretization error which can easily be estimated by using an embedded formula of inferior order. They are also defined by (3.3)–(3.4) but with different coefficients $\hat{b}_i$. Then, local error estimates are provided without further computation by

$$le_n^t := \| \sum_{i=1}^{s} (\hat{b}_i - b_i) V_i^n \|.$$

The coefficients b_i, $\hat{b}_i$, c_{ij}, a_{ij}, and γ can be chosen to gain the desired order, high consistency and good stability properties like A- or L-stability [16, 20].

The equations (3.4) are linear elliptic problems which can be solved consecutively. This is done by means of an adaptive multilevel finite element method. The main idea of the multilevel technique consists of replacing the solution space by a sequence of discrete spaces with successively increasing dimensions to improve the approximation property. After computing approximations of the intermediate values V_i^n, $i = 1, \dots, s$, a posteriori error estimates can be utilized to give specific assessment of the error distribution [11]. To estimate the local spatial errors, we make use of the hierarchical basis concept. The errors are estimated by solving local Dirichlet problems on smaller subdomains as described in Ref. [20]. As long as a weighted sum of error estimators of all finite elements is greater than $\mathrm{tol}_{x,y}$, the spatial grid is refined where local estimators are too big. If local estimators are unreasonable small, the grid is locally coarsened as well. At the end, the spatial grid at each discrete time node is adapted in such a way, that $le_i^x \le tol_{x,y}$, where le_i^x denotes the spatial error estimator computed at $t = t_i$. Interpolation errors of the initial values $y_k(t_{i-1})$, resulting from changing grids, and the dynamic behavior of the new solution $y_k(t_i)$ are taken into account.

The linear algebraic systems resulting from the above-described discretization are solved by the stabilized bi-conjugated gradient method BiCGstab preconditioned with an incomplete LU-factorization.

This fully adaptive environment gives us the possibility to allow for appropriate independently refined spatial grids in each point of time, independently for each PDAE. The grids, estimated within the state solve, are used as an initial grid for the adjoint computations. However, it is important to control spatial error estimates in the backwards solve as well. This is due to the fact that a certain accuracy of the adjoint variables has to be guaranteed to ensure global convergence of the optimization algorithm. Furthermore, an independent adjoint mesh allows to fulfill the desired accuracy by using no more mesh points than necessary, in particular when type and behavior of state and adjoint PDAE systems differ.

Generally, the application of non-adjoint time integration schemes leads to inconsistency between the reduced gradient and the minimization problem itself. Therefore it is essential to control global discretization errors in state and adjoint system to ensure a sufficient consistency between discrete and continuous problem (see Algorithm 2.2 and equation (2.17)).

3.2. Global error estimation and multilevel strategy

Because of the semi-discretization the global error $y_k^{\mathrm{ex}} - y_k$ can be split into two parts, the temporal error $ge_k^t = y_k^{\mathrm{ex}} - Y_k^{\mathrm{ex}}$ and the spatial error $ge_k^x = Y_k^{\mathrm{ex}} - y_k$, where Y_k^{ex} denotes the linear interpolant in time of the exact solutions of the stationary PDAEs obtained after applying time integration. As usual, we bound the global error by controlling spatial and temporal discretizations separately, which yields $\|y_k^{\mathrm{ex}} - y_k\|_Y \le \|ge_k^t\|_Y + \|ge_k^x\|_Y$.

In what follows, we will exemplify our global error control in the case $Y = V = L^2(0, t_e; B)$, where B is an appropriate function space. This setting is used in our application with $B = H^1(\Omega)$, see Section 4. Suppose the time integrator has delivered discrete solutions at the time nodes $t_0 < t_1 < \cdots < t_{m_k - 1} < t_{m_k} = t_e$. Then, defining global errors in time, $e_i^t = \|y_k^{\mathrm{ex}}(t_i) - Y_k^{\mathrm{ex}}(t_i)\|_B$, we can approximate the overall temporal error ge_k^t through a weighted l_2-norm of the vector $e^t = (e_0^t, e_1^t, \ldots, e_{m_k}^t)$, defined by

$$\|e^t\|_{l_2} = \left(\sum_{k=0}^{m_k-1} \frac{\tau_k}{2} \left((e_k^t)^2 + (e_{k+1}^t)^2 \right) \right)^{1/2}.$$

Note the first component e_0^t equals zero, because exact initial values are used. Since we use approximate local time errors to steer the step size selection with respect to the equidistribution principle, we have to bridge the gap between local and global time errors. Let $Y_k(t_i)$ be the solution after one time step, starting with the exact initial value $y_k^{\text{ex}}(t_{i-1})$. The local time error is then defined by $le_i^t = \|y_k^{\text{ex}}(t_i) - Y_k(t_i)\|_B$. The error-per-step strategy we use ensures $le_i^t \le \text{tol}_t$ in each time step. It is a known fact that in this case the global time errors e_i^t are proportional to $(\text{tol}_t)^{p/(p+1)}$, where p is the order of the time integrator. This gives us the opportunity to reduce the global time error through downsizing the local tolerance tol_t. Based on the local error estimators in time, $le^t = (0, le_1^t, \ldots, le_{m_k}^t)$, we compute for the state variables y the scaled global quantity

$$ge_{y,\text{est}}^t = (t_e - t_0)^{-\frac{1}{2}} \|le^t\|_{l_2}. \tag{3.5}$$

Observe that having $le_i^t \le \text{tol}_t$ we get $ge_{y,\text{est}}^t \approx \text{tol}_t$.

Based on the temporary local hierarchical error estimators in space $le^x = (0, le_1^x, \ldots, le_{m_k}^x)$, see Section 3.1 and Ref. [20], we compute for the state variables y the scaled global quantity

$$ge_{y,\text{est}}^x = (t_e - t_0)^{-\frac{1}{2}} \|le^x\|_{l_2}. \tag{3.6}$$

Again, having $le_i^x \le \text{tol}_{x,y}$ we get $ge_{y,\text{est}}^x \approx \text{tol}_{x,y}$.

Clearly, the spatial and temporal accuracy of the adjoint variables can be controlled analogously. Let $ge_{\lambda,\text{est}}^t$ and $ge_{\lambda,\text{est}}^x$ be the corresponding scaled global quantities for the adjoint variables λ.

We then implement (2.12a) and (2.12b) by the checkable conditions

$$ge_{y,\text{est}}^t + ge_{y,\text{est}}^x \le c_y^* \, \|P_{U_{\text{ad}}^h - u_k}(-g_k)\|_U \tag{3.7a}$$

$$ge_{\lambda,\text{est}}^t + ge_{\lambda,\text{est}}^x \le c_\lambda^* \, \|P_{U_{\text{ad}}^h - u_k}(-g_k)\|_U \tag{3.7b}$$

with new constants c_y^* and c_λ^*. We emphasize that changing the constants bears no restrictions for the multilevel optimization since c_y and c_λ have to be prescribed by the user anyway. The new constants only take into account the hidden proportionality between local and global errors.

To balance temporal and spatial errors, we divide the constants c_y^* and c_λ^* into a time and a spatial component $c_y^t + c_y^x := c_y^*$ and $c_\lambda^t + c_\lambda^x := c_\lambda^*$, respectively. Appropriately reducing the local tolerances $\text{tol}_{x,y}$, $\text{tol}_{x,\lambda}$ and tol_t with respect to given constants c_y^t, c_y^x, c_λ^t, c_λ^x, we are able to meet (3.7a)–(3.7b) and eventually (2.11a)–(2.11b) with unknown, but well-defined constants c_y and c_λ. In general only one grid refinement is necessary to achieve the desired accuracy. Even though, success after the first refinement can not be guaranteed, the numerical observations made in [20, 8, 24] show that relying on tolerance proportionality is a reliable way to bridge the gap between local and global estimates.

3.3. Control update and inner iteration

To determine a new control update $u_{k+1} = u_k + s_u$ we have to solve the current local SQP problem (2.4). Here, reduced gradient and reduced Hessian are derived in infinite-dimensional vector spaces and computed by solving one or two additional systems of PDAEs. The reduced gradient, given by

$$\hat{J}'(u) = J_u(y,u) + e_u^*(y,u)\lambda, \tag{3.8}$$

is evaluated with the solution λ of the adjoint system

$$e_y^*(y,u)\lambda = -J_y(y,u). \tag{3.9}$$

The reduced Hessian applied to the direction s_u is defined as

$$\hat{J}''(u)s_u = J_{uu}(y,u)s_u + \lambda e_{uu}(y,u)s_u + e_u^*(y,u)w + J_{uy}(y,u)s_y + \lambda e_{uy}(y,u)s_y, \tag{3.10}$$

with second adjoint state w. It can be evaluated by solving two additional systems, the tangential state system

$$e_y(y,u)s_y = -e_u(y,u)s_u, \tag{3.11}$$

and the second adjoint system

$$e_y^*(y,u)w = -J_{yy}(y,u)s_y - \lambda e_{yy}(y,u)s_y - J_{yu}(y,u)s_u - \lambda e_{yu}(y,u)s_u. \tag{3.12}$$

So far, we do not consider control constraints within the KARDOS based optimization environment. However, test problems for Algorithm 2.2 with and without control constraints that use finite elements in space and Rosenbrock schemes in time together with uniform spatial and adaptive time refinements were computed in a MATLAB implementation with promising results [38].

Now, let $g_k \approx \hat{J}'(u_k)$ and $\hat{H}_k \approx \hat{J}''(u_k)$, then minimizing (2.4) coincides with solving the linear equation

$$\hat{H}_k s_u = -g_k, \tag{3.13}$$

additionally controlling the curvature of $\hat{H}_k$ and considering the trust region Δ_k. To solve (3.13) we apply the stabilized biconjugate gradient method BiCGstab [34]. As all Krylov methods, BiCGstab does not require the reduced Hessian $\hat{H}_k$ itself but only an application of it. In contrast to the standard CG, BiCGstab can deal with unsymmetric matrices as well. Even though the reduced Hessian is symmetric in theory, this is not necessarily the case when applying independent forward and backward discretizations. By controlling discretization errors in state and adjoint variables we automatically control the errors in symmetry. Therefore, in our setting the reduced Hessian is sufficiently close to being symmetric in every iteration. Nevertheless, our experience shows, that in real engineering applications BiCGstab works more reliable than a CG-method. Following the method of Steihaug [30] we augment BiCGstab by a verification of negative curvature and a step restriction to the trust region. Furthermore, we terminate this inner iteration, if the current inner iterate $s_u^{(j)}$ solves the system accurate enough. Here, we couple the inner accuracy with the outer accuracy by the exponent 1.5 which mirrors the order of the outer SQP-iteration [10]. In addition, we still force the generalized Cauchy decrease condition to hold.

Even though the computation of the reduced Hessian requires four PDAE solves per inner iteration, our experience shows that the additional work pays off in comparison to an SQP method using, for example, BFGS-updates.

4. Application to glass cooling problem

The presented multilevel generalized SQP method has been implemented within the framework of a newly developed all-in-one optimization environment based on the PDAE solver KARDOS. In this section we present numerical experiments and their results of the application to a space, time- and frequency-dependent radiative heat transfer problem modelling the cooling of hot glass. The state system $e(y, u)$ is described by a semilinear partial differential algebraic equation with high non-linearities in control and state due to energy exchange caused by radiation.

4.1. The SP_1 N-band model for radiative heat transfer

Because the high dimension of the phase space makes the numerical solution of the full radiative heat transfer equation very expensive, especially for optimization purposes, we use a first-order approximation of spherical harmonics including a practically relevant frequency bands model with N bands. This SP_1-approximation has been tested fairly extensively for various radiation transfer problems in glass and has proven to be an efficient way to improve the classical diffusion approximations [25]. It results in the following space-time-dependent system of semilinear partial differential algebraic equations of mixed parabolic-elliptic type in $N+1$ components:

$$\partial_t T - \nabla \cdot (k_c \nabla T) = \sum_{i=1}^{N} \nabla \cdot \left(\frac{1}{3(\sigma_i + \kappa_i)} \nabla \phi_i \right), \tag{4.1}$$

$$-\epsilon^2 \nabla \cdot \left(\frac{1}{3(\sigma_i + \kappa_i)} \nabla \phi_i \right) + \kappa_i \phi_i = 4\pi \kappa_i B^{(i)}(T), \quad i = 1, \dots, N, \tag{4.2}$$

with boundary and initial conditions

$$k_c n \cdot \nabla T + \sum_{i=1}^{N} \frac{1}{3(\sigma_i + \kappa_i)} n \cdot \nabla \phi_i \tag{4.3}$$

$$= \frac{h_c}{\epsilon}(u - T) + \frac{\alpha \pi}{\epsilon} \left(\frac{n_a}{n_g} \right)^2 \left(B^{(0)}(u) - B^{(0)}(T) \right) + \frac{c_1}{\epsilon} \sum_{i=1}^{N} \left(4\pi B^{(i)}(u) - \phi_i \right),$$

$$\frac{\epsilon^2}{3(\sigma_i + \kappa_i)} n \cdot \nabla \phi_i = c_1 \epsilon \left(4\pi B^{(i)}(u) - \phi_i\right), \quad i = 1, \ldots, N, \tag{4.4}$$

$$T(x, 0) = T_0(x). \tag{4.5}$$

The state y consists of the glass temperature $T(x,t)$ and the mean radiative intensities $\phi_i(x,t)$, $i = 1, \ldots, N$. Splitting up the continuous frequency spectrum into N bands $[\nu_{i-1}, \nu_i]$, $i = 0, \ldots, N$, with $\nu_N = \infty$ and $\nu_{-1} := 0$ they are defined as

$$\phi_i(x,t) := \int_{\nu_{i-1}}^{\nu_i} \int_{\mathbb{S}^2} I(x,t,\nu,s) \mathrm{d}s \mathrm{d}\nu, \; i = 1, \ldots, N, \tag{4.6}$$

with space, time, frequency and directional dependent radiative intensity

$$I(x,t,\nu,s).$$

Furthermore, the frequency-dependent absorption and scattering coefficients κ_ν and σ_ν and the Planck function

$$B(T,\nu) = \frac{n_g^2}{c_0^2} \frac{2h_p \nu^3}{e^{h_p \nu/(k_b T)} - 1}, \tag{4.7}$$

with Planck constant h_b, Boltzmann constant k_b and the speed of light in vacuum c_0, are set constant with respect to the frequency on each of the bands,

$$\kappa_\nu = \kappa_i, \quad \sigma_\nu = \sigma_i, \; \text{for } \nu \in [\nu_{i-1}, \nu_i], \qquad i = 1, \ldots, N, \tag{4.8}$$

$$B^{(i)}(v) := \int_{\nu_{i-1}}^{\nu_i} B(v,\nu) \mathrm{d}\nu, \qquad i = 1, \ldots, N. \tag{4.9}$$

The system is control by the furnace temperature $u(t)$ which is reduced from values close to the initial glass temperature T_0 towards room temperature. Further quantities are the heat conductivity k, the heat convection h, the optical thickness ϵ, the ratio of the speed of light in vacuum and in glass n_g, the index of refraction of air n_a and the mean hemispheric surface emissivity α in the opaque spectral region $[0, \nu_0]$, where glass strongly absorbs radiation. Division by proper reference quantities brings the system to the above-presented dimensionless form.

4.1.1. The objective functional. In the optimization of glass cooling processes, it is important for the quality of the glass that its temperature distribution follows a desired profile to control the chemical reactions in the glass. Typically, there are time-dependent reference values provided by engineers for the furnace temperature as well, which should be considered within the objective. To allow for a different control value at final time than the given reference value, we additionally include a tracking of the glass temperature at final time. Without this term the optimal control at the final time would coincide with the reference control even though there might be a control with smaller objective value which differs only at final time, as discussed in [5]. The above-described requirements can be summarized in the following objective functional:

$$J(T,u) := \frac{1}{2} \int_0^{t_e} \|T - T_d\|_{L^2(\Omega)}^2 \, \mathrm{d}t + \frac{\delta_e}{2} \|(T - T_d)(t_e)\|_{L^2(\Omega)}^2 + \frac{\delta_u}{2} \int_0^{t_e} (u - u_d)^2 \mathrm{d}t,$$

with the desired glass temperature distribution $T_d(x,t)$ and a guideline for the control $u_d(t)$. The final time of the cooling process is denoted by t_e. The positive weights δ_e and δ_u are used to steer the influence of the corresponding components.

4.2. Numerical experiments

In the following we will interpret the glass as an infinitely long block. It is therefore sufficient to consider a two-dimensional cross section as computational domain. The model and problem specific quantities used are summarized in Table 1. We

Ω	comp. domain	$[-3,3]\times[-1,1]$	c_y^t	state time error const.	$5.0e-3$
t_e	final time	$1.0e-1$	c_y^x	state space error const.	$5.0e-2$
$T_0(x)$	initial glass temp.	$9.0e+2$	c_λ^t	adjoint time error const.	$5.0e-3$
			c_λ^x	adjoint space error const.	$1.0e-1$
$T_d(t)$	desired glass temp.	$T_0\cdot e^{\frac{-\log(\frac{T_0}{300})t}{t_e}}$	N	number of frequency bands	8
			k	conductivity coefficient	$1.0e+0$
$u_0(t)$	initial control	$T_d(t)$	h	convection coefficient	$1.0e-3$
$u_d(t)$	desired control	$T_d(t)$	κ	absorption coefficient	$1.0e+0$
δ_e	final value weight	$5.0e-2$	ϵ	optical thickness coeff.	$1.0e+1$
δ_u	control reg. weight	$1.0e-1$	a	radiated energy coeff.	$1.8e-8$

TABLE 1. Problem and Model Specific Qualities

start the optimization on quite coarse grids, with 9 points in time and a number of spatial nodes between 181 and 413 forced by the local tolerances $\text{tol}_t = \text{tol}_{x,y} = 5e-2$ and $\text{tol}_{x,\lambda} = 1e-1$. For the time discretization we apply ROS3PL, which is an L-stable order three method for nonlinear stiff differential algebraic equations of index one and is especially designed to avoid order reduction for parabolic problems [22, 23].

The optimal control iterates, the corresponding reduced gradients and the resulting glass temperature in the interior point $(1,1)$ are presented in Figure 1. In each subfigure, the lightest line represents the initial quality and the darkest the final or optimal quality, additionally labelled with its iteration number.

A schematic representation showing the performance of the optimization and the simultaneous grid refinement is presented in Figure 2. During the optimization the reduced gradient is reduced significantly (see Figure 2). After 3 iterations the scaled state time error $ge^t_{y,\text{est}}/c^t_y$ and the reduced gradient norm intersect. This means, criteria (3.7a) is violated and the grid has to be refined. We adjust the local tolerance tol_t with respect to the ratio of violating error estimate and gradient norm. Reading from left to right each intersection of the gradient norm with any of the four scaled global error estimates correlates with a grid refinement. The increasing slope of the reduced gradient decrease throughout the optimization process mirrors the superlinear or even quadratic convergence of the optimization method. We terminate the optimization if the reduced gradient norm is less then 10^{-2} which implies a reduction by more than 4 decimal orders. Since we refine the

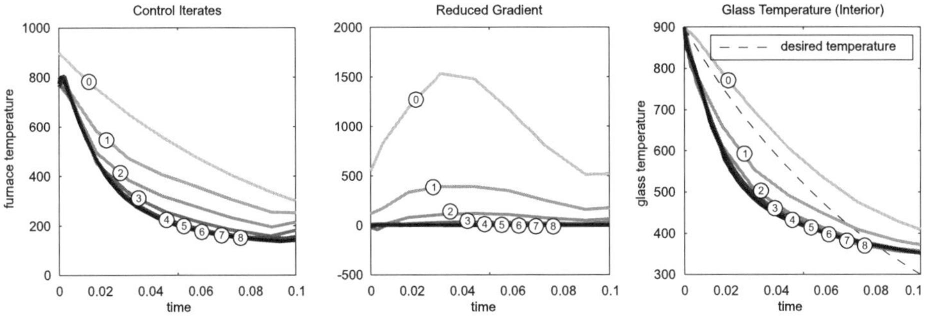

FIGURE 1. Optimization Results (Multilevel SQP Method)

Optimal control iterates, corresponding reduced gradients and resulting glass temperature in the interior point $(1,1)$ determined by the presented SQP method. The lightest line (0) represents the initial quality and the darkest (8) the final or optimal quality.

meshes as the discrete reduced gradient approaches towards zero, it can be shown that the infinite-dimensional gradient converges towards zero. These observations coincide with the theoretical results presented in Section 2 and Ref. [38].

Additional information about the optimization progress and the grid refinement is presented in Table 2. Each horizontal line and the row below represent a grid refinement. After each grid refinement the reduced gradient norm increases,

opt it	target value	gradient norm	time nodes	space nodes	BiCGstab loops	term. reason	overall cpu time
0	2.19932e+4	3.5060e+2	9	181-413			7.57e+1
1	1.31374e+4	9.6134e+1	9	181-413	2	res. small	6.62e+2
2	1.17238e+4	2.7958e+1	9	181-413	2	res. small	1.37e+3
3	1.14556e+4	7.5189e+0	9	181-413	2	res. small	2.07e+3
3	1.16706e+4	7.7070e+0	13	181-413			2.17e+3
4	1.16387e+4	2.3535e+0	13	181-413	3	max no it.	3.07e+3
4	1.16517e+4	2.1034e+0	16	181-413			3.17e+3
5	1.16484e+4	3.4187e−1	16	181-413	3	max no it.	4.16e+3
5	1.15196e+4	1.0511e+0	30	181-441			4.40e+3
6	1.15182e+4	3.8596e−1	30	181-441	3	max no it.	6.97e+3
6	1.13976e+4	8.1932e−1	35	181-629			7.31e+3
7	1.13970e+4	5.8765e−2	35	181-629	3	max no it.	1.09e+4
7	1.15477e+4	3.0908e+0	93	413-3435			1.71e+4
8	1.15434e+4	2.2688e−1	93	413-3435	3	max no it.	7.45e+4
9	1.15432e+4	7.3990e−3	93	413-3435	3	max no it.	1.32e+5

TABLE 2. Optimization Protocol (Multilevel SQP Method)

due to marginal differences between the adjoint values on old and new grid. Nevertheless, after just one or two optimization iteration the reduced gradient is significantly reduced in comparison to the last iterate on the old grid. Column 6 and 7 show the number of BiCGstab iterations within the inner iteration and the reason to terminate. Our experience shows that a maximum iteration number of 3 gives satisfactory results. In column 8 the overall computing time up to the corresponding optimization iteration is given.

Clearly, carrying out all optimization iterations on the same level requires less optimization iterations, since there is no increase of the reduced gradient after a grid refinement. But having a look at the overall computing time in Table 2 it can be seen that the optimization iteration on the finest level does take more than

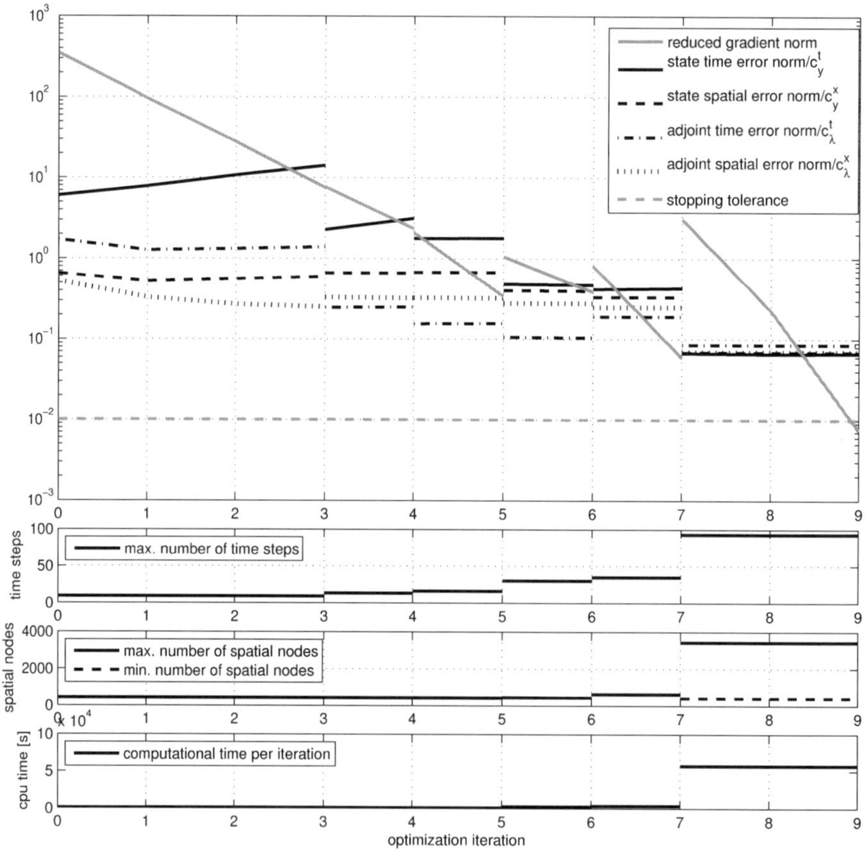

FIGURE 2. Performance of Multilevel SQP Method

Error estimators via number of optimization steps. Each intersection of the reduced gradient norm with one of the scaled global error estimators gives rise to a grid refinement.

three times longer than the whole optimization took so far. This means that only two additional iteration on the finest level would almost double the computational effort.

To study the computational benefit of the presented multilevel approach we consider a less time consuming gray scale model approximating the same underlying radiative heat transfer problem [6]. Comparing the results presented above to those of an SQP method with fixed local tolerances comparable to those of the finest level we generally observe a difference in computing time by at least a factor 5.

It is also interesting to verify whether the additional effort for the computation of the reduced Hessian and the iterative solving of system (3.13) really pays off. Therefore, we compare the multilevel SQP method to a multilevel gradient method with Armijo line search [5, 21]. Even though one optimization iteration of the gradient method is significantly faster than one of the SQP method, there are also significantly more iterations necessary to reduce the scaled gradient under the predefined tolerance which all in all leads to a multiplication of computing time by a factor 5.

If the number of spatial nodes differs significantly, like it does on the finest level (see Table 2), it is also of great computational benefit to allow independent spatial grids at different points of time and for different PDAEs. Numerical experiments show that if all computations would be carried out on the same spatial mesh, one optimization iteration serving the highest accuracy would take significantly longer than in the adaptive case.

Summing up, we can say that both, the higher order of convergence and the multilevel strategy, each save approximately 80% of the computing time. This means computational savings of approximately 96% in comparison to an optimization method with linear order of convergence and without multilevel strategy. In addition we can observe substantial savings in computation time due to independent and fully adaptive space-time grids.

5. Conclusions

We have presented an adaptive multilevel generalized SQP method to solve PDAE-constrained optimal control problems which allows the use of independent discretization schemes to integrate state and adjoint equations. The algorithm controls the accuracy of the state and adjoint solver in such a way that the solution of the infinite-dimensional problem is approached efficiently by using a hierarchy of adaptively refined discretizations. In particular the inconsistencies of the reduced gradient resulting from an independent discretization of state equation and adjoint equation are controlled in a rigorous way to ensure sufficient progress of the optimization iteration. The accuracy requirements of the multilevel generalized SQP method can be implemented by a posteriori error estimators and provide a modular interface to existing solvers with a posteriori error estimation. In this work we use global error estimators based on local strategies and the principle of tolerance proportionality in time and hierarchical basis techniques in space.

The problem solving environment of the PDAE solver KARDOS was used to implement our adaptive solution strategy and to apply it to a two-dimensional radiative heat transfer problem governed by a complex nonlinear system of time-dependent partial differential algebraic equations.

Acknowledgement

The authors gratefully acknowledge the support of the German Research Foundation (DFG) within the Priority Program 1253 "Optimization with Partial Differential Equations" under grants LA1372/6-2 and UL158/7-2. Furthermore, we thank the referee for the constructive comments.

References

[1] R. Becker, M. Braack, D. Meidner, R. Rannacher, and B. Vexler. Adaptive finite element methods for PDE-constrained optimal control problems. In *Reactive flows, diffusion and transport*, pages 177–205. Springer, Berlin, 2007.

[2] A. Borzì. Smoothers for control- and state-constrained optimal control problems. *Comput. Vis. Sci.*, 11(1):59–66, 2008.

[3] A. Borzì and K. Kunisch. A multigrid scheme for elliptic constrained optimal control problems. *Comput. Optim. Appl.*, 31(3):309–333, 2005.

[4] R.H. Byrd. Robust trust region methods for constrained optimization. In *Third SIAM Conference on Optimization, Houston, TX*, 1987.

[5] D. Clever and J. Lang. Optimal control of radiative heat transfer in glass cooling with restrictions on the temperature gradient. Preprint 2595, Technische Universität Darmstadt, 2009.

[6] D. Clever, J. Lang, S. Ulbrich, and J.C. Ziems. Combination of an adaptive multilevel SQP method and a space-time adaptive PDAE solver for optimal control problems. Appears in Procedia Computer Science, 2010.

[7] Andrew R. Conn, N.I.M. Gould, and P.L. Toint. *Trust-region methods.* MPS/SIAM Series on Optimization. Society for Industrial and Applied Mathematics (SIAM), Philadelphia, PA, 2000.

[8] K. Debrabant and J. Lang. On global error estimation and control for parabolic equations. Technical Report 2512, Technische Universität Darmstadt, 2007.

[9] J.E. Dennis, Jr., M. El-Alem, and M.C. Maciel. A global convergence theory for general trust-region-based algorithms for equality constrained optimization. *SIAM J. Optim.*, 7(1):177–207, 1997.

[10] P. Deuflhard. Global inexact Newton methods for very large scale nonlinear problems. In *IMPACT of Computing in Science and Engineering*, pages 366–393, 1991.

[11] P. Deuflhard, P. Leinen, and H. Yserentant. Concepts of an adaptive hierarchical finite element code. *Impact of Comput. in Sci. and Engrg.*, 1:3–35, 1989.

[12] T. Dreyer, B. Maar, and V. Schulz. Multigrid optimization in applications. *J. Comput. Appl. Math.*, 120(1-2):67–84, 2000. SQP-based direct discretization methods for practical optimal control problems.

[13] B. Erdmann, J. Lang, and R. Roitzsch. KARDOS-User's Guide. Manual, Konrad-Zuse-Zentrum Berlin, 2002.

[14] S. Gratton, M. Mouffe, Ph.L. Toint, and M. Weber-Mendonca. A recursive trust-region method in infinity norm for bound-constrained nonlinear optimization. *IMA Journal of Numerical Analysis*, 2008 (to appear).

[15] S. Gratton, A. Sartenaer, and P.L. Toint. Recursive trust-region methods for multiscale nonlinear optimization. *SIAM J. Optim.*, 19(1):414–444, 2008.

[16] E. Hairer and G. Wanner. *Solving Ordinary Differential Equations II, Second Revised Edition.* Springer Verlag, 2002.

[17] M. Heinkenschloss and L.N. Vicente. Analysis of inexact trust-region SQP algorithms. *SIAM J. Optim.*, 12(2):283–302, 2001/02.

[18] M. Hinze, R. Pinnau, M. Ulbrich, and S. Ulbrich. *Optimization with PDE constraints.* Mathematical Modelling: Theory and Applications (23). Springer, 2008.

[19] C.T. Kelley. *Iterative Methods for Optimization.* Frontiers in Applied Mathematics. SIAM, 1999.

[20] J. Lang. *Adaptive Multilevel Solution of Nonlinear Parabolic PDE Systems. Theory, Algorithm, and Applications*, volume 16 of *Lecture Notes in Computational Science and Engineering.* Springer Verlag, 2000.

[21] J. Lang. Adaptive computation for boundary control of radiative heat transfer in glass. *Journal of Computational and Applied Mathematics*, 183:312–326, 2005.

[22] J. Lang and D. Teleaga. Towards a fully space-time adaptive FEM for magnetoquasistatics. *IEEE Transactions on Magnetics*, 44,6:1238–124, 2008.

[23] J. Lang and J. Verwer. ROS3P – an accurate third-order Rosenbrock solver designed for parabolic problems. *BIT*, 41:730–737, 2001.

[24] J. Lang and J. Verwer. On global error estimation and control for initial value problems. *SIAM J. Sci. Comput.*, 29:1460–1475, 2007.

[25] E.W. Larsen, G. Thömmes, A. Klar, M. Seaïd, and T. Götz. Simplified P_N approximations to the equations of radiative heat transfer and applications. *Journal of Computational Physics*, 183:652–675, 2002.

[26] R.M. Lewis and S.G. Nash. Model problems for the multigrid optimization of systems governed by differential equations. *SIAM J. Sci. Comput.*, 26(6):1811–1837 (electronic), 2005.

[27] D. Meidner and B. Vexler. Adaptive space-time finite element methods for parabolic optimization problems. *SIAM J. Control Optim.*, 46(1):116–142 (electronic), 2007.

[28] E.O. Omojokun. Trust region algorithms for optimization with nonlinear equality and inequality constraints. PhD thesis, University of Colorado, Boulder, Colorado, USA, 1989.

[29] L. Petzold. Adjoint sensitivity analysis for time-dependent partial differential equations with adaptive mesh refinement. Journal of Computational Physics (198), 310-325, 2004.

[30] T. Steihaug. The conjugate gradient method and trust regions in large scale optimization. *SIAM, Journal of Numerical Analysis*, 20,1:626–637, 1983.

[31] F. Tröltzsch. *Optimale Steuerung partieller Differentialgleichungen.* Vieweg, 2005.

[32] S. Ulbrich. Generalized SQP methods with "parareal" time-domain decomposition for time-dependent PDE-constrained optimization. In *Real-time PDE-constrained optimization*, volume 3 of *Comput. Sci. Eng.*, pages 145–168. SIAM, Philadelphia, PA, 2007.

[33] M. Vallejos and A. Borzì. Multigrid optimization methods for linear and bilinear elliptic optimal control problems. *Computing*, 82(1):31–52, 2008.

[34] H.A. van der Vorst. Bi-CGSTAB: A fast and smoothly converging variant of bi-cg for the solution of nonsymmetric linear systems. *SIAM Journal on Scientific and Statistical Computing*, 13:631–644, 1992.

[35] R. Verfürth. A posteriori error estimators for convection-diffusion equations. *Numer. Math.*, 80(4):641–663, 1998.

[36] R. Verfürth. A posteriori error estimates for linear parabolic equations. Preprint, Ruhr-Universität Bochum, Fakultät für Mathematik, Bochum, Germany, 2004.

[37] R. Verfürth. A posteriori error estimates for non-linear parabolic equations. Preprint, Ruhr-Universität Bochum, Fakultät für Mathematik, Bochum, Germany, 2004.

[38] J.C. Ziems. *Adaptive Multilevel SQP-methods for PDE-constrained optimization.* Dissertation, Fachbereich Mathematik, Technische Universität Darmstadt. Verlag Dr. Hut, München, 2010.

[39] J.C. Ziems and S. Ulbrich. Adaptive multilevel inexact SQP methods for PDE-constrained optimization. *SIAM J. Optim.*, 21(1):1–40, 2011.

Debora Clever, Jens Lang and Stefan Ulbrich
Technische Universität Darmstadt
Department of Mathematics and
Graduate School of Computational Engineering
Dolivostr. 15
D-64293 Darmstadt, Germany
e-mail: clever@mathematik.tu-darmstadt.de
lang@mathematik.tu-darmstadt.de
ulbrich@mathematik.tu-darmstadt.de

Carsten Ziems
Technische Universität Darmstadt
Department of Mathematics
Dolivostr. 15
D-64293 Darmstadt, Germany
e-mail: ziems@mathematik.tu-darmstadt.de

International Series of Numerical Mathematics, Vol. 160, 75–98

Projection Based Model Reduction for Optimal Design of the Time-dependent Stokes System

Thomas Franke, Ronald H.W. Hoppe, Christopher Linsenmann and Achim Wixforth

Abstract. The optimal design of structures and systems described by partial differential equations (PDEs) often gives rise to large-scale optimization problems, in particular if the underlying system of PDEs represents a multi-scale, multi-physics problem. Therefore, reduced order modeling techniques such as balanced truncation model reduction, proper orthogonal decomposition, or reduced basis methods are used to significantly decrease the computational complexity while maintaining the desired accuracy of the approximation. In particular, we are interested in such shape optimization problems where the design issue is restricted to a relatively small portion of the computational domain. In this case, it appears to be natural to rely on a full order model only in that specific part of the domain and to use a reduced order model elsewhere. A convenient methodology to realize this idea consists in a suitable combination of domain decomposition techniques and balanced truncation model reduction. We will consider such an approach for shape optimization problems associated with the time-dependent Stokes system and derive explicit error bounds for the modeling error.

As an application in life sciences, we will be concerned with the optimal design of capillary barriers as part of a network of microchannels and reservoirs on microfluidic biochips that are used in clinical diagnostics, pharmacology, and forensics for high-throughput screening and hybridization in genomics and protein profiling in proteomics.

Mathematics Subject Classification (2000). Primary 65K10; Secondary 49M05; 49M15; 65M55; 65M60; 76Z99; 90C06.

Keywords. Projection based model reduction, shape optimization, time-dependent Stokes system; domain decomposition, balanced truncation, biochips.

The authors acknowledge support by the German National Science Foundation DFG within the DFG Priority Program SPP 1253 'Optimierung mit partiellen Differentialgleichungen'.

1. Introduction

The numerical solution of optimization problems governed by time-dependent partial differential equations (PDEs) can be computationally very expensive with respect to both storage and CPU times. Therefore, serious attempts have been undertaken to achieve a significant reduction of the computational costs based on Reduced Order Models (ROMs). ROMs determine a subspace that contains the 'essential' dynamics of the time-dependent PDEs and project these PDEs onto the subspace. If the subspace is small, the original PDEs in the optimization problem can be replaced by a small system of ordinary differential equations and the resulting approximate optimization problem can be solved efficiently. Among the most commonly used ROM techniques are balanced truncation model reduction, Krylov subspace methods, proper orthogonal decomposition (POD), and reduced basis methods (see, e.g., the books and survey articles [8, 10, 11, 13, 21, 45, 52] and the references therein).

Among the challenges one has to overcome when one wants to apply ROM techniques for optimization problems are the efficient computation of ROMs for use in optimization and the derivation of error estimates for ROMs. Some aspects of these questions have been addressed in [3, 4, 16, 17, 18, 31, 32, 33, 36, 37, 38, 41, 42, 49, 50, 55, 56, 57]. In most of these applications, estimates for the error between the solution of the original optimization problem and the optimization problem governed by the reduced order model are not available; if they exist, then only for a restricted class of problems. Different types of error estimates for some ROM approaches have been discussed in, e.g., [3, 4, 24, 25, 29, 31, 32, 48, 55].

In this contribution, which is based on [2, 4], we will apply balanced truncation model reduction (BTMR) to the optimal design of systems whose operational behavior is described by the time-dependent Stokes equations. Since BTMR is essentially restricted to linear time-invariant systems, whereas optimal design problems are genuinely nonlinear in nature due to the nonlinear dependence on the design variables, we consider a semi-discretization in space of the time-dependent Stokes equations (Section 2) and focus on such problems where the design is restricted to a relatively small part of the computational domain. Following the exposition in [4], we use a combination of domain decomposition and BTMR in the sense that we keep the full order model for the subdomain subject to the design and use BTMR for the rest of the domain (Sections 3 and 4). It turns out that the reduced optimality system represents the necessary optimality conditions of a reduced optimal design problem featuring a reduced objective functional. This observation is the key for an a priori analysis of the modeling error (Section 5). It should be emphasized that the full order model refers to the semi-discretized Stokes system. Hence, the error induced by the discretization in space is not taken into account. The main result states that under some assumptions on the objective functional the error between the optimal design for the full order model and the reduced order model is bounded by a constant times the sum of those Hankel singular values in the associated singular value decomposition involving the controllability and observability Gramians that are not used for the derivation of the ROM.

The combination of domain decomposition and balanced truncation model reduction (DDBTMR) is applied to the optimal design of surface acoustic wave driven microfluidic biochips which are used for high-throughput screening and hybridization in genomics and protein profiling in proteomics (see [2]). In particular, we consider the shape optimization of capillary barriers between the channels on top of the biochips and reservoirs where the chemical analysis of the probes is performed (Section 6).

2. Optimal design of the Stokes equations and semi-discretization in space

Let $\Omega(\theta) \subset \mathbb{R}^2$ be a bounded domain that depends on design variables $\theta = (\theta_1, \ldots, \theta_d)^T \in \Theta$, where $\theta_i, 1 \leq i \leq d$, are the Bézier control points of a Bézier curve representation of the boundary $\partial\Omega(\theta)$ and Θ stands for the convex set

$$\Theta := \{\theta_i \in \mathbb{R} \mid \theta_{\min}^{(i)} \leq \theta_i \leq \theta_{\max}^{(i)}, 1 \leq i \leq d\},$$

with $\theta_{\min}^{(i)}, \theta_{\max}^{(i)}, 1 \leq i \leq d$, being given. We assume that the boundary $\partial\Omega(\theta)$ consists of an inflow boundary $\Gamma_{\text{in}}(\theta)$, an outflow boundary $\Gamma_{\text{out}}(\theta)$, and a lateral boundary $\Gamma_{\text{lat}}(\theta)$ such that $\partial\Omega(\theta) = \overline{\Gamma}_{\text{in}}(\theta) \cup \overline{\Gamma}_{\text{out}}(\theta) \cup \overline{\Gamma}_{\text{lat}}(\theta)$ with pairwise disjoint $\Gamma_{\text{in}}(\theta), \Gamma_{\text{out}}(\theta), \Gamma_{\text{lat}}(\theta)$. We set $Q(\theta) := \Omega(\theta) \times (0,T), \Sigma(\theta) := \partial\Omega(\theta) \times (0,T), \Sigma_{\text{in}}(\theta) := \Gamma_{\text{in}}(\theta) \times (0,T), \Sigma_{\text{lat}}(\theta) := \Gamma_{\text{lat}}(\theta) \times (0,T), T > 0$. Denoting by $v = (v_1, v_2)^T$ and p the velocity and the pressure of a fluid with viscosity ν in $\overline{\Omega}(\theta)$ and by $\ell(v,p,x,t,\theta)$ a given function of the velocity, the pressure, the independent variables x, t, and the design variable θ, we consider optimal design problems associated with the time-dependent Stokes system of the form

$$\inf_{\theta \in \Theta} J(\theta) := \frac{1}{2} \int_0^T \int_{\Omega(\theta)} \ell(v,p,x,t,\theta)\, dxdt, \tag{2.1a}$$

subject to the Stokes flow

$$\frac{\partial v}{\partial t} - \nu\Delta v + \nabla p = f \quad \text{in } Q(\theta), \tag{2.1b}$$

$$\nabla \cdot v = 0 \quad \text{in } Q(\theta), \tag{2.1c}$$

$$v = v^{\text{in}} \quad \text{on } \Sigma_{\text{in}}(\theta), \tag{2.1d}$$

$$v = 0 \quad \text{on } \Sigma_{\text{lat}}(\theta), \tag{2.1e}$$

$$(\nu\nabla v - pI)n = 0 \quad \text{on } \Sigma_{\text{out}}(\theta), \tag{2.1f}$$

$$v(\cdot, 0) = v^{(0)} \quad \text{in } \Omega(\theta), \tag{2.1g}$$

where f is a given forcing term, v^{in} stands for a prescribed velocity at the inflow boundary, n in (2.1f) is the unit exterior normal, and $v^{(0)}$ is the velocity distribution at initial time $t = 0$ with $\nabla \cdot v^{(0)} = 0$.

The discretization in space and time of the optimality system associated with (2.1a)–(2.1g) gives rise to a large-scale nonlinear optimization problem whose numerical solution requires considerable computational efforts, even if efficient solvers such as those based on 'all-at-once methods' are used. A significant reduction of the computational complexity can be achieved by projection based model reduction applied to a semi-discretization in space of the time-dependent Stokes equations, e.g., by stable continuous elements such as the Taylor-Hood P2-P1 element with respect to a simplicial triangulation of the spatial domain (cf., e.g., [14, 15, 22]). Then, the semi-discrete optimization problem reads

$$\inf_{\theta \in \Theta} \mathbf{J}(\theta) := \int_0^T \boldsymbol{\ell}(\mathbf{v}, \mathbf{p}, \theta, t)\, dt. \tag{2.2a}$$

The integrand $\boldsymbol{\ell}$ in (2.2a) stems from the semidiscretization of the inner integral in (2.1a), and the pair $(\mathbf{v}, \mathbf{p})$ is assumed to solve the Hessenberg index 2 system

$$\begin{pmatrix} \mathbf{M}(\theta) & 0 \\ 0 & 0 \end{pmatrix} \frac{d}{dt} \begin{pmatrix} \mathbf{v}(t) \\ \mathbf{p}(t) \end{pmatrix} = - \begin{pmatrix} \mathbf{A}(\theta) & \mathbf{B}^T(\theta) \\ \mathbf{B}(\theta) & 0 \end{pmatrix} \begin{pmatrix} \mathbf{v}(t) \\ \mathbf{p}(t) \end{pmatrix} + \begin{pmatrix} \mathbf{K}(\theta) \\ \mathbf{L}(\theta) \end{pmatrix} \mathbf{f}(t), \quad t \in (0, T], \tag{2.2b}$$

$$\mathbf{M}(\theta)\mathbf{v}(0) = \mathbf{v}^0, \tag{2.2c}$$

where $\mathbf{M}(\theta), \mathbf{A}(\theta) \in \mathbb{R}^{n\times n}$ stand for the mass and stiffness matrices, $\mathbf{B}(\theta) \in \mathbb{R}^{m\times n}$, $m < n$, refers to the discrete divergence operator, and $\mathbf{K}(\theta) \in \mathbb{R}^{n\times k}, \mathbf{L}(\theta) \in \mathbb{R}^{m\times k}, \mathbf{f}(t) \in \mathbb{R}^k, t \in (0, T)$. A Hessenberg index 2 system is an index 2 differential algebraic system where the algebraic variable is absent from the algebraic equation.

Under some assumptions on the matrices $\mathbf{M}(\theta), \mathbf{A}(\theta)$, and $\mathbf{B}(\theta)$, which are satisfied when using stable continuous elements, we can show continuous dependence of the solution of (2.2b), (2.2c) on the data. For a more detailed discussion we refer to [4]. This result will play a significant role in the a priori analysis of the modeling error in Section 5. For ease of notation we drop the dependence on the design variable θ.

Theorem 2.1. *Assume that $\mathbf{M} \in \mathbb{R}^{n\times n}$ is symmetric positive definite, $\mathbf{A} \in \mathbb{R}^{n\times n}$ is symmetric positive definite on Ker $\mathbf{B}$, i.e.,*

$$\mathbf{v}^T \mathbf{A} \mathbf{v} \geq \alpha\, \|\mathbf{v}\|^2 \quad , \quad \mathbf{v} \in Ker\, \mathbf{B}, \tag{2.3}$$

and $\mathbf{B} \in \mathbb{R}^{m\times n}$ has full row rank m. Then, there exist positive constants C_1 and C_2 such that

$$\begin{pmatrix} \|\mathbf{v}\|_{L^2} \\ \|\mathbf{p}\|_{L^2} \end{pmatrix} \leq C_1\, \|\mathbf{v}^0\| + C_2 \begin{pmatrix} \|\mathbf{f}\|_{\mathbf{L^2}} \\ \|\mathbf{f}\|_{L^2} + \|\frac{d}{dt}\mathbf{f}\|_{L^2} \end{pmatrix}. \tag{2.4}$$

Proof. Following [43], we introduce $\mathbf{\Pi} := \mathbf{I}-\mathbf{B}^T(\mathbf{B}\mathbf{M}^{-1}\mathbf{B}^T)^{-1}\mathbf{B}\mathbf{M}^{-1}$ as an oblique projection onto Ker $\mathbf{B}^T$ along Im $\mathbf{B}$ and split $\mathbf{v}(t) = \mathbf{v}_H(t)+\mathbf{v}_P(t)$, where $\mathbf{v}_H(t) \in$ Ker $\mathbf{B}$ and

$$\mathbf{v}_P(t) := \mathbf{M}^{-1}\mathbf{B}^T(\mathbf{B}\mathbf{M}^{-1}\mathbf{B}^T)^{-1}\mathbf{L}\mathbf{f}(t) \tag{2.5}$$

is a particular solution of the second equation of the semi-discrete Stokes system (2.2b), (2.2c). The semi-discrete Stokes system transforms to

$$\mathbf{\Pi}\mathbf{M}\mathbf{\Pi}^T \frac{d}{dt}\mathbf{v}_H(t) = -\mathbf{\Pi}\mathbf{A}\mathbf{\Pi}^T \mathbf{v}_H(t) + \mathbf{\Pi}\widetilde{\mathbf{K}}\mathbf{f}(t) \quad , \quad t \in (0,T], \tag{2.6a}$$

$$\mathbf{\Pi}\mathbf{M}\mathbf{\Pi}^T \mathbf{v}_H(0) = \mathbf{\Pi}\mathbf{v}^0. \tag{2.6b}$$

The pressure $\mathbf{p}$ can be recovered according to

$$\mathbf{p}(t) = (\mathbf{B}\mathbf{M}^{-1}\mathbf{B}^T)^{-1}\Big(\mathbf{B}\mathbf{M}^{-1}\Big(-\mathbf{A}\mathbf{v}_H(t) + \widetilde{\mathbf{K}}\mathbf{f}(t)\Big) - \mathbf{L}\frac{d}{dt}\mathbf{f}(t)\Big), \tag{2.7}$$

where $\widetilde{\mathbf{K}} := \mathbf{K} - \mathbf{A}\mathbf{M}^{-1}\mathbf{B}^T(\mathbf{B}\mathbf{M}^{-1}\mathbf{B}^T)^{-1}\mathbf{L}$. The assertion follows from Gronwall's lemma applied to (2.6a), (2.6b) and from (2.5), (2.7). □

3. Balanced truncation model reduction of the semi-discretized Stokes optimality system

We assume that the integrand $\boldsymbol{\ell}$ in (2.2a) is of the form

$$\boldsymbol{\ell}(\mathbf{v},\mathbf{p},\theta,t) := \frac{1}{2}\,|\mathbf{C}(\theta)\mathbf{v}(t) + \mathbf{D}(\theta)\mathbf{p}(t) + \mathbf{F}(\theta)\mathbf{f}(t) - \mathbf{d}(t)|^2, \tag{3.1}$$

where $\mathbf{C}(\theta) \in \mathbb{R}^{q\times n}$ and $\mathbf{D}(\theta) \in \mathbb{R}^{q\times m}$ are observation matrices, $\mathbf{F}(\theta) \in \mathbb{R}^{q\times k}$, is a feedthrough matrix, and $\mathbf{d}(t) \in \mathbb{R}^q, t \in (0,T)$. Dropping again the dependence on θ for ease of notation, the semi-discretized Stokes optimality system consists of the state equations

$$\begin{pmatrix} \mathbf{M} & \mathbf{0} \\ \mathbf{0} & \mathbf{0} \end{pmatrix} \frac{d}{dt} \begin{pmatrix} \mathbf{v}(t) \\ \mathbf{p}(t) \end{pmatrix} = -\begin{pmatrix} \mathbf{A} & \mathbf{B}^T \\ \mathbf{B} & \mathbf{0} \end{pmatrix} \begin{pmatrix} \mathbf{v}(t) \\ \mathbf{p}(t) \end{pmatrix} + \begin{pmatrix} \mathbf{K} \\ \mathbf{L} \end{pmatrix} \mathbf{f}(t), \tag{3.2a}$$

$$\mathbf{z}(t) = \mathbf{C}\mathbf{v}(t) + \mathbf{D}\mathbf{p}(t) + \mathbf{F}\mathbf{f}(t), \tag{3.2b}$$

$$\mathbf{M}\mathbf{v}(0) = \mathbf{v}^0, \tag{3.2c}$$

and the adjoint equations

$$-\begin{pmatrix} \mathbf{M} & \mathbf{0} \\ \mathbf{0} & \mathbf{0} \end{pmatrix} \frac{d}{dt} \begin{pmatrix} \boldsymbol{\lambda}(t) \\ \boldsymbol{\kappa}(t) \end{pmatrix} = -\begin{pmatrix} \mathbf{A} & \mathbf{B}^T \\ \mathbf{B} & \mathbf{0} \end{pmatrix} \begin{pmatrix} \boldsymbol{\lambda}(t) \\ \boldsymbol{\kappa}(t) \end{pmatrix} + \begin{pmatrix} \mathbf{C}^T \\ \mathbf{D}^T \end{pmatrix} \mathbf{z}(t), \tag{3.3a}$$

$$\mathbf{q}(t) = \mathbf{K}^T\boldsymbol{\lambda}(t) + \mathbf{L}^T\boldsymbol{\kappa}(t) + \mathbf{F}^T\mathbf{z}(t), \tag{3.3b}$$

$$\mathbf{M}\boldsymbol{\lambda}(T) = \boldsymbol{\lambda}^{(T)}. \tag{3.3c}$$

For the realization of the BTMR, we compute the controllability and observability Gramians $\mathbf{P}, \mathbf{Q} \in \mathbb{R}^{n \times n}$ as the solutions of the matrix Lyapunov equations

$$\bar{\mathbf{A}}\mathbf{P}\bar{\mathbf{M}} + \bar{\mathbf{M}}\mathbf{P}\bar{\mathbf{A}} + \bar{\mathbf{K}}\bar{\mathbf{K}}^T = 0, \tag{3.4a}$$

$$\bar{\mathbf{A}}\mathbf{Q}\bar{\mathbf{M}} + \bar{\mathbf{M}}\mathbf{Q}\bar{\mathbf{A}} + \bar{\mathbf{C}}^T\bar{\mathbf{C}} = 0, \tag{3.4b}$$

where

$$\bar{\mathbf{A}} := -\mathbf{\Pi}\mathbf{A}\mathbf{\Pi}^T, \quad \bar{\mathbf{M}} := \mathbf{\Pi}\mathbf{M}\mathbf{\Pi}^T, \quad \bar{\mathbf{K}} := \mathbf{\Pi}\widetilde{\mathbf{K}},$$
$$\bar{\mathbf{C}} := \mathbf{\Pi}\widetilde{\mathbf{C}}, \quad \widetilde{\mathbf{C}} := \mathbf{C} - \mathbf{F}(\mathbf{B}\mathbf{M}^{-1}\mathbf{B}^T)^{-1}\mathbf{B}\mathbf{M}^{-1}\mathbf{A},$$

and $\mathbf{\Pi}$ refers to the oblique projection from the proof of Theorem 2.1. The Lyapunov equations (3.4a), (3.4b) can be solved approximately by multishift ADI techniques (cf., e.g., [12, 27, 39, 46]). We factorize $\mathbf{P} = \mathbf{U}\mathbf{U}^T, \mathbf{Q} = \mathbf{E}\mathbf{E}^T$ and perform the singular value decomposition

$$\mathbf{U}^T\mathbf{M}\mathbf{E} = \mathbf{Z}\mathbf{S}_n\mathbf{Y}^T \ , \ \mathbf{S}_n := \mathrm{diag}(\sigma_1, \ldots, \sigma_n) \ , \ \sigma_i > \sigma_{i+1} \ , \ 1 \le i \le n-1. \tag{3.5}$$

We compute $\mathbf{V}, \mathbf{W}$ according to

$$\mathbf{V} := \mathbf{U}\mathbf{Z}_p\mathbf{S}_p^{-1/2} \quad , \quad \mathbf{W} := \mathbf{E}\mathbf{Y}_p\mathbf{S}_p^{-1/2}, \tag{3.6}$$

where $1 \le p \le n$ is chosen such that $\sigma_{p+1} < \tau\sigma_1$ for some threshold $\tau > 0$ and $\mathbf{Y}_p, \mathbf{Z}_p$ are the matrices built by the leading p columns of $\mathbf{Y}, \mathbf{Z}$.
The projection matrices satisfy

$$\mathbf{V} = \mathbf{\Pi}^T\mathbf{V} \quad , \quad \mathbf{W} = \mathbf{\Pi}^T\mathbf{W} \quad , \quad \mathbf{W}^T\mathbf{M}\mathbf{V} = \mathbf{I}.$$

Multiplying the state equations by $\mathbf{W}^T$ and the adjoint equations by $\mathbf{V}^T$ and using the preceding equations results in a reduced order optimality system, where the reduced order state equations turn out to be

$$\frac{d}{dt}\widehat{\mathbf{v}}_H(t) = -\widehat{\mathbf{A}}\widehat{\mathbf{v}}_H(t) + \widehat{\mathbf{K}}\mathbf{f}(t) \quad , \quad t \in (0, T], \tag{3.7a}$$

$$\widehat{\mathbf{z}}(t) = \widehat{\mathbf{C}}\widehat{\mathbf{v}}_H(t) + \widehat{\mathbf{G}}\mathbf{f}(t) - \widehat{\mathbf{H}}\frac{d}{dt}\mathbf{f}(t) \quad , \quad t \in (0, T], \tag{3.7b}$$

$$\widehat{\mathbf{v}}_H(0) = \widehat{\mathbf{v}}_H^0, \tag{3.7c}$$

whereas the reduced order adjoint state equations are given according to

$$-\frac{d}{dt}\widehat{\boldsymbol{\lambda}}_H(t) = -\widehat{\mathbf{A}}^T\widehat{\boldsymbol{\lambda}}_H(t) + \widehat{\mathbf{C}}^T\widehat{\mathbf{z}}(t) \quad , \quad t \in [0, T), \tag{3.8a}$$

$$\widehat{\mathbf{q}}(t) = \widehat{\mathbf{K}}^T\widehat{\boldsymbol{\lambda}}_H(t) + \widehat{\mathbf{G}}^T\widehat{\mathbf{z}}(t) - \widehat{\mathbf{H}}\frac{d}{dt}\widehat{\mathbf{z}}(t) \quad , \quad t \in [0, T), \tag{3.8b}$$

$$\widehat{\boldsymbol{\lambda}}_H(T) = \widehat{\boldsymbol{\lambda}}^{(T)}, \tag{3.8c}$$

with appropriately defined $\widehat{\mathbf{A}}, \widehat{\mathbf{C}}, \widehat{\mathbf{G}}, \widehat{\mathbf{H}}$, and $\widehat{\mathbf{K}}$. Due to the stability of $\mathbf{W}^T\mathbf{A}\mathbf{V}$, the classical BTMR estimate for the error in the observations and the outputs can be shown to hold true.

Theorem 3.1. *Let* $\mathbf{z}(t), \mathbf{q}(t), t \in [0,T]$, *and* $\widehat{\mathbf{z}}(t), \widehat{\mathbf{q}}(t), t \in [0,T]$, *be the observations and outputs of the full order and the reduced order optimality system as given by* (3.2b), (3.3b) *and* (3.7b), (3.8b), *and let* $\sigma_i, 1 \le i \le n$, *be the Hankel singular values from the singular value decomposition* (3.5). *Moreover, suppose that* $\mathbf{v}_H(0) = 0$ *and* $\boldsymbol{\lambda}_H(T) = 0$. *Then, there holds*

$$\|\mathbf{z} - \widehat{\mathbf{z}}\|_{L^2} \le 2\ \|\mathbf{f}\|_{L^2} \Big(\sigma_{p+1} + \cdots + \sigma_n\Big), \tag{3.9a}$$

$$\|\mathbf{q} - \widehat{\mathbf{q}}\|_{L^2} \le 2\ \|\widehat{\mathbf{z}}\|_{L^2} \Big(\sigma_{p+1} + \cdots + \sigma_n\Big). \tag{3.9b}$$

Proof. We refer to Section 7 in [30]. □

4. Domain decomposition and balanced truncation model reduction

For optimal design problems associated with linear state equations, where the design only effects a relatively small part of the computational domain, the nonlinearity is thus restricted to that part and motivates to consider a combination of domain decomposition and BTMR. Let us consider a domain $\Omega(\theta)$ such that

$$\overline{\Omega(\theta)} = \overline{\Omega}_1 \cup \overline{\Omega_2(\theta)}, \quad \Omega_1 \cap \Omega_2(\theta) = \emptyset, \quad \Gamma(\theta) := \overline{\Omega}_1 \cap \overline{\Omega_2(\theta)}, \tag{4.1}$$

where the local area of interest is $\Omega_2(\theta)$, whereas the design variables θ do not apply to the rest Ω_1 of the computational domain. The fine-scale model results from a spatial discretization by P2-P1 Taylor-Hood elements with respect to a simplicial triangulation of the computational domain which aligns with the decomposition of the spatial domain. We have to make sure that the solutions of the Stokes subdomain problems associated with Ω_1 and $\Omega_2(\theta)$ are the restrictions of the solution of the global problem to the subdomains. To this end, the subdomain pressures $\mathbf{p}_i, 1 \le i \le 2$, are split into a constant $\mathbf{p}_{0,i}$ and a pressure with zero spatial average. While the latter is uniquely determined as the solution of the subdomain problem, $\mathbf{p}_0 = (\mathbf{p}_{0,1}, \mathbf{p}_{0,2})^T$ is determined through the coupling of the subdomain problems via the interface. The fine-scale model is used only in the local area of interest, whereas a reduced order model based on balanced truncation is used for the rest of the domain. The objective functional

$$\mathbf{J}(\mathbf{v},\mathbf{p},\theta) := \mathbf{J}_1(\mathbf{v},\mathbf{p}) + \mathbf{J}_2(\mathbf{v},\mathbf{p},\theta), \tag{4.2}$$

$$\mathbf{J}_1(\mathbf{v},\mathbf{p}) := \int_0^T |\mathbf{C}_1\mathbf{v}_1(t) + \mathbf{D}_1\mathbf{p}_1(t) + \mathbf{F}_1\mathbf{f}(t) - \mathbf{d}(t)|^2\ dt,$$

$$\mathbf{J}_2(\mathbf{v},\mathbf{p},\theta) := \int_0^T \ell(\mathbf{v}_2(t), \mathbf{p}_2(t), \mathbf{v}_\Gamma(t), \mathbf{p}_0(t), t, \theta)\ dt,$$

is assumed to consist of an objective functional $\mathbf{J}_1$ of tracking type for subdomain Ω_1, depending only on the velocity and the pressure in Ω_1, and an objective functional $\mathbf{J}_2$ for subdomain $\Omega_2(\theta)$, depending on the velocities $\mathbf{v}_2$ in $\Omega_2(\theta)$ and $\mathbf{v}_\Gamma$ on the interface $\Gamma(\theta)$ as well as on the pressure $\mathbf{p}_2$ in $\Omega_2(\theta)$ and on $\mathbf{p}_0$.

Grouping the state variables according to $\mathbf{x}_i := (\mathbf{v}_i, \mathbf{p}_i)^T, 1 \leq i \leq 2$, and $\mathbf{x}_\Gamma := (\mathbf{v}_\Gamma, \mathbf{p}_0)$, the semi-discretized domain decomposed Stokes system can be written in block structured form according to

$$\begin{pmatrix} \mathbf{P}_1\mathbf{x}_1 \\ \mathbf{P}_2(\theta)\mathbf{x}_2 \\ \mathbf{P}_\Gamma(\theta)\mathbf{x}_\Gamma \end{pmatrix} := \begin{pmatrix} \mathbf{E}_1 & \mathbf{0} & \mathbf{0} \\ \mathbf{0} & \mathbf{E}_2(\theta) & \mathbf{0} \\ \mathbf{0} & \mathbf{0} & \mathbf{E}_\Gamma(\theta) \end{pmatrix} \frac{d}{dt} \begin{pmatrix} \mathbf{x}_1 \\ \mathbf{x}_2 \\ \mathbf{x}_\Gamma \end{pmatrix} \tag{4.3}$$
$$+ \begin{pmatrix} \mathbf{S}_{11} & \mathbf{0} & \mathbf{S}_{1\Gamma} \\ \mathbf{0} & \mathbf{S}_{22}(\theta) & \mathbf{S}_{2\Gamma}(\theta) \\ \mathbf{S}_{1\Gamma}^T & \mathbf{S}_{2\Gamma}^T(\theta) & \mathbf{S}_{\Gamma\Gamma}(\theta) \end{pmatrix} \begin{pmatrix} \mathbf{x}_1 \\ \mathbf{x}_2 \\ \mathbf{x}_\Gamma \end{pmatrix} = \begin{pmatrix} \mathbf{N}_1 \\ \mathbf{N}_2(\theta) \\ \mathbf{N}_\Gamma(\theta) \end{pmatrix} \mathbf{f}.$$

Here, the singular block matrices $\mathbf{E}_1, \mathbf{E}_2(\theta)$ and $\mathbf{E}_\Gamma(\theta)$ are given by

$$\mathbf{E}_1 := \begin{pmatrix} \mathbf{M}_1 & \mathbf{0} \\ \mathbf{0} & \mathbf{0} \end{pmatrix}, \ \mathbf{E}_2(\theta) := \begin{pmatrix} \mathbf{M}_2(\theta) & \mathbf{0} \\ \mathbf{0} & \mathbf{0} \end{pmatrix}, \ \mathbf{E}_\Gamma(\theta) := \begin{pmatrix} \mathbf{M}_\Gamma(\theta) & \mathbf{0} \\ \mathbf{0} & \mathbf{0} \end{pmatrix},$$

whereas $\mathbf{S}_{11}, \mathbf{S}_{22}(\theta)$ and $\mathbf{S}_{\Gamma\Gamma}(\theta)$ are the Stokes matrices associated with the subdomains $\Omega_1, \Omega_2(\theta)$ and the interface $\Gamma(\theta)$

$$\mathbf{S}_{11} := \begin{pmatrix} \mathbf{A}_{11} & \mathbf{B}_{11}^T \\ \mathbf{B}_{11} & \mathbf{0} \end{pmatrix}, \quad \mathbf{S}_{22}(\theta) := \begin{pmatrix} \mathbf{A}_{22}(\theta) & \mathbf{B}_{22}^T(\theta) \\ \mathbf{B}_{22}(\theta) & \mathbf{0} \end{pmatrix},$$
$$\mathbf{S}_{\Gamma\Gamma}(\theta) := \begin{pmatrix} \mathbf{A}_{\Gamma\Gamma}(\theta) & \mathbf{B}_0^T \\ \mathbf{B}_0 & \mathbf{0} \end{pmatrix},$$

and $\mathbf{S}_{1\Gamma}, \mathbf{S}_{2\Gamma}(\theta)$ are of the form

$$\mathbf{S}_{1\Gamma} := \begin{pmatrix} \mathbf{A}_{11} & \mathbf{B}_{11}^T \\ \mathbf{B}_{11} & \mathbf{0} \end{pmatrix}, \quad \mathbf{S}_{2\Gamma}(\theta) := \begin{pmatrix} \mathbf{A}_{2\Gamma}(\theta) & \mathbf{0} \\ \mathbf{B}_{2\Gamma}(\theta) & \mathbf{0} \end{pmatrix}.$$

Finally, $\mathbf{N}_1, \mathbf{N}_2(\theta)$ and $\mathbf{N}_\Gamma(\theta)$ are given by

$$\mathbf{N}_1 := \begin{pmatrix} \mathbf{K}_1 \\ \mathbf{L}_1 \end{pmatrix}, \quad \mathbf{N}_2(\theta) := \begin{pmatrix} \mathbf{K}_2(\theta) \\ \mathbf{L}_2(\theta) \end{pmatrix}, \quad \mathbf{N}_\Gamma(\theta) := \begin{pmatrix} \mathbf{K}_\Gamma(\theta) \\ \mathbf{L}_0(\theta) \end{pmatrix}.$$

Introducing Lagrange multipliers $\boldsymbol{\lambda}_1(t), \boldsymbol{\lambda}_2(t), \boldsymbol{\lambda}_\Gamma(t)$ and $\boldsymbol{\kappa}_1(t), \boldsymbol{\kappa}_2(t), \boldsymbol{\kappa}_0(t)$, and partitioning them by means of $\boldsymbol{\mu}_i(t) := (\boldsymbol{\lambda}_i(t), \boldsymbol{\kappa}_i(t))^T, 1 \leq i \leq 2, \boldsymbol{\mu}_\Gamma(t) := (\boldsymbol{\lambda}_\Gamma(t), \boldsymbol{\kappa}_0(t))^T$, the Lagrangian associated with (4.2),(4.3) is given by

$$\mathcal{L}(\mathbf{x}, \boldsymbol{\mu}, \theta) := \mathbf{J}(\mathbf{v}, \mathbf{p}, \theta) + \int_0^T \begin{pmatrix} \boldsymbol{\mu}_1(t) \\ \boldsymbol{\mu}_2(t) \\ \boldsymbol{\mu}_\Gamma(t) \end{pmatrix} \cdot \begin{pmatrix} \mathbf{P}_1\mathbf{x}_1(t) - \mathbf{N}_1\mathbf{f} \\ \mathbf{P}_2(\theta)\mathbf{x}_2(t) - \mathbf{N}_2(\theta)\mathbf{f} \\ \mathbf{P}_\Gamma(\theta)\mathbf{x}_\Gamma(t) - \mathbf{N}_\Gamma(\theta)\mathbf{f} \end{pmatrix} dt. \tag{4.4}$$

We now focus on the optimality system associated with subdomain Ω_1.

Lemma 4.1. *The optimality system associated with subdomain* Ω_1 *consists of the state equations*

$$\begin{pmatrix} \mathbf{M}_1 & \mathbf{0} \\ \mathbf{0} & \mathbf{0} \end{pmatrix} \frac{d}{dt} \begin{pmatrix} \mathbf{v}_1(t) \\ \mathbf{p}_1(t) \end{pmatrix} \tag{4.5a}$$

$$= - \begin{pmatrix} \mathbf{A}_{11} & \mathbf{B}_{11}^T \\ \mathbf{B}_{11} & \mathbf{0} \end{pmatrix} \begin{pmatrix} \mathbf{v}_1(t) \\ \mathbf{p}_1(t) \end{pmatrix} - \begin{pmatrix} \mathbf{A}_{1\Gamma} & \mathbf{0} \\ \mathbf{B}_{1\Gamma} & \mathbf{0} \end{pmatrix} \begin{pmatrix} \mathbf{v}_\Gamma(t) \\ \mathbf{p}_0(t) \end{pmatrix} + \begin{pmatrix} \mathbf{K}_1 \\ \mathbf{L}_1 \end{pmatrix} \mathbf{f}(t),$$

$$\mathbf{z}_1(t) = \mathbf{C}_1 \mathbf{v}_1(t) + \mathbf{D}_1 \mathbf{p}_1(t) + \mathbf{D}_0 \mathbf{p}_0(t) + \mathbf{F}_1 \mathbf{f}(t) - \mathbf{d}(t), \tag{4.5b}$$

and the adjoint state equations

$$- \begin{pmatrix} \mathbf{M}_1 & \mathbf{0} \\ \mathbf{0} & \mathbf{0} \end{pmatrix} \frac{d}{dt} \begin{pmatrix} \boldsymbol{\lambda}_1(t) \\ \boldsymbol{\kappa}_1(t) \end{pmatrix} \tag{4.6a}$$

$$= - \begin{pmatrix} \mathbf{A}_{11} & \mathbf{B}_{11}^T \\ \mathbf{B}_{11} & \mathbf{0} \end{pmatrix} \begin{pmatrix} \boldsymbol{\lambda}_1(t) \\ \boldsymbol{\kappa}_1(t) \end{pmatrix} - \begin{pmatrix} \mathbf{A}_{1\Gamma} & \mathbf{0} \\ \mathbf{B}_{1\Gamma} & \mathbf{0} \end{pmatrix} \begin{pmatrix} \boldsymbol{\lambda}_\Gamma(t) \\ \boldsymbol{\kappa}_0(t) \end{pmatrix} - \begin{pmatrix} \mathbf{C}_1^T \\ \mathbf{F}_1^T \end{pmatrix} \mathbf{z}_1(t),$$

$$\mathbf{q}_1(t) = \mathbf{K}_1^T \boldsymbol{\lambda}_1(t) + \mathbf{L}_1^T \boldsymbol{\kappa}_1(t) + \mathbf{D}_1^T \mathbf{z}_1(t). \tag{4.6b}$$

Proof. The proof follows readily from (4.2) and (4.3). □

The optimality system for $\Omega_2(\theta)$ and the interface $\Gamma(\theta)$ can be derived likewise. We further have a variational inequality due to the constraints on θ.
We see that (4.5a), (4.5b) and (4.6a), (4.6b) have exactly the form which we considered before in Section 3. Hence, it is directly amenable to the application of BTMR.

Lemma 4.2. *There exist projection matrices* $\mathbf{V}_1, \mathbf{W}_1$ *such that the reduced state equations associated with subdomain* Ω_1 *are of the form*

$$\frac{d}{dt}\,\widehat{\mathbf{v}}_1(t) = - \mathbf{W}_1^T \mathbf{A}_{11} \mathbf{V}_1 \widehat{\mathbf{v}}_1(t) - \mathbf{W}_1^T \widetilde{\mathbf{B}}_1 \begin{pmatrix} \widehat{\mathbf{v}}_\Gamma(t) \\ \widehat{\mathbf{p}}_0(t) \\ \mathbf{f}(t) \end{pmatrix}, \tag{4.7a}$$

$$\begin{pmatrix} \widehat{\mathbf{z}}_{v,\Gamma}(t) \\ \widehat{\mathbf{z}}_{p,\Gamma}(t) \\ \widehat{\mathbf{z}}_1(t) \end{pmatrix} = \widetilde{\mathbf{C}}_1 \mathbf{V}_1 \widehat{\mathbf{v}}_1(t) + \widetilde{\mathbf{D}}_1 \begin{pmatrix} \widehat{\mathbf{v}}_\Gamma(t) \\ \widehat{\mathbf{p}}_0(t) \\ \mathbf{f}(t) \end{pmatrix} - \widetilde{\mathbf{H}}_1 \frac{d}{dt} \begin{pmatrix} \widehat{\mathbf{v}}_\Gamma(t) \\ \widehat{\mathbf{p}}_0(t) \\ \mathbf{f}(t) \end{pmatrix}, \tag{4.7b}$$

whereas the reduced adjoint state equations are given by

$$-\frac{d}{dt}\,\widehat{\boldsymbol{\lambda}}_1(t) = - \mathbf{V}_1^T \mathbf{A}_{11} \mathbf{W}_1 \widehat{\boldsymbol{\lambda}}_1(t) + \mathbf{V}_1^T \widetilde{\mathbf{C}}_1 \begin{pmatrix} \widehat{\boldsymbol{\lambda}}_1(t) \\ \widehat{\boldsymbol{\kappa}}_0(t) \\ -\widehat{\mathbf{z}}_1(t) \end{pmatrix}, \tag{4.8a}$$

$$\begin{pmatrix} \widehat{\mathbf{q}}_{v,\Gamma}(t) \\ \widehat{\mathbf{q}}_{p,\Gamma}(t) \\ \widehat{\mathbf{q}}_1(t) \end{pmatrix} = - \widetilde{\mathbf{B}}_1^T \mathbf{W}_1 \widehat{\boldsymbol{\lambda}}_1(t) + \widetilde{\mathbf{D}}_1^T \begin{pmatrix} \widehat{\boldsymbol{\lambda}}_1(t) \\ \widehat{\boldsymbol{\kappa}}_0(t) \\ -\widehat{\mathbf{z}}_1(t) \end{pmatrix} + \widetilde{\mathbf{H}}_1^T \frac{d}{dt} \begin{pmatrix} \widehat{\boldsymbol{\lambda}}_1(t) \\ \widehat{\boldsymbol{\kappa}}_0(t) \\ -\widehat{\mathbf{z}}_1(t) \end{pmatrix}. \tag{4.8b}$$

Since we neither apply BTMR to subdomain $\Omega_2(\theta)$ nor to the interface $\Gamma(\theta)$, the corresponding state and adjoint state equations can be derived in a straightforward way.

Lemma 4.3. *The state and the adjoint state equations associated with the subdomain $\Omega_2(\theta)$ read as follows*

$$\begin{pmatrix} \mathbf{M}_2(\theta) & \mathbf{0} \\ \mathbf{0} & \mathbf{0} \end{pmatrix} \frac{d}{dt} \begin{pmatrix} \widehat{\mathbf{v}}_2(t) \\ \widehat{\mathbf{p}}_2(t) \end{pmatrix} = - \begin{pmatrix} \mathbf{A}_{22}(\theta) & \mathbf{B}_{22}^T(\theta) \\ \mathbf{B}_{22}(\theta) & \mathbf{0} \end{pmatrix} \begin{pmatrix} \widehat{\mathbf{v}}_2(t) \\ \widehat{\mathbf{p}}_2(t) \end{pmatrix} \tag{4.9a}$$

$$- \begin{pmatrix} \mathbf{A}_{2\Gamma}(\theta) & \mathbf{0} \\ \mathbf{B}_{2\Gamma}(\theta) & \mathbf{0} \end{pmatrix} \begin{pmatrix} \widehat{\mathbf{v}}_\Gamma(t) \\ \widehat{\mathbf{p}}_0(t) \end{pmatrix} + \begin{pmatrix} \mathbf{K}_2(\theta) \\ \mathbf{L}_2(\theta) \end{pmatrix} \mathbf{f}(t),$$

$$- \begin{pmatrix} \mathbf{M}_2(\theta) & \mathbf{0} \\ \mathbf{0} & \mathbf{0} \end{pmatrix} \frac{d}{dt} \begin{pmatrix} \widehat{\boldsymbol{\lambda}}_2(t) \\ \widehat{\boldsymbol{\kappa}}_2(t) \end{pmatrix} = - \begin{pmatrix} \mathbf{A}_{22}(\theta) & \mathbf{B}_{22}^T(\theta) \\ \mathbf{B}_{22}(\theta) & \mathbf{0} \end{pmatrix} \begin{pmatrix} \widehat{\boldsymbol{\lambda}}_2(t) \\ \widehat{\boldsymbol{\kappa}}_2(t) \end{pmatrix} \tag{4.9b}$$

$$- \begin{pmatrix} \mathbf{A}_{2\Gamma}(\theta) & \mathbf{0} \\ \mathbf{B}_{2\Gamma}(\theta) & \mathbf{0} \end{pmatrix} \begin{pmatrix} \widehat{\boldsymbol{\lambda}}_\Gamma(t) \\ \widehat{\boldsymbol{\kappa}}_0(t) \end{pmatrix} - \begin{pmatrix} \nabla_{\widehat{v}_2} \boldsymbol{\ell}(\widehat{\mathbf{v}}_2, \widehat{\mathbf{p}}_2, \widehat{\mathbf{v}}_\Gamma, \widehat{\mathbf{p}}_0, t, \theta) \\ \nabla_{\widehat{\mathbf{p}}_2} \boldsymbol{\ell}(\widehat{\mathbf{v}}_2, \widehat{\mathbf{p}}_2, \widehat{\mathbf{v}}_\Gamma, \widehat{\mathbf{p}}_0, t, \theta) \end{pmatrix}.$$

The state and the adjoint state equations associated with the interface $\Gamma(\theta)$ are given by

$$\begin{pmatrix} \mathbf{M}_\Gamma(\theta) & \mathbf{0} \\ \mathbf{0} & \mathbf{0} \end{pmatrix} \frac{d}{dt} \begin{pmatrix} \widehat{\mathbf{v}}_\Gamma(t) \\ \widehat{\mathbf{p}}_0(t) \end{pmatrix} = - \begin{pmatrix} \mathbf{A}_{\Gamma\Gamma}(\theta) & \mathbf{B}_0^T(\theta) \\ \mathbf{B}_0(\theta) & \mathbf{0} \end{pmatrix} \begin{pmatrix} \widehat{\mathbf{v}}_\Gamma(t) \\ \widehat{\mathbf{p}}_0(t) \end{pmatrix} \tag{4.10a}$$

$$+ \begin{pmatrix} \widehat{\mathbf{z}}_{v,\Gamma}(t) \\ \widehat{\mathbf{z}}_{p,\Gamma}(t) \end{pmatrix} - \begin{pmatrix} \mathbf{A}_{2\Gamma}^T(\theta) & \mathbf{B}_{2\Gamma}^T(\theta) \\ \mathbf{0} & \mathbf{0} \end{pmatrix} \begin{pmatrix} \widehat{\mathbf{v}}_2(t) \\ \widehat{\mathbf{p}}_2(t) \end{pmatrix} + \begin{pmatrix} \mathbf{K}_\Gamma(\theta) \\ \mathbf{L}_0(\theta) \end{pmatrix} \mathbf{f}(t),$$

$$- \begin{pmatrix} \mathbf{M}_\Gamma(\theta) & \mathbf{0} \\ \mathbf{0} & \mathbf{0} \end{pmatrix} \frac{d}{dt} \begin{pmatrix} \widehat{\boldsymbol{\lambda}}_\Gamma(t) \\ \widehat{\boldsymbol{\kappa}}_0(t) \end{pmatrix} = - \begin{pmatrix} \mathbf{A}_{\Gamma\Gamma}(\theta) & \mathbf{B}_0^T(\theta) \\ \mathbf{B}_0(\theta) & \mathbf{0} \end{pmatrix} \begin{pmatrix} \widehat{\boldsymbol{\lambda}}_\Gamma(t) \\ \widehat{\boldsymbol{\kappa}}_0(t) \end{pmatrix} \tag{4.10b}$$

$$+ \begin{pmatrix} \widehat{\mathbf{q}}_{v,\Gamma}(t) \\ \widehat{\mathbf{q}}_{p,\Gamma}(t) \end{pmatrix} - \begin{pmatrix} \mathbf{A}_{2\Gamma}^T(\theta) & \mathbf{B}_{2\Gamma}^T(\theta) \\ \mathbf{0} & \mathbf{0} \end{pmatrix} \begin{pmatrix} \widehat{\boldsymbol{\lambda}}_2(t) \\ \widehat{\boldsymbol{\kappa}}_2(t) \end{pmatrix}$$

$$- \begin{pmatrix} \nabla_{\widehat{\mathbf{v}}_\Gamma} \boldsymbol{\ell}(\widehat{\mathbf{v}}_2, \widehat{\mathbf{p}}_2, \widehat{\mathbf{v}}_\Gamma, \widehat{\mathbf{p}}_0, t, \theta) \\ \nabla_{\widehat{\mathbf{p}}_0} \boldsymbol{\ell}(\widehat{\mathbf{v}}_2, \widehat{\mathbf{p}}_2, \widehat{\mathbf{v}}_\Gamma, \widehat{\mathbf{p}}_0, t, \theta) \end{pmatrix}.$$

The equations (4.9a), (4.9b) *and* (4.10a), (4.10b) *are complemented by the variational inequality*

$$\int_0^T \nabla_\theta \boldsymbol{\ell}(\mathbf{v}_2, \mathbf{p}_2, \mathbf{v}_\Gamma, \mathbf{p}_0, t, \theta)\, dt \tag{4.11}$$

$$+ \int_0^T \begin{pmatrix} \widehat{\boldsymbol{\mu}}_2(t) \\ \widehat{\boldsymbol{\mu}}_\Gamma(t) \end{pmatrix}^T \begin{pmatrix} (\mathbf{D}_\theta \mathbf{P}_2(\theta)(\tilde{\theta} - \theta)\widehat{\mathbf{x}}_2(t) - (\mathbf{D}_\theta \mathbf{N}_2(\theta)(\tilde{\theta} - \theta)\mathbf{f}(t) \\ (\mathbf{D}_\theta \mathbf{P}_\Gamma(\theta)(\tilde{\theta} - \theta)\widehat{\mathbf{x}}_\Gamma(t) - (\mathbf{D}_\theta \mathbf{N}_\Gamma(\theta)(\tilde{\theta} - \theta)\mathbf{f}(t) \end{pmatrix} dt \geq 0,$$

which is supposed to hold true for all $\tilde{\theta} \in \Theta$. Here, $\widehat{\mathbf{x}}_2 := (\widehat{\mathbf{v}}_2, \widehat{\mathbf{p}}_2), \widehat{\mathbf{x}}_\Gamma := (\widehat{\mathbf{v}}_\Gamma, \widehat{\mathbf{p}}_0)$. Moreover, $\mathbf{N}_2(\theta), \mathbf{N}_\Gamma(\theta)$ and $\mathbf{P}_2(\theta), \mathbf{P}_\Gamma(\theta)$ are given as in (4.3).

The following result can be verified by straightforward computation.

Theorem 4.4. *The reduced order optimality system* (4.7a)–(4.11) *represents the first-order necessary optimality conditions for the reduced order optimization problem*

$$\inf_{\theta\in\Theta} \widehat{\mathbf{J}}(\theta), \quad \widehat{\mathbf{J}}(\theta) := \widehat{\mathbf{J}}_1(\widehat{\mathbf{v}}_1, \widehat{\mathbf{p}}_1) + \widehat{\mathbf{J}}_2(\widehat{\mathbf{v}}_2, \widehat{\mathbf{p}}_2, \widehat{\mathbf{v}}_\Gamma, \widehat{\mathbf{p}}_0, \theta), \tag{4.12}$$

where the reduced order functionals $\widehat{\mathbf{J}}_1$ *and* $\widehat{\mathbf{J}}_2$ *are given by*

$$\widehat{\mathbf{J}}_1(\widehat{\mathbf{v}}_1, \widehat{\mathbf{p}}_1) := \frac{1}{2}\int_0^T |\widehat{\mathbf{z}}_1|^2 dt, \quad \widehat{\mathbf{J}}_2(\widehat{\mathbf{v}}_2, \widehat{\mathbf{p}}_2, \widehat{\mathbf{v}}_\Gamma, \widehat{\mathbf{p}}_0, \theta) := \int_0^T \boldsymbol{\ell}(\widehat{\mathbf{v}}_2, \widehat{\mathbf{p}}_2, \widehat{\mathbf{v}}_\Gamma, \widehat{\mathbf{p}}_0, t, \theta) dt.$$

5. A priori estimates of the modeling error

We now focus our attention on an a priori analysis of the modeling error due to the approximation of the full order model by the reduced order model obtained by the application of the combined domain decomposition and balanced truncation model reduction approach. We will show that under some assumptions the error in the optimal design can be bounded from above by the sum of the remaining Hankel singular values, i.e., we are able to derive an upper bound of the same form as in the standard BTMR estimates:

$$\|\theta^* - \widehat{\theta}^*\| \ \le \ C \left(\sigma_{p+1} + \cdots + \sigma_n\right).$$

One of these assumptions is to require the objective functional $\mathbf{J}$ to be strongly convex.

($\mathbf{A_1}$) There exists a constant $\kappa > 0$ such that for all $\widehat{\theta}, \theta \in \Theta$ there holds

$$\left(\nabla\mathbf{J}(\widehat{\theta}) - \nabla\mathbf{J}(\theta)\right)^T (\widehat{\theta} - \theta) \ \ge \ \kappa \, \|\widehat{\theta} - \theta\|^2.$$

Then, it is easy to see that the error in the optimal design is bounded from above by the difference of the gradients of the objective functional for the reduced optimization problem and the gradient of the objective functional for the full optimization problem at optimality $\widehat{\theta}^*$.

Lemma 5.1. *Assume that the objective functional J satisfies* ($\mathbf{A_1}$). *Then, if* $\theta^* \in \Theta$ *and* $\widehat{\theta}^* \in \Theta$ *are the solutions of the full order and the reduced order optimization problem, there holds*

$$\|\theta^* - \widehat{\theta}^*\| \ \le \ \kappa^{-1} \, \|\nabla\widehat{\mathbf{J}}(\widehat{\theta}^*) - \nabla\mathbf{J}(\widehat{\theta}^*)\|. \tag{5.1}$$

Proof. Obviously, we have

$$\left.\begin{aligned} \nabla\mathbf{J}(\theta^*)^T(\theta - \theta^*) \ &\ge \ 0 \\ \nabla\widehat{\mathbf{J}}(\widehat{\theta}^*)^T(\theta - \widehat{\theta}^*) \ &\ge \ 0 \end{aligned}\right\} \quad \Longrightarrow \quad \left(\nabla\mathbf{J}(\theta^*) - \nabla\widehat{\mathbf{J}}(\widehat{\theta}^*)\right)^T (\widehat{\theta}^* - \theta^*) \ \ge \ 0. \tag{5.2}$$

Combining (5.2) with the strong convexity of $\mathbf{J}$ allows to deduce (5.1). □

Lemma 5.1 tells us that we have to provide an upper bound for the difference of these gradients. To this end, we further have to require that the objective functional $\mathbf{J}_1$ associated with the subdomain Ω_1 does not explicitly depend on the pressure. Moreover, as far as the objective functional $\mathbf{J}_2$ associated with the subdomain $\Omega_2(\theta)$ is concerned, we assume that the derivatives with respect to the state variables $\mathbf{v}_2, \mathbf{v}_\Gamma, \mathbf{p}_2$ and $\mathbf{p}_0$ are Lipschitz continuous uniformly in the design variable θ.

($\mathbf{A_2}$) The objective functional $\mathbf{J}_1$ does not explicitly depend on the pressure, i.e., it is supposed to be of the form

$$\mathbf{J}_1(\mathbf{v}_1) = \frac{1}{2}\int_0^T |\mathbf{C}_1\mathbf{v}_1(t) + \mathbf{F}_1\mathbf{f}(t) - \mathbf{d}(t)|^2 \, dt.$$

($\mathbf{A_3}$) Treating the states $\mathbf{x}_2 := (\mathbf{v}_2, \mathbf{p}_2)^T$ and $\mathbf{x}_\Gamma := (\mathbf{v}_\Gamma, \mathbf{p}_0)^T$ in the integrand $\boldsymbol{\ell}$ of the objective functional

$$\mathbf{J}_2(\mathbf{x}_2, \mathbf{x}_\Gamma, \theta) = \frac{1}{2}\int_0^T \boldsymbol{\ell}(\mathbf{x}_2, \mathbf{x}_\Gamma, t, \theta) \, dt,$$

as implicit functions of θ, we assume that for some positive constant L_1 the Lipschitz condition

$$\|\nabla_\theta \boldsymbol{\ell}(\mathbf{x}_2, \mathbf{x}_\Gamma, t, \theta) - \nabla_\theta \boldsymbol{\ell}(\mathbf{x}_2', \mathbf{x}_\Gamma', t, \theta)\| \le L_1\Big(\|\mathbf{x}_2 - \mathbf{x}_2'\|^2 + \|\mathbf{x}_\Gamma - \mathbf{x}_\Gamma'\|^2\Big)^{1/2}$$

is satisfied uniformly in $\theta \in \Theta$ and $t \in [0, T]$.

($\mathbf{A_4}$) There exists a constant $L_2 > 0$ such that for all $\theta \in \Theta$ and all θ' with $\|\theta'\| \le 1$ there holds

$$\max\{\|\mathbf{D}_\theta \mathbf{P}_2(\theta)\theta'\|, \|\mathbf{D}_\theta \mathbf{P}_\Gamma(\theta)\theta'\|, \|\mathbf{D}_\theta \mathbf{N}_2(\theta)\theta'\|, \|\mathbf{D}_\theta \mathbf{N}_\Gamma(\theta)\theta'\|\} \le L_2.$$

Under these requirements, it follows that the difference in the gradients can be bounded from above by the L^2-norms of the differences between the full and the reduced states as well as the full and reduced adjoint states.

Lemma 5.2. *Assume that* ($\mathbf{A_2}$), ($\mathbf{A_3}$), ($\mathbf{A_4}$) *hold true. Then, there exists a constant* $C > 0$, *depending on* L_1, L_2 *in assumptions* ($\mathbf{A_3}$), ($\mathbf{A_4}$), *such that for* $\theta \in \Theta$

$$\|\nabla \mathbf{J}(\theta) - \nabla \widehat{\mathbf{J}}(\theta)\| \le C\left(\left\|\begin{pmatrix} \mathbf{x}_2 - \widehat{\mathbf{x}}_2 \\ \mathbf{x}_\Gamma - \widehat{\mathbf{x}}_\Gamma \end{pmatrix}\right\|_{L^2} + \left\|\begin{pmatrix} \boldsymbol{\mu}_2 - \widehat{\boldsymbol{\mu}}_2 \\ \boldsymbol{\mu}_\Gamma - \widehat{\boldsymbol{\mu}}_\Gamma \end{pmatrix}\right\|_{L^2}\right), \tag{5.3}$$

where $\mathbf{x}_2 - \widehat{\mathbf{x}}_2, \mathbf{x}_\Gamma - \widehat{\mathbf{x}}_\Gamma$ *and* $\boldsymbol{\mu}_2 - \widehat{\boldsymbol{\mu}}_2, \boldsymbol{\mu}_\Gamma - \widehat{\boldsymbol{\mu}}_\Gamma$ *are given by*

$$\mathbf{x}_2 - \widehat{\mathbf{x}}_2 = \begin{pmatrix} \mathbf{v}_2 - \widehat{\mathbf{v}}_2 \\ \mathbf{p}_2 - \widehat{\mathbf{p}}_2 \end{pmatrix}, \quad \mathbf{x}_\Gamma - \widehat{\mathbf{x}}_\Gamma = \begin{pmatrix} \mathbf{v}_\Gamma - \widehat{\mathbf{v}}_\Gamma \\ \mathbf{p}_0 - \widehat{\mathbf{p}}_0 \end{pmatrix},$$

$$\boldsymbol{\mu}_2 - \widehat{\boldsymbol{\mu}}_2 = \begin{pmatrix} \boldsymbol{\lambda}_2 - \widehat{\boldsymbol{\lambda}}_2 \\ \boldsymbol{\kappa}_2 - \widehat{\boldsymbol{\kappa}}_2 \end{pmatrix}, \quad \boldsymbol{\mu}_\Gamma - \widehat{\boldsymbol{\mu}}_\Gamma = \begin{pmatrix} \boldsymbol{\lambda}_\Gamma - \widehat{\boldsymbol{\lambda}}_\Gamma \\ \boldsymbol{\kappa}_0 - \widehat{\boldsymbol{\kappa}}_0 \end{pmatrix}.$$

Proof. Expressing the objective functional $\mathbf{J}$ in terms of the associated Lagrangian $\mathcal{L}$ according to (4.4), we find that for $\tilde{\theta} \in \Theta$ there holds

$$\nabla \mathbf{J}(\theta)^T \tilde{\theta} = \int_0^T (\nabla_\theta \boldsymbol{\ell}(\mathbf{x}_2, \mathbf{x}_\Gamma, t, \theta))^T \tilde{\theta} \, dt \tag{5.4}$$

$$+ \int_0^T \begin{pmatrix} \boldsymbol{\mu}_2(t) \\ \boldsymbol{\mu}_\Gamma(t) \end{pmatrix}^T \begin{pmatrix} (\mathbf{D}_\theta \mathbf{P}_2(\theta)\tilde{\theta})\mathbf{x}_2(t) - (\mathbf{D}_\theta \mathbf{N}_2(\theta)\tilde{\theta})\mathbf{f}(t) \\ (\mathbf{D}_\theta \mathbf{P}_\Gamma(\theta)\tilde{\theta})\mathbf{x}_\Gamma(t) - (\mathbf{D}_\theta \mathbf{N}_\Gamma(\theta)\tilde{\theta})\mathbf{f}(t) \end{pmatrix} dt.$$

Proceeding analogously for the reduced objective functional $\widehat{\mathbf{J}}$, we have

$$\nabla \widehat{\mathbf{J}}(\theta)^T \tilde{\theta} = \int_0^T (\nabla_\theta \boldsymbol{\ell}(\widehat{\mathbf{x}}_2, \widehat{\mathbf{x}}_\Gamma, t, \theta))^T \tilde{\theta} \, dt \tag{5.5}$$

$$+ \int_0^T \begin{pmatrix} \widehat{\boldsymbol{\mu}}_2(t) \\ \widehat{\boldsymbol{\mu}}_\Gamma(t) \end{pmatrix}^T \begin{pmatrix} (\mathbf{D}_\theta \mathbf{P}_2(\theta)\tilde{\theta})\widehat{\mathbf{x}}_2(t) - (\mathbf{D}_\theta \mathbf{N}_2(\theta)\tilde{\theta})\mathbf{f}(t) \\ (\mathbf{D}_\theta \mathbf{P}_\Gamma(\theta)\tilde{\theta})\widehat{\mathbf{x}}_\Gamma(t) - (\mathbf{D}_\theta \mathbf{N}_\Gamma(\theta)\tilde{\theta})\mathbf{f}(t) \end{pmatrix} dt.$$

Subtracting (5.5) from (5.4) and using $(\mathbf{A_3}), (\mathbf{A_4})$ gives the assertion. □

Consequently, it remains to estimate the modeling error in those states and adjoint states associated with subdomain $\Omega_2(\theta)$ and the interface $\Gamma(\theta)$. For this purpose, we suppose stability of the semi-discrete Stokes system and the subsystem associated with subdomain Ω_1.

$(\mathbf{A_5})$ The matrix $\mathbf{A}(\theta) \in \mathbb{R}^{n \times n}$ is symmetric positive definite and the matrix $\mathbf{B}(\theta) \in \mathbb{R}^{m \times n}$ has rank m. The generalized eigenvalues of $(\mathbf{A}(\theta), \mathbf{M}(\theta))$ have positive real part.
The matrix $\mathbf{A}_{11}(\theta) \in \mathbb{R}^{n_1 \times n_1}$ is symmetric positive definite and the matrix $\mathbf{B}_{11}(\theta) \in \mathbb{R}^{m_1 \times n_1}$ has rank m_1. The generalized eigenvalues of $(\mathbf{A}_{11}(\theta), \mathbf{M}_{11}(\theta))$ have positive real part.

For the modeling errors in the velocities and the pressures we will show that they can be bounded from above by the sum of the remaining Hankel singular values. The same holds true for the errors in the observations in Ω_1 and on the interface.

Lemma 5.3. *Let* $\mathbf{x} = (\mathbf{x}_1, \mathbf{x}_2, \mathbf{x}_\Gamma)^T$ *with* $\mathbf{x}_i = (\mathbf{v}_i, \mathbf{p}_i)^T, 1 \le i \le 2, \mathbf{x}_\Gamma = (\mathbf{v}_\Gamma, \mathbf{p}_\Gamma)^T$ *and* $\widehat{\mathbf{x}} = (\widehat{\mathbf{x}}_1, \widehat{\mathbf{x}}_2, \widehat{\mathbf{x}}_\Gamma)^T$ *with* $\widehat{\mathbf{x}}_1 = \widehat{\mathbf{v}}_1, \widehat{\mathbf{x}}_2 = (\widehat{\mathbf{v}}_2, \widehat{\mathbf{p}}_2)^T, \widehat{\mathbf{x}}_\Gamma = (\widehat{\mathbf{v}}_\Gamma, \widehat{\mathbf{p}}_0)^T$*, be the states satisfying the optimality systems associated with the full order and the reduced order model. Then, under assumption* $(\mathbf{A_5})$ *and for* $\mathbf{v}_1^{(0)} = 0$ *there exists* $C > 0$ *such that*

$$\left\| \begin{pmatrix} \mathbf{v}_2 - \widehat{\mathbf{v}}_2 \\ \mathbf{v}_\Gamma - \widehat{\mathbf{v}}_\Gamma \end{pmatrix} \right\|_{L^2} \le C \left\| \begin{pmatrix} \mathbf{f} \\ \widehat{\mathbf{x}}_\Gamma \end{pmatrix} \right\|_{L^2} \Big(\sigma_{p+1} + \cdots + \sigma_n \Big), \tag{5.6a}$$

$$\left\| \begin{pmatrix} \mathbf{p}_2 - \widehat{\mathbf{p}}_2 \\ \mathbf{p}_0 - \widehat{\mathbf{p}}_0 \end{pmatrix} \right\|_{L^2} \le C \left\| \begin{pmatrix} \mathbf{f} \\ \widehat{\mathbf{x}}_\Gamma \end{pmatrix} \right\|_{L^2} \Big(\sigma_{p+1} + \cdots + \sigma_n \Big), \tag{5.6b}$$

$$\left\| \begin{pmatrix} \mathbf{z}_1 - \widehat{\mathbf{z}}_1 \\ \mathbf{z}_{v,\Gamma} - \widehat{\mathbf{z}}_{v,\Gamma} \\ \mathbf{z}_{p,\Gamma} - \widehat{\mathbf{z}}_{p,\Gamma} \end{pmatrix} \right\|_{L^2} \leq C \left\| \begin{pmatrix} \mathbf{f} \\ \widehat{\mathbf{x}}_\Gamma \end{pmatrix} \right\|_{L^2} \Big(\sigma_{p+1} + \cdots + \sigma_n\Big). \tag{5.6c}$$

Proof. Since the full order optimality equations for subdomain Ω_1 differ from the reduced order optimality system by the inputs from the interface Γ, we construct an auxiliary full order system with velocity $\widetilde{\mathbf{v}}_1$, pressure $\widetilde{\mathbf{p}}_1$, and observations $\widetilde{\mathbf{z}}$ that has the same inputs as the reduced order system.

$$\begin{aligned} \mathbf{E}_1 \frac{d}{dt} \begin{pmatrix} \tilde{\mathbf{v}}_1(t) \\ \tilde{\mathbf{p}}_1(t) \end{pmatrix} &= -\mathbf{S}_{11} \begin{pmatrix} \tilde{\mathbf{v}}_1(t) \\ \tilde{\mathbf{p}}_1(t) \end{pmatrix} - \mathbf{S}_{1\Gamma} \begin{pmatrix} \widehat{\mathbf{v}}_\Gamma(t) \\ \widehat{\mathbf{p}}_0(t) \end{pmatrix} + \begin{pmatrix} \mathbf{K}_1 \\ \mathbf{L}_1 \end{pmatrix} \mathbf{f}(t), \\ \tilde{\mathbf{z}}_1(t) &= \mathbf{C}_1\tilde{\mathbf{v}}_1(t) + \mathbf{F}_1\tilde{\mathbf{p}}_1(t) + \mathbf{F}_0\widehat{\mathbf{p}}_0(t) + \mathbf{D}_1\mathbf{f}(t) - \mathbf{d}(t), \\ \begin{pmatrix} \tilde{\mathbf{z}}_{v,\Gamma}(t) \\ \tilde{\mathbf{z}}_{p,\Gamma}(t) \end{pmatrix} &= -\mathbf{S}_{1\Gamma}^T \begin{pmatrix} \tilde{\mathbf{v}}_1(t) \\ \tilde{\mathbf{p}}_1(t) \end{pmatrix}, \\ \mathbf{M}_1\tilde{\mathbf{v}}_1(0) = \mathbf{v}_1^{(0)} \quad &, \quad \mathbf{L}_1\mathbf{f}(0) = \mathbf{B}_{11}\mathbf{M}_1^{-1}\mathbf{v}_1^{(0)} + \mathbf{B}_{1\Gamma}\mathbf{M}_\Gamma(\theta)^{-1}\mathbf{v}_\Gamma^{(0)}(\theta). \end{aligned}$$

Then, the standard BT error bound tells us that the error in the observations between the auxiliary full order system and the reduced order system can be bounded from above by the sum of the remaining Hankel singular values.

$$\left\| \begin{pmatrix} \tilde{\mathbf{z}}_1 - \widehat{\mathbf{z}}_1 \\ \tilde{\mathbf{z}}_{v,\Gamma} - \widehat{\mathbf{z}}_{v,\Gamma} \\ \tilde{\mathbf{z}}_{p,\Gamma} - \widehat{\mathbf{z}}_{p,\Gamma} \end{pmatrix} \right\|_{L^2} \leq 2 \Big(\sigma_{p+1} + \cdots + \sigma_n\Big) \left\| \begin{pmatrix} \mathbf{f} \\ \widehat{\mathbf{v}}_\Gamma \\ \widehat{\mathbf{p}}_0 \end{pmatrix} \right\|_{L^2}. \tag{5.7}$$

Considering the errors in the states with $\widehat{\mathbf{v}}_1$ and $\widehat{\mathbf{p}}_1$ replaced with the velocity $\widetilde{\mathbf{v}}_1$ and the pressure $\widetilde{\mathbf{p}}_1$ from the auxiliary system

$$\mathbf{e}_v := (\mathbf{v}_1 - \tilde{\mathbf{v}}_1, \mathbf{v}_2 - \widehat{\mathbf{v}}_2, \mathbf{v}_\Gamma - \widehat{\mathbf{v}}_\Gamma)^T \quad , \quad \mathbf{e}_p := (\mathbf{p}_1 - \tilde{\mathbf{p}}_1, \mathbf{p}_2 - \widehat{\mathbf{p}}_2, \mathbf{p}_0 - \widehat{\mathbf{p}}_0)^T,$$

we see that the errors satisfy an index 2 differential algebraic equation with the forcing term being the difference in the observations at the interface.

$$\begin{aligned} \mathbf{E}(\theta) \frac{d}{dt} \begin{pmatrix} \mathbf{e}_v(t) \\ \mathbf{e}_p(t) \end{pmatrix} &= -\mathbf{S}(\theta) \begin{pmatrix} \mathbf{e}_v(t) \\ \mathbf{e}_p(t) \end{pmatrix} + \begin{pmatrix} \mathbf{g}_1(t) \\ \mathbf{0} \end{pmatrix}, \quad t \in (0,T], \\ \mathbf{M}(\theta)\mathbf{e}_v(0) &= \mathbf{0}. \end{aligned}$$

Here, $\mathbf{g}_1(t) := (\mathbf{0}, \mathbf{0}, \tilde{\mathbf{z}}_{\mathbf{v},\mathbf{\Gamma}} - \widehat{\mathbf{z}}_{\mathbf{v},\mathbf{\Gamma}})^\mathbf{T}$. Theorem 2.1 implies

$$\left\| \begin{pmatrix} \mathbf{v}_1 - \tilde{\mathbf{v}}_1 \\ \mathbf{v}_2 - \widehat{\mathbf{v}}_2 \\ \mathbf{v}_\Gamma - \widehat{\mathbf{v}}_\Gamma \end{pmatrix} \right\|_{L^2} \leq C \, \|\tilde{\mathbf{z}}_{v,\Gamma} - \widehat{\mathbf{z}}_{v,\Gamma}\|_{L^2}, \tag{5.8a}$$

$$\left\| \begin{pmatrix} \mathbf{p}_1 - \tilde{\mathbf{p}}_1 \\ \mathbf{p}_2 - \widehat{\mathbf{p}}_2 \\ \mathbf{p}_0 - \widehat{\mathbf{p}}_0 \end{pmatrix} \right\|_{L^2} \leq C \, \|\tilde{\mathbf{z}}_{v,\Gamma} - \widehat{\mathbf{z}}_{v,\Gamma}\|_{L^2}. \tag{5.8b}$$

Using (5.7) in (5.8a) and (5.8b) results in (5.6a) and (5.6b). □

For the error in the adjoint states, under assumptions $(\mathbf{A_2})$, $(\mathbf{A_3})$, and $(\mathbf{A_4})$, we can derive a similar upper bound.

Lemma 5.4. *Let* $\mathbf{x}, \mathbf{x}_\Gamma$ *be as in Lemma 5.3 and assume that* $\boldsymbol{\mu} := (\boldsymbol{\mu}_1, \boldsymbol{\mu}_2, \boldsymbol{\mu}_\Gamma)^T$ *with* $\boldsymbol{\mu}_i := (\boldsymbol{\lambda}_i, \boldsymbol{\kappa}_i)^T, 1 \le i \le 2, \boldsymbol{\mu}_\Gamma := (\boldsymbol{\lambda}_\Gamma, \boldsymbol{\kappa}_0)^T$ *and* $\widehat{\boldsymbol{\mu}} := (\widehat{\boldsymbol{\mu}}_1, \widehat{\boldsymbol{\mu}}_2, \widehat{\boldsymbol{\mu}}_\Gamma)^T$ *with* $\widehat{\boldsymbol{\mu}}_1 := \widehat{\boldsymbol{\lambda}}_1, \widehat{\boldsymbol{\mu}}_2 := (\widehat{\boldsymbol{\lambda}}_2, \widehat{\boldsymbol{\kappa}}_2)^T, \widehat{\boldsymbol{\mu}}_\Gamma := (\widehat{\boldsymbol{\lambda}}_\Gamma, \widehat{\boldsymbol{\kappa}}_0)^T$ *satisfy the optimality systems associated with the full order and the reduced order model. Then, under assumptions* $(\mathbf{A_2}), (\mathbf{A_3}), (\mathbf{A_4})$ *and for* $\boldsymbol{\lambda}_1^{(T)} = 0$ *there exists* $C > 0$ *such that*

$$\left\| \begin{pmatrix} \boldsymbol{\lambda}_2 - \widehat{\boldsymbol{\lambda}}_2 \\ \boldsymbol{\lambda}_\Gamma - \widehat{\boldsymbol{\lambda}}_\Gamma \end{pmatrix} \right\|_{L^2} \le C \left(\left\| \begin{pmatrix} \mathbf{f} \\ \widehat{\mathbf{x}}_\Gamma \end{pmatrix} \right\|_{L^2} + \left\| \begin{pmatrix} \widehat{\mathbf{z}}_1 \\ \widehat{\boldsymbol{\mu}}_\Gamma \end{pmatrix} \right\|_{L^2} \right) \Big(\sigma_{p+1} + \cdots + \sigma_n \Big),$$

$$\left\| \begin{pmatrix} \boldsymbol{\kappa}_2 - \widehat{\boldsymbol{\kappa}}_2 \\ \boldsymbol{\kappa}_0 - \widehat{\boldsymbol{\kappa}}_0 \end{pmatrix} \right\|_{L^2} \le C \left(\left\| \begin{pmatrix} \mathbf{f} \\ \widehat{\mathbf{x}}_\Gamma \end{pmatrix} \right\|_{L^2} + \left\| \begin{pmatrix} \widehat{\mathbf{z}}_1 \\ \widehat{\boldsymbol{\mu}}_\Gamma \end{pmatrix} \right\|_{L^2} \right) \Big(\sigma_{p+1} + \cdots + \sigma_n \Big).$$

Proof. Using the same reasoning as in the proof of Lemma 5.3, we construct an appropriate auxiliary system for subdomain Ω_1 which has the same inputs on the interface as the reduced order system in the adjoint states. For details we refer to Lemma 6.2 in [4]. □

Combining the results of the previous lemmas, we obtain the desired a priori estimate of the modeling error.

Theorem 5.5. *Under assumptions* $(\mathbf{A_1})$–$(\mathbf{A_5})$ *let* θ^* *and* $\widehat{\theta}^*$ *be the optimal designs obtained by the solution of the full order and the reduced order optimization problem. Then, there exists* $C > 0$ *such that*

$$\|\theta^* - \widehat{\theta}^*\| \le C \Big(\sigma_{p+1} + \cdots + \sigma_n \Big). \tag{5.10}$$

6. Application to surface acoustic wave driven microfluidic biochips

We apply the model reduction based optimization of the Stokes system to the optimal design of capillary barriers for surface acoustic wave driven microfluidic biochips. Microfluidic biochips are used in pharmaceutical, medical, and forensic applications for high throughput screening, genotyping, and sequencing in genomics, protein profiling in proteomics, and cytometry in cell analysis [47, 53]. They provide a much better sensitivity and a greater flexibility than traditional approaches. More importantly, they give rise to a significant speed-up of the hybridization processes and allow the in-situ investigation of these processes at an extremely high time resolution. This can be achieved by integrating the fluidics on top of the chip by means of a lithographically produced network of channels and reservoirs (cf. Fig. 1 (left)). The idea is to inject a DNA or protein containing probe and to transport it in the fluid to a reservoir where a chemical analysis is performed. The fluid flow is induced by surface acoustic waves (SAW) generated by interdigital transducers placed on a piezoelectric substrate which allows a much more accurate control of the flow than conventional external pumps [20, 28, 58, 59, 60]. In order to guarantee the filling of the reservoirs with a precise amount of the probe, pressure driven capillary barriers are placed between the channels and the reservoirs (cf. Fig. 1 (right)). As long as the pressure is above

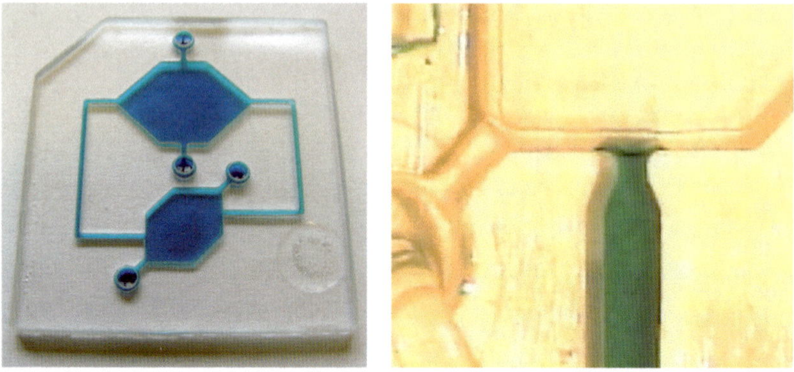

FIGURE 1. Microfluidic biochip (left) and capillary barrier (right)

a certain threshold, there is fluid flow from the channel into the reservoir (flow mode). Otherwise, there is no inflow, i.e., the barrier is in the stopping mode. One of the optimization issues is to design the capillary barrier such that the velocity and the pressure in the flow mode is as close as possible to a prescribed velocity and pressure profile.

SAW driven microfluidic biochips are modeled by a system of PDEs consisting of the linearized equations of piezoelectricity coupled with the compressible Navier-Stokes equations (see, e.g., [1, 2]). We consider the SAW induced fluid flow in the fluidic network on top of the biochip and denote by $\Omega \subset \mathbb{R}^2$ the domain occupied by the fluid. Its boundary is split according to $\Gamma = \overline{\Gamma}_D \cup \overline{\Gamma}_N$, $\Gamma_D \cap \Gamma_N = \emptyset$, where Γ_D stands for the part where the SAW enter the fluid filled microchannels. We further denote by $\mathbf{v}$ and p the velocity and the pressure, and we refer to ρ, η, and ξ as the density of the fluid and the standard and bulk viscosities. The pair $(\mathbf{v}, p)$ satisfies the following initial-boundary value problem

$$\rho\Big(\frac{\partial \mathbf{v}}{\partial t} + \mathbf{v}\cdot\nabla\mathbf{v}\Big) = \nabla\cdot\boldsymbol{\sigma} \qquad \text{in } \Omega,\ t \in (0,T], \tag{6.1a}$$

$$\frac{\partial \rho}{\partial t} + \nabla\cdot(\rho\mathbf{v}) = 0 \qquad \text{in } \Omega,\ t \in (0,T], \tag{6.1b}$$

$$\mathbf{v}(\cdot + \mathbf{u}(\cdot,t),t) = \frac{\partial \mathbf{u}}{\partial t}(\cdot,t) \qquad \text{on } \Gamma_D,\ t \in (0,T] \tag{6.1c}$$

$$\boldsymbol{\sigma}\mathbf{n} = 0 \qquad \text{on } \Gamma_N,\ t \in (0,T], \tag{6.1d}$$

$$\mathbf{v}(\cdot,0) = \mathbf{v}_0, \quad p(\cdot,0) = p_0 \qquad \text{in } \Omega(0), \tag{6.1e}$$

where $\boldsymbol{\sigma} = (\boldsymbol{\sigma}_{ij})_{i,j=1}^2$, $\boldsymbol{\sigma}_{ij} := -p\,\delta_{ij} + 2\eta\varepsilon_{ij}(\mathbf{v}) + \delta_{ij}(\xi - 2\eta/3)\nabla\cdot\mathbf{v}$ and $\mathbf{u}$ in (6.1c) stands for the deflection of the walls of the microchannels caused by the SAW. We note that $\mathbf{u}$ can be computed by the solution of the linearized equations of piezoelectricity (see, e.g., [23]) and that we have neglected the time-dependence of the domain, since the deflection of the walls of the microchannels by the SAW (approximately $10^{-9}\,m$) is small compared to the lengths, widths, and heights of the microchannels (μm to mm).

The SAW induced fluid flow occurs on two different time scales. When the SAW enter the fluid filled microchannels, sharp jets are created which is a process that happens within nanoseconds. Then, the SAW propagate along the channels and experience a significant damping resulting in a flow pattern called acoustic streaming. This relaxation process happens on a time scale of milliseconds. We are thus faced with a multiscale fluid flow which can be appropriately handled by homogenization. Following [1, 35], we introduce a scale parameter $0 < \varepsilon \ll 1$ and consider the asymptotic expansions

$$\begin{aligned} \rho &= \rho_0 + \varepsilon\, \rho' + \varepsilon^2\, \rho'' + O(\varepsilon^3)\,, \\ \mathbf{v} &= \mathbf{v}_0 + \varepsilon\, \mathbf{v}' + \varepsilon^2\, \mathbf{v}'' + O(\varepsilon^3)\,, \\ p &= p_0 + \varepsilon\, p' + \varepsilon^2\, p'' + O(\varepsilon^3)\,. \end{aligned}$$

Here, $\rho_0, \mathbf{v}_0, p_0$ are constant in time and space and represent the known equilibrium state without SAW excitation, whereas $\rho\prime, \rho\prime\prime$ etc. are functions of space and time. We insert the expansion into (6.1a)–(6.1e) and collect all terms of order $O(\varepsilon)$. Setting $\rho_1 = \varepsilon\rho', \mathbf{v}_1 := \varepsilon\mathbf{v}', p_1 := \varepsilon p'$, we find that the triple $(\rho_1, \mathbf{v}_1, p_1)$ satisfies the following time-periodic initial-boundary value problem for the linear compressible Navier-Stokes equations

$$\rho_0 \frac{\partial \mathbf{v}_1}{\partial t} - \nabla \cdot \boldsymbol{\sigma}_1 = 0 \qquad \text{in } \Omega \times (0, T_1], \tag{6.2a}$$

$$\frac{\partial \rho_1}{\partial t} + \rho_0 \nabla \cdot \mathbf{v}_1 = 0 \qquad \text{in } \Omega \times (0, T_1], \tag{6.2b}$$

$$v_1 = \mathbf{g}_1 \qquad \text{on } \Gamma_D \times (0, T_1], \tag{6.2c}$$

$$\boldsymbol{\sigma}_1 \mathbf{n} = 0 \qquad \text{on } \Gamma_N \times (0, T_1], \tag{6.2d}$$

$$\mathbf{v}_1(\cdot, 0) = 0, \quad p_1(\cdot, 0) = 0 \qquad \text{in } \Omega, \tag{6.2e}$$

where $T_1 := 2\pi/\omega$ with ω being the angular frequency of the time harmonic SAW excitation, $\mathbf{g}_1 := \partial u/\partial t$ in (6.2c), and

$$\boldsymbol{\sigma}_1 = ((\boldsymbol{\sigma}_1)_{ij})_{i,j=1}^2\,, \quad (\boldsymbol{\sigma}_1)_{ij} := -p_1\, \delta_{ij} + 2\eta\varepsilon_{ij}(\mathbf{v}_1) + \delta_{ij}(\xi - 2\eta/3)\nabla \cdot \mathbf{v}_1.$$

Moreover, p_1 and ρ_1 are related by the constitutive equation $p_1 = c_0^2\, \rho_1$ in $\Omega \times (0, T_1]$, where c_0 stands for the small signal sound speed in the fluid. The system (6.2a)–(6.2e) describes the propagation and damping of the acoustic waves in the microchannels.

Now, we collect all terms of order $O(\varepsilon^2)$. We set $\rho_2 := \varepsilon^2\rho'', \mathbf{v}_2 := \varepsilon^2\mathbf{v}'', p_2 := \varepsilon^2 p''$. Performing the time-averaging $\langle w\rangle := T_1^{-1}\int_{t_0}^{t_0+T_1} w dt$ allows to eliminate ρ_2 from the equations. We thus arrive at the compressible Stokes system

$$\rho_0 \frac{\partial \mathbf{v}_2}{\partial t} - \nabla \cdot \boldsymbol{\sigma}_2 = \langle -\rho_1 \frac{\partial \mathbf{v}_1}{\partial t} - \rho_0 (\nabla \mathbf{v}_1)\mathbf{v}_1 \rangle \qquad \text{in } \Omega \times (0, T], \tag{6.3a}$$

$$\rho_0 \nabla \cdot \mathbf{v}_2 = \langle -\nabla \cdot (\rho_1 \mathbf{v}_1) \rangle \qquad \text{in } \Omega \times (0, T], \tag{6.3b}$$

$$\mathbf{v}_2 = \mathbf{g}_2 \qquad \text{on } \Gamma_D \times (0,T], \tag{6.3c}$$

$$\boldsymbol{\sigma}_2 \mathbf{n} = 0 \qquad \text{on } \Gamma_N \times (0,T], \tag{6.3d}$$

$$\mathbf{v}_2(\cdot,0) = 0, \; p_2(\cdot,0) = 0 \qquad \text{in } \Omega, \tag{6.3e}$$

where $\mathbf{g}_2 := -\langle(\nabla \mathbf{v}_1)\mathbf{u}\rangle$ in (6.3c) and

$$\boldsymbol{\sigma}_2 = ((\boldsymbol{\sigma}_2)_{ij})_{i,j=1}^2 \;, \; (\boldsymbol{\sigma}_2)_{ij} := -p_2\, \delta_{ij} + 2\eta\varepsilon_{ij}(\mathbf{v}_2) + \delta_{ij}(\xi - 2\eta/3)\nabla \cdot \mathbf{v}_2.$$

The density ρ_2 can be obtained via the constitutive equation $p_2 = c_0^2\, \rho_2$ in $\Omega \times (0,T]$. We use the compressible Stokes system (6.3a)–(6.3e) as a model for the acoustic streaming. For a theoretical justification of (6.2a)–(6.2e) and (6.3a)–(6.3e) and a model validation based on experimental data we refer to [2].

For the optimal design of a capillary barrier, we consider acoustic streaming as described by (6.3a)–(6.3e) in a network of microchannels and reservoirs on top of a microfluidic biochip with a capillary barrier between a channel and a reservoir (cf. Figure 2). The computational domain Ω is decomposed into subdomains $\Omega_1 = \Omega \setminus \Omega_2$, and $\Omega_2 = (1.5, 2.5) \times (9, 10)$ mm^2. The boundary $\partial\Omega$ is split into $\Gamma_{\text{in}} = \{0\} \times (9.4, 10), \Gamma_{\text{out}} = \{10\} \times (0,1)$ mm^2, and $\Gamma_{\text{lat}} = \partial\Omega \setminus (\Gamma_{\text{in}} \cup \Gamma_{\text{out}})$. We assume that an interdigital transducer of width 6mm is placed at Γ_{in} and that the input velocity profile $\mathbf{u} = (\mathbf{u}_1, \mathbf{u}_2)$ is given by

$$\mathbf{u}_1(t, x_1) = 0.6\epsilon \sin(2\pi(-\hat{k}x_1 + ft)),$$
$$\mathbf{u}_2(t, x_1) = -\epsilon \cos(2\pi(-\hat{k}x_1 + ft))$$

with appropriately chosen constants $\epsilon, \hat{k}$ and f. We further choose a constant velocity profile $\mathbf{v}_{\text{in}}(x_1, x_2)$ on $\Gamma_{\text{in}} \times (0,T)$, outflow boundary conditions on $\Gamma_{\text{out}} \times (0,T)$, and no-slip conditions on $\Gamma_{\text{lat}} \times (0,T)$. The objective is to design the shape of the top $\Gamma_{2,T}$ and the bottom $\Gamma_{2,B}$ of $\partial\Omega_2$ in such a way that a prescribed velocity profile $\mathbf{v}^d$ is achieved in $\Omega_2 \times (0,T)$ and that the vorticity is minimized in Ω_{obs} (the

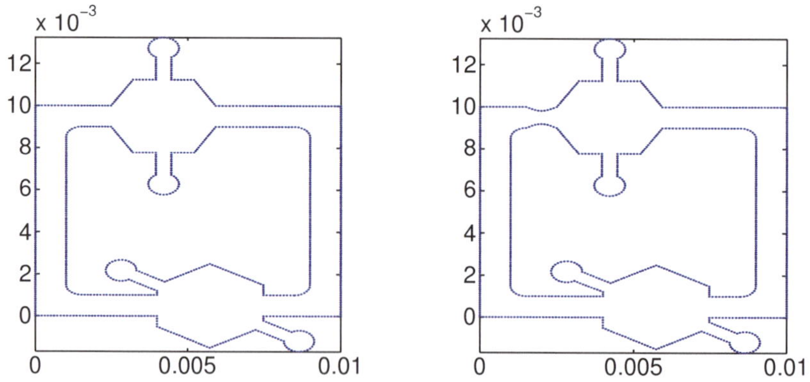

FIGURE 2. Optimal design of a capillary barrier: The reference domain Ω_{ref} (left, in [m]) and the optimal domain (right, in [m]).

two bulb shaped structures associated with the lower reservoir in Figure 2). The subdomain Ω_2 is parameterized by representing the top and bottom boundary by Bézier curves with $d_{\text{top}} = 6$ and $d_{\text{bot}} = 6$ control points, respectively. This leads to a parametrization $\Omega_2(\theta)$ of Ω_2 with parameters $\theta \in \mathbb{R}^{d_{\text{top}}+d_{\text{bot}}}$.

The shape optimization problem amounts to the minimization of

$$J(\theta) = \int_0^T \int_{\Omega_{\text{obs}}} |\nabla \times \mathbf{v}(x,t)|^2 dx dt + \int_0^T \int_{\Omega_2(\theta)} |\mathbf{v}(x,t) - \mathbf{v}^d(x,t)|^2 dx dt \tag{6.4}$$

subject to (6.3a)–(6.3e) and the design parameter constraints

$$\theta^{\min} \leq \theta \leq \theta^{\max}.$$

The final time T is $T = 0.1$ ms, and the bounds $\theta^{\min}, \theta^{\max}$ on the design parameters are chosen such that the design constraints are never active in this example. The optimal domain $\Omega(\theta^*)$ is shown in Figure 2 (right).

We consider a geometrically conforming simplicial triangulation $\mathcal{T}_h(\Omega)$ of Ω that aligns with the decomposition into the subdomains Ω_1 and Ω_2 as well as the respective boundaries. The semi-discretization in space is performed by Taylor-Hood P2-P1 elements. We denote by $N_v^{(1)}$, $N_v^{(2)}$, N_v^Γ the number of velocity degrees of freedom in the subdomains $\overline{\Omega}_1 \setminus \Gamma, \overline{\Omega}_2 \setminus \Gamma$ and in Γ, respectively, and set $N_v = N_v^{(1)} + N_v^{(2)} + N_v^\Gamma$. Similarly, $N_p^{(1)}$, $N_p^{(2)}$ stand for the numbers of pressure degrees of freedom in the subdomains $\overline{\Omega}_1, \overline{\Omega}_2$ and $N_p = N_p^{(1)} + N_p^{(2)}$ is the total number of pressure degrees of freedom.

The semi-discretized optimization problems is solved by a projected BFGS method with Armijo line search [34], and the optimization algorithm is terminated when the norm of the projected gradient is less than $2 \cdot 10^{-8}$. We use automatic differentiation [26, 51] to compute the derivatives with respect to the design variables.

We have applied the combination of domain decomposition and balanced truncation model reduction (DDBTMR) to the semi-discretized Stokes system (6.3a)–(6.3e) using four different finite element meshes. Figure 3 (left) displays the convergence of the multi-shift ADI algorithm from [27] for the computation of the controllability Gramian $\mathbf{P}$ and the observability Gramian $\mathbf{Q}$, and Figure 3 (right) shows the computed Hankel singular values for the finest grid problem. The constant C in the estimate (5.10) for the error between the optimal design parameters, computed by the full and the reduced order problems, depends on quantities like α in (2.3) of Theorem 2.1, the derivatives of $\mathbf{A}(\theta)$ with respect to θ, etc. Numerical experiments indicate that for the current scaling of the problem, the constant C in the estimate (5.10) is large. Therefore, we require a rather small truncation level of $\sigma_{p+1} < 10^{-12}\sigma_1$ for the Hankel singular values.

Table 1 shows the sizes $N_v^{(1)}$, N_v of the full order models on the four grids as well as the sizes $N_{\widehat{v}}^{(1)}$, $N_{\widehat{v}}$ of the reduced order models in subdomain Ω_1 and in Ω. For the finest grid, DDBTMR reduced the size of the Ω_1 subproblem from $N_v^{(1)} = 48324$ to $N_{\widehat{v}}^{(1)} = 766$. The velocity degrees of freedom in $\Omega_2 \cup \Gamma$ are not

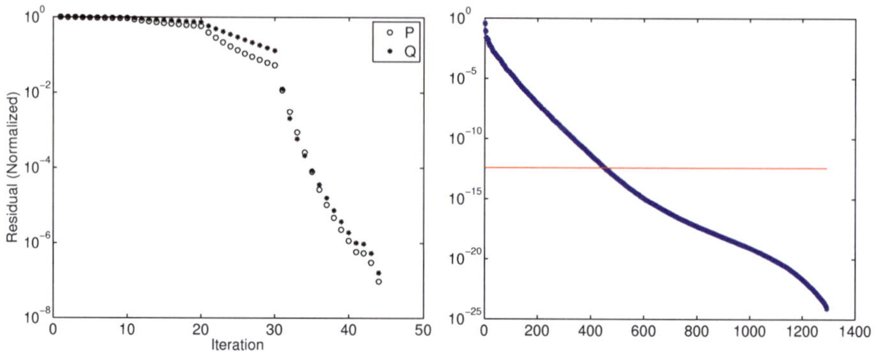

FIGURE 3. Convergence of the multishift ADI (left); the largest Hankel singular values and the threshold $10^{-12}\sigma_1$ (right).

reduced. On the finest grid these are $N_v^{(2)} + N_v^{\Gamma} = 914$. Therefore, the reduced order problem has $N_{\widehat{v}} = 914 + 766 = 1680$ degrees of freedom.

grid	m	$N_v^{(1)}$	$N_{\widehat{v}}^{(1)}$	N_v	$N_{\widehat{v}}$
1	167	7482	351	7640	509
2	195	11442	370	11668	596
3	291	16504	451	16830	777
4	802	48324	766	49238	1680

TABLE 1. The number m of observations, the numbers $N_v^{(1)}$, N_v of velocity degrees of freedom in subdomain Ω_1 and in Ω for the full order model, and the numbers $N_{\widehat{v}}^{(1)}$, $N_{\widehat{v}}$ of velocity degrees of freedom in subdomain Ω_1 and in Ω for the reduced order model.

The optimal shape parameters θ^* and $\widehat{\theta}^*$ computed by minimizing the full and the reduced order model, respectively, are shown in Table 2. For the finest grid, the error between the full and the reduced order model solutions is $\|\theta^* - \widehat{\theta}^*\| = 3.9165 \cdot 10^{-5}$.

θ^*	$(9.8833, 9.7467, 9.7572, 9.8671, 9.1336, 9.2015, 9.1971, 9.1310)\times 10^{-3}$
$\widehat{\theta}^*$	$(9.8694, 9.7374, 9.7525, 9.8628, 9.1498, 9.2044, 9.1895, 9.1204)\times 10^{-3}$

TABLE 2. Optimal shape parameters θ^* and $\widehat{\theta}^*$ computed by minimizing the full and the reduced order model.

Acknowledgement. The results presented in this paper have been obtained in co-operation with Harbir Antil and Matthias Heinkenschloss from Rice University. We are thankful to our colleagues for their most valuable contributions.

References

[1] H. Antil, A. Gantner, R.H.W. Hoppe, D. Köster, K.G. Siebert, and A. Wixforth, *Modeling and simulation of piezoelectrically agitated acoustic streaming on microfluidic biochips*, in *Domain Decomposition Methods in Science and Engineering XVII*, U. Langer et al., eds., Lecture Notes in Computational Science and Engineering, Vol. 60, Springer, Berlin-Heidelberg-New York, 2008, pp. 305–312.

[2] H. Antil, R. Glowinski, R.H.W. Hoppe, C. Linsenmann, T.W. Pan, and A. Wixforth, *Modeling, simulation, and optimization of surface acoustic wave driven microfluidic biochips.* J. Comp. Math. **28**, 149–169, 2010.

[3] H. Antil, M. Heinkenschloss, R.H.W. Hoppe, and D.C. Sorensen, *Domain decomposition and model reduction for the numerical solution of PDE constrained optimization problems with localized optimization variables.* Comp. Visual. Sci. **13**, 249–264, 2010.

[4] H. Antil, M. Heinkenschloss, and R.H.W. Hoppe, *Domain decomposition and balanced truncation model reduction for shape optimization of the Stokes system.* Optimization Methods & Software, DOI:10.1080/10556781003767904, 2010.

[5] H. Antil, R.H.W. Hoppe, and C. Linsenmann, *Path-following primal-dual interior-point methods for shape optimization of stationary flow problems.* J. Numer. Math. **11**, 81–100, 2007.

[6] H. Antil, R.H.W. Hoppe, and C. Linsenmann, *Adaptive path following primal dual interior point methods for shape optimization of linear and nonlinear Stokes flow problems.* In: Lecture Notes in Computer Science, Vol. 4818, pp. 259–266, Springer, Berlin-Heidelberg-New York, 2008.

[7] H. Antil, R.H.W. Hoppe, and C. Linsenmann, *Optimal design of stationary flow problems by path-following interior-point methods.* Control and Cybernetics **37**, 771–796, 2008.

[8] A.C. Antoulas, *Approximation of Large-Scale Systems.* SIAM, Philadelphia, 2005.

[9] A.C. Antoulas, M. Heinkenschloss, and T. Reis, *On balanced truncation for inhomogeneously initialized systems.* Technical Report TR09-29, Department of Computational and Applied Mathematics, Rice University, 2009.

[10] Z. Bai, P. Dewilde, and R. Freund, *Reduced order modeling.* Handbook of Numerical Analysis, Vol. XIII (W. Schilders and E.J.W. ter Maten; eds.), pp. 825–895, North-Holland/Elsevier, Amsterdam, 2005.

[11] P. Benner, R.W. Freund, D. Sorensen, and A. Varga (eds.), *Special issue on 'order reduction of large-scale systems'.* Linear Algebra and its Applications **415** (2-3), 2006.

[12] P. Benner, J.-R. Li, and T. Penzl, *Numerical solution of large Lyapunov equations, Riccati equations, and linear-quadratic control problems.* Numer. Lin. Alg. Appl. **15**, 755–777, 2008.

[13] P. Benner, V. Mehrmann, and D.C. Sorensen (eds.), *Dimension Reduction of Large-Scale Systems.* Lecture Notes in Computational Science and Engineering, Vol. 45, Springer, Berlin-Heidelberg-New York, 2005.

[14] D. Braess, *Finite elements. Theory, Fast Solvers, and Applications in Elasticity Theory. 3rd Ed.* Cambridge University Press, Cambridge, 2007.

[15] F. Brezzi and M. Fortin, *Mixed and Hybrid Finite Element Methods.* Springer, Berlin-Heidelberg-New York, 1991.

[16] T. Bui-Thanh, K. Willcox, and O. Ghattas, *Model reduction for large-scale systems with high-dimensional parametric input space.* SIAM J. Sci. Comp. **30**, 3270–3288, 2008.

[17] T. Bui-Thanh, K. Willcox, O. Ghattas, and B. van Bloemen Wanders, *Goal-oriented, model-constrained optimization for reduction of large-scale systems.* J. Comp. Phys. **224**, 880–896, 2007.

[18] M. Fahl and E. Sachs, *Reduced order modelling approaches to PDE-constrained optimization based on proper orthogonal decomposition.* In: Large-Scale PDE-Constrained Optimization (L.T. Biegler et al.; eds.), Lect. Notes in Comput. Sci. Engrg., Vol. 30, Springer, Berlin-Heidelberg-New York, 2003.

[19] M.A. Fallah, *SAW induced acoustic streaming in microchannels of different geometry.* Master's Thesis. Institute of Physics, University of Augsburg, 2008.

[20] T. Franke and A. Wixforth, *Microfluidics for miniaturized laboratories on a chip.* ChemPhysChem **9**, 2140–2156, 2008.

[21] R.W. Freund, *Model reduction methods based on Krylov subspaces.* Acta Numerica, 267–319, 2003.

[22] V. Girault and P.-A. Raviart, *Finite Element Methods for Navier-Stokes Equations. Theory and Algorithms.* Springer, Berlin-Heidelberg-New York, 1986.

[23] A. Gantner, R.H.W. Hoppe, D. Köster, K.G. Siebert, and A. Wixforth, *Numerical simulation of piezoelectrically agitated surface acoustic waves on microfluidic biochips*, Comp. Visual. Sci. **10**, 145–161, 2007.

[24] M.A. Grepl and A.T. Patera, *A posteriori error bounds for reduced-basis approximations of parametrized parabolic partial differential equations.* ESAIM: M2AN, **39**, 157–181, 2005.

[25] M.A. Grepl, Y. Maday, N.C. Nguyen, and A.T. Patera, *Efficient reduced-basis treatment of nonaffine and nonlinear partial differential equations.* ESAIM: M2AN, **41**, 575–605, 2007.

[26] A. Griewank and A. Walther, *Evaluating Derivatives. Principles and Techniques of Algorithmic Differentiation. 2nd Ed.* SIAM, Philadelphia, 2008.

[27] S. Gugercin, D.C. Sorensen, and A.C. Antoulas, *A modified low-rank Smith method for large scale Lyapunov equations.* Numer. Algorithms **32**, 27–55, 2003.

[28] Z. Guttenberg, H. Muller, H. Habermuller, A. Geisbauer, J. Pipper, J. Felbel, M. Kielpinski, J. Scriba, and A. Wixforth, *Planar chip device for PCR and hybridization with surface acoustic wave pump.* Lab Chip **5**, 308–317, 2005.

[29] B. Haasdonk, M. Ohlberger, and G. Rozza, *A reduced basis method for evolution schemes with parameter-dependent explicit operators.* Electronic Transactions on Numerical Analysis **32**, 145–161, 2008.

[30] M. Heinkenschloss, D.C. Sorensen, and K. Sun, *Balanced truncation model reduction for a class of descriptor systems with application to the Oseen equations.* SIAM Journal on Scientific Computing, **30**, 1038–1063, 2008.

[31] M. Hinze and S. Volkwein, *Proper orthogonal decomposition surrogate models for nonlinear dynamical systems: Error estimates and suboptimal control.* In: Dimension Reduction of Large-Scale Systems (P. Benner, V. Mehrmann, and D.S. Sorensen; eds.), pages 261–306, Lecture Notes in Computational Science and Engineering, Vol. 45, Springer, Berlin-Heidelberg-New York, 2005.

[32] M. Hinze and S. Volkwein, *Error estimates for abstract linear-quadratic optimal control problems using proper orthogonal decomposition.* Comp. Optim. Appl. **39**, 319–345, 2008.

[33] K. Ito and S.S. Ravindran, *A reduced order method for simulation and control of fluid flows.* J. Comp. Phys. **143**, 403–425, 1998.

[34] C.T. Kelley, *Iterative Methods for Optimization.* SIAM, Philadelphia, 1999.

[35] D. Köster, *Numerical simulation of acoustic streaming on SAW-driven biochips,* SIAM J. Comp. Sci. **29**, 2352–2380, 2007.

[36] K. Kunisch and S. Volkwein, *Galerkin proper orthogonal decomposition methods for parabolic problems.* Numer. Math. **90**, 117–148, 2001.

[37] K. Kunisch and S. Volkwein, *Galerkin proper orthogonal decomposition methods for a general equation in fluid dynamics.* SIAM J. Numer. Anal. **40**, 492–515, 2002.

[38] K. Kunisch and S. Volkwein, *Proper orthogonal decomposition for optimality systems.* ESAIM: M2AN, **42**, 1–23, 2008.

[39] J.-R. Li and J. White, *Low rank solution of Lyapunov equations.* SIAM J. Matrix Anal. Appl. **24**, 260–280, 2002.

[40] J.L. Lions and E. Magenes, *Non-homogeneous Boundary Value Problems and Applications. Vol. I, II.* Springer, Berlin-Heidelberg-New York, 1972.

[41] H.V. Ly and H.T. Tran, *Modelling and control of physical processes using proper orthogonal decomposition.* Math. and Comput. Modelling **33**, 223–236, 2001.

[42] H.V. Ly and H.T. Tran, *Proper orthogonal decomposition for flow calculations and optimal control in a horizontal CVD reactor.* Quart. Appl. Math. **60**, 631–656, 2002.

[43] R. März, *Canonical projectors for linear differential algebraic equations.* Comput. Math. Appl. **31**, 121–135, 1996.

[44] V. Mehrmann and T. Stykel, *Balanced truncation model reduction for large-scale systems in descriptor form.* In: Dimension Reduction of Large-Scale Systems (P. Benner, V. Mehrmann, and D.C. Sorensen; eds.), Lecture Notes in Computational Science and Engineering, Vol. 45, pp. 83–115, Springer, Berlin-Heidelberg-New York, 2005.

[45] A.T. Patera and G. Rozza, *Reduced Basis Methods and A Posteriori Error Estimation for Parametrized Partial Differential Equations.* (c) M.I.T. 2006-08. MIT Pappalardo Graduate Monographs in Mechanical Engineering, MIT, Boston, in progress.

[46] T. Penzl, *Eigenvalue decay bounds for solutions of Lyapunov equations: the symmetric case.* Syst. Control Lett. **40**, 139–144, 2000.

[47] J. Pollard and B. Castrodale, *Outlook for DNA microarrays: emerging applications and insights on optimizing microarray studies*, Report. Cambridge Health Institute, Cambridge 2003.

[48] T. Reis and M. Heinkenschloss, *Model reduction for a class of nonlinear electrical circuits.* In: Proc. 48th IEEE Conference on Decision and Control, Shanghai, P.R. China, December 16–18, 2009, pp. 5376–5383, IEEE, 2010.

[49] C.W. Rowley, *Model reduction for fluids, using balanced proper orthogonal decomposition.* Int. J. on Bifurcation and Chaos **15**, 997–1013, 2005.

[50] C.W. Rowley, T. Colonius, and R.M. Murray, *Model reduction for compressible fluids using POD and Galerkin projection.* Phys. D **189**, 115–129, 2004.

[51] S.M. Rump, *INTLAB – INTerval LABoratory.* In: Tibor Csendes, editor, *Developments in Reliable Computing*, pages 77–104. Kluwer Academic Publishers, Dordrecht, 1999.

[52] W.H.A. Schilders, H.A. van der Vorst, and J. Rommes (eds.), *Model Order Reduction: Theory, Research Aspects, and Applications.* Mathematics in Industry, Vol. 13, Springer, Berlin-Heidelberg-New York, 2008.

[53] H.M. Shapiro, *Practical flow cytometry.* Wiley-Liss, New York, 2003.

[54] T. Stykel, *Balanced truncation model reduction for semidiscretized Stokes equation.* Linear Algebra Appl. **415**, 262–289, 2006.

[55] F. Tröltzsch and S. Volkwein, *POD a posteriori error estimates for linear-quadratic optimal control problems.* Comp. Opt. and Appl. **44**, 83–115, 2009.

[56] S. Volkwein, *Optimal control of a phase-field model using proper orthogonal decomposition.* ZAMM **81**, 83–97, 2001.

[57] K. Willcox and J. Peraire, *Balanced model reduction via the proper orthogonal decomposition.* AIAA **40**, 2323–2330, 2002.

[58] A. Wixforth, *Acoustically driven programmable microfluidics for biological and chemical applications.* JALA **11**, 399-405, 2006.

[59] A. Wixforth, J. Scriba, and G. Gauer, *Flatland fluidics.* mst news **5**, 42-43, 2002.

[60] L.Y. Yeo and J.R. Friend, *Ultrafast microfluidics using surface acoustic waves.* Biomicrofluidics **3**, 012002–012023, 2009.

Thomas Franke and Achim Wixforth
Institute of Physics
Universität Augsburg
D-86159 Augsburg, Germany
e-mail: `franketh@physik.uni-augsburg.de`
`achim.wixforth@physik.uni-augsburg.de`

Ronald H.W. Hoppe
Institute of Mathematics
Universität Augsburg
D-86159 Augsburg, Germany

and

Department of Mathematics
University of Houston
Houston, TX 77204-3008, USA
e-mail: `hoppe@math.uni-augsburg.de`
`rohop@math.uh.edu`

Christopher Linsenmann
Institute of Mathematics
Universität Augsburg
D-86159 Augsburg, Germany
e-mail: `christopher.linsenmann@math.uni-augsburg.de`

International Series of Numerical Mathematics, Vol. 160, 99–122

Automated Extension of Fixed Point PDE Solvers for Optimal Design with Bounded Retardation

Nicolas Gauger, Andreas Griewank, Adel Hamdi, Claudia Kratzenstein, Emre Özkaya and Thomas Slawig

Abstract. We study PDE-constrained optimization problems where the state equation is solved by a pseudo-time stepping or fixed point iteration. We present a technique that improves primal, dual feasibility and optimality simultaneously in each iteration step, thus coupling state and adjoint iteration and control/design update. Our goal is to obtain bounded retardation of this coupled iteration compared to the original one for the state, since the latter in many cases has only a Q-factor close to one. For this purpose and based on a doubly augmented Lagrangian, which can be shown to be an exact penalty function, we discuss in detail the choice of an appropriate control or design space preconditioner, discuss implementation issues and present a convergence analysis. We show numerical examples, among them applications from shape design in fluid mechanics and parameter optimization in a climate model.

Mathematics Subject Classification (2000). Primary 90C30; Secondary 99Z99.

Keywords. Nonlinear optimization, fixed point solvers, exact penalty function, augmented Lagrangian, automatic differentiation, aerodynamic shape optimization, parameter optimization, climate model.

1. Introduction

Design optimization or control problems with PDEs may be distinguished from general nonlinear programming problems by the fact that the vector of variables is partitioned into a state vector $y \in Y$ and control or design variables $u \in \mathcal{A} \subset U$. For applications of this scenario in Computational Fluid Dynamics (CFD) see for example [22, 27, 28]. In this paper, we aim to solve an optimization problem

$$\min_{y,u} f(y,u) \quad \text{s.t.} \quad c(y,u) = 0, \tag{1.1}$$

where the constraint or state equation is solved by an iterative process. Usually, the admissible set $\mathcal{A}$ is a closed convex subset of the design space U. For simplicity, we assume that Y and U are Hilbert spaces. When Y and U have finite dimensions $n = \dim(Y)$ and $m = \dim(U)$, their elements may be identified by coordinate vectors in $\mathbb{R}^n$ and $\mathbb{R}^m$ with respect to suitable Hilbert bases. This convention allows us to write duals as transposed vectors and inner products as the usual scalar products in Euclidean space.

The problem of augmenting fixed point solvers for PDEs with sensitivity and optimization has been considered by various authors during the last few years [13, 17, 19, 18, 11, 24]. In [17], the author studied a *One-Shot* approach involving preconditioned design corrections to solve design optimization problems. It is an approach that aims at attaining feasibility and optimality simultaneously (see [17, 11]). In fact, within one step the primal, dual and design variables are updated simultaneously. Using automatic differentiation [14, 16] this requires only one simultaneous evaluation of the function value with one directional and one adjoint derivative. The focus in [17] was about the derivation of an "ideal" preconditioner that ensures local contractivity of the three coupled iterative processes. From analyzing the eigenvalues of the associated Jacobian, the author derived necessary but not sufficient conditions to bound those eigenvalues below 1 in modulus.

Deriving a preconditioner that ensures even local convergence of the three coupled iteration seems to be quite difficult. Instead, we study in this paper the introduction of an exact penalty function of doubly augmented Lagrangian type (see [6, 7, 8, 9]) to coordinate the three iterative processes. This penalty function is defined from the Lagrangian of the optimization problem augmented by weighted primal and dual residuals. The approach should be useful for any combination and sequencing of steps for improving primal, dual feasibility and optimality. In this work we firstly analyze the dual retardation behavior that means the slow down in the overall convergence when the adjoint is coupled with the primal iteration. Section 3 is devoted to derive conditions on the involved weighting coefficients in view of a consistent reduction of the considered exact penalty function. In Section 4, we establish reasonable choices for the weighting coefficients. Then we elaborate a line search procedure that does not require the computation of any second derivatives of the original problem and propose a suitable preconditioner in Section 5. We show a global convergence result in Section 6 and present three examples in Sections 7–9.

1.1. Problem statement

We suppose that the constraint $c(y,u) = 0$ is solved by a fixed point iteration and can be thus be equivalently written as $y = G(y,u)$. We assume that f and G are $C^{2,1}$ functions on the closed convex set $Y \times \mathcal{A}$, and that the Jacobian $G_y := \partial G/\partial y$ has a spectral radius $\rho < 1$. Then, for a suitable inner product norm $\|.\|$, we have

$$\|G_y(y,u)\| = \|G_y^\top(y,u)\| \leq \rho < 1. \tag{1.2}$$

Hence, the mean value theorem implies on any convex subdomain of Y that G is a contraction. By Banach fixed point theorem, for fixed u, the sequence $y_{k+1} = G(y_k, u)$ converges to a unique limit $y^* = y^*(u)$. The Lagrangian associated to the constrained optimization problem is

$$L(y, \bar{y}, u) \quad = \quad f(y, u) + (G(y, u) - y)^\top \bar{y} \; = \; N(y, \bar{y}, u) - y^\top \bar{y},$$

where we introduced is the shifted Lagrangian N as

$$N(y, \bar{y}, u) \quad := \quad f(y, u) + G(y, u)^\top \bar{y}.$$

Furthermore, according to the first-order necessary condition [1], a KKT point $(y^*, \bar{y}^*, u^*)$ of the optimization problem (1.1) must satisfy

$$\left.\begin{array}{rcllll} y^* & = & G(y^*, u^*) & & \\ \bar{y}^* & = & N_y(y^*, \bar{y}^*, u^*)^\top & = & f_y(y^*, u^*)^\top + G_y(y^*, u^*)^\top \bar{y}^* \\ 0 & = & N_u(y^*, \bar{y}^*, u^*)^\top & = & f_u(y^*, u^*)^\top + G_u(y^*, u^*)^\top \bar{y}^*. \end{array}\right\} \tag{1.3}$$

Denoting by $\mathcal{F} := \{z = (y, u) \in Y \times \mathcal{A} : y = G(y, u)\}$ the feasible set, any z in the tangent plane $\mathcal{T}$ can be represented by the Implicit Function Theorem as $z = Z\tilde{z}$ where $\tilde{z} \in \mathbb{R}^m$ and

$$Z = \begin{bmatrix} (I - G_y)^{-1} G_u \\ I \end{bmatrix}.$$

In view of (1.2), we have that $I - G_y$ is invertible. Therefore, $\mathcal{F}$ is a smooth manifold of dimension $\dim(u) = m$ with tangent space spanned by the columns of Z. According to the second-order necessary condition, the reduced Hessian

$$H = Z^\top N_{xx} Z \qquad \text{where} \quad N_{xx} = \begin{bmatrix} N_{yy} & N_{yu} \\ N_{uy} & N_{uu} \end{bmatrix}, \tag{1.4}$$

must be positive semi-definite at a local minimizer. We will make the slightly stronger assumption of second-order sufficiency, i.e., H is positive definite.

1.2. One-shot strategy

Motivated by (1.3), one can use the following coupled full step iteration, called *One-shot strategy*, to reach a KKT point (see [17, 11]):

$$\left.\begin{array}{rcl} y_{k+1} & = & G(y_k, u_k), \\ \bar{y}_{k+1} & = & N_y(y_k, \bar{y}_k, u_k)^\top, \\ u_{k+1} & = & u_k - B_k^{-1} N_u(y_k, \bar{y}_k, u_k)^\top. \end{array}\right\} \tag{1.5}$$

Here, B_k is a design space preconditioner which must be selected to be symmetric positive definite. The contractivity (1.2) ensures that the first equation in the coupled full step (1.5) converges ρ-linearly for fixed u. Although the second equation exhibits a certain time-lag, it converges with the same asymptotic R-factor (see [19]). As far as the convergence of the coupled iteration (1.5) is concerned, the goal is to find B_k that ensures that the spectral radius of the coupled iteration (1.5) stays below 1 and as close as possible to ρ. We use the following notations:

$$\begin{aligned} \Delta y_k &:= G(y_k, u_k) - y_k, \\ \Delta \bar{y}_k &:= N_y(y_k, \bar{y}_k, u_k)^\top - \bar{y}_k, \\ \Delta u_k &:= -B_k^{-1} N_u(y_k, \bar{y}_k, u_k)^\top. \end{aligned}$$

2. Dual retardation

Since the solution $\bar{y}^*$ of the second equation introduced in (1.3) depends on the primal solution y^*, it cannot be computed accurately as long as y^* is not known, and the dual iterates $\bar{y}_k$ computed from the second equation will be affected by the remaining inaccuracy in the primal iterates y_k. Actually, the dual step corrections typically lag a little bit behind as the perturbations caused by the errors in the primal iterates tend to accumulate initially. We refer to this delay of convergence relative to the primal iterates as dual retardation. Nevertheless, asymptotically the dual correction steps tend to be no larger than the primal correction steps.

Theorem 2.1. *For u fixed let f, G be once Lipschitz continuously differentiable with respect to y near the fixed point $y^* = y^*(u)$ of G, and let $y_k \to y^*$ such that*

$$\lim_{k\to\infty} \frac{\|\Delta y_k\|}{\|\Delta y_{k-1}\|} = \rho^* := \|G_y(y^*, u)\|.$$

Now define for any $\varepsilon, \alpha, \beta > 0$ the smallest pair of integers $(\ell_p^\varepsilon, \ell_d^\varepsilon)$ such that

$$\sqrt{\alpha}\|\Delta y_{\ell_p^\varepsilon}\| \le \varepsilon \quad \text{and} \quad \sqrt{\beta}\|\Delta \bar{y}_{\ell_d^\varepsilon}\| \le \varepsilon.$$

Then, we have

$$\limsup_{\varepsilon\to 0} \frac{\ell_d^\varepsilon}{\ell_p^\varepsilon} \le 1. \tag{2.1}$$

Proof. See [21, Theorem 2.1]. □

One can show that this result is sharp considering the scalar case with the cost function

$$f(y) = \frac{\theta}{2} y^2 + \nu y, \quad y \in \mathbb{R},\ \nu \in \mathbb{R},\ \theta \in \mathbb{R}_+^*,$$

and $G(y, u) = \rho y + u$ with $0 < \rho < 1$. The coupled primal and dual iteration is

$$\begin{aligned} y_{k+1} &= \rho y_k + u, \\ \bar{y}_{k+1} &= \rho \bar{y}_k + \theta y_k + \nu. \end{aligned}$$

Then, we get the limit in (2.1) is 1. For details see again [21].

As we will see later, a more or less rather natural choice for the weights α, β is $\sqrt{\beta/\alpha} = (1-\rho)/\theta$. If we use $|\Delta y_0| = 1$, $\sqrt{\beta/\alpha} = (1-\rho)/\theta$, we can show that in this example $|\Delta y_k| = \rho^k$ and $\sqrt{\beta/\alpha}|\Delta \bar{y}_k| = k(1-\rho)\rho^{k-1}$.

3. Doubly augmented Lagrangian

The asymptotic rate of convergence of the coupled iteration (1.5) to a limit point $(y^*, \bar{y}^*, u^*)$ is determined by the spectral radius $\hat{\rho}(J^*)$ of the block Jacobian:

$$J^* = \left.\frac{\partial(y_{k+1}, \bar{y}_{k+1}, u_{k+1})}{\partial(y_k, \bar{y}_k, u_k)}\right|_{(y^*,\bar{y}^*,u^*)} = \begin{bmatrix} G_y & 0 & G_u \\ N_{yy} & G_y^\top & N_{yu} \\ -B^{-1}N_{uy} & -B^{-1}G_u^\top & I - B^{-1}N_{uu} \end{bmatrix}.$$

In [17], the author proved that unless they happen to coincide with those of G_y, the eigenvalues of J^* solve the following nonlinear eigenvalues problem:

$$\det[(\lambda - 1)B + H(\lambda)] = 0,$$

where

$$H(\lambda) = Z(\lambda)^\top N_{xx} Z(\lambda) \quad \text{and} \quad Z(\lambda) = \begin{bmatrix} (\lambda I - G_y)^{-1} G_u \\ I \end{bmatrix}.$$

As discussed in [17], although the conditions $B = B^\top \succ 0$ and $B \succ \frac{1}{2}H(-1)$ ensure that real eigenvalues of J^* stay less than 1, they are just necessary but not sufficient to exclude real eigenvalues less than -1. In addition, no constructive condition to also bound complex eigenvalues below 1 in modulus has been found. However, these do arise even when the underlying primal solver is Jacobi's method on the elliptic boundary value problem $y'' = 0$ in one space dimension. Therefore, deriving a design space preconditioner that ensures even local convergence of the coupled full step iteration (1.5) seems to be quite difficult.

Instead, we base our analysis on the following penalty or merit function of doubly augmented Lagrangian type (see [6, 7]), defined for $\alpha, \beta > 0$:

$$L^a(y, \bar{y}, u) := \frac{\alpha}{2}\|G(y,u) - y\|^2 + \frac{\beta}{2}\|N_y(y, \bar{y}, u)^\top - \bar{y}\|^2 + N(y, \bar{y}, u) - \bar{y}^\top y. \quad (3.1)$$

Now, we aim to solve the optimization problem (1.1) by looking for descent on L^a.

3.1. Gradient of L^a

In the remainder, we use the notation $\Delta G_y = I - G_y$. Note that ΔG_y is invertible. By an elementary calculation, we obtain:

Proposition 3.1. *The gradient of L^a is given by*

$$\nabla L^a(y, \bar{y}, u) = -Ms(y, \bar{y}, u), \quad \textit{where} \quad M = \begin{bmatrix} \alpha \Delta G_y^\top & -I - \beta N_{yy} & 0 \\ -I & \beta \Delta G_y & 0 \\ -\alpha G_u^\top & -\beta N_{yu}^\top & B \end{bmatrix}, \quad (3.2)$$

and s is the step increment vector associated with the iteration (1.5)

$$s(y, \bar{y}, u) = \begin{bmatrix} G(y,u) - y \\ N_y(y, \bar{y}, u)^\top - \bar{y} \\ -B^{-1} N_u(y, \bar{y}, u)^\top \end{bmatrix}. \quad (3.3)$$

The gradient ∇L^a involves vector derivatives as well as matrix derivatives where the complexity of their computations may grow with respect to the dimension of u. To avoid that dependence, we propose an economical computation of ∇L^a using Automatic Differentiation (AD), see [16]. Actually, to compute vector derivatives we can use the *reverse mode* of the package ADOL-C developed at Dresden University of Technology [15]. Furthermore, we present two options to compute terms in ∇L^a involving matrix derivatives namely $\Delta\bar{y}^\top N_{yy}$ and $\Delta\bar{y}^\top N_{yu}$. The first option consists on using one reverse sweep of *Second Order Adjoint* (SOA) by employing some (AD) tools, like ADOL-C [16] that ensures a cost proportional to the cost of (f, G) evaluation and independent of dimensions. Whereas the second option consists on simply using the definition

$$\frac{\partial}{\partial t}(N_y(y + t\Delta\bar{y}, \bar{y}, u))\Big|_{t=0} = N_{yy}(y, \bar{y}, u)\Delta\bar{y},$$

to approximate $\Delta\bar{y}^\top N_{yy}$. In fact, for $t \neq 0$, we have

$$\Delta\bar{y}^\top N_{yy}(y, \bar{y}, u) = \frac{N_y(y + t\Delta\bar{y}, \bar{y}, u)^\top - N_y(y, \bar{y}, u)^\top}{t} + o(t), \tag{3.4}$$

and thus the $\Delta\bar{y}^\top N_{yy}$ (and similarly $\Delta\bar{y}^\top N_{yu}$) can be approximated using (3.4).

4. Conditions on the weights α, β

Here we derive conditions on the weights α, β which in turn influence the merit function L^a to be an exact penalty function and the step increment vector associated to iteration (1.5) to yield descent on it.

4.1. Correspondence conditions

The first condition characterizes the correspondence between stationary points of L^a and zero increments s of the one-shot iteration, i.e., stationary points of 99.

Corollary 4.1 (Correspondence condition). *There is a one-to-one correspondence between the stationary points of L^a and the roots of s, defined in* (3.3), *wherever*

$$\det[\alpha\beta\Delta G_y^\top \Delta G_y - I - \beta N_{yy}] \neq 0,$$

which is implied by the correspondence condition

$$\alpha\beta(1-\rho)^2 > 1 + \beta\theta, \tag{4.1}$$

where $\theta = \|N_{yy}\|$.

Proof. See [21, Corollary 3.2]. □

The correspondence condition (4.1) now implies that the merit function L^a introduced in (3.1) is an exact penalty function:

Corollary 4.2. *If the condition*

$$\alpha\beta\Delta G_y^\top \Delta G_y > I + \beta N_{yy},$$

holds, then the penalty function L^a *introduced in* (3.1) *has a positive definite Hessian at a stationary point of the optimization problem* (1.1) *if and only if the reduced Hessian* H *introduced in* (1.4) *is positive definite at that point.*

Proof. See [21, Corollary 3.3]. □

4.2. Descent properties of the step increment

Here we derive conditions under which the step increment vector s introduced in (3.3) yields descent on the exact penalty function L^a.

Proposition 4.3 (Descent condition). *The step increment vector* s *yields descent on* L^a *for all large positive* B *if*

$$\alpha\beta\Delta\bar{G}_y \quad > \quad \left(I + \frac{\beta}{2}N_{yy}\right)(\Delta\bar{G}_y)^{-1}\left(I + \frac{\beta}{2}N_{yy}\right), \tag{4.2}$$

where $\Delta\bar{G}_y = \frac{1}{2}(\Delta G_y + \Delta G_y^\top)$. *Moreover,* (4.2) *is implied by the condition*

$$\sqrt{\alpha\beta}(1-\rho) \quad > \quad 1 + \frac{\beta}{2}\theta. \tag{4.3}$$

Proof. See [21, Proposition 3.4]. □

The design corrections given by the third component in s involve the inverse of B. Thus, provided (4.3) holds, a pure feasibility step (with fixed design) yields also descent on L^a. Considering a base point

$$(y_{k-1}, \bar{y}_{k-1}, u)$$

where $k \geq 1$ and analyzing L^a at the current point

$$(y_k = G(y_{k-1}, u), \bar{y}_k = N_y(y_{k-1}, \bar{y}_{k-1}, u), u),$$

we can establish a condition that leads to reduction on the exact penalty function L^a using a full pure feasibility step.

Theorem 4.4 (Full step descent condition). *Let* $\alpha, \beta > 0$ *satisfy*

$$\alpha \quad > \quad \frac{\theta(\beta\theta + 2)}{1-\rho^2} + \frac{(5 + \rho(1+\beta\theta))^2}{\beta(1-\rho^2)^2}. \tag{4.4}$$

Then a full pure feasibility step yields descent in L^a, *i.e.,*

$$L^a(y_k, \bar{y}_k, u) - L^a(y_{k-1}, \bar{y}_{k-1}, u) \quad < \quad 0.$$

Proof. See [21, Theorem 3.6] □

Note that the descent condition (4.3) is a bit stronger than the correspondence condition (4.1). However, the condition (4.4) is stronger than (4.3).

4.3. Bounded level sets of L^a

In order to establish a global convergence result, we show that under reasonable assumptions all level sets of the doubly augmented Lagrangian L^a are bounded.

Theorem 4.5. *Let $f \in C^{1,1}(Y \times U)$ be radially unbounded and satisfy*

$$\liminf_{\|y\|+\|u\|\to\infty} \frac{f}{\|\nabla_y f\|^2} \quad > \quad 0. \tag{4.5}$$

Then there exists (α, β) fulfilling (4.3) such that

$$\lim_{\|y\|+\|\bar{y}\|+\|u\|\to\infty} L^a(y, \bar{y}, u) \quad = \quad +\infty. \tag{4.6}$$

If the limit in (4.5) is equal to infinity, the assertion (4.6) holds without any additional restriction on (α, β).

Proof. See [20, Theorem 2.1]. □

Assumption (4.5) requires that f grows quadratically or slower as a function of $\|y\| + \|u\|$. If for example f is quadratic, i.e.,

$$f(x) = \frac{1}{2} x^\top A x + b^\top x \qquad \text{where} \quad A \in \mathbb{R}^{n,n}, A^\top = A \succ 0, \quad b, x \in \mathbb{R}^n,$$

we have (with $\lambda_{\min}, \lambda_{\max}$ the smallest and biggest eigenvalue, respectively, of A)

$$\lim_{\|x\|\to\infty} \frac{f(x)}{\|\nabla f(x)\|^2} \quad = \quad \frac{1}{2} \lim_{\|x\|\to\infty} \frac{x^\top A x}{\|Ax\|^2} \geq \frac{\lambda_{\min}}{2\|A\|_2^2} \quad = \quad \frac{\lambda_{\min}}{2\lambda_{\max}^2} > 0.$$

4.4. Particular choice of α and β

To ensure consistent reduction on L^a, a rather large primal weight α may be necessary, which severely restricts the step size and slows down the convergence. We present two options to compute the α, β based on selecting α as small as possible which is still fulfilling at least the condition (4.3) or (4.4) as an equality.

Option 1: Deriving α, β from (4.3) and minimizing α as a function of β leads to

$$\beta_1 = \frac{2}{\theta} \quad \text{and} \quad \alpha_1 = \frac{2\theta}{(1-\rho)^2}. \tag{4.7}$$

Option 2: Deriving α, β from the condition (4.4):
Fulfilling (4.4) as an equality implies

$$\alpha = \frac{\theta^2\beta^2 + 2\theta(1+5\rho)\beta + 5(5+2\rho) + \rho^2}{\beta(1-\rho^2)^2}. \tag{4.8}$$

Minimizing the right-hand side with respect to β gives

$$\beta_2 = \frac{\sqrt{5(5+2\rho)+\rho^2}}{\theta} \quad \text{and} \quad \alpha_2 = \frac{2\theta(1+5\rho+\theta\beta_2)}{(1-\rho^2)^2}. \tag{4.9}$$

If as usually $\rho \approx 1$, we obtain

$$\beta_2 \approx \frac{6}{\theta} \quad \text{and} \quad \alpha_2 \approx \frac{24\theta}{(1+\rho)^2(1-\rho)^2} \approx \frac{6\theta}{(1-\rho)^2}. \tag{4.10}$$

4.5. Estimation of problem parameters

To compute the weights we need estimates for ρ and θ. Neglecting the change in u, they can be obtained from the primal and dual iterates. From (1.2) we have

$$\|G(y_k, u) - G(y_{k-1}, u)\| \le \rho\|y_k - y_{k-1}\| \quad \text{for all } k \in \mathbb{N}.$$

Therefore, starting with an initial value ρ_0, we may update ρ using

$$\rho_{k+1} = \max\{\frac{\|\Delta y_k\|}{\|\Delta y_{k-1}\|}, \tau\rho_k\},$$

where $\tau \in (0,1)$. To estimate the value of θ, we use the approximation

$$\begin{bmatrix}\Delta y_{k+1}\\ \Delta\bar{y}_{k+1}\end{bmatrix} \approx \begin{bmatrix} G_y(y_k, u_k) & 0\\ N_{yy}(y_k, \bar{y}_k, u_k) & G_y(y_k, u_k)^\top\end{bmatrix}\begin{bmatrix}\Delta y_k\\ \Delta\bar{y}_k\end{bmatrix}.$$

Then, we obtain

$$\begin{bmatrix}\Delta\bar{y}_k\\ -\Delta y_k\end{bmatrix}^\top \begin{bmatrix}\Delta y_{k+1}\\ \Delta\bar{y}_{k+1}\end{bmatrix} \approx -(\Delta y_k)^\top N_{yy}(y_k, \bar{y}_k, u_k)\Delta y_k,$$

and

$$(\Delta y_k)^\top N_{yy}(y_k, \bar{y}_k, u_k)\Delta y_k \approx (\Delta y_k)^\top \Delta\bar{y}_{k+1} - (\Delta\bar{y}_k)^\top\Delta y_{k+1}.$$

Thus, one can approximate the value of θ as follows:

$$\theta_{k+1} \approx \max\{\frac{|(\Delta y_k)^\top\Delta\bar{y}_{k+1} - (\Delta\bar{y}_k)^\top\Delta y_{k+1}|}{\|\Delta y_k\|^2}, \tau\theta_k\}.$$

As far as numerical experiments are concerned, we obtained the best results when $N_{yy} = I$ which theoretically can be attained by using a coordinate transformation provided that $N_{yy} \succ 0$. In fact, let $N_{yy} = N_{yy}^{\frac{\top}{2}} N_{yy}^{\frac{1}{2}}$ be the Cholesky factorization of N_{yy}. Then, by considering $\tilde{y} = N_{yy}^{\frac{1}{2}}\, y$ and $\tilde{\bar{y}} = N_{yy}^{-\frac{\top}{2}}\,\bar{y}$ we obtain

$$\tilde{L}^a(\tilde{y}, \tilde{\bar{y}}, u) = \frac{\alpha}{2}\|\tilde{G}(\tilde{y}, u) - \tilde{y}\|^2 + \frac{\beta}{2}\|\tilde{N}_{\tilde{y}}(\tilde{y}, \tilde{\bar{y}}, u) - \tilde{\bar{y}}\|^2 + \tilde{N}(\tilde{y}, \tilde{\bar{y}}, u) - \tilde{\bar{y}}^\top\tilde{y}.$$

where $\tilde{G}(\tilde{y}, u) = N_{yy}^{\frac{1}{2}} G(N_{yy}^{-\frac{1}{2}}\tilde{y}, u)$, $\tilde{f}(\tilde{y}, u) = f(N_{yy}^{-\frac{1}{2}}\tilde{y}, u)$ and $\tilde{N}_{\tilde{y}}(\tilde{y}, \tilde{\bar{y}}, u) = \tilde{f}(\tilde{y}, u) + \tilde{G}(\tilde{y}, u)^\top\tilde{\bar{y}}$. Thus, we get $\nabla\tilde{L}^a(\tilde{y}, \tilde{\bar{y}}, u) = -\tilde{M}\tilde{s}(\tilde{y}, \tilde{\bar{y}}, u)$ where

$$\tilde{M} = \begin{bmatrix}\alpha\Delta\tilde{G}_{\tilde{y}}^\top & -(1+\beta)I & 0\\ -I & \beta\Delta\tilde{G}_{\tilde{y}} & 0\\ -\alpha\tilde{G}_u^\top & -\beta\tilde{N}_{\tilde{y}u}^\top & B\end{bmatrix}, \quad \tilde{s}(\tilde{y}, \tilde{\bar{y}}, u) = \begin{bmatrix}\tilde{G}(\tilde{y}, u) - \tilde{y}\\ \tilde{N}_{\tilde{y}}(\tilde{y}, \tilde{\bar{y}}, u)^\top - \tilde{\bar{y}}\\ -B^{-1}\tilde{N}_u(\tilde{y}, \tilde{\bar{y}}, u)^\top\end{bmatrix}.$$

With $\Delta\tilde{G}_{\tilde{y}} = I - N_{yy}^{\frac{1}{2}}\, G_y\, N_{yy}^{-\frac{1}{2}}$, $\tilde{G}_u = N_{yy}^{\frac{1}{2}}\, G_u$ and $\tilde{N}_{\tilde{y}u} = N_{yy}^{-\frac{\top}{2}}\, N_{yu}$. Furthermore, $\rho(\tilde{G}_{\tilde{y}}) = \rho(G_y) = \rho$. Therefore, if we have a state space preconditioner $P \approx N_{yy}^{\frac{1}{2}}$ for which P and P^{-1} can be evaluated at reasonable cost, then we may use

$$\tilde{L}^a(y, \bar{y}, u) = \frac{\alpha}{2}\|P(G(y, u) - y)\|^2 + \frac{\beta}{2}\|P^{-\top}(N_y(y, \bar{y}, u) - \bar{y})\|^2 + N(y, \bar{y}, u) - \bar{y}^\top y.$$

Hence, by the same techniques used to prove Proposition 4.3, we obtain the descent condition associated to the transformed coordinates given by

$$\sqrt{\alpha\beta}(1-\rho) > 1 + \frac{\beta\tilde{\theta}}{2} \quad \text{where} \quad \tilde{\theta} = \|P^{-1}N_{yy}^{\frac{1}{2}}\|$$

and the following approximations:

$$\beta_1 = \frac{2}{\tilde{\theta}},\ \alpha_1 = \frac{2\tilde{\theta}}{(1-\rho)^2} \quad \text{and} \quad \beta_2 = \frac{6}{\tilde{\theta}},\ \alpha_2 = \frac{6\tilde{\theta}}{(1-\rho)^2}.$$

5. Choice of the preconditioner B

Here, we derive a design space preconditioner B that results in a step s that yields descent on L^a. We assume that α, β are chosen such that (4.1) holds.

5.1. Explicit condition on B

In this subsection, we derive an explicit condition that leads to a first choice for the design space preconditioner. Since $s^\top Ms = \frac{1}{2}s^\top(M + M^\top)s$ and using (3.2), (3.3), we compute the symmetric matrix M_S defined as follows:

$$M_S := \frac{1}{2}(M^\top + M) = \begin{bmatrix} \alpha\Delta\bar{G}_y & -I - \frac{\beta}{2}N_{yy} & -\frac{\alpha}{2}G_u \\ -I - \frac{\beta}{2}N_{yy} & \beta\Delta\bar{G}_y & -\frac{\beta}{2}N_{yu} \\ -\frac{\alpha}{2}G_u^\top & -\frac{\beta}{2}N_{yu}^\top & B \end{bmatrix}. \tag{5.1}$$

Here $\Delta\bar{G}_y$ is the symmetric matrix given in Proposition 4.3. Therefore, we obtain

$$s^\top \nabla L^a = -s^\top M_S s. \tag{5.2}$$

Let $B^{\frac{1}{2}}$ be a Cholesky factor of B. Rescaling $u = B^{-\frac{1}{2}}\tilde{u}$, we find a result similar to (5.1) involving $\tilde{s}$ and $\tilde{M}_S$ where $\tilde{s}$ is obtained from the increment vector s by replacing its third component $\Delta u = -B^{-1}{N_u}^\top$ by

$$\Delta\tilde{u} = B^{\frac{1}{2}}\Delta u = -B^{-\frac{\top}{2}}{N_u}^\top = -{N_{\tilde{u}}}^\top$$

and the matrix $\tilde{M}_S$ is derived from M_S by substituting B with I and all derivatives with respect to the design u with $G_{\tilde{u}} = G_u B^{-\frac{1}{2}}$, $N_{\tilde{u}} = N_u B^{-\frac{1}{2}}$ and $N_{y\tilde{u}} = N_{yu}B^{-\frac{1}{2}}$. Thus, we get

$$\tilde{M}_S = \begin{bmatrix} \alpha\Delta\bar{G}_y & -I - \frac{\beta}{2}N_{yy} & -\frac{\alpha}{2}G_{\tilde{u}} \\ -I - \frac{\beta}{2}N_{yy} & \beta\Delta\bar{G}_y & -\frac{\beta}{2}N_{y\tilde{u}} \\ -\frac{\alpha}{2}G_{\tilde{u}}^\top & -\frac{\beta}{2}N_{y\tilde{u}}^\top & I \end{bmatrix}.$$

and $\tilde{M}_S$ is obtained from the matrix M_S as follows:

$$\tilde{M}_S = \operatorname{diag}(I, I, B^{-\frac{\top}{2}})\ M_S\ \operatorname{diag}(I, I, B^{-\frac{1}{2}}). \tag{5.3}$$

The aim now is to derive explicit conditions on B that ensure positive definiteness of $\tilde{M}_S$ which in view of (5.2), (5.3) implies that the increment vector s introduced in (3.3) yields descent on L^a. In fact it suffices to show positive-definiteness of a much simpler, real matrix in order to get the desired result for $\tilde{M}_S$:

Proposition 5.1. *Let* $\theta = \|N_{yy}\|$ *and* D_C *be the matrix defined by*

$$D_C = \begin{bmatrix} \alpha(1-\rho) & -1-\frac{\beta}{2}\theta & -\frac{\alpha}{2}\|G_{\tilde{u}}\| \\ -1-\frac{\beta}{2}\theta & \beta(1-\rho) & -\frac{\beta}{2}\|N_{y\tilde{u}}\| \\ -\frac{\alpha}{2}\|G_{\tilde{u}}\| & -\frac{\beta}{2}\|N_{y\tilde{u}}\| & 1 \end{bmatrix}. \tag{5.4}$$

Then, we have for all $v_1, v_2 \in \mathbb{R}^n$ *and* $v_3 \in \mathbb{R}^m$,

$$\begin{bmatrix} v_1 \\ v_2 \\ v_3 \end{bmatrix}^\top \tilde{M}_S \begin{bmatrix} v_1 \\ v_2 \\ v_3 \end{bmatrix} \geq \begin{bmatrix} \|v_1\| \\ \|v_2\| \\ \|v_3\| \end{bmatrix}^\top D_C \begin{bmatrix} \|v_1\| \\ \|v_2\| \\ \|v_3\| \end{bmatrix}.$$

Proof. See [20, Proposition 3.1]. □

Now we get a condition on B that ensures positive definiteness of D_C:

Proposition 5.2. *Let* $\theta = \|N_{yy}\|$ *and* α, β *satisfy* (4.3). *If*

$$\left(\frac{\sqrt{\alpha}}{2}\|G_{\tilde{u}}\| + \frac{\sqrt{\beta}}{2}\|N_{y\tilde{u}}\|\right)^2 \leq (1-\rho) - \frac{(1+\frac{\theta}{2}\beta)^2}{\alpha\beta(1-\rho)}, \tag{5.5}$$

then D_C *introduced in* (5.4) *is a positive definite matrix.*

Proof. See [20, Proposition 3.2] □

To get explicit conditions on B that ensure (5.5), we note that

$$\frac{1}{2}(\sqrt{\alpha}\|G_{\tilde{u}}\| + \sqrt{\beta}\|N_{y\tilde{u}}\|) \leq \left\| \begin{matrix} \sqrt{\alpha}G_{\tilde{u}} \\ \sqrt{\beta}N_{y\tilde{u}} \end{matrix} \right\|_2 = \left\| \begin{pmatrix} \sqrt{\alpha}G_u \\ \sqrt{\beta}N_{yu} \end{pmatrix} B^{-\frac{1}{2}} \right\|_2, \tag{5.6}$$

and, using a QR decomposition on the right-hand side obtain

$$\left\| \begin{pmatrix} \sqrt{\alpha}G_u \\ \sqrt{\beta}N_{yu} \end{pmatrix} B^{-\frac{1}{2}} \right\|_2^2 = \|RB^{-\frac{1}{2}}\|_2^2 = \|RB^{-1}R^\top\|_2.$$

As design corrections involve B^{-1}, the aim is to chose it as large as possible. The largest B_0^{-1} for which $\|RB^{-1}R^\top\|_2$ is equal to some $\sigma > 0$ is $RB_0^{-1}R^\top = \sigma I$, i.e., according to (5.6), all preconditioners B satisfying

$$B = B^\top \succeq B_0 = \frac{1}{\sigma}R^\top R = \frac{1}{\sigma}(\alpha G_u^\top G_u + \beta {N_{yu}}^\top N_{yu}). \tag{5.7}$$

lead to $D_C \succ 0$, and thus to a descent on L^a by the increment vector s. Here, σ must be chosen such that Proposition 5.2 applies, i.e.,

$$\sigma = 1 - \rho - \frac{(1+\frac{\theta}{2}\beta)^2}{\alpha\beta(1-\rho)} > 0. \tag{5.8}$$

5.2. Particular choice of weighting coefficients

Now we define weighting coefficients α, β fulfilling (4.3) and independent of all linear transformation in the design space. If we assume that the rectangular matrix $G_u \in \mathbb{R}^{n,m}$ has full column rank and denote by C a Cholesky factor such that $G_u^\top G_u = C^\top C \succ 0$. Then we can show (see [20, Section 3.2]) that

$$\|C^{-\top} B_0 C^{-1}\| \;\leq\; \varphi(\alpha,\beta) \;:=\; \frac{\alpha + q\beta}{\psi(\alpha,\beta)}, \tag{5.9}$$

where

$$\begin{aligned} \psi(\alpha,\beta) \;&:=\; 1 - \rho - \frac{(1+\frac{\theta}{2}\beta)^2}{\alpha\beta(1-\rho)} \;>\; 0, \\ q \;&:=\; \|C^{-\top} N_{yu}^\top N_{yu} C^{-1}\|_2 = \|N_{yu} C^{-\top}\|_2^2 \;=\; \max_{0\neq z\in U} \frac{\|N_{yu} z\|_2^2}{\|G_u^\top z\|_2^2}. \end{aligned} \tag{5.10}$$

The ratio q quantifies the perturbation of the adjoint equation $N_y = 0$ caused by a design variation z relative to that in the primal equation. Since the aim is to maximize B^{-1} in order to make significant design corrections, we define optimal penalty weights α, β which satisfy (4.3) and realize a minimum of the function φ.

Proposition 5.3. *For $q > 0$ the function φ in* (5.9) *reaches its minimum for*

$$\beta = \frac{3}{\sqrt{\theta^2 + 3q(1-\rho)^2} + \frac{\theta}{2}} \quad \text{and} \quad \alpha = q\frac{\beta(1+\frac{\theta}{2}\beta)}{1-\frac{\theta}{2}\beta}. \tag{5.11}$$

Proof. See [20, Proposition 3.3 and Appendix]. □

For $q = 0$ this directly gives $\beta = 2/\theta$. Combining both equations in (5.11) (with q kept in there) and setting $q = 0$ afterwards gives

$$\alpha = \frac{4\theta}{(1-\rho)^2}. \tag{5.12}$$

5.3. Suitable B and relation to $\nabla_{uu} L^a$

Here, using B_0 derived in (5.7) we define a suitable B and establish its relation to the Hessian of L^a with respect to the design. We consider Δu such that

$$\min_{\Delta u} \; L^a(y + \Delta y, \bar{y} + \Delta\bar{y}, u + \Delta u).$$

Using a quadratic approximation of L^a, assuming $\nabla_{uu} L^a \succ 0$ and $N_{uu} \succeq 0$, we define a suitable design space preconditioner B from (5.7) and (5.8) such that

$$B \;=\; B_0 + \frac{1}{\sigma} N_{uu} \;=\; \frac{1}{\sigma}\left(\alpha G_u^\top G_u + \beta N_{yu}^\top N_{yu} + N_{uu}\right). \tag{5.13}$$

In view of (5.7) the increment vector s obtained using the preconditioner B introduced in (5.13) yields descent on L^a. In addition, we have $B \approx \nabla_{uu} L^a$. This approximation turns to an equality at primal and dual feasibility. Besides, as L^a is an exact penalty function, we have $\nabla^2 L^a \succ 0$ in a neighborhood of a local minimizer and then in particular $\nabla_{uu} L^a = B \succ 0$.

5.4. BFGS update to an approximation of B

As the suitable preconditioner B derived in (5.13) involves matrix derivatives which may be costly evaluated, numerically we use the BFGS method to update its approximation H_k. In view of the relation $B \approx \nabla_{uu} L^a$ established in the previous subsection, we use the following secant equation in the update of H_k:

$$H_{k+1} R_k = \Delta u_k, \quad R_k := \nabla_u L^a(y_k, \bar{y}_k, u_k + \Delta u_k) - \nabla_u L^a(y_k, \bar{y}_k, u_k).$$

Imposing to the step multiplier η to satisfy the second Wolfe's condition

$$\Delta u_k{}^\top \nabla_u L^a(y_k, \bar{y}_k, u_k + \eta \Delta u_k) \geq c_2 \Delta u_k{}^\top \nabla_u L^a(y_k, \bar{y}_k, u_k) \quad \text{with } c_2 \in [0, 1],$$

implies the necessary curvature condition

$$R_k{}^\top \Delta u_k > 0. \tag{5.14}$$

A simpler procedure could skip the update whenever (5.14) does not hold by either setting H_{k+1} to identity or to the last iterate H_k. Provided (5.14) holds, we use

$$H_{k+1} = (I - r_k \Delta u_k R_k{}^\top) H_k (I - r_k R_k \Delta u_k{}^\top) + r_k \Delta u_k \Delta u_k{}^\top \text{ with } r_k = \frac{1}{R_k{}^\top \Delta u_k}.$$

5.5. Alternating approach

Each BFGS update of H_k needs to make a pure design step (step with fixed primal and dual variables) in order to compute the coefficient R_k. The approach presented here aims to achieve minimization of L^a using alternating between pure design and pure feasibility steps. For several applications, design corrections may be costly evaluated especially where each design update implies a modification of the geometry which requires to remesh and update the data structure (see [22, 27]. Thus it could be more convenient to perform only *significant* design corrections. If the suggested change in the design variable u is small, we directly improve feasibility, leaving u unchanged. Actually, we perform a design correction only if

$$\Delta u^\top \nabla_u L^a < 0 \quad \text{and} \quad \tau\, \Delta u^\top \nabla_u L^a < \Delta y^\top \nabla_y L^a + \Delta \bar{y}^\top \nabla_{\bar{y}} L^a, \tag{5.15}$$

where $\tau \in]0, 1]$ is a percent which may be fixed by the user. We suppose there exists $\bar{B}$ such that for all iteration k we have

$$B(y, \bar{y}, u) \leq B_k \leq \bar{B} \qquad \text{for all } (y, \bar{y}, u) \in \mathcal{N}_0, \tag{5.16}$$

where $\mathcal{N}_0$ is a level set of L^a. And thus $\|B_k\|$ is finite for all iterations. An algorithm realizing this approach is presented in details in [20, Section 4].

5.6. Line search procedures

With the preconditioner B derived in (5.13), we expect full step convergence near a local minimizer of L^a. To enforce convergence in the earlier stage of iterations, we briefly sketch two backtracking line search procedures based on two slightly different quadratic forms: The first one consists in applying a standard backtracking line search on a quadratic interpolation Q of L^a (see [1, 5]):

$$Q(\eta) = \xi_2 \eta^2 + \xi_1 \eta + \xi_0 \qquad \text{for} \quad \eta \in [0, \eta_c],$$

where

$$\xi_0 = L^a(y_k, \bar{y}_k, u_k), \qquad \xi_1 = \nabla L^a(y_k, \bar{y}_k, u_k)^\top s_k < 0,$$
$$\xi_2 = \frac{1}{{\eta_c}^2}\left(L^a(y_k + \eta_c \Delta y_k, \bar{y}_k + \eta_c \Delta \bar{y}_k, u_k + \eta_c \Delta u_k) - \xi_1 \eta_c - \xi_0\right).$$

Here, $\xi_1 < 0$ is implied by the fact that the increment vector yields descent on L^a.

The second procedure does not require the computation of ∇L^a which may save calculation cost. Linear interpolations P, D of the primal and dual residuals, respectively, and a standard parabolic interpolation q based on the initial descent and two function values for the unpenalized Lagrangian lead to the approximation

$$\tilde{Q}(\eta) = \frac{\alpha}{2}\|P(\eta)\|_2^2 + \frac{\beta}{2}\|D(\eta)\|_2^2 + q(\eta), \quad \eta \in [0, \eta_c],$$

of L^a. Here η_c is a tentative step size. If η^* denotes the (explicitly computable) stationary point of $\tilde{Q}$ multiplied by the sign of its second-order term, we accept η_c only if $\eta^* \geq \frac{2}{3}\eta_c$ which ensures $\tilde{Q}(\eta_c) < \tilde{Q}(0)$ and thus

$$L^a(y_k + \eta_c \Delta y_k, \bar{y}_k + \eta_c \Delta \bar{y}_k, u_k + \eta_c \Delta u_k) < L^a(y_k, \bar{y}_k, u_k).$$

As long as $\eta^* \geq \frac{2}{3}\eta_c$ is violated, we set

$$\eta_c = \operatorname{sign}(\eta^*) \max\{0.2|\eta_c|, \min\{0.8|\eta_c|, |\eta^*|\}\}$$

and recompute η^*. For the acceptance of the initial step multiplier $\eta_c = 1$, we also require $\eta^* \leq \frac{4}{3}\eta_c$. Failing this, η_c is once increased to $\eta_c = \eta^*$ and then always reduced until the main condition $\eta^* \geq \frac{2}{3}\eta_c$ is fulfilled. In both cases, $\frac{\eta^*}{\eta_c} \geq \frac{2}{3}$ is fulfilled after a finite number of steps and the line search procedure terminates.

6. Global convergence

If the assumptions of Theorem 4.5 apply and the line search ensures a monotonic decrease of L^a, all iterates during the optimization lie in the bounded level set

$$\mathcal{N}_0 := \{(y, \bar{y}, u) : L^a(y, \bar{y}, u) \leq L^a(y_0, \bar{y}_0, u_0)\}.$$

We can show that the search directions s are gradient related:

Proposition 6.1. *If Theorem* 4.5 *applies and* $N_{uu} \succeq 0$, *then there exists* $C > 0$ *with*

$$\cos\gamma = -\frac{s^\top \nabla L^a}{\|\nabla L^a\|\|s\|} \geq C > 0 \qquad \text{for all} \quad (y, \bar{y}, u) \in \mathcal{N}_0,$$

where the step s is computed with the preconditioner B introduced in (5.13).

Proof. See [20, Proposition 5.1]. □

The alternating approach does not affect the above result. Actually, we employ a pure design step only if (5.15) holds and thus

$$\frac{-\Delta u^\top \nabla_u L^a}{\|\nabla L^a\|\|\Delta u\|} \geq \frac{1}{(1+\tau)} \frac{-s^\top \nabla L^a}{\|\nabla L^a\|\|s\|}.$$

We use a pure feasibility step if $\tau\ \Delta u^\top \nabla_u L^a \geq \Delta y^\top \nabla_y L^a + \Delta \bar{y}^\top \nabla_{\bar{y}} L^a$ which gives

$$-(\Delta y^\top \nabla_y L^a + \Delta \bar{y}^\top \nabla_{\bar{y}} L^a) \geq -\frac{\tau}{(1+\tau)}\ s^\top \nabla L^a.$$

In addition, since Theorem 4.5 applies all level sets of the continuous function L^a are bounded which implies that L^a is bounded below. Therefore, using the well-known effectiveness of the line search procedure based on a standard backtracking [1] and the gradient relatedness result established in Proposition 6.1, we obtain

$$\lim_{k\to\infty} \|\nabla L^a(y_k, \bar{y}_k, u_k)\| = 0.$$

7. Numerical experiment: The Bratu problem

The Bratu problem is frequently used in combustion modeling:

$$\begin{array}{rcll} \Delta y(x) + e^{y(x)} & = & 0 & x = (x_1, x_2) \in [0,1]^2 \\ y(0, x_2) & = & y(1, x_2) & x_2 \in [0,1] \\ y(x_1, 0) & = & \sin(2\pi x_1) & x_1 \in [0,1] \\ y(x_1, 1) & = & u(x_1) & x_1 \in [0,1]. \end{array}$$

The function u is a boundary control that can be varied to minimize the objective

$$f(y,u) = \int_0^1 (\partial_{x_2} y(x_1, 1) - 4 - \cos(2\pi x_1))^2\, dx_1 + \sigma \int_0^1 \left(u^2 + u'^2\right) dx.$$

We use $\sigma = 0.001$, an initial control is $u(x_1) = 2.2$ (see [17]), a five point central difference scheme with $h = 1/10$, and Jacobi's method.

To solve the minimization problem, we use power iterations to compute the spectral radius $\rho_{N_{yy}}$ of the matrix N_{yy} and $\rho_{G_y^*}$ of $G_y^\top G_y$. Then, we update $\theta = \rho_{N_{yy}}$ and $\rho = \sqrt{\rho_{G_y^*}}$. We update the ratio q introduced in (5.10) from

$$q_k = \max\{q_{k-1}, \frac{\|N_y(y_k, \bar{y}_k, u_k + \Delta u_k) - N_y(y_k, \bar{y}_k, u_k)\|_2^2}{\|G(y_k, u_k + \Delta u_k) - G(y_k, u_k)\|_2^2}\},$$

and used α, β, σ as in (5.11), (5.8). We compared the number of iterations N_{opt} needed for the optimization with the alternating approach, i.e., to reach

$$\alpha\|G(y_k, u_k) - y_k\|_2^2 + \beta\|N_y(y_k, \bar{y}_k, u_k) - \bar{y}_k\|_2^2 + \|\Delta u_k\|_2^2 \leq \varepsilon := 10^{-4},$$

and the number of iterations N_f required to reach feasibility with fixed u:

$$\|G(y_k, u) - y_k\|_2^2 + \|N_y(y_k, \bar{y}_k, u) - \bar{y}_k\|_2^2 \leq \varepsilon,$$

We used a mesh size $h \in [0.055, 0.125]$. The behaviors with respect to h of N_{opt}, N_f and of the ratio $R = N_{\text{opt}}/N_f$ are depicted in Figure 1. It shows that the number of iterations N_{opt} needed to solve the optimization problem is always bounded by a reasonable factor (here 4.6 at maximum) times the number of iterations N_f required to reach feasibility: bounded retardation. Although both numbers grow while decreasing h, the ratio $R = N_{\text{opt}}/N_f$ in Figure 2 seems reaching some limit slightly bigger than 2 for small values of h. In the case of L^a, the two line search procedures give results that are numerically indistinguishable.

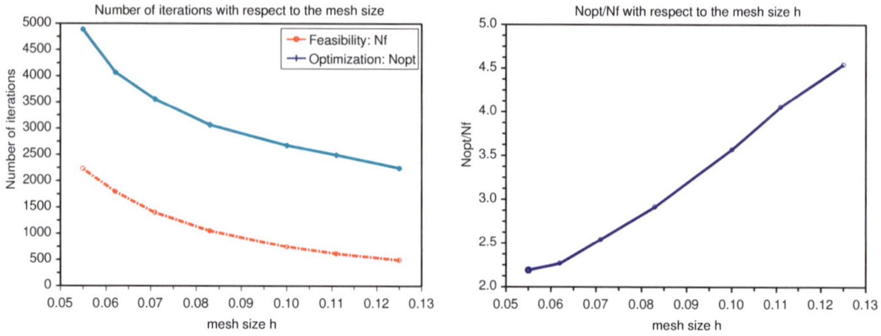

FIGURE 1. Iterations (left) and retardation factor w.r.t. mesh size h

8. Application: Aerodynamic shape optimization

As a first realistic application, we consider shape optimization of a RAE2822 transonic airfoil whose aerodynamic properties are calculated by a structured Euler solver with quasi-unsteady formulation based on pseudo time steps. The objective is to reduce the inviscid shock that is present on the initial airfoil and therefore to minimize the drag. The adjoint solver which calculates the necessary sensitivities for the optimization is based on discrete adjoints and derived by using reverse mode of automatic differentiation. A detailed information about the presented work can be found in [29].

8.1. Governing equations and boundary conditions

Since we are interested in drag reduction in transonic flow regime, the compressible Euler equations are an appropriate choice. They are capable of describing the (inviscid) shocks, which are the main sources of the pressure drag.

Even though the flow steady, the solution is obtained by integrating the (quasi-)unsteady Euler equations in time until a steady state is reached. These time steps do not have any physical meaning and are called pseudo-time steps.

For 2D flow, the compressible Euler equations in Cartesian coordinates read:

$$\frac{\partial w}{\partial t} + \frac{\partial f}{\partial x} + \frac{\partial g}{\partial y} = 0 \text{ with } f = \begin{bmatrix} \rho u \\ \rho u^2 + p \\ \rho uv \\ \rho uH \end{bmatrix} \text{ and } g = \begin{bmatrix} \rho v \\ \rho vu \\ \rho v^2 + p \\ \rho vH \end{bmatrix}, \tag{8.1}$$

where w is the vector of conserved variables $\{\rho,\ \rho u,\ \rho v,\ \rho E\}$ (ρ is the density, u and v are the velocity components and E denotes the energy).

As boundary conditions, we assume the Euler slip condition on the wall ($\vec{n}^T\vec{v} = 0$) and free stream conditions at the farfield. For a perfect gas holds

$$p = (\gamma - 1)\rho(E - \frac{1}{2}(u^2 + v^2)) \tag{8.2}$$

$$\rho H = \rho E + p \tag{8.3}$$

for pressure p and enthalpy H. The pressure and drag coefficients are defined as

$$C_p := \frac{2(p - p_\infty)}{\gamma M_\infty^2 p_\infty}, \qquad C_d := \frac{1}{C_{ref}} \int_C C_p(n_x \cos\alpha + n_y \cos\alpha) dl \ . \tag{8.4}$$

8.2. Shape parameterization

In aerodynamic shape optimization, there are mainly two ways of doing the shape updates: Either parameterizing the shape itself or parameterizing shape deformations. In [27] these possibilities are investigated in detail. In the following, we take the second approach, such that an initial airfoil shape is deformed by some set of basis functions that are scaled by certain design parameters. Here, the basic idea of shape deformation is to evaluate these basis functions scaled with certain design parameters and to deform the camberline of the airfoil accordingly. Then, the new shape is simply obtained by using the deformed camberline and the initial thickness distribution. The result is a surface deformation that maintains the airfoil thickness.

We have chosen Hicks-Henne functions, which are widely used in airfoil optimization. These function have the positive property that they are defined in the interval $[0,1]$ with a peak position at a and they are analytically smooth at zero and one. The normalized airfoil shape is deformed by using Hicks-Henne functions multiplied by the design parameters u_i:

$$\Delta \,\text{camber}(x) = \sum u_i h\left\{a, b\right\}(x) \text{ and } \text{camber}(x) += \Delta \,\text{camber}(x) \ . \tag{8.5}$$

After deforming the airfoil geometry, a difference vector is calculated and finally performs a mesh deformation by using this difference vector. This approach is also very advantageous in terms of gradient computations, since we have to differentiate only the simple structured mesh and shape deformation tools, instead of complex mesh generators.

The numerical solution of (8.1) is computed by the TAUij code, which is a structured quasi 2D version of the TAU code, developed at the German Aerospace Center (DLR). For the spatial discretization the MAPS+ [31] scheme is used. For the pseudo time stepping, a fourth-order Runge-Kutta scheme is applied. To accelerate the convergence, local time stepping, explicit residual smoothing and a multigrid methods are used. The code TAUij is written in C and comprises approximately 6000 lines of code distributed over several files.

8.3. Gradient computation and implementation issues

One of the key points in aerodynamic shape optimization with gradient-based methods is the computation of the derivatives. For this study, we generate the adjoint codes in a semi-automatic fashion by using reverse mode of automatic differentiation. The reverse mode of AD allows to generate discrete adjoint codes in which computational cost is independent from number of optimization parameters. The freeware AD tool ADOL-C [15] gives the possibility of applying reverse AD. Since ADOL-C is based on operator overloading strategy, for the reverse mode it is usually necessary to tape all operations that are done on the active variables (the variables which are to differentiate with respect to the selected independent parameters) on memory or disk. Because in our case the primal iteration is a fixed-point iteration, a complete taping of the primal iteration is not necessary. Since we use a one-shot approach rather than a hierarchical approach, we need to tape only one pseudo-time step in each iteration instead of the whole time-stepping. This is of course very advantageous, since the tape sizes would be extremely large, even for the case of rather coarse meshes, because the tape size of a primal iterate would be multiplied by the number of pseudo-time steps. Nevertheless, in [32] it is demonstrated how to overcome this kind of drawbacks in cases of hierarchical approaches by the so-called reverse accumulation of adjoints [3].

For the coupled iteration, we need to evaluate several derivative vectors

$$\nabla_u L^a = \alpha \Delta y^T G_u + \beta \Delta \overline{y}^T N_{yu} + N_u \ , \tag{8.6}$$

in order to update the design vectors u. Furthermore, we need to evaluate the terms N_y for the update of the adjoint states $\overline{y}$.

Note, that all expressions in (8.6) are either vectors or matrix vector products. Several subroutines of ADOL-C allow us to calculate these matrix vector products easily by using the reverse mode of AD for the first-order terms and reverse on tangent for the second-order term.

For the differentiation, we simply set the independent vector as $[u; y]$, the dependent vector as $[N; y]$ and correspondingly calculate N inside the routine that returns the goal functional C_d. It should also be mentioned that, apart from N_y, the derivatives with respect to the design parameters u are propagated within the design chain in the reverse order as vector matrix products. In addition to the flow solver, the other programs of the design chain, namely **meshdefo, difgeo, defgeo**, have to be differentiated, too. In [10], this reverse propagation of the adjoint vectors is covered in detail, and comparisons of the resulting adjoint sensitivities versus finite differences are also illustrated.

8.4. Numerical results

The numerical tests are done on the transonic shape optimization problem of a RAE2822 airfoil that is introduced previously. The number of Hicks-Henne functions for the shape parameterization are chosen to be 20. The single-step one-shot method is applied in the sense that full steps are taken in the design update. As a stopping criteria, we choose $|\Delta u| < \epsilon$, where ϵ is a user defined tolerance. For our particular application we have chosen $\epsilon = 0.0001$.

As flow conditions, we have an inflow Mach number of $M_\infty = 0.73$ and an angle of attack of $\alpha = 2°$. Within the first 30 iterations, in order to smooth out

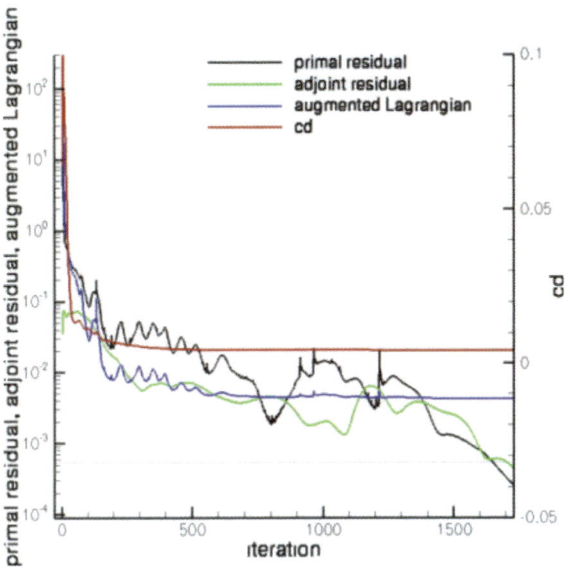

FIGURE 2. Optimization history

possible oscillatory effects, caused by the initialization of the flow field, we do only updates of the state and the adjoint state, without changes of the airfoil geometry. After these smoothing iterations, we do one-shot iterations. Figure 2 shows the optimization histories of the augmented Lagrangian, the cost functional C_d, the primal as well as the adjoint state residual. We observe, that after approximately 1600 iterations, the coupled iteration converges and the drag coefficient is reduced drastically. Consequently, we just measure a deterioration factor of 4 from the simulation to the one-shot optimization.

9. Application: Parameter optimization in a climate model

Here we present a second real-world example, this time from climate modeling, which is in detail described in [25, 26]. Parameter optimization is an important task in all kind of climate models. Many processes are not well known, some are too small-scaled in time or space, and others are just beyond the scope of the model. Here, parameters of a simplified model of the north Atlantic thermohaline circulation (THC) are optimized to fit the results to data given by a more detailed climate model of intermediate complexity.

The 4-box model of the Atlantic THC described in [34] simulates the flow rate of the Atlantic Ocean known as the 'conveyor belt', carrying heat northward and having a significant impact on climate in northwestern Europe. Temperatures T_i and salinity differences S_i in four different boxes, namely the southern, northern,

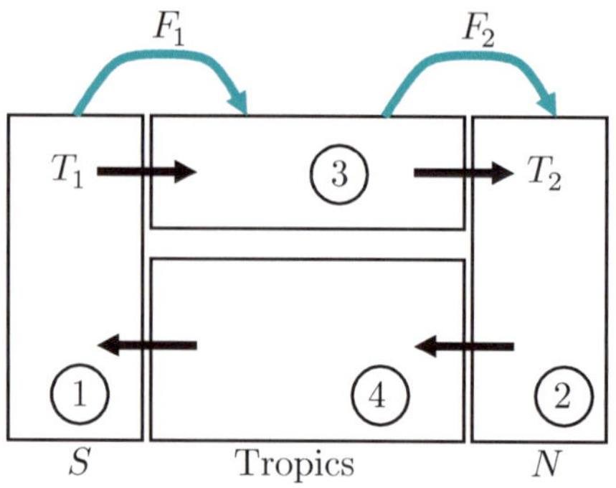

FIGURE 3. Rahmstorf box model, flow direction shown for $m > 0$.

tropical and the deep Atlantic, are the characteristics inducing the flow rate. The surface boxes exchange heat and freshwater with the atmosphere, which causes a pressure-driven circulation, compare Figure 3.

In [33] a smooth coupling of the two possible flow directions is proposed. The resulting ODE system, e.g., for boxes $i = 1, 2$, reads:

$$\begin{aligned}
\dot{T}_1 &= \lambda_1(T_1^* - T_1) + \frac{m^+}{V_1}(T_4 - T_1) + \frac{m^-}{V_1}(T_3 - T_1) \\
\dot{T}_2 &= \lambda_2(T_2^* - T_2) + \frac{m^+}{V_2}(T_3 - T_2) + \frac{m^-}{V_2}(T_4 - T_2) \\
\dot{S}_1 &= \frac{S_0 f_1}{V1} + \frac{m^+}{V_1}(S_4 - S_1) + \frac{m^-}{V_1}(S_3 - S_1) \\
\dot{S}_2 &= -\frac{S_0 f_2}{V_2} + \frac{m^+}{V_2}(S_3 - S_2) + \frac{m^-}{V_2}(S_4 - S_2).
\end{aligned}$$

Here $m = k(\beta(S_2 - S_1) - \alpha(T_2 - T_1))$ is the meridional volume transport or overturning. For boxes $i = 3, 4$ there are similar, also coupled equations. Several model parameters are involved, the most important being the freshwater flux f_1 containing atmospheric water vapor transport and wind-driven oceanic transport; they are used to simulate global warming in the model. The parameter a in $m^+ := \frac{m}{1-e^{-am}}, m^- := \frac{-m}{1-e^{am}}$ allows to use the model for both flow directions.

9.1. The optimization problem

Given fresh water fluxes $(f_{1,i})_{i=1}^n$ (with $n = 68$ here), the aim is to fit the values $m_i = m(f_{1,i})$ obtained by the model to data $m_{d,i}$ from a more complex model *Climber* 2, see [30], while $u = (T_1^*, T_2^*, T_3^*, \Gamma, k, \alpha)$ are the control parameters. If $F(y, u)$ denotes the right-hand side of the ODE system of the model, we get

$$\begin{aligned}
&\min_{y,u} J(y, u) := \frac{1}{2}\|m - m_d\|_2^2 + \frac{\alpha_w}{2}\|u - u_0\|_2^2, \quad m = (m_i)_{i=1}^n, m_d = (m_{d,i})_{i=1}^n, \\
&\text{s.t.} \quad \dot{y}(f_{1,i}) = F(y(f_{1,i}), u), \quad i = 1, \ldots, n.
\end{aligned}$$

The regularization term incorporates a prior guess u_0 for the parameters. The ODE system is solved by an explicit Euler method, thus G defined in Section 1.1 here represents one Euler step, but operating on all parameters f_{1i} together, i.e., for fixed u we have $G(\cdot, u) : \mathbb{R}^{8n} \to \mathbb{R}^{8n}$. Contractivity of G is not given in general, i.e., ρ in (1.2) exceeds 1 for several steps, but in average is less than 1. Also the assumption $\partial c/\partial y$ being always invertible is violated. Nevertheless, in practice G converges for fixed u but different starting values y_0 to the same stationary y^*. About 400 to 11,000 steps are needed to reach $\|y_{k+1} - y_k\| < \varepsilon = 10^{-6}$ for iteration index k.

9.2. One-shot method for the box model

We calculate the preconditioner B defined in (5.13) in every iteration including all first- and second-order derivatives. To compute α, β and σ, we set the (iteration-dependent) contraction factor of the Euler time stepping to $\rho = 0.9$. We determine $\|G_u\|_2$ and $\|N_{yu}\|_2$ computing the Eigenvalues of $G_u^\top G_u, N_{yu}^\top N_{yu} \in \mathbb{R}^{6\times 6}$ directly, whereas for those of $N_{yy}^\top N_{yy}$ we apply a power iteration.

The forward mode of TAF [12] is used for G_u, the reverse mode for $\bar{y}^\top G_y$, and for $\bar{y}^\top G_{yu}$ and $\bar{y}^\top G_{yy}$ first the reverse and then the forward mode is applied. With only 6 parameters in our optimization problem, the reverse mode is only slightly cheaper than the forward mode for this example, and therefore is not mandatory.

For the calculation of necessary matrix-vector products (i.e., directional derivatives) we determine $\bar{y}^\top G_{yy} b, b \in \mathbb{R}^{8n}$, with a TAF generated subroutine and $J_{yy} b$ by hand. A second call of the TAF subroutine computes $N_{yy}^\top N_{yy} b = (J_{yy} + \bar{y}^\top G_{yy})^\top N_{yy} b$. In our testings, the dominant part of $N_{yy}^\top N_{yy}$ is the constant matrix $J_{yy}^\top J_{yy}$ and thus $\|N_{yy}\|_2$ does not change significantly from iteration to iteration. As one can see in table 1, an update performed only after several time-steps does not significantly influence the optimization. In our calculations β becomes very

update of N_{yy}	#iterations	time (min)	$J(y^*, u^*)$	data fit
every 10000 iterations	1,037,804	6.174	14.879	14.053
every 1000 iterations	1,010,011	6.148	14.879	14.053
every iteration	1,015,563	10.394	14.879	14.053

TABLE 1. Effect on the optimization of rare update of N_{yy}, $\alpha_w = 0.1$.

small ($\approx 10^{-5}$) whereas α is large ($\approx 10^5$). Since $N_{yu}^\top N_{yu}$ contains quite large values and $G_u^\top G_u$ only small ones, we assume that α, β are well chosen.

9.3. Comparison between BFGS and One-shot strategy, bounded retardation

We compared the One-shot approach with a standard BFGS method. For the latter, in each iteration the model has to be run into a steady state for all $f_{1,i}$.

Without any regularization, the One-shot method does not converge, and BFGS finds optimal values u^* with $\frac{\partial J}{\partial u}(u^*) = 0$ being far away from reality. Generally, the smaller α_w the better the fit of the data becomes. The optimization

	$\alpha_w = 10$		$\alpha_w = 0.1$		$\alpha_w = 0.001$	
	One-shot	BFGS	One-shot	BFGS	One-shot	BFGS
$J(y^*, u^*)$	25.854	26.221	14.879	15.926	12.748	11.411
data fit	0.269	0.277	0.206	0.213	0.183	0.166
# iterations	1,269,019	20	1,010,011	28	10,678,000	65
# Euler steps	1,269,019	1,285,203	1,010,011	1,808,823	10,678,000	4,236,481

TABLE 2. Results of the optimization

results by both methods differ only slightly, see Table 2. The total number of needed time steps for the One-shot method was smaller compared to BFGS for $\alpha_w \geq 0.1$. Concerning computational time, the BFGS method is a little bit faster (since the Euler steps are cheaper due to the smaller system size), and only in the case where $\alpha_w = 0.001$ even significantly with a relation of $1 : 5.5$.

The most promising point of this study is that the One-shot strategy is much faster *close* to the optimal pair (y^*, u^*) than the BFGS method, which is actually the motivation for the One-shot approach, and one could save iterations mitigating the stopping criterion. We refer to [26] for details.

Finally we remark that it is typical in climate models that less theoretical analysis can be provided because of the complexity of real world models. For the optimization strategy, its quality and usefulness is even more convincing if good results are achieved even though convergence assumptions are not fulfilled.

References

[1] J.F. Bonnans, J. Charles Gilbert, C. Lemaréchal, C.A. Sagastizábal,*Numerical Optimization Theoretical and Practical Aspects*, Springer Berlin Heidelberg (2003).

[2] A. Brandt, *Multi-Grid techniques:* 1984 *guide with applications to fluid dynamics*, GMD-Studien. no. 85, St. Augustin, Germany (1984).

[3] Christianson B., *Reverse accumulation and attractive fixed points*, Optimization Methods and Software, **3** (1994), 311–326.

[4] Christianson, B., Reverse Accumulation and Implicit Functions. *Optimization Methods and Software* **9**:4 (1998), 307–322.

[5] J.E. Dennis, Jr. and R.B. Schnabel, *Numerical Methods for unconstrained optimization and Nonlinear Equations*, Prentice-Hall (1983).

[6] G. Di Pillo and L. Grippo. *A Continuously Differentiable Exact Penalty Function for Nonlinear Programming Problems with Inequality Constraints.* SIAM J. Control Optim. **23** (1986), 72–84.

[7] G. Di Pillo, *Exact penalty methods, in E. Spedicato (ed.), Algorithms for continuous optimization: the state of the art*, Kluwer Academic Publishers (1994), 209–253.

[8] L.C.W. Dixon, *Exact Penalty Function Methods in Nonlinear Programming.* Report NOC, The Hatfield Polytechnic, **103** (1979).

[9] R. Fontecilla, T. Steihaug, R.A. Tapia, *A Convergence Theory for a Class of Quasi-Newton Methods for Constrained Optimization.* SIAM Num. An. **24** (1987), 1133–1151.

[10] N.R. Gauger, A. Walther, C. Moldenhauer, M. Widhalm, *Automatic Differentiation of an Entire Design Chain for Aerodynamic Shape Optimization*, Notes on Numerical Fluid Mechanics and Multidisciplinary Design **96** (2007), 454–461.

[11] N.R. Gauger, A. Griewank, J. Riehme, *Extension of fixed point PDE solvers for optimal design by one-shot method – with first applications to aerodynamic shape optimization*, European J. Computational Mechanics (REMN), **17** (2008), 87–102.

[12] R. Giering, T. Kaminski, T. Slawig, Generating efficient derivative code with TAF: Adjoint and tangent linear Euler flow around an airfoil. *Future Generation Computer Systems* **21**:8 (2005), 1345–1355.

[13] I. Gherman, V. Schulz, *Preconditioning of one-shot pseudo-timestepping methods for shape optimization*, PAMM **5**:1(2005), 741–759.

[14] J.C. Gilbert, *Automatic Differentiation and iterative Processes*, Optimization Methods and Software **1** (1992), 13–21.

[15] A. Griewank, D. Juedes, J. Utke, *ADOL-C: A package for the automatic differentiation of algorithms written in C/C++.* ACM Trans. Math. Softw. **22** (1996), 131–167.

[16] A. Griewank, *Evaluating Derivatives: Principles and Techniques of Algorithmic Differentiation.* No. 19 in Frontiers in Appl. Math. SIAM, Philadelphia, PA, (2000).

[17] A. Griewank, *Projected Hessians for Preconditioning in One-Step Design Optimization*, Large Scale Nonlinear Optimization, Kluwer Academic Publ. (2006), 151–171.

[18] A. Griewank, C. Faure, *Reduced Functions, Gradients and Hessians from Fixed Point Iteration for state Equations*, Numerical Algorithms **30,2** (2002), 113–139.

[19] A. Griewank, D. Kressner, *Time-lag in Derivative Convergence for Fixed Point Iterations*, ARIMA Numéro spécial CARI'04 (2005), 87–102.

[20] A. Hamdi, A. Griewank, *Reduced quasi-Newton method for simultaneous design and optimization*, Comput. Optim. Appl. online `www.springerlink.com`

[21] A. Hamdi, A. Griewank, *Properties of an augmented Lagrangian for design optimization*, Optimization Methods and Software **25**:4 (2010), 645–664.

[22] A. Jameson, *Optimum aerodynamic design using CFD and control theory*, In 12th AIAA Computational Fluid Dynamics Conference, AIAA paper 95-1729, San Diego, CA, 1995. American Institute of Aeronautics and Astronautics (1995).

[23] T. Kaminski, R. Giering, M. Vossbeck, Efficient sensitivities for the spin-up phase. In *Automatic Differentiation: Applications, Theory, and Implementations*, H.M. Bücker, G. Corliss, P. Hovland, U. Naumann, and B. Norris, eds., Springer, New York. Lecture Notes in Computational Science and Engineering **50** (2005), 283–291.

[24] C.T. Kelley, *Iterative Methods for optimizations*, Society for Industrial and Applied Mathematics (1999).

[25] C. Kratzenstein, T. Slawig, One-shot parameter optimization in a box-model of the North Atlantic thermohaline circulation. *DFG Preprint SPP*1253-11-03 (2008).

[26] C. Kratzenstein, T. Slawig, One-shot Parameter Identification – Simultaneous Model Spin-up and Parameter Optimization in a Box Model of the North Atlantic Thermohaline Circulation. *DFG Preprint SPP*1253-082 (2009).

[27] B. Mohammadi, O. Pironneau. *Applied Shape Optimization for Fluids. Numerical Mathematics and scientific Computation*, Cladenen Press, Oxford (2001).

[28] P.A. Newman, G.J.-W. Hou, H.E. Jones, A.C. Taylor. V.M. Korivi. *Observations on computational methodologies for use in large-scale, gradient-based, multidisciplinary design incorporating advanced CFD codes.* Technical Memorandum 104206, NASA Langley Research Center. AVSCOM Technical Report 92-B-007 (1992).

[29] E. Özkaya, N. Gauger, Single-step one-shot aerodynamic shape optimization. *DFG Preprint SPP*1253-10-04 (2008).

[30] S. Rahmstorf, V. Brovkin, M. Claussen, C. Kubatzki, Climber-2: A climate system model of intermediate complexity. Part ii. *Clim. Dyn.* **17** (2001), 735–751.

[31] C.C. Rossow, *A flux splitting scheme for compressible and incompressible flows*, Journal of Computational Physics, **164** (2000), 104–122.

[32] S. Schlenkrich, A. Walther, N.R. Gauger, R. Heinrich, *Differentiating fixed point iterations with ADOL-C: Gradient calculation for fluid dynamics*, in H.G. Bock, E. Kostina, H.X. Phu, R. Rannacher, editors, Modeling, Simulation and Optimization of Complex Processes – Proceedings of the Third International Conference on High Performance Scientific Computing 2006 (2008), 499–508.

[33] S. Titz, T. Kuhlbrodt, S. Rahmstorf, U. Feudel, On freshwater-dependent bifurcations in box models of the interhemispheric thermohaline circulation. *Tellus A* **54** (2002), 89–98.

[34] K. Zickfeld, T. Slawig, T., S. Rahmstorf, A low-order model for the response of the Atlantic thermohaline circulation to climate change. *Ocean Dyn.* **54** (2004), 8–26.

Nicolas R. Gauger and Emre Özkaya
RWTH Aachen University
Department of Mathematics and CCES
Schinkelstr. 2
D-52062 Aachen, Germany
e-mail: gauger@mathcces.rwth-aachen.de
ozkaya@mathcces.rwth-aachen.de

Andreas Griewank and Adel Hamdi
Institut für Mathematik
Fakultät Math.-Nat. II
Humboldt-Universität
Unter den Linden 6
D-10099 Berlin, Germany
e-mail: Andreas.Griewank@math.hu-berlin.de
adel.hamdi@insa-rouen.fr

Claudia Kratzenstein and Thomas Slawig
Christian-Albrechts-Universität zu Kiel
Institut für Informatik and DFG Cluster of Excellence *The Future Ocean*
D-24098 Kiel, Germany
e-mail: ctu@informatik.uni-kiel.de
ts@informatik.uni-kiel.de

International Series of Numerical Mathematics, Vol. 160, 123–146

Well-posedness of Networked Hyperbolic Systems of Balance Laws

Martin Gugat, Michael Herty, Axel Klar, Günther Leugering and Veronika Schleper

Abstract. We present an overview on recent results concerning hyperbolic systems on networks. We present a summary of theoretical results on existence, uniqueness and stability. The established theory extends previously known results on the Cauchy problem for nonlinear, 2×2 hyperbolic balance laws. The proofs are based on Wave-Front Tracking and therefore we present detailed results on the Riemann problem first.

Mathematics Subject Classification (2000). 35L65, 49J20.

Keywords. Hyperbolic conservation laws on networks, optimal control of networked systems, management of fluids in pipelines.

1. Introduction

We are interested in analytical properties of hyperbolic balance laws on networks. We will recall recent results on these kind of problems using wave front tracking. We start giving some overview on existing literature on this topic before collecting recent results.

From a mathematical point of view, there exist a variety of publications concerning flows on networks. These publications range from application in data networks (e.g., [31, 32]) and traffic flow (e.g., [61, 8, 19]) over supply chains (e.g., [11, 7, 42]) to flow of water in canals (e.g., [54, 56, 46, 60, 33]).

In the engineering community, gas flow in pipelines is in general modeled by steady state or simple transient models, see, e.g., [15, 53, 64, 65, 38] and the publications of the pipeline simulation interest group [66]. More detailed models based on partial differential equations can be found, e.g., in [10, 69, 22]. Therein, the flow of gas inside the pipes is modeled by a system of balance laws. At junctions of two or more pipes, algebraic conditions couple the solution inside the pipeline segments and yield a network solution. For more details on the modeling of junctions see also [9, 21, 51, 17, 43, 28]. The recent literature offers several results on the modeling of systems governed by conservation laws on networks. For instance, in [10, 9, 22] the modeling of a network of gas pipelines is considered, the basic model being the p-system. The key problem in these papers is the description of the evolution of fluid at a junction between two or more pipes. A different physical problem, leading to a similar analytical framework, is that of the flow of water in open channels, considered, for example, in [56]. Starting with the classical work, several other recent papers deal with the control of smooth solutions, see, for instance, [34, 44, 48, 45, 47, 56, 59, 58, 60]. In all these cases an assumption on the C^1-norm of initial and boundary data is necessary. Under this very strong assumption no shocks can appear. The theoretical results in the cited papers are based on results on classical solutions for quasi-linear systems. This approach also allows for an explicit construction of the optimal control. However, these results **cannot** be extended to include the shock phenomena. Other approaches are based on numerical discretizations, as in [37, 62, 63, 67]. This approach is completely different from the existing approaches in the sense that no underlying function space theory is provided or could be derived from the equations. The main purpose of this approach is to compute numerically optimal controls. Whether or not these controls converge when the discretization mesh refines was unclear. The current literature offers several results on the well-posedness of systems of conservation laws, with various variations on this theme; see [1, 2, 3, 22, 24, 27]. However, none of these papers deals with control or optimization problems, as is done, for instance, in [4, 5, 6, 13, 14, 16, 20, 40, 68].

Here, we summarize recent results on well-posedness of networked system of hyperbolic balance laws as published in [26, 10, 9, 27]. We will refer to the proofs at the corresponding paragraphs in the document.

2. The model

The finest model discussed in this work is a nonlinear PDE model based on the Euler equations. Since the pipeline cross section is in general very small compared to the length of the pipeline segments (pipes) we neglect variations of pressure and velocity over the cross sectional area and use a model with only one space dimension. A second common assumption in gas pipeline simulation is the thermodynamical equilibrium, which yields a constant temperature of the gas along the pipe. Furthermore, the main forces inside the pipes are due to friction and inclination of the pipe. This all together leads to the isothermal Euler equations in one space dimension and an additional nonlinear source term. In one single pipe we have therefore the following set of equations:

$$\partial_t \rho + \partial_x (\rho v) = 0$$
$$\partial_t (\rho v) + \partial_x \left(\rho v^2 + p(\rho)\right) = g_1(\rho, v) + g_2(x, \rho),$$

where $\rho(x,t)$ is the mass density of the gas, $v(x,t)$ denotes the gas velocity and $p(\rho)$ is a pressure law satisfying the following assumption

(P) $p \in \mathbf{C}^2(\mathbb{R}^+;\mathbb{R}^+)$ with $p(0) = 0$ and $p'(\rho) > 0$ and $p''(\rho) \geq 0$ for all $\rho \in \mathbb{R}^+$.

In the context of gas pipelines, $p(\rho)$ is often chosen as

$$p(\rho) = \frac{zRT}{M_g}\rho = a^2\rho, \tag{2.1}$$

where z is the natural gas compressibility factor, R the universal gas constant, T the temperature and M_g the molecular weight of the gas. Under the assumption of constant gas temperature, the resulting constant a can be viewed as the speed of sound in the gas, depending on gas type and temperature.
Furthermore, $g_1(\rho, v)$ and $g_2(x, \rho)$ are source terms modeling the influence of friction and inclination respectively. For a pipe of diameter D, we use

$$g_1(\rho, v) = -\frac{\nu}{2D}\rho v|v|$$

for the (steady state) friction, see [62, 69]. The friction factor ν is calculated for example using Chen's formula [18]

$$\frac{1}{\sqrt{\nu}} := -2\log\Big(\frac{\varepsilon/D}{3.7065} - \frac{5.0452}{N_{\mathrm{Re}}}\log\big(\frac{1}{2.8257}\big(\frac{\varepsilon}{D}\big)^{1.1098} + \frac{5.8506}{0.8981 N_{\mathrm{Re}}}\big)\Big),$$

where N_{Re} is the Reynolds number $N_{\mathrm{Re}} = \rho v D/\mu$, μ the gas dynamic viscosity and ϵ the pipeline roughness. Furthermore, we assume that the wall extension or contraction under pressure load is negligible, such that the pipe diameter D is constant.

The inclination of the pipe is modeled by

$$g_2(x, \rho) = -\mathrm{g}\rho\sin\alpha(x),$$

where g is the gravity and $\alpha(x)$ is the slope of the pipe.
To simplify the notation, we define the flux variable $q = \rho v$ and call

$$\partial_t\rho + \partial_x q = 0 \tag{2.2a}$$

$$\partial_t q + \partial_x\left(\frac{q^2}{\rho} + a^2\rho\right) = g(x, \rho, q) \tag{2.2b}$$

the fine or isothermal Euler model for the gas flow in one pipe, where g is a combination of the source terms g_1 and g_2. We make now two additional assumptions, which simplify the mathematical treatment.

($\mathbf{A_1}$) There are no vacuum states, i.e., $\rho > 0$.
($\mathbf{A_2}$) All states are subsonic, i.e., $\frac{q}{\rho} < a$

The use of the first restriction can be justified by the fact that the atmospheric pressure is a natural lower bound for the pressure in the pipes. Lower pressures occur only due to waves travelling through the pipe. However, waves of a size that creates vacuum states would lead to exploding or imploding pipes and do not occur in practice. The second assumption is justified by the fact that pipelines are operated at high pressure (40 to 60 bar) and the gas velocity is in general very low (< 10 m/s), whereas the speed of sound in natural gas is around 370 m/s. In standard pipeline operation conditions, all states are therefore sufficiently far away from the sonic states that even travelling waves will not create sonic or supersonic states. In the following, we denote by $y = (\rho, q)$ the state variables and by

$$f(\rho, q) = \begin{pmatrix} q \\ \frac{q^2}{\rho} + a^2\rho \end{pmatrix}$$

the flux function of the fine model (2.2) and recall the basic properties of the above model.

For $\rho \neq 0$, the system matrix

$$f'(\rho, q) = \begin{pmatrix} 0 & 1 \\ a^2 - \frac{q^2}{\rho^2} & 2\frac{q}{\rho} \end{pmatrix}$$

has two real distinct eigenvalues

$$\lambda_1(\rho, q) = \frac{q}{\rho} - a, \qquad \lambda_2(\rho, q) = \frac{q}{\rho} + a$$

provided that $\frac{q}{\rho} \neq a$. Clearly, assumptions $(\mathbf{A_1})$ and $(\mathbf{A_2})$ insure that the model (2.2) remains strictly hyperbolic.
Furthermore, the right eigenvectors are given by

$$r_1 = \begin{pmatrix} \rho \\ q - a\rho \end{pmatrix}, \qquad r_2 = \begin{pmatrix} \rho \\ q + a\rho \end{pmatrix}.$$

This implies that both characteristic fields are genuinely nonlinear, since we have $\nabla\lambda_1 \cdot r_1 = -a$ and $\nabla\lambda_2 \cdot r_2 = a$.
In the following, we denote by

$$\mathcal{L}_1(\rho; \bar{\rho}, \bar{q}) = \begin{cases} \frac{\bar{q}}{\bar{\rho}}\rho - a\rho\log\left(\frac{\rho}{\bar{\rho}}\right) & \rho < \bar{\rho} \\ \bar{q} + (\rho - \bar{\rho})\, s_1(\rho, \bar{\rho}, \bar{q}) & \rho > \bar{\rho} \end{cases}$$

the union of the 1-rarefaction and the 1-shock curve through the (left) state $\bar{y} = (\bar{\rho}, \bar{q})$. The union of the 2-shock and 2-rarefaction curve through $(\bar{\rho}, \bar{q})$ will be denoted by

$$\mathcal{L}_2(\rho; \bar{\rho}, \bar{q}) = \begin{cases} \bar{q} + (\rho - \bar{\rho})\, s_2(\rho, \bar{\rho}, \bar{q}) & \rho < \bar{\rho} \\ \frac{\bar{q}}{\bar{\rho}}\rho + a\rho\log\left(\frac{\rho}{\bar{\rho}}\right) & \rho > \bar{\rho} \end{cases},$$

where the shock speeds s_1 and s_2 are given by

$$s_1(\rho; \bar{\rho}, \bar{q}) = \frac{\bar{q}}{\bar{\rho}} - a\sqrt{\frac{\rho}{\bar{\rho}}}, \qquad s_2(\rho; \bar{\rho}, \bar{q}) = \frac{\bar{q}}{\bar{\rho}} + a\sqrt{\frac{\rho}{\bar{\rho}}}.$$

These curves are called the (forward) 1 and 2 Lax curves [55, 57, 30]. Analogously we can define the backward 1 and 2 Lax curves through a (right) state.

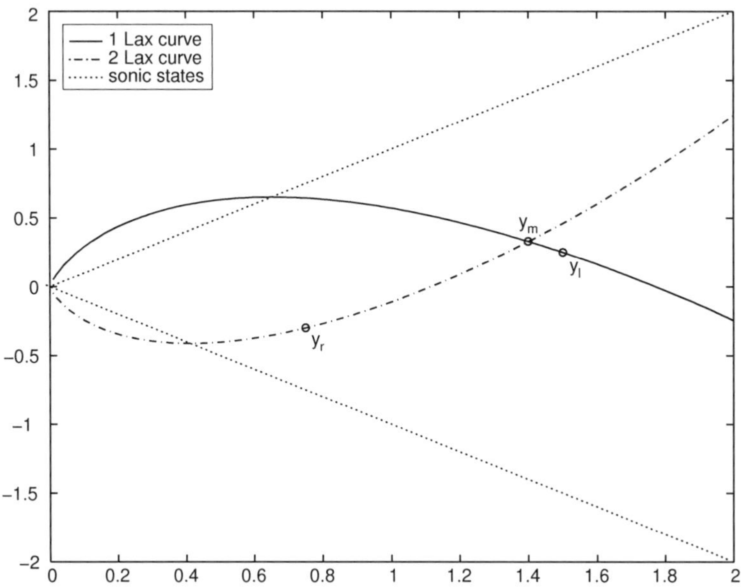

FIGURE 1. Construction of the solution to a Riemann problem. The initial states are y_l and y_r, the solution of the Riemann problem consists of an additional state y_m.

2.1. Extension to a network model

In the previous section, we derived a model for the gas flow in one single pipe. Extending this to networks of pipes, we want to keep the property of one spatial dimension. To this end, we introduce conditions at the nodes (junctions) of the network, coupling the single pipe equations (2.2).

Using the notation introduced in [9], we define a network of pipes as a finite directed graph $(\mathcal{J}, V)$. Both sets are supposed to be non-empty sets of indices. Each element (edge) $j \in \mathcal{J}$ corresponds to a pipe, parameterized by a finite interval $\mathcal{I}_l := \left[x_a^{(l)}, x_b^{(l)}\right]$. Each element (node) $v \in \mathcal{V}$ corresponds to a single intersection of pipes. For a fixed vertex $v \in \mathcal{V}$, we define δ_v^- and δ_v^+ as the sets of all indices of pipes ingoing to and outgoing from the node v respectively and denote by $\delta_v = \delta_v^- \cup \delta_v^+$ the union of in- and outcoming pipes at the node v. Furthermore, we define the degree of a vertex $v \in \mathcal{V}$ as the number of connected pipes. The degree of a vertex v is denoted by $|\cdot|$.

We divide the set of nodes $\mathcal{V}$ according to their physical interpretation. Any node of degree one, i.e., with $|\delta_v^+ \cup \delta_v^-| = 1$ is either an inflow ($\delta_v^- = \emptyset$) or an outflow ($\delta_v^+ = \emptyset$) boundary node. The sets of these nodes are denoted by $\mathcal{V}_I$ and $\mathcal{V}_O$ respectively and can be interpreted as gas providers and costumers. Furthermore, some nodes v of degree 2 are controllable vertices (e.g., compressor stations, valves,...). This subset is denoted by $\mathcal{V}_C \subset \mathcal{V}$. The remaining nodes $\mathcal{V}_P = \mathcal{V} \backslash (\mathcal{V}_I \cup \mathcal{V}_O \cup \mathcal{V}_C)$ are simple pipe to pipe intersections, also called standard junctions. Additionally to the above assumptions $(\mathbf{A_1})$ and $(\mathbf{A_2})$, we claim that

$(\mathbf{A_3})$ All pipes have the same diameter D. The cross sectional area is given by $A = \frac{D^2}{4}\pi$. As for a single pipe, the wall extension or contraction under pressure load is negligible.

$(\mathbf{A_4})$ The friction factor ν is the same for all pipes.

It remains now to model the different types of nodes. Clearly, the value at a node v depends only on the flow in the in- and outgoing pipes and possibly some time-dependent controls. Mathematically this yields coupling conditions of the form

$$\Psi\left(\rho^{(l_1)}, q^{(l_1)}, \ldots, \rho^{(l_n)}, q^{(l_n)}\right) = \Pi(t), \qquad \delta_v = \{l_1, \ldots, l_n\},$$

where Ψ is a possibly nonlinear function of the states and Π is a control function, depending only on time. Summarizing, we have the following network model for gas flow in pipelines consisting of m pipes:

$$\left.\begin{aligned} &\partial_t \rho^{(l)} + \partial_x q^{(l)} = 0 \\ \partial_t q^{(l)} + \partial_x &\left(\frac{\left(q^{(l)}\right)^2}{\rho^{(l)}} + a^2 \rho^{(l)}\right) = g(x, \rho^{(l)}, q^{(l)}) \end{aligned}\right. \quad \begin{aligned} &\forall l \in \mathcal{J} = \{1, \ldots, m\}, \\ &x \in \mathcal{I}_l, t \geq 0 \end{aligned} \tag{2.3a}$$

$$\Psi(\rho^{(l_1)}, q^{(l_1)}, \ldots, \rho^{(l_n)}, q^{(l_n)}) = \Pi(t) \qquad \begin{aligned} &\forall v \in \mathcal{V}, \\ &\delta_v = \{l_1, \ldots, l_n\} \end{aligned} \tag{2.3b}$$

We discuss now the different types of junctions and the resulting coupling conditions.

Note first that the in- and outflow nodes $v \in \mathcal{V}_I \cup \mathcal{V}_O$ can be identified with boundary conditions for the equations (2.3a), modeling for example the (time-dependent) inflow pressure or the outgoing amount of gas. The formulation of boundary conditions for systems of balance laws is well known so that we omit a detailed discussion here. At standard junctions $v \in \mathcal{V}_P$, no external forces are applied so that the amount of gas is conserved at pipe to pipe intersections, i.e.,

$$\sum_{l \in \delta_v^+} q^{(l)} = \sum_{\bar{l} \in \delta_v^-} q^{(\bar{l})}.$$

Furthermore, we claim that the pressure is equal in all pipes, i.e.,

$$p(\rho^{(l)}) = p(\rho^{(\bar{l})}) \qquad \forall l, \bar{l} \in \delta_v^+ \cup \delta_v^-.$$

These coupling conditions were introduced in [10] and analyzed in [9]. The conservation of mass at junctions is unquestioned in the literature, the pressure condition however is up for discussion, since it is not clear how the pressure tables of the engineering community can be represented in mathematical coupling conditions. For a different pressure condition at the junction see, e.g., [21]. However, the differences in the solutions obtained by different pressure coupling is quite small, as shown in [27]. We therefore restrict to the above-presented condition of equal pressure.

All together, the coupling conditions at a standard junction are given by

$$\Psi(\rho^{(l_1)}, q^{(l_1)}, \dots, \rho^{(l_n)}, q^{(l_n)}) = \begin{pmatrix} \sum_{l\in\delta_v^+} q^{(l)} - \sum_{\bar{l}\in\delta_v^-} q^{(\bar{l})} \\ p(\rho^{(l_1)}) - p(\rho^{(l_2)}) \\ \dots \\ p(\rho^{(l_1)}) - p(\rho^{(l_n)}) \end{pmatrix} = 0, \tag{2.4}$$

for $\delta_v^- \cup \delta_v^+ = \{l_1, \dots, l_n\}$.

Inspired by the construction of solutions to the standard Riemann problem, we can view the problem of finding a solution at a vertex of the network as a so-called *half Riemann problem*, see, e.g., [50]. For a rigorous definition of half Riemann problems see [21, 9]. Given a junction of n pipes coupled through the above coupling conditions (2.4), in each of the n pipes, we assume a constant subsonic state $(\bar{\rho}^{(l)}, \bar{q}^{(l)})$. The states $(\bar{\rho}^{(1)}, \bar{q}^{(1)}), \dots, (\bar{\rho}^{(n)}, \bar{q}^{(n)})$ do in general not fulfill the coupling condition (2.4), but as for the standard Riemann problem, we can use the Lax curves to construct the intermediate states that arise at the junction. Note that solutions to this problem are only feasible if the occurring waves travel with negative speed in incoming pipes and with positive speed in outgoing pipes, which implies that the coupling conditions do not provide feasible solutions to all choices of states $(\bar{\rho}^{(1)}, \bar{q}^{(1)}), \dots, (\bar{\rho}^{(n)}, \bar{q}^{(n)})$. For more details see section 3, Proposition 3.5.

To simplify the notation in the construction, we identify the incoming pipes of the junction with the indices $1, \dots, j$ and the outgoing pipes with $j+1, \dots, n$. Since we claim that the pressure law $p(\rho)$ depends only on the mass density ρ, the pressure conditions in (2.4) imply

$$\rho^{(1)} = \rho^{(2)} = \dots = \rho^{(n)} =: \tilde{\rho}$$

at the junction.

Furthermore, we know that the states $(\tilde{\rho}, q^{(1)}), \dots, (\tilde{\rho}, q^{(j)})$ in the incoming pipes must be connected to the 1-Lax curves through $(\bar{\rho}^{(1)}, \bar{q}^{(1)}), \dots, (\bar{\rho}^{(j)}, \bar{q}^{(j)})$ to create (subsonic) waves of negative speed only. Analogously, the states

$$(\tilde{\rho}, q^{(j+1)}), \dots, (\tilde{\rho}, q^{(n)})$$

in the outgoing pipes must be connected to the (backward) 2-Lax curves through $(\bar{\rho}^{(j+1)}, \bar{q}^{(j+1)}), \dots, (\bar{\rho}^{(m)}, \bar{q}^{(m)})$ to create (subsonic) waves of positive speed only. The conservation of mass in (2.4) implies therefore

$$\sum_{l=1}^{j} \mathcal{L}_1(\rho, \bar{\rho}^{(l)}, \bar{q}^{(l)}) = \sum_{\bar{l}=j+1}^{n} \bar{\mathcal{L}}_2(\rho, \bar{\rho}^{(\bar{l})}, \bar{q}^{(\bar{l})}).$$

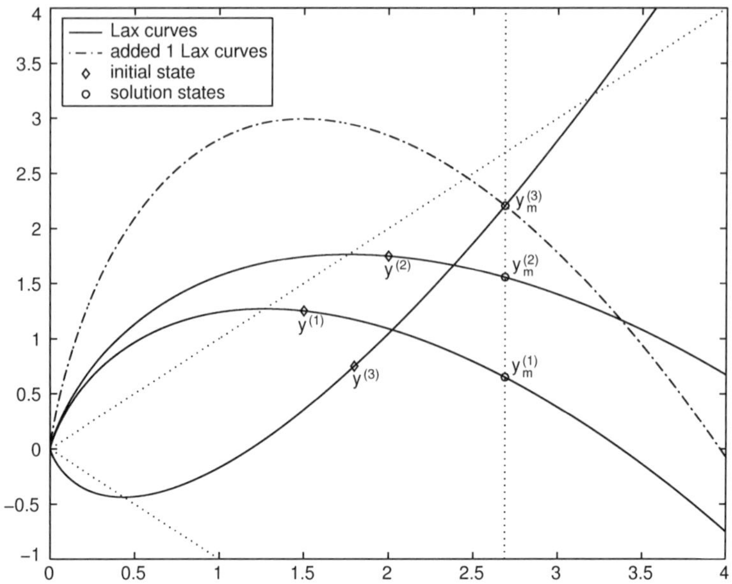

FIGURE 2. Construction of a solution to a half Riemann problem at a junction of 2 incoming and 1 outgoing pipes. The initial states are marked by a diamond, the solution of the half Riemann problem by circles.

Graphically this means that the intersection of the curve $\left(\rho, \sum_{l=1}^{j} \mathcal{L}_1(\rho, \bar{\rho}^{(l)}, \bar{q}^{(l)})\right)$ with the curve $\left(\rho, \sum_{\bar{l}=j+1}^{n} \bar{\mathcal{L}}_2(\rho, \bar{\rho}^{(\bar{l})}, \bar{q}^{(\bar{l})})\right)$ defines the mass density $\tilde{\rho}$ at the junction. The corresponding fluxes $q^{(l)}$ at the junction are then given by the value of the Lax-curves at $\tilde{\rho}$, i.e., the solution at the junction is given by

$$y^{(l)} = \left(\tilde{\rho}, \mathcal{L}_1(\tilde{\rho}, \bar{\rho}^{(l)}, \bar{q}^{(l)})\right), \qquad l = 1, \ldots, j$$

$$y^{(l)} = \left(\tilde{\rho}, \bar{\mathcal{L}}_2(\tilde{\rho}, \bar{\rho}^{(l)}, \bar{q}^{(l)})\right), \qquad l = j+1, \ldots, n$$

see also Figure 2.

Note furthermore that the parameterization of the network does not influence the solution at the junction. To illustrate this, consider one single pipe of the network. As an ingoing pipe, it is parametrized by the interval $[x_a, x_b]$ and the flow inside this pipe has density ρ and velocity v. Since this pipe is ingoing, the junction is located at x_b. If we reparametrize the network to make this pipe an outgoing one, the junction will be located at the beginning of the pipe, i.e., at $\tilde{x}_a$. As the direction of the flow does not change in our system this means that the flow inside the pipe is now given by the density ρ and the velocity $-v$. For an illustration see also Figure 3.

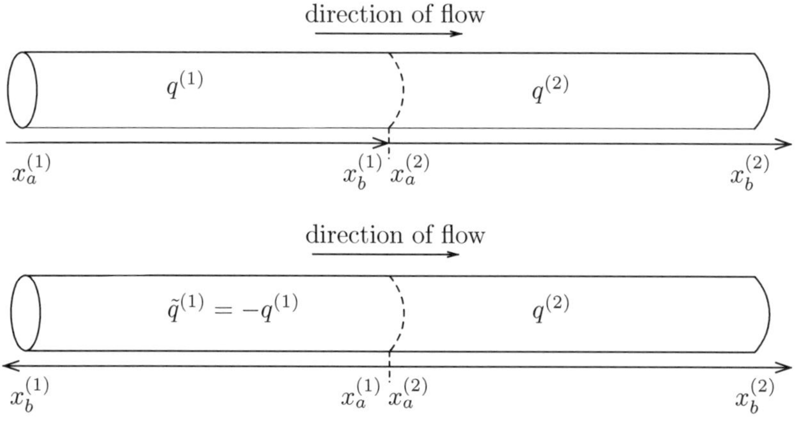

FIGURE 3. Illustration of a reparametrization of a network.

Recalling that $q = \rho v$, a close look at the definition of the Lax curves shows now that

$$\mathcal{L}_1(\rho, \bar{\rho}, \bar{q}) = \bar{\mathcal{L}}_2(\rho, \bar{\rho}, -\bar{q}).$$

The result of the above construction will therefore not change due to a redefinition of incoming and outgoing arcs in the network. This property will be of importance in section 3 where we consider a prototype network of only one node with arcs parameterized such that the junction is at $x = 0$ for all pipes, i.e., the node has only outgoing arcs.

At controllable junctions $v \in \mathcal{V}_C$, a compressor acts as a *directed* external force, in other words the compressor can pump gas only in one direction. A by-pass pipe circumvents the arise of back pressure so that the controllable junction becomes a standard junction in case of reversed flow or compressor maintenance. Note further that any controllable junction is of degree 2.

For simplicity we denote the ingoing quantities by a superscript $v-$ and the outgoing ones by a superscript $v+$. We assume that the compressors are powered using external energy only. Therefore the amount of gas is conserved at controllable nodes, i.e.,

$$q^{(v-)} = q^{(v+)}.$$

As proposed, e.g., in [17] and used in wide parts of the community (see [52, 67, 49]), the effect of the compressor is modeled in the second coupling condition, given by

$$P_v(t) = q^{(v-)} \left(\left(\frac{p(\rho^{(v+)})}{p(\rho^{(v-)})} \right)^{\kappa} - 1 \right).$$

Hereby κ is defined by $\kappa = \frac{\gamma-1}{\gamma}$ using the isentropic coefficient $\gamma \in \left\{\frac{5}{3}, \frac{7}{5}\right\}$ of the gas. $P_v(t)$ is proportional to the time-dependent compressor power applied at the junction. Note that $q^{(v-)} > 0$ by assumption. As mentioned above, the coupling for $q^{(v-)} \leq 0$ is given by the standard junction conditions.

All together, the coupling conditions at a controllable vertex are given by

$$\Psi(\rho^{(v-)}, q^{(v-)}, \rho^{(v+)}, q^{(v+)}) = \begin{pmatrix} q^{(v-)} - q^{(v+)} \\ q^{(v-)} \left(\left(\frac{p(\rho^{(v+)})}{p(\rho^{(v-)})} \right)^{\kappa} - 1 \right) \end{pmatrix} = \begin{pmatrix} 0 \\ P_v(t) \end{pmatrix}. \quad (2.5)$$

Note that the second condition in (2.5) reduces to $p(\rho^{(v+)}) = p(\rho^{(v-)})$ if the compressor is turned off ($P_v = 0$). The compressor coupling conditions are therefore a generalization of the coupling conditions at standard junctions of degree 2.

Due to the nonlinearity of the second coupling condition, it is not possible to state a direct construction rule as for standard junctions. However, the Lax curves can be used to illustrate how the compressor works, see also Figure 4.

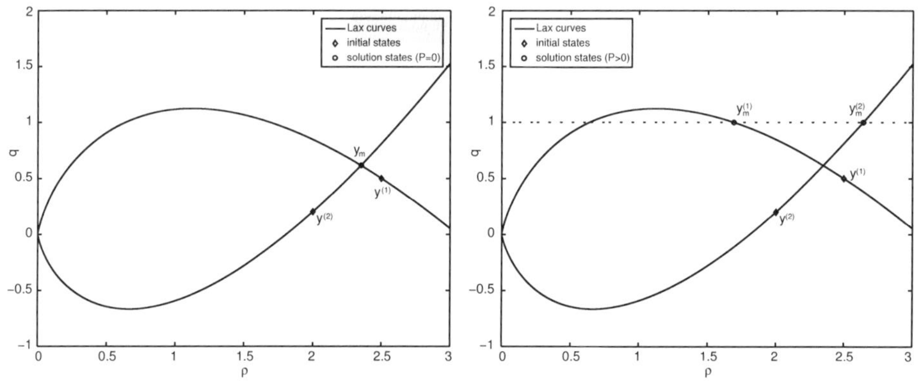

FIGURE 4. Construction of a solution to a Riemann problem with left state y_l and right state y_r (left) and illustration of the effect of a non zero compressor power at a controllable node (right).

As for standard junctions, the wave in the incoming pipe $(v-)$ can have only negative speed, whereas the wave in the outgoing pipe $(v+)$ must have positive speed. Assume that $y_m = (\rho_m, q_m)$ is the solution of a Riemann problem at the junction for $P = 0$. For $P > 0$, the resulting states $y_m^{(v+)}$ and $y_m^{(v-)}$ will lie above y_m in the sense that $q_m^{(v+)} = q_m^{(v-)} > q_m$. Since $q_m^{(v-)}$ must be part of the 1-Lax curve and $q_m^{(v+)}$ is part of the 2-Lax curve, we have furthermore $\rho_m^{(v-)} < \rho_m^{(v+)}$. This is consistent with the expected behavior, since the compressor increases the pressure in the outgoing pipe at the expense of the pressure in the incoming pipe. Figure 4 also shows that the pressure in the outgoing pipe cannot be increased arbitrarily, since the maximum flux at the junction is bounded due to the boundedness from above of the 1-Lax curve.

3. Analytical properties of the fine model

In this part, we consider a more general initial value problem consisting of the 2×2 systems of balance laws

$$\partial_t y^{(l)}(t,x) + \partial_x f_l\left(y^{(l)}(t,x)\right) = g\left(t,x,y^{(l)}(t,x)\right) \tag{3.1}$$

with $l = 1,\dots,n$, $t \in [0,+\infty]$ and $x \in [0,+\infty]$ along n pipes. At $x = 0$ the pipes are coupled through the possibly nonlinear coupling conditions

$$\Psi\left(y^{(1)}(t,0+),\dots,y^{(n)}(t,0+)\right) = \Pi(t). \tag{3.2}$$

In other words we deal with an initial boundary value problem for systems of balance laws coupled at $x = 0$ through nonlinear boundary conditions.

The well-posedness of systems of conservation or balance laws is widely studied and the literature offers a variety of papers on this subject, see, e.g., [3, 24, 29, 35]. Furthermore, in [19, 21, 22, 39, 51], the well-posedness of network problems involving hyperbolic systems (without source terms) is studied. Also results on initial-boundary value problems for systems of conservation laws are available, see, e.g., [1, 2, 25, 36]. The present work in this section (Section 3.1 to 3.3) deals with 2×2 hyperbolic systems on a network, also allowing the presence of source terms. The result includes a general condition for the well-posedness of coupling conditions as well as an existence result for the Cauchy problem on networks.

The treatment of the source in most publications on systems of balance laws uses the classical time splitting technique, see, e.g., [23, 25]. This method is used also in Section 3.3 to include the effect of friction and gravity. However, it often requires rather strong assumptions on the source term.

This section is organized as follows. In a first step we derive conditions such that the Riemann problem at the junction located at $x = 0$ is well posed. Using this result, we prove that the Cauchy problem for (3.1) has a unique solution, provided that certain conditions are satisfied. Furthermore, the solution depends Lipschitz continuously on the initial data and the coupling conditions. Throughout this section, we refer to [12] and [30] for the general theory of hyperbolic systems of conservation laws.

3.1. Basic notations, definitions and assumptions

For convenience, we recall the following basic definitions and notations. More details on the basic theoretical aspects of conservation laws can be found for example in [12] or [30]. We assume in the following:

(H) The system

$$\partial_t y + \partial_x f(y) = 0, \tag{3.3}$$

is strictly hyperbolic with smooth coefficients, defined on the open set Ω. For each $i = 1,\dots,n$, the ith characteristic family is either genuinely nonlinear or linearly degenerate.

We denote by $R_i(\sigma)(y_o)$ the parametrized *i-rarefaction curve* through y_o. Furthermore, we define the i-shock curves $S_i(\sigma)(y_o)$ using the Ranking-Hugoniot jump condition, i.e.,

$$f\left(S_i(\sigma)(y_o)\right) - f(y_o) = \boldsymbol{\lambda}_i(\sigma)\left(S_i(\sigma)(y_o) - y_o\right),$$

for some scalar function $\boldsymbol{\lambda}_i(\sigma)$, where the parametrization is chosen such that $\left\|\frac{dS_i}{d\sigma}\right\| \equiv 1$. If the ith characteristic family is linearly degenerate, the i-shock and i-rarefaction curves coincide. In this case, the shock and rarefaction curves are parameterized by arc-length. If the ith characteristic family is genuinely nonlinear, r_i is chosen such that $D\lambda_i\left(y\right)r_i\left(y\right) \equiv 1$ We can chose the parameterization, such that S_i and R_i are twice continuously differentiable at $\sigma = 0$ and we define

$$\mathcal{L}_i(\sigma)(y_o) = \begin{cases} R_i(\sigma)(y_o) & \text{if } \sigma \geq 0 \\ S_i(\sigma)(y_o) & \text{if } \sigma < 0 \end{cases}. \tag{3.4}$$

Here and in what follows, $\mathbb{R}^+ = [0, +\infty[$ and $\mathring{\mathbb{R}}^+ =]0, +\infty[$. Furthermore, let $\Omega_l \in \mathbb{R}^2$ be a nonempty open set and define $\Omega = \Omega_1 \times \Omega_2 \times \cdots \times \Omega_n$. Let $f \equiv (f_1, \ldots, f_n)$ and $y \equiv (y^{(1)}, \ldots, y^{(n)})$. For later use, with a slight abuse of notation, we denote by

$$\begin{aligned} \|y\| &= \sum_{l=1}^{n} \left\|y^{(l)}\right\| && \text{for } y \in \Omega, \\ \|y\|_{\mathbf{L}^1} &= \int_{\mathbb{R}^+} \|y(x)\|\,dx && \text{for } y \in \mathbf{L}^1\left(\mathbb{R}^+;\Omega\right), \\ \mathrm{TV}(y) &= \sum_{l=1}^{n} \mathrm{TV}(y^{(l)}) && \text{for } y \in \mathbf{BV}\left(\mathbb{R}^+;\Omega\right). \end{aligned}$$

Assume that the flows f_l satisfy the following assumption at an n-tuple $\bar{y} \in \Omega$:

(F): For $l = 1, \ldots, n$, the flow f_l is in $C^4\left(\Omega_l;\mathbb{R}^2\right)$, $Df_l\left(\bar{y}^{(l)}\right)$ admits a strictly negative eigenvector $\lambda_1^{(l)}\left(\bar{y}^{(l)}\right)$, a strictly positive one $\lambda_2^{(l)}\left(\bar{y}^{(l)}\right)$ and each characteristic field is either genuinely nonlinear or linearly degenerate.

This condition ensures that for $g_l \equiv 0$ and $x \in \mathbb{R}$ each system of (3.1) generates a standard Riemann semigroup [12, Section 8]. The gas flow models introduced clearly fulfill condition **(F)** due to assumptions $(\mathbf{A_1})$ and $(\mathbf{A_2})$.

In the following, the constant states $\bar{y} \in \Omega$, $\bar{\Pi} \in \mathbb{R}$ are fixed. Throughout, we also fix a time $\hat{T} \in]0, +\infty]$ and a positive $\hat{\delta}$. For all $\delta \in \left]0, \hat{\delta}\right]$, we set

$$\mathcal{Y}_\delta = \left\{y \in \bar{y} + \mathbf{L}^1\left(\mathbb{R}^+;\Omega\right) : \mathrm{TV}(y) \leq \delta\right\}.$$

As usual in the context of initial boundary value problems (see [1, 2]), we consider the metric space $X = \left(\bar{y} + \mathbf{L}^1(\mathbb{R}^+;\Omega)\right) \times \left(\bar{\Pi} + \mathbf{L}^1\left(\mathbb{R}^+;\mathbb{R}^n\right)\right)$ equipped with the $\mathbf{L}^1$ distance. The extended variable is denoted by $\mathbf{p} \equiv (y, \Pi)$ with $y \in \bar{y} + \mathbf{L}^1(\mathbb{R}^+;\Omega)$

and $\Pi \in \bar{\Pi} + \mathbf{L}^1(\mathbb{R}^+; \mathbb{R}^n)$. Correspondingly, denote for all $(y, \Pi), (\tilde{y}, \tilde{\Pi}) \in X$

$$\begin{aligned} d_X\left((y,\Pi),(\tilde{y},\tilde{\Pi})\right) &= \left\|(y,\Pi)-(\tilde{y},\tilde{\Pi})\right\|_X = \|y-\tilde{y}\|_{\mathbf{L}^1} + \left\|\Pi-\tilde{\Pi}\right\|_{\mathbf{L}^1} \\ \mathrm{TV}(y,\Pi) &= \mathrm{TV}(y) + \mathrm{TV}(\Pi) + \|\Psi(y(0+)) - \Pi(0+)\| \\ \mathcal{D}^\delta &= \{\mathbf{p} \in X : \mathrm{TV}(\mathbf{p}) \le \delta\} \end{aligned} \tag{3.5}$$

Note that $\mathcal{Y}_\delta$ denotes the set where the functions y vary, while $\mathcal{D}^\delta$ contains pairs (y, Π).

Furthermore, $\mathcal{O}(1)$ denotes a sufficiently large constant dependent only on f restricted to a neighborhood of the initial states.

On the source term $g \equiv (g_1, \ldots, g_n)$ we require that if G is defined by

$$(G(t,y))(x) = \left(g_1\left(t,x,y^{(1)}(x)\right), \ldots, g_n\left(t,x,y^{(n)}(x)\right)\right),$$

then G satisfies:

(G): $G\colon [0,\hat{T}] \times \mathcal{Y}_{\hat{\delta}} \to \mathbf{L}^1(\mathbb{R}^+; \mathbb{R}^{2n})$ is such that there exist positive L_1, L_2 and for all $t, s \in [0, \hat{T}]$

$$\begin{aligned} \|G(t,y_1) - G(s,y_2)\|_{\mathbf{L}^1} &\le L_1 \cdot (\|y_1 - y_2\|_{\mathbf{L}^1} + |t-s|) && \forall y_1, y_2 \in \mathcal{Y}_{\hat{\delta}}, \\ \mathrm{TV}(G(t,y)) &\le L_2 && \forall y \in \mathcal{Y}_{\hat{\delta}}. \end{aligned}$$

Note that **(G)** also comprises non-local terms, see, e.g., [24]. This rather weak assumption is sufficient to prove the well-posedness of the Cauchy problem at junctions. Examples of sources g such that the corresponding G satisfies **(G)** are provided by the next proposition, which comprises the source terms of the gas flow models presented.

Proposition 3.1. *Assume that the map $g\colon \mathbb{R}^+ \times \Omega \to \mathbb{R}^{2n}$ satisfies:*

1. *there exists a state $\bar{y}$ and a compact subset $\bar{\mathcal{K}}$ of $\mathbb{R}^+$ such that $g(x, \bar{y}) = 0$ for all $x \in \mathbb{R}^+ \setminus \bar{\mathcal{K}}$;*
2. *there exists a finite positive measure μ such that for all $x_1, x_2 \in \mathbb{R}^+$ with $x_1 \le x_2$, and all $y \in \Omega^l$,*
$$\|g_l(x_2+, y) - g_l(x_1-, y)\| \le \mu([x_1, x_2]);$$
3. *there exists a positive $\hat{L}$ such that for all $y_1, y_2 \in \Omega$, for all $x \in \mathbb{R}^+$,*
$$\|g(x,y_1) - g(x,y_2)\| \le \hat{L} \cdot \|y_1 - y_2\|.$$

Then, the autonomous source term $G(t, y) = g(y)$ satisfies condition **(G)**.

Provided we have positive density $\rho \ge \epsilon > 0$ the following observations can be made. The source term of the gas flow models $y = (\rho, \rho v)$ based on partial differential equations introduced is given by

$$g(x, \rho, \rho v) = -\frac{\nu}{2D}\frac{\rho v|\rho v|}{\rho} - \mathrm{g}\rho \sin\alpha(x).$$

We show now, that this source term fulfills the three assumptions of Proposition 3.1. Since each pipe has a finite length, $\alpha(x) \equiv 0$ outside an interval $[x_a, x_b]$, condition 1 is satisfied by choosing $\bar{\mathcal{K}} = [x_a, x_b]$ and $\bar{y} = (y_1, 0)$. Furthermore, conditions 2 and 3 are fulfilled due to the Lipschitz continuity of the source term.

3.2. Well-posedness of the coupling conditions

In this section, we show that the Riemann problem at the junction is well posed. To this end, we consider the homogeneous problem given by (3.1) with $g \equiv 0$. The role of the source term can be neglected in the discussion of the Riemann problem, as the source is included later through operator splitting techniques, see Section 3.3.

We start with two basic definitions concerning the Riemann problem at a junction.

Definition 3.2. By *Riemann problem at the junction* we mean the problem

$$\begin{cases} \partial_t y^{(l)} + \partial_x f_l(y^{(l)}) = 0, \\ \Psi\left(y(t,0+)\right) = \Pi, \\ y^{(l)}(0,x) = y_o^{(l)}, \end{cases} \qquad \begin{array}{ll} t \in \mathbb{R}^+, & l \in \{1,\dots,n\}, \\ x \in \mathbb{R}^+, & y^{(l)} \in \Omega \end{array} \tag{3.6}$$

where $y_o^{(1)}, \dots, y_o^{(n)}$ are constant states in Ω and $\Pi \in \mathbb{R}^n$ is also constant.

Definition 3.3. Fix a map $\Psi \in \mathbf{C}^1(\Omega^n; \mathbb{R}^n)$ and $\Pi \in \mathbb{R}$. A *Ψ-solution* to the Riemann problem (3.6) is a function $y\colon \mathbb{R}^+ \times \mathbb{R}^+ \mapsto \Omega^n$ such that the following hold:

(L) For $l = 1, \dots, n$, the function $(t,x) \mapsto y^{(l)}(t,x)$ is self-similar and coincides with the restriction to $x \in \mathring{\mathbb{R}}^+$ of the Lax solution to the standard Riemann problem

$$\begin{cases} \partial_t y^{(l)} + \partial_x f_l(y^{(l)}) = 0, \\ y^{(l)}(0,x) = \begin{cases} y_o^{(l)} & \text{if } x > 0, \\ y^{(l)}(1,0+) & \text{if } x < 0. \end{cases} \end{cases}.$$

(Ψ) The trace $y(t,0+)$ of y at the junction satisfies (3.2) for a.e. $t > 0$.

Note that the (arbitrary) choice of $t = 1$ in **(L)** above is justified by the self similarity of the solutions to Riemann problems.

Furthermore, denote by $\sigma \mapsto \mathcal{L}_i^{(l)}(y_o^{(l)}, \sigma)$ the ith Lax curve through $y_o^{(l)}$ for $i = 1, 2$ corresponding to f_l, as defined in Section 3.1.

Lemma 3.4. *For sufficiently small σ we have*

$$\frac{1}{C}|\sigma| \le \left\| \mathcal{L}_i^{(l)}\left(y^{(l)}, \sigma\right) - y^{(l)} \right\| \le C|\sigma|.$$

Note that contrary to the previous definition of a junction in the modeling part, we have parametrized the junction in this section such that pipes are only outgoing from a junction. This makes the notations much clearer in the following sections, as only 1-waves can hit a junction, whereas only 2-waves can leave the junction. Therefore, in the following proposition, which yields the continuous dependence of the solution to the Riemann problem from the initial state, from the coupling condition Ψ and from the control term Π, the eigenvectors $r_2^{(l)}$ of the second family play a decisive role.

Proposition 3.5. *Let $n \in \mathbb{N}$ with $n \geq 2$. Fix $\Psi \in \mathbf{C}^1(\Omega; \mathbb{R}^n)$ and a constant $\bar{\mathbf{p}} = (\bar{y}, \bar{\Pi})$ such that $\bar{\Pi} = \Psi(\bar{y})$ and f satisfies* **(F)** *at $\bar{y}$. If*

$$\det\left[D_1\Psi(\bar{y})r_2^{(1)}\left(\bar{y}^{(1)}\right) \quad D_2\Psi(\bar{y})r_2^{(2)}\left(\bar{y}^{(2)}\right) \quad \ldots \quad D_n\Psi(\bar{y})r_2^{(n)}\left(\bar{y}^{(n)}\right)\right] \neq 0 \tag{3.7}$$

where $D_l\Psi = D_{y^{(l)}}\Psi$, then there exist positive δ, K such that

1. *for all $\mathbf{p} \equiv (y_o, \Pi)$ with $\|\mathbf{p} - \bar{\mathbf{p}}\| < \delta$, the Riemann problem (3.6) admits a unique self-similar solution $(t, x) \mapsto \left(\boldsymbol{\mathcal{R}}^{\Psi}(\mathbf{p})\right)(t, x)$ in the sense of Definition 3.3;*
2. *let $\mathbf{p}, \tilde{\mathbf{p}}$ satisfy $\|\mathbf{p} - \bar{\mathbf{p}}\| < \delta$ and $\|\tilde{\mathbf{p}} - \bar{\mathbf{p}}\| < \delta$. Then, the traces at the junction of the corresponding solutions to (3.6) satisfy*
$$\left\|\left(\boldsymbol{\mathcal{R}}^{\Psi}(\mathbf{p})\right)(t, 0+) - \left(\boldsymbol{\mathcal{R}}^{\Psi}(\tilde{\mathbf{p}})\right)(t, 0+)\right\| \leq K \cdot \|\mathbf{p} - \tilde{\mathbf{p}}\|; \tag{3.8}$$
3. *for any $\tilde{\Psi} \in \mathbf{C}^1(\Omega; \mathbb{R}^n)$ with $\left\|\Psi - \tilde{\Psi}\right\|_{\mathbf{C}^1} < \delta$, $\tilde{\Psi}$ also satisfies (3.7) and for all $\mathbf{p} \in \Omega$ satisfying $\|\mathbf{p} - \bar{\mathbf{p}}\| < \delta$,*
$$\left\|\left(\boldsymbol{\mathcal{R}}^{\tilde{\Psi}}(\mathbf{p})\right)(t) - \left(\boldsymbol{\mathcal{R}}^{\Psi}(\mathbf{p})\right)(t)\right\|_{\mathbf{L}^1} \leq K \cdot \left\|\tilde{\Psi} - \Psi\right\|_{\mathbf{C}_1} \cdot t. \tag{3.9}$$
4. *let $\mathbf{p}, \tilde{\mathbf{p}}$ satisfy $\|\mathbf{p} - \bar{\mathbf{p}}\| < \delta$ and $\|\tilde{\mathbf{p}} - \bar{\mathbf{p}}\| < \delta$ and call $\Sigma(\mathbf{p})$ the n-vector of the total sizes of the 2-waves in the solution to (3.6). Then,*
$$\|\Sigma(\mathbf{p}) - \Sigma(\tilde{\mathbf{p}})\| \leq K \cdot \|\mathbf{p} - \tilde{\mathbf{p}}\|. \tag{3.10}$$

Application to gas flow

To prove the well-posedness of Riemann problems for the gas flow models, we have to show that the coupling conditions introduced fulfill the assumptions of Proposition 3.5. Note that Proposition 3.5 allows for different flux functions $f^{(l)}$ at a junction. The coupling conditions remain therefore well posed also for mixed junctions, i.e., for junctions where pipes modeled by the isothermal Euler equations are connected to pipes modeled by the wave equation. This is especially useful in large pipeline systems, where only parts have to be resolved as accurate as possible.

Standard junctions of n pipes:

Due to the special parameterization of the junction, the coupling condition for a standard junction of n pipes is given by

$$\Psi(\rho^{(l_1)}, q^{(l_1)}, \ldots, \rho^{(l_n)}, q^{(l_n)}) = \begin{pmatrix} \sum_{l=1}^{n} q^{(l)} \\ \rho^{(l_1)} - \rho^{(l_2)} \\ \cdots \\ \rho^{(l_1)} - \rho^{(l_n)} \end{pmatrix} = 0.$$

The determinant in Proposition 3.5 reads then

$$\det \begin{bmatrix} \lambda_2(\bar y^{(1)}) & \lambda_2(\bar y^{(2)}) & \lambda_2(\bar y^{(3)}) & \cdots & \lambda_2(\bar y^{(n)}) \\ 1 & -1 & 0 & \cdots & 0 \\ 1 & 0 & -1 & & \vdots \\ \vdots & \vdots & & \ddots & \\ 1 & 0 & \cdots & & -1 \end{bmatrix} = (-1)^{n-1} \prod_{l=1}^{n} \lambda_2(\bar y^{(l)}).$$

Since $\lambda_2 > 0$ holds for the isothermal Euler equations (see assumption $(\mathbf{A_2})$) as well as for the wave equation, the determinant condition (3.7) is satisfied and the Riemann problem at the junction is well posed.

Compressor junctions:
Recall that compressor junctions are restricted to 2 pipes, so that we can simplify the notation and write the coupling conditions as

$$\Psi(\rho^{(1)}, q^{(1)}, \rho^{(2)}, q^{(2)}) = \begin{pmatrix} q^{(1)} + q^{(2)} \\ q^{(2)} \left(\left(\frac{\rho^{(2)}}{\rho^{(1)}} \right)^{\kappa} - 1 \right) \end{pmatrix} = \begin{pmatrix} 0 \\ P(t) \end{pmatrix}.$$

The determinant condition (3.7) reads then

$$\det \begin{bmatrix} \lambda_2(\bar y^{(1)}) & \lambda_2(\bar y^{(2)}) \\ -\kappa \frac{\bar q^{(2)}}{\bar\rho^{(1)}} \left(\frac{\bar\rho^{(2)}}{\bar\rho^{(1)}} \right)^{\kappa} & \kappa \frac{\bar q^{(2)}}{\bar\rho^{(1)}} \left(\frac{\bar\rho^{(2)}}{\bar\rho^{(1)}} \right)^{\kappa} + \left(\left(\frac{\bar\rho^{(2)}}{\bar\rho^{(1)}} \right)^{\kappa} - 1 \right) \lambda_2(\bar y^{(2)}) \end{bmatrix}$$

$$= \kappa \frac{\bar q^{(2)}}{\bar\rho^{(2)}} \left(\frac{\bar\rho^{(2)}}{\bar\rho^{(1)}} \right)^{\kappa} \left(\lambda_2 \left(\bar y^{(1)} \right) + \lambda_2 \left(\bar y^{(2)} \right) \right) + \lambda_2(\bar y^{(1)}) \lambda_2(\bar y^{(2)}) \left(\left(\frac{\bar\rho^{(2)}}{\bar\rho^{(1)}} \right)^{\kappa} - 1 \right).$$

This determinant is nonzero for isothermal Euler equations that $P > 0$ and that the compressor is directed. This implies that $q^{(2)} > 0$ and $\rho^{(2)} > \rho^{(1)}$, which makes the determinant strictly positive.

3.3. Existence of solutions to the Cauchy problem at the junction

In this Section, we prove the existence and uniqueness of solutions to the Cauchy problem at a junction. To this end, we start with the definition of a weak Ψ-solution.

Definition 3.6. Fix the maps $\Psi \in \mathbf{C}^1(\Omega; \mathbb{R}^n)$ and $\Pi \in \mathbf{BV}(\mathbb{R}^+; \mathbb{R}^n)$. A *weak solution* on $[0, T]$ to

$$\begin{cases} \partial_t y^{(l)} + \partial_x f_l(y^{(l)}) = g_l(t, x, y^{(l)}), & t \in \mathbb{R}^+, \quad l \in \{1, \ldots, n\} \\ \Psi(y(t, 0)) = \Pi(t), & x \in \mathbb{R}^+, \quad y_o \in \bar y + \mathbf{L}^1(\mathbb{R}^+; \Omega) \\ y(0, x) = y_o(x), & \end{cases} \tag{3.11}$$

is a map $y \in \mathbf{C}^0([0,T]; \bar y + \mathbf{L}^1(\mathbb{R}^+; \Omega))$ such that for all $t \in [0, T]$, $y(t) \in \mathbf{BV}(\mathbb{R}^+; \Omega)$ and

(W): $y(0) = y_o$ and for all $\phi \in \mathbf{C}_c^\infty\left(]0,T[\times \mathring{\mathbb{R}}^+;\mathbb{R}\right)$ and for $l = 1,\dots,n$

$$\int_0^T \int_{\mathbb{R}^+} \left(y^{(l)}\,\partial_t\phi + f_l(y^{(l)})\,\partial_x\phi\right)\,dx\,dt + \int_0^T \int_{\mathbb{R}^+} \phi(t,x)\,g_l(t,x,y^{(l)})\,dx\,dt = 0\,.$$

(Ψ): The condition at the junction is met: for a.e. $t \in \mathbb{R}^+$, $\Psi\left(y(t,0+)\right) = \Pi(t)$.

The weak solution y is an entropy solution if for any convex entropy-entropy flux pair (η_l, q_l), for all $\phi \in \mathbf{C}_c^\infty\left(]0,T[\times \mathring{\mathbb{R}}^+;\mathbb{R}^+\right)$ and for $l = 1,\dots,n$

$$\int_0^T \int_{\mathbb{R}^+} \left(\eta_l\left(y^{(l)}\right)\partial_t\phi + q_l\left(y^{(l)}\right)\partial_x\phi\right)dx\,dt + \int_0^T \int_{\mathbb{R}^+} D\eta_l\left(y^{(l)}\right)g(t,x,y)\,\phi\,dx\,dt \geq 0.$$

3.3.1. The homogeneous Cauchy problem. In this section we construct a solution to the homogeneous version of the above-defined initial-boundary value problem by means of the front tracking method. An overview of the different steps necessary in this construction can be found in [12]. To keep the presentation well arranged, we present here only the changes in the standard algorithm required for the treatment of the initial-boundary problem. Below, we denote by $\boldsymbol{\mathcal{T}}_t$ the right translation, i.e., $(\boldsymbol{\mathcal{T}}_t\Pi)(s) = \Pi(t+s)$. Using this notation, we prove the following theorem, yielding the existence and stability of solutions to the homogeneous Cauchy problem. We denote by $\boldsymbol{\mathcal{S}}_t$ the semigroup solutions to the homogeneous Cauchy problem (3.12) for $g \equiv 0$. The map $t \mapsto \Upsilon$ and $t \mapsto \Phi$ are a straightforward extensions to the same map introduced in [12]. More details can be found in [26].

Theorem 3.7. *Let $n \in \mathbb{N}$, $n \geq 2$ and f satisfy* **(F)** *at $\bar{y}$. Let $\bar{\Pi} = \Psi(\bar{y})$. Then, there exist positive δ, L and a semigroup $\boldsymbol{\mathcal{P}} : [0,+\infty[\times \mathcal{D} \to \mathcal{D}$ such that:*

1. $\mathcal{D} \supseteq \overline{\mathcal{D}^\delta}$;
2. *For all $(y,\Pi) \in \mathcal{D}$, $\boldsymbol{\mathcal{P}}_t(y,\Pi) = (\boldsymbol{\mathcal{S}}_t(y,\Pi), \boldsymbol{\mathcal{T}}_t\Pi)$, with $\boldsymbol{\mathcal{P}}_0\mathbf{p} = \mathbf{p}$ and for $s,t \geq 0$, $\boldsymbol{\mathcal{P}}_s\boldsymbol{\mathcal{P}}_t\mathbf{p} = \boldsymbol{\mathcal{P}}_{s+t}\mathbf{p}$;*
3. *For all $(y_o,\Pi) \in \mathcal{D}$, the map $t \to \boldsymbol{\mathcal{S}}_t(y_o,\Pi)$ solves*

$$\begin{cases} \partial_t y + \partial_x f(y) = 0 \\ \Psi\left(y(0+,t)\right) = \Pi(t) \\ y(0,x) = y_o(x) \end{cases} \tag{3.12}$$

   according to Definition 3.6;

4. *For $\mathbf{p}, \tilde{\mathbf{p}} \in \mathcal{D}$ and $t, \tilde{t} \geq 0$*

$$\begin{aligned} \|\boldsymbol{\mathcal{S}}_t\mathbf{p} - \boldsymbol{\mathcal{S}}_t\tilde{\mathbf{p}}\|_{\mathbf{L}^1(\mathbb{R}^+)} &\leq L \cdot \left(\|y - \tilde{y}\|_{\mathbf{L}^1(\mathbb{R}^+)} + \left\|\Pi - \tilde{\Pi}\right\|_{\mathbf{L}^1([0,t])}\right) \\ \|\boldsymbol{\mathcal{S}}_t\mathbf{p} - \boldsymbol{\mathcal{S}}_{\tilde{t}}\mathbf{p}\|_{\mathbf{L}^1(\mathbb{R}^+)} &\leq L \cdot |t - \tilde{t}|\,. \end{aligned}$$

   Furthermore, if $\tilde{\Psi}$ fulfills (3.7), then for all $\mathbf{p} \in \mathcal{D}$,

$$\left\|\boldsymbol{\mathcal{S}}_t^{\tilde{\Psi}}\mathbf{p} - \boldsymbol{\mathcal{S}}_t^{\Psi}\mathbf{p}\right\|_{\mathbf{L}^1} \leq L \cdot \left\|\tilde{\Psi} - \Psi\right\|_{\mathbf{C}^1} \cdot t. \tag{3.13}$$

5. *If* $\mathbf{p} \in \mathcal{D}$ *is piecewise constant then, for* $t > 0$ *sufficiently small,* $\boldsymbol{\mathcal{S}}_t\mathbf{p}$ *coincides with the juxtaposition of the solutions to Riemann problems centered at the points of jumps or at the junction.*
6. *For all* $\mathbf{p} \in \mathcal{D}$*, the map* $t \mapsto \Upsilon(\boldsymbol{\mathcal{P}}_t\mathbf{p})$ *is non increasing.*
7. *for all* $\mathbf{p}, \tilde{\mathbf{p}} \in \mathcal{D}$*, the map* $t \mapsto \Phi(\boldsymbol{\mathcal{P}}_t\mathbf{p}, \boldsymbol{\mathcal{P}}_t\tilde{\mathbf{p}})$ *is non increasing.*
8. *There exist constants* C*,* $\eta > 0$ *such that for all* $t > 0$*, for all* $\mathbf{p}, \tilde{\mathbf{p}} \in \mathcal{D}$ *and* v *with* $v(t,.) \in \mathbf{L}^1(\mathbb{R}^+;\Omega)$ *and* $\mathrm{TV}(v(t,.)) < \eta$ *for all* t

$$\|\boldsymbol{\mathcal{S}}_t\mathbf{p} - \boldsymbol{\mathcal{S}}_t\tilde{\mathbf{p}} - v(t,.)\|_{\mathbf{L}^1(\mathbb{R}^+)} \leq L\left(\|y - \tilde{y} - v(s,.)\|_{\mathbf{L}^1(\mathbb{R}^+)} + \left\|\Pi - \tilde{\Pi}\right\|_{\mathbf{L}^1([0,t])}\right) + C \cdot t \cdot \mathrm{TV}(v(s,.)) .$$

As mentioned before, the Cauchy problem at the junction can also be seen as an initial-boundary value problem of $2n$ conservation laws on $\mathbb{R}^+$ with a nonlinear boundary condition at $x = 0$. This is very convenient, since there are already several basic results in the existing literature, which can be applied also in our case, see for instance [1, 2, 25, 36, 41].

3.3.2. The source term. In the previous section, we showed the existence of a semigroup of solutions to the homogeneous Cauchy problem (3.12) for initial data with sufficiently small total variation. Now, we have to include the source term, to prove the existence of solutions to the inhomogeneous Cauchy problem (3.11). To this end, we use an operator splitting technique, presented, e.g., in [23] in the framework of differential equations in metric spaces. The following theorem states the existence of solutions to the inhomogeneous Cauchy problem.

Theorem 3.8. *Let* $n \in \mathbb{N}$*,* $n \geq 2$ *and assume that* f *satisfies* **(F)** *at* $\bar{u}$ *and* G *satisfies* **(G)***. Fix a map* $\Psi \in \mathbf{C}^1(\Omega;\mathbb{R}^n)$ *that satisfies condition* (3.7)*.*
Then, there exist positive δ*,* δ'*,* L*,* T*, domains* $\mathcal{D}_t$*, for* $t \in [0,T]$*, and a map*

$$\begin{aligned} \boldsymbol{\mathcal{E}} \colon \ \{(\tau, t_o, \mathbf{p}) \colon t_o \in [0,T[\,, \tau \in [0, T - t_o], \mathbf{p} \in \mathcal{D}_{t_o}\} &\to \mathcal{D}^\delta \\ (\tau, t_o, \mathbf{p}) &\mapsto \boldsymbol{\mathcal{E}}(\tau, t_o)\mathbf{p} \end{aligned}$$

such that:
1. $\overline{\mathcal{D}^{\delta'}} \subseteq \mathcal{D}_t \subseteq \overline{\mathcal{D}^{\delta}}$ *for all* $t \in [0,T]$*;*
2. *for all* $t_o \in [0,T]$ *and* $\mathbf{p} \in \mathcal{D}_{t_o}$*,* $\boldsymbol{\mathcal{E}}(0,t_o)\mathbf{p} = \mathbf{p}$*;*
3. *for all* $t_o \in [0,T]$ *and* $\tau \in [0, T - t_o]$*,* $\boldsymbol{\mathcal{E}}(\tau, t_o)\mathcal{D}_{t_o} \subseteq \mathcal{D}_{t_o+\tau}$*;*
4. *for all* $t_o \in [0,T]$*,* $\tau_1, \tau_2 \geq 0$ *with* $\tau_1 + \tau_2 \in [0, T - t_o]$*,*
$$\boldsymbol{\mathcal{E}}(\tau_2, t_o + \tau_1) \circ \boldsymbol{\mathcal{E}}(\tau_1, t_o) = \boldsymbol{\mathcal{E}}(\tau_2 + \tau_1, t_o);$$
5. *for all* $(y_o, \Pi) \in \mathcal{D}_{t_o}$*, we have* $\boldsymbol{\mathcal{E}}(t,t_o)(y_o,\Pi) = (y(t), \boldsymbol{\mathcal{T}}_t\Pi)$ *where* $t \mapsto y(t)$ *is the entropy solution to the Cauchy Problem* (3.11) *according to Definition* 3.6 *while* $t \mapsto \boldsymbol{\mathcal{T}}_t\Pi$ *is the right translation;*
6. $\boldsymbol{\mathcal{E}}$ *is tangent to Euler polygonal, in the sense that for all* $t_o \in [0,T]$*, for all* $(y_o, \Pi) \in \mathcal{D}_{t_o}$*, setting* $\boldsymbol{\mathcal{E}}(t,t_o)(y_o,\Pi) = (y(t), \boldsymbol{\mathcal{T}}_t\Pi)$*,*
$$\lim_{t \to 0} (1/t) \|y(t) - (\boldsymbol{\mathcal{S}}_t(y_o, \Pi) + tG(t_o, y_o))\|_{\mathbf{L}^1} = 0$$
where $\boldsymbol{\mathcal{S}}$ *is the semigroup defined in Theorem* 3.7*;*

7. *for all* $t_o \in [0,T]$, $\tau \in [0, T-t_o]$ *and for all* $\mathbf{p}, \tilde{\mathbf{p}} \in \mathcal{D}_{t_o}$,

$$\|\boldsymbol{\mathcal{E}}(\tau,t_o)\mathbf{p} - \boldsymbol{\mathcal{E}}(\tau,t_o)\tilde{\mathbf{p}}\|_{\mathbf{L}^1} \leq L \cdot \|y - \tilde{y}\|_{\mathbf{L}^1} + L \cdot \int_{t_o}^{t_o+\tau} \left\|\tilde{\Pi}(t) - \Pi(t)\right\| dt. \tag{3.14}$$

8. *call* $\tilde{\boldsymbol{\mathcal{E}}}$ *the analogous map generated by* $\tilde{\Psi}$, *then for all* $t_o \in [0,T]$, $\tau \in [0, T-t_o]$ *and for all* $\mathbf{p} \in \mathcal{D}_{t_o}$,

$$\left\|\tilde{\boldsymbol{\mathcal{E}}}(\tau,t_o)\mathbf{p} - \boldsymbol{\mathcal{E}}(\tau,t_o)\mathbf{p}\right\|_{\mathbf{L}^1} \leq L \cdot \left\|\tilde{\Psi} - \Psi\right\|_{\mathbf{C}^1} \cdot \tau.$$

The proof of this theorem is based on the theoretical framework of [23], which relies on an operator splitting technique in metric spaces. Note that X together with the $\mathbf{L}^1$-distance defined in (3.5) defines a metric space. In this metric space, we define the map

$$F_{t,t_o}(y,\Pi) = \left(\boldsymbol{\mathcal{S}}_t\,(y,\Pi) + tG\left(t_o, \boldsymbol{\mathcal{S}}_t\,(y,\Pi)\right), \boldsymbol{\mathcal{T}}_t\Pi\right),$$

which approximates the solution of the inhomogeneous Cauchy problem in the interval $[t_o, t_o + t]$. However, we use as construction in the proof a wave or front tracking algorithm. We therefore need a time-discrete version of the previous functional. To this end we divide a time step t in further fixed time-steps of length ϵ and approximate thereby F_{t,t_o}. The operator F is then the composition of these operators, see [23]. The combination of these operators are called Euler-ε-polygonal on the interval $[t_o, t_o + t]$ and are defined by

$$F^{\varepsilon}\left(t,t_o\right)\mathbf{p} = F_{t-k\varepsilon,t_o+k\varepsilon} \circ \bigcirc_{h=0}^{k} F_{\varepsilon,t_o+h\varepsilon}\mathbf{p},$$

where $k = \max\left\{h \in \mathbb{N} : h\varepsilon < t_o + t\right\}$ and $\bigcirc_{h=0}^{k} F_{\varepsilon,t_o+h\varepsilon}\mathbf{p} = F_{\varepsilon,t_o+k\varepsilon} \circ \ldots \circ F_{\varepsilon,t_o}\mathbf{p}$. Essentially, this means that we approximate the solution to the inhomogeneous Cauchy problem by the solution to the homogeneous Cauchy problem, corrected by an Euler-approximation of the source term after every time step of size ε. This mimics a first-order Euler scheme for the numerical solution. Several estimates on F^{ε} and the influence of the source term on the solution assure than the existence of the limit semigroup $\boldsymbol{\mathcal{E}}$.

3.3.3. Optimal control of networked hyperbolic balance laws. We show now a result, proving the existence of optimal controls for a small network of two pipes and one compressor. Herby, we use the existence result of Theorem 3.8 for a general class of 2×2 balance laws. Contrary to [40] where the one-dimensional isentropic Euler equations are controlled to a given final state $y(T,x)$ by boundary controls, we allow only one single control function acting at the junction and influencing the state variables indirectly through a (possibly nonlinear) condition. Furthermore, the demand is not given by a final state, but by a cost functional involving the control function as well as parts of the solution of the state system. For a rigorous definition of the cost functional see Proposition 3.9.

To prove the existence of optimal controls, we recall first 7. in Theorem 3.8. Let y_o and $\tilde{y}_o$ be two suitable initial states and Π, $\tilde{\Pi}$ be two control functions with sufficiently small total variation. We denote by $\mathcal{E}(\tau,0)(y_o,\Pi)$ and $\mathcal{E}(\tau,0)(\tilde{y}_o,\tilde{\Pi})$ for $\tau \in [0,T]$ the solutions of the state system with initial condition y_o and $\tilde{y}_o$ and control functions Π and $\tilde{\Pi}$ respectively. Then, the following estimate holds true:

$$\|\mathcal{E}(\tau,0)(y_o,\Pi) - \mathcal{E}(\tau,0)(\tilde{y}_o,\tilde{\Pi})\| \le L \cdot \|y_o - \tilde{y}_o\|_{\mathbf{L}^1} + L \cdot \int_0^\tau \|\Pi(t) - \tilde{\Pi}(t)\| dt.$$

Recall further the following definitions, where $\bar{y}$ and $\bar{\Pi}$ are constant states.

$$\begin{aligned} X &= \left(\bar{y} + \mathbf{L}^1\left(\mathbb{R}^+;\Omega\right)\right) \times \left(\bar{\Pi} + \mathbf{L}^1\left(\mathbb{R}^+;\mathbb{R}^n\right)\right) \\ \mathrm{TV}(y,\Pi) &= \mathrm{TV}(y) + \mathrm{TV}(\Pi) + \|\Psi(y(0+)) - \Pi(0+)\| \\ \mathcal{D}^\delta &= \{(y,\Pi) \in X : \mathrm{TV}(y,\Pi) \le \delta\} \\ \mathcal{D}_0 &\subseteq \mathcal{D}^\delta \end{aligned}$$

From the above estimate (7. in Theorem 3.8), we immediately obtain the following existence result for an optimal control function Π to the nonlinear constrained optimization problem.

Proposition 3.9. *Let $n \in \mathbb{N}$, $n \ge 2$. Assume that f satisfies* **(F)** *at $\bar{y}$ and G satisfies* **(G)**. *Fix a map $\Psi \in \mathbf{C}^1(\Omega;\mathbb{R}^n)$ satisfying* (3.7) *and let $\bar{\Pi} = \Psi(\bar{y})$. With the notation in Section* 3.3 *and especially in Theorem* 3.8, *for a fixed $y_o \in \mathcal{Y}_\delta$, assume that*

$$\begin{aligned} &J_o\colon \left\{\Pi_{|[0,T]}\colon \Pi \in \left(\bar{\Pi} + \mathbf{L}^1(\mathbb{R}^+;\mathbb{R}^n)\right) \text{ and } (y_o,\Pi) \in \mathcal{D}^\delta\right\} \to \mathbb{R} \\ &J_1\colon \mathcal{D}^\delta \to \mathbb{R} \end{aligned}$$

are non negative and lower semicontinuous with respect to the $\mathbf{L}^1$ norm. Then, the cost functional

$$\mathcal{J}(\Pi) = J_o(\Pi) + \int_0^T J_1\left(\boldsymbol{\mathcal{E}}(\tau,0)(y_o,\Pi)\right) d\tau \tag{3.15}$$

admits a minimum on $\left\{\Pi_{|[0,T]}\colon \Pi \in \left(\bar{\Pi} + \mathbf{L}^1(\mathbb{R}^+;\mathbb{R}^n)\right) \text{ and } (y_o,\Pi) \in \mathcal{D}_0\right\}$.

Acknowledgment

This work has been supported by DFG LE 595/22-1, KL1105/16-1, HE5386/6-1, HE5386/7-1 and DAAD D/08/11076.

References

[1] D. Amadori. Initial-boundary value problems for nonlinear systems of conservation laws. *NoDEA Nonlinear Differential Equations Appl.*, 4(1):1–42, 1997.

[2] D. Amadori and R.M. Colombo. Continuous dependence for 2 × 2 conservation laws with boundary. *J. Differential Equations*, 138(2):229–266, 1997.

[3] D. Amadori and G. Guerra. Uniqueness and continuous dependence for systems of balance laws with dissipation. *Nonlinear Anal.*, 49(7, Ser. A: Theory Methods):987–1014, 2002.

[4] F. Ancona and G.M. Coclite. On the attainable set for Temple class systems with boundary controls. *SIAM J. Control Optim.*, 43(6):2166–2190 (electronic), 2005.

[5] F. Ancona and A. Marson. On the attainable set for scalar nonlinear conservation laws with boundary control. *SIAM J. Control Optim.*, 36:290–312, 1998.

[6] F. Ancona and A. Marson. Asymptotic stabilization of systems of conservation laws by controls acting at a single boundary point. *Control Methods in PDE-Dynamical Systems (Amer. Math. Soc. Providence)*, 426, 2007.

[7] D. Armbruster, P. Degond, and C. Ringhofer. A model for the dynamics of large queuing networks and supply chains. *SIAM J. Appl. Math.*, 66:896–920, 2006.

[8] A. Aw and M. Rascle. Resurrection of second-order models of traffic flow? *SIAM J. Appl. Math.*, 60:916–944, 2000.

[9] M.K. Banda, M. Herty, and A. Klar. Coupling conditions for gas networks governed by the isothermal Euler equations. *Networks and Heterogeneous Media*, 1(2):275–294, 2006.

[10] M.K. Banda, M. Herty, and A. Klar. Gas flow in pipeline networks. *Networks and Heterogeneous Media*, 1(1):41–56, 2006.

[11] S. Bolch, G.and Greiner, H. de Meer, and K. Trivedi. *Queuing Networks and Markov Chains.* John Wiley & Sons, New York, 1998.

[12] A. Bressan. *Hyperbolic systems of conservation laws: The one-dimensional Cauchy problem*, volume 20 of *Oxford Lecture Series in Mathematics and its Applications.* Oxford University Press, Oxford, 2000.

[13] A. Bressan and G.M. Coclite. On the boundary control of systems of conservation laws. *SIAM J. Control Optim.*, 41:607–622, 2002.

[14] A. Bressan and A. Marson. A maximum principle for optimally controlled systems of conservation laws. *Rend. Sem. Mat. Univ. Padova*, 94:79–94, 1995.

[15] I. Cameron. Using an excel-based model for steady-state and transient simulation. Technical report, TransCanada Transmission Company, 2000.

[16] C. Castro, F. Palacios, and E. Zuazua. An alternating descent method for the optimal control of the inviscid Burgers equation in the presence of shocks. *Math. Models Methods Appl. Sci.*, 18:369–416, 2008.

[17] K.S. Chapman and M. Abbaspour. Non-isothermal Compressor Station Transient Modeling. Technical report, PSIG, 2003.

[18] N.H. Chen. An explicit equation for friction factor in pipe. *Ind. Eng. Chem. Fund.*, 18(3):296–297, 1979.

[19] G. Coclite, M. Garavello, and B. Piccoli. Traffic flow on a road network. *SIAM J. Math. Anal.*, 36(6):1862–1886, 2005.

[20] R. Colombo and A. Groli. Minimising stop and go waves to optimise traffic flow. *Appl. Math. Lett.*, 17(6):697–701, 2004.

[21] R.M. Colombo and M. Garavello. A well-posed Riemann problem for the p-system at a junction. *Networks and Heterogeneous Media*, 1:495–511, 2006.

[22] R.M. Colombo and M. Garavello. On the Cauchy problem for the p-system at a junction. *SIAM J. Math. Anal.*, 39:1456–1471, 2008.

[23] R.M. Colombo and G. Guerra. Differential equations in metric spaces with applications. *Discrete and Continuous Dynamical Systems*, 2007. To appear. Preprint http://arxiv.org/abs/0712.0560.

[24] R.M. Colombo and G. Guerra. Hyperbolic balance laws with a non local source. *Communications in Partial Differential Equations*, 32(12), 2007.

[25] R.M. Colombo and G. Guerra. On general balance laws with boundary. Preprint, 2008.

[26] R.M. Colombo, G. Guerra, M. Herty, and V. Schleper. Optimal control in networks of pipes and canals. *SIAM Journal on Control and Optimization*, 48(3):2032–2050, 2009.

[27] R.M. Colombo, M. Herty, and V. Sachers. On 2×2 conservation laws at a junction. *SIAM J. Math. Anal.*, 40(2):605–622, 2008.

[28] R.M. Colombo and F. Marcellini. Smooth and discontinuous junctions in the p-system. to appear in J. Math. Anal. Appl., 2009.

[29] G. Crasta and B. Piccoli. Viscosity solutions and uniqueness for systems of inhomogeneous balance laws. *Discrete Contin. Dyn. Syst.*, 3(4):477–502, 1997.

[30] C.M. Dafermos. *Hyperbolic Conservation Laws in Continuum Physics*, volume 325 of *A Series of Comprehensive Studies in Mathematics*. Springer Verlag, Berlin, Heidelberg, New York, 2005. Second edition.

[31] C. D'Apice and R. Manzo. Calculation of predicted average packet delay and its application for flow control in data network. *J. Inf. Optimization Sci.*, 27:411–423, 2006.

[32] C. D'Apice, R. Manzo, and B. Piccoli. Packet flow on telecommunication networks. *SIAM J. Math. Anal.*, 38(3):717–740, 2006.

[33] J. de Halleux, C. Prieur, J.-M. Coron, B. d'Andréa Novel, and G. Bastin. Boundary feedback control in networks of open channels. *Automatica J. IFAC*, 39(8):1365–1376, 2003.

[34] J. de Halleux, C. Prieur, J.-M. Coron, B. d'Andréa Novel, and G. Bastin. Boundary feedback control in networks of open channels. *Automatica J. IFAC*, 39(8):1365–1376, 2003.

[35] R.J. DiPerna. Global existence of solutions to nonlinear hyperbolic systems of conservation laws. *J. Differ. Equations*, 20:187–212, 1976.

[36] C. Donadello and A. Marson. Stability of front tracking solutions to the initial and boundary value problem for systems of conservation laws. *NoDEA Nonlinear Differential Equations Appl.*, 14(5-6):569–592, 2007.

[37] K. Ehrhardt and M. Steinbach. Nonlinear gas optimization in gas networks. In H. Bock, E. Kostina, H. Pu, and R. Rannacher, editors, *Modeling, Simulation and Optimization of Complex Processes*. Springer Verlag, Heidelberg, 2005.

[38] I.R. Ellul, G. Saether, and M.E. Shippen. The modeling of multiphase systems under steady-state and transient conditions a tutorial. Technical report, PSIG, 2004.

[39] M. Garavello and B. Piccoli. Traffic flow on a road network using the aw-rascle model. *Commun. Partial Differ. Equations*, 31(1-3):243–275, 2006.

[40] O. Glass. On the controllability of the 1-D isentropic Euler equation. *J. Eur. Math. Soc. (JEMS)*, 9(3):427–486, 2007.

[41] J. Goodman. *Initial Boundary Value Problems for Hyperbolic Systems of Conservation Laws*. PhD thesis, California University, 1982.

[42] S. Göttlich, M. Herty, and A. Klar. Modelling and optimization of supply chains on complex networks. *Communications in Mathematical Sciences*, 4(2):315–330, 2006.

[43] G. Guerra, F. Marcellini, and V. Schleper. Balance laws with integrable unbounded sources. *SIAM Journal of Mathematical Analysis*, 41(3), 2009.

[44] M. Gugat. Nodal control of conservation laws on networks. Sensitivity calculations for the control of systems of conservation laws with source terms on networks. Cagnol, John (ed.) et al., Control and boundary analysis. Selected papers based on the presentation at the 21st conference on system modeling and optimization, Sophia Antipolis, France, July 21–25, 2003. Boca Raton, FL: Chapman & Hall/CRC. Lecture Notes in Pure and Applied Mathematics 240, 201–215 (2005), 2005.

[45] M. Gugat and G. Leugering. Global boundary controllability of the de St. Venant equations between steady states. *Annales de l'Institut Henri Poincaré, Nonlinear Analysis*, 20(1):1–11, 2003.

[46] M. Gugat and G. Leugering. Global boundary controllability of the saint-venant system for sloped canals with friction. *Annales de l'Institut Henri Poincaré, Nonlinear Analysis*, 26(1):257–270, 2009.

[47] M. Gugat, G. Leugering, and E.J.P. Georg Schmidt. Global controllability between steady supercritical flows in channel networks. *Math. Methods Appl. Sci.*, 27(7):781–802, 2004.

[48] M. Gugat, G. Leugering, K. Schittkowski, and E.J.P.G. Schmidt. Modelling, stabilization, and control of flow in networks of open channels. In *Online optimization of large scale systems*, pages 251–270. Springer, Berlin, 2001.

[49] M. Herty. Modeling, simulation and optimization of gas networks with compressors. *Networks and Heterogenous Media*, 2:81, 2007.

[50] M. Herty and M. Rascle. Coupling conditions for a class of second-order models for traffic flow. *SIAM Journal on Mathematical Analysis*, 38, 2006.

[51] H. Holden and N.H. Risebro. Riemann problems with a kink. *SIAM J. Math. Anal.*, 30:497–515, 1999.

[52] C. Kelling, K. Reith, and E. Sekirnjak. A Practical Approach to Transient Optimization for Gas Networks. Technical report, PSIG, 2000.

[53] J. Králink, P. Stiegler, Z. Vostrý, and J. Závorka. Modelle für die Dynamik von Rohrleitungsnetzen – theoretischer Hintergrund. *Gas-Wasser-Abwasser*, 64(4):187–193, 1984.

[54] C.D. Laird, L.T. Biegler, B.G. van Bloemen Waanders, and R.A. Barlett. Contamination source determination for water networks. *A.S.C.E. Journal of Water Resources Planning and Management*, 131(2):125–134, 2005.

[55] P.D. Lax. Hyperbolic systems of conservation laws. {II}. *Comm. Pure Appl. Math.*, 10:537–566, 1957.

[56] G. Leugering and E.J.P.G. Schmidt. On the modeling and stabilization of flows in networks of open canals. *SIAM J. Control Opt.*, 41(1):164–180, 2002.

[57] R.J. LeVeque. *Numerical Methods for Conservation Laws*. Lectures in Mathematics ETH Zürich. Birkhäuser Verlag, 2006. Third edition.

[58] T.-T. Li. Exact controllability for quasilinear hyperbolic systems and its application to unsteady flows in a network of open canals. *Mathematical Methods in the Applied Sciences*, 27:1089–1114, 2004.

[59] T.-T. Li and B.-P. Rao. Exact boundary controllability for quasi-linear hyperbolic systems. *SIAM J. Control Optim.*, 41(6):1748–1755, 2003.

[60] T.-T. Li and B.-P. Rao. Exact boundary controllability of unsteady flows in a tree-like network of open canals. *Methods Appl. Anal.*, 11(3):353–365, 2004.

[61] M.J. Lighthill and G.B. Whitham. On kinematic waves. *Proc. Royal Soc. Edinburgh*, 229:281–316, 1955.

[62] A. Martin, M. Möller, and S. Moritz. Mixed integer models for the stationary case of gas network optimization. *Math. Programming.*

[63] A. Osiadacz. Optimization of high pressure gas networks using hierarchical systems theory. Technical report, Warsaw University, 2002.

[64] A.J. Osiadacz. *Simulation and analysis of gas networks.* Gulf Publishing Company, Houston, 1989.

[65] A.J. Osiadacz. Different transient models – limitations, advantages and disadvantages. Technical report, PSIG, 1996.

[66] Pipeline Simulation Interest Group. www.psig.org.

[67] M. Steinbach. On PDE solution in transient optimization of gas networks. *J. Comput. Appl. Math.*, 203(2):345–361, 2007.

[68] S. Ulbrich. Adjoint-based derivative computations for the optimal control of discontinuous solutions of hyperbolic conservation laws. *Syst. Control Lett,*, 48(3–4):313–328, 2003.

[69] J. Zhou and M.A. Adewumi. Simulation of transients in natural gas pipelines using hybrid tvd schemes. *Int. J. Numer. Meth. Fluids*, 32:407–437, 2000.

Michael Gugat and Günther Leugering
Department of Mathematics, University of Erlangen–Nürnberg
D-91058 Erlangen, Germany
e-mail: gugat@am.uni-erlangen.de
leugering@am.uni-erlangen.de

Michael Herty
Department of Mathematics, RWTH Aachen University
D-52056 Aachen, Germany
e-mail: herty@mathc.rwth-aachen.de

Axel Klar
Department of Mathematics, TU Kaiserslautern
D-67663 Kaiserslautern, Germany
e-mail: klar@mathematik.uni-kl.de

Veronika Schleper
Department of Mathematics, University of Stuttgart
D-70569 Stuttgart, Germany
e-mail: veronika.schleper@mathematik.uni-stuttgart.de

International Series of Numerical Mathematics, Vol. 160, 147–170

A Space-Time Multigrid Method for Optimal Flow Control

Michael Hinze, Michael Köster and Stefan Turek

Abstract. We present a hierarchical solution concept for optimization problems governed by the time-dependent Navier–Stokes system. Discretisation is carried out with finite elements in space and a one-step-θ-scheme in time. By combining a Newton solver for the treatment of the nonlinearity with a space-time multigrid solver for linear subproblems, we obtain a robust solver whose convergence behaviour is independent of the refinement level of the discrete problem. A set of numerical examples analyses the solver behaviour for various problem settings with respect to efficiency and robustness of this approach.

Mathematics Subject Classification (2000). 35Q30, 49K20, 49M05, 49M15, 49M29, 65M55, 65M60, 76D05, 76D55.

Keywords. Distributed control, finite elements, time-dependent Navier–Stokes, Newton, space-time multigrid, optimal control.

1. Introduction

Active flow control plays a central role in many practical applications such as, e.g., control of crystal growth processes [9, 16, 15, 17], where the flow in the melt has a significant impact on the quality of the crystal. Optimal control of the flow by electro-magnetic fields and/or boundary temperatures leads to optimisation problems with PDE constraints, which are frequently governed by the time-dependent Navier–Stokes system.

The mathematical formulation is a minimisation problem with PDE constraints. By exploiting the special structure of the first-order necessary optimality conditions, the so-called Karush-Kuhn-Tucker (KKT)-system, we are able to develop a hierarchical solution approach for the optimisation of the Stokes and Navier–Stokes equations which satisfies

$$\frac{\text{effort for optimisation}}{\text{effort for simulation}} \leq C, \tag{1}$$

for a constant $C > 0$ of moderate size, independent of the refinement level. Tests show a factor of $20 - 30$ for the Navier–Stokes equations. Here, the effort for the simulation is assumed to be optimal in that sense, that a solver needs $O(N)$ operations, N denoting the total number of unknowns for a given computational mesh in space and time. This can be achieved by utilising appropriate multigrid techniques for the linear subproblems in space. Because of (1), the developed solution approach for the optimal control problem has also complexity $O(N)$; this is achieved by a combination of a space-time Newton approach for the nonlinearity and a space-time multigrid approach for linear subproblems. The complexity of this algorithm distinguishes our method from adjoint-based steepest descent methods used to solve optimisation problems in many practical applications, which in general do not satisfy this complexity requirement. A related approach can be found, e.g., in [4] where multigrid methods for the numerical solution of optimal control problems for parabolic PDEs are developed based on Finite Difference techniques for the discretisation. In [6] a space-time multigrid method for the corresponding *integral equation approach* of [10] is developed, compare also [8].

The paper is organised as follows: In Section 2 we describe the discretisation of a flow control problem and give an introduction to the ingredients needed to design a multigrid solver. The discretisation is carried out with finite elements in space and finite differences in time. In Section 3 we propose the basic algorithms that are necessary to construct our multigrid solver for linear and a Newton solver for nonlinear problems. Finally, Section 4 is devoted to numerical examples which we present to confirm the predicted behaviour.

2. Problem formulation and discretisation

We consider the optimal control problem

$$J(y,u) := \frac{1}{2}||y-z||^2_{L^2(Q)} + \frac{\alpha}{2}||u||^2_{L^2(Q)} + \frac{\gamma}{2}||y(T)-z(T)||^2_{L^2(\Omega)} \quad \longrightarrow \quad \text{min!} \tag{2}$$

$$\text{s.t.} \qquad \begin{aligned} y_t - \nu\Delta y + y\nabla y + \nabla p &= u && \text{in } Q,\\ -\text{div}\; y &= 0 && \text{in } Q,\\ y(0,\cdot) &= y^0 && \text{in } \Omega,\\ y &= g && \text{at } \Sigma. \end{aligned}$$

Here, $\Omega \subset \mathbb{R}^d$ $(d = 2, 3)$ denotes an open bounded domain, $\Gamma := \partial\Omega$, $T > 0$ defines the time horizon, and $Q = (0,T) \times \Omega$ denotes the corresponding space-time cylinder with space-time boundary $\Sigma := (0,T) \times \Gamma$. The function $g : \Sigma \to \mathbb{R}^d$ specifies some Dirichlet boundary conditions, u denotes the control, y the velocity vector, p the pressure and $z : Q \to \mathbb{R}^d$ a given target velocity field for y. Finally, $\gamma \geq 0$, $\alpha > 0$ denote constants. For simplicity, we do not assume any restrictions on the control u.

The first-order necessary optimality conditions are then given through the so-called *Karush-Kuhn-Tucker* system

$$
\begin{array}{rcll}
y_t - \nu\Delta y + y\nabla y + \nabla p &=& u & \text{in } Q\\
-\text{div } y &=& 0 & \text{in } Q\\
y &=& g & \text{at } \Sigma\\
y(0,\cdot) &=& y^0 & \text{in } \Omega\\[1ex]
-\lambda_t - \nu\Delta\lambda - y\nabla\lambda + (\nabla y)^t\lambda + \nabla\xi &=& y - z & \text{in } Q\\
-\text{div } \lambda &=& 0 & \text{in } Q\\
\lambda &=& 0 & \text{at } \Sigma\\
\lambda(T) &=& \gamma(y(T) - z(T)) & \text{in } \Omega\\[2ex]
u &=& -\frac{1}{\alpha}\lambda & \text{in } Q,
\end{array}
$$

where λ denotes the dual velocity and ξ the dual pressure. We eliminate u in the KKT system, and (ignoring boundary conditions at the moment), we obtain

$$
\begin{array}{rcl}
y_t - \nu\Delta y + y\nabla y + \nabla p &=& -\dfrac{1}{\alpha}\lambda,\\
-\text{div } y &=& 0,\\
y(0,\cdot) &=& y_0,
\end{array}
\tag{3}
$$

$$
\begin{array}{rcl}
-\lambda_t - \nu\Delta\lambda - y\nabla\lambda + (\nabla y)^t\lambda + \nabla\xi &=& y - z,\\
-\text{div } \lambda &=& 0,\\
\lambda(T) &=& \gamma(y(T) - z(T))
\end{array}
\tag{4}
$$

where we call (3) the *primal* and (4) the *dual* equation.

Coupled discretisation in time. For the discretisation of the system, we follow the approach *first optimise, then discretise*: We semi-discretise the KKT-system in time. For stability reasons (cf. [21]) we prefer implicit time stepping techniques. This allows us to be able to choose the timestep size only depending on the accuracy demands, independent of any stability constraints. The approach is demonstrated on the standard 1st order backward Euler scheme as a representative of implicit schemes. Higher-order schemes like Crank-Nicolson are possible as well but lead to a more complicated matrix structure (with a couple of nonlinear terms also on the offdiagonals) and, depending on the time discretisation, some less straightforward time prolongation/restriction.

The time discretisation of (3) yields

$$
\begin{array}{rcl}
\dfrac{y_{k+1} - y_k}{\Delta t} - \nu\Delta y_{k+1} + y_{k+1}\nabla y_{k+1} + \nabla p_{k+1} &=& -\dfrac{1}{\alpha}\lambda_{k+1}\\
-\text{div } y_{k+1} &=& 0\\
y_0 &=& y^0
\end{array}
\tag{5}
$$

where $N \in \mathbb{N}$, $k = 0, \ldots, N-1$ and $\Delta t = 1/N$. To (3), (4) we apply the discretisation recipe from [3]. For this purpose, we define the following operators:

$\mathcal{A}v := -\nu\Delta v$, $\mathcal{I}v := v$, $\mathcal{G}q := \nabla q$, $\mathcal{D}v := -\text{div } v$, $\mathcal{K}_n v := \mathcal{K}(y_n)v := (y_n\nabla)v$, $\overline{\mathcal{K}}_n v := \overline{\mathcal{K}}(y_n)v := (v\nabla)y_n$ and $\mathcal{C}_n v := \mathcal{C}(y_n)v := \mathcal{A}v + \mathcal{K}(y_n)v$ for all velocity vectors v and and pressure functions q in space, $n \in \mathbb{N}$.

As the initial solution y^0 may not be solenoidal, we realise the initial condition $y_0 = y^0$ by the following solenoidal projection,

$$\begin{aligned} \frac{1}{\Delta t} y_0 - \nu\Delta y_0 + y_0\nabla y_0 + \nabla p_0 &= \frac{1}{\Delta t} y^0 - \nu\Delta y^0 + y^0\nabla y^0, \\ -\text{div } y_0 &= 0. \end{aligned}$$

This projection operator uses the same operations like the backward Euler scheme which allows for a more consistent notation. Using $x := (y_0, p_0, y_1, p_1, \ldots, y_N, p_N)$, this yields the nonlinear system of the primal equation,

$$\begin{aligned} \mathcal{H}x &:= \mathcal{H}(x)x \\ &= \left(\begin{array}{c|c|c|c} \begin{matrix} \frac{\mathcal{I}}{\Delta t} + \mathcal{C}_0 & \mathcal{G} \\ \mathcal{D} & \end{matrix} & & & \\ \hline -\frac{\mathcal{I}}{\Delta t} & \begin{matrix} \frac{\mathcal{I}}{\Delta t} + \mathcal{C}_1 & \mathcal{G} \\ \mathcal{D} & \end{matrix} & & \\ \hline & \ddots & \ddots \quad \ddots & \\ \hline & & -\frac{\mathcal{I}}{\Delta t} & \begin{matrix} \frac{\mathcal{I}}{\Delta t} + \mathcal{C}_N & \mathcal{G} \\ \mathcal{D} & \end{matrix} \end{array}\right) \begin{pmatrix} y_0 \\ p_0 \\ y_1 \\ p_1 \\ \vdots \\ y_N \\ p_N \end{pmatrix} \\ &= \left(\left(\frac{\mathcal{I}}{\Delta t} + \mathcal{C}_0\right) y^0, 0, -\frac{\lambda_1}{\alpha}, 0, \ldots, -\frac{\lambda_N}{\alpha}, 0, \right) \end{aligned}$$

which is equivalent to (5) if y^0 is solenoidal. In the second step, we focus on the Fréchet derivative of the Navier–Stokes equations. For a vector $(\bar{y}, \bar{p})$ the Fréchet derivative in (y, p) reads

$$\mathcal{F}(y,p) \begin{pmatrix} \bar{y} \\ \bar{p} \end{pmatrix} := \begin{pmatrix} \bar{y}_t - \nu\Delta\bar{y} + (\bar{y}\nabla y + y\nabla\bar{y}) + \nabla\bar{p} \\ -\text{div } \bar{y} \end{pmatrix}.$$

We again carry out the time discretisation as above. For vectors $x := (y_0, p_0, y_1, p_1, \ldots, y_N, p_N)$ and $\bar{x} := (\bar{y}_0, \bar{p}_0, \bar{y}_1, \bar{p}_1, \ldots, \bar{y}_N, \bar{p}_N)$ this results in the scheme

$$\begin{aligned} \mathcal{M}\bar{x} &:= \mathcal{M}(x)\bar{x} \\ &= \left(\begin{array}{c|c|c|c} \begin{matrix} \frac{\mathcal{I}}{\Delta t} + \mathcal{N}_0 & \mathcal{G} \\ \mathcal{D} & \end{matrix} & & & \\ \hline -\frac{\mathcal{I}}{\Delta t} & \begin{matrix} \frac{\mathcal{I}}{\Delta t} + \mathcal{N}_1 & \mathcal{G} \\ \mathcal{D} & \end{matrix} & & \\ \hline & \ddots & \ddots \quad \ddots & \\ \hline & & -\frac{\mathcal{I}}{\Delta t} & \begin{matrix} \frac{\mathcal{I}}{\Delta t} + \mathcal{N}_N & \mathcal{G} \\ \mathcal{D} & \end{matrix} \end{array}\right) \begin{pmatrix} \bar{y}_0 \\ \bar{p}_0 \\ \bar{y}_1 \\ \bar{p}_1 \\ \vdots \\ \bar{y}_N \\ \bar{p}_N \end{pmatrix} \end{aligned}$$

with the additional operator $\mathcal{N}_n := \mathcal{N}(y_n) := \mathcal{A} + \mathcal{K}(y_n) + \overline{\mathcal{K}}(y_n)$. The time discretisation of the dual equation corresponding to $\mathcal{H}$ is now defined as the adjoint $\mathcal{M}^*$ of $\mathcal{M}$,

$$(\mathcal{M}\bar{x}, \lambda) = (\bar{x}, \mathcal{M}^*\lambda),$$

where $\lambda := (\lambda_0, \xi_0, \lambda_1, \xi_1, \dots, \lambda_N, \xi_N)$. With $\mathcal{N}_n^* := \mathcal{N}^*(y_n) = \mathcal{A} - \mathcal{K}(y_n) + \overline{\mathcal{K}}^*(y_n)$, $\overline{\mathcal{K}}^*(y_n)v = (\nabla y_n)^t v$ for all velocity vectors v, this reads

$$\mathcal{M}^*\lambda = \mathcal{M}^*(x)\lambda$$
$$= \left(\begin{array}{cc|cc|c|cc} \frac{\mathcal{I}}{\Delta t} + \mathcal{N}_0^* & \mathcal{G} & -\frac{\mathcal{I}}{\Delta t} & & & & \\ \mathcal{D} & & & & & & \\ \hline & & \frac{\mathcal{I}}{\Delta t} + \mathcal{N}_1^* & \mathcal{G} & -\frac{\mathcal{I}}{\Delta t} & & \\ & & \mathcal{D} & & & & \\ \hline & & & & \ddots & \ddots & \ddots \\ \hline & & & & & \frac{\mathcal{I}}{\Delta t} + \mathcal{N}_N^* & \mathcal{G} \\ & & & & & \mathcal{D} & \end{array}\right) \lambda$$
$$= \left(y_0 - z_0, 0,\ y_1 - z_1, 0, \dots, \left(1 + \frac{\gamma}{\Delta t}\right)(y_N - z_N), 0\right)^T$$

where the right-hand side and terminal condition is chosen in such a way that the *optimise-then-discretise* approach we are using here commutes with the *discretise-then-optimise* approach. This corresponds to the time discretisation scheme

$$\begin{aligned} \frac{\lambda_k - \lambda_{k+1}}{\Delta t} - \nu\Delta\lambda_k - y_k\nabla\lambda_k + (\nabla y_k)^t\lambda_k + \nabla\xi_k &= y_k - z_k \\ -\text{div}\ \lambda_k &= 0 \\ \lambda_N &= \gamma(y_N - z_N). \end{aligned} \tag{6}$$

of (4). Here we have used $\mathcal{D}^* = \mathcal{G}$ and $\mathcal{A}^* = \mathcal{A}$. Now let us define $w_n := (y_n, p_n, \lambda_n, \xi_n)$ and

$$w := (w_0, w_1, \dots) := (y_0, \lambda_0, p_0, \xi_0, y_1, \lambda_1, p_1, \xi_1, y_2, \lambda_2, p_2, \xi_2, \dots).$$

After shifting the terms with λ_{k+1} and y_k in (5) and (6) from the right-hand side to the left-hand side and mixing the two matrices stemming from $\mathcal{H}$ and $\mathcal{M}^*$, we obtain a semi-discrete system

$$G(w)w = f. \tag{7}$$

The right-hand side is given by

$$f = \Big(\underbrace{(\mathcal{I}/\Delta t + \mathcal{C}_0)y^0, -z_0, 0, 0}, \underbrace{0, -z_1, 0, 0},$$
$$\dots, \underbrace{0, -z_{N-1}, 0, 0}, \underbrace{0, -(1 + \gamma/\Delta t)z_N, 0, 0}\Big)$$

and the operator reads

$$G = G(w) = \left(\begin{array}{c|c|c|c|c} \mathcal{G}_0 & \hat{\mathcal{I}}_0 & & & \\ \hline \tilde{\mathcal{I}}_1 & \mathcal{G}_1 & \hat{\mathcal{I}}_1 & & \\ \hline & \tilde{\mathcal{I}}_2 & \mathcal{G}_2 & \hat{\mathcal{I}}_2 & \\ \hline & & \ddots & \ddots & \ddots \\ \hline & & & \tilde{\mathcal{I}}_N & \mathcal{G}_N \end{array}\right) \tag{8}$$

with

$$\mathcal{G}_0 = \begin{pmatrix} \frac{\mathcal{I}}{\Delta t} + \mathcal{C}_0 & 0 & \mathcal{G} & 0 \\ -\mathcal{I} & \frac{\mathcal{I}}{\Delta t} + \mathcal{N}_0^* & 0 & \mathcal{G} \\ \mathcal{D} & 0 & 0 & 0 \\ 0 & \mathcal{D} & 0 & 0 \end{pmatrix}, \quad \mathcal{G}_i = \begin{pmatrix} \frac{\mathcal{I}}{\Delta t} + \mathcal{C}_i & \frac{\mathcal{I}}{\alpha} & \mathcal{G} & 0 \\ -\mathcal{I} & \frac{\mathcal{I}}{\Delta t} + \mathcal{N}_i^* & 0 & \mathcal{G} \\ \mathcal{D} & 0 & 0 & 0 \\ 0 & \mathcal{D} & 0 & 0 \end{pmatrix}$$

for $i = 1, \ldots, N-1$,

$$\tilde{\mathcal{I}}_{i+1} = \begin{pmatrix} -\frac{\mathcal{I}}{\Delta t} & & & \\ & 0 & & \\ & & 0 & \\ & & & 0 \end{pmatrix}, \quad \hat{\mathcal{I}}_i = \begin{pmatrix} 0 & & & \\ & -\frac{\mathcal{I}}{\Delta t} & & \\ & & 0 & \\ & & & 0 \end{pmatrix}$$

for $i = 0, \ldots, N-1$ and

$$\mathcal{G}_N = \begin{pmatrix} \frac{\mathcal{I}}{\Delta t} + \mathcal{C}_N & \frac{\mathcal{I}}{\alpha} & \mathcal{G} & 0 \\ -(1 + \frac{\gamma}{\Delta t})\mathcal{I} & \frac{\mathcal{I}}{\Delta t} + \mathcal{N}_N^* & 0 & \mathcal{G} \\ \mathcal{D} & 0 & 0 & 0 \\ 0 & \mathcal{D} & 0 & 0 \end{pmatrix}.$$

At this point, we discretise in space with a finite element approach. The fully discrete version of the KKT system is defined by replacing the operators $\mathcal{I}$, $\mathcal{A}$, $\mathcal{D}$, ... by their finite element versions $\mathcal{I}^h$, $\mathcal{A}^h$, $\mathcal{D}^h$, ... and by incorporating boundary conditions into the right-hand side f. We finally end up with the nonlinear system

$$G^h(w^h)w^h = f^h \tag{9}$$

with the vector $w^h := (w_0^h, w_1^h, \ldots)$ and $w_n^h := (y_n^h,\ \lambda_n^h,\ p_n^h,\ \xi_n^h)$. Note that the system matrix is a block tridiagonal matrix of the form

$$G^h = G^h(w_h) = \left(\begin{array}{c|c|c|c} G_0 & \hat{M}_0 & & \\ \hline \tilde{M}_1 & G_1 & \hat{M}_1 & \\ \hline & \ddots & \ddots & \ddots \\ \hline & & \tilde{M}_N & G_N \end{array}\right) \tag{10}$$

where $N \in \mathbb{N}$ denotes the number of time intervals. This way, the solver for the optimal control problem reduces to a solver for a sparse block tridiagonal system where the diagonal blocks $G_n = G_n(w_h)$ correspond to the timesteps of the fully coupled KKT system. This system does not have to be set up in memory in its complete form: Utilising defect correction algorithms reduces the solution process to a sequence of matrix vector multiplications in space and time. A matrix-vector multiplication of a solution w^h with the space-time matrix G^h on the other hand reduces to $3N+1$ local matrix-vector multiplications, three in each timestep with subsequent $\tilde{M}_n$, G_n and $\hat{M}_n$.

Discretisation of the Newton system associated to (7). The Newton algorithm in space and time can be written in defect correction form as follows:

$$w_{i+1} := w_i + F(w_i)^{-1}(f - G(w_i)w_i), \qquad i \in \mathbb{N}$$

with $F(w)$ being the Fréchet derivative of the operator $w \mapsto G(w)w$ which is given by the Newton matrix

$$F(w) = \begin{pmatrix} \mathcal{F}_0 & \hat{\mathcal{I}}_0 & & & \\ \tilde{\mathcal{I}}_1 & \mathcal{F}_1 & \hat{\mathcal{I}}_1 & & \\ & \tilde{\mathcal{I}}_2 & \mathcal{F}_2 & \hat{\mathcal{I}}_2 & \\ & & \ddots & \ddots & \ddots \\ & & & \tilde{\mathcal{I}}_N & \mathcal{F}_N \end{pmatrix}$$

with

$$\mathcal{F}_0 = \begin{pmatrix} \frac{\mathcal{I}}{\Delta t} + \mathcal{N}_0 & 0 & \mathcal{G} & 0 \\ -\mathcal{I} + \mathcal{R}_0 & \frac{\mathcal{I}}{\Delta t} + \mathcal{N}_0^* & 0 & \mathcal{G} \\ \mathcal{D} & 0 & 0 & 0 \\ 0 & \mathcal{D} & 0 & 0 \end{pmatrix}, \quad \mathcal{F}_i = \begin{pmatrix} \frac{\mathcal{I}}{\Delta t} + \mathcal{N}_i & \frac{1}{\alpha}\mathcal{I} & \mathcal{G} & 0 \\ -\mathcal{I} + \mathcal{R}_i & \frac{\mathcal{I}}{\Delta t} + \mathcal{N}_i^* & 0 & \mathcal{G} \\ \mathcal{D} & 0 & 0 & 0 \\ 0 & \mathcal{D} & 0 & 0 \end{pmatrix}$$

for $i = 1, \ldots, N-1$ and

$$\mathcal{F}_N = \begin{pmatrix} \frac{\mathcal{I}}{\Delta t} + \mathcal{N}_N & \frac{1}{\alpha}\mathcal{I} & \mathcal{G} & 0 \\ -(1 + \frac{\gamma}{\Delta t})\mathcal{I} + \mathcal{R}_N & \frac{\mathcal{I}}{\Delta t} + \mathcal{N}_N^* & 0 & \mathcal{G} \\ \mathcal{D} & 0 & 0 & 0 \\ 0 & \mathcal{D} & 0 & 0 \end{pmatrix}.$$

Here, we use the additional operator $\mathcal{R}_n v := \mathcal{R}(\lambda_n)v := -(v\nabla)\lambda_n + (\nabla v)^t \lambda_n$ for all velocity vectors v.

3. The Newton-multigrid solver

The KKT-system represents a boundary value problem in the space-time domain. It is shown, e.g., in [6] that, assuming sufficient regularity on the state (y, p) and the adjoint state (λ, ξ), it can equivalently be rewritten as higher-order elliptic equation in the space-time domain for either the state or the adjoint state. This indicates that multigrid can be used as efficient solver for the (linearised) KKT system as it is an ideal solver for elliptic PDEs.

We formally define the solution approach as outer nonlinear loop that has to solve a linear subproblem in each nonlinear step.

3.1. The outer defect correction/Newton loop

To treat the nonlinearity in the underlying Navier–Stokes equations, we use a standard nonlinear fixed point iteration as well as a space-time Newton iteration. Both algorithms can be written down as fully discrete preconditioned defect correction loop,

$$\begin{aligned} 1.)\quad C(w_i^h)d_i &= d_i := \left(f^h - G^h(w_i^h)w_i^h\right) \\ 2.)\quad w_{i+1}^h &:= w_i^h + d_i. \end{aligned}$$

For the fixed point method, we choose $C(w^h) := G^h(w^h)$ as preconditioner, while the space-time Newton method is characterised by $C(w^h) := F^h(w^h)$ with $F^h(w^h)$ being the discrete analogon to $F(w)$ from Section 2. The solution of the auxiliary problem $C(w_i^h)d_i = d_i$ is obtained by applying the following space-time multigrid method.

3.2. The inner multigrid solver

Let $\Omega_1, \ldots, \Omega_{\mathrm{NLMAX}}$ be a conforming hierarchy of triangulations of the domain Ω in the sense of [7]. Ω_1 is a basic coarse mesh and Ω_{l+1} stems from a regular refinement of Ω_l (i.e., new vertices, cells and edges are generated by connecting opposite midpoints of edges). We use $V_1, \ldots, V_{\mathrm{NLMAX}}$ to refer to the different Finite Element spaces in space built upon these meshes. Furthermore, let $T_1, \ldots, T_{\mathrm{NLMAX}}$ be a hierarchy of decompositions of the time interval $[0, T]$, where each T_{l+1} stems from T_l by bisecting each time interval. For each l, the above discretisation in space and time yields a solution space $W_l = V_l \times T_l$ and a space-time system

$$G^l w^l = f^l, \qquad l = 1, \ldots, \mathrm{NLMAX}$$

of the form (9). $f^{\mathrm{NLMAX}} = f^h$, $w^{\mathrm{NLMAX}} = w^h$ and $G^{\mathrm{NLMAX}} = G^h(w^h)$ identify the discrete right-hand side, the solution and system operator on the finest level, respectively.

Algorithm 1 Space-time multigrid

```
function SpaceTimeMultigrid(w;f;l)
    if (l = 1) then
        return (G^1)^{-1} f                                  ▷ coarse grid solver
    end if
    while (not converged) do
        w ← S(G^l, w, f, NSMpre)                             ▷ presmoothing
        d ← R(f − G^l w)                                     ▷ restricion of the defect
        w ← w + P(SpaceTimeMultigrid(0; d; l − 1))           ▷ coarse grid correction
        w ← S(G^l, w, f, NSMpost)                            ▷ postsmoothing
    end while
    return w                                                 ▷ solution
end function
```

Let us denote by $P : W_l \to W_{l+1}$ a prolongation operator and by $R : W_l \to W_{l-1}$ the corresponding restriction. Furthermore, let $S : W_l \to W_l$ define a *smoothing* operator (see the following sections for a definiton of these operators). Let us denote with NSMpre, NSMpost the numbers of pre- and postsmoothing steps, respectively. With these components and definitions, Algorithm 1 implements a basic multigrid V-cycle. For variations of this algorithm which use the W- or F-cycle, see [2, 11, 27]. The algorithm is called on the maximum level by

$$w^{\mathrm{NLMAX}} := \mathrm{SpaceTimeMultigrid}\left(0; f^{\mathrm{NLMAX}}; \mathrm{NLMAX}\right)$$

and implicitely uses the matrices $G^1, \ldots, G^{\mathrm{NLMAX}}$.

3.3. Prolongation/restriction

Our discretisation is based on Finite Differences in time and Finite Elements in space. The operators for exchanging solutions and right-hand side vectors between the different levels therefore decompose into a time prolongation/restriction and space prolongation/restriction. Let k be the space level, $P_S : V_k \to V_{k+1}$ the prolongation operator in space and $R_S : V_{k+1} \to V_k$ the corresponding restriction. The prolongation for a space-time vector $w^l = (w_0^l, \ldots, w_N^l)$ on space-time level l can be written as:

$$P(w^l) := \left(P_S(w_0^l),\ \frac{P_S(w_0^l) + P_S(w_1^l)}{2},\ P_S(w_1^l),\ \frac{P_S(w_1^l) + P_S(w_2^l)}{2},\ \ldots,\ P_S(w_N^l)\right)$$

and is a composition of the usual Finite Difference prolongation in time (see also [12]) and Finite Element prolongation in space. The corresponding restriction for a defect vector $d^{l+1} = (d_0^{l+1}, \ldots, d_{2N}^{l+1})$ follows from the 1D multigrid theory [11, 12]:

$$\begin{aligned} R(d^{l+1}) := \Bigg(&R_S\Big(\frac{1}{4}(2d_0^{l+1} + d_1^{l+1})\Big),\ R_S\Big(\frac{1}{4}(d_1^{l+1} + 2d_2^{l+1} + d_3^l)\Big),\\ &\ldots, R_S\Big(\frac{1}{4}(d_{2N-1}^{l+1} + 2d_{2N}^{l+1})\Big)\Bigg). \end{aligned}$$

Our numerical tests in Section 4 are carried out with the nonconforming $\tilde{Q}_1/Q_0$ Finite Element pair in space. For these elements, we use the standard prolongation/restriction operators which can be found, e.g., in [18, 21].

3.4. Smoothing operators and coarse grid solver

The special matrix structure of the global space-time matrix (9) allows to define iterative smoothing operators based on defect correction. Note that usually, every smoother can also be used as coarse grid solver to solve the equation $(G^1)^{-1} f$ in the first step of the algorithm; for that purpose, one has to replace the fixed number of iterations by a stopping criterion depending on the residuum.

Let us first introduce some notations. The space-time matrix G^l at level l can be decomposed into the block submatrices as follows with diagonal submatrices

G_i^l for the timesteps $i = 1, \ldots, N$:

$$G^l = \left(\begin{array}{c|c|c|c} G_0^l & \hat{M}_0^l & & \\ \hline \tilde{M}_1^l & G_1^l & \hat{M}_1^l & \\ \hline & \ddots & \ddots & \ddots \\ \hline & & \tilde{M}_N^l & G_N^l \end{array}\right), \quad G_i^l =: \begin{pmatrix} A_i^{\text{primal}} & M_i^{\text{dual}} & B & 0 \\ M_i^{\text{primal}} & A_i^{\text{dual}} & 0 & B \\ B^T & 0 & 0 & 0 \\ 0 & B^T & 0 & 0 \end{pmatrix}$$

A_i^{primal}, A_i^{dual} are velocity submatrices, M_i^{primal} and M_i^{dual} coupling matrices between the primal and dual velocity and B and B^T clustering the gradient/divergence matrices which are independent of the timestep i. For simplicity, we dropped the index l here. Furthermore, we assume the decompositions $x_i = (x_i^y, x_i^\lambda, x_i^p, x_i^\xi)$ and $d_i = (d_i^y, d_i^\lambda, d_i^p, d_i^\xi)$ of vectors into primal/dual subvectors.

We introduce three iterative block smoothing algorithms. Let $\omega, \omega_1, \omega_2 \in \mathbb{R}$ be damping parameters. The special matrix structure suggests the use of a Block-Jacobi method in the form of Algorithm 2.

Similar to a Block-Jacobi algorithm, it is possible to design a forward-backward block SOR algorithm for smoothing, see Algorithm 3. (For the sake of notation, we define $x_{-1} := x_{N+1} := 0$, $\tilde{M}_0 := \hat{M}_N := 0$.) In contrast to Block-Jacobi, this algorithm exploits basic coupling in time without significant additional costs. The algorithm allows to specify an additional parameter NSMinner which defines how many forward-backward-sweeps are calculated before updating the flow; in our computations however, we always use NSMinner=1.

Above smoothers always treat the primal and dual solution in a coupled way. On the other hand, one can also decouple these solution parts and perform a forward simulation for the primal solution, followed by a backward simulation for the dual solution, see Algorithm 4. This type of algorithm, which we call 'forward-backward simulation algorithm' is rather natural and was used in a similar form by other authors before as a solver (see, e.g., [9, 24]). It is expected to be a compromise in speed and stability: Fully coupled systems in space are avoided, so

---

Algorithm 2 Space-time Block-Jacobi smoother

---

function JacSmoother(G^l,w,f,NSM)
  for $j = 0$ to NSM **do**
    $d \leftarrow f - G^l w$ ▷ Defect
    for $i = 0$ to N **do**
      $d_i \leftarrow (G_i^l)^{-1} d_i$ ▷ Block-Jacobi preconditioning
    end for
    $w \leftarrow w + \omega d$
  end for
  return w ▷ Solution
end function

Algorithm 3 Forward-Backward Block-SOR smoother

```
function FBSORSmoother(G^l, w, f, NSM)
  for istep = 1 to NSM do
    r ← f − G^l w                                          ▷ Defect
    x ← 0                                                   ▷ correction vector
    for istepinner = 1 to NSMinner do
      x^old ← x
      for i = 0 to N do                                     ▷ Forward in time
        d_i ← r_i − M̃_i x_{i−1} − M̂_i x^old_{i+1}           ▷ Defect in time
        x_i ← (1 − ω_1) x^old_i + ω_1 (G^l_i)^{−1} d_i
      end for
      x^old ← x
      for i = N downto 0 do                                 ▷ Backward in time
        d_i ← r_i − M̃_i x^old_{i−1} − M̂_i x_{i+1}           ▷ Defect in time
        x_i ← (1 − ω_1) x^old_i + ω_1 (G^l_i)^{−1} d_i
      end for
    end for
    w ← w + ω_2 x                                           ▷ Correction
  end for
  return w                                                  ▷ Solution
end function
```

the computation of each timestep is faster. On the other hand, due to the reduced coupling, the convergence speed of the overall smoother might be reduced.

Note that the key feature of all algorithms is the solution of saddle point subsystems of the form

$$\begin{pmatrix} A_V & B \\ B^T & 0 \end{pmatrix} \begin{pmatrix} c_1 \\ c_2 \end{pmatrix} = \begin{pmatrix} d_1 \\ d_2 \end{pmatrix} \quad \Leftrightarrow: \quad A_{\mathrm{sp}} c = d$$

in one time step. A_V contains here all velocity submatrices and B, B^T all gradient/divergence submatrices. The full space-time algorithm therefore reduces to an algorithm in space. All space-time operations (e.g., matrix-vector multiplications) can therefore be carried out without setting up the whole space time matrix in memory. The system $A_{\mathrm{sp}} c = d$ is a coupled saddle point problem for primal and/or dual velocity and pressure. It can be solved, e.g., by using direct solvers (as long as the number of unknowns in space is not too large) or sophisticated techniques from computational fluid dynamics, namely a spatial multigrid with Pressure-Schur-Complement based smoothers. A typical approach is presented in the next section.

3.5. Coupled multigrid solvers in space

Systems of the form $A_{\mathrm{sp}} c = d$ for subproblems in space can efficiently be solved with a multigrid solver in space. For a proper description, we have to formu-

Algorithm 4 Forward-Backward simulation smoother

```
function FBSimSmoother(G^l,w,f,NSM)
    for istep = 1 to NSM do
        r ← f − G^l w                                             ▷ Defect
        x ← 0                                                     ▷ correction vector
        for i = 0 to N do                                         ▷ Forward in time
            d_i ← r_i − G^l_i x_i − M̃_i x_{i−1}
            (x_i^y; x_i^p) ← (x_i^y; x_i^p) + ω_1 (A_i^primal B; B^T 0)^{-1} (d_i^y; d_i^p)
        end for
        for i = N downto 0 do                                     ▷ Backward in time
            d_i ← r_i − G^l_i x_i − M̂_i x_{i+1}
            (x_i^λ; x_i^ξ) ← (x_i^λ; x_i^ξ) + ω_1 (A_i^dual B; B^T 0)^{-1} (d_i^λ; d_i^ξ)
        end for
        w ← w + ω_2 x                                             ▷ Correction
    end for
    return w                                                      ▷ Solution
end function
```

late prolongation, restriction and smoothing operators. Prolongation and restriction operators based on the applied Finite Element spaces are standard and well known (see, e.g., [2, 5, 11, 27, 21]). Smoothing operators acting simultaneously on the primal and dual variables can be constructed, e.g., using the pressure Schur complement ('PSC') approach for CFD problems (see also [20, 25, 26]). We shortly describe this approach here.

We first introduce some notations. In each timestep, a linear system $Ax = b$ has to be solved; in our case, this system can be written in the form

$$\begin{pmatrix} A_V & B \\ B^T & 0 \end{pmatrix} \begin{pmatrix} x_1 \\ x_2 \end{pmatrix} = \begin{pmatrix} b_1 \\ b_2 \end{pmatrix}$$

which is a typical saddle point problem for primal and/or dual variables, with A_V being a velocity submatrix and B and B^T clustering the gradient/divergence matrices. Depending on the type of the underlying space-time smoother, this system contains either only primal or dual variables, or it represents a combined system of primal and dual variables. In the latter case, A_V, B and B^T decompose into proper 2×2 block submatrices.

Let iel denote the number of an arbitrary element in the mesh. Furthermore, let $I(iel)$ identify a list of all degrees of freedom that can be found on element iel, containing numbers for the primal and/or dual velocity vectors in all spatial dimensions and the primal and/or dual pressure. With this index set, we define $A_{I(iel)}$ to be a (rectangular) matrix containing only those rows from A identified by the index set $I(iel)$. In the same way, let $x_{I(iel)}$ and $b_{I(iel)}$ define the subvectors of x and b containing only the entries identified by $I(iel)$. Furthermore we define $A_{I(iel),I(iel)}$ to be the (square) matrix that stems from extracting only those rows and columns from A identified by $I(iel)$.

Algorithm 5 PSC-Smoother for smoothing an approximate solution to $Ax = b$

```
function PSCSmoother(A,x,b,NSM)
    for ism = 1,NSM do                                          ▷ NSM smoothing sweeps
        for iel = 1 to NEL do                                   ▷ Loop over the elements
            x_I(iel) ← x_I(iel) + ω C_iel^{-1} (b_I(iel) − A_I(iel) x)   ▷ Local Correction
        end for
    end for
    return x                                                    ▷ Solution
end function
```

This notation allows to formulate the basic PSC smoother in space, see Algorithm 5; $\omega \in \mathbb{R}$ is used here as a damping parameter with default value $\omega = 1$. Of course, this formulation is not yet complete, as it is lacking a proper definition of the local preconditioner C_{iel}^{-1} which is a small square matrix with as many unknowns as indices in $I(iel)$.

There are two basic approaches for this preconditioner. The first approach, which we entitle by PSCSmootherFull, results in the simple choice of $C_{iel} := A_{I(iel),I(iel)}$ and calculating C_{iel}^{-1} by invoking a LU decomposition, e.g., with the LAPACK package [19]. That approach is rather robust and still feasible as the system is small; for the $\tilde{Q}_1/Q_0$ space that is used in our discretisation (see [21]), the system has 18 unknowns.

The second approach, which we call PSCSmootherDiag, results in taking a different subset of the matrix A for forming $C_{I(iel)}$. To describe this approach, we define

$$\hat{A} := \begin{pmatrix} \operatorname{diag}(A_V) & B \\ B^T & 0 \end{pmatrix}$$

where diag(·) refers to the operator taking only the diagonal elements of a given matrix. The local preconditioner can then be formulated as $C_{iel} := \hat{A}_{I(iel),I(iel)}$. If the local system is a combined system of primal and dual variables, this approach decouples the primal from the dual variables. Applying $\hat{A}_{I(iel),I(iel)}^{-1}$ then decomposes into two independent subproblems which is much faster but leads to reduced stability. Most of the numerical tests in the later sections were carried out using PSCSmootherDiag except where noted. We note that it is even possible to increase the stability by applying this approach to patches of cells (cf. [20]) but we do not apply this approach here.

4. Numerical examples

In this section we numerically analyse the proposed solver strategies with respect to robustness and efficiency. The nonlinearity is captured by a space-time fixed point/Newton iteration, both preconditioned by the proposed space-time multigrid.

4.1. The Driven-Cavity example

Example 4.2 (Driven-Cavity configuration). Let a domain $\Omega = (0,1)^2$ be given. On the four boundary edges $\Gamma_1 := \{0\} \times (0,1), \Gamma_2 := [0,1] \times \{0\}, \Gamma_3 := \{1\} \times (0,1)$, $\Gamma_4 := [0,1] \times \{1\}$ we describe Dirichlet boundary conditions as $y(x,t) = (0,0)$ for $x \in \Gamma_1 \cup \Gamma_2 \cup \Gamma_3$ and $y(x,t) = (1,0)$ for $x \in \Gamma_4$. The coarse grid consists of only one quadratic element. The time interval for this test case is $[0,T]$ with $T = 1$, the viscosity parameter of the uncontrolled and controlled flow is set to $\nu = 1/400$. The initial flow y^0 is the stationary fully developed Navier–Stokes flow at $\nu = 1/400$, while the target flow z is chosen as the fully developed Stokes-flow.

A stationary analogon of this example was analysed in [1] and in [24] the authors analyse this problem under constraints, see also [13, 14]. Figure 1 shows a picture of the streamlines of the target flow and the initial flow with the corresponding velocity magnitude in the background. For better visualisation, we took a different resolution of the positive and negative streamlines. Figure 2 depicts the controlled flow and the control at $t = 0.075$, $t = 0.25$ and $t = 0.5$. One identifies two main vortices which 'push' the Navier–Stokes flow to the Stokes-Flow at the beginning of the time interval. Table 1 lists the number of unknowns for a pure forward simulation and the optimisation.

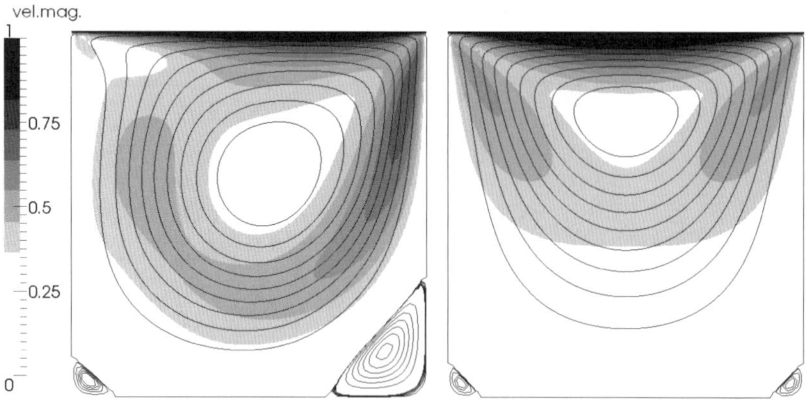

FIGURE 1. *Driven-Cavity* example, Streamlines of the stationary Navier–Stokes (initial) flow (left) and stationary Stokes (target-) flow (right). Velocity magnitude in the background.

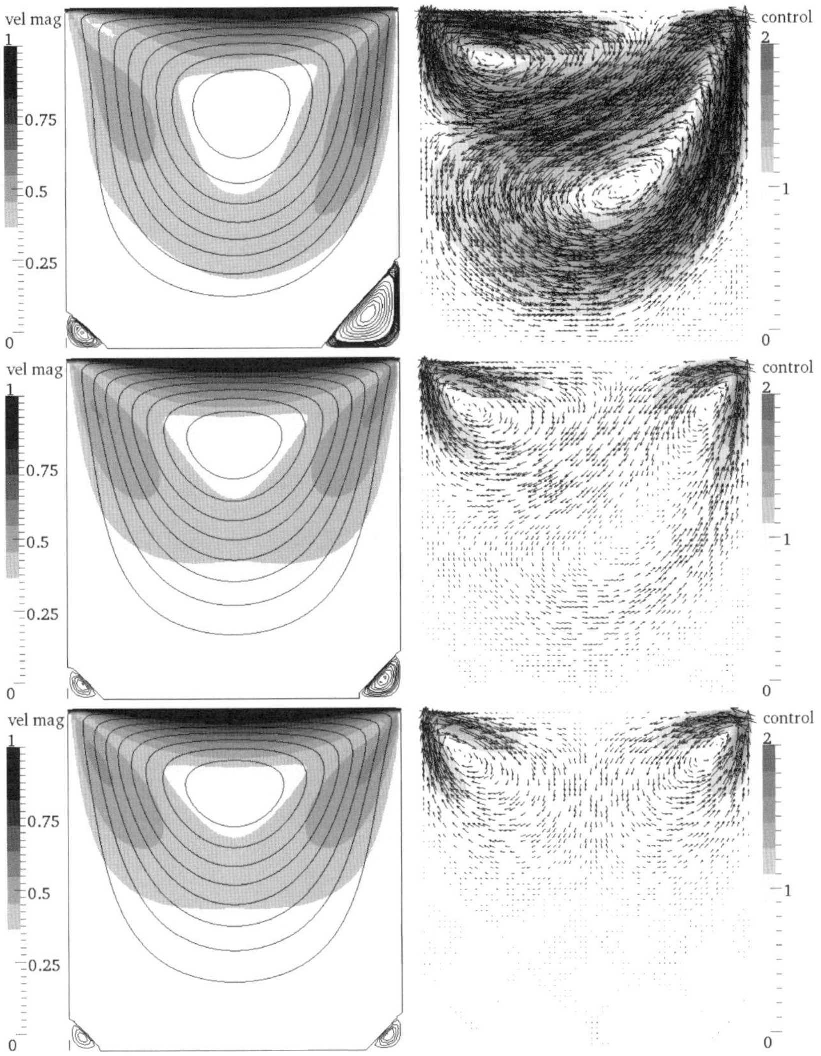

FIGURE 2. *Driven-Cavity* example, controlled flow at $t = 0.075$ (top), $t = 0.25$ (center) and $t = 0.5$ (bottom). Left: Streamlines with primal velocity magnitude in the background. Right: Control.

In the first couple of tests we analyse the behaviour of the solver if being applied to the Driven-Cavity example. For our tests, we prototypically choose the regularisation parameters in the KKT-system to be $\alpha = 0.01$ and $\gamma = 0$ and start with defining a basic coarse mesh in space and time. We choose $\Delta t_{\text{coarse}} = 1/10$ and $h_{\text{coarse}} = 1/4$, although any other relation between Δt_{coarse} and h_{coarse} would

		simulation		optimisation	
Δt	h	#DOF space	#DOF total	#DOF space	#DOF total
1/20	1/8	352	7 392	704	14 784
1/40	1/16	1 344	55 104	2 688	110 208
1/80	1/32	5 248	425 088	10 496	850 176
1/160	1/64	20 736	3 338 496	41 472	6 676 992

TABLE 1. *Driven-Cavity* example, problem size. Number of degrees of freedom in space ('#DOF space') and on the whole space-time domain including the initial condition ('#DOF total').

be suitable as well. This mesh is simultaneously refined by regular refinement in space and time. On each space-time level, we perform the following tests:

1) We calculate an optimal control problem with the target flow as specified above. The nonlinear space-time solver damps the norm of the residual by $\varepsilon_{\text{OptNL}} = 10^{-5}$, the linear space-time multigrid in each nonlinear iteration by $\varepsilon_{\text{OptMG}}$. The convergence criterion of the innermost spatial multigrid solver in each timestep was set to damp the norm of the residual by $\varepsilon_{\text{SpaceMG}}$.
2) We calculate a pure simulation with a fully implicit Navier–Stokes solver in time, using the control computed in 1.) as right-hand side. In each timestep the norm of the residual is damped by $\varepsilon_{\text{SimNL}} = 10^{-5}$. The linear multigrid subsolver in each nonlinear iteration damps the norm of the residual by $\varepsilon_{\text{SimMG}}$.

General tests. In the first couple of tests we analyse the behaviour of the nonlinear space-time solver for optimal control. We fix the space-time mesh to $\Delta t = 1/40$ and $h = 1/16$ which is already fine enough for a qualitative analysis. The convergence criterion of the innermost solver is set to $\varepsilon_{\text{SpaceMG}} = 10^{-2}$. The smoother in space is PSCSmootherDiag, the space-time smoother FBSORSmoother($\omega_1 = 0.8$, $\omega_2 = 1$). From Figure 3 one can see linear convergence for the fixed point iteration and quadratic convergence of the Newton iteration. Note that because of $\varepsilon_{\text{OptNL}} = 10^{-5}$ the impact of the parameter $\varepsilon_{\text{OptMG}}$ to Newton is rather low, it does not (yet) influence the number of iterations.

The next test analyses the influence of the innermost stopping criterion $\varepsilon_{\text{SpaceMG}}$. For the space-time smoothers JACSmoother($\omega = 0.7$, NSM $= 4$), FBSimSmoother($\omega_1 = 0.8, \omega_2 = 0.5$, NSM $= 4$) and FBSORSmoother($\omega_1 = 0.8, \omega_2 = 1$, NSM $= 1$) we fix the convergence criterion of the space-time multigrid to $\varepsilon_{\text{OptMG}} = 10^{-2}$ (Table 2). Then, we calculate the fixed point and Newton iteration for different settings of $\varepsilon_{\text{SpaceMG}}$. As one can see from the table, the solver behaves very robust against $\varepsilon_{\text{SpaceMG}}$, so we choose $\varepsilon_{\text{SpaceMG}} = 10^{-2}$ for all later tests. Furthermore one can see that the efficiency of FBSimSmoother with four smoothing steps is comparable to FBSORSmoother with one smoothing step. JACSmoother on the other hand is the least efficient smoother of these three, so we omit further investigations of it.

Table 3 reveals that for a reasonable convergence criterion of $\varepsilon_{\text{OptNL}} = 10^{-5}$, the number of linear and nonlinear iterations is rather independent of the convergence criterion $\varepsilon_{\text{OptMG}}$ of the space-time multigrid solver. (The smoother in these computations is FBSORSMOOTHER($\omega_1 = 0.8, \omega_2 = 1$, NSM $= 1$).) Furthermore, the number of linear and nonlinear iterations stays almost constant upon increasing the resolution of the space-time mesh which confirms the linear complexity of the

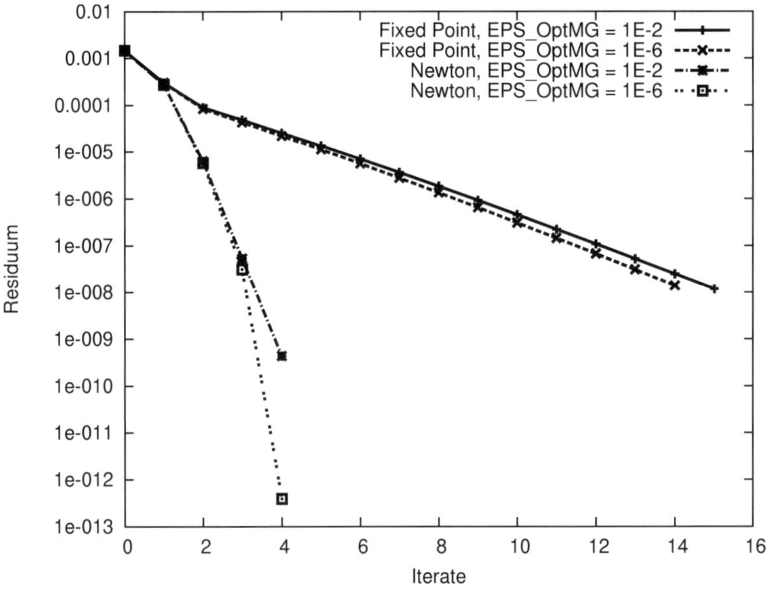

FIGURE 3. *Driven-Cavity* example, convergence of the fixed point and Newton algorithm.

		JACSmoother			FBSimSmoother			FBSORSmoother		
Slv.	$\varepsilon_{\text{SpMG}}$	#NL	#MG	Time	#NL	#MG	Time	#NL	#MG	Time
FP	10^{-1}	15	134	811	15	60	350	15	75	310
	10^{-2}	15	134	846	15	60	352	15	75	319
	10^{-6}	15	134	2405	15	60	650	15	75	775
N	10^{-1}	4	50	449	4	16	138	4	25	150
	10^{-2}	4	50	467	4	16	138	4	25	160
	10^{-6}	4	50	1181	4	16	251	4	25	349

TABLE 2. The influence of $\varepsilon_{\text{SpaceMG}}$ to the Fixed point ('FP') and Newton ('N') solver. $\varepsilon_{\text{OptMG}} = 10^{-2}$. '#NL'=number of iterations, nonlinear solver. '#MG'=number of iterations, space-time multigrid. 'Time'= comp. time in sec.; *Driven-Cavity* example.

$\varepsilon_{\text{OptMG}}$:		10^{-2}				10^{-6}			
Nonl. Solv.:		fixed point		Newton		fixed point		Newton	
Δt	h	#NL	#MG	#NL	#MG	#NL	#MG	#NL	#MG
1/40	1/16	15	75	4	25	14	252	4	80
1/80	1/32	8	40	4	25	8	158	4	87
1/160	1/64	6	33	4	27	6	132	3	68

TABLE 3. The influence of $\varepsilon_{\text{OptMG}}$. '#NL'=number of iterations, nonlinear solver. '#MG'=number of iterations, space-time multigrid. *Driven-Cavity* example.

algorithm. More precisely, the number of iterations even reduces with increasing space-time level – an effect which was also observed and proven in [6].

We note that the impact of the setting of $\varepsilon_{\text{OptMG}}$ would be stronger if $\varepsilon_{\text{OptNL}}$ is set smaller; for such cases, one would prefer an inexact Newton algorithm which adaptively determins the stopping criterion of the linear space-time solver; for details see, e.g., [14]. In all further tests, we take $\varepsilon_{\text{OptMG}} = 10^{-2}$.

Optimisation vs. Simulation. In the following tests we compare the behaviour of the solver for the optimal control problem with a pure simulation. As convergence criterion for the solver we choose $\varepsilon_{\text{SimNL}} = \varepsilon_{\text{OptNL}} = 10^{-5}$ and $\varepsilon_{\text{SimMG}} = \varepsilon_{\text{OptMG}} = 10^{-2}$. Table 4 depicts the result of a set of forward simulations for various settings of Δh and t. We tested both nonlinear solvers, the simple nonlinear fixed point iteration (entitled by 'FP') as well as the Newton iteration (entitled by 'N'). The linear subproblems in space were solved by a multigrid solver in space with a PSCSmootherDiag-type smoother, the space-time multigrid used an FBSORSmoother($\omega_1 = 0.8, \omega_2 = 1$, NSM = 1)-smoother.

The columns ⊘#NL and ⊘#MG in this table describe the average number of linear/nonlinear iterations per timestep in the simulation, which is comparable to the number of nonlinear/linear iterations of the optimisation (columns #NL and #MG, compare also Table 3). #MG in the optimisation task differs from ⊘#MG in the simulation by a factor of approx. 2–3, independent of the level, which means that the effort for both, the simulation and the optimisation, grows with same complexity when increasing the problem size.

Table 4 furthermore compares the different execution times of the simulation and optimisation solver. Using the Newton method clearly shows that the execution time of the optimisation is a bounded multiple of the execution time of the simulation. One can see a factor of $C \approx 50 - 60$ for this example[1], even being

[1] We note here that the large factors for 'small' mesh resolutions ($h \leq 1/16$) are merely an effect of acceleration due to cache effects on the side of the simulation and can be expected: A simulation with only some hundreds of unknowns is able to rather completely run in the cache, whereas the optimisation of a fully nonstationary PDE in space and time usually does not fit into the computer cache anymore. A more detailed analysis and exploitation of the computer cache and its effects can be found in [22, 23].

Slv.	Δt	h	simulation ⊘#NL	⊘#MG	optimisation #NL	#MG	T_{sim}	T_{opt}	$\frac{T_{\text{opt}}}{T_{\text{sim}}}$
FP	1/40	1/16	4	15	15	75	3	316	124
	1/80	1/32	4	16	8	40	21	1414	68
	1/160	1/64	3	16	6	33	207	10436	51
N	1/40	1/16	3	9	4	25	3	158	63
	1/80	1/32	3	10	4	25	21	1359	63
	1/160	1/64	2	11	4	27	219	11359	52
N	1/16	1/16	3	12	4	13	1.2	46	38
	1/32	1/32	3	13	4	13	10.4	330	32
	1/64	1/64	3	13	4	12	109.9	2275	21

TABLE 4. *Driven-Cavity*-example, optimisation and simulation. Execution time as well as mean number of iterations for a pure forward simulation and the corresponding optimisation.

	coarse mesh		fine mesh		FBSORSmoother			FBSimSmoother		
Slv.	Δt	h	Δt	h	#NL	#MG	T_{opt}	#NL	#MG	T_{opt}
FP	1/4	1/4	1/16	1/16	15	45	99	div	div	div
	1/6	1/4	1/24	1/16	15	59	164	15	46	176
	1/8	1/4	1/32	1/16	15	60	197	15	59	279
N	1/4	1/4	1/16	1/16	4	13	45	div	div	div
	1/6	1/4	1/24	1/16	4	17	74	div	div	div
	1/8	1/4	1/32	1/16	4	21	117	4	15	109

TABLE 5. *Driven-Cavity*-example. Smoother robustness.

better for higher mesh resolutions. This factor depends strongly on the anisotropy of the space-time coarse mesh: The lower part of the table shows the results for a time coarse mesh with $\Delta t_{\text{coarse}} = 1/4$ instead of $\Delta t_{\text{coarse}} = 1/10$, which is fully isotropic in space and time. For this mesh, we obtain factors of $C \approx 20 - 30$.

Table 5 compares the FBSIMSMOOTHER against the FBSORSMOOTHER in terms of robustness. Both algorithms have a similar efficiency, but the solver does not converge anymore for large Δt_{coarse} if FBSIMSMOOTHER is used. Furthermore one can see, that the number of space-time multigrid steps #MG (and thus the solver time T_{opt}) increases with finer Δt_{coarse}. This behaviour is typical for SOR-like algorithms and depends on the anisotropy in the space-time mesh.

4.3. Tests with the Flow-around-Cylinder example

In a similar way as above, we now carry out a set of tests for the more complicated *Flow-around-Cylinder* problem, which is a modification of a well-known CFD benchmark problem in [21]:

Example 4.4 (Flow-around-Cylinder configuration). As spatial domain, we prescribe a rectangle without an inner cylinder $\Omega := (0, 2.2) \times (0, 0.41) \setminus \overline{B_r(0.2, 0.2)}$, $r = 0.05$. We decompose the boundary of this domain into five parts: $\Gamma_1 := \{0\} \times [0, 0.41], \Gamma_2 := (0, 2.2] \times \{0\}, \Gamma_3 := \{2.2\} \times (0, 0.41), \Gamma_4 := [0, 2.2] \times \{0.41\}$ and $\Gamma_5 := \partial B_r(0.2, 0.2)$. Boundary conditions are: $y(x, t) := (0, 0)$ for $x \in \Gamma_2 \cup \Gamma_4 \cup \Gamma_5$, do-nothing boundary conditions on Γ_3 and a parabolic inflow profile with maximum velocity $U_{\max} := 0.3$ on Γ_1. The time interval for this test case is $[0, T]$ with $T = 1$. Similar to the previous *Driven-Cavity*-example, our initial flow is the stationary Navier–Stokes-flow at $\nu = 1/500$ while the target flow is the stationary Stokes-flow. The viscosity parameter in the optimisation is set to $\nu = 1/500$ as well (resulting in a Re=10 optimisation).

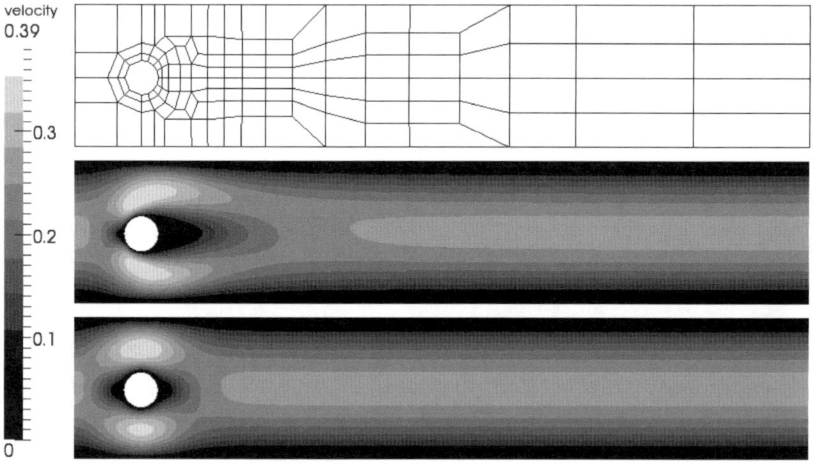

FIGURE 4. *Flow-around-Cylinder* example. Velocity magnitude. Top: Coarse mesh. Center: Initial Navier–Stokes flow. Bottom: Target Stokes flow.

Figure 4 depicts the basic mesh, the initial flow and the target flow. For the time discretisation, we choose $N = 10$ timesteps on time level 1 which leads to 20, 40 and 80 timesteps on space-time level 2, 3 and 4, resp. In Table 6 the corresponding problem size for the simulation and optimisation can be seen. The convergence criteria for the solvers are defined as $\varepsilon_{\text{SimNL}} = \varepsilon_{\text{OptNL}} = 10^{-5}$ and $\varepsilon_{\text{SimMG}} = \varepsilon_{\text{OptMG}} = \varepsilon_{\text{SpaceMG}} = 10^{-2}$. We again focus on the difference in the execution time between the simulation and a corresponding optimisation and proceed as in the last section: We first compute a control with the optimisation and then execute a pure simulation with the computed control as right-hand side.

Table 7 compares the mean number of nonlinear and linear iterations per timestep in the simulation to the number of nonlinear and linear iterations in the optimisation. We use FBSIMSMOOTHER($\omega_1 = 0.8$, $\omega_2 = 0.5$, NSM $= 4$) here as

Δt	Space-Lv.	simulation #DOF space	 #DOF total	optimisation #DOF space	 #DOF total
1/20	2	2 704	29 744	5 408	59 488
1/40	3	10 608	222 768	21 216	445 536
1/80	4	42 016	1 722 656	84 032	3 445 312

TABLE 6. *Flow-around-Cylinder* example, problem size. Number of degrees of freedom in space ('#DOF space') and on the whole space-time domain including the initial condition ('#DOF total').

			Simulation		Optimisation				
Slv.	Δt	Space-Lvl.	⊘#NL	⊘#MG	#NL	#MG	T_{sim}	T_{opt}	$\frac{T_{\text{opt}}}{T_{\text{sim}}}$
FP	1/20	2	4	19	4	14	3	111	37
	1/40	3	4	19	4	17	37	972	26
	1/80	4	3	20	4	19	324	9383	29
N	1/20	2	3	12	3	10	3	112	37
	1/40	3	3	15	3	13	61	924	15
	1/80	4	3	14	3	14	387	9410	24

TABLE 7. *Flow-around-Cylinder* example. Simulation and optimisation.

space-time smoother. As in the Driven-Cavity example, the number of nonlinear iterations for the optimisation is comparable to the mean number of nonlinear iterations in the simulation, and so it does for the number of linear iterations. The execution time of the simulation and the optimisation[2] differs by only a constant factor which is typically decreasing for higher levels of refinement. The table indicates a factor $C \approx 20 - 30$ for reasonable levels.

5. Conclusions

Optimal control of the time-dependent Navier–Stokes equations can be carried out with iterative nonlinear and linear solution methods that act on the whole space-time domain. As nonlinear solver Newton's method is used. Because of the special structure of the system matrix a space-time multigrid algorithm can be formulated for the linear subproblems in the Newton iteration. All matrix vector multiplications and smoothing operations can be reduced to local operations in

[2]In the table, the execution time for the simulation using the fixed point algorithm is usually lower than the execution time with the Newton algorithm. This stems from the fact that if the Newton iteration is used, the effort for solving the spatial system in every timestep is much higher, and due to the convergence criterion $\varepsilon_{\text{OptNL}} < 10^{-5}$ the Newton and the fixed point method need approximately the same number of linear/nonlinear iterations. This situation usually changes if a higher Re number is used as this implies a stronger influence of the nonlinearity.

space, thus avoiding the necessity of storing the whole space-time matrix in memory. Problems in space can be tackled by efficient spatial multigrid and Pressure-Schur-Complement techniques from Computational Fluid Dynamics. The overall solver works with optimal complexity, the numerical effort growing linearly with the problem size. The execution time necessary for the optimisation is a bounded multiple of the execution time necessary for a 'similar' simulation, where numerical tests indicate a factor of $C \approx 20-30$ for reasonable configurations. Being based on Finite Elements, the solver can be applied to rather general computational meshes.

This article concentrated on the basic ingredients of the solver. For simplicity, we restricted to first-order implicit Euler discretisation in time and ignored any restrictions on the controls. Nevertheless, higher-order schemes like Crank-Nicolson are possible and preferable for larger timesteps, but the additional coupling leads to a more complicated matrix structure. Similarly, bounds on the control, other finite element spaces for higher accuracy and stabilisation techniques which are necessary to compute with higher Reynolds numbers are topics which have to be addressed in the future.

References

[1] G.V. Alekseyev and V.V. Malikin. Numerical analysis of optimal boundary control problems for the stationary Navier–Stokes equations. *Computational Fluid Dynamics Journal*, 3(1):1–26, 1994.

[2] R.E. Bank and T.F. Dupond. An optimal order process for solving finite element equations. *Math. Comput.*, 36(153):35–51, 1981.

[3] G. Bärwolff and M. Hinze. Optimization of semiconductor melts. *Zeitschrift für Angewandte Mathematik und Mechanik*, 86:423–437, 2006.

[4] A. Borzi. Multigrid methods for parabolic distributed optimal control problems. *Journal of Computational and Applied Mathematics*, 157:365–382, 2003.

[5] S.C. Brenner. An optimal-order multigrid method for P_1 nonconforming finite elements. *Math. Comput.*, 52(185):1–15, 1989.

[6] G. Büttner. *Ein Mehrgitterverfahren zur optimalen Steuerung parabolischer Probleme.* PhD thesis, Fakultät II – Mathematik und Naturwissenschaften der Technischen Universität Berlin, 2004.
`http://edocs.tu-berlin.de/diss/2004/buettner_guido.pdf`.

[7] Ph.G. Ciarlet. *The finite element method for elliptic problems.* Studies in mathematics and its applications, Vol. 4. North-Holland Publishing Company, Amsterdam, New-York, Oxford, 1978. ISBN 0444850287.

[8] H. Goldberg and F. Tröltzsch. On a SQP-multigrid technique for nonlinear parabolic boundary control problems. In W.W. Hager and P.M. Pardalos, editors, *Optimal Control: Theory, Algorithms, and Applications*, pages 154–174. Kluwer, 1998.

[9] M. Gunzburger, E. Ozugurlu, J. Turner, and H. Zhang. Controlling transport phenomena in the czochralski crystal growth process. *Journal of Crystal Growth*, 234:47–62, 2002.

[10] W. Hackbusch. Fast solution of elliptic optimal control problems. *J. Opt. Theory and Appl.*, 31(4):565–581, 1980.

[11] W. Hackbusch. *Multi-Grid Methods and Applications.* Springer, Berlin, 1985. ISBN 3-540-12761-5.

[12] W. Hackbusch. Multigrid methods for FEM and BEM applications. In E. Stein, R. de Borst, and Th.J.R. Hughes, editors, *Encyclopedia of Computational Mechanics*, chapter 20. John Wiley & Sons Ltd., 2004.

[13] M. Hintermüller and M. Hinze. A SQP-semi-smooth Newton-type algorithm applied to control of the instationary Navier–Stokes system subject to control constraints. *Siam J. Optim.*, 16:1177–1200, 2006.

[14] M. Hinze. Optimal and instantaneous control of the instationary Navier–Stokes equations. Institut für Numerische Mathematik, Technische Universität Dresden, 2000. Habilitation.

[15] M. Hinze and S. Ziegenbalg. Optimal control of the free boundary in a two-phase Stefan problem. *J. Comput. Phys.*, 223:657–684, 2007.

[16] M. Hinze and S. Ziegenbalg. Optimal control of the free boundary in a two-phase Stefan problem with flow driven by convection. *Z. Angew. Math. Mech.*, 87:430–448, 2007.

[17] M. Hinze and S. Ziegenbalg. Optimal control of the phase interface during solidification of a GaAs melt. *Proc. Appl. Math. Mech.*, 311(8):2501–2507, 2008.

[18] M. Köster. Robuste Mehrgitter-Krylowraum-Techniken für FEM-Verfahren, 2007. Diplomarbeit, Universität Dortmund, Diplomarbeit, `http://www.mathematik.tu-dortmund.de/lsiii/static/schriften_eng.html`.

[19] NETLIB. LAPACK – Linear Algebra PACKage, 1992. `http://www.netlib.org/lapack/`.

[20] R. Schmachtel. *Robuste lineare und nichtlineare Lösungsverfahren für die inkompressiblen Navier–Stokes-Gleichungen.* PhD thesis, TU Dortmund, June 2003. `http://www.mathematik.tu-dortmund.de/lsiii/static/schriften_eng.html`.

[21] S. Turek. *Efficient Solvers for Incompressible Flow Problems: An Algorithmic and Computational Approach.* Springer, Berlin, 1999. ISBN 3-540-65433-X.

[22] S. Turek, Ch. Becker, and S. Kilian. Hardware-oriented numerics and concepts for PDE software. *Future Generation Computer Systems*, 22(1-2):217–238, 2006. doi: 10.1016/j.future.2003.09.007.

[23] S. Turek, D. Göddeke, Ch. Becker, S.H.M. Buijssen, and H. Wobker. FEAST – realisation of hardware-oriented numerics for HPC simulations with finite elements. *Concurrency and Computation: Practice and Experience*, 2010. Special Issue Proceedings of ISC 2008, accepted.

[24] M. Ulbrich. Constrained optimal control of Navier–Stokes flow by semismooth Newton methods. *Systems Control Lett.*, 48:297–311, 2003.

[25] S.P. Vanka. Block-implicit multigrid solution of Navier–Stokes equations in primitive variables. *Journal of Computational Physics*, 65:138–158, 1986.

[26] H. Wobker and S. Turek. Numerical studies of Vanka-type smoothers in computational solid mechanics. *Advances in Applied Mathematics and Mechanics*, 1(1):29–55, 2009.

[27] H. Yserentant. Old and new convergence proofs for multigrid methods. *Acta Numerica*, pages 1–44, 1992.

Michael Hinze
Department of Mathematics
University of Hamburg
Bundesstrasse 55
D-20146 Hamburg, Germany
e-mail: `michael.hinze@uni-hamburg.de`

Michael Köster (corresponding author) and Stefan Turek
Institut für Angewandte Mathematik
Technische Universität Dortmund
Vogelpothsweg 87
D-44227 Dortmund, Germany
e-mail: `michael.koester@mathematik.tu-dortmund.de`
`stefan.turek@mathematik.tu-dortmund.de`

International Series of Numerical Mathematics, Vol. 160, 171–182

A Globalized Semi-smooth Newton Method for Variational Discretization of Control Constrained Elliptic Optimal Control Problems

Michael Hinze and Morten Vierling

Abstract. When combining the numerical concept of variational discretization introduced in [Hin03, Hin05] and semi-smooth Newton methods for the numerical solution of pde constrained optimization with control constraints [HIK03, Ulb03] special emphasis has to be placed on the implementation, convergence and globalization of the numerical algorithm. In the present work we address all these issues following [HV]. In particular we prove fast local convergence of the algorithm and propose a globalization strategy which is applicable in many practically relevant mathematical settings. We illustrate our analytical and algorithmical findings by numerical experiments.

Mathematics Subject Classification (2000). 49J20, 49K20, 49M15.

Keywords. Variational discretization, semi-smooth Newton method, primal-dual active set strategy, Elliptic optimal control problem, control constraints, error estimates.

1. Introduction and mathematical setting

We are interested in the numerical treatment of the control problem

$$(\mathbb{P}) \quad \begin{cases} \min_{(y,u)\in Y\times U_{\text{ad}}} J(y,u) := \frac{1}{2}\|y-z\|^2_{L^2(\Omega)} + \frac{\alpha}{2}\|u\|^2_{L^2(\Omega)} \\ \text{s.t.} \\ \quad -\Delta y = \imath u \quad \text{in } \Omega, \\ \quad\quad y = 0 \quad \text{on } \partial\Omega\,. \end{cases}$$

Here, $\alpha > 0$ and $\Omega \subset \mathbb{R}^d$ $(d = 2, 3)$ denotes an open, bounded convex polyhedral domain, and $U_{\text{ad}} := \{v \in L^2(\Omega); a \le v \le b\}$ with $a, b \in \mathbb{R}$, $a < b$. Furthermore, $\imath : L^2(\Omega) \to H^{-1}(\Omega)$ denotes the canonical injection, so that the state space for the Poisson problem in $(\mathbb{P})$ is $Y := H^1_0(\Omega)$. We note that also additional state constraints could be included into our problem setting, see, e.g., in [DH07a, DH07b], and also more general (linear) elliptic or parabolic state equations may be considered. However, all structural issues discussed in the present work are induced by the control constraints, hence to keep the exposition as simple as possible state constraints are not considered here.

Problem ($\mathbb{P}$) admits a unique solution $(y, u) \in Y \times U_{\text{ad}}$, and can equivalently be rewritten as the optimization problem

$$\min_{u \in U_{\text{ad}}} \hat{J}(u) \tag{1.1}$$

for the reduced functional $\hat{J}(u) := J(y(u), u) \equiv J(S\imath\, u, u)$ over the set U_{ad}, where $S : Y^* \to L^2(\Omega)$ denotes the (compact) prolongated solution operator associated with $-\Delta$ and homogeneous Dirichlet boundary conditions. We further know that the first-order necessary (and here also sufficient) optimality conditions take the form

$$\langle \hat{J}'(u), v - u \rangle_{L^2(\Omega)} \geq 0 \text{ for all } v \in U_{\text{ad}} \tag{1.2}$$

where $\hat{J}'(u) = \alpha(u, \cdot)_{L^2(\Omega)} + \imath^* S^*(S\imath\, u - z) \equiv \alpha(u, \cdot)_{L^2(\Omega)} + \imath^* p$, with $p := S^*(S\imath\, u - z)$ denoting the adjoint variable. From now on, we omit the operators $\imath, \imath^*$ for notational convenience, considering S a continuous endomorphism of $L^2(\Omega)$. The function p in our setting satisfies

$$\begin{aligned} -\Delta p &= y - z && \text{in } \Omega, \\ p &= 0 && \text{on } \partial\Omega. \end{aligned} \tag{1.3}$$

For the numerical treatment of problem ($\mathbb{P}$) it is convenient to rewrite (1.2) for $\sigma > 0$ arbitrary in form

$$u = P_{U_{\text{ad}}}\left(u - \sigma \nabla \hat{J}(u)\right) \overset{\sigma = 1/\alpha}{\equiv} P_{U_{\text{ad}}}\left(-\frac{1}{\alpha} R^{-1} p\right),$$

with the Riesz isomorphism $R : L^2(\Omega) \to (L^2(\Omega))^*$, the gradient $\nabla \hat{J}(u) = R^{-1}\hat{J}'(u)$ and $P_{U_{\text{ad}}}$ denoting the orthogonal projector onto U_{ad}.

2. Finite element discretization

To discretize ($\mathbb{P}$) we concentrate on a Finite Element approach based on a quasi-uniform sequence of triangulations $\mathcal{T}_h$ of the polyhedral domain Ω, where h denotes the mesh parameter. For $k \in \mathbb{N}$ we set

$$Y_h := \{v \in C^0(\bar{\Omega}); v_{|\partial\Omega} = 0; v_{|T} \in \mathbb{P}_k(T) \text{ for all } T \in \mathcal{T}_h\} \subset Y\,.$$

Now we approximate problem ($\mathbb{P}$) by its variational discretization (cf. [Hin05])

$$(\mathbb{P}_h) \quad \begin{cases} \min_{(y_h, u) \in Y_h \times U_{\text{ad}}} J(y_h, u) := \frac{1}{2}\|y_h - z\|^2_{L^2(\Omega)} + \frac{\alpha}{2}\|u\|^2_{L^2(\Omega)} \\ \text{s.t.} \\ a(y_h, v_h) = (u, v_h)_{L^2(\Omega)} \quad \text{for all } v_h \in Y_h, \end{cases}$$

where $a(y,v) := \int_\Omega \nabla y \nabla v dx$ denotes the bilinear form associated with $-\Delta$. Problem $(\mathbb{P}_h)$ admits a unique solution $(y_h, u_h) \in Y_h \times U_{\rm ad}$ and, as above, can equivalently be rewritten as the optimization problem

$$\min_{u \in U_{\rm ad}} \hat{J}_h(u) \tag{2.1}$$

for the discrete reduced functional $\hat{J}_h(u) := J(y_h(u), u) \equiv J(S_h u, u)$ over the set $U_{\rm ad}$, where $S_h : L^2(\Omega) \to Y_h \subset L^2(\Omega)$ denotes the solution operator associated with the finite element discretization of $-\Delta$. The first-order necessary (and here also sufficient) optimality conditions take the form

$$\langle \hat{J}_h'(u_h), v - u_h \rangle_{L^2(\Omega)} \geq 0 \text{ for all } v \in U_{\rm ad}$$

where $\hat{J}_h'(v) = \alpha(v, \cdot)_{L^2(\Omega)} + S_h^*(S_h v - z) \equiv \alpha(v, \cdot)_{L^2(\Omega)} + p_h$, with $p_h := S_h^*(S_h v - z)$ denoting the adjoint variable. The function p_h in our setting satisfies

$$a(v_h, p_h) = (y_h - z, v_h)_{L^2(\Omega)} \text{ for all } v_h \in Y_h. \tag{2.2}$$

Analogously to (1.2), for $\sigma > 0$ arbitrary, omitting R we have

$$u_h = P_{U_{\rm ad}}\left(u_h - \sigma \nabla \hat{J}_h(u_h)\right) \overset{\sigma=1/\alpha}{\equiv} P_{U_{\rm ad}}\left(-\frac{1}{\alpha} p_h\right). \tag{2.3}$$

Remark 2.1. Problem $(\mathbb{P}_h)$ is still infinite-dimensional in that the control space is not discretized. This is reflected through the appearance of the projector $P_{U_{\rm ad}}$ in (2.3). The numerical challenge now consists in designing numerical solution algorithms for problem $(\mathbb{P}_h)$ which are implementable, and which reflect the infinite-dimensional structure of the *discrete* problem $(\mathbb{P}_h)$, see also [Hin03, Hin05].

For the error $\|u - u_h\|_{L^2(\Omega)}$ between the solutions u of (1.1) and u_h of (2.1) we have from, e.g., [HV] and [HPUU09, Chapt. 3]

Theorem 2.2. *Let u denote the unique solution of* (1.1), *and u_h the unique solution of* (2.1). *Then there holds*

$$\begin{aligned} &\alpha\|u - u_h\|_{L^2(\Omega)}^2 + \frac{1}{2}\|y(u) - y_h\|_{L^2(\Omega)}^2 \\ &\leq \langle p(u) - \tilde{p}_h(u), u_h - u \rangle_{L^2(\Omega)} + \frac{1}{2}\|y(u) - y_h(u)\|_{L^2(\Omega)}^2, \end{aligned} \tag{2.4}$$

where $\tilde{p}_h(u) := S_h^*(SBu - z)$, $y_h(u) := S_h Bu$, and $y(u) := SBu$.

It follows from (2.4) that the error $\|u - u_h\|_{L^2(\Omega)}$ between the solution u of problem (1.1) and u_h of (2.1) is completely determined by the approximation properties of the discrete solution operator S_h and its adjoint S_h^*.

3. Semi-smooth Newton algorithm

Problem ($\mathbb{P}$) and its variational discretization ($\mathbb{P}_h$) can now be expressed by means of

$$G(v) := v - P_{[a,b]}\left(-\frac{1}{\alpha}p(y(v))\right), \quad \text{and} \quad G_h(v) := v - P_{[a,b]}\left(-\frac{1}{\alpha}p_h(y_h(v))\right), \tag{3.1}$$

where $P_{[a,b]}$ is the pointwise projection onto the interval $[a,b]$, and for given $v \in L^2(\Omega)$ the functions p, p_h are defined through (1.3) and (2.2), respectively. As discussed in the previous sections,

$$G(u),\ G_h(u_h) = 0 \text{ in } L^2(\Omega). \tag{3.2}$$

These equations will be shown to be amenable to semi-smooth Newton methods as proposed in [HIK03] and [Ulb03]. Following [Ulb03], we begin with formulating

Algorithm 3.1. (Semi-smooth Newton algorithm for (3.2))

Start with $v \in L^2(\Omega)$ given. Do until convergence
Choose $M \in \partial G_h(v)$.
Solve $M\delta v = -G_h(v)$, $v := v + \delta v$,

where we use Clarke's generalized differential ∂. If we choose generalized Jacobians $M \in \partial G_h(v)$ with $\|M^{-1}\|$ uniformly bounded throughout the iteration, and at the solution u_h the function G_h is ∂G_h-semi-smooth of order μ, this algorithm is locally superconvergent of order $1+\mu$, [Ulb03]. Although Algorithm 3.1 works on the infinite-dimensional space $L^2(\Omega)$, it is possible to implement it numerically, as is shown subsequently.

3.1. Semismoothness

To apply the Newton algorithm, we need to ensure that the discretized operator G_h is indeed semi-smooth. This is shown in [HV], using [Ulb03, Theorem 5.2].

Theorem 3.2. *The function G_h defined in* (3.1) *is ∂G_h-semi-smooth of order $\mu < \frac{1}{3}$. There holds*

$$\partial G_h(v)w = w + \frac{1}{\alpha}\partial P_{[a,b]}\left(-\frac{1}{\alpha}p_h(y_h(v))\right) \cdot \left(S_h^* S_h w\right),$$

where the application of the differential

$$\partial P_{[a,b]}(x) = \begin{cases} 0 & \text{if } x \notin [a,b] \\ 1 & \text{if } x \in (a,b) \\ [0,1] & \text{if } x = a \text{ or } x = b \end{cases}.$$

and the multiplication by $S_h^* S_h w$ are pointwise operations a.e. in Ω.

Remark 3.3. In [HU04] the mesh independence of the superlinear convergence is stated. Recent results from [Ulb09] indicate semismoothness of G of order $\frac{1}{2}$ as well as mesh independent q-superlinear convergence of the Newton algorithm of order $\frac{3}{2}$, if for example the modulus of the slope of $-\frac{1}{\alpha}p(y(u))$ is bounded away from zero on the border of the active set, and if the mesh parameter h is reduced appropriately.

3.2. Newton algorithm

In order to implement Algorithm 3.1, we have to choose $M \in \partial G_h(v)$. The set-valued function $\partial P_{[a,b]}\big(-\frac{1}{\alpha}p_h(y_h(v))\big)$ contains the characteristic function $\chi_{\mathcal{I}(v)}$ of the inactive set

$$\mathcal{I}(v) = \left\{\omega \in \Omega \,\middle|\, \big(-\frac{1}{\alpha}p_h(y_h(v))\big)(\omega) \in (a,b)\right\}.$$

By χ^v we will denote synonymously the characteristic function

$$\chi_{\mathcal{I}(v)} = \begin{cases} 1 \text{ on } \mathcal{I}(v) \\ 0 \text{ everywhere else} \end{cases}$$

as well as the self-adjoint endomorphism in $L^2(\Omega)$ given by the pointwise multiplication with $\chi_{\mathcal{I}(v)}$. With $\chi^v \in \partial P_{[a,b]}\left(-\frac{1}{\alpha}p_h(y_h(v))\right)$ the Newton-step in Algorithm 3.1 takes the form

$$\Big(I + \frac{1}{\alpha}\chi^v S_h^* S_h\Big)\delta v = -v + P_{[a,b]}\left(-\frac{1}{\alpha}p_h(y_h(v))\right). \tag{3.3}$$

To obtain an impression of the structure of the next iterate $v^+ = v + \delta v$ we rewrite (3.3) as

$$v^+ = P_{[a,b]}\left(-\frac{1}{\alpha}p_h(y_h(v))\right) - \frac{1}{\alpha}\chi^v S_h^* S_h \delta v.$$

Since the range of S_h^* is Y_h, the first addend is continuous and piecewise polynomial (of degree k) on a refinement $\mathcal{K}_h$ of $\mathcal{T}_h$. The partition $\mathcal{K}_h$ is obtained from $\mathcal{T}_h$ by inserting nodes and edges along the boundary between the inactive set $\mathcal{I}(v)$ and the according active set. The inserted edges are level sets of polynomials of degree $\le k$, since we assume $a, b \in \mathbb{R}$. Thus, in the case $k \ge 2$ edges in general are curved.

The second addend, involving the cut-off function χ^v, is also piecewise polynomial of degree k on $\mathcal{K}_h$ but may jump along the edges not contained in $\mathcal{T}_h$.

Finally v^+ is an element of the following finite-dimensional affine subspace

$$Y_h^+ = \left\{\chi^v \varphi + (1-\chi^v)P_{[a,b]}\left(-\frac{1}{\alpha}p_h(y_h(v))\right) \,\middle|\, \varphi \in Y_h\right\}$$

of $L^2(\Omega)$. The iterates generated by the Newton algorithm can be represented exactly with about constant effort, since the number of inserted nodes varies only mildly from step to step, once the algorithm begins to converge. Furthermore the number of inserted nodes is bounded, see [Hin03, Hin05].

Since the Newton increment δv may have jumps along the borders of both the new and the old active and inactive sets, it is advantageous to compute v^+ directly, because v^+ lies in Y_h^+. To achieve an equation for v^+ we add $G'_h(v)v$ on both sides of (3.3) to obtain

$$\Big(I + \frac{1}{\alpha}\chi^v S_h^* S_h\Big)v^+ = P_{[a,b]}\Big(-\frac{1}{\alpha}p_h(y_h(v))\Big) + \frac{1}{\alpha}\chi^v S_h^* S_h v\,. \tag{3.4}$$

This leads to

Algorithm 3.4 (Newton algorithm).

$v \in U$ given. Do until convergence
Solve (3.4) for v^+, $v := v^+$.

3.3. Computing the Newton step

Since v^+ defined by (3.4) is known on the active set $\mathcal{A}(v) := \Omega \setminus \mathcal{I}(v)$ it remains to compute v^+ on the inactive set. So we rewrite (3.4) in terms of the unknown $\chi^v v^+$ by splitting v^+ as

$$v^+ = (1-\chi^v)v^+ + \chi^v v^+.$$

This yields

$$\begin{aligned}&\Big(I + \frac{1}{\alpha}\chi^v S_h^* S_h\Big)\chi^v v^+\\ &\quad= P_{[a,b]}\Big(-\frac{1}{\alpha}p_h(y_h(v))\Big) + \frac{1}{\alpha}\chi^v S_h^* S_h v - \Big(I + \frac{1}{\alpha}\chi^v S_h^* S_h\Big)(1-\chi^v)v^+\,.\end{aligned}$$

Since $(1-\chi^v)v^+$ is already known, we can restrict the latter equation to the inactive set $\mathcal{I}(v)$

$$\Big(\chi^v + \frac{1}{\alpha}\chi^v S_h^* S_h \chi^v\Big)v^+ = \frac{1}{\alpha}\chi^v S_h^* z - \frac{1}{\alpha}\chi^v S_h^* S_h(1-\chi^v)v^+\,. \tag{3.5}$$

On the left-hand side of (3.5) now a continuous, self-adjoint operator on $L^2(\mathcal{I}(v))$ appears, which is positive definite, because it is the restriction of the positive definite operator $\big(I + \frac{1}{\alpha}\chi^v S_h^* S_h \chi^v\big)$ to $L^2(\mathcal{I}(v))$.

Hence we are in the position to apply a CG-algorithm to solve (3.5). Moreover under the assumption of the initial guess living in

$$Y_h^+\big|_{\mathcal{I}^v} = \{\chi^v\varphi \mid \varphi \in Y_h\}\,,$$

as it holds for $\chi^v v^+$, the algorithm does not leave this space because of

$$\Big(I + \frac{1}{\alpha}\chi^v S_h^* S_h \chi^v\Big)\,Y_h^+\big|_{\mathcal{I}(v)} \subset Y_h^+\big|_{\mathcal{I}(v)},$$

and all CG-iterates are elements of $Y_h^+\big|_{\mathcal{I}(v)}$. These considerations lead to the following

Algorithm 3.5 (Solving (3.4)).

1. Compute the active and inactive sets $\mathcal{A}(v)$ and $\mathcal{I}(v)$.

2. $\forall w \in \mathcal{A}(v)$ set

$$v^+(w) = P_{[a,b]}\left(-\frac{1}{\alpha}p_h(y_h(v))(w)\right).$$

3. Solve (3.5) for $\chi^v v^+$ by CG-iteration. By choosing an initial iterate in $Y_h^+|_{\mathcal{I}(v)}$ one ensures that all iterates lie inside $Y_h^+|_{\mathcal{I}(v)}$.
4. $v^+ = (1-\chi^v)v^+ + \chi^v v^+$.

We note that the use of this procedure in Algorithm 3.4 coincides with the active set strategy proposed in [HIK03].

3.4. Globalization

Globalization of Algorithm 3.4 may require a damping step of the form

$$v_\lambda^+ = v + \lambda(v^+ - v)$$

with some $\lambda > 0$. According to the considerations above, we have

$$v_\lambda^+ = (1-\lambda)v + \lambda\Big(P_{[a,b]}\left(-\frac{1}{\alpha}p_h(y_h(v))\right) - \frac{1}{\alpha}\chi^v S_h^* S_h \delta v\Big).$$

Unless $\lambda = 1$ the effort of representing v_λ^+ will in general grow with every iteration of the algorithm, due to the jumps introduced in each step. This problem can be bypassed by focusing on the adjoint state $p_h(v)$ instead of the control v. In fact the function χ^v, now referred to as χ^p, and thus also Equation (3.4), do depend on v only indirectly through the adjoint $p = p_h(v) = S_h^*(S_h v - z)$

$$\Big(I + \frac{1}{\alpha}\chi^p S_h^* S_h\Big)v^+ = P_{[a,b]}\left(-\frac{1}{\alpha}p\right) + \frac{1}{\alpha}\chi^p(p + S_h^* z). \tag{3.6}$$

Now in each iteration the next full-step iterate v^+ is computed from (3.6). If damping is necessary, one computes $p_\lambda^+ = p_h(v_\lambda^+)$ instead of v_λ^+. In our (linear) setting the adjoint state p_λ^+ simply is a convex combination of $p = p_h(v)$ and $p^+ = p_h(v^+)$

$$p_\lambda^+ = \lambda p^+ + (1-\lambda)p,$$

and unlike v_λ^+ the adjoint state p_λ^+ lies in the finite element space Y_h. Thus only a set of additional nodes according to the jumps of the most recent full-step iterate v^+ have to be managed, exactly as in the undamped case.

Algorithm 3.6 (Damped Newton algorithm). $v \in U$ given.
Do until convergence

Solve Equation (3.6) for v^+.
Compute $p^+ = p_h(y_h(v^+))$.
Choose the damping-parameter λ. (for example by Armijo line search)
Set $p := p_\lambda^+ = \lambda p^+ + (1-\lambda)p$.

Algorithm 3.4 is identical to Algorithm 3.6 without damping ($\lambda = 1$).

Remark 3.7. The above algorithm is equivalent to a damped Newton algorithm applied to the equation

$$p_h = S_h^* S_h P_{[a,b]}\left(-\frac{1}{\alpha}p_h\right) - S_h^* z\,, \qquad u := P_{[a,b]}\left(-\frac{1}{\alpha}p_h\right)\,.$$

3.5. Global convergence of the undamped Newton algorithm

Since orthogonal projections are non-expansive, it is not difficult to see that the fixed-point equation for problem ($\mathbb{P}_h$)

$$u_h = P_{[a,b]}\left(-\frac{1}{\alpha}S_h^*(S_h u_h - z)\right)$$

can be solved by simple fixed-point iteration that converges globally for $\alpha > \|S_h\|^2_{L^2(\Omega),L^2(\Omega)}$, see [Hin03, Hin05]. A similar global convergence result holds for the undamped Newton algorithm 3.4, see [Vie07].

Lemma 3.8. *For sufficiently small $h > 0$, the Newton algorithm 3.4 converges globally and mesh independently if $\alpha > \frac{4}{3}\|S\|^2$.*

4. Numerical examples

We end this paper by illustrating our theoretical findings by numerical examples. We apply the globalized Algorithm 3.6 with the following Armijo line search strategy. The merit function

$$\Phi_h(p) = \left\| p - S_h^* S_h P_{[a,b]}\left(-\frac{1}{\alpha}p\right) + S_h^* z \right\|^2_{L^2(\Omega)}\,,$$

is chosen to govern the step size. It presents the mean-square residual of the optimality system in the adjoint-reduced formulation.

Algorithm 4.1 (Armijo). Start with $\lambda = 1$. If

$$\Phi_h(p_\lambda^+) \le \Phi_h(p) + 0.02\,\underbrace{\langle \Phi_h'(p), p_\lambda^+ - p\rangle}_{\le 0}, \tag{4.1}$$

accept p_λ^+. If not, redefine $\lambda := 0.7\,\lambda$ and apply test (4.1) again.

As stopping criterion we use $\|P_{[a,b]}(-\frac{1}{\alpha}p_\lambda^+) - u_h\|_{L^2(\Omega)} < 10^{-11}$ in Algorithm 3.6, using the a posteriori bound for admissible $v \in U_{\mathrm{ad}}$

$$\|v - u_h\|_{L^2(\Omega)} \le \frac{1}{\alpha}\|\zeta\|_{L^2(\Omega)}\,, \qquad \zeta(\omega) = \begin{cases} [\alpha v + p_h(v)]_- & \text{if } v(\omega) = a \\ [\alpha v + p_h(v)]_+ & \text{if } v(\omega) = b \\ \alpha v + p_h(v) & \text{if } a < v(\omega) < b \end{cases}\,,$$

presented in [KR08] and [VT09]. Clearly, this estimate applies to $v = P_{[a,b]}(-\frac{1}{\alpha}p_\lambda^+)$, whereas in general it does not hold for the iterates v^+ generated by Algorithm 3.4 or 3.6 that need not lie in U_{ad}.

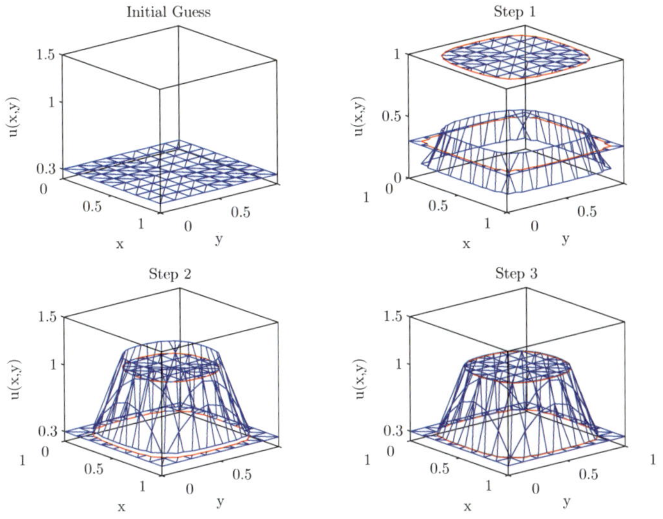

FIGURE 1. The first four Newton-iterates for Example 4.2 (Dirichlet) with parameter $\alpha = 0.001$

For the first example, Algorithm 3.6 reduces to Algorithm 3.4, i.e., the algorithm works with full Newton steps ($\lambda = 1$), thus reflecting the global convergence property from Lemma 3.8. The 2nd example involves a small parameter $\alpha = 10^{-7}$ and the undamped Algorithm 3.4 would not converge in this case.

Example 4.2 (Dirichlet). We consider problem ($\mathbb{P}$) on the unit square $\Omega = (0,1)^2$ with $a = 0.3$ and $b = 1$. Further we set

$$z = 4\pi^2 \alpha \sin(\pi x) \sin(\pi y) + (S \circ \imath) r \ ,$$

where

$$r = \min\left(1, \max\left(0.3, 2\sin(\pi x)\sin(\pi y)\right)\right) .$$

The choice of parameters implies a unique solution $u = r$ to the continuous problem ($\mathbb{P}$).

Throughout this section, solutions to the state equation are approximated by continuous, piecewise linear finite elements on a regular quasi-uniform triangulation $\mathcal{T}_h$ with maximal edge length $h > 0$. The meshes are generated through regular refinement starting from the coarsest mesh.

Problem ($\mathbb{P}_h$) admits a unique solution u_h. In the setting of the present example we have

$$\|u_h - u\|_{L^2(\Omega)} = O(h^2)$$

and

$$\|u_h - u\|_{L^\infty(\Omega)} = O(|\log(h)|^{\frac{1}{2}} h^2)$$

mesh param. h	ERR	ERR_∞	EOC	EOC_∞	Iterations	$\|\zeta\|/\alpha$
$\sqrt{2}/16$	2.5865e-03	1.2370e-02	1.95	1.79	4	2.16e-15
$\sqrt{2}/32$	6.5043e-04	3.2484e-03	1.99	1.93	4	2.08e-15
$\sqrt{2}/64$	1.6090e-04	8.1167e-04	2.02	2.00	4	2.03e-15
$\sqrt{2}/128$	4.0844e-05	2.1056e-04	1.98	1.95	4	1.99e-15
$\sqrt{2}/256$	1.0025e-05	5.3806e-05	2.03	1.97	4	1.69e-15
$\sqrt{2}/512$	2.5318e-06	1.3486e-05	1.99	2.00	4	1.95e-15

TABLE 1. L^2- and L^∞-error development for Example 4.2 (Dirichlet)

for domains $\Omega \subset \mathbb{R}^2$, see [Hin03, Hin05]. Both convergence rates are confirmed in Table 1, where the L^2- and the L^∞-errors for Example 4.2 are presented, together with the corresponding experimental orders of convergence

$$EOC_i = \frac{\ln ERR(h_{i-1}) - \ln ERR(h_i)}{\ln(h_{i-1}) - \ln(h_i)}.$$

Lemma 3.8 ensures global convergence of the undamped Algorithm 3.4 only for $\alpha > 1/(3\pi^4) \simeq 0.0034$, but it is still observed for $\alpha = 0.001$. The algorithm is initialized with $v_0 \equiv 0.3$. The resulting number of Newton steps as well as the value of $\|\zeta\|/\alpha$ for the computed solution are also given in Table 1. Figure 1 shows the Newton iterates. Active and inactive sets are very well distinguishable, and jumps along their frontiers can be observed.

To demonstrate Algorithm 3.6 with damping we again consider Example 4.2, this time with $\alpha = 10^{-7}$, again using the same stopping criterion as in the previous examples. Table 2 shows errors, the number of iterations and the maximal number of Armijo damping steps performed for different mesh parameters h. To compare the number of iterations we choose a common initial guess $u_0 \equiv 1$. The number of iterations appears to be independent of h, while the amount of damping necessary seems to decrease with decreasing h. Note that small steps during damping might be meaningful due to the smallness of α.

Algorithm 3.4 has also been implemented successfully for parabolic control problems with a discontinuous Galerkin approximation of the states, as well as for elliptic control problems with Lavrentiev-regularized state constraints, see [HM08], [Vie07].

Acknowledgement

The first author gratefully acknowledges the support of the DFG Priority Program 1253 entitled Optimization With Partial Differential Equations. We also thank Andreas Günther for some fruitful discussions.

mesh param. h	ERR	ERR_∞	EOC	EOC_∞	Iterations	max#Armijo
$\sqrt{2}/2$	1.1230e-01	3.0654e-01	-	-	11	14
$\sqrt{2}/4$	3.8398e-02	1.4857e-01	1.55	1.04	22	11
$\sqrt{2}/8$	9.8000e-03	4.4963e-02	1.97	1.72	16	10
$\sqrt{2}/16$	1.7134e-03	1.2316e-02	2.52	1.87	18	10
$\sqrt{2}/32$	4.0973e-04	2.8473e-03	2.06	2.11	30	9
$\sqrt{2}/64$	8.2719e-05	6.2580e-04	2.31	2.19	15	7
$\sqrt{2}/128$	2.0605e-05	1.4410e-04	2.01	2.12	15	6
$\sqrt{2}/256$	4.7280e-06	4.6075e-05	2.12	1.65	15	6
$\sqrt{2}/512$	1.1720e-06	1.0363e-05	2.01	2.15	15	6

TABLE 2. Development of the error in Example 4.2 (Dirichlet) for $\alpha = 10^{-7}$.

References

[DH07a] Klaus Deckelnick and Michael Hinze. Convergence of a finite element approximation to a state-constrained elliptic control problem. *SIAM J. Numer. Anal.*, 45(5):1937–1953, 2007.

[DH07b] Klaus Deckelnick and Michael Hinze. A finite element approximation to elliptic control problems in the presence of control and state constraints. *Preprint, Hamburger Beitr. z. Ang. Math.* 2007-01, 2007.

[HIK03] M. Hintermüller, K. Ito, and K. Kunisch. The primal-dual active set strategy as a semismooth Newton method. *SIAM J. Optim.*, 13(3):865–888, 2003.

[Hin03] Michael Hinze. A generalized discretization concept for optimal control problems with control constraints. Technical Report Preprint MATH-NM-02-2003, Institut für Numerische Mathematik, Technische Universität Dresden, 2003.

[Hin05] Michael Hinze. A variational discretization concept in control constrained optimization: The linear-quadratic case. *Comput. Optim. Appl.*, 30(1):45–61, 2005.

[HM08] Michael Hinze and Christian Meyer. Numerical analysis of Lavrentiev-regularized state-constrained elliptic control problems. *Comp. Optim. Appl.*, 2008.

[HPUU09] M. Hinze, R. Pinnau, M. Ulbrich, and S. Ulbrich. *Optimization with PDE constraints.* Mathematical Modelling: Theory and Applications 23. Dordrecht: Springer., 2009.

[HU04] Michael Hintermüller and Michael Ulbrich. A mesh-independence result for semismooth Newton methods. *Mathematical Programming*, 101:151–184, 2004.

[HV] Michael Hinze and Morten Vierling. A globalized semi-smooth Newton method for variational discretization of control constrained elliptic optimal control problems. In *Constrained Optimization and Optimal Control for Partial Differential Equations*. Birkhäuser, 2011, to appear.

[KR08] Klaus Krumbiegel and Arnd Rösch. A new stopping criterion for iterative solvers for control constrained optimal control problems. *Archives of Control Sciences*, 18(1):17–42, 2008.

[Ulb03] Michael Ulbrich. Semismooth Newton methods for operator equations in function spaces. *SIAM J. Optim.*, 13(3):805–841, 2003.

[Ulb09] Michael Ulbrich. A new mesh-independence result for semismooth Newton methods. *Oberwolfach Rep.*, 4:78–81, 2009.

[Vie07] Morten Vierling. Ein semiglattes Newtonverfahren für semidiskretisierte steuerungsbeschränkte Optimalsteuerungsprobleme. Master's thesis, Universität Hamburg, 2007.

[VT09] Stefan Volkwein and Fredi Tröltzsch. Pod a posteriori error estimates for linear-quadratic optimal control problems. *Computational Optimization and Applications*, 44(1):83–115, 2009.

Michael Hinze and Morten Vierling
Schwerpunkt Optimierung und Approximation
Universität Hamburg
Bundesstraße 55
D-20146 Hamburg, Germany
e-mail: michael.hinze@uni-hamburg.de
morten.vierling@math.uni-hamburg.de

International Series of Numerical Mathematics, Vol. 160, 183–196

Structure Exploiting Adjoints for Finite Element Discretizations

Denise Holfeld, Philipp Stumm and Andrea Walther

Abstract. This paper presents some details for the development, analysis, and implementation of efficient numerical optimization algorithms using algorithmic differentiation (AD) in the context of partial differential equation (PDE) constrained optimization. This includes an error analysis for the discrete adjoints computed with AD and a systematic structure exploitation including efficient checkpointing routines, especially multistage and online checkpointing approaches.

Mathematics Subject Classification (2000). 65Y20, 90C30, 49N90, 68W40.

Keywords. Optimal control, calculation of adjoints, algorithmic differentiation, structure exploitation, checkpointing.

1. Introduction

In time-dependent flow control as well as in the framework of goal-oriented a posteriori error control, the calculation of adjoint information forms a basic ingredient to generate the required derivatives of the cost functional (see, e.g., [4]). However, the corresponding computations may become extremely tedious if possible at all because of the sheer size of the resulting discretized problem as well as its possibly nonlinear character, which imposes a need to keep track of the complete forward solution to be able to integrate the corresponding adjoint differential equation backwards. Due to the resulting memory requirement, several checkpointing techniques have been developed. Here, only a few intermediate states are stored as checkpoints. Subsequently, the required forward information is recomputed piece by piece from the checkpoints according to the requests of the adjoint calculation. If the number of time steps for integrating the differential equation describing the state is known a priori, one very popular checkpointing strategy is to distribute the checkpoints equidistantly over the time interval. However, it was shown in [11] that this approach is not optimal. One can compute optimal so-called binomial

This work was completed with the support of the DFG SPP 1253.

checkpointing schedules in advance to achieve for a given number of checkpoints an optimal, i.e., minimal, runtime increase [3]. This procedure is referred to as offline checkpointing and implemented in the package revolve [3].

However, in the context of flow control, the PDEs to be solved are usually stiff, and the solution process relies therefore on some adaptive time stepping procedure. Since in these cases the number of time steps performed is known only after the complete integration, an offline checkpointing strategy is not applicable. Instead, one may apply a straightforward checkpointing by placing a checkpoint each time a certain number of time steps has been executed. This transforms the uncertainty in the number of time steps to a uncertainty in the number of checkpoints needed. This approach is used by CVODES [5]. However, when the amount of memory per checkpoint is very high, one certainly wants to determine the number of checkpoints required a priori. For that purpose, we propose a new procedure for online checkpointing that distributes a given number of checkpoints on the fly during the integration procedure. This new approach yields a time-optimal adjoint computation for a given number of checkpoints for a wide range of applications.

Additionally, we present concepts how to employ AD to integrate the adjoint PDE backwards in time. The state equation to be solved has to be discretized in time and space in order to be adjoined by AD with the reverse mode. We present a new approach for the computation of adjoints by AD by exploiting spatial and time structure of the state equation. We show that this approach yields a considerable runtime reduction [9]. Furthermore, we derived an error analysis for the AD adjoints [7].

2. Checkpointing

The calculation of adjoints of time-dependent nonlinear PDEs often requires the full forward solution to be stored. This may be impossible due to lack of storage capacities. The checkpointing technique is an approach to memory reduction by storing only parts of the forward solution in combination with the execution of additional time steps. Furthermore, checkpointing techniques may reduce the overall runtime even in cases where the full forward trajectory could be stored due to the different access times of the available memory layers. For a given number of checkpoints c and a given number of time steps l, the binomial checkpointing minimizes the number of time steps to be performed for the adjoint calculation. By defining $\beta(c,r) \equiv \binom{c+r}{c}$, the minimal number of time steps is given by $t(c,l) = rl - \beta(c+1, r-1)$, where r is the unique integer satisfying $\beta(c,r-1) < l \leq \beta(c,r)$. r is called repetition number and denotes the maximum number of time step evaluations. An overview to $\beta(c,r)$ for repetition numbers $r = 2,3$ is illustrated in Tab. 1. The binomial checkpointing approach has been extended to multistage checkpointing, i.e., the checkpoints may be stored on different memory levels and to online checkpointing, where the number of time steps is a priori unknown.

2.1. Multistage checkpointing

If all checkpoints are held in the main memory, the access costs to a checkpoint are negligible. This changes if the checkpoints are stored at different levels of the memory hierarchy. In this case one has to take the memory access cost of the different checkpoints into account. We determine the checkpoint write and read counts for a binomial checkpointing approach and keep checkpoints with a high number of read and write counts in the main memory and the other ones on disc. The following subsections are a summary of [6], where the proofs of the theorems are given.

2.1.1. Determining the write counts. The function $w_i(l,c)$ is called write count for the checkpoint i for a binomial checkpoint distribution with l time steps and c checkpoints. To compute the value of $w_i(l,c)$, one has to distinguish the ranges $\beta(c,r-1) \le l \le \beta(c,r-1)+\beta(c-1,r-1)$ analyzed in Theorem 2.1 and $\beta(c,r-1)+\beta(c-1,r-1) \le l \le \beta(c,r)$ considered in Theorem 2.3. The write counts for the first range can be calculated as follows.

Theorem 2.1. *Assume $\beta(c,r-1) \le l \le \beta(c,r-1)+\beta(c-1,r-1)$. Then one has for the write counts $w_i = \beta(i,r-2)$.*

Before determining the number of write counts for the other range, one has to consider the write counts on special lengths l_k^w in the second range.

Theorem 2.2. *Let k be given with $0 \le k < c$. If $\beta(c,r-1)+\beta(c-1,r-1) \le l_k^w \le \beta(c,r)$, one has*

$$w_i(l_k^w,c) = \begin{cases} \beta(i,r-2) & \text{if } k=0 \text{ or } i=0 \\ \beta(i,r-1) & \text{if } i<k \\ \beta(i,r-1)-\beta(i-k-1,r-1) & \text{else} \end{cases}$$

where $l_k^w = \sum_{j=0}^{k+1} \beta(c-j,r-1)$.

Using Theorem 2.2, we derive the write counts for an arbitrary number l of time steps for the second range.

Theorem 2.3. *Let $\beta(c,r-1)+\beta(c-1,r-1) \le l \le \beta(c,r)$. Then one has*

$$w_i(l,c) = \begin{cases} w_i(l_k^w,c) & \text{if } i<k \\ w_i(l_k^w,c)+w_{i-k}(\tilde{l},c-k)-\beta(i-k,r-3) & \text{else} \end{cases}$$

where k is the unique integer satisfying

$$\sum_{j=0}^{k+1} \beta(c-j,r-1) \le l < \sum_{j=0}^{k+2} \beta(c-j,r-1)$$

and $\tilde{l} = l - l_k^w + \beta(c-k,r-2)+\beta(c-k-1,r-2)$.

2.1.2. Determining the read counts. The function $r_i(l,c)$ is called read count for the checkpoint i for a binomial checkpoint distribution with l time steps and c checkpoints. Once again, for the computation of $r_i(l,c)$ one has to consider the two ranges $\beta(c,r-1)+\beta(c-1,r-1) \leq l \leq \beta(c,r)$ (Theorem 2.4) and $\beta(c,r-1) \leq l \leq \beta(c,r-1)+\beta(c-1,r-1)$ (Theorem 2.5). If l lies the the first range, the read counts can be calculated using the write counts.

Theorem 2.4. *Let $\beta(c,r-1)+\beta(c-1,r-1) \leq l \leq \beta(c,r)$. Then one obtains for the read counts $r_i(l,c) = w_i(l,c) + \beta(i+1,r-2)$.*

The calculation of the read counts for the second range of time step numbers is given in the next theorem.

Theorem 2.5. *Let $\beta(c,r-1) \leq l \leq \beta(c,r-1)+\beta(c-1,r-1)$. Then one has for $c-i \leq k+1$ that*

$$r_i(l,c) = \beta(i+1,r-2) + \beta(i+k-c,r-2) + r_{i-k+1-c}(\tilde{l},k+1)$$

where $\tilde{l} = l - l_k^r + \beta(k+1,r-2)$ for the unique integer k satisfying

$$\beta(c,r-1) + \sum_{j=0}^{k-1} \beta(j,r-2) < l \leq \beta(c,r-1) + \sum_{j=0}^{k} \beta(j,r-2).$$

with $l_k^r = \beta(c,r-1) + \sum_{j=0}^{k} \beta(j,r-2)$.

The multistage checkpointing technique was integrated into the checkpointing software revolve (`www2.math.uni-paderborn.de/index.php?id=12067&L=1`).

2.2. Online checkpointing

The following subsection is a summary of [8], that also contains the proofs. If the number of time steps is unknown a priori the time integration process is given by Algorithm 1 where for each time step has to be tested if the termination criterion is fulfilled. As soon as it is fulfilled, the adjoint calculation starts. We developed

Algorithm 1 Adaptive Time Integration

```
for i = 0, 1, ... do
  u_{i+1} = F_i(u_i)
  if Termination criterion fulfilled then
    Start adjoint calculation.
  end if
end for
```

two algorithms for an online checkpointing procedure that determines the checkpoint distributions on the fly. We proved that these approaches yield checkpointing distributions that are either optimal or almost optimal with only a small gap to optimality. The first algorithm can be applied for a number l of time steps with

c	10	20	40	80	160	320
$\beta(c,2)$	66	231	861	3321	13041	51681
$\beta(c,3)$	286	1771	12341	91881	708561	5564321

TABLE 1. Maximal number of time steps for algorithms 2 and 3

$l \leq \beta(c,2)$ and the second one for $\beta(c,2) \leq l \leq \beta(c,3)$. The maximum number of time steps that can be handled with these algorithms is illustrated in Tab. 1. If l exceeds the upper bound for these two algorithms an alternative checkpointing strategy explained later can be applied.

Algorithm 2 Optimal Online Checkpointing Algorithm for $l \leq \beta(c,2)$

$\mathbf{C} = \{t | t = (j+1)c - \frac{1}{2}j(j-1) \; j = 0, \ldots, c\}$
for $l = 0, 1, \ldots, \beta(c,2)$ **do**
  Evaluate $u_{l+1} = F_l(u_l)$
  if Termination criterion fulfilled **then**
    Start adjoint calculation.
  end if
  if $l \notin \mathbf{C}$ **then**
    Store state u_l in a checkpoint.
  end if
end for

Algorithm 2 ensures optimal checkpointing schedules if $l \leq \beta(c,2)$. If the number of time steps exceeds $\beta(c,2)$, one has to switch to Algorithm 3. This algorithm is more complicated to analyze. For this purpose, a general checkpoint distribution for c checkpoints is illustrated in Fig. 1, where d_{c-i} defines the distance between the $(i+1)$-st and i-th checkpoint.

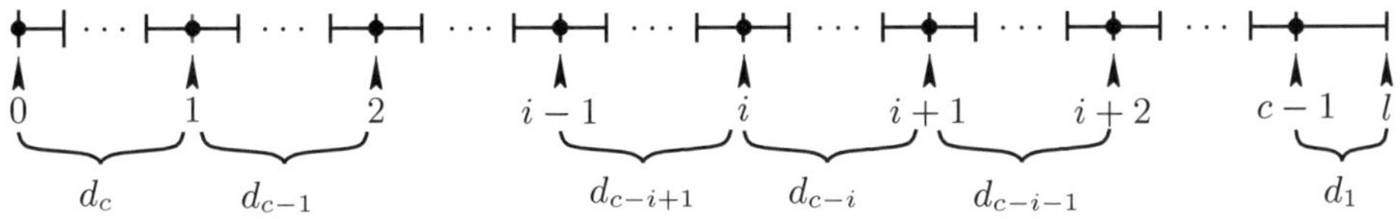

FIGURE 1. General checkpoint distribution

d_1^s denotes the distance between the last checkpoint and l and d_i^s the distance between the $(c-i)$th and $(c-i+1)$th checkpoint for $l = \beta(c,2)$. One has that

$$d_1^s = 3\,, \quad d_i^s = i+2 \quad \text{for} \quad 2 \leq i \leq c.$$

Denoting with d_i^e the corresponding distances for $l = \beta(c,3)$ gives

$$d_1^e = 4\,, \quad d_2^e = 6\,, \quad d_i^e = d_{i-1}^e + i + 2 \quad \text{for} \quad 2 \le i \le c.$$

An optimal checkpoint distribution cannot be found for all values of l in this range. However if the gap to optimality is one, one can overwrite a current state in a checkpoint as described in the next theorem and the distribution is then optimal for the next two time steps. For this purpose, we define $\hat{t}(c,l)$ denoting the number of time steps to be performed for the adjoint calculation for a given checkpoint distribution. As above, $t(c,l) = rl - \beta(c+1, r-1)$ denotes the number of time steps to be reversed for the binomial checkpointing approach, hence the minimal number of time step executions.

Theorem 2.6 (replacement condition). *Let a checkpoint distribution be given such that* $\hat{t}(c,l) = t(c,l) + 1$. *Applying one of the rules*

1. *the content of a checkpoint* $i = 1, \ldots, c-2$ *is replaced by* x_{l+1} *in the time integration from* x_l *to* x_{l+3} *if*

$$d_{c-i+1} + d_{c-i} \le \; d_{c-i+1}^e \quad \textit{and} \quad d_j > \; d_j^s \quad \textit{for } j = 2, \ldots, c-i-1$$

   holds, or

2. *the content of the last checkpoint* $c-1$ *is replaced by* x_{l+1} *if* $d_2 = 3$

ensures $\hat{t}(c, l+3) = t(c, l+3) + 1$.

We derive the following Algorithm 3 for online checkpointing if $\beta(c,2) \le l \le \beta(c,3)$ based on Theorem 2.6.

Algorithm 3 Online Checkpointing Algorithm for $\beta(c,2) = l \le \beta(c,3)$

```
Set f = 0 and D = {d_i^e | i = 1,...,c}.
for l = β(c,2),...,β(c,3) do
  Evaluate u_{l+1} = F_l(u_l).
  if Termination criterion fulfilled then
    Start reversal.
  end if
  if f = 0 then
    Store state u_l in a checkpoint with Theorem 2.6.
    if l + 1 ∈ D then
      f=0
    else
      f=2
    end if
  end if
end for
```

If $l = \beta(c,3)$, Algorithm 3 stops. Since the checkpointing distribution of Algorithm 3 coincides with the dynamical checkpointing algorithm presented in [12] for $l = \beta(c, 3)$, one may proceed with this algorithm that minimizes at least the repetition number. But does not yield an optimal or almost optimal checkpointing scheme for all time steps for a general r. However, it allows the application of checkpointing schedules for an unlimited number of time steps. For this purpose, we proceed with this algorithm. The online checkpointing technique is also integrated in the checkpointing software revolve, hence revolve now provides the binomial, the multistage and the online checkpointing technique in one tool.

3. Structure exploiting adjoints

When solving PDE constrained optimal control problems, the corresponding adjoint PDE has a special time and spatial structure. Our main contribution here is the derivation of the adjoint equation with an efficient algorithm using AD by exploiting the structure in space and time and the error analysis of the AD adjoint to the adjoint of the discretize-then-optimize approach. The following aspects are a summary of [7]. We consider a PDE constrained optimal control problem of the form

$$J(u, q) = j(u(T)) + R(q) \to \min! \tag{3.1}$$

$$s.t.\ \partial_t u + \nabla \cdot f(u) = S(q) \quad \text{in } \Omega \tag{3.2}$$

$$f(u) \cdot n = f^b(u) \quad \text{on } \Gamma$$

$$u(0, x) = u_0(x)$$

with the nonlinear flux $f : \mathbb{R} \to \mathbb{R}$ and the state $u : [0, T] \times \Omega \subset [0, T] \times \mathbb{R}^d \to \mathbb{R}$. The normal direction vector is denoted by n and the boundary of Ω by Γ. The distributed control is determined by $q : [0, T] \to \mathbb{R}$. Regularization terms are represented by $R(q)$.

The PDE is transformed into an ODE by a dG-method. We denote the state discretized in space by u_h. Then the required integrals are replaced by quadrature rules. The resulting discretized state is denoted by $\tilde{u}_h$. Finally, a Runge Kutta method is applied for the time discretization yielding the fully discretized state $\mathbf{u}$ which is then differentiated by AD yielding $\bar{\mathbf{u}}$. We derived an error analysis for the AD adjoint and the adjoints of the discretize-then-optimize approach without/with replacement of the integrals (denoted by $\lambda_h/\tilde{\lambda}_h$).

3.1. Structure exploiting AD

To employ AD in optimal control problems, we use the reverse mode since one has for a distributed control problem as considered here a large number of inputs and only one output, the objective value. There exist at least three approaches to the application of AD in optimal control problems – black box, TSE and TSSE approach. Using the black box approach means that we consider the whole time integration as a black box to compute our adjoints. This method may result in a massive memory requirement for large problems. Second, in the TSE approach every time step of the Runge-Kutta integration is considered as a black box. This yields that in every time step one adjoins one Runge-Kutta step. This means that if Φ denotes the increment function, one obtains $\mathbf{u}_{j+1} = \Phi(\mathbf{u}_j)$ in the time integration step $j \to j+1$. Hence, one obtains the adjoint by $\bar{\mathbf{u}}_j = (\Phi_u)^T \bar{\mathbf{u}}_{j+1}$ in the adjoint time step $j+1 \to j$. The memory requirement for this approach is much lower compared to the black box approach. Third, for the TSSE approach the increment function $\tilde{\Phi}$ is adjoined on each finite element K separately and accumulated appropriately. The TSSE approach reduces the memory requirement and the runtime as shown for a numerical example [9].

For the theoretical analysis, we need the following notation. The domain Ω is subdivided into shape-regular meshes $T_h = \{K_k\}_{k=1,\dots,N}$ where N denotes the number of elements. For simplicity, we use K instead of K_k for one element of T_h. The increment function Φ can be decomposed into $\Phi = (\Phi^1, \dots, \Phi^N)$ for the elements $k = 1, \dots, N$. The adjoint for the adjoint integration step $j+1 \to j$ is determined by $\bar{\mathbf{u}}_j^K = (\Phi_u^K)^T \bar{\mathbf{u}}_{j+1}^K$ for all elements K of the triangulation T_h and then accumulated to $\bar{\mathbf{u}}_j = (\bar{\mathbf{u}}_j^1, \dots, \bar{\mathbf{u}}_j^N)$. Therefore, we consider Φ^K for the finite element K as a black box. This technique leads to a lower memory requirement compared to the TSE approach which may also result in a lower runtime [9].

3.2. Error analysis

We provide error estimates for $\|u_h(t_j) - \mathbf{u}_j\|$ and $\|\lambda_h(t_j) - \bar{\mathbf{u}}_j\|$. Both errors can be split into two terms yielding

$$\begin{aligned}
\|u_h(t_j) - \mathbf{u}_j\| &\leq \|u_h(t_j) - \tilde{u}_h(t_j)\| + \|\tilde{u}_h(t_j) - \mathbf{u}_j\| \\
\|\lambda_h(t_j) - \bar{\mathbf{u}}_j\| &\leq \|\lambda_h(t_j) - \tilde{\lambda}_h(t_j)\| + \|\tilde{\lambda}_h(t_j) - \bar{\mathbf{u}}_j\|.
\end{aligned}$$

Due to the properties of the Runge-Kutta method, $\|\tilde{u}_h(t_j) - \mathbf{u}_j\| \leq C_T h_T^\nu$ for $C_T > 0$ and $\|\tilde{\lambda}_h(t_j) - \bar{\mathbf{u}}_j\| \leq D_T h_T^\nu$ for $D_T > 0$ [10] if an s-stage Runge-Kutta method of order ν is used. The errors $\|u_h(t) - \tilde{u}_h(t)\|$ and $\|\lambda_h(t) - \tilde{\lambda}_h(t)\|$ are caused by the quadrature rules. For the error estimate, we introduce the definition

$$\begin{aligned}
h_d &= \max_{K \in T_h} \{h_K | h_K \text{ is the diameter of K}\}, \\
h_l &= \max_{K \in T_h} \{h_e | h_e \text{ is the length of the edge of K}\}, \text{ and} \\
h_T &= \max_{i=1,\dots,\hat{N}} \Delta t \\
H &= (h_d, h_l)
\end{aligned}$$

and the assumption

Assumption 3.1. *Assume that Ω is finite and there exits a regular triangulation T_h of Ω. Furthermore, let the quadrature rule on $K \in T_h$ be of order k and on the edge of $K \in T_h$ of order l.*

We use a discontinuous Galerkin method for the spatial discretization. Therefore, the flux function f in 3.2 has to be replaced by a numerical flux function h^f on the edge of each element K.

Theorem 3.2. *Let Assumption* 3.1 *be fulfilled. Let the flux function f, the numerical flux function h^f, and f^b be Lipschitz-continuous. Then the following error estimate*

$$\|u_h(t) - \tilde{u}_h(t)\| \leq \rho_1 \exp\left(L_1(H)(t - t_0)\right) + \frac{\epsilon_1(H)}{L_1(H)} \left(\exp\left(L_1(H)(t - t_0)\right) - 1\right)$$

holds where

$$\epsilon_1(H) = C_1 h_d^{k+1} + C_2 h_l^{l+1} \quad \text{and} \quad \rho_1 = \|u_h(t_0) - \tilde{u}_h(t_0)\|$$

for positive constants L_1, C_1 and C_2, where L_1 depends on H and C_1 and C_2 are independent of H.

Theorem 3.3. *Let Assumption* 3.1 *be fulfilled and let the flux function f, the numerical flux function h^f, and f^b be continuously differentiable. Then we have the following error estimate*

$$\|\lambda_h(t) - \tilde{\lambda}_h(t)\| \leq \rho_2 \exp\left(L_2(H)(T - t)\right) + \frac{\epsilon_2(H)}{L_2(H)} \left(\exp\left(L_2(H)(T - t)\right) - 1\right)$$

where

$$\epsilon_2(H) = D_1 h_d^{k+1} + D_2 h_l^{l+1} \quad \text{and} \quad \rho_2 = \|\lambda_h(T) - \tilde{\lambda}_h(T)\|$$

for positive constants L_2, D_1 and D_2, where L_2 depends on H and D_1 and D_2 are independent of H.

Since the Lipschitz-constants L_1 and L_2 are $O(h^{-1})$, $\tilde{u}_h$ and $\tilde{\lambda}_h$ may not converge to u_h and λ_h as shown for a numerical example in [7]. However, the resulting curve of the error motivates to determine an optimal diameter h_d.

Theorem 3.4. *Let the quadrature rules on K be of order k and on ∂K be exact. Let f, h^f, and f^b be Lipschitz-continuous and let $u(t_0) = \tilde{u}(t_0)$. Let $\epsilon(h_d) = \epsilon_1(h_d)/L_1(h_d)$. Then, the optimal grid size h_{opt} satisfies*

$$\epsilon'(h_{\text{opt}}) + L'(h_{\text{opt}})\epsilon(h_{\text{opt}})(t - t_0) = 0$$

and is given by

$$h_{\text{opt}} = \frac{P_1(t - t_0)}{k + 2}$$

where $L(h_d) = P_1/h_d$ for an appropriate constant P_1.

4. Numerical results

For the three presented aspects of optimal control for time-dependent problems, numerical tests were performed.

4.1. Multistage checkpointing

The numerical example provided by the Institute for Aerospace Engineering, Technical University Dresden, is motivated by a plasma spraying procedure. The PDEs to be solved are the compressible Navier-Stokes equations given in vector form by

$$\partial_t U + \nabla \cdot \vec{F} = \nabla \cdot \vec{D} + S(q)$$

with the preservation variable $U = (\rho, \rho v_x, \rho v_y, \rho e)$, the advective fluxes F, the diffusive fluxes D and the source term S that depends on a control q. These PDEs are discretized by a discontinuous Galerkin method with symmetric interior penalization. The objective is the determination of a heat source distribution q such that $T[q] = T_0$, i.e.,

$$J(q) = \int_0^{t_F} \int_\Omega (T[q] - T_0)^2 \mathrm{d}\Omega \ \mathrm{d}t \longrightarrow 0$$

where $\Omega = (0, a) \times (-a, a)$. For our numerical example, we set $a = 0.001$ and $T_0 = 100$. The adjoints for one time step are determined by AD. We compared the normalized runtime, i.e., $\mathrm{TIME}(\nabla J)/\mathrm{TIME}(J)$, for one gradient evaluation for different grid sizes. The main result is surprising because the minimal runtime is not observed if the complete forward trajectory is stored. This runtime ratio is given by the straight line in Fig. 2. In addition, the least runtime depends on the grid, respectively on the size of the checkpoints. The runtime behavior can be explained this way. First, the number of recomputations decreases. Second, the access cost to disc increases the more checkpoints are placed on disc. The second effect outweighs the first one the more checkpoints are saved onto disc yielding that the overall normalized runtime increases at a special number of checkpoints.

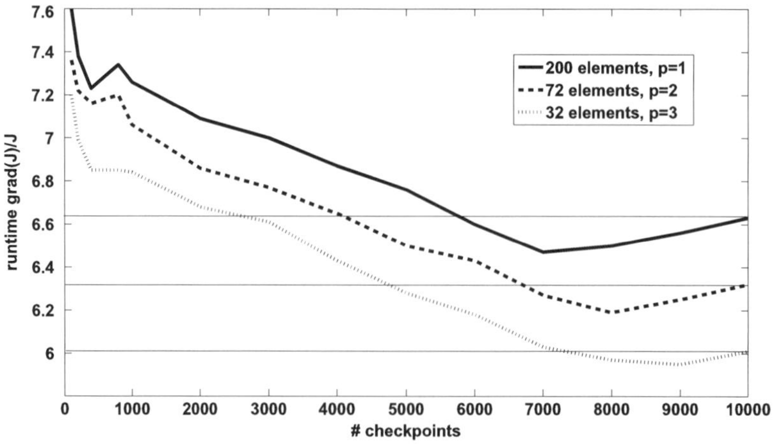

FIGURE 2. Comparison runtime optimal flow control with 10000 time steps for 400 checkpoints in RAM and different grids

4.2. Online checkpointing

Here, we consider a laminar flow over a flat plate. Small disturbance waves, so-called Tollmien-Schlichting waves, extend the boundary layer during time. Hence, Tollmien-Schlichting waves cause the transition from a laminar to a turbulent boundary layer. The transition point is determined by the critical Reynolds number. Electromagnetic forces may be used to achieve a flow profile where the transition point is delayed considerably. That is, a larger Reynolds number has to be achieved for which the flow is still stable.

A direct numerical software called Semtex [2] is used to simulate the flow using spectral elements. Semtex was extended by a modulated Lorentz force for the optimal control problem considered here. To couple the numerical simulation with flow control approaches an efficient provision of derivatives is indispensable. For this purpose, a calculation of adjoints was added to SEMTEX including hand-coded and AD-generated adjoints [1].

For a laminar flow, an optimization based on a variable time step number and an open loop approach turned out to be most efficiently [1]. In this case, the number of time steps for one gradient calculation is determined on the basis of an appropriate stopping criterion of the time integration. Therefore, online checkpointing has to be used.

We compare the normalized runtime for one gradient evaluation by the online checkpointing presented here versus the optimal offline checkpointing both implemented in revolve. Fig. 3 illustrates the runtime behavior and the number of recomputations of the two checkpointing approaches for a number of 13700 time steps to perform and up to 300 checkpoints yielding a repetition rate of $r \leq 3$. Hence, Algorithm I is used for 150 and more checkpoints and Algorithms I and II for 40 up to 150 checkpoints. For less than 40 checkpoints the presented algorithm in [12] is used additionally. We observe the typical behavior of checkpointing approaches, i.e., the runtime and the number of recomputations decrease when the number of checkpoints increases. One can see that the runtime of optimal offline checkpointing and the online checkpointing approach almost coincide for a wide range. Only for less than 40 checkpoints the difference between the two approaches is marginal greater, which might be caused by the non-optimal checkpointing algorithm presented in [12] that has to be used.

4.3. Structure exploitation

Here, we consider once more the numerical example described in Sec. 4.1 with a number of time steps varying in the range $100 - 5000$ where $\Delta t = 5 * 10^{-8}$ for one time step. We use linear test and trial functions over a grid containing 200 elements and compare the TSE approach and the TSSE approach. In Fig. 4 the dashed lines represent the results for the memory requirement in MByte and the solid lines the results for the runtime in seconds. Evidently, the memory reduction of the TSSE approach is enormous. Up to 80% memory reduction compared to the TSE approach can be obtained leading to a reduction of 50% of runtime in average for the considered numbers of time steps. This reduction would be even

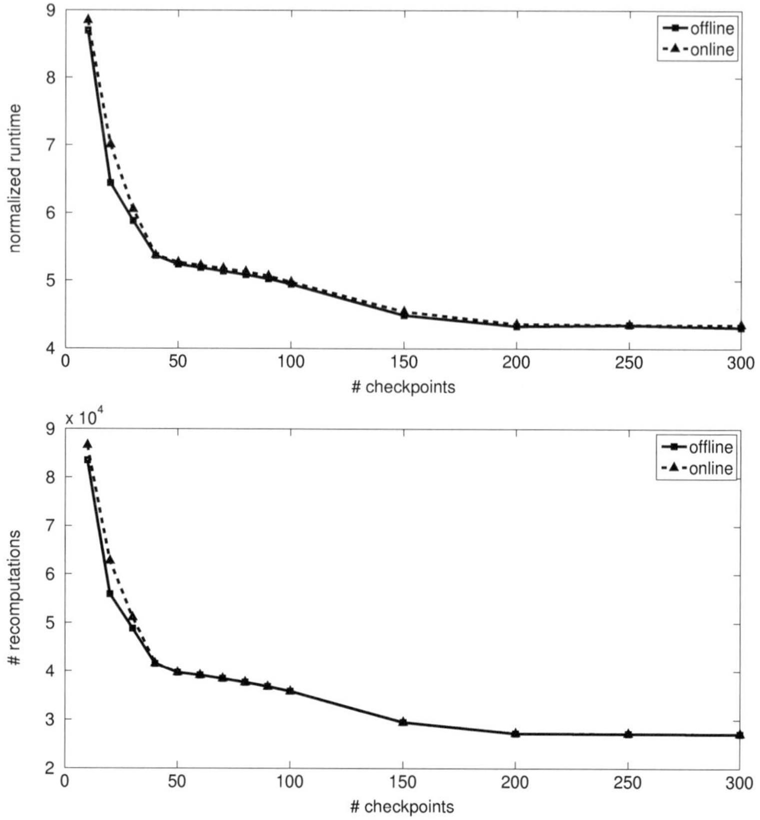

FIGURE 3. Comparison of normalized runtime and number of recomputations required by optimal offline checkpointing and (almost) optimal online checkpointing

greater for more time steps since more checkpoints can be stored using the TSSE approach. Note that the runtime increases linearly in both cases.

In [7] we provide a detailed analysis for the error caused by AD adjoints and by quadrature rules. We show that applying an inexact quadrature rule may lead to divergence behavior.

5. Conclusion

In this project, we provided a systematic way how to employ AD in PDE constrained optimal control problems, especially concentrating on the spatial and time structure exploitation to achieve good runtime results and a low memory requirement. Additionally, we derived an error estimate for the AD adjoint and

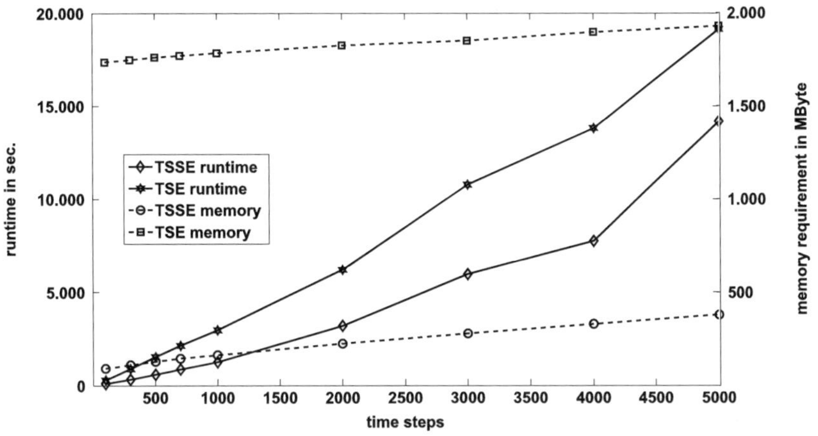

FIGURE 4. Runtime and memory requirement for the TSE and TSSE approach

showed that the AD adjoint converges to the adjoint of the discretize-then-optimize approach if quadrature rules are used that determine the integrals correctly. In order to reduce the memory requirement, we extended the binomial checkpointing approach to a multistage approach, such that states may be stored on disc too and to an online checkpointing approach such that the distributions almost coincide with these of the binomial checkpointing by providing a theoretical analysis. A very surprising result is that using multistage checkpointing may result in a lower runtime compared to storing all states. Also, our online checkpointing approach leads to runtimes that almost coincides with binomial checkpointing approaches that cannot be applied for adaptive time stepping schemes.

References

[1] T. Albrecht, H. Metskes, J. Stiller, D. Holfeld, and A. Walther. Adjoint-based optimization of fluids flows. Technical report, TU Dresden, 2010.

[2] H.M. Blackburn and S.J. Sherwin. Formulation of a Galerkin spectral element-fourier method for three-dimensional incompressible flows in cylindrical geometries. *J Comput Phys*, 197(2):759-778, 2004.

[3] A. Griewank and A. Walther. Revolve: An implementation of checkpointing for the reverse or adjoint mode of computational differentiation. *ACM Transactions on Mathematical Software*, 26:19–45, 2000.

[4] M.D. Gunzburger. *Perspectives in flow control and optimization.* Advances in Design and Control 5. Philadelphia, PA: Society for Industrial and Applied Mathematics (SIAM). xiv, 261 p., 2003.

[5] R. Serban and A.C. Hindmarsh. CVODES: An ODE solver with sensitivity analysis capabilities. UCRL-JP-20039, LLNL, 2003.

[6] P. Stumm and A. Walther. Multistage approaches for optimal offline checkpointing. *SIAM Journal on Scientific Computing*, 31(3):1946–1967, 2009.

[7] P. Stumm and A. Walther. Error estimation for structure exploiting adjoints. Technical Report SPP1253-103, TU Dresden, 2010.

[8] P. Stumm and A. Walther. New algorithms for optimal online checkpointing. *SIAM Journal on Scientific Computing*, 32(2):836–854, 2010.

[9] P. Stumm, A. Walther, J. Riehme, and U. Naumann. Structure-exploiting automatic differentiation of finite element discretizations. In Christian H. Bischof and H. Martin Bücker, editors, *Advances in Automatic Differentiation*, pages 339–349. Springer, 2008.

[10] A. Walther. Automatic differentiation of explicit Runge-Kutta methods for optimal control. *Comput. Optim. Appl.*, 36(1):83–108, 2007.

[11] A. Walther and A. Griewank. Advantages of binomial checkpointing for memory-reduced adjoint calculations. In M. Feistauer, V. Dolejší, P. Knobloch, and K. Najzar, editors, *Numerical Mathematics and Advanced Applications, ENUMATH* 2003. Springer, 2004.

[12] Q. Wang, P. Moin, and G. Iaccarino. Minimal repetition dynamic checkpointing algorithm for unsteady adjoint calculation. *SIAM Journal on Scientific Computing*, 31(4):2549–2567, 2009.

Denise Holfeld and Andrea Walther
Universität Paderborn
Institut für Mathematik
Warburgerstr. 100
D-33098 Paderborn, Germany
e-mail: denise.holfeld@uni-paderborn.de
andrea.walther@uni-paderborn.de

Philipp Stumm
TU Dresden
Fachrichtung Mathematik
Institut für Wissenschaftliches Rechnen
D-01062 Dresden, Germany
e-mail: Philipp.Stumm@tu-dresden.de

International Series of Numerical Mathematics, Vol. 160, 197–212

Computing Covariance Matrices for Constrained Nonlinear Large Scale Parameter Estimation Problems Using Krylov Subspace Methods

Ekaterina Kostina and Olga Kostyukova

Abstract. In the paper we show how, based on the preconditioned Krylov subspace methods, to compute the covariance matrix of parameter estimates, which is crucial for efficient methods of optimum experimental design.

Mathematics Subject Classification (2000). Primary 65K10; Secondary 15A09, 65F30.

Keywords. Constrained parameter estimation, covariance matrix of parameter estimates, optimal experimental design, nonlinear equality constraints, iterative matrix methods, preconditioning.

1. Introduction

Parameter estimation and optimal design of experiments are important steps in establishing models that reproduce a given process quantitatively correctly. The aim of parameter estimation is to reliably and accurately identify model parameters from sets of noisy experimental data. The "accuracy" of the parameters, i.e., their statistical distribution depending on data noise, can be estimated up to first order by means of a covariance matrix approximation and corresponding confidence regions. In practical applications, however, one often finds that the experiments performed to obtain the required measurements are expensive, but nevertheless do not guarantee satisfactory parameter accuracy or even well-posedness of the parameter estimation problem. In order to maximize the accuracy of the parameter estimates additional experiments can be designed with optimal experimental settings or controls (e.g., initial conditions, measurement devices, sampling times,

This work was completed with the support of ESF Activity "Optimization with PDE Constraints" (Short visit grant 2990).

temperature profiles, feed streams etc.) subject to constraints. As an objective functional a suitable function of the covariance matrix can be used. The possible constraints in this problem describe costs, feasibility of experiments, domain of models etc.

The methods for optimum experimental design that have been developed over the last few years can handle processes governed by differential algebraic equations (DAE), but methods for processes governed by partial differential equations (PDE) are still in their infancy because of the extreme complexity of models and the optimization problem. The aim of this paper is to show how to compute covariance matrices using Krylov subspace methods and thus to make a step towards efficient numerical methods for optimum experimental design for processes described by systems of non-stationary PDEs.

2. Problems of parameter estimation (PE) and optimum experimental design (OED) for dynamic processes

2.1. Parameter estimation problems in dynamic processes

The parameter estimation problem can be described as follows. Let the dynamics of the model be described by a system of differential equations which depends on a finite number of unknown parameters. It is assumed that there is a possibility to measure a signal of an output device that writes at given points the output signal of the dynamic system with some errors. According to a common approach, in order to determine the unknown parameters an optimization problem is solved in which a special functional is minimized under constraints that describe the specifics of the model. The optimization functional describes the deviation of model response and known data which are output signals measured by the output device in a suitable norm. The choice of an adequate norm usually depends on the statistical properties of the measurement errors. The traditional choice is a weighted least-squares-type functional. An abstract (simplified) parameter estimation problem maybe formally written as follows:

$$\begin{aligned} \min \quad & J(q,u) := ||(S(u(t_i,x_i)) - \bar{S}_i, i = 1,\dots,N)||_Z^2, \qquad (2.1) \\ \text{s.t.} \quad & \mathcal{F}(t,x,q,u,\frac{\partial u}{\partial t},\frac{\partial u}{\partial x},\frac{\partial^2 u}{\partial x^2}) = 0, t \in T, x \in X; \\ & \text{some initial and boundary conditions,} \end{aligned}$$

where $t \in T \subset \mathbb{R}$ is time, $x \in X \subset \mathbb{R}^d$ are coordinate variables of a d-dimensional space, T and X bounded, the state variables $u = u(t,x)$ are determined in an appropriate space V, q are unknown parameters in an appropriate space Q, $\mathcal{F}$ is a vector function, $S(u)$ is an observation operator that maps the state variable u to the appropriate space of measurements, $\bar{S}_i$, $i = 1,\dots,N$, is a given set of output measurements, Z is a given positive definite matrix; $||z||_Z^2 = z^T Zz$.

The numerical methods for parameter estimation for processes described by systems of differential-algebraic equations are well established. They are based on the so-called "all-at-once" approach, see, e.g., [3, 7, 8, 13, 28]. The basic principle of this approach consists in discretizing the dynamic model including possible boundary conditions like a boundary value problem – independent of its specific nature – and in treating the resulting discretized model as nonlinear constraint in the optimization problem. This discretization is accomplished by finite differences, collocation, see, e.g., Ascher [3], Schulz [28] and Biegler [13], or multiple shooting methods, see, e.g., [7, 10].

The discretization of the dynamic model in (2.1) yields a finite-dimensional, large-scale, nonlinear constrained approximation problem which can be formally written as

$$\min_{z\in\mathbb{R}^n} \quad ||F_1(z)||_2^2, \text{s.t. } F_2(z) = 0. \tag{2.2}$$

Note, that the equality conditions $F_2(z) = 0$, include the discretized dynamic model. We assume that the functions $F_i : D \subset R^n \to R^{m_i}$, $i = 1, 2$, are twice-continuously differentiable. The vector $z \in \mathbb{R}^n$ combines parameters and all variables resulted from discretization of model dynamics. The problem (2.2) is best solved by a generalized Gauss-Newton method, which iteration scheme is (basically) given by

$$z^{k+1} = z^k + t^k \Delta z^k, \qquad 0 < t^k \leq 1, \tag{2.3}$$

where the increment Δz^k is the solution of the linearized problem at $z = z^k$

$$\min_{\Delta z\in\mathbb{R}^n} \quad ||F_1(z) + J_1(z)\Delta z||_2^2, \text{s.t. } F_2(z) + J_2(z)\Delta z = 0, \tag{2.4}$$

Here we use notations $J_i(z) = \dfrac{\partial F_i(z)}{\partial z}$, $i = 1, 2$. The Gauss-Newton method almost shows performance of a second-order method requiring only provision of the first-order derivatives. The method usually makes use of a tailored linear algebra, to exploit the special structures arising from the all-at-once-approach discretization of the dynamics. At each iteration of the Gauss-Newton method one solves a linear least-squares constrained approximation problem (2.4).

Under certain regularity assumptions on the Jacobians $J_i(z^k)$, $i = 1, 2$, a linearized problem (2.4) has a unique solution Δz^k which can be formally written with the help of a solution operator J^+

$$\Delta z^k = -J^+ F(z^k), \; F(z) = \begin{pmatrix} F_1(z) \\ F_2(z) \end{pmatrix}.$$

The solution operator J^+ is a generalized inverse, that is it satisfies $J^+ J J^+ = J^+$ with $J = J(z) = \begin{pmatrix} J_1(z) \\ J_2(z) \end{pmatrix}$.

It is important for parameter estimation problems to compute not only parameters but also a statistical assessment of the accuracy of these parameter estimates. This can be done by means of the covariance matrix. A representation of the covariance matrix for constrained nonlinear parameter estimation problems can be found in [8, 9] and can be briefly summarized as follows. If the experimental data is normally distributed ($\mathcal{N}(0, \mathbb{I}_{m_1})$), where $\mathbb{I}_s$ denotes the $s \times s$ identity matrix then the computed parameter estimates are also random variables which are normally distributed in the first order with the (unknown) true value as expected values and the variance-covariance matrix $C \in \mathbb{R}^{n \times n}$ given by

$$C = J^+ \begin{pmatrix} \mathbb{I}_{m_1} & 0_{m_1 \times m_2} \\ 0_{m_2 \times m_1} & 0_{m_2 \times m_2} \end{pmatrix} J^{+T}.$$

Here $0_{k \times s}$ denotes the $k \times s$ zero matrix.

The variance-covariance matrix describes the confidence ellipsoid which is an approximation of the nonlinear confidence region of the estimated variables. For details see [8, 9].

2.2. Optimum experimental design problems

It often happens in practice that data evaluated under different experimental conditions leads to estimations of parameters with good or poor confidence regions depending on the experimental conditions. One would like to find those experiments that result in the best statistical quality for the estimated parameters and at the same time satisfy additional constraints, e.g., experimental costs, safety, feasibility of experiments, validity of the model etc. Thus the aim of optimum experimental design is to construct N_{ex} complementary experiments by choosing appropriate experimental variables and experimental controls $\xi_1, \ldots, \xi_{N_{\text{ex}}}$ in order to maximize the statistical reliability of the unknown variables under estimation. For this purpose one solves a special optimization problem. Since the "size" of a confidence region is described by the covariance matrix C a suitable function of matrix C may be taken as a cost functional. One possible choice is $\Phi(C) = \text{trace}(C)$.

The optimum experimental design approach leads to an optimal control problem in dynamic systems

$$\begin{array}{ll}
\min\limits_{u,\xi} & \Phi(C(u, \xi, q)), \\
\text{where} & C \text{ is the covariance matrix of the underlying PE problem,} \\
& u = (u_1, \ldots, u_{N_{\text{ex}}}), \xi = (\xi_1, \ldots, \xi_{N_{\text{ex}}}) \\
\text{s.t.} & c_i(t, u_i(t,x), \xi_i(t), q) \geq 0,\ t \in T_i,\ \text{a.e. },\ i = 1, \ldots, N_{\text{ex}}, \\
& u_i = (u_i(t,x),\ t \in T_i,\ x \in X_i),\ q \in Q,\ \xi_i = (\xi_i(t),\ t \in T_i) \\
& T_i \subset \mathbb{R}, X_i \subset \mathbb{R}^d,\ i = 1, \ldots, N_{\text{ex}}, \\
& T_i, X_i \text{ are bounded,} \\
& \text{satisfy model dynamics, e.g.,} \\
& \mathcal{F}_i(t, x, q, u_i, \dfrac{\partial u_i}{\partial t}, \dfrac{\partial u_i}{\partial x}, \dfrac{\partial^2 u_i}{\partial x^2}) = 0, t \in T_i, x \in X_i, \\
& \text{and initial and boundary conditions.}
\end{array}$$

The decision variables in this optimization problem are the control profiles (e.g., temperature profiles of cooling/heating or inflow profiles in reaction kinetics experiments, or force and torque inputs in mechanical multibody systems) and the time-independent control variables (e.g., initial concentrations or reactor volume in reaction kinetics, or properties of the experimental device) including integer variables describing whether a possible data evaluation should be carried out or not. The state and control constraints $c_i(\cdot) \geq 0, i = 1, \ldots, N_{\text{ex}}$, describe limitations of experimental costs, safety requirements, feasibility of experiments, validity of the model etc in each experiment. The experimental design optimization problem is a nonlinear constrained optimal control problem. The main difficulty lies in the non-standard objective function due to its implicit definition on the sensitivities of the underlying parameter estimation problem, i.e., on the derivatives of the solution of the differential equation system with respect to the parameters and initial values.

Since the early pioneering works of Kiefer and Wolfowitz in the middle of the last century, the theory of experimental design for linear systems has been further developed and now may be considered as well-established. A comprehensive overview can be found in the monograph of Pukelsheim [24]. Comparable state-of-the-art comprehensive methods for nonlinear models do not exist yet. Beyond theoretical studies of Fedorov and Atkinson [1, 15] there exist few numerical methods. The applications described in the literature are concentrated mainly on very simple model problems [1, 2, 26]. The main reasons for this are the complexity of the optimization problem and the high computational costs in applications to realistic processes.

Numerical methods for design of experiments for nonlinear DAE systems have been only developed in the last years in Heidelberg, see [5, 6, 19, 20]. These numerical methods for optimum experimental design in DAE are based on the *direct* optimization approach, according to which the control functions are parametrized on an appropriate grid by a special discretization approach, and the solution of the DAE systems and the state constraints are discretized. As a result, one obtains a finite-dimensional constrained nonlinear optimization problem, which is solved by an SQP method. The main effort for the solution of the optimization problem by the SQP method is spent on the calculation of the values of the objective function and the constraints as well as their gradients. Efficient methods for derivative computations combining internal numerical differentiation [8] of the discretized solution of the DAE and automatic differentiation of the model functions [16] have been developed, e.g., in [20]. Note that second-order derivatives are required, since the objective function is based on first derivatives already. The methods allow simultaneous optimization of controls and design parameters as well as sampling design. Methods for the design of robust optimal experiments are developed in [19]. They allow to compute the worst-case designs for the case that the parameter values are only known to lie in a possibly large confidence region. The treatment of many applications in particular in chemistry and biology show that drastic improvements of the quality of parameters can be achieved.

First steps towards optimum experimental design methods in the context of PDE are described in [11, 12, 17]. The main numerical issues addressed in [11, 12] are a posteriori error estimates and development of adaptive finite element methods. Haber et al [17] discuss numerical methods for experimental design for ill-posed linear inverse problem when regularization techniques are used.

The aim of the paper is another aspect of numerical OED methods, namely an efficient computation of covariance matrices and their derivatives. So far numerical methods for parameter estimation and optimal design of experiments in dynamic processes have been based on *direct linear algebra methods*. On the other hand, for very large scale constrained systems with sparse matrices of special structure, e.g., originating from discretization of PDEs, the direct linear algebra methods are not competitive with *iterative linear algebra* methods even for forward models. Generally, the covariance matrix can be calculated via a corresponding generalized inverse matrix of Jacobians of parameter estimation problem. But basically the generalized inverse can not be computed explicitly by iterative methods, hence the statistical assessment of the parameter estimate can not be provided by the standard procedures of such methods. Hence, in case of PE and OED in PDE models, generalizations of iterative linear algebra methods to the computation of the covariance matrix and its derivatives are crucial for practical applications.

3. Covariance matrix and its numerical computation using Krylov subspace methods

As in [9], we consider the constrained nonlinear parameter estimation problem (2.2), which results from problem (2.1) after corresponding discretization. Here the vector z includes unknown parameters and variables resulting from discretization of a PDE.

To solve problem (2.2) we apply a generalized Gauss-Newton method. At each iteration of the Gauss-Newton algorithm we solve a linear least-squares problem, which can be written in the form (see also (2.4))

$$\min_y ||Ay - \bar{b}||_2^2, \text{ s.t. } By = c, \tag{3.1}$$

where

$$\bar{b} = -F_1(z), \quad c = -F_2(z), \quad A = A(z) = \tfrac{\partial F_1(z)}{\partial z} \in \mathbb{R}^{m_1 \times n},$$
$$B = B(z) = \tfrac{\partial F_2(z)}{\partial z} \in \mathbb{R}^{m_2 \times n}, \quad \bar{m} = n - m_2 > 0,$$

z is given, and

$$\text{rank} B = m_2, \quad \text{rank} \begin{pmatrix} A \\ B \end{pmatrix} = n. \tag{3.2}$$

Let $\bar{y}$ be a vector satisfying $B\bar{y} = c$. Then the problem (3.1) is equivalent to the following one

$$\min_x ||Ax - b||_2^2, \text{ s.t. } Bx = 0, \tag{3.3}$$

with $b = \bar{b} - A\bar{y}$. Having the solution $x^\star$ of the problem (3.3), the solution $y^\star$ of the problem (3.1) can be computed as $y^\star = \bar{y} + x^\star$. Hence in what follows without loss of generality we consider the problem (3.3).

It is shown in [9] how to compute covariance matrices when the underlying finite-dimensional constrained linearized parameter estimation problem (3.3) is solved using a conjugate gradient technique. One of the intriguing results of [9] is that solving linear constrained least squares problems by conjugate gradient methods we get as a by-product the covariance matrix and confidence intervals as well as their derivatives. The results are generalized for LSQR methods and are numerically tested in [27, 21] and are briefly summarized in the following.

Under the regularity assumptions (3.2), the optimal solution x^* and the Lagrange vector λ^* satisfy the KKT system

$$\mathcal{K}\begin{pmatrix} x^* \\ \lambda^* \end{pmatrix} = \begin{pmatrix} A^T b \\ 0 \end{pmatrix}, \quad \mathcal{K} := \begin{pmatrix} A^T A & B^T \\ B & 0_{m_2 \times m_2} \end{pmatrix},$$

and can be explicitly written using the linear operator A^+ as follows

$$x^* = A^+ \begin{pmatrix} b \\ 0 \end{pmatrix}, \quad A^+ = \begin{pmatrix} \mathbb{I}_n & 0_{n\times m_2} \end{pmatrix} \mathcal{K}^{-1} \begin{pmatrix} A^T & 0_{n\times m_2} \\ 0_{m_2\times m_1} & \mathbb{I}_{m_2} \end{pmatrix}.$$

The approximation of the variance-covariance matrix can be expressed as the following matrix [9]

$$C = A^+ \begin{pmatrix} \mathbb{I}_{m_1} & 0_{m_1\times m_2} \\ 0_{m_2\times m_1} & 0_{m_2\times m_2} \end{pmatrix} A^{+T}. \tag{3.4}$$

which we will call in the following, for the sake of brevity, a covariance matrix. It is shown in [9] that C satisfies a linear system of equations.

Lemma 3.1. *The covariance matrix C (3.4) is equal to the sub-matrix X of the matrix $\mathcal{K}^{-1}$*

$$\mathcal{K}^{-1} = \begin{pmatrix} X & W \\ S & T \end{pmatrix}$$

and satisfies the following linear equation system with respect to variables $C \in \mathbb{R}^{n\times n}$ and $S \in \mathbb{R}^{m_2\times n}$

$$\mathcal{K}\begin{pmatrix} C \\ S \end{pmatrix} = \begin{pmatrix} \mathbb{I}_n \\ 0_{m_2\times n} \end{pmatrix}. \tag{3.5}$$

Another presentation of the covariance matrix is given in [27]

Lemma 3.2. *The covariance matrix C (3.4) is equal to*

$$C = Z(Z^T A^T A Z)^{-1} Z^T,$$

where the columns of the orthogonal matrix $Z \in \mathbb{R}^{n\times\bar{m}}$ span the null space of B.

Let us present formulas for the computation of the derivatives of the covariance matrix $C = C(A,B)$ and the matrix $S = S(A,B)$, as the functions of the matrices A and B. These derivatives are needed in numerical methods for design of optimal nonlinear experiments in case the trace of C is chosen as a criterion.

Let $A(t) = A + t\Delta A$ and $B(\mu) = B + \mu\Delta B$. Then

$$\left.\frac{\partial \mathrm{tr} C(A(t), B)}{\partial t}\right|_{t=0} = -\sum_{i=1}^{n} C^{(i)T}(\Delta A^T A + A^T \Delta A)C^{(i)},$$

$$\left.\frac{\partial \mathrm{tr} C(A, B(\mu))}{\partial \mu}\right|_{\mu=0} = -2\sum_{i=1}^{n} C^{(i)T} \Delta B S^{(i)},$$

where $C^{(i)}$ and $S^{(i)}$ denote ith columns of matrices C and S respectively. A column $S^{(i)}$ of the matrix S satisfies $B^T S^{(i)} = (e_i - A^T A C^{(i)})$.

Let us show how to compute the covariance matrix using LSQR, an iterative method for solving large linear systems or least-squares problems which is known to be numerically more reliable than other conjugate-gradient methods [22, 23].

Algorithm LSQR($\mathcal{A}$, b) (for solving $\min_x ||\mathcal{A}x - b||_2^2$)

Step 1: (*Initialization*)
$\beta_1 u_1 = b$, $\alpha_1 v_1 = \mathcal{A}^T u_1$, $p_1 = v_1$, $x_0 = 0$,
$\bar{\phi}_1 = \beta_1$, $\bar{\rho}_1 = \alpha_1$

Step 2: For $k = 1, 2, 3, \ldots$ repeat steps **2.1**–**2.4**.

2.1: (*Continue the bidiagonalization*)
1. $\beta_{k+1} u_{k+1} = \mathcal{A} v_k - \alpha_k u_k$
2. $\alpha_{k+1} v_{k+1} = \mathcal{A}^T u_{k+1} - \beta_{k+1} v_k$

2.2: (*Construct and apply next orthogonal transformation*)
1. $\rho_k = (\bar{\rho}_k^2 + \beta_{k+1}^2)^{\frac{1}{2}}$
2. $c_k = \bar{\rho}_k / \rho_k$
3. $s_k = \beta_{k+1} / \rho_k$
4. $\theta_{k+1} = s_k \alpha_{k+1} / \rho_k$
5. $\bar{\rho}_{k+1} = -c_k \alpha_{k+1}$
6. $\phi_k = c_k \bar{\phi}_k$
7. $\bar{\phi}_{k+1} = s_k \bar{\phi}_k$.

2.3: (*Update*)
1. $d_k = (1/\rho_k) p_k$
2. $x_k = x_{k-1} + (\phi_k / \rho_k) p_k$
3. $p_{k+1} = v_{k+1} - \theta_{k+1} p_k$

2.4: (*Test for convergence*)
If $\frac{||\mathcal{A}^T r_k||}{||\mathcal{A}|| ||r_k||} \leq \mathsf{Tol}$, $r_k := \mathcal{A}x_k - b$, Tol is some tolerance, then STOP.

In this algorithm, the parameters β_i and α_i are computed such that $||u_i||_2 = ||v_i||_2 = 1$.

To solve problem (3.3) we apply the Algorithm LSQR($\mathcal{A}$, b) with $\mathcal{A} = A\mathcal{P}$, where $\mathcal{P}$ is an orthogonal projector onto the null space of B. In Step 2.1 we have to compute vectors $A\mathcal{P}v_k$ and $\mathcal{P}A^T u_{k+1}$. Since $\mathcal{P}v_k = v_k$ by construction (see [27]), then $A\mathcal{P}v_k = Av_k$, and we need to compute only one projection $\mathcal{P}A^T u_{k+1}$ of a vector $A^T u_{k+1}$ onto the null space of the matrix B. This projection is not calculated explicitly, but a corresponding unconstrained least squares problem (lower level problem) is solved, i.e., to compute $\mathcal{P}w$ we solve $q^* = \arg\min_q ||B^T q - w||_2^2$ and set $\mathcal{P}w = w - B^T q^*$.

In further theoretical analysis we assume that all computations are performed in exact arithmetic. Under this assumption it is easy to show that the Algorithm LSQR($A\mathcal{P}$, b) terminates in at most $\bar{m}$ iterations independently of the tolerance value Tol≥ 0.

Theorem 3.3. *Assume, that the premature termination does not occur, i.e., the* Algorithm LSQR($A\mathcal{P}$, b) *results after* $\bar{m}$ *iterations in the vectors* $x_{\bar{m}}, d_1, \ldots, d_{\bar{m}}$. *The exact solution* x^* *of the problem* (3.3) *and the corresponding exact covariance matrix* C *can be computed as* $x^* = x_{\bar{m}}$, $C = DD^T$ *with* $D = (d_1, \ldots, d_{\bar{m}})$.

Proof. The first assertion follows from the theory of Krylov-type methods. We prove that $C = DD^T$.

From properties of LSQR [22], we get $D^T \mathcal{P} A^T A \mathcal{P} D = \mathbb{I}$, where the projection matrix $\mathcal{P}$ can be expressed as $\mathcal{P} = ZZ^T$ with orthogonal matrix Z spanning the null space of B. Hence

$$D^T ZZ^T A^T AZZ^T D = (D^T Z)(Z^T A^T AZ)(Z^T D) = \mathbb{I}. \tag{3.6}$$

We show, that the matrix $Z^T D$ is nonsingular. Indeed, by LSQR [22] we have $D = VR^{-1}$, where R is an upper bi-diagonal matrix, $V = (v_1, \ldots, v_{\bar{m}})$, $\operatorname{rank} V = \bar{m}$. Since $v_i \in \operatorname{Ker} B$ and $\operatorname{rank} V = \bar{m}$, it follows that $V = ZN$ with a nonsingular matrix N. Summing up, $Z^T D = Z^T ZNR^{-1} = NR^{-1}$ is nonsingular as a product of nonsingular matrices and we get from (3.6) that $(Z^T A^T AZ)^{-1} = (Z^T D)(D^T Z)$. Hence

$$C = Z(Z^T A^T AZ)^{-1} Z^T = Z(Z^T D)(D^T Z)Z^T = DD^T$$

since $ZZ^T D = ZZ^T ZNR^{-1} = ZNR^{-1} = VR^{-1} = D$. □

Remark 3.4. With exact arithmetic all algorithms considered in the paper converge in at most $\bar{m}$ iterations. If the algorithm is stopped after k ($k < \bar{m}$) iterations because of stopping criteria then x_k is an approximate solution of the problem (3.3), and the matrix $C_k = \sum_{j=1}^{k} d_j d_j^T$ is an approximation of the covariance matrix C. In this case we may continue the process to get "complete" set of the vectors $d_1, \ldots, d_{\bar{m}}$ and the exact covariance matrix $C = C_{\bar{m}}$. Notice that approximations C_k, $k = 1, 2, \ldots$, are "monotonically increasing" in the sense that $C_{k+1} \succeq C_k$. As usual, for two symmetric matrices M and N, $M \succeq N$ denotes $M - N \succeq 0$ and $M \succeq 0$ denotes that M is positive semi-definite.

Remark 3.5. In parameter estimation problem we need only few diagonal elements of the covariance matrix corresponding to the parameters. These values of interest

$\sigma_{ii} := e_i^T C e_i$ for some i can be approximated by monotonically increasing estimates

$$\sigma_{ii}^{(k)} = e_i^T \sum_{j=1}^{k} d_j d_j^T e_i = \sum_{j=1}^{k} d_{ij}^2, k = 1, \ldots, \bar{m}.$$

Preliminary numerical results [21, 27] show, as expected, that the iterative process needs a proper preconditioning in order to achieve a reasonable efficiency and accuracy in the elements of the covariance matrix. Our aim is to accelerate the solution process by applying appropriate preconditioners [4, 14, 25]. The question to be answered further in this paper is how to compute the linear operator A^+ and the covariance matrix in case the KKT systems are solved iteratively with preconditioning.

4. Computing covariance matrix using preconditioned Krylov subspace methods

Solving problem (3.3) by a Krylov subspace method is equivalent to solving the following system of linear equations

$$\mathcal{P}^T A^T A \mathcal{P} x = \mathcal{P}^T A^T b. \tag{4.1}$$

Here, $\mathcal{P}$ as before is an orthogonal projector onto null space of B. Preconditioning consists in an equivalent reformulation of the original linear system (4.1)

$$\bar{\tilde{A}} x = \bar{b} \tag{4.2}$$

where the new matrix $\bar{\tilde{A}}$ has "better" properties than the matrix $\tilde{A} = \mathcal{P}^T A^T A \mathcal{P}$ from the original system (4.1).

Suppose, that we apply preconditioning, that is we change the variables x using a nonsingular matrix $W \in \mathbb{R}^{n \times n}$

$$\bar{x} = W^{-1} x.$$

Then the problem (3.3) is equivalent to the following problem

$$\min_{\bar{x}} ||\bar{A}\bar{x} - b||_2^2, \text{ s.t. } \bar{B}\bar{x} = 0, \tag{4.3}$$

in the sense, that if $\bar{x}^*$ solves the problem (4.3) then $x^* = W\bar{x}^*$ solves the problem (3.3). Here $\bar{A} = AW$, $\bar{B} = BW$. Furthermore, the solution of problem (4.3) is equivalent to solving the linear system

$$\mathcal{M}^T \bar{A}^T \bar{A} \mathcal{M} y = \mathcal{M}^T \bar{A}^T b \tag{4.4}$$

where $\mathcal{M}$ is a projector onto Ker $\bar{B}$: if y^* is a solution of system (4.4) then $\bar{x}^* = \mathcal{M} y^*$ solves the problem (4.3) and $x^* = W\bar{x}^*$ is a solution to problem (3.3).

We consider two types of projectors $\mathcal{M}$ onto null space of $\bar{B}$: $\mathcal{M} = \bar{\mathcal{P}}$ and $\mathcal{M} = W^{-1}\mathcal{P}W$ where $\mathcal{P}$ and $\bar{\mathcal{P}}$ are orthogonal projection operators onto subspaces Ker B and Ker $\bar{B}$ respectively. These types of projector operators generate two types of preconditioning for system (4.1):

- *Preconditioning of type I* corresponds to solving system (4.4) with $\mathcal{M} = \bar{\mathcal{P}}$, i.e., the system

$$\bar{\mathcal{P}}^T W^T A^T A W \bar{\mathcal{P}} y = \bar{\mathcal{P}}^T W^T A^T b. \tag{4.5}$$

- *Preconditioning of type II* corresponds to solving system (4.4) with $\mathcal{M} = W^{-1}\mathcal{P}W$, i.e., the system

$$W^T \mathcal{P}^T A^T A \mathcal{P} W y = W^T \mathcal{P}^T b. \tag{4.6}$$

4.1. Preconditioning of type I

Setting $\mathcal{A} = \bar{A}\bar{\mathcal{P}}$, where $\bar{\mathcal{P}}$ is an orthogonal projector onto null space of the matrix $\bar{B}$ we may apply the Algorithm LSQR($\mathcal{A}$, b) directly to solve the problem (4.3). Note, that now in the step 2.1 $\mathcal{A}v_k = \bar{A}\bar{\mathcal{P}}v_k = \bar{A}v_k$ and $\mathcal{A}^T u_{k+1} = \bar{\mathcal{P}}\bar{A}^T u_{k+1} = \bar{A}^T u_{k+1} - \bar{B}^T q^*$, where $q^* = \arg\min_q ||\bar{B}^T q - \bar{A}^T u_{k+1}||$.

Suppose, that the premature termination does not occur, i.e., the Algorithm LSQR($\mathcal{A}$, b) with $\mathcal{A} = \bar{A}\bar{\mathcal{P}}$ results after $\bar{m}$ iterations in the vectors $x_{\bar{m}}, d_1, \ldots, d_{\bar{m}}$. Then the solution of the problem (4.3) and the corresponding covariance matrix are as follows $\bar{x}^* = x_{\bar{m}}$, $\bar{C} = \bar{D}\bar{D}^T$ where $\bar{D} = (d_1, d_2, \ldots, d_{\bar{m}})$.

Lemma 4.1. *The solution x^* of the problem* (3.3) *and the corresponding covariance matrix C is computed as $x^* = W\bar{x}^*$, $C = W\bar{C}W^T$.*

Proof. It follows from Lemma 3.1 that the matrix $\bar{C}$ solves the system

$$\begin{aligned}
& \begin{pmatrix} \bar{A}^T\bar{A} & \bar{B}^T \\ \bar{B} & 0 \end{pmatrix} \begin{pmatrix} \bar{C} \\ \bar{S} \end{pmatrix} = \begin{pmatrix} \mathbb{I} \\ 0 \end{pmatrix} \\
\Rightarrow \quad & \begin{pmatrix} W^T & 0 \\ 0 & \mathbb{I} \end{pmatrix} \mathcal{K} \begin{pmatrix} W & 0 \\ 0 & \mathbb{I} \end{pmatrix} \begin{pmatrix} \bar{C} \\ \bar{S} \end{pmatrix} = \begin{pmatrix} \mathbb{I} \\ 0 \end{pmatrix} \\
\Rightarrow \quad & \begin{pmatrix} W^T & 0 \\ 0 & \mathbb{I} \end{pmatrix} \mathcal{K} \begin{pmatrix} W\bar{C} \\ \bar{S} \end{pmatrix} = \begin{pmatrix} \mathbb{I} \\ 0 \end{pmatrix}
\end{aligned}$$

while the covariance C solves the system (3.5). Hence

$$\begin{pmatrix} W\bar{C} \\ \bar{S} \end{pmatrix} = \mathcal{K}^{-1} \begin{pmatrix} W^T & 0 \\ 0 & \mathbb{I} \end{pmatrix}^{-1} \begin{pmatrix} \mathbb{I} \\ 0 \end{pmatrix} = \mathcal{K}^{-1} \begin{pmatrix} \mathbb{I} \\ 0 \end{pmatrix} W^{-T} = \begin{pmatrix} C \\ S \end{pmatrix} W^{-T}.$$

It follows from the last equation that $W\bar{C} = CW^{-T}$, $\bar{S} = SW^{-T}$. Thus, $C = W\bar{C}W^T$, $S = \bar{S}W^T$. □

The Algorithm LSQR($\mathcal{A}$, b) for problem (4.3) makes use of the matrix products AW and BW, which is not always reasonable, see [18]. It is not difficult to rewrite the Algorithm LSQR($\mathcal{A}$, b) without carrying out the variable transformation explicitly.

Algorithm Preconditioned LSQR-I(A, B, b, $Q = WW^T$) (for solving problem (3.3))

Step 1: (*Initialization*)
$\beta_1 u_1 = b$, $\alpha_1 v_1 = A^T u_1 - B^T q_0$,
where q_0 solves the problem $\min_q ||B^T q - A^T u_1||_Q^2$,
$p_1 = Qv_1$, $x_0 = 0$, $\bar{\phi}_1 = \beta_1$, $\bar{\rho}_1 = \alpha_1$

Step 2: For $k = 1, 2, 3, \ldots$ repeat steps **2.1**–**2.4**.

2.1: (*Continue the bidiagonalization*)
1. $\beta_{k+1} u_{k+1} = AQv_k - \alpha_k u_k$
2. $\alpha_{k+1} v_{k+1} = A^T u_{k+1} - B^T q_k - \beta_{k+1} v_k$, where q_k solves the problem $\min_q ||B^T q - A^T u_{k+1}||_Q^2$

2.2: *as in* Algorithm LSQR($\mathcal{A}$, b)

2.3: (*Update*)
1. $d_k = (1/\rho_k) p_k$
2. $x_k = x_{k-1} + (\phi_k / \rho_k) p_k$
3. $p_{k+1} = Qv_{k+1} - \theta_{k+1} p_k$.

2.4: (*Test for convergence*)
If $\frac{||\mathcal{A}^T r_k - B^T q(r_k)||_Q}{||\mathcal{A}|| ||r_k||} \le$ Tol, then STOP. Here $r_k = Ax_k - b$,
$q(r)$ solves the problem $\min_q ||B^T q - A^T r||_Q^2$.

In this algorithm, the parameters β_i and α_i are computed in such a way, that $||u_i||_2 = ||v_i||_Q = 1$. Here $||s||_2^2 = s^T s$, $||s||_Q^2 := s^T Q s$.

Lemma 4.2 (Computation of covariance matrix). *Suppose, that the* Algorithm PLSQR-I(A, B, b, Q) *converges after $\bar{m}$ iterations and we get the vectors $x_{\bar{m}}$ and $d_1, \ldots, d_{\bar{m}}$. Then the solution x^* of the problem* (3.3) *and the corresponding covariance matrix are given by*

$$x^* = x_{\bar{m}}, C = DD^T, D = (d_1, \ldots, d_{\bar{m}}).$$

4.2. Preconditioning of type II

Setting $\mathcal{A} = A\mathcal{P}W$, where $\mathcal{P}$ is an orthogonal projector onto null space of the matrix B we may apply the Algorithm LSQR($\mathcal{A}$, b) directly to solve the system (4.6). Note, that now in the step 2.1 we need to compute $\mathcal{A}v_k = A\mathcal{P}Wv_k$, where $\mathcal{P}Wv_k = Wv_k - B^T q^*$, $q^* = \arg\min_q ||B^T q - Wv_k||$, and $\mathcal{A}^T u_{k+1} = W^T \mathcal{P} A^T u_{k+1}$, where $\mathcal{P} A^T u_{k+1} = A^T u_{k+1} - B^T q^*$, where $q^* = \arg\min_q ||B^T q - A^T u_{k+1}||$, which means that we have to solve twice the lower level problem.

Suppose that after applying the Algorithm LSQR($A\mathcal{P}W$, b) we get vectors $x_{\bar{m}}, d_1, \ldots, d_{\bar{m}}; v_1, \ldots, v_{\bar{m}}$. Then the solution of the system (4.6) and the corresponding covariance matrix have the form $y^* = x_{\bar{m}}$, $\tilde{C} = \tilde{D}\tilde{D}^T$, $\tilde{D} = (d_1, \ldots, d_{\bar{m}})$. Let us show, how we can use this information in order to compute the solution x^* of the problem (3.3) and the corresponding matrix C.

By construction, y^* solves the system (4.6). Then $\bar{x}^* = \mathcal{M}y^*$ with $\mathcal{M} = W^{-1}\mathcal{P}W$ solves the problem (4.3) and $x^* = W\bar{x}^*$ is a solution to problem (3.3).

Hence,

$$x^* = W\mathcal{M}y^* = \mathcal{P}Wy^* = \sum_{k=1}^{\bar{m}} \frac{\phi_k}{\rho_k}\mathcal{P}Wp_k.$$

Similarly,

$$C = \mathcal{P}W\tilde{C}W^T\mathcal{P} = \mathcal{P}W\tilde{D}\tilde{D}^TW^T\mathcal{P}$$

As $\mathcal{P}Wd_k = \frac{1}{\rho_k}\mathcal{P}Wp_k$ in order to compute x^* and C we need vectors

$$\mathcal{P}Wp_k, \quad k = 1, \ldots, \bar{m}. \tag{4.7}$$

Since $p_1 = v_1$, $p_k = v_k - \theta_k p_{k-1}$, $k = 2, 3, \ldots$, and the vectors $\mathcal{P}Wv_k$ are have calculated in the algorithm (see step 2.1 and remarks at the beginning of this subsection), it is easy to modify the Algorithm LSQR($A\mathcal{P}W$, b) in order to compute recursively the vectors (4.7) and the vector x^*. Further, we modify the algorithm such that it makes use of the matrix $Q = WW^T$:

Algorithm Preconditioned LSQR-II(A, B, b, Q) (for solving problem (3.3))

Step 1: (*Initialization*)
$\beta_1 u_1 = b$, $\alpha_1 v_1 = A^T u_1 - B^T q_0$,
where q_0 solves $\min_q ||B^T q - A^T u_1||_2^2$,
$p_1 = Qv_1$, $g_0 = 0$, $x_0 = 0$,
$\bar{\phi}_1 = \beta_1$, $\bar{\rho}_1 = \alpha_1$, $\theta_1 = 0$

Step 2: For $k = 1, 2, 3, \ldots$ repeat steps **2.1**–**2.4**.

2.1: (*Continue the bidiagonalization*)

1. $\hat{v}_k = Qv_k - B^T\hat{q}_k$, where $\hat{q}_k$ solves $\min_q ||B^T q - Qv_k||_2^2$,
2. $\beta_{k+1}u_{k+1} = A\hat{v}_k - \alpha_k u_k$
3. $\alpha_{k+1}v_{k+1} = A^T u_{k+1} - B^T q_k - \beta_{k+1}v_k$, where q_k solves $\min_q ||B^T q - A^T u_{k+1}||_2^2$,
4. $g_k = \hat{v}_k - \theta_k g_{k-1}$

2.2: *as in* Algorithm LSQR($\mathcal{A}$, b)

2.3: (*Update*)

1. $d_k = (1/\rho_k)g_k$
2. $x_k = x_{k-1} + (\phi_k/\rho_k)g_k$
3. $p_{k+1} = Qv_{k+1} - \theta_{k+1}p_k$.

2.4: If $\frac{||A^T r_k - B^T q(r_k)||_Q}{||\mathcal{A}|| \, ||r_k||} \leq \mathsf{Tol}$, then STOP. Here $r_k = Ax_k - b$, $q(r)$ solves the problem $\min_q ||B^T q - A^T r||_2^2$.

In this algorithm, the parameters β_i and α_i are computed in such a way, that $||u_i||_2 = ||v_i||_Q = 1$. Note, that in Step 2.1 we have to solve twice the lower level problem: to compute $\mathcal{P}Qv_k$ and $\mathcal{P}A^T u_{k+1}$.

Lemma 4.3 (Computation of covariance matrix). *Suppose, that the* Algorithm PLSQR-II(A, B, b, Q) *converges after* $\bar{m}$ *iterations and we get the vectors* $x_{\bar{m}}$ *and* $d_1, \ldots, d_{\bar{m}}$. *Then the solution* x^* *of the problem* (3.3) *and the corresponding covariance matrix are given by*

$$x^* = x_{\bar{m}}, C = DD^T, D = (d_1, \ldots, d_{\bar{m}})$$

Remark 4.4. In the Algorithms PLSQR-I(A, B, b, Q) and PLSQR-II(A, B, b, Q), matrix products like AW and BW are never explicitly performed. Only the action of applying the preconditioner solver operation $Q := WW^T$ to a given vector need be computed. This property is important to systems resulted from PDE discretization. In this sense the Algorithms PLSQR-I(A, B, b, Q) and PLSQR-II(A, B, b, Q) are more preferable than the corresponding Algorithms LSQR ($\bar{A}\bar{\mathcal{P}}$, b) and LSQR-($A\mathcal{P}W$, b).

Remark 4.5. Obviously, the vectors $x_k, k = 1, 2, \ldots$ generated by the Algorithm LSQR ($\bar{A}\bar{\mathcal{P}}$, b) (or by the Algorithm LSQR($A\mathcal{P}W$, b)) in an exact arithmetic are connected with the vectors $\bar{x}_k = x_k, k = 1, 2, \ldots$ generated by the Algorithm PLSQR-I(A, B, b, Q) (or by the Algorithm PLSQR-II(A, B, b, Q)) as follows $\bar{x}_k = Wx_k, k = 1, 2, \ldots$. Thus, these vector sequences maybe considered as the same except for a multiplication with W. But we can show that the Algorithms LSQR-I(A, B, b, Q) and LSQR-II(A, B, b, Q) generate in general completely different vector sequences.

Remark 4.6. The vector sequence $x_k, k = 1, 2, \ldots$ generated by the Algorithm PLSQR-I(A, B, b, Q) is the same (in exact arithmetic) as the vector sequence $x_k, k = 1, 2, \ldots$ generated by the Algorithm 3.1 (preconditioned CG in expanded form) from [14]. But the Algorithm PLSQR-I(A, B, b, Q) is based on LSQR methods, which are known to be more numerically stable compared to CG methods for least squares problems.

5. Conclusions

For solving constraint parameter estimation and optimal design problems, we need the knowledge of covariance matrix of the parameter estimates and its derivatives. Hence, development of effective methods for presentation and computation of the covariance matrix and its derivatives, based on iterative methods, are crucial for practical applications. In the paper, we have shown that solving nonlinear constrained least squares problems by Krylov subspace methods we get as a by-product practically for free these matrices. The forthcoming research will be devoted to numerical aspects including derivation of an estimation for the error $||C - C^k||$ in a suitable norm, choice of effective preconditioners and effective implementation of the described methods for parameter estimation and design of optimal parameters in processes defined by partial differential equations.

References

[1] A.C. Atkinson and A.N. Donev. *Optimum Experimental Designs.* Oxford University Press, 1992.

[2] S.P. Asprey and S. Macchietto. Statistical tools for optimal dynamic model building. *Computers and Chemical Engineering*, 24:1261–1267, 2000.

[3] U. Ascher. Collocation for two-point boundary value problems revisited. *SIAM Journal on Numerical Analysis*, 23(3):596–609, 1986.

[4] A. Battermann and E.W. Sachs. Block preconditioners for KKT systems in PDE-governed optimal control problems. In K.H. Hoffman, R.H.W. Hoppe, and V. Schulz, editors, *Fast solution of discretized optimization problems*, 1–18. ISNM, Int. Ser. Numer. Math. 138, 2001.

[5] I. Bauer, H.G. Bock, S. Körkel, and J.P. Schlöder. Numerical methods for initial value problems and derivative generation for DAE models with application to optimum experimental design of chemical processes. In F. Keil, W. Mackens, H. Voss, and J. Werther, editors, *Scientific Computing in Chemical Engineering II*, volume 2, 282–289. Springer, Berlin-Heidelberg, 1999.

[6] I. Bauer, H.G. Bock, S. Körkel, and J.P. Schlöder. Numerical methods for optimum experimental design in DAE systems. *Journal of Computational and Applied Mathematics*, 120:1–25, 2000.

[7] H.G. Bock. Numerical treatment of inverse problems in chemical reaction kinetics. In K.-H. Ebert, P. Deuflhard, and W. Jäger, editors, *Modelling of Chemical Reaction Systems*, volume 18 of *Springer Series in Chemical Physics*, pages 102–125. Springer Verlag, 1981.

[8] H.G. Bock. *Randwertproblemmethoden zur Parameteridentifizierung in Systemen nichtlinearer Differentialgleichungen*, volume 183 of *Bonner Mathematische Schriften*. University of Bonn, 1987.

[9] H.G. Bock, E.A. Kostina, and O.I. Kostyukova. Conjugate gradient methods for computing covariance matrices for constrained parameter estimation problems. *SIAM Journal on Matrix Analysis and Application*, 29:626–642, 2007.

[10] R. Bulirsch. Die Mehrzielmethode zur numerischen Lösung von nichtlinearen Randwertproblemen und Aufgaben der optimalen Steuerung. Technical report, Carl-Cranz-Gesellschaft, 1971.

[11] T. Carraro. *Parameter estimation and optimal experimental design in flow reactors.* Phd thesis, University of Heidelberg, 2005.

[12] T. Carraro, V. Heuveline, and R. Rannacher. Determination of kinetic parameters in laminar flow reactors. I. Numerical aspects. In W. Jäger, R. Rannacher, and J. Warnatz, editors, *Reactive Flows, Diffusion and Transport. From Experiments via Mathematical Modeling to Numerical Simulation and Optimization.* Springer, 2007.

[13] A. Cervantes and L.T. Biegler. Large-scale DAE optimization using a simultaneous NLP formulation. *AIChE Journal*, 44(5):1038–1050, 2004.

[14] H.S. Dollar, and A.J. Wathen. Approximate factorization constraint preconditioners for saddle-point matrices. *Siam J. Sci. Comput.*, 27(5): 1555-1572, 2006.

[15] V.V. Fedorov. *Theory of Optimal Experiments.* Probability and Mathematical Statistics. Academic Press, London, 1972.

[16] A. Griewank. *Evaluating Derivatives. Principles and Techniques of Algorithmic Differentiation.* Frontiers in Applied Mathematics. SIAM, 2000.

[17] E. Haber, L. Horesh, and L. Tenorio. Numerical methods for optimal experimental design of large-scale ill-posed problems. *Inverse Problems.* 24(5), 2008.

[18] C.T. Kelley, *Iterative Methods for Linear and Nonlinear Equations*, SIAM, Philadelphia, 1995.

[19] S. Körkel and E.A. Kostina. Numerical methods for nonlinear experimental design. In H.G. Bock, E.A. Kostina, H.X. Phu, and R. Rannacher, editors, *Modeling, Simulation and Optimization of Complex Processes, Proceedings of the International Conference on High Performance Scientific Computing,* 2003, *Hanoi, Vietnam.* Springer, 2004.

[20] S. Körkel. *Numerische Methoden für Optimale Versuchsplanungsprobleme bei nichtlinearen DAE-Modellen.* Phd thesis, Universität Heidelberg, 2002.

[21] E. Kostina, M. Saunders, and I. Schierle. Computation of covariance matrices for constrained parameter estimation problems using LSQR. Technical Report, Department of Mathematics and Computer Science, U Marburg, 2008.

[22] C.C. Paige and M.A. Saunders, LSQR: An algorithm for sparse linear equations and sparse least-squares, *ACM Trans. Math. Softw.*, 8(1):43 – 71, 1982.

[23] C.C. Paige and M.A. Saunders, LSQR: Sparse linear equations and least-squares, *ACM Trans. Math. Softw.*, 8(2):195–209, 1982.

[24] F. Pukelsheim. *Optimal Design of Experiments.* John Wiley & Sons, Inc., New York, 1993.

[25] T. Rees, H.S. Dollar, and A.J. Wathen. Optimal solvers for PDE-constrained optimization. Technical report RAL-TR-2008-018, Rutherford Appleton Laboratory, 2008.

[26] P.E. Rudolph and G. Herrendörfer. Optimal experimental design and accuracy of parameter estimation for nonlinear regression models used in long-term selection. *Biom. J.*, 37(2):183–190, 1995.

[27] I. Schierle. *Computation of Covariance Matrices for Constrained Nonlinear Parameter Estimation Problems in Dynamic Processes Using iterative Linear Algebra Methods.* Diploma thesis, Universität Heidelberg, 2008.

[28] V.H. Schulz. *Ein effizientes Kollokationsverfahren zur numerischen Behandlung von Mehrpunktrandwertaufgaben in der Parameteridentifizierung und Optimalen Steuerung.* Diploma thesis, Universität Augsburg, 1990.

Ekaterina Kostina
University of Marburg
Hans-Meerwein-Strasse
D-35032 Marburg, Germany
e-mail: kostina@mathematik.uni-marburg.de

Olga Kostyukova
Institute of Mathematics
Belarus Academy of Sciences
Surganov Str. 11
220072 Minsk, Belarus
e-mail: kostyukova@im.bas-net.by

Part II

Shape and Topology Optimization

Introduction to Part II
Shape and Topology Optimization

In this part, novel results in the field of shape and topology optimization achieved within the special priority program are presented.

Pradeep Atwal, Sergio Conti, Benedict Geihe, Martin Pach, Martin Rumpf and Rüdiger Schultz present, in *On Shape Optimization with Stochastic Loadings*, algorithmic approaches for topology optimization problems within a stochastic framework.

Luise Blank, Harald Garcke, Lavinia Sarbu, Tarin Srisupattarawanit, Vanessa Styles and Axel Voigt compare, in *Phase-field Approaches to Structural Topology Optimization*, Cahn-Hilliard and Allen-Cahn formulations of topology optimization problems.

Christian Brandenburg, Florian Lindemann, Michael Ulbrich and Stefan Ulbrich present, in *Advanced Numerical Methods for PDE Constrained Optimization with Application to Optimal Design in Navier Stokes Flow*, an approach to parametric shape optimization which is based on transformation to a reference domain with continuous adjoint computations and applied to the instationary Navier-Stokes equations.

Karsten Eppler and Helmut Harbrecht solve, in *Shape Optimization for Free Boundary Problems – Analysis and Numerics*, a Bernoulli type free boundary problem by means of shape optimization and characterize well-posed and ill-posed problems by analyzing the shape Hessian.

Nicolas Gauger, Caslav Ilic, Stephan Schmidt and Volker Schulz derive, in *Non-parametric Aerodynamic Shape Optimization*, very efficient numerical schemes by exploiting the arising structures in the shape optimization problems – in particular by approximating arising shape Hessians.

Volker Schulz

International Series of Numerical Mathematics, Vol. 160, 215–243

On Shape Optimization with Stochastic Loadings

Pradeep Atwal †, Sergio Conti, Benedict Geihe, Martin Pach, Martin Rumpf and Rüdiger Schultz

Abstract. This article is concerned with different approaches to elastic shape optimization under stochastic loading. The underlying stochastic optimization strategy builds upon the methodology of two-stage stochastic programming. In fact, in the case of linear elasticity and quadratic objective functionals our strategy leads to a computational cost which scales linearly in the number of *linearly independent* applied forces, even for a large set of realizations of the random loading. We consider, besides minimization of the expectation value of suitable objective functionals, also two different risk averse approaches, namely the *expected excess* and the *excess probability*. Numerical computations are performed using either a level set approach representing implicit shapes of general topology in combination with composite finite elements to resolve elasticity in two and three dimensions, or a collocation boundary element approach, where polygonal shapes represent geometric details attached to a lattice and describing a perforated elastic domain. Topology optimization is performed using the concept of topological derivatives. We generalize this concept, and derive an analytical expression which takes into account the interaction between neighboring holes. This is expected to allow efficient and reliable optimization strategies of elastic objects with a large number of geometric details on a fine scale.

Mathematics Subject Classification (2000). 90C15, 74B05, 65N30, 65N38, 34E08, 49K45.

Keywords. Shape optimization in elasticity, two-stage stochastic programming, risk averse optimization, level set method, boundary element method, topological derivative.

1. Introduction

Data uncertainty is a critical feature of many real-world shape optimization problems. When optimizing elastic structures, it has to be taken into account that volume and in particular surface loadings typically vary over time. Often, these variations can be captured by probability distributions. In order to study the resulting random optimization problem it is crucial to observe that decisions on the shape must be nonanticipative, i.e., be made before applying the stochastic forcing. In terms of optimization this implies that, for a fixed candidate shape, each realization of the stochastic force determines a value of the optimality criterion. Hence, the shape is assigned a random variable whose realizations are the values of the optimality criterion induced by the data realizations.

From this point of view, shape optimization under (stochastic) uncertainty amounts to finding "optimal" shapes in sets of random variable shapes. The quotes shall indicate that different concepts for ranking random variables may apply. In the present paper we discuss risk neutral as well as risk averse models. While in the former the ranking is done by comparing the expectation values, in the latter statistical parameters reflecting different perceptions of risk are applied. This bears striking conceptual similarity with finite-dimensional two-stage stochastic programming.

The paper is organized as follows. After a brief review of related work in Section 2, we introduce in Section 3 the elastic state equation, define suitable cost functionals, and formulate an initial random shape optimization model with stochastic loading. Starting from basic paradigms of finite-dimensional two-stage stochastic programming, Section 4 serves to introduce general classes of risk neutral and risk averse shape optimization models. Computation of shape derivatives is addressed in Section 5. In Section 6 we present our methods for describing shapes, namely, the level set approach and parametrized perforated domains. Section 7 describes our numerical algorithms. A selection of case studies is presented in Section 8, to demonstrate the validity of our conceptual approach and the effectiveness of the algorithm. Section 9 discusses a new analytical result on the hole-hole interaction in topological derivatives, and outlines its possible application to a reduced model for the optimization of perforated geometries.

2. Related work

Deterministic shape optimization. Shape optimization under deterministic loading is a well-developed field within PDE-constrained optimization; see for example the books [59, 17, 3]. We shall not review here the details, but simply remark that the usual strategy of performing iteratively modifications of the boundary, following the shape derivative, does not permit to optimize the topology of the shape, and that the homogenization strategy delivers shapes with microstructures at many different scales, which may be difficult to realize in practice.

Topology optimization. The optimal shape determined using just the shape derivative may strongly depend on the choice of the initial topology (cf. [5, 8, 26]). This is not only due to the discretization, but also to the descent strategy itself. Indeed, even with a level set description which in principle entails no restriction on the topology, the evolution of a level set based on a gradient flow is able to merge existing holes, but not capable of creating new holes. The possibility of creating new holes is the key idea behind the so-called "bubble method" or *topological sensitivity*. The method is based on considering variations of the domain corresponding to the insertion of small holes of prescribed shape and computing an asymptotic expansion depending on the radius ρ. After its introduction by Schumacher [51], the method was generalized to a wider class of shape functionals by Sokołowski and Żokowski [56] and applied to 3D elasticity in [57]. In [58], the approach is extended to the case of finitely many circular holes, combining topology variations with boundary variations simultaneously. Using an adjoint method and a truncation technique, Garreau et al. [32] computed the topological sensitivity for general objective functionals and arbitrarily-shaped holes. The topological derivative has been incorporated into the level set method, e.g., in [18], and also combined with the shape derivative in that context (cf., e.g., [5, 10, 36]). The topological derivative permits arbitrary changes in the topology, but remains a local modification. In particular, even in this case there is no guarantee to reach the global minimum, although one produces a local minimum which is stable under a wider class of variations and includes in particular topology changes (cf. Fig. 14).

Finite-dimensional stochastic programming. The principal modeling approaches in optimization under uncertainty differ in their requirements on data information. Worst-case models, as in online or robust optimization [2, 15], merely ask for ranges of the uncertain parameters. Applications to shape optimization can be found in [13, 37, 27, 22, 21]. In stochastic optimization, see [53] for a recent textbook, data uncertainty is captured by probability distributions. In finite dimension, there exists a rich theory and methodology of linear models [47, 53] and, to lesser extent, linear mixed-integer or nonlinear models [14, 44, 48, 60]. Stochastic optimization in continuous time has been analyzed in stochastic dynamic programming and stochastic control, see [31, 20]. Stochastic shape optimization in the design of mechanical or aerodynamic structures, has been addressed in [41, 50]. The difference of these contributions to the present paper is in the decision spaces. In [41, 50] these are Euclidean spaces, while our design decisions are allowed to be shapes (open sets) in suitable working domains.

Multiload shape optimization. Most shape optimization approaches are based on considering one given, typical, load configuration. This is not always realistic, and indeed multiload approaches have been developed, in which a fixed (usually small) number of different loading configurations is considered and optimization refers to this set of configurations, see, e.g., [6, 33, 63, 11] and references therein. These approaches are based on evaluating the objective functional separately for each of the possible values of the loading, which renders them infeasible if the set of possible forces is large, as for example is the case when one aims at approximating a continuous distribution of forces. In case of additional geometrical assumptions a more efficient method was derived in [9], where optimization of the expected compliance is shown to be equivalent to a convex problem, and hence efficiently solvable. Our approach is instead based on the solution of the elasticity PDE only for the basis modes for the surface and volume loading. Additionally, besides an optimization of the expected value of the cost we investigate also the optimization of nonlinear risk measures acting on the cost functional. A robust probabilistic approach for the optimization of simple beam models is discussed in [1], whereas in [42] structural reliability is discussed for beam geometries with uncertain load magnitude. Worst-case situations in a multiload context have also been considered, see, e.g., [16]. Structural optimization under incomplete information is also addressed in [13]. The authors investigate different types of uncertainties and follow a non-probabilistic, robust worst case approach. They work out effective techniques to transform optimal design problems with uncertainties into conventional structural optimization problems. In [46] a worst case analysis for given bounds on uncertain loads has been implemented based on a boundary element approach. A worst case compliance optimization is investigated in [7] based on a level set description of shapes and a semi-definite programming approach in the minimization algorithm. Schulz et al. [50] discussed shape optimization under uncertainty in aerodynamic design. Uncertainty is captured by Euclidean parameters following truncated normal distributions and leading to optimization models of the type of finite-dimensional nonlinear programs. Due to the distributional assumptions, there are explicit formulas for relevant probability functionals.

3. Shape optimization model with linear elasticity and random loading

The state equation. For a fixed elastic domain $\mathcal{O}$ and given stochastic loading we take into account linearized elasticity to determine a displacement $u : \mathcal{O} \to \mathbb{R}^d$ as the elastic response to applied forces. Hereby, we assume $\mathcal{O}$ to be a sufficiently regular subset of a fixed, bounded working domain $D \subset \mathbb{R}^d$. The boundary of $\mathcal{O}$ is assumed to be decomposed into a Dirichlet boundary Γ_D on which we prescribe homogeneous Dirichlet boundary conditions $u = 0$, a Neumann boundary Γ_N on which the surface loading is applied, and the remaining homogeneous Neumann boundary $\partial\mathcal{O} \setminus (\Gamma_D \cup \Gamma_N)$, whose shape is to be optimized according to some objective functional, to be discussed later. Neither Γ_D nor Γ_N will be subject to the shape optimization. For details on the concrete handling of the boundary conditions in the shape optimization we refer to [26]. Concerning the loading, we are in particular interested in stochastic loads. Thus, we assume a random volume force $f(\omega) \in L^2(D;\mathbb{R}^d)$ and a random surface force $g(\omega) \in L^2(\Gamma_N;\mathbb{R}^d)$ to be given. Here ω denotes a realization on an abstract probability space $(\Omega, \mathcal{A}, \wp)$. Based on this setup u is given as the solution of the following elliptic boundary value problem of linearized elasticity

$$\begin{aligned} -\mathrm{div}\,(\mathbf{C}\varepsilon(u)) &= f(\omega) && \text{in } \mathcal{O}\,, \\ u &= 0 && \text{on } \Gamma_D\,, \\ (\mathbf{C}\varepsilon(u))n &= g(\omega) && \text{on } \Gamma_N\,, \\ (\mathbf{C}\varepsilon(u))n &= 0 && \text{on } \partial\mathcal{O}\setminus\Gamma_N\setminus\Gamma_D\,. \end{aligned} \tag{1}$$

Here, $\varepsilon(u) = \frac{1}{2}(\nabla u + \nabla u^\top)$ is the linearized strain tensor and $\mathbf{C} = (\mathbf{C}_{ijkl})_{ijkl}$ denotes the elasticity tensor. For the sake of simplicity, we restrict ourselves here to isotropic materials, i.e., $\mathbf{C}_{ijkl} = 2\mu\delta_{ik}\delta_{jl} + \lambda\delta_{ij}\delta_{kl}$, where δ_{ij} denotes the Kronecker symbol and μ, λ the positive Lamé constants of the material. For any open, connected set $\mathcal{O}$ with Lipschitz boundary and any fixed realization ω of the random loading, there exists a unique weak solution $u = u(\mathcal{O},\omega) \in H^1(\mathcal{O};\mathbb{R}^d)$ of (1) [23, 40]. This solution can equivalently be characterized as the unique minimizer of the quadratic functional

$$E(\mathcal{O},u,\omega):=\frac{1}{2}a(\mathcal{O},u,u) - l(\mathcal{O},u,\omega) \quad \text{with} \tag{2}$$

$$a(\mathcal{O},\psi,\theta):=\int_{\mathcal{O}} \mathbf{C}_{ijkl}\varepsilon_{ij}(\psi)\varepsilon_{kl}(\theta)\,\mathrm{d}x\,, \tag{3}$$

$$l(\mathcal{O},\theta,\omega):=\int_{\mathcal{O}} f_i(\omega)\theta_i\,\mathrm{d}x + \int_{\Gamma_N} g_i(\omega)\theta_i\,\mathrm{d}\mathcal{H}^{d-1} \tag{4}$$

on $H^1_{\Gamma_D}(\mathcal{O};\mathbb{R}^d):=\{u \in H^1(\mathcal{O};\mathbb{R}^d)\,|\, u = 0 \text{ on } \Gamma_D \text{ in the sense of traces}\}$. Here and below, we use the summation convention and implicitly sum over repeated indices.

The random shape optimization model. Next, we formulate the actual shape optimization problem and take into account an objective functional J which depends on both the shape $\mathcal{O}$ and the resulting elastic displacement $u(\mathcal{O},\omega)$. Let us assume that J is of the form

$$J(\mathcal{O},\omega) = \int_{\mathcal{O}} j\left(u\left(\mathcal{O},\omega\right)\right)\,\mathrm{d}x + \int_{\Gamma_N} k\left(u\left(\mathcal{O},\omega\right)\right)\,\mathrm{d}\mathcal{H}^{d-1}\,. \tag{5}$$

To allow for a subsequent efficient numerical realization in the context of a high number of stochastic scenarios of the loading we confine ourselves to linear or quadratic functions $j(.)$ and $k(.)$. Finally, we obtain the random shape optimization model with stochastic loading

$$\min\left\{J\left(\mathcal{O},\omega\right) :\, \mathcal{O}\in\mathcal{U}_{\mathrm{ad}}\right\} \tag{6}$$

where $\mathcal{U}_{\mathrm{ad}}$ is a suitable set of admissible shapes. The potential energy of the applied loading for given displacement

$$J_1\left(\mathcal{O},\omega\right) := \int_{\mathcal{O}} f\cdot u(\mathcal{O},\omega)\,\mathrm{d}x + \int_{\Gamma_N} g\cdot u(\mathcal{O},\omega)\,\mathrm{d}\mathcal{H}^{d-1}\,, \tag{7}$$

which is denoted the compliance, is the most common objective functional, which we pick up here also in the stochastic context. Alternatively, one might consider the least square error compared to a target displacement u_0

$$J_2(\mathcal{O},\omega) := \frac{1}{2}\int_{\mathcal{O}} |u(\mathcal{O},\omega) - u_0|^2 \,\mathrm{d}x \tag{8}$$

or alternatively

$$J_3(\mathcal{O},\omega) := \frac{1}{2}\int_{\partial\mathcal{O}} |u(\mathcal{O},\omega) - u_0|^2 \,\mathrm{d}\mathcal{H}^{d-1}, \tag{9}$$

which is used in the context of parametrized perforated domains. For the level set approach we usually include in the objective functional a penalty for the object volume $\alpha\int_{\mathcal{O}} \mathrm{d}x$ with $\alpha \geq 0$. Having defined the objective functional we are in a position to consider the optimization of the shape with respect to a fixed realization of the randomness. Starting from this deterministic case we will then introduce general classes of risk neutral (cf. Fig. 1) and risk averse stochastic cost functionals, where a different (nonlinear) weighting is applied on the probability space $(\Omega, \mathcal{A}, \wp)$.

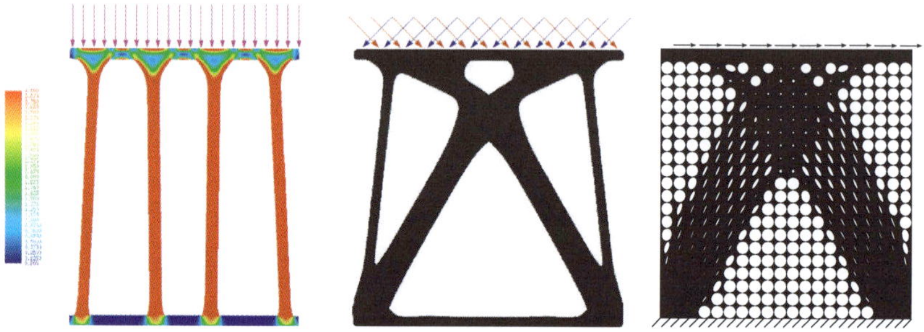

FIGURE 1. A direct comparison of two-stage stochastic optimization and deterministic optimization for an averaged load is shown. In the middle, a stochastically optimal shape is rendered together with the two underlying load scenarios ω_1 and ω_2 on the upper plate, with surface loads $g(\omega_1)$ and $g(\omega_2)$ both with probability $\frac{1}{2}$. On the left the optimal shape color-coded with the von Mises stress is drawn for a deterministic load $\frac{1}{2}g(\omega_1) + \frac{1}{2}g(\omega_2)$ (both reprinted from [26, Fig. 4.2]). An optimal shape for a perforated domain under deterministic shear load is shown on the right.

4. Stochastic shape optimization models

Two-stage stochastic programming in finite dimension. Finite-dimensional linear stochastic programs serve as blueprints for our stochastic shape optimization models. For a quick review consider the initial random optimization problem

$$\min\{c^\top x + q^\top y \,:\, Tx + Wy = z(\omega),\ x \in X,\ y \in Y\}, \tag{10}$$

for polyhedra X, Y in Euclidean spaces, together with the information constraint

$$\text{decide } x \mapsto \text{ observe } z(\omega) \mapsto \text{ decide } y = y(x, z(\omega)). \tag{11}$$

Hence, problem (10) is accompanied by a two-stage scheme of alternating decision and observation, where the first-stage decision x must not anticipate future information on the random data $z(\omega)$. The second-stage decision $y = y(x, z(\omega))$ often is interpreted as a recourse action. It is taken as an optimal solution to the second-stage problem, i.e., the minimization problem remaining after x and $z(\omega)$ were fixed. The overall aim is to find an x which is "optimal" under these circumstances. Assume the minimum in (10) exists and rewrite

$$\begin{aligned} &\min_x \left\{c^\top x + \min_y\{q^\top y \, : \, Wy = z(\omega) - Tx, \; y \in Y\} \; : \; x \in X\right\} \\ &\quad = \min\{c^\top x + \Phi(z(\omega) - Tx) \; : \; x \in X\} \\ &\quad = \min\{\mathsf{j}(x, \omega) \; : \; x \in X\} \end{aligned} \tag{12}$$

where $\Phi(v) := \min\{q^\top y \, : \, Wy = v, y \in Y\}$ is the value function of a linear program with parameters in the right-hand side. In many cases, including the situation in elasticity we shall focus on below, the operator W is invertible, hence y can in principle be written as a linear functional of v. In the infinite-dimensional case of interest here, however, one does not have an explicit representation of the inverse operator, and it is convenient to keep the variational structure of the problem.

The representation (12) now gives rise to understanding the search for an "optimal" x as the search for a "minimal" member in the family of random variables $\{\mathsf{j}(x, \omega) \, : \, x \in X\}$. Different modes of ranking random variables then lead to different types of stochastic programs. In a risk neutral setting the ranking is done by taking the expectation $\mathbb{E}_\omega$, leading to

$$\min\{Q_{\mathrm{EV}}(x) := \mathbb{E}_\omega(\mathsf{j}(x, \omega)) \; : \; x \in X\}$$

which is a well-defined optimization problem under mild conditions. With risk aversion, the expectation is replaced by statistical parameters reflecting some perception of risk (risk measures). As examples let us mention minimization of the expected excess of some preselected target $\eta \in \mathbb{R}$

$$\min\{Q_{\mathrm{EE}_\eta}(x) := \mathbb{E}_\omega(\max\{\mathsf{j}(x, \omega) - \eta, 0\}) \; : \; x \in X\},$$

and minimization of the excess probability of some preselected $\eta \in \mathbb{R}$

$$\min\{Q_{\mathrm{EP}_\eta}(x) := \wp_\omega(\mathsf{j}(x, \omega) > \eta) \; : \; x \in X\}.$$

Besides these pure expectation and risk models there are mean-risk models where weighted sums of expected value and risk expressions are minimized, for further details see [45, 49, 53].

Two-stage stochastic programming formulation of shape optimization. In the optimization problem (6) there is a natural information constraint stating that first, and independently of the realizations of the forces $f(\omega)$, $g(\omega)$, the shape $\mathcal{O}$ has to be selected. Then, after observation of $f(\omega)$, $g(\omega)$, (1) determines the displacement field $u = u(\mathcal{O}, \omega)$, leading to the objective value $J(\mathcal{O}, \omega)$. This manifests the interpretation of (6) as a two-stage random optimization problem: In the outer optimization, or first stage, the nonanticipative decision on $\mathcal{O}$ has to be taken. After observation of $f(\omega)$, $g(\omega)$ the second-stage optimization problem is the variational problem of linearized elasticity, where for a fixed elastic domain $\mathcal{O}$ and random state ω one seeks a displacement u minimizing the energy $E(\mathcal{O}, u, \omega)$ defined in (2).

This second-stage optimization process is neither associated with further stochastic parameters nor with the optimization of additional material properties. In fact, it consists of the determination of the elastic displacement, which in turn is required for the computation of the elastic energy and the cost functional. Even though there is no additional decision making involved, the variational structure of the elasticity problem we are solving gives an obvious analogy to the second-stage problem in finite-dimensional stochastic programming.

As counterpart to (11), the information constraint

$$\text{decide } \mathcal{O} \mapsto \text{ observe } \omega \mapsto \text{ compute } u = u(\mathcal{O}, \omega).$$

must be fulfilled. The role of $\mathrm{j}(x, \omega)$ in (12) then is taken by $J(\mathcal{O}, \omega)$ from (5). Altogether, the random shape optimization problem (6)

$$\min \{J(\mathcal{O}, \omega) : \mathcal{O} \in \mathcal{U}_{\mathrm{ad}}\}$$

arises as the analogon to (12). It amounts to finding a "minimal" member in the family of random variables

$$\{J(\mathcal{O}, \omega) : \mathcal{O} \in \mathcal{U}_{\mathrm{ad}}\}.$$

As in the finite-dimensional case, now different ranking modes for random variables give rise to risk neutral and risk averse stochastic shape optimization models. Taking the expectation yields the risk neutral problem

$$\min \{\mathbf{Q}_{\mathrm{EV}}(\mathcal{O}) := \mathbb{E}_\omega(J(\mathcal{O}, \omega)) : \mathcal{O} \in \mathcal{U}_{\mathrm{ad}}\}. \tag{13}$$

Of the variety of risk averse problems arising with the expectation replaced by some risk measure, we study in more detail those given by the expected excess over some preselected target $\eta \in \mathbb{R}$

$$\min \left\{\mathbf{Q}_{\mathrm{EE}_\eta}(\mathcal{O}) := \mathbb{E}_\omega(\max\{J(\mathcal{O}, \omega) - \eta, 0\}) : \mathcal{O} \in \mathcal{U}_{\mathrm{ad}}\right\}, \tag{14}$$

and by the excess probability over $\eta \in \mathbb{R}$

$$\min \left\{\mathbf{Q}_{\mathrm{EP}_\eta}(\mathcal{O}) := \wp_\omega(J(\mathcal{O}, \omega) > \eta) : \mathcal{O} \in \mathcal{U}_{\mathrm{ad}}\right\}. \tag{15}$$

In our applications we use smooth approximations of the max-function in (14) and of the discontinuity caused by the probability in (15). For the expected excess this leads to

$$\mathbf{Q}^\epsilon_{\mathrm{EE}_\eta}(\mathcal{O}) = \mathbb{E}_\omega\left(q^\epsilon(J(\mathcal{O}, \omega))\right),$$

where $q^\epsilon(t) := \frac{1}{2}\left(\sqrt{(t-\eta)^2+\epsilon}+(t-\eta)\right)$, $\epsilon > 0$. For the excess probability we obtain

$$\mathbf{Q}^\epsilon_{\mathrm{EP}_\eta}(\mathcal{O}) = \mathbb{E}_\omega\left(H^\epsilon(J(\mathcal{O},\omega))\right)$$

with $H^\epsilon(t) := \left(1+e^{-\frac{2(t-\eta)}{\epsilon}}\right)^{-1}$, $\epsilon > 0$. If the random variable ω follows a discrete distribution with scenarios ω_i and probabilities $\pi_i, i = 1,\ldots,N_s$ for an in general large number N_s of scenarios, then

$$\mathbf{Q}_{\mathrm{EV}}(\mathcal{O}) = \sum_{i=1}^{N_s} \pi_i J\left(\mathcal{O},\omega_i\right), \tag{16}$$

and accordingly for $\mathbf{Q}^\epsilon_{\mathrm{EE}_\eta}(\mathcal{O})$ and $\mathbf{Q}^\epsilon_{\mathrm{EP}_\eta}(\mathcal{O})$.

5. Shape derivatives

Shape derivatives in the deterministic case. To compute the shape derivative, we consider variations $\mathcal{O}_v = (\mathrm{Id}+v)(\mathcal{O})$ of a smooth elastic domain $\mathcal{O}$ for a smooth vector field v defined on the working domain D. For the deterministic objective functional

$$\mathbf{J}\left(\mathcal{O}\right) := J\left(\mathcal{O},u\left(\mathcal{O}\right)\right) = \int_{\mathcal{O}} j\left(u\left(\mathcal{O}\right)\right)\,\mathrm{d}x + \int_{\Gamma_N} k\left(u\left(\mathcal{O}\right)\right)\,\mathrm{d}\mathcal{H}^{d-1} \tag{17}$$

the shape derivative [28] in the direction v initially takes the form $\mathbf{J}'(\mathcal{O})(v) = J_{,\mathcal{O}}(\mathcal{O},u(\mathcal{O}))(v) + J_{,u}(\mathcal{O},u(\mathcal{O}))(u'(\mathcal{O})(v))$. To avoid an evaluation of $u'(\mathcal{O})(v)$ for any test vector field v, one introduces the dual or adjoint problem. In fact, the dual solution $p = p(\mathcal{O}) \in H^1_{\Gamma_D}\left(\mathcal{O};\mathbb{R}^d\right)$ is defined as the minimizer of the dual functional

$$E_{\mathrm{dual}}\left(\mathcal{O},p\right) := \frac{1}{2}a\left(\mathcal{O},p,p\right) + l_{\mathrm{dual}}\left(\mathcal{O},p\right),$$

with $l_{\mathrm{dual}}(\mathcal{O},p) = \int_{\mathcal{O}} j'(u)p\,\mathrm{d}x + \int_{\Gamma_N} k'(u)p\,\mathrm{d}\mathcal{H}^{d-1}$. For the (deterministic version of the) compliance objective (7) we observe that $p = -u$. Given p for fixed u and $\mathcal{O}$ we can evaluate the shape derivative of the deterministic cost functional as follows:

$$\begin{aligned}\mathbf{J}'(\mathcal{O})(v) &= J_{,\mathcal{O}}(\mathcal{O},u(\mathcal{O}))(v) - l_{,\mathcal{O}}(\mathcal{O},p(\mathcal{O}))(v) + a_{,\mathcal{O}}(\mathcal{O},u(\mathcal{O}),p(\mathcal{O}))(v) \\ &= \int_{\partial\mathcal{O}\setminus(\Gamma_D\cup\Gamma_N)} (v\cdot n)\left[j(u(\mathcal{O})) - f p(\mathcal{O}) + \mathbf{C}_{ijkl}e_{ij}(u(\mathcal{O}))e_{kl}(p(\mathcal{O}))\right]\mathrm{d}\mathcal{H}^{d-1}.\end{aligned} \tag{18}$$

Let us remark that in a general situation the shape calculus is more subtle. For a discussion of the appropriate differentiation of boundary integrals we refer to [28, 59]. Here, we confine ourselves, as in [26, 25], to the simpler case that homogeneous Neumann boundary conditions are assumed on the part of the boundary which is optimized. Furthermore, let us emphasize that the above classical shape derivative is only admissible in case of additional regularity for the primal solution u and the dual solution p, which holds under sufficiently strong regularity assumptions on loads and geometry of the considered shapes.

Shape derivatives of expectation and risk measures. With a discrete probability distribution as in (16) we compute the shape derivative of the expectation as

$$\mathbf{Q}'_{\mathrm{EV}}(\mathcal{O})(v) = \sum_{i=1}^{N_s} \pi_i \, J'\left(\mathcal{O}, \omega_i\right)(v) \, . \tag{19}$$

For the approximated expected excess and excess probability the chain rule yields

$$\left(\mathbf{Q}^{\epsilon}_{\mathrm{EE}_\eta}\right)'(\mathcal{O})\,(v) = \sum_{i=1}^{N_s} \frac{\pi_i}{2} J'\left(\mathcal{O}, \omega_i\right)(v) \left(\frac{J\left(\mathcal{O}, \omega_i\right) - \eta}{\sqrt{\left(J\left(\mathcal{O}, \omega_i\right) - \eta\right)^2 + \epsilon}} + 1 \right) ,$$

$$\left(\mathbf{Q}^{\epsilon}_{\mathrm{EP}_\eta}\right)'(\mathcal{O})\,(v) = \sum_{i=1}^{N_s} \frac{2}{\epsilon} \pi_i J'\left(\mathcal{O}, \omega_i\right)(v)\, e^{-\frac{2}{\epsilon}(J(\mathcal{O},\omega_i)-\eta)} \left(1 + e^{-\frac{2}{\epsilon}(J(\mathcal{O},\omega_i)-\eta)}\right)^{-2} .$$

Efficient evaluation of primal and dual solutions for realizations of the randomness. The preceding conceptual discussion concerned an arbitrary probability space $(\Omega, \mathcal{A}, \wp)$. When it comes to computations, however, further assumptions are needed to enable calculation of the integrals behind the relevant expected values and probabilities. To this end, we approximate the probability distribution with a discrete one, and assume that $(\Omega, \mathcal{A}, \wp)$ is finite, in the sense that there are finitely many realizations ω. The number of realizations will be denoted by N_s.

Then, it seems that for every realization one has to compute a primal solution $u(\mathcal{O}, \omega_i)$ and a dual solution $p(\mathcal{O}, \omega_i)$. Under our assumption that $j(\cdot)$ and $k(\cdot)$ are linear or quadratic functions there is a significant algorithm shortcut at our disposal for $N_s \gg 1$, which we will recall here. For details we refer to [26]. Let us assume that there is a small number of (deterministic) basis volume forces $f_1, \ldots, f_{K_1}$, and of (deterministic) basis surface loads $g_1, \ldots, g_{K_2}$ and that the actual loads $f(\omega)$ and $g(\omega)$ arise as linear combinations of these deterministic basis loads, $f(\omega) = \sum_{i=1}^{K_1} c_i^f(\omega) f_i$ and $g(\omega) = \sum_{j=1}^{K_2} c_j^g(\omega) g_j$, with random coefficients $c_i^f(\omega) \in \mathbb{R}$, $i = 1, \ldots, K_1$, and $c_j^g(\omega) \in \mathbb{R}$, $j = 1, \ldots, K_2$. The advantage of this approach is that we do not need to solve the elasticity PDE for each of the N_s different realizations of the stochastic variable ω, but only for the different basis forces. This will reduce the computing cost significantly in case $N_s \gg K_1 + K_2$. Suppose $u^{(i,0)}$ is the elastic displacement for given volume force $f := f_i$ and surface load $g := 0$, for all $i = 1, \ldots, K_1$. Similarly, let $u^{(0,j)}$ be the displacement corresponding to the volume force $f := 0$ and the surface load $g := g_j$, for all $j = 1, \ldots, K_2$. Then,

$$\bar{u}\left(\mathcal{O}; \omega\right) := \sum_{i=1}^{K_1} c_i^f(\omega) u^{(i,0)} + \sum_{j=1}^{K_2} c_j^g(\omega) u^{(0,j)} \tag{20}$$

is the minimizer of (2) with volume force $f := f(\omega)$, and surface load $g := g(\omega)$. A similar relation holds for the adjoint state $\bar{p}\left(\mathcal{O}; \omega\right)$ in scenario ω, if we additionally assume that $\sum_{i=1}^{K_1} c_i^f(\omega) + \sum_{j=1}^{K_2} c_j^g(\omega) = 1$, which can always be achieved by choosing the basis forces appropriately. In that case, and if we denote the adjoint states for the individual basis forces by $p^{(i,0)}$ and $p^{(0,j)}$,

$$\bar{p}\,(\mathcal{O};\omega) := \sum_{i=1}^{K_1} c_i^f(\omega)p^{(i,0)} + \sum_{j=1}^{K_2} c_j^g(\omega)p^{(0,j)} \tag{21}$$

is the adjoint state belonging to the state $\bar{u}\,(\mathcal{O};\omega)$.

6. Description of two different classes of admissible shapes

In this section, we will investigate two different shape representations and discuss how to evaluate cost functionals and derivatives with respect to a modification of the shape. On the one hand we will employ a level set method to flexibly describe shapes given as implicit surfaces. On the other hand, geometric details constituting an elastic domain will be described by polygonal models determined by a small number of parameters each.

Implicit shapes described via level sets. To be flexible with respect to the topology of the optimal shape we formulate the optimization problem in terms of shapes described by level sets – let us in particular refer to the approach by Allaire and coworkers [3, 4, 5, 8] and to [29, 39, 52]. Explicitly, we assume the elastic body $\mathcal{O}$ to be represented by a level set function ϕ, that is $\mathcal{O} = \{\phi < 0\}:=\{x \in D \,|\, \phi(x) < 0\}$. The boundary $\partial\mathcal{O}$ corresponds to the zero level set of ϕ, i.e., $D \cap \partial\mathcal{O} = \{\phi = 0\}$. Essential is the observation that the (normal) variation $\delta x \cdot n$ of the domain boundary $\partial\mathcal{O}$ can be identified with the variation $\delta\phi$ of the level set function ϕ via the level set equation $\delta\phi + |\nabla\phi|\,\delta x \cdot n = 0$, where $n = \frac{\nabla\phi}{|\nabla\phi|}$ are the outer normals on the level sets. Hence, for the objective functional rephrased in terms of the level set function as $\mathcal{J}(\phi):=J(\{\phi < 0\})$ one obtains for the shape derivative

$$\mathcal{J}'(\phi)(\delta\phi) = J'(\{\phi < 0\})\left(-\delta\phi\,|\nabla\phi|^{-1}n\right)\,. \tag{22}$$

This representation of the shape derivative can then be used in a relaxation scheme for the objective functional. We emphasize that the usual level set approach allows only simplifications of the topology during a descent scheme for the cost functionals. New holes cannot be created.

Perforated domains with parametrized geometric details. At variance with the usual description of shapes as volumetric macroscopic objects with a piecewise smooth boundary let us now consider a fixed domain D perforated with a large number of holes. In fact, we do not aim to optimize the outer boundary of D but the geometry of the holes inside the domain. Thereby, we perform an optimization of the fine-scale geometric structure of the elastic object. In the application section, we will comment on the relation of this type of shape optimization to homogenization approaches in shape optimization. As a structural restriction let us assume that the fine scale geometric details are centered at the nodes of a lattice covering the domain D. The considered fine scale geometry is parametrized over a low-dimensional set of parameters. Furthermore, we restrict ourselves here to N elliptical holes $\mathcal{B}(c_i, \alpha_i, a_i, b_i)$ $(1 \le i \le N)$, parametrized by a rotation α_i and two scaling factors a_i and b_i for the two semi-axes, i.e.,

$$x(s) = c_i + h \begin{pmatrix} a_i \cos(\alpha_i)\cos(s) - b_i \sin(\alpha_i)\sin(s) \\ a_i \sin(\alpha_i)\cos(s) + b_i \cos(\alpha_i)\sin(s) \end{pmatrix}$$

where h is the grid size of the lattice and the center points are characterized by $\frac{c_i}{h} + \frac{1}{2} \in \mathbb{Z}$. To avoid overlapping of adjacent or completely vanishing holes we require the scaling parameters to be bounded, i.e., we require $\epsilon \le a_i,\, b_i \le \frac{1}{2} - \epsilon$ with $\epsilon > 0$ being a small additional offset to prevent numerical instability. Finally, we impose a constraint on the total 2D volume of the elastic object $\mathcal{O} := D \setminus (\bigcup_{i=1,\dots,N} \mathcal{B}(c_i, \alpha_i, a_i, b_i))$ and require $|D| - \sum_{i=1}^{N} \pi h^2 a_i b_i = V$. If we assume that the ellipses are attached to a fixed set of cell centers c_i in D, then the resulting shape optimization problem turns into a finite-dimensional optimization problem in $\mathbb{R}^{3N}$ with inequality constraints for the a_i and b_i and one equality constraint due to the prescribed total 2D volume. The associated cost function is given by

$$\mathcal{J}((\alpha_i, a_i, b_i)_{i=1,\dots,N}) := J\Big(D \setminus (\bigcup_{i=1,\dots,N} \mathcal{B}(c_i, \alpha_i, a_i, b_i))\Big),$$

for which one easily derives the shape gradient

$$\mathcal{J}'((\alpha_i, a_i, b_i)_{i=1,\dots,N}) = \Big((\partial_{\alpha_j}\mathcal{J}, \partial_{a_j}\mathcal{J}, \partial_{b_j}\mathcal{J})((\alpha_i, a_i, b_i)_{i=1,\dots,N})\Big)_{j=1,\dots,N}$$

as a vector in $\mathbb{R}^{3N}$. In this context, we are in particular interested in the interaction of different holes with respect to the shape optimization.

7. A sketch of the numerical algorithms

In this section we detail the numerical algorithms in two space dimensions ($d = 2$). On the one hand we discuss a finite element approach for the level set-based shape optimization and in particular describe suitable finite element spaces for the primal and the dual solutions. Furthermore, we discuss how to combine the topological derivative and the regularized gradient descent approaches in a robust discrete-energy minimization. On the other hand we review the boundary element method as a tool to calculate the cost functional and its gradient in case of perforated domains with fine scale geometries.

Finite element discretization of the level set approach. For the ease of implementation we restrict ourselves to the working domain $D = [0,1]^2$, and use a dyadic hierarchy of regular simplicial grids and discrete level set functions Φ in the corresponding space of piecewise affine functions $\mathcal{V}_h$. To avoid inaccurate evaluations of the objective functional and the shape derivative and a complicated regular remeshing of the boundary [30, 43] we resort to composite finite elements introduced by Hackbusch and Sauter [35] for the efficient multigrid solution of PDEs on complicated domains. In fact, we define the space $\mathcal{V}_h^{\mathrm{cfe}}$ spanned by the vector-valued composite finite element basis function $\Theta_{ij}^{\text{cfe}}(x) = e_j\, \chi_{\mathcal{O}_h}(x)\, \Theta_i(x)$, where Θ_i is one of the usual nodal basis functions of $\mathcal{V}_h$ and $e_1 = (1,0)$, $e_2 = (0,1)$, and $\chi_{[\Phi<0]}$ the characteristic function of the elastic object. Notice that to keep the Dirichlet boundary and the inhomogeneous Neumann boundary fixed we freeze the level set function Φ on a small neighborhood of $\Gamma_D \cup \Gamma_N$ (cf. the problem set up in Section 3). For a given basis load the discrete primal solution is defined as the finite element function $U^{i,j} \in \mathcal{V}_h^{\text{cfe}}$ solving $A(\mathcal{O}_h, U^{i,j}, \Theta) = l^{i,j}(\mathcal{O}_h, \Theta)$ for all $\Theta \in \mathcal{V}_h^{\text{cfe}}$, where $l^{i,0}(\mathcal{O}_h, \Theta) := \int_{\mathcal{O}_h} f_i \cdot \Theta \,\mathrm{d}x$ and $l^{0,j}(\mathcal{O}_h, \Theta) := \int_{\partial\mathcal{O}_h} g_j \cdot \Theta \,\mathrm{d}\mathcal{H}^{d-1}$ for $1 \le i \le K_1$ and $1 \le j \le K_2$. The corresponding set of dual solutions are those functions $P^{i,j} \in \mathcal{V}_h^{\text{cfe}}$, for which $A(\mathcal{O}_h, \Theta, P^{i,j}) = -J_{,u}(\mathcal{O}_h, U^{i,j})(\Theta)$ for all $\Theta \in \mathcal{V}_h^{\text{cfe}}$. Due to the assumption on $J(\mathcal{O}, \cdot)$ the integrand is at most quadratic and can be integrated exactly using a Gauss quadrature rule. Finally, the primal and dual solutions for the actual set of realizations ω_i for $i = 1, \dots, N_S$ are composed of the discrete primal and dual solutions $U^{i,j}$ and $P^{i,j}$, respectively, based on the discrete analog of (20) and (21), respectively (cf. [26, 38]).

For the relaxation of the shape functional we consider a time discrete, regularized gradient descent for the risk measure $\mathbf{Q}$ defined on shapes now described by level set functions. Explicitly, given a level set function Φ^k we compute Φ^{k+1} in the next iteration solving

$$\mathcal{G}(\Phi^{k+1} - \Phi^k, \Psi) = -\tau \mathbf{Q}'(\Phi^k)(\Psi) \tag{23}$$

for all $\Psi \in \mathcal{V}^h$, where $\mathcal{G}(\zeta, \xi) = \int_D \zeta\xi + \frac{\rho^2}{2}\nabla\zeta \cdot \nabla\xi \,\mathrm{d}x$ is a H^1-type metric on variations of the level set function ϕ related to a Gaussian filtering of the L^2 gradient with the filter width ρ. Furthermore, we consider Armijo's rule as a step size control. For an overview on optimal design based on level sets and suitable energy descent methods we refer to a recent survey by Burger and Osher [19]. Finally, after a fixed number of gradient descent steps we use a thresholding with respect to the topological derivative to enable the creation of new holes in the elastic domain. This relaxation algorithm is now applied in a coarse to fine manner. Starting on coarse grids, we successively solve the minimization and prolongate on the next finer grid level until the a priori given finest grid resolution is reached.

Boundary element approach in the context of perforated domains. We will now briefly review the boundary element method which is used to solve the primal and dual problem in the context of perforated domains. Main ingredient is a fundamental solution for the PDE of linearized elasticity [24]. Identifying points $x \in \mathbb{R}^2$ with points $z(x) = x_1 + p\,x_2$ on the complex plane, where p is a complex constant, which is used in order to transform balls into ellipses, the fundamental solution u_{ki}^* and its normal derivative v_{ki}^* are given by

$$u_{ki}^* = \frac{1}{2\pi} \sum\nolimits_m \Re\Big\{ \sum\nolimits_n A_{kn} N_{nj}\ \ln\left(z_n(x-y)\right) \Big\}\, d_{mi}\,,$$

$$v_{ki}^* = -\frac{1}{2\pi} \sum\nolimits_{j,m} \Re\left\{ \sum\nolimits_n \frac{D_{ijn} N_{nm}}{z_n(x-y)} \right\} d_{mi}\, n_j\,.$$

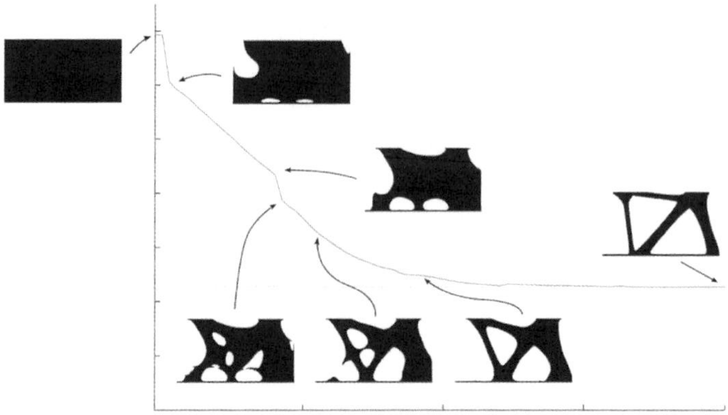

FIGURE 2. The energy decay is depicted and some intermediate time steps are shown for the successive shape relaxation via the level set / CFE method. The impact of the topological derivative can be seen two times, first directly at the start of the algorithm and second associated with a jump of the energy during the relaxation process.

Here n denotes the outer normal and A_{kn}, N_{nj},D_{ijn}, d_{mi} are appropriately chosen constants. We now consider equation (2), apply Green's formula, substitute ϑ by the fundamental solutions u_i^* and arrive at

$$u_{ki}(x) = \int_{\partial\mathcal{O}} (\mathbf{C}\varepsilon(u)n)\cdot u_{ki}^* \, \mathrm{d}\mathcal{H}^1(y) - \int_{\partial\mathcal{O}} (\mathbf{C}\varepsilon(u_{ki}^*)n)\cdot u \, \mathrm{d}\mathcal{H}^1(y) + \int_{\mathcal{O}} f\cdot u_{ki}^* \, \mathrm{d}y. \tag{24}$$

From this formula one now derives in the usual way a collocation type boundary element method (for details we refer to [34]). Under the assumption that $f \equiv 0$, for u on the boundary and $\sigma_n = \mathbf{C}\varepsilon(u)n$ denoting the normal tensions at the boundary we obtain from (24) the integral equation

$$u = U[\sigma_n] - V[u]\,, \tag{25}$$

where $U[\sigma_n]$ is a *single layer operator* and $V[u]$ a *double layer operator*.

For the discretization we restrict ourselves to polygonal domains $\mathcal{O}_h$, where the ellipses and the outer boundary are replaced by polygons with a fixed number of vertices, which are considered as *collocation points*. Effectively, these points depend on the parameters c_i, α_i, a_i, and b_i of the ellipses. Furthermore, we approximate u and σ_n on the boundary via piecewise affine, continuous functions. Equation (25) must now hold for every collocation point leading to a linear system of equations for the values of the displacement and the normal tensions at each point. To evaluate the objective functionals observe that (7) and (9) have already been phrased as boundary integrals. Shape derivatives of the objectives are also expressed in terms of boundary integrals in (18). For the evaluation of these integrals full gradients of the primal and dual solution are required whereas we only have (discrete approximations) of the normal stresses at our disposal so

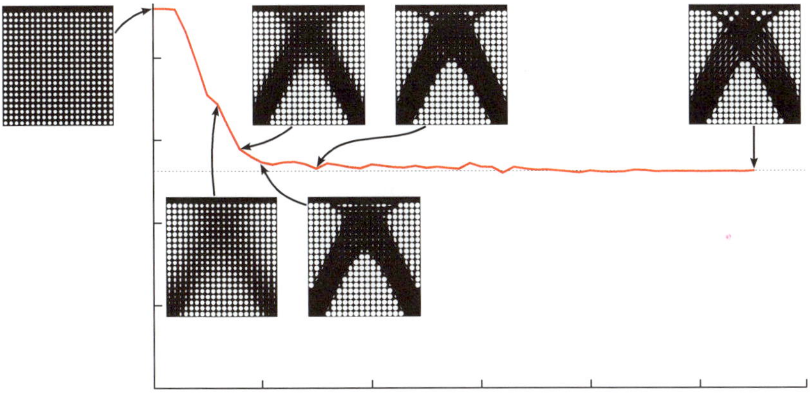

FIGURE 3. The energy during optimization of a perforated square domain under shear load by `Ipopt` is plotted with the final value indicated by the dashed line.

far. However one can reconstruct them approximately using finite differences on the polygonal boundary for the tangential derivatives of the displacement. Thus, we are able to write suitable approximations of the objective functional and the shape derivative in terms of the vector of free parameters $(\alpha_i, a_i, b_i)_{i=1,\dots,N}$. The treatment of the resulting finite-dimensional constrained optimization problems is a classical and well-developed field. For our problem we rely on the software package `Ipopt` [61, 62] which implements a primal-dual interior point filter line search algorithm. Fig. 3 shows a plot of the energy during optimization by `Ipopt` together with some intermediate results.

8. Applications of stochastic risk measure optimization

In this section we present several applications of the methods described before. We discuss the stochastic optimization in the context of shapes described by level sets and by parametrized geometry (and also shortly compare the latter with results derived by the homogenization method). Furthermore, we give a detailed description of the results obtained by using risk averse stochastic optimization.

Stochastic optimization of a 2D carrier plate. As a first application we look at a 2D carrier plate, where the supporting construction between a floor slap, whose lower boundary is assumed to be the Dirichlet boundary, and an upper plate, on which forces act, is optimized. The working domain is $D = [0,1]^2$.

Fig. 1 (left and center) shows a straightforward comparison between a stochastic optimization as described before and an optimization for a deterministic load corresponding to the expected value of the stochastic loads. Also the optimal shape (for a deterministic load) with fine scale parametrized geometry is shown on the right. For comparison Fig. 4 depicts additional results in the context

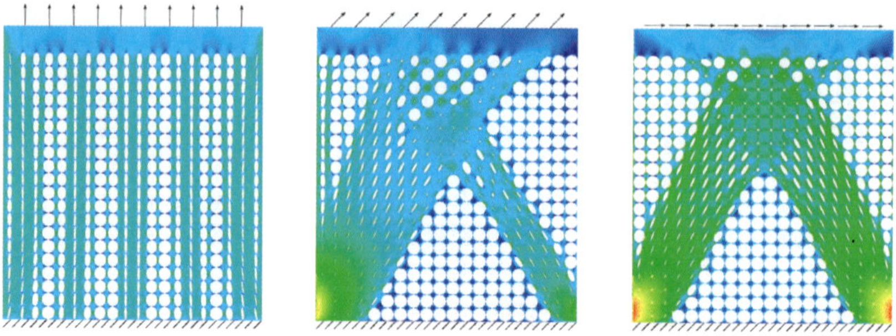

FIGURE 4. Different loads and corresponding optimal shapes color-coded by the von Mises stress.

of perforated domains under similar loading conditions color-coded by the von Mises stress. One observes a rather sharp interface between regions with high and low volume fraction of the elliptical holes. Furthermore, the resulting shapes are roughly similar to the ones obtained in the level set context but have additionally developed fine trusses aligned with the main loading directions. For the image generation in Fig. 4 a triangulation of the computational domain was created using the software `Triangle` [54, 55] and at each interior vertex the linearized strain tensor was computed using (24).

Our next case study is a more elaborate example for stochastic optimization in the context of level sets. Fig. 5 shows a sketch of the stochastic loading on the upper plate and the (single) stochastically optimal shape obtained by the level set based stochastic optimization algorithm. Furthermore, the von Mises stress distribution is rendered for the different load scenarios. Since each realization of the stochastic load is chosen to be spatially uniform on the upper plate, the realizations only differ by the direction of the force. Hence, the load space containing all realizations of the stochastic loads can be represented by the span of two base loads g^1 and g^2. In the notation of Section 5, we have $K_2 = 2$ and $N_S = 20$.

2D and 3D cantilever. As a second application, we discuss the shape optimization of a 2D cantilever. We choose $D = [0, 2] \times [0, 1]$ and model a cantilever that is fixed on the left side and where a deterministic downward pointing force is applied on a small plate on the right. Three methods, namely homogenization (by Allaire [3], first with a composite solution and then with a penalization of composite regions), level set based optimization, and optimization of a perforated geometry are used to create the results shown in Fig. 6 (rotated by 90 degrees). Also, for the latter two methods, the von Mises stress is shown by color-coding.

As a generalization of the application before, we now consider a 3D cantilever. Here again the cantilever is fixed on one side by having a Dirichlet condition on a disk-shaped plate. Whereas on the other side we prescribe a Neumann boundary load on a small rectangular plate opposite to the center of the disk. Applying eight

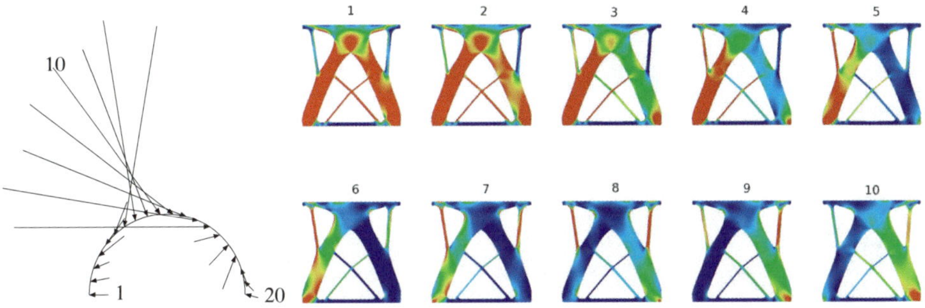

FIGURE 5. Stochastic shape optimization based on 20 scenarios is depicted. On the left the different loads $g(\omega_\sigma)$ with probabilities π_σ are sketched. Each arrow represents one scenario where the arrow length is determined by the corresponding force intensity weighted with the probability π_σ of the corresponding scenario. On the right the von Mises stress distribution is color-coded on the optimal shape for 10 out of the 20 realizations of the stochastic loading. Due to the nonsymmetric loading configuration the resulting shape is nonsymmetric as well. In particular, the right carrier is significantly thicker than the left one, whereas the connecting diagonal stray pointing up right is thinner than the one pointing down left. (Figure reprinted from [26, Fig. 4.3])

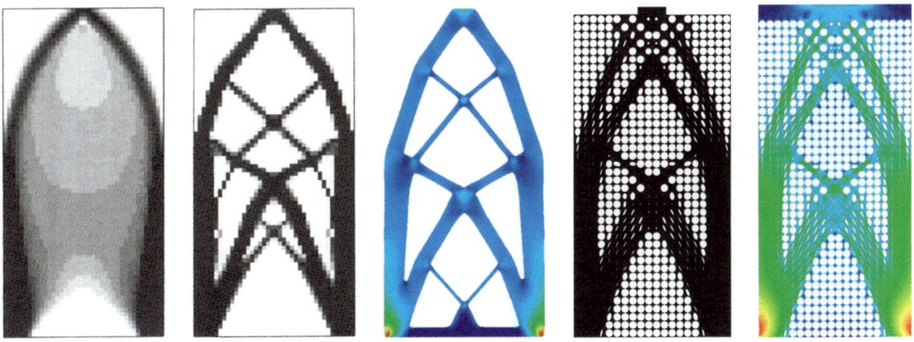

FIGURE 6. Comparison between optimal shapes for the cantilever scenario obtained by the homogenization method by Allaire (first two panels, reprinted from [3, Fig. 5.8, p. 370]), level set based optimization (central panel, reprinted from [26, Fig. 4.5]), and optimization of perforated geometries.

stochastic loading scenarios with equally probable and equally distributed forces in a plane parallel to the disk-shaped plate results in the optimal shape shown in Fig. 7. Again a colorcoding of the von Mises stress distribution is displayed.

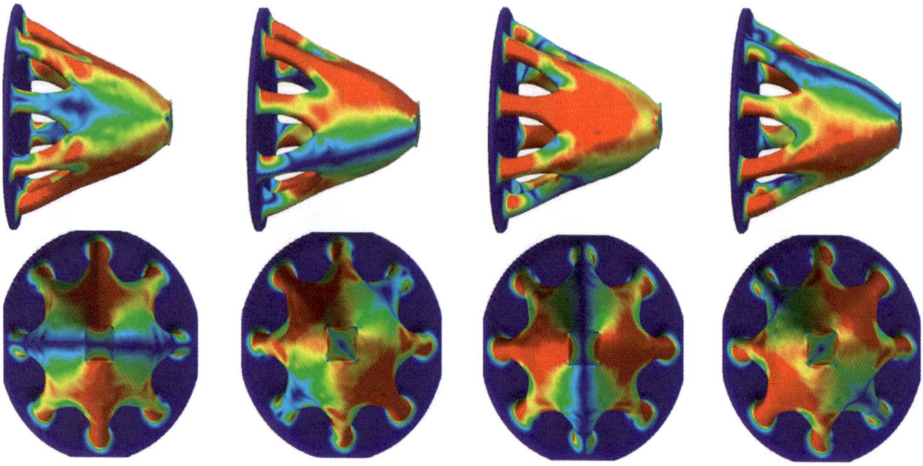

FIGURE 7. The optimal design in the case of stochastic shape optimization for the cantilever problem with eight scenarios is depicted. From left to right four scenarios are color-coded with the von Mises stress in a consecutive clockwise ordering with respect to the attacking loads. The upper and the lower row show the shape geometry under different perspectives. (Figure reprinted from [26, Fig. 4.8])

2D bridge. For the sake of completeness also a typical bridge-type scenario was investigated in the context of perforated domains. Here the computational domain was chosen as $D = [0,3] \times [0,1]$, the lower corners were kept fixed by imposing Dirichlet boundary values and the remainder of the lower boundary was subject to downward pointing forces. Fig. 8 shows the obtained result color-coded with the von Mises stress and again in comparison to results of the homogenization method.

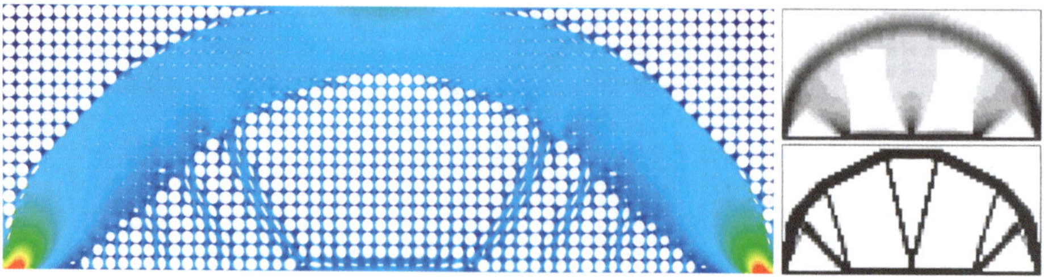

FIGURE 8. Comparison between optimal shapes for the bridge scenario obtained by Allaire with the homogenization method (right panel, reprinted from [3, Fig. 5.28, p. 399]) and optimization of perforated geometries.

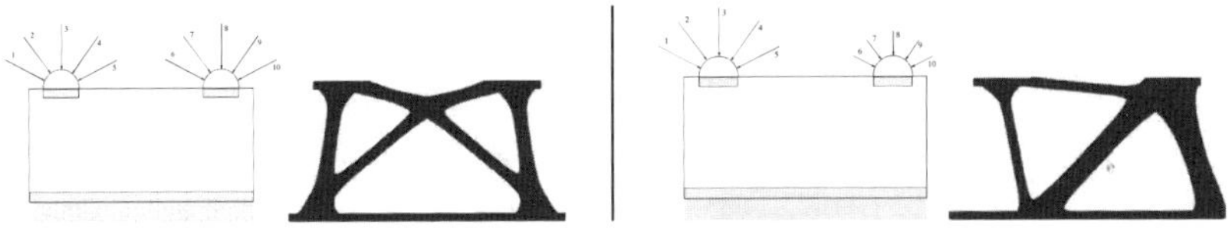

FIGURE 9. A symmetric stochastic load configuration (left) leads to symmetric support jibs (left panel), whereas a nonsymmetric stochastic loading (right panel) favors a correspondingly nonsymmetric truss construction. (Figure reprinted from [25, Fig. 4])

Trusses underneath two fixed bearings. In this section we will discuss an application with the focus on risk averse stochastic optimization and study the impact of different risk measures on the optimal shape. First, we discuss the optimization of the expected value objective. We consider two instances with 10 loading scenarios each (shown in Fig. 9). In the first instance (plotted on the left) all scenarios have the same probabilities and all scenario loads have the same magnitude. In the second instance (plotted on the right) we deal with an asymmetric case: loads acting on the left bearing are twice as strong as those acting on the right bearing. But the left loads have a much lower probability of 0.01 whereas each load on the right is active with probability 0.19. Hence, on average, the forces acting on the left bearing play a minor role. In both instances, the number of base forces is 4 (i.e., $K_2 = 4$), since two of them have their support only at the left bearing, whereas the other two apply only on top of the right bearing. The shapes shown in Fig. 9 minimize the expected value of the compliance cost functional. It can be observed that the symmetric load configuration leads to a symmetric truss construction, whereas for nonsymmetric loading nonsymmetric outriggers minimize the expected value of the cost. Fig. 2 shows the energy decay and selected snapshots of the evolving shape for the level set based method. The intermediate results illustrate the interplay between hole creation by thresholding based on the topological derivative and the optimization of the boundary by the level set / CFE method.

Expected excess. As mentioned above, one of the main objectives of this section is to compare the different risk measures and we will discuss this in the context of the second (nonsymmetric) load configuration. Fig. 10 shows a family of optimal shapes minimizing the expected excess for varying excess parameter η.

Although we start each computation with the same initial solid bulk domain $\mathcal{O}_0 = D$, we observe a continuous evolution of the geometry with η. Due to the nonanticipativity of our stochastic shape optimization model, the actual loading on an optimal shape, but not the shape itself, depends on the load scenario.

Is is important to point out that in the minimization of the expected excess all the scenarios ω where $J(\mathcal{O}, \omega) \leq \eta$ are irrelevant. In practice, since we use the smooth approximation $\mathbf{Q}^{\epsilon}_{\mathrm{EE}_\eta}$ (see Section 4) to regularize the stochastic cost functional, these scenarios still have a small effect. Hence, the optimization tries to choose $\mathcal{O}$ such that the objective $J(\mathcal{O}, \omega)$ stays below the risk parameter η in the

FIGURE 10. A sequence of results for the optimization with respect to the expected excess for $\eta = 0.1, 0.2, 0.3, 0.4, 0.6, 0.8, 1.0, 1.5$. The underlying loading is shown in Fig. 9 (right panel). (Figure reprinted from [25, Fig. 5])

high-probability, and therefore "expensive" scenarios, which correspond to load acting on the right bearings. In fact, the higher the value of η is, the easier it is to keep the objective below η. This leads to the effect that less material is needed on the right and therefore more material can be used to improve the situation for the scenarios with the strong but low-probability loading on the left bearing.

An analogous computation was done in the framework of perforated domains. In Fig. 11 one can observe qualitatively the same behavior as seen in the level set based computation. In particular, one observes that for high η the initial configuration remains unchanged because there is no need for improvement to stay below the excess parameter (cf. Fig. 12). We have to remark that because of the different scaling of the attacking forces the values of the parameter η in the two computations are not directly comparable.

Excess probability. We now turn to optimizing the excess probability. Albeit somewhat related, strikingly different shapes are being realized. Looking at Fig. 13 and

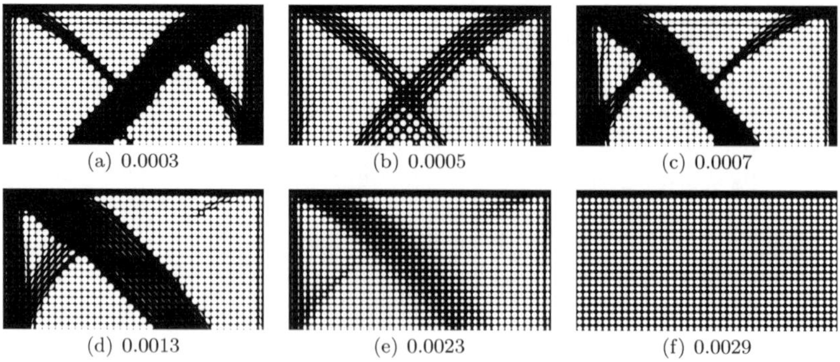

FIGURE 11. A sequence of results for the optimization with respect to the expected excess for $\eta = 0.0003, 0.0005, 0.0007, 0.0013, 0.0023, 0.0029$. The underlying loading is shown in Fig. 9 (right panel).

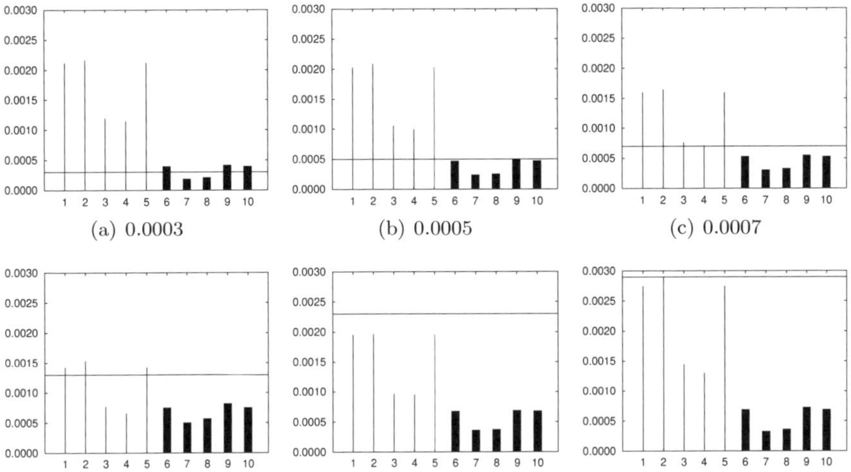

(a) 0.0003 (b) 0.0005 (c) 0.0007

(d) 0.0013 (e) 0.0023 (f) 0.0029

FIGURE 12. Objective values for each of the 10 scenarios are rendered with bar charts for the optimal shapes in Fig. 11. The bar thickness is chosen proportional to the probability.

Fig. 14 one observes a complete "mass shift" towards the right bearing. The important difference to the expected excess is that the amount of excess is irrelevant here. Therefore huge objective values may be acceptable for the five (low-probability) scenarios on the left as long as it is possible to push the corresponding values for the (high-probability) scenarios on the right bearing below η (cf. Fig. 15 for the corresponding cost diagram). Consequently trusses for the left bearing are allowed to become thinner and thinner. This can be seen in Fig. 13(a) but even better for the combined level set and topological shape optimization approach in Fig. 14 where finally the narrow trusses are removed via a topology optimization step.

Let us conclude with a final review of the three considered risk measures. Fig. 15 shows objective values for each loading scenario at the optimum for the

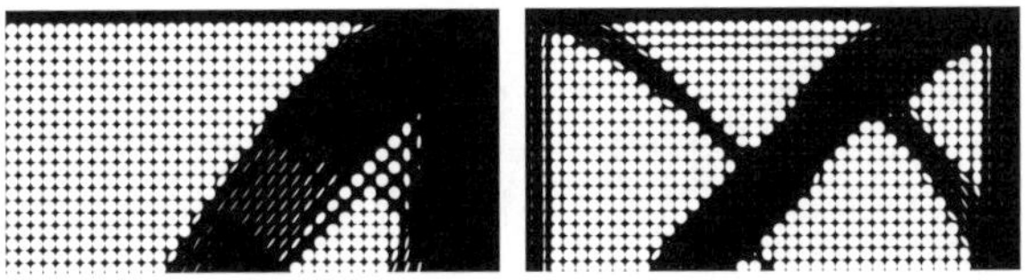

(a) 0.0001 (b) 0.0003

FIGURE 13. Two results for the optimization with respect to the excess probability for $\eta = 0.0001, 0.0003$. The underlying loading is shown in Fig. 9 (right panel).

FIGURE 14. In the optimization of the excess probability for $\eta = 0.4$ decreasingly thin trusses are realized on the intermediate grey shape on the left until the left bearing is completely truncated via a topology optimization step (cf. Fig. 9 for the load configuration). The final result is shown on the right. (Figure reprinted from [25, Fig. 7])

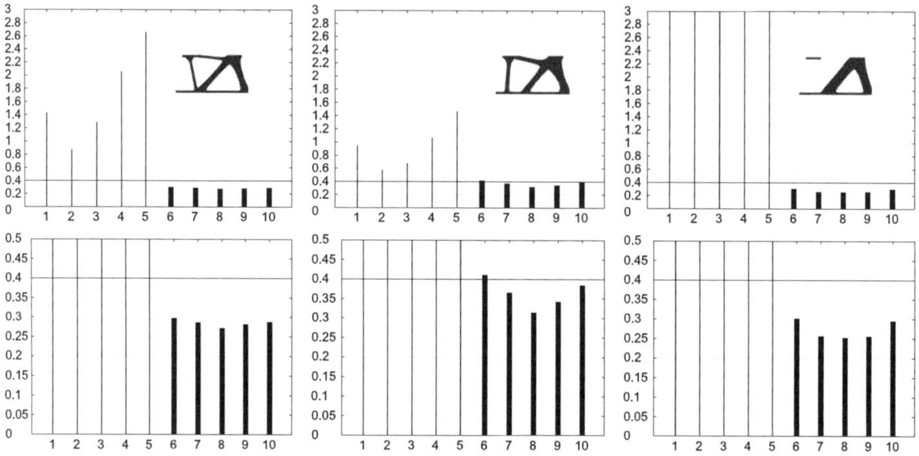

FIGURE 15. Objective values for each of the 10 scenarios are rendered with bar charts for the optimal shapes corresponding to the stochastic cost functionals $\mathbf{Q}_{\mathrm{EV}}$, $\mathbf{Q}^{\epsilon}_{\mathrm{EE}_\eta}$, $\mathbf{Q}^{\epsilon}_{\mathrm{EP}_\eta}$ (from left to the right) for $\eta = 0.4$. The second row shows a zoom into the diagrams. The bar thickness is chosen proportional to the probability. (Figure reprinted from [25, Fig. 8])

different risk measures. The expected value causes a weighted average over all single-scenario realizations. Therefore objective values for high-probability scenarios are reduced as much as possible whereas rather big values are accepted for the low-probability scenarios. For the expected excess the main difference is that there is no need to further optimize scenarios whose objective values have already been pushed below the threshold. The gained flexibility may instead be used to reduce the cost of scenarios above η more aggressively. Finally the excess probability reduces the process of assigning an objective value to the question of how likely scenarios exceeding the threshold occur. The consequence is ignorance of objective values for scenarios significantly above the threshold, which may be considered to be "lost" anyway. A compliance tending to infinity is indeed accepted when the left bearing is left floating. The arising flexibility is then used to further reduce maximal (weighted) scenarios whose costs are slightly bigger than the threshold. As to the built-in regularization in $\mathbf{Q}^{\epsilon}_{\mathrm{EE}_\eta}$, $\mathbf{Q}^{\epsilon}_{\mathrm{EP}_\eta}$ one has to remark that the discussed threshold behavior is smeared out at η, thereby favoring cost reduction of single scenarios in the vicinity of the threshold.

9. Hole-hole interaction in the topological derivative

The case studies presented above demonstrate the applicability of the parametric approach with a large number of elliptical holes to stochastic shape optimization problems. The explicit analysis of the limiting case of a large number of small holes is however numerically very demanding. Therefore we turn to a two-scale approach. In order to solve the cell problem, containing a periodic arrangement of holes, we need to understand hole-hole interactions. This is best done in the framework of topology derivatives, which we now briefly review.

Introducing topological derivatives. As discussed in the introduction, classical shape optimization methods are based on iteratively improving a given shape $\mathcal{O}$, by determining at each step an optimal small modification to the position of the boundary $\partial\mathcal{O}$, and are therefore unable to change the topology of the shape. The method of topology optimization is based on a determination of the gradient of the functional with respect to changes in topology of $\mathcal{O}$ [56, 32]. Precisely, for $\bar{x} \in \mathcal{O}$ one considers the family of domains $\mathcal{O} \setminus B(\bar{x}, \rho)$. The topology derivative at $\bar{x}$ is then defined as the limit $\lim_{\rho\to 0}(J[\mathcal{O} \setminus B(\bar{x}, \rho)] - J[\mathcal{O}])/\rho^n$, where J is the target functional. As usual in PDE-constrained optimization the target functional depends indirectly on $\mathcal{O}$, through a function u which is obtained by solving a PDE, which in our case corresponds to linear elasticity. Therefore an analytic evaluation of the topological derivatives requires a determination of the leading-order correction to the elastic deformation arising from the formation of a small hole [32]. This is done by a separation of length scales, between the one where the hole acts and the macroscopic one of $\mathcal{O}$; the interaction between the two can be represented by a suitable Dirichlet-to-Neumann operator acting on the new (fictitious) boundary. Thus, the inner problem can be analytically solved, depending on the local values of the strain coming from the exterior problem. The topological derivative is then evaluated numerically, and used to modify the computational domain $\mathcal{O}$.

Interaction between holes. As explained above the topological derivative describes the behavior of the cost functional with respect to the removal of a small hole. In practice, it is often used to remove a complete area, determined as the set where the derivative is below a threshold (see Section 7). This and also the observed hole structures in Fig 16(a) motivate the study of an asymptotic expansion with respect to multiple holes on a lattice as in

$$J\Big(\mathcal{O} \setminus \bigcup_{z\in\mathbb{Z}^n} \delta z + \epsilon\omega\Big) = J\left(\mathcal{O}\right) + \cdots \tag{26}$$

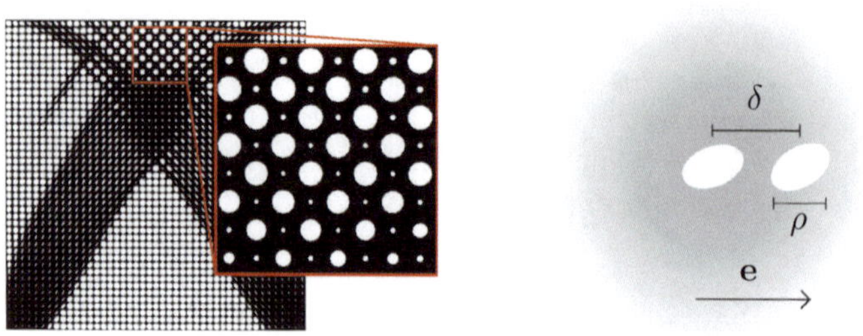

(a) (b)

FIGURE 16. Fig. (a) shows some hole structures observed in the optimization of parametrized domains (see Section 6) and Fig. (b) gives a small sketch of the two-hole problem dealt with in Theorem 1.

We start by studying the interaction between two holes (see Fig. 16(b) for a sketch). In order to make the notation easier we deal with the scalar case, by considering the Poisson equation on the set $\mathcal{O}$ (note that one can also apply an analogous result to the case of linear elasticity). Here $a_0(u, v)$ is the corresponding bilinear form and u_0 the solution to the equation. Let $\omega_1, \omega_2 \subset \mathbb{R}^2$ be the shapes of the two holes and for convenience assume that both contain the origin. We also define the sets $\mathcal{O}_R = \mathcal{O} \setminus B(\bar{x}, R)$, the new fictitious boundary (see above) $\Gamma_R = \partial B(\bar{x}, R)$ and the perforated domain $\mathcal{O}_\rho = \mathcal{O} \setminus (\bar{x} + (\rho\omega_1 \cup (\delta\mathbf{e} + \rho\omega_2)))$. Also let u_ρ be the solution to the Poisson equation on $\mathcal{O}_\rho$ with free Neumann conditions on the boundaries of the holes. Then we have the following statement giving an asymptotic expansion of the cost function with respect to ρ:

Theorem 1 (From [12]). *Let $j(\rho) = J(u_\rho)$ be a cost function where J does not explicitly depend on ρ and $DJ(u)$ is continuous and linear on $\mathcal{V}_R = \{u \in H^1(\mathcal{O}_R),\ u = 0 \text{ on } \Gamma_R\}$. Let $v_0 \in \mathcal{V}_R$ be the solution to the adjoint equation*

$$a_0(w, v_0) = -DJ(u_0)\, w \quad \forall w \in \mathcal{V}_R,$$

and let

$$\mathfrak{d}^{(1)}_{\text{topo}} := \int_{\Gamma_R} \nabla_x \left(S^{(1),1}_{\omega_1}(x, \nabla u_0(\bar{x})) + S^{(1),1}_{\omega_2}(x, \nabla u_0(\bar{x})) \right) \mathbf{n}(x)\, v_0(x)\, \mathrm{d}\gamma(x)$$

$$\mathfrak{d}^{(1)}_{\text{int},1} := \int_{\Gamma_R} \nabla_x \left(\widetilde{S}^{(1),1}_{\omega_1}(x, \nabla u_0(\bar{x})) + \widetilde{S}^{(1),1}_{\omega_2}(x, \nabla u_0(\bar{x})) \right) \mathbf{n}(x)\, v_0(x)\, \mathrm{d}\gamma(x)$$

$$\mathfrak{d}^{(1)}_{\text{int},2} := \int_{\Gamma_R} \nabla_x \left(\widehat{S}^{(1),1}_{\omega_1}\left(x, \nabla^2 u_0(\bar{x})\right) + \widehat{S}^{(1),1}_{\omega_2}\left(x, \nabla^2 u_0(\bar{x})\right) \right) \mathbf{n}(x)\, v_0(x)\, \mathrm{d}\gamma(x)$$
$$+ \int_{\Gamma_R} \mathbf{e} \nabla^2_x \left(S^{(1),1}_{\omega_2}(x, \nabla u_0(\bar{x})) \right) \mathbf{n}(x)\, v_0(x)\, \mathrm{d}\gamma(x)$$

$$\mathfrak{d}^{(2)}_{\text{topo}} := \int_{\Gamma_R} \nabla_x \left(S^{(1),2}_{\omega_1} (x, \nabla u_0(\bar{x})) + S^{(2),1}_{\omega_1} \left(x, \nabla^2 u_0(\bar{x})\right)\right) \mathbf{n}(x)\, v_0(x)\, \mathrm{d}\gamma(x)$$
$$+ \int_{\Gamma_R} \nabla_x \left(S^{(1),2}_{\omega_2} (x, \nabla u_0(\bar{x})) + S^{(2),1}_{\omega_2} \left(x, \nabla^2 u_0(\bar{x})\right)\right) \mathbf{n}(x)\, v_0(x)\, \mathrm{d}\gamma(x).$$

Then the function has the following asymptotic expansion:

$$j(\rho) = j(0) + \left\{\mathfrak{d}^{(1)}_{topo} + \left(\frac{\rho}{\delta}\right)^n \mathfrak{d}^{(1)}_{int,1} + \delta \mathfrak{d}^{(1)}_{int,2}\right\} \rho^n + \left\{\mathfrak{d}^{(2)}_{topo}\right\} \rho^{n+1} + o\left(\rho^{n+1}\right).$$

Here we assume that $\delta = \rho^\alpha$, *with* $\alpha \in (\frac{1}{2}, \frac{3}{4})$.

The terms $S^{(i),j}_{\omega_k}$, $\widetilde{S}^{(i),j}_{\omega_k}$ and $\widehat{S}^{(i),j}_{\omega_k}$ are based on approximations of single layer potentials associated to exterior Neumann problems, reflecting the behavior of the inner problem. In particular, the terms in $\mathfrak{d}^{(1)}_{\text{topo}}$ are the same that appear when computing the topological derivative for each individual hole (see also [58]) and the terms in $\mathfrak{d}^{(2)}_{\text{topo}}$ are reflecting the second-order topological derivatives of each hole. The interaction is described by $\mathfrak{d}^{(1)}_{\text{int},1}$and $\mathfrak{d}^{(1)}_{\text{int},2}$. Here, $\mathfrak{d}^{(1)}_{\text{int},1}$ is the first expansion term (with respect to the scaling parameter $(\frac{\rho}{\delta})$) approximating the interaction between two holes in the aforementioned exterior Neumann problem. On the other hand, $\mathfrak{d}^{(1)}_{\text{int},2}$ is the leading order correction term for the fact that the topology derivative for ω_2 should be computed at $\bar{x} + \delta$ but is instead computed only in $\bar{x}$.

Outlook: a reduced model for the optimization of perforated geometries. Theorem 1 gives an explicit expression for the hole-hole interaction in the topological derivative. Controlling the pairwise interaction permits to obtain in a natural way an estimate for the interaction terms in a lattice of holes, as was indicated in (26) and in Fig. 16(a). Indeed, much as in atomic lattices, one can approximate the total energy of the lattice by the sum of the pairwise interactions of neighbors. This strategy will permit us to estimate the influence of lattice-like microstructures on the objective functional, either in a single-scale approach as in Section 6 or in a two-scale approach where hole structures of ellipses on a periodic lattice are used to describe microstructures. An efficient solution of the cell problem will then make a study of macroscopic stochastic loadings possible. This class of microstructures is markedly different from the laminates typically used in two-scale shape optimization [3]. It remains an open problem to understand if this difference in the microstructure will result in a significant difference at the macroscopic level.

Acknowledgment

The authors thank Martin Lenz for the cooperation on the boundary element approach applied to shape optimization. This work was supported by the Deutsche Forschungsgemeinschaft through the Schwerpunktprogramm 1253 *Optimization with Partial Differential Equations.*

Dedication: The first author, Pradeep Atwal, died tragically while this paper was under review. We, the co-authors, dedicate this paper to the memory of Pradeep Atwal.

References

[1] Adali, S., J.C. Bruch, J., Sadek, I., and Sloss, J. Robust shape control of beams with load uncertainties by optimally placed piezo actuators. *Structural and Multidisciplinary Optimization* 19, 4 (2000), 274–281.

[2] Albers, S. Online algorithms: a survey. *Mathematical Programming* 97 (2003), 3–26.

[3] Allaire, G. *Shape Optimization by the Homogenization Method*, vol. 146. Springer Applied Mathematical Sciences, 2002.

[4] Allaire, G., Bonnetier, E., Francfort, G., and Jouve, F. Shape optimization by the homogenization method. *Numerische Mathematik* 76 (1997), 27–68.

[5] Allaire, G., de Gournay, F., Jouve, F., and Toader, A.-M. Structural optimization using topological and shape sensitivity via a level set method. *Control and Cybernetics* 34 (2005), 59–80.

[6] Allaire, G., and Jouve, F. A level-set method for vibration and multiple loads structural optimization. *Comput. Methods Appl. Mech. Engrg.* 194, 30-33 (2005), 3269–3290.

[7] Allaire, G., Jouve, F., and de Gournay, F. Shape and topology optimization of the robust compliance via the level set method. *ESAIM Control Optim. Calc. Var.* 14 (2008), 43–70.

[8] Allaire, G., Jouve, F., and Toader, A.-M. Structural optimization using sensitivity analysis and a level-set method. *Journal of Computational Physics* 194, 1 (2004), 363–393.

[9] Alvarez, F., and Carrasco, M. Minimization of the expected compliance as an alternative approach to multiload truss optimization. *Struct. Multidiscip. Optim.* 29 (2005), 470–476.

[10] Amstutz, S., and Andrä, H. A new algorithm for topology optimization using a level-set method. *Journal of Computational Physics* 216 (2006), 573–588.

[11] Atwal, P. Continuum limit of a double-chain model for multiload shape optimization. *Journal of Convex Analysis* 18, 1 (2011).

[12] Atwal, P. *Hole-hole interaction in shape optimization via topology derivatives.* PhD thesis, University of Bonn, in preparation.

[13] Banichuk, N.V., and Neittaanmäki, P. On structural optimization with incomplete information. *Mechanics Based Design of Structures and Machines* 35 (2007), 75–95.

[14] Bastin, F., Cirillo, C., and Toint, P. Convergence theory for nonconvex stochastic programming with an application to mixed logit. *Math. Program.* 108 (2006), 207–234.

[15] Ben-Tal, A., El-Ghaoui, L., and Nemirovski, A. *Robust Optimization.* Princeton University Press, Princeton and Oxford, 2009.

[16] Ben-Tal, A., Kočvara, M., Nemirovski, A., and Zowe, J. Free material design via semidefinite programming: the multiload case with contact conditions. *SIAM J. Optim.* 9 (1999), 813–832.

[17] Bendsøe, M.P. *Optimization of structural topology, shape, and material.* Springer-Verlag, Berlin, 1995.

[18] Burger, M., Hackl, B., and Ring, W. Incorporating topological derivatives into level set methods. *J. Comp. Phys.* 194 (2004), 344–362.

[19] Burger, M., and Osher, S.J. A survey on level set methods for inverse problems and optimal design. *European Journal of Applied Mathematics* 16, 2 (2005), 263–301.

[20] Chang, F.R. *Stochastic Optimization in Continuous Time.* Cambridge University Press, Cambridge, 2004.

[21] Cherkaev, A., and Cherkaev, E. Stable optimal design for uncertain loading conditions. In *Homogenization*, V. B. et al., ed., vol. 50 of *Series on Advances in Mathematics for Applied Sciences.* World Scientific, Singapore, 1999, pp. 193–213.

[22] Cherkaev, A., and Cherkaev, E. Principal compliance and robust optimal design. *Journal of Elasticity* 72 (2003), 71–98.

[23] Ciarlet, P.G. *Mathematical Elasticity Volume I: Three-Dimensional Elasticity*, vol. 20. Studies in Mathematics and its Applications, North-Holland, 1988.

[24] Clements, D., and Rizzo, F. A Method for the Numerical Solution of Boundary Value Problems Governed by Second-order Elliptic Systems. *IMA Journal of Applied Mathematics* 22 (1978), 197–202.

[25] Conti, S., Held, H., Pach, M., Rumpf, M., and Schultz, R. Risk averse shape optimization. *Siam Journal on Control and Optimization* (2009). Submitted.

[26] Conti, S., Held, H., Pach, M., Rumpf, M., and Schultz, R. Shape optimization under uncertainty – a stochastic programming perspective. *SIAM J. Optim.* 19 (2009), 1610–1632.

[27] de Gournay, F., Allaire, G., and Jouve, F. Shape and topology optimization of the robust compliance via the level set method. *ESAIM: Control, Optimisation and Calculus of Variations* 14 (2007), 43–70.

[28] Delfour, M.C., and Zolésio, J. *Geometries and Shapes: Analysis, Differential Calculus and Optimization.* Adv. Des. Control 4. SIAM, Philadelphia, 2001.

[29] Dias, G.P., Herskovits, J., and Rochinha, F.A. Simultaneous shape optimization and nonlinear analysis of elastic solids. *Computational Mechanics* (1998).

[30] Du, Q., and Wang, D. Tetrahedral mesh generation and optimization based on centroidal Voronoi tesselations. *International Journal for Numerical Methods in Engineering* 56 (2003), 1355–1373.

[31] Fleming, W.H., and Rishel, R.W. *Deterministic and Stochastic Optimal Control.* Springer, New York, 1975.

[32] Garreau, S., Guillaume, P., and Masmoudi, M. The topological asymptotic for PDE systems: The elasticity case. *SIAM J. Control Optim.* 39, 6 (2001), 1756–1778.

[33] Guedes, J.M., Rodrigues, H.C., and Bendsøe, M.P. A material optimization model to approximate energy bounds for cellular materials under multiload conditions. *Struct. Multidiscip. Optim.* 25 (2003), 446–452.

[34] Hackbusch, W. *Integral Equations*, vol. 120 of *International Series of Numerical Mathematics.* Birkhäuser Verlag, Basel, Boston, Berlin, 1995.

[35] Hackbusch, W., and Sauter, S. Composite finite elements for the approximation of PDEs on domains with complicated micro-structures. *Numerische Mathematik* 75 (1997), 447–472.

[36] He, L., Kao, C.-Y., and Osher, S. Incorporating topological derivatives into shape derivatives based level set methods. *Journal of Computational Physics* 225, 1 (2007), 891–909.

[37] Huyse, L. *Free-form airfoil shape optimization under uncertainty using maximum expected value and second-order second-moment strategies.* ICASE report; no. 2001-18. ICASE, NASA Langley Research Center Available from NASA Center for Aerospace Information, Hampton, VA, 2001.

[38] Liehr, F., Preusser, T., Rumpf, M., Sauter, S., and Schwen, L.O. Composite finite elements for 3D image based computing. *Computing and Visualization in Science* 12 (2009), 171–188.

[39] Liu, Z., Korvink, J.G., and Huang, R. Structure topology optimization: Fully coupled level set method via femlab. *Structural and Multidisciplinary Optimization* 29 (June 2005), 407–417.

[40] Marsden, J., and Hughes, T. *Mathematical foundations of elasticity.* Dover Publications Inc., New York, 1993.

[41] Marti, K. *Stochastic Optimization Methods.* Springer, Berlin, 2005.

[42] Melchers, R. Optimality-criteria-based probabilistic structural design. *Structural and Multidisciplinary Optimization* 23, 1 (2001), 34–39.

[43] Owen, S.J. A survey of unstructured mesh generation technology. In *Proceedings of the 7th International Meshing Roundtable* (Dearborn, Michigan, 1998), Sandia National Laboratories, pp. 239–267.

[44] Pennanen, T. Epi-convergent discretizations of multistage stochastic programs. *Mathematics of Operations Research* 30 (2005), 245–256.

[45] Pflug, G.C., and Römisch, W. *Modeling, Measuring and Managing Risk.* World Scientific, Singapore, 2007.

[46] Rumigny, N., Papadopoulos, P., and Polak, E. On the use of consistent approximations in boundary element-based shape optimization in the presence of uncertainty. *Comput. Methods Appl. Mech. Engrg.* 196 (2007), 3999–4010.

[47] Ruszczyński, A., and Shapiro, A., eds. *Handbooks in Operations Research and Management Sciences, 10: Stochastic Programming.* Elsevier, Amsterdam, 2003.

[48] Schultz, R. Stochastic programming with integer variables. *Mathematical Programming* 97 (2003), 285–309.

[49] Schultz, R., and Tiedemann, S. Risk aversion via excess probabilities in stochastic programs with mixed-integer recourse. *SIAM J. on Optimization* 14, 1 (2003), 115–138.

[50] Schulz, V., and Schillings, C. On the nature and treatment of uncertainties in aerodynamic design. *AIAA Journal* 47 (2009), 646–654.

[51] Schumacher, A. *Topologieoptimierung von Bauteilstrukturen unter Verwendung von Lochpositionierungskriterien.* PhD thesis, Universität – Gesamthochschule Siegen, 1996.

[52] Sethian, J.A., and Wiegmann, A. Structural boundary design via level set and immersed interface methods. *Journal of Computational Physics* 163, 2 (2000), 489–528.

[53] Shapiro, A., Dentcheva, D., and Ruszczyński, A. *Lectures on Stochastic Programming.* SIAM-MPS, Philadelphia, 2009.

[54] Shewchuk, J. Triangle: Engineering a 2d quality mesh generator and Delaunay triangulator. *Applied Computational Geometry: Towards Geometric Engineering* 1148 (1996), 203–222.

[55] SHEWCHUK, J. Delaunay refinement algorithms for triangular mesh generation. *Computational Geometry: Theory and Applications* 22 (2002), 21–74.

[56] SOKOŁOWSKI, J., AND ŻOCHOWSKI, A. On the topological derivative in shape optimization. *SIAM J. Control Optim.* 37, 4 (1999), 1251–1272.

[57] SOKOŁOWSKI, J., AND ŻOCHOWSKI, A. Topological derivatives of shape functionals for elasticity systems. *Mech. Struct. & Mach.* 29, 3 (2001), 331–349.

[58] SOKOŁOWSKI, J., AND ŻOCHOWSKI, A. Optimality conditions for simultaneous topology and shape optimization. *SIAM Journal on Control and Optimization* 42, 4 (2003), 1198–1221.

[59] SOKOŁOWSKI, J., AND ZOLÉSIO, J.-P. *Introduction to Shape Optimization: Shape Sensitivity Analysis.* Springer, 1992.

[60] STEINBACH, M. Tree-sparse convex programs. *Mathematical Methods of Operations Research* 56 (2002), 347–376.

[61] WÄCHTER, A. *An Interior Point Algorithm for Large-Scale Nonlinear Optimization with Applications in Process Engineering.* Phd thesis, Carnegie Mellon University, 2002.

[62] WÄCHTER, A., AND BIEGLER, L. On the implementation of a primal-dual interior point filter line search algorithm for large-scale nonlinear programming. *Mathematical Programming* 106, 1 (2006), 25–57.

[63] ZHUANG, C., XIONG, Z., AND DING, H. A level set method for topology optimization of heat conduction problem under multiple load cases. *Comput. Methods Appl. Mech. Engrg.* 196 (2007), 1074–1084.

Pradeep Atwal † and Sergio Conti
Institut für Angewandte Mathematik
Universität Bonn
Endenicher Allee 60
D-53115 Bonn, Germany
e-mail: sergio.conti@uni-bonn.de

Benedict Geihe and Martin Rumpf
Institut für Numerische Simulation
Universität Bonn
Endenicher Allee 60
D-53115 Bonn, Germany
e-mail: benedict.geihe@ins.uni-bonn.de
martin.rumpf@ins.uni-bonn.de

Martin Pach and Rüdiger Schultz
Fakultät für Mathematik
Universität Duisburg-Essen
Forsthausweg 2
D-47057 Duisburg, Germany
e-mail: pach@math.uni-duisburg.de
schultz@math.uni-duisburg.de

International Series of Numerical Mathematics, Vol. 160, 245–256

Phase-field Approaches to Structural Topology Optimization

Luise Blank, Harald Garcke, Lavinia Sarbu, Tarin Srisupattarawanit, Vanessa Styles and Axel Voigt

Abstract. The mean compliance minimization in structural topology optimization is solved with the help of a phase field approach. Two steepest descent approaches based on L^2- and H^{-1}-gradient flow dynamics are discussed. The resulting flows are given by Allen-Cahn and Cahn-Hilliard type dynamics coupled to a linear elasticity system. We finally compare numerical results obtained from the two different approaches.

Mathematics Subject Classification (2000). 74P15, 74P05, 74S03, 35K99.

Keywords. Structural topology optimization, phase-field approximation, Allen-Cahn model, Cahn-Hilliard model, elasticity, gradient flow.

1. Introduction

Structural topology optimization denotes problems of finding optimal material distributions in a given design domain subject to certain criteria. It has become a standard tool of engineering design, in particular in structural mechanics, see [4] and the literature therein for more details. There are two different problems of importance: (a) the maximization of material stiffness at given mass, and (b) the minimization of mass while keeping a certain stiffness. We consider only the first approach which is known as the minimal compliance problem and is today well understood with respect to its mathematical formulation, see [1] for an overview. Various successful numerical techniques have been proposed, which rely on sensitivity analysis, mathematical programming, homogenization, see [4] for an overview, or more recently on level-set and phase-field methods [2, 33]. The connection to level-set and phase-field methods is best seen using a relation to image processing.

This work was supported by DFG within SPP 1253 "Optimization with partial differential equations" under BL433/2-1 and Vo899/5-1 and by the Vielberth foundation. Also the fifth author was supported by the EPSRC grant EP/D078334/1.

In [9] the analogy between basic concepts of image segmentation and structural topology optimization is clearly illustrated. While level-set methods have become an accepted tool in structural topology optimization, the use of phase-field methods in this field has not yet become popular. There are only a few approaches considered, see [33, 11, 29, 10]. Some approaches are based on the fourth-order Cahn-Hilliard equation and hence require a high computational cost. We will here consider an approach which reduces the cost by replacing the Cahn-Hilliard equation by a volume conserved second-order Allen-Cahn equation. Finally, let us point out that phase field approaches have the advantage that topology changes can easily be handled, see Figures 2 and 3.

The outline of the paper is as follows: In Section 2 we describe the phase field approach. In Section 3 the discretization of the Allen-Cahn and the Cahn-Hilliard equations with elasticity are discussed. In Section 4 numerical results for both approaches are shown and compared with each other, and in Section 5 we draw conclusions.

2. Phase-field approach

We consider a structural topology optimization problem of a statically loaded linear elastic structure. The goal is to compute the material distribution in a given bounded design domain $\Omega \subset \mathbb{R}^d$.

We will describe the material distribution with the help of a phase field variable φ. The phase field φ will take values close to 1 in the void and values close to -1 if material is present. In phase field approaches the interface between material and void is described by a diffuse interfacial layer of a thickness which is proportional to a small length scale parameter ε and at the interface the phase field φ rapidly but smoothly changes its value. We can prescribe a given mass by requiring $\fint_\Omega \varphi = m$ where $m \in (-1,1)$ and $\fint_\Omega \varphi$ is the mean value of φ. We now assume a linear elastic material with an elasticity tensor $\mathcal{C}_1$ and we model the void with a very small elasticity tensor $\mathcal{C}_2$ where we later choose $\mathcal{C}_2 = \varepsilon^2 \mathcal{C}_1$ but other choices are possible. In the interfacial region we interpolate the elastic properties and set

$$\mathcal{C}(\varphi) = \mathcal{C}_1 + \tfrac{1}{2}(1+\varphi)(\mathcal{C}_2 - \mathcal{C}_1)\,.$$

We now denote by $\mathbf{u} : \Omega \to \mathbb{R}^d$ the displacement vector and by $\mathcal{E}(\mathbf{u}) := \frac{1}{2}(\nabla \mathbf{u} + \nabla \mathbf{u}^t)$ the strain tensor. Assuming that the outer forces are given by a linear functional F on the Sobolev space $H^1(\Omega, \mathbb{R}^d)$ the goal in classical structural topology optimization is to minimize the mean compliance $F(\mathbf{u})$ subject to $\fint_\Omega \varphi(x)dx = m$ and

$$\langle \mathcal{E}(\mathbf{u}), \mathcal{E}(\boldsymbol{\eta}) \rangle_{\mathcal{C}(\varphi)} = F(\boldsymbol{\eta}) \tag{2.1}$$

which has to hold for all $\boldsymbol{\eta} \in H^1(\Omega, \mathbb{R}^d)$ such that $\boldsymbol{\eta} = \mathbf{0}$ on a given non-empty Dirichlet boundary Γ_D. Here we use the notation

$$\langle \mathcal{A}, \mathcal{B} \rangle_{\mathcal{C}} := \int_\Omega \mathcal{A} : \mathcal{C}\mathcal{B}$$

where the :-product of matrices $\mathcal{G}$ and $\mathcal{H}$ is given as $\mathcal{G} : \mathcal{H} := \sum_{i,j=1}^d \mathcal{G}_{ij}\mathcal{H}_{ij}$.

The outer forces F can be given for example by a boundary traction on $\Gamma_F \subset \partial\Omega \setminus \Gamma_D$ and in this case we have

$$F(\boldsymbol{\eta}) = \int_{\Gamma_F} \mathbf{f} \cdot \boldsymbol{\eta} \tag{2.2}$$

where $\mathbf{f} : \Gamma_F \to \mathbb{R}^d$ describes outer forces acting on the structure. The strong formulation of (2.1) with F of the form (2.2) is now given as

$$\begin{aligned} -\nabla \cdot [\mathcal{C}(\varphi)\mathcal{E}(\mathbf{u})] &= \mathbf{0} && \text{in } \Omega\,, \\ \mathbf{u} &= \mathbf{0} && \text{on } \Gamma_D\,, \\ [\mathcal{C}(\varphi)\mathcal{E}(\mathbf{u})] \cdot n &= \mathbf{f} && \text{on } \Gamma_F\,, \\ [\mathcal{C}(\varphi)\mathcal{E}(\mathbf{u})] \cdot n &= \mathbf{0} && \text{on } \partial\Omega \setminus (\Gamma_D \cup F_F)\,, \end{aligned}$$

where n is the outer unit normal to $\partial\Omega$. In the above formulation the problem is ill posed and unwanted checkerboard patterns and mesh dependencies are well-known phenomena, see [28].

A possible regularization is to add a perimeter penalization to the functional which penalizes length for $d = 2$ and area if $d = 3$ for the interface between material and void. This regularization in particular avoids checkerboard patterns if spatial discretization parameters tend to zero, see [19, 24].

In phase field approaches such a penalization can be modeled with the help of a Ginzburg-Landau energy

$$E(\varphi) := \int_\Omega \left(\tfrac{\gamma\varepsilon}{2}|\nabla\varphi|^2 + \tfrac{\gamma}{\varepsilon}\psi(\varphi)\right)dx$$

where γ is a parameter related to the interfacial energy density. The potential function $\psi : \mathbb{R} \to \mathbb{R}_0^+ \cup \{\infty\}$ is assumed to have two global minima at the points ± 1. Examples are $\psi(\varphi) = \psi_1(\varphi) := c_1(1-\varphi^2)^2$ with $c_1 \in \mathbb{R}^+$ or the obstacle potential

$$\psi(\varphi) = \begin{cases} \psi_0(\varphi) & \text{if } \quad |\varphi| \le 1\,, \\ \infty & \text{if } \quad |\varphi| > 1 \end{cases} \tag{2.3}$$

with, e.g., $\psi_0(\varphi) := \frac{1}{2}(1-\varphi^2)$. It is well known that the energy E converges to a scalar multiple of the perimeter functional, see [23].

We now want to solve

$$\min J(\varphi, \mathbf{u}) := E(\varphi) + F(\mathbf{u}) \tag{2.4}$$

subject to (2.1) and $\fint_\Omega \varphi(x)dx = m$. For a given φ we can compute a unique $\mathbf{u}(\varphi)$ with $\mathbf{u}(\varphi) = \mathbf{0}$ on Γ_D which solves (2.1). We can hence consider the reduced problem

$$\min \hat{J}(\varphi) \quad \text{subject to} \quad \fint_\Omega \varphi(x)dx = m \tag{2.5}$$

with the reduced functional

$$\hat{J}(\varphi) := J(\varphi, \mathbf{u}(\varphi)) .$$

In order to compute the first variation of the reduced functional $\hat{J}$ we apply a formal Lagrange approach, see, e.g., Hinze et al. [20]. We therefore introduce the adjoint variable $\mathbf{p} : \Omega \to \mathbb{R}^d$ and define the Lagrangian

$$L(\varphi, \mathbf{u}, \mathbf{p}) := E(\varphi) + F(\mathbf{u}) - \langle \mathcal{E}(\mathbf{u}), \mathcal{E}(\mathbf{p}) \rangle_{\mathcal{C}(\varphi)} + F(\mathbf{p}) .$$

We now seek stationary states $(\varphi, \mathbf{u}, \mathbf{p})$ of L. If the first variation for $(\varphi, \mathbf{u}, \mathbf{p})$ vanishes we observe that $\mathbf{u}$ and $\mathbf{p}$ both solve (2.1). Assuming $\Gamma_D \neq \emptyset$ we obtain that (2.1) has a unique solution with Dirichlet data on Γ_D and we hence conclude $\mathbf{u} \equiv \mathbf{p}$. Using this we get

$$\tfrac{\delta \hat{J}}{\delta \varphi}(\varphi) = \tfrac{\delta L}{\delta \varphi}(\varphi, \mathbf{u}, \mathbf{p}) = \tfrac{\delta E}{\delta \varphi}(\varphi) - \langle \mathcal{E}(\mathbf{u}), \mathcal{E}(\mathbf{u}) \rangle_{\mathcal{C}'(\varphi)} ,$$

where $\frac{\delta \hat{J}}{\delta \varphi}$, $\frac{\delta L}{\delta \varphi}$ and $\frac{\delta E}{\delta \varphi}$ denote the first variation with respect to φ and $\mathbf{u}$ solves (2.1).

We now want to use a steepest descent approach in order to find (local) minima of (2.5). We choose a gradient flow dynamics with an artificial time variable and this leads to a pseudo time stepping approach. Given an inner product $\langle .,. \rangle$ the gradient flow for (2.5) with respect to $\langle .,. \rangle$ is given as

$$\begin{aligned} \langle \partial_t \varphi, \zeta \rangle &= -\tfrac{\delta \hat{J}}{\delta \varphi}(\varphi) = -\tfrac{\delta L}{\delta \varphi}(\varphi, \mathbf{u}, \mathbf{p})(\zeta) \\ &= -\int_\Omega [\gamma \varepsilon \nabla \varphi \cdot \nabla \zeta + \tfrac{\gamma}{\varepsilon} \psi'(\varphi) \zeta - \mathcal{E}(\mathbf{u}) : \mathcal{C}'(\varphi) \mathcal{E}(\mathbf{u}) \zeta] \end{aligned}$$

where $\mathbf{u}$ solves (2.1). Of course the steepest descent method should take the constraint on the total mass given as $\fint_\Omega \varphi(x) = m$ into account. Furthermore, the steepest descent direction is given by the gradient and the gradient of course depends on the inner product chosen. As inner product we either choose the L^2-inner product which results in an Allen-Cahn type dynamics or the mass conserving H^{-1}-inner product leading to a modified Cahn-Hilliard problem.

In the following we briefly discuss how we obtain the Allen-Cahn dynamics and a modified Cahn-Hilliard equation as gradient flows. For further details we refer to [30] and [5].

We first formulate the problem in the case that $\langle .,. \rangle$ is given by a scaled L^2-inner product $(.,.)$ leading to an Allen-Cahn type dynamics, where also the mass constraint $\fint_\Omega \varphi = m$ has to be enforced. Using the obstacle potential (2.3) we obtain on an arbitrary time interval $(0, T)$ (see [6] for further details):

($\mathbf{P}_1$) *Find* $\varphi \in H^1(\Omega_T)$ *and* $\mathbf{u} \in L^\infty(0,T;H^1(\Omega,\mathbb{R}^d))$ *such that*

$$\fint_\Omega \varphi(x,t)dx = m, \varphi(.,0) = \varphi_0, |\varphi| \le 1 \text{ a.e. in } \Omega_T, \mathbf{u} = \mathbf{0} \text{ a.e. on } \Gamma_D \times (0,T),$$
$$(\varepsilon\partial_t\varphi + \tfrac{\gamma}{\varepsilon}\psi_0'(\varphi), \chi - \varphi) + \gamma\varepsilon(\nabla\varphi, \nabla(\chi-\varphi)) \ge \tfrac{1}{2}\langle\mathcal{E}(\mathbf{u}), \mathcal{E}(\mathbf{u})(\chi-\varphi)\rangle_{\mathcal{C}_2-\mathcal{C}_1},$$
$$\langle\mathcal{E}(\mathbf{u}), \mathcal{E}(\boldsymbol{\eta})\rangle_{\mathcal{C}(\varphi)} = F(\boldsymbol{\eta})$$

which has to hold for almost all t *and all* $\chi \in H^1(\Omega)$ *with* $|\chi| \le 1$ *and* $\fint_\Omega \chi = m$ *and all* $\boldsymbol{\eta} \in H^1(\Omega,\mathbb{R}^d)$ *such that* $\boldsymbol{\eta} = \mathbf{0}$ *on the Dirichlet boundary* Γ_D.

Since we consider the L^2-gradient flow of $\hat{J}$ one obtains directly an a priori estimate on $\partial_t\varphi$ in $L^2(\Omega_T)$ and hence we require $\varphi \in H^1(\Omega_T)$ in ($\mathbf{P}_1$), see also [6].

We now discuss the mass conserving H^{-1}-gradient flow which leads to the Cahn-Hilliard type dynamics. For functions v_1, v_2 with mean value zero we define the inner product

$$(v_1, v_2)_{-1} := \int_\Omega \nabla(-\Delta)^{-1}v_1 \cdot \nabla(-\Delta)^{-1}v_2$$

where $y = (-\Delta)^{-1}v$ is the weak solution of $-\Delta y = v$ in Ω with $\fint_\Omega y = 0$ and $\frac{\partial y}{\partial n} = 0$ on $\partial\Omega$. The H^{-1}-gradient flow

$$(\partial_t\varphi, \chi)_{-1} = -\tfrac{\delta L}{\delta\varphi}(\varphi, \mathbf{u}, \mathbf{p})(\chi)$$

can now be rewritten by introducing the chemical potential $w = -(-\Delta)^{-1}\partial_t\varphi + \overline{w}$ where $\overline{w}$ is an appropriate constant. Considering the smooth potential $\psi = \psi_1$ and a variable diffusivity B, we obtain the following problem which is a modified Cahn-Hilliard equation, see also [15, 21].

($\mathbf{P}_2$) *Find sufficiently regular* $(\varphi, w, \mathbf{u})$ *such that*

$$\varphi(.,0) = \varphi_0, |\varphi| \le 1 \text{ a.e.}, \ \mathbf{u} = \mathbf{0} \text{ a.e. on } \Gamma_D \times (0,T) \text{ and}$$
$$\partial_t\varphi = \nabla\cdot(B(\varphi)\nabla w) \text{ in the distributional sense}, \tag{2.6}$$
$$\tfrac{\partial w}{\partial n} = 0, \tfrac{\partial\varphi}{\partial n} = 0 \text{ on } \partial\Omega \times (0,T),$$
$$w = -\gamma\varepsilon\Delta\varphi + \tfrac{\gamma}{\varepsilon}\psi_1'(\varphi) - \mathcal{E}(\mathbf{u}) : \mathcal{C}'(\varphi)\mathcal{E}(\mathbf{u}) \text{ in the distributional sense}$$

together with (2.1).

Strictly speaking we obtain ($\mathbf{P}_2$) as the gradient flow of $(.,.)_{-1}$ only in the case that the mobility function B in (2.6) is equal to one. We refer to Taylor and Cahn [30] who discuss how the definition of $(.,.)_{-1}$ has to be modified for a variable mobility. With this modification we obtain (2.6) also for a variable mobility. We also remark that (2.6) together with the Neumann boundary conditions on w imply that the mass $\int_\Omega \varphi$ is preserved. For further information on elastically modified Cahn-Hilliard models we refer to the overview [16].

Stationary states of ($\mathbf{P}_1$) and ($\mathbf{P}_2$) respectively fulfil the first-order necessary conditions for (2.4). In the following section we describe how we numerically solve ($\mathbf{P}_1$) and ($\mathbf{P}_2$) and in Section 4 we will compare numerical results of ($\mathbf{P}_1$) and ($\mathbf{P}_2$).

3. Discretization

In this section we present finite element approximations of $(\mathbf{P}_1)$ and $(\mathbf{P}_2)$.

3.1. Notation

For simplicity we assume that Ω is a polyhedral domain. Let $\mathcal{T}_h$ be a regular triangulation of Ω into disjoint open simplices, i.e., $\Omega = \cup_{T\in\mathcal{T}_h}\overline{T}$. Furthermore, we define $h := \max_{T\in\mathcal{T}_h}\{\text{diam } T\}$ the maximal element size of $\mathcal{T}_h$ and we set $\mathcal{J}$ to be the set of nodes of $\mathcal{T}_h$ and $\{p_j\}_{j\in\mathcal{J}}$ to be the coordinates of these nodes. Associated with $\mathcal{T}_h$ is the piecewise linear finite element space

$$S_h := \left\{\eta \in C^0(\overline{\Omega}) \middle| \eta_{|_T} \in P_1(T)\ \forall\, T \in \mathcal{T}_h\right\} \subset H^1(\Omega),$$

where we denote by $P_1(T)$ the set of all affine linear functions on T. Furthermore, we denote the standard nodal basis functions of S_h by χ_j for all $j \in \mathcal{J}$. Then φ_j for $j = 1,\ldots,\mathcal{J}$ denote the coefficients of the basis representation of φ_h in S_h which is given by $\varphi_h = \sum_{j\in\mathcal{J}}\varphi_j\chi_j$.

In order to derive a discretization of $(\mathbf{P}_1)$ we define

$$\mathcal{K}_h^m := \{\eta \in S_h |\, |\eta(x)| \le 1 \text{ for all } x \in \Omega, \fint_\Omega \eta dx = m\}.$$

We introduce also the lumped mass scalar product $(f,g)_h = \int_\Omega I_h(fg)$ instead of (f,g), where $I_h : C^0(\overline{\Omega}) \to S_h$ is the standard interpolation operator such that $(I_h f)(p_j) = f(p_j)$ for all nodes $j \in \mathcal{J}$. In addition, we employ a quadrature formula $\langle\mathcal{A},\mathcal{B}\rangle_{\mathcal{C}}^h$ in place of $\langle\mathcal{A},\mathcal{B}\rangle_{\mathcal{C}}$, with the property that $\langle\mathcal{A},\mathcal{B}\rangle_{\mathcal{C}}^h = \langle\mathcal{A},\mathcal{B}\rangle_{\mathcal{C}}$ for piecewise affine linear integrands $\mathcal{A} : \mathcal{C}\mathcal{B}$.

3.2. Finite element approximation of the Allen-Cahn approach with mass conservation and obstacle potential

Taking a fixed time step $\tau = t_n - t_{n-1}$ we obtain the following finite element approximation of $(\mathbf{P}_1)$:

$(\mathbf{P}_1^h)$ *Given* $\varphi_h^{n-1} \in \mathcal{K}_h^m$ *find* $(\varphi_h^n, \mathbf{u}_h^n) \in \mathcal{K}_h^m \times (S_h)^d$ *such that*

$$\mathbf{u}_h^n = \mathbf{0} \text{ on } \mathbf{\Gamma}_D,$$

$$\langle\mathcal{E}(\mathbf{u}_h^n),\mathcal{E}(\boldsymbol{\eta})\rangle_{\mathcal{C}(\varphi_h^{n-1})}^h = F(\boldsymbol{\eta})\ \forall\, \boldsymbol{\eta} \in (S_h)^d \text{ with } \boldsymbol{\eta} = \mathbf{0} \text{ on } \Gamma_D, \tag{3.1}$$

$$\begin{aligned}&(\tfrac{\varepsilon}{\tau}(\varphi_h^n - \varphi_h^{n-1}) - \tfrac{\gamma}{\varepsilon}\varphi_h^n, \chi - \varphi_h^n)_h + \gamma\varepsilon(\nabla\varphi_h^n, \nabla(\chi - \varphi_h^n))\\ &\quad\ge \tfrac{1}{2}\langle\mathcal{E}(\mathbf{u}_h^n),\mathcal{E}(\mathbf{u}_h^n)(\chi - \varphi_h^n)\rangle_{\mathcal{C}_2-\mathcal{C}_1}^h\ \ \forall \chi \in \mathcal{K}_h^m.\end{aligned} \tag{3.2}$$

As (3.1) is independent of φ_h^n we use a preconditioned conjugate gradient solver to compute $\mathbf{u}_h^n$ from this equation, see also [18, 17]. Due to the use of piecewise linear finite elements and mass lumping the reformulation of (3.2) with Lagrange multipliers $\mu_h \in S_h$ and $\lambda \in \mathbb{R}$ can be stated as follows, see [6]:

Given $(\varphi_h^{n-1}, \mathbf{u}_h^n) \in \mathcal{K}_h^m \times (S_h)^d$, *find* $\varphi_h^n \in \mathcal{K}_h^m$, $\mu_h \in S_h$ *and* $\lambda \in \mathbb{R}$ *such that*

$$(\tfrac{\varepsilon^2}{\tau} - \gamma)(\varphi_h^n, \chi)_h + \gamma\varepsilon^2(\nabla\varphi_h^n, \nabla\chi) + (\mu_h, \chi)_h - \lambda(1, \chi) = \tfrac{\varepsilon^2}{\tau}(\varphi_h^{n-1}, \chi)_h + \tfrac{\varepsilon}{2}\langle\mathcal{E}(\mathbf{u}_h^n), \mathcal{E}(\mathbf{u}_h^n)\chi\rangle_{\mathcal{C}_2-\mathcal{C}_1}^h \ \forall\chi \in S_h, \tag{3.3}$$

$$\fint_\Omega \varphi_h^n = m, \tag{3.4}$$

$$(\mu_j)_- \geq 0, \ (\mu_j)_+ \geq 0, \ |\varphi_j| \leq 1, \tag{3.5}$$

$$(\varphi_j + 1)(\mu_j)_- = (\varphi_j - 1)(\mu_j)_+ = 0 \ \forall \ j \in \mathcal{J}, \tag{3.6}$$

where $(.)_+$ *and* $(.)_-$ *are the positive and negative parts of a quantity in the brackets.*

To solve (3.3)–(3.6) we apply the PDAS-method presented in [6], yielding the following algorithm:

Primal-Dual Active Set Algorithm (PDAS):

0. *Set* $k = 0$ *and initialize* $\mathcal{A}_0^\pm$.
1. *Define* $\mathcal{I}_k = \mathcal{J} \setminus (\mathcal{A}_k^+ \cup \mathcal{A}_k^-)$.
   Set $\varphi_j^k = \pm 1$ *for* $j \in \mathcal{A}_k^\pm$ *and* $\mu_j^k = 0$ *for* $j \in \mathcal{I}_k$.
2. *Solve the discretized PDE* (3.3) *with the non-local constraint* (3.4) *to obtain* φ_j^k *for* $j \in \mathcal{I}_k$ *and* $\lambda^k \in \mathbb{R}$.
3. *Determine* μ_j^k *for* $j \in \mathcal{A}_k^\pm$ *using* (3.3).
4. *Set* $\mathcal{A}_{k+1}^+ := \{j \in \mathcal{J} : \varphi_j^k + \frac{\mu_j^k}{c} > 1\}$, $\mathcal{A}_{k+1}^- := \{j \in \mathcal{J} : \varphi_j^k + \frac{\mu_j^k}{c} < -1\}$.
5. *If* $\mathcal{A}_{k+1}^\pm = \mathcal{A}_k^\pm$ *stop, otherwise set* $k = k + 1$ *and goto* 1.

Remark 3.1. We solve the system arising from Step 2 using MINRES, see [8].

The Allen-Cahn variational inequality with volume constraint is implemented using the adaptive finite element toolbox Alberta 1.2 [27].

3.3. Finite element approximation of the Cahn-Hilliard approach with smooth potential

For the case of a fixed time step $\tau = t_n - t_{n-1}$ we obtain the following finite element approximation of $(\mathbf{P}_2)$:

$(\mathbf{P}_2^h)$ *Given* $\varphi_h^{n-1} \in S_h$ *find* $(\varphi_h^n, w_h^n, \mathbf{u}_h^n) \in S_h \times S_h \times (S_h)^d$ *such that*

$$\mathbf{u}_h^n = \mathbf{0} \ on \ \mathbf{\Gamma}_D, \tag{3.7}$$

$$\langle\mathcal{E}(\mathbf{u}_h^n), \mathcal{E}(\boldsymbol{\eta})\rangle_{\mathcal{C}(\varphi_h^{n-1})}^h = F(\boldsymbol{\eta}) \ \forall \ \boldsymbol{\eta} \in (S_h)^d \ with \ \boldsymbol{\eta} = \mathbf{0} \ on \ \mathbf{\Gamma}_D, \tag{3.8}$$

$$(\tfrac{\varepsilon}{\tau}(\varphi_h^n - \varphi_h^{n-1}), \chi)_h + (B(\varphi_h^{n-1})\nabla w_h^n, \nabla\chi)_h = 0 \ \forall \ \chi \in S_h, \tag{3.9}$$

$$(w_h^n, \chi)_h = \gamma\epsilon(\nabla\varphi_h^n, \nabla\chi)_h + \tfrac{\gamma}{\epsilon}(\psi_1'(\varphi_h^{n-1}) + \psi_1''(\varphi_h^{n-1})(\varphi_h^n - \varphi_h^{n-1}), \chi)_h - \tfrac{1}{2}\langle\mathcal{E}(\mathbf{u}_h^n), \mathcal{E}(\mathbf{u}_h^n)\chi\rangle_{\mathcal{C}_2-\mathcal{C}_1}^h \ \ \forall\chi \in S_h, \tag{3.10}$$

where $B(\varphi) = \frac{9}{4}(1 - \varphi^2)^2$.

We solve equation (3.8) as in $(\mathbf{P}_1)$. Equations (3.9) and (3.10) on the other hand, define a system of two discretized second-order equations for φ_h^n and the chemical potential w_h^n. The derivative of the double well potential was linearized

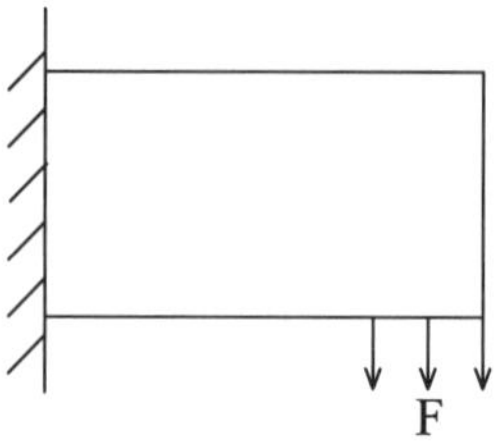

FIGURE 1. The design domain for a cantilever beam

as $\psi_1'(\varphi_h^n) \approx \psi_1'(\varphi_h^{n-1}) + \psi_1''(\varphi_h^{n-1})(\varphi_h^n - \varphi_h^{n-1})$ with $\psi_1'(\varphi) = \frac{9}{8}(1-\varphi^2)^2$, see (3.10). The resulting linear system is solved using BiCGStab, see [26, 32] for details.

The Cahn-Hilliard equation is implemented using the adaptive finite element toolbox AMDiS [32].

4. Numerics

In this section we present some numerical results for both the Allen-Cahn and the Cahn-Hilliard approach.

Since the interfacial thickness in both approaches is proportional to ε we need to choose $h \ll \varepsilon$ in order to resolve the interfacial layer (see [12, 13] for details). Away from the interface h can be chosen larger and hence adaptivity in space can heavily speed up computations. We use the same mesh refinement strategy as in Barrett, Nürnberg and Styles [3], i.e., a fine mesh is constructed where $|\varphi_h^{n-1}| < 1$ with a coarser mesh present in the bulk regions $\varphi_h^{n-1} = \pm 1$. We set the interfacial parameters $\varepsilon = \frac{1}{16\pi}$ and $\gamma = 1$ and we take the minimal diameter of an element $h_{\min} = 7.81 \times 10^{-3}$ and the maximal diameter $h_{\max} = 6.25 \cdot 10^{-2}$. The time step is chosen as $\tau = 6.25 \cdot 10^{-6}$ for the Allen-Cahn approach. In the Cahn-Hilliard case an adaptive time step is used.

We use a cantilever beam geometry, see Figure 1, where we pose Dirichlet boundary conditions on the left boundary Γ_D and a vertical force is acting at the bottom of its free vertical edge. We take $\Omega = (-1,1) \times (0,1)$, and hence $\Gamma_D = \{(-1,y) \in \mathbb{R}^2 : y \in [0,1]\}$. The force F is acting on $\Gamma_F := \{(x,0) \in \mathbb{R}^2 : x \in [0.75,1]\}$ and is defined by $\mathbf{f}(x) = (0,250)^t$ for $x \in \Gamma_F$. In our computations we use an isotropic elasticity tensor $\mathcal{C}_1$ of the form $\mathcal{C}_1\mathcal{E} = 2\mu_1\mathcal{E} + \lambda_1(tr\mathcal{E})I$ with the Lamé constants $\lambda_1 = \mu_1 = 5000$ and choose $\mathcal{C}_2 = \varepsilon^2\mathcal{C}_1$ in the void. We initialize the order parameter φ with random values between -0.1 and 0.1 for the Allen-Cahn approach and -0.2 and 0.2 for the Cahn-Hilliard approach. In both cases the random field ensures that we approximately have the same proportion of material and void, i.e., $m \approx 0$.

Figure 2 shows the results obtained using the Allen-Cahn variational inequality with volume constraint, where the state at $t = 0.160$ appears to be a numerical steady state.

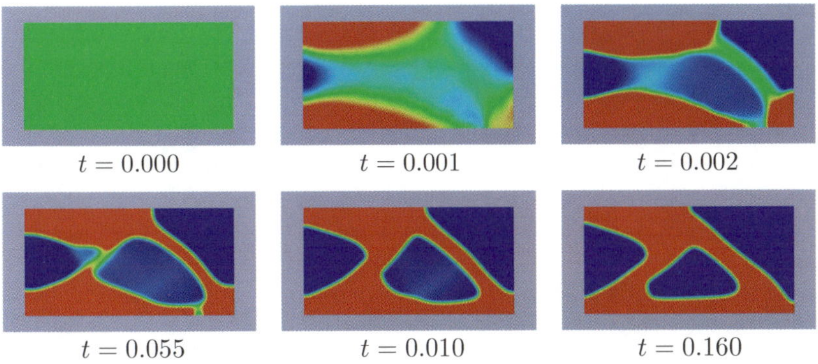

FIGURE 2. Allen-Cahn results for the cantilever beam computation at various times; material in red and void in blue

Figure 3 shows the results obtained using the Cahn-Hilliard equation with a variable mobility. Again the state at $t = 0.168$ appears to be a numerical steady state.

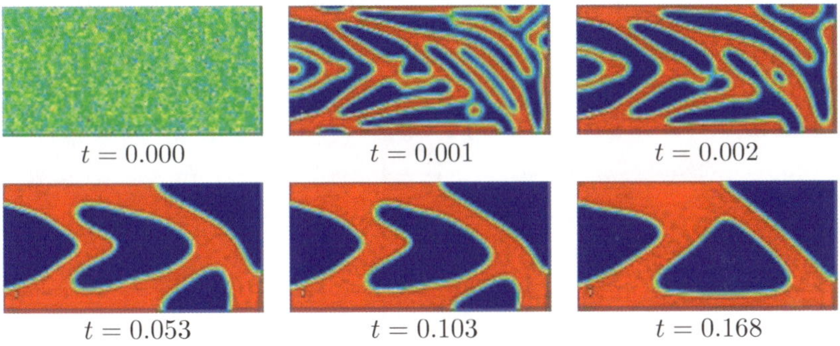

FIGURE 3. Cahn-Hilliard results for the cantilever beam computation at various times; material in red and void in blue

A comparison of both simulations gives two results: First the obtained optimal shape is almost the same, and second the evolution towards this shape is very different. Within the Allen-Cahn approach the final structure evolved directly from the random initial state within the same spatial scale. Also "new material" can be formed during the evolution in regions which previously have been occupied by void material. Within the Cahn-Hilliard approach such forming of "new material" was never observed. Instead the evolution always follows a coarsening process from fine scale structures, as a result of the spinodal decomposition in the early evolution, to coarser scales.

5. Conclusions

The use of phase-field methods in structural topology optimization has been limited due to high computational cost, associated with solving the underlying fourth-order Cahn-Hilliard equation. We have demonstrated on a simple example that a volume conserved Allen-Cahn equation can be used instead, which reduces the computational cost and thus makes the phase-field approach more efficient. We also point out that an obstacle potential together with the primal-dual active set approach allows us to compute for the phase field only in the interfacial region which reduces the total size of the problem, see also [5, 6]. We also mention that phase field approaches can be generalized for multimaterial structural topology optimization, see [33] for the Cahn-Hilliard case and [7] for the Allen-Cahn case.

We further want to point out, that the use of phase-field methods might also allow structural topology optimization to be extended to other fields besides structural mechanics. Due to the flexibility of the phase-field approach it can easily be coupled with other fields, such as flow, temperature or concentration fields. In [25, 22, 31, 14] a method is described which allows it to solve general partial differential equations with general boundary conditions in evolving geometries, which are implicitly described using a phase-field function. Allowing the phase-field function to evolve in order to minimize an objective function, which depends on the variables of the partial differential equation to be solved, will lead to new structural topology optimization problems.

References

[1] G. Allaire, *Optimization by the Homogenization Method.* Springer, Berlin 2002.

[2] G. Allaire, F. Jouve, A.-M. Toader, *Structural optimization using sensitivity analysis and a level-set method.* J. Comput. Phys. **194** (2004), 363–393.

[3] J.W. Barrett, R. Nürnberg, V. Styles, *Finite element approximation of a phase field model for void electromigration.* SIAM J. Numer. Anal. **46** (2004), 738–772.

[4] M.P. Bendsøe, O. Sigmund, *Topology Optimization.* Springer, Berlin 2003.

[5] L. Blank, M. Butz, H. Garcke, *Solving the Cahn-Hilliard variational inequality with a semi-smooth Newton method.* To appear in ESAIM Control Optim. Calc. Var., online: DOI 10.1051/cocv/2010032.

[6] L. Blank, H. Garcke, L. Sarbu, V. Styles, *Primal-dual active set methods for Allen-Cahn variational inequalities with non-local constraints.* DFG priority program 1253 "Optimization with PDEs", preprint no. 1253-09-01.

[7] L. Blank, H. Garcke, L. Sarbu, V. Styles, *Non-local Allen-Cahn systems: Analysis and primal dual active set methods.* Preprint No. 02/2011, Universtität Regensburg.

[8] L. Blank, L. Sarbu, M. Stoll, *Preconditioning for Allen-Cahn variational inequalities with non-local constraints.* Preprint Nr.11/2010, Universität Regensburg, Mathematik.

[9] B. Bourdin, A. Chambolle, *Design-dependent loads in topology optimization.* ESAIM Control Optim. Calc. Var. **9** (2003), 19–48.

[10] B. Bourdin, A. Chambolle, *The phase-field method in optimal design.* IUTAM Symposium on Topological Design Optimization of Structures, Machines and Materials (M.P. Bendsøe, N.Olhoff, and O. Sigmund eds.), Solid Mechanics and its Applications **137** (2006), 207–216.

[11] M. Burger, R. Stainko, *Phase-field relaxation of topology optimization with local stress constraints.* SIAM J. Control Optim. **45** (2006), 1447–1466.

[12] C.M. Elliott, *Approximation of curvature dependent interface motion.* State of the art in Numerical Analysis, IMA Conference Proceedings **63**, Clarendon Press, Oxford, 1997, 407–440.

[13] C.M. Elliott, V. Styles, *Computations of bi-directional grain boundary dynamics in thin films.* J. Comput. Phys. **187** (2003), 524–543.

[14] C.M. Elliott, B. Stinner, V. Styles, R. Welford, *Numerical computation of advection and diffusion on evolving diffuse interfaces.* IMA J. Num. Anal. **31**(3) (2011), 786–812, doi:10.1093/imanum/drq005.

[15] H. Garcke, *On Cahn-Hilliard systems with elasticity.* Proc. Roy. Soc. Edinburgh Sect. A **133** (2003), no. 2, 307–331.

[16] H. Garcke, *Mechanical effects in the Cahn-Hilliard model: A review on mathematical results.* In "Mathematical Methods and Models in phase transitions", ed.: Alain Miranville, Nova Science Publ. (2005), 43–77.

[17] H. Garcke, R. Nürnberg, V. Styles, *Stress and diffusion induced interface motion: Modelling and numerical simulations.* European Journal of Applied Math. **18** (2007), 631–657.

[18] H. Garcke, U. Weikard, *Numerical approximation of the Cahn-Larché equation.* Numer. Math. **100** (2005) 639–662.

[19] R.B. Haber, C.S. Jog, M.P. Bendsøe, *A new approach to variable topology shape design using a constraint on perimeter.* Struct. Multidisc. Optim. **11** (1996), 1–12.

[20] M. Hinze, R. Pinnau, M. Ulbrich, S. Ulbrich *Optimization with PDE Constraints.* Mathematical Modelling: Theory and Applications **23**, Springer (2008).

[21] F.C. Larché, J.W. Cahn, *The effect of self-stress on diffusion in solids*, Acta Metall. **30** (1982) 1835–1845.

[22] X. Li, J. Lowengrub, A. Rätz, A. Voigt, *Solving PDEs in complex geometries- a diffuse domain approach.* Comm. Math. Sci. **7** (2009), 81–107.

[23] L. Modica, *The gradient theory of phase transitions and the minimal interface criterion.* Arch. Rat. Mech. Anal. **98** (1987), 123–142.

[24] J. Petersson, *Some convergence results in perimeter-controlled topology optimization.* Computer Meth. Appl. Mech. Eng. **171** (1999), 123–140.

[25] A. Rätz, A. Voigt, *PDEs on surfaces – a diffuse interface approach.* Comm. Math. Sci. **4** (2006), 575–590.

[26] A. Rätz, A. Ribalta, A. Voigt, *Surface evolution of elastically stressed films under deposition by a diffuse interface model.* J. Comput. Phys. **214** (2006), 187–208.

[27] A. Schmidt, K.G. Siebert, *Design of adaptive finite element software. The finite element toolbox ALBERTA*, Lecture Notes in Computational Science and Engineering 42. Springer-Verlag, Berlin, 2005.

[28] O. Sigmund, J. Petersson, *Numerical instabilities in topology optimization: A survey on procedures dealing with checkerboards, mesh-dependencies and local minima.* Struct. Multidisc. Optim. **16** (1998), 68–75.

[29] A. Takezawa, S. Nishiwaki, M. Kitamura, *Shape and topology optimization based on the phase field method and sensitivity analysis.* Journal of Computational Physics **229** (7) (2010), 2697–2718.

[30] J.E. Taylor, J.W. Cahn, *Linking anisotropic sharp and diffuse surface motion laws via gradient flows.* J. Statist. Phys. **77** (1994), no. 1-2, 183–197.

[31] K.E. Teigen, X. Li, J. Lowengrub, F. Wang, A. Voigt, *A diffuse-interface approach for modeling transport, diffusion and adsorption/desorption of material quantities on a deformable interface.* Comm. Math. Sci. **7** (2009), 1009–1037.

[32] S. Vey, A. Voigt, *AMDiS – adaptive multidimensional simulations.* Comput. Vis. Sci. **10** (2007), 57–67.

[33] M.Y. Wang, S.W. Zhou, *Phase field: A variational method for structural topology optimization.* Comput. Model. Eng. Sci. **6** (2004), 547–566.

Luise Blank and Harald Garcke
Fakultät für Mathematik
Universität Regensburg
D-93040 Regensburg, Germany
e-mail: luise.blank@mathematik.uni-regensburg.de
harald.garcke@mathematik.uni-regensburg.de

Lavinia Sarbu and Vanessa Styles
Department of Mathematics
University of Sussex
Brighton BN1 9RF, UK
e-mail: ls99@sussex.ac.uk
v.styles@sussex.ac.uk

Tarin Srisupattarawanit and Axel Voigt
Institut für Wissenschaftliches Rechnen
TU Dresden
D-01062 Dresden, Germany
e-mail: tarin.srisupattarawanit@tu-dresden.de
axel.voigt@tu-dresden.de

International Series of Numerical Mathematics, Vol. 160, 257–275

Advanced Numerical Methods for PDE Constrained Optimization with Application to Optimal Design in Navier Stokes Flow

Christian Brandenburg, Florian Lindemann, Michael Ulbrich and Stefan Ulbrich

Abstract. We present an approach to shape optimization which is based on transformation to a reference domain with continuous adjoint computations. This method is applied to the instationary Navier-Stokes equations for which we discuss the appropriate setting and discuss Fréchet differentiability of the velocity field with respect to domain transformations. Goal-oriented error estimation is used for an adaptive refinement strategy. Finally, we give some numerical results.

Mathematics Subject Classification (2000). Primary 76D55; Secondary 49K20.

Keywords. Shape optimization, Navier-Stokes equations, PDE-constrained optimization, goal-oriented error estimation.

1. Introduction

This paper serves as the final report for the project "Advanced Numerical Methods for PDE Constrained Optimization with Application to Optimal Design and Control of a Racing Yacht in the America's Cup" in the DFG priority program 1253, *Optimization with Partial Differential Equations*. It covers some of the major results that were achieved in the course of the first funding period. The main task was to develop and investigate advanced numerical methods for finding the optimal shape of a body B, subject to constraints on the design, that is exposed to instationary incompressible Navier-Stokes flow. As an objective function the drag

We gratefully acknowledge the support of the Schwerpunktprogramm 1253 sponsored by the German Research Foundation (DFG). Part of the numerical computations were performed on a Linux cluster at TUM supported by DFG grant INST 95/919-1 FUGG. The work of the first and the fourth author was in parts supported by the International Research Training Group 1529 "Mathematical Fluid Dynamics" of the DFG.

of the body was chosen in the numerical computations. The resulting optimization problems are complex and highly nonlinear. As the developed techniques are quite general, they can, without significant conceptual difficulties, be used to address a wide class of shape optimization problems.

Shape optimization is an important and active field of research with many engineering applications; detailed accounts of its theory and applications can be found in, e.g., [8, 9, 15, 24, 28]. We use the approach of transformation to a reference domain, as originally introduced by Murat and Simon [26], see also [4, 14], which makes optimal control techniques readily applicable. Furthermore, as observed in [14] in the context of linear elliptic equations, and discussed in the present setting in [6], discretization and optimization can be made commutable. This allows to circumvent the tedious differentiation of finite element code with respect to the position of the vertices of the mesh.

We apply this approach to shape optimization problems governed by the instationary Navier-Stokes equations. On one hand we characterize the appropriate function spaces for domain transformations in this framework and give a theoretical result about the Fréchet differentiability of the design-to-state operator, which is done in Section 2. On the other hand, we focus on the practical implementation of shape optimization methods. In Section 3 we present the discretization and stabilization techniques we use to solve the Navier-Stokes equations numerically. Section 4 covers the use of goal-oriented error estimation and adaptivity for efficiently solving shape optimization problems. We then present numerical results obtained for some model problems in Section 5, followed by some conclusions in Section 6.

2. Shape optimization for the Navier-Stokes equations

2.1. Shape Optimization in the abstract setting

We use the approach of transformation to a reference domain to formulate the optimization problem in a functional analytical setting. The idea of using transformations to describe varying domains was suggested by Murat and Simon, see [26], and forms an excellent basis for deriving rigorous Fréchet differentiability results with respect to domain variations [4, 14]. This approach provides a flexible framework that can be used for many types of transformations (e.g., boundary displacements, free form deformation).

We consider a reference domain Ω_{ref} and interpret admissible domains $\Omega \in \mathcal{O}_{\text{ad}}$ as images of Ω_{ref} under suitable transformations τ. Then the abstract optimization problem is given as:

$$\min \quad J(y,\tau) \quad \text{s.t.} \quad E(y,\tau) = 0, \quad \tau \in T_{\text{ad}}. \tag{2.1}$$

We minimize an objective functional $J : Y(\Omega_{\text{ref}}) \times T(\Omega_{\text{ref}}) \to \mathbb{R}$ where the state $y \in Y(\Omega_{\text{ref}})$ and the transformation $\tau \in T(\Omega_{\text{ref}})$ are coupled by the *state equation* $E(y,\tau) = 0$ with $E : Y(\Omega_{\text{ref}}) \times T(\Omega_{\text{ref}}) \to Z(\Omega_{\text{ref}})$. Here, $T_{\text{ad}} \subset T(\Omega_{\text{ref}})$ is the set of admissible transformations corresponding to the set of admissible domains $\mathcal{O}_{\text{ad}}$. $T(\Omega_{\text{ref}}) \subset \{\tau : \Omega_{\text{ref}} \to \mathbb{R}^d\}$, $d = 2$ or 3, is assumed to be a Banach space of bicontinuous transformations of Ω_{ref}. In this context, $Y(\Omega)$ is a Banach space of functions defined on $\Omega \subset \mathbb{R}^d$ and we assume that

$$\left.\begin{aligned} Y(\Omega_{\text{ref}}) = \{\tilde{y} \circ \tau : \tilde{y} \in Y(\tau(\Omega_{\text{ref}}))\} \\ \tilde{y} \in Y(\tau(\Omega_{\text{ref}})) \mapsto y := \tilde{y} \circ \tau \in Y(\Omega_{\text{ref}}) \text{ is a homeomorphism} \end{aligned}\right\} \quad \forall \tau \in T_{\text{ad}}. \quad \text{(A)}$$

If the state equation is given in variational form defined on the domain $\tau(\Omega_{\text{ref}})$, then usually an equivalent variational form E defined on the reference domain can be obtained by using the transformation rule for integrals. This will be demonstrated for the instationary Navier-Stokes equations in the following. By convention, we denote all quantities on the physical domains $\tau(\Omega_{\text{ref}})$ by $\tilde{}$.

2.2. The Navier-Stokes equations

We apply the presented approach to shape optimization problems governed by the instationary Navier-Stokes equations for a viscous, incompressible fluid on a bounded domain $\Omega = \tau(\Omega_{\text{ref}}) \subset \mathbb{R}^d$ with Lipschitz boundary. To avoid technicalities in the formulation of the equations, we consider homogeneous Dirichlet boundary conditions. We arrive at the problem

$$\begin{aligned} \tilde{\boldsymbol{v}}_t - \nu\Delta\tilde{\boldsymbol{v}} + (\tilde{\boldsymbol{v}} \cdot \nabla)\tilde{\boldsymbol{v}} + \nabla\tilde{p} = \tilde{\boldsymbol{f}} \quad &\text{on } \Omega \times I, \quad \text{div } \tilde{\boldsymbol{v}} = 0 \quad \text{on } \Omega \times I \\ \tilde{\boldsymbol{v}} = \boldsymbol{0} \quad &\text{on } \partial\Omega \times I, \ \tilde{\boldsymbol{v}}(\cdot, 0) = \tilde{\boldsymbol{v}}_0 \quad \text{on } \Omega \end{aligned}$$

where $\tilde{\boldsymbol{v}} : \Omega \times I \to \mathbb{R}^d$ denotes the velocity and $\tilde{p} : \Omega \times I \to \mathbb{R}$ the pressure of the fluid. Here $I = (0, T), T > 0$ is the time interval and $\nu > 0$ is the kinematic viscosity.

We introduce the spaces

$$\mathcal{V}(\Omega) := \{\tilde{\boldsymbol{v}} \in C_0^\infty(\Omega)^d : \text{ div } \tilde{\boldsymbol{v}} = 0\}, \quad V(\Omega) := \text{cl}_{H_0^1}(\mathcal{V}(\Omega)),$$

$$H(\Omega) := \text{cl}_{L^2}(\mathcal{V}(\Omega)), \quad L_0^2(\Omega) := \{\tilde{p} \in L^2(\Omega) : \int_\Omega \tilde{p} = 0\},$$

the corresponding Gelfand triple $V(\Omega) \hookrightarrow H(\Omega) \hookrightarrow V(\Omega)^*$, and define

$$W(I; V(\Omega)) := \{\tilde{\boldsymbol{v}} \in L^2(I; V(\Omega)) : \ \tilde{\boldsymbol{v}}_t \in L^2(I; V(\Omega)^*)\}.$$

In the same way, the space $W(I; H_0^1(\Omega)^d)$ used later is defined.

Now let $\tilde{\boldsymbol{f}} \in L^2(I; H^{-1}(\Omega)^d), \tilde{\boldsymbol{v}}_0 \in H(\Omega)$. It is well known that so far the question of existence and uniqueness is answered satisfactorily only in the case $d \leq 2$. In fact, under these assumptions with $d = 2$, there exists a unique solution $(\tilde{\boldsymbol{v}}, \tilde{p})$ with $\tilde{\boldsymbol{v}} \in W(I; V(\Omega))$ where $\tilde{p}$ is a $L_0^2(\Omega)$-valued distribution, see [30]. Assuming that the data $\tilde{\boldsymbol{f}}$ and $\tilde{\boldsymbol{v}}_0$ are sufficiently regular, the solution has further regularity as stated by

Lemma 2.1. *Let $d = 2$ and assume*

$$\tilde{\boldsymbol{f}}, \tilde{\boldsymbol{f}}_t \in L^2(I; H^{-1}(\Omega)^2), \tilde{\boldsymbol{f}}(\cdot, 0) \in H, \tilde{\boldsymbol{v}}_0 \in V(\Omega) \cap H^2(\Omega)^2. \tag{2.2}$$

Then the solution $(\tilde{\boldsymbol{v}}, \tilde{p})$ *of the Navier-Stokes equations satisfies*

$$\tilde{\boldsymbol{v}} \in C(I; V(\Omega)),\ \tilde{\boldsymbol{v}}_t \in L^2(I, V(\Omega)) \cap L^\infty(I; H(\Omega)),\ \tilde{p} \in L^\infty(I; L_0^2(\Omega)) \tag{2.3}$$

The proof can be found for example in [30, Ch. III].

2.3. Weak formulation and transformation to the reference domain

Under the assumptions of Lemma 2.1, we consider a weak velocity-pressure formulation of the problem where the divergence-freeness of the velocity is not included in the trial and test spaces: Find $(\tilde{\boldsymbol{v}}, \tilde{p})$ such that

$$\begin{aligned}
&\langle \tilde{\boldsymbol{v}}_t(\cdot, t), \tilde{\boldsymbol{w}} \rangle_{H^{-1}, H_0^1} + \int_\Omega \tilde{\boldsymbol{v}}(x,t)^T \nabla \tilde{\boldsymbol{v}}(x,t) \tilde{\boldsymbol{w}}(x)\, dx \\
&\quad + \int_\Omega \nu \nabla \tilde{\boldsymbol{v}}(x,t) : \nabla \tilde{\boldsymbol{w}}(x)\, dx - \int_\Omega \tilde{p}(x,t)\ \operatorname{div}\ \tilde{\boldsymbol{w}}(x)\, dx \\
&\quad = \int_\Omega \tilde{\boldsymbol{f}}(x,t)^T \tilde{\boldsymbol{w}}(x)\, dx \quad \forall \tilde{\boldsymbol{w}} \in H_0^1(\Omega)^d \text{ for a.a. } t \in I \\
&\int_\Omega \tilde{q}(x)\ \operatorname{div}\ \tilde{\boldsymbol{v}}(x,t)\, dx = 0 \quad \forall \tilde{q} \in L_0^2(\Omega) \text{ for a.a. } t \in I \\
&\tilde{\boldsymbol{v}}(\cdot, 0) = \tilde{\boldsymbol{v}}_0.
\end{aligned} \tag{2.4}$$

In the following we assume that

(T) Ω_{ref} is a bounded Lipschitz domain and $\Omega' \supset \bar{\Omega}_{\text{ref}}$ is open and bounded with Lipschitz boundary. Moreover $T_{\text{ad}} \subset W^{2,\infty}(\Omega')^d$ is bounded such that for all $\tau \in T_{\text{ad}}$ the mappings $\tau : \bar{\Omega}_{\text{ref}} \to \tau(\bar{\Omega}_{\text{ref}})$ satisfy $\tau^{-1} \in W^{2,\infty}(\tau(\bar{\Omega}_{\text{ref}}))^d$ and $\det(\tau') \geq \delta > 0$, with a constant $\delta > 0$. Here, $\tau'(x) = \nabla \tau(x)^T$ denotes the Jacobian of τ.
Moreover, the data $\tilde{\boldsymbol{v}}_0, \tilde{\boldsymbol{f}}$ are given such that

$$\begin{aligned}
&\tilde{\boldsymbol{f}} \in L^\infty(I; C^1(\Omega)^d),\ \tilde{\boldsymbol{f}}_t \in L^2(I; H^{-1}(\Omega)^d), \\
&\tilde{\boldsymbol{f}}(0) \in H(\Omega),\ \tilde{\boldsymbol{v}}_0 \in V(\Omega) \cap H^2(\Omega)^d \cap C^1(\Omega)^d
\end{aligned}$$

for all $\Omega \in \mathcal{O}_{\text{ad}} = \{\tau(\Omega_{\text{ref}}) \ :\ \tau \in T_{ad}\}$, i.e., the data $\tilde{\boldsymbol{v}}_0, \tilde{\boldsymbol{f}}_0$ are used on all $\Omega \in \mathcal{O}_{\text{ad}}$.

Assumption (T) ensures in particular (2.2) and assumption (A) holds in the following obvious version for time-dependent problems, where the transformation acts only in space.

Lemma 2.2. *Let* T_{ad} *satisfy assumption* (T). *Then the space* $W(I; H_0^1(\Omega)^d)$ *for the velocity satisfies assumption* (A), *more precisely, we have for all* $\tau \in T_{ad}$

$$W(I; H_0^1(\Omega_{\text{ref}})^d) = \{\tilde{\boldsymbol{v}}(\tau(\cdot)) : \tilde{\boldsymbol{v}} \in W(I; H_0^1(\tau(\Omega_{\text{ref}}))^d)\},$$
$$\tilde{\boldsymbol{v}} \in W(I; H_0^1(\tau(\Omega_{\text{ref}}))^d) \mapsto \boldsymbol{v} := \tilde{\boldsymbol{v}}(\tau(\cdot)) \in W(I; H_0^1(\Omega_{\text{ref}})^d) \textit{ is a homeom.}$$

A proof of this result is beyond the scope of this paper and will be given in [20]. A similar result can be shown for the pressure if we define a topology on the pressure space like $L^p(I; L_0^2(\Omega))$ as guaranteed by Lemma 2.1.

Given the weak formulation of the Navier-Stokes equations on a domain $\tau(\Omega_{\text{ref}})$ we can apply the transformation rule for integrals to obtain a variational formulation based on the domain Ω_{ref} which is equivalent to (2.4): Find $(\boldsymbol{v}, p)$ such that

$$\begin{aligned}
&\int_{\Omega_{\text{ref}}} \boldsymbol{v}_t^T \boldsymbol{w} \det \tau' \, dx + \sum_{i=1}^{d} \int_{\Omega_{\text{ref}}} \nu \nabla v_i^T \tau'^{-1} \tau'^{-T} \nabla w_i \det \tau' \, dx \\
&+ \int_{\Omega_{\text{ref}}} \boldsymbol{v}^T \tau'^{-T} \nabla \boldsymbol{v} \, \boldsymbol{w} \det \tau' \, dx - \int_{\Omega_{\text{ref}}} p \operatorname{tr}(\tau'^{-T} \nabla \boldsymbol{w}) \det \tau' \, dx \\
&- \int_{\Omega_{\text{ref}}} \tilde{\boldsymbol{f}}(\tau(x), t)^T \boldsymbol{w} \det \tau' \, dx = 0 \quad \forall \boldsymbol{w} \in H_0^1(\Omega_{\text{ref}})^d \text{ for a.a. } t \in I \\
&\int_{\Omega_{\text{ref}}} q \operatorname{tr}(\tau'^{-T} \nabla \boldsymbol{v}) \det \tau' \, dx = 0 \quad \forall q \in L_0^2(\Omega_{\text{ref}}) \text{ for a.a. } t \in I \\
&\boldsymbol{v}(\cdot, 0) = \tilde{\boldsymbol{v}}_0(\tau(\cdot)).
\end{aligned} \tag{2.5}$$

2.4. Differentiability of the design-to-state operator

Very recently, we succeeded in proving Fréchet differentiability of the velocity field with respect to domain variations in 2D under reasonable assumptions. In the following, we consider the case $d = 2$. A paper on these results is in preparation [20]. For the stationary Navier-Stokes equations, a corresponding investigation can be found in [4]. Significant additional complications for the time-dependent case are caused by the fact that in the standard $W(I; V(\Omega))$-setting the time regularity of the pressure is very low. In fact, the $\nabla \tilde{p}$ term takes care of those parts of the residual that are not seen when tested with solenoidal functions. Now $L^2(I; V(\Omega)^*)$ is a weaker space than $L^2(I; H^{-1}(\Omega)^2)$, since $V(\Omega) \subsetneq H_0^1(\Omega)^2$ is a closed subspace strictly smaller than $H_0^1(\Omega)^2$. Therefore, the fact that $\tilde{\boldsymbol{v}}_t \in L^2(I; V(\Omega)^*)$ cannot be used to derive time regularity results for the pressure. This causes difficulties since after transformation to the reference domain the velocity field is no longer solenoidal. Thus, the pressure cannot be eliminated and the skew symmetry of the trilinear convection form cannot be used since it only holds if the first argument is solenoidal. For achieving that the required regularity properties of solutions are maintained under transformation, we need the requirement $\tau \in T_{\text{ad}} \subset W^{2,\infty}(\Omega')^2$. This especially concerns the regularity of the time derivative. As already discussed, in 2D, the Navier-Stokes equations on the domain $\Omega = \tau(\Omega_{\text{ref}})$ have a unique solution $\tilde{\boldsymbol{v}} \in W(I; V(\Omega))$. Under the assumptions stated in Lemma 2.1, there also holds

$$\tilde{\boldsymbol{v}} \in C(I; V(\Omega)), \quad \tilde{\boldsymbol{v}}_t \in L^2(I; V(\Omega)) \cap L^\infty(I; H(\Omega)). \tag{2.6}$$

This implies the pressure regularity $\tilde{p} \in L^\infty(I; L_0^2(\Omega))$. The latter regularity properties are maintained under transformation to the reference domain Ω_{ref}. After transformation of the Navier-Stokes equations, the condition $\operatorname{div} \tilde{\boldsymbol{v}} = 0$ becomes $\operatorname{tr}(g(\tau')\nabla \boldsymbol{v}) = 0$, where $g(M) = \det M \, M^{-T}$. To proceed further, it is crucial to require and exploit the additional regularity of $\boldsymbol{v}$. As demonstrated in [4, 6], it is sufficient to consider the differentiability of $\boldsymbol{v}(\tau)$ at $\tau = \text{id}$, since $\tau(\Omega_{\text{ref}})$ can be taken as the current reference domain and the derivative w.r.t. variations of this domain can be transformed to derivatives w.r.t. domain variations of Ω_{ref}. We build on the uniform boundedness of $\boldsymbol{v}(\tau)$ in the sense of (2.6) for τ sufficiently close to id. In a first step, we prove continuity results for the solution operator $\tau \mapsto \boldsymbol{v}(\tau)$ in $L^\infty(I; L^2(\Omega)^2) \cap L^2(I; H_0^1(\Omega)^2)$. From this and the boundedness in (2.6), continuity in stronger spaces can be proved, e.g., by interpolation.

For further presentation, we write the equations (2.5) on Ω_{ref} schematically as

$$E(\boldsymbol{v}, p, \tau) = 0.$$

Formal linearization of (2.5) about $(\bar{\boldsymbol{v}}, \bar{p}, \bar{\tau}) = (\boldsymbol{v}(\text{id}), p(\text{id}), \text{id})$ yields

$$A_1(\bar{\boldsymbol{v}}, \text{id})\delta\boldsymbol{v} + A_2(\text{id})\delta p + A_3(\bar{\boldsymbol{v}}, \bar{p}, \text{id})\delta\tau = 0.$$

As mentioned, $\delta\boldsymbol{v}$ is not solenoidal, but satisfies $\operatorname{div}(\delta\boldsymbol{v}) = -\operatorname{tr}(g'(I)\delta\tau'\nabla\bar{\boldsymbol{v}})$ with $g(M)$ as defined above. To apply standard theory, we use the existence of a right inverse $B : L_0^2(\Omega_{\text{ref}}) \to H_0^1(\Omega_{\text{ref}})^2$ of the div operator that, as shown in [13], can be chosen such that it also is a bounded linear operator from $H^1(\Omega_{\text{ref}})^*$ to $L_0^2(\Omega_{\text{ref}})^2$. We set $\boldsymbol{e}_0 = -B \operatorname{tr}(g'(I)\delta\tau'\nabla\bar{\boldsymbol{v}})$ and obtain a splitting $\delta\boldsymbol{v} = \boldsymbol{e}_0 + \boldsymbol{e}$, where now $\boldsymbol{e}$ is solenoidal. Note that $\boldsymbol{e}_0 = A_4(\nabla\bar{\boldsymbol{v}}, \delta\tau')$ is bilinear. Inserting $\delta\boldsymbol{v} = \boldsymbol{e}_0 + \boldsymbol{e}$ into the linearized equation gives a linearized Navier-Stokes equation for the solenoidal function $\boldsymbol{e}$. The right-hand side generated by $\boldsymbol{e}_0$ can be carefully estimated to obtain linear bounds for $\delta\boldsymbol{v}$ in terms of $\delta\tau$. Clearly, $\delta\boldsymbol{v}$ is the candidate for the derivative $\boldsymbol{v}_\tau(\text{id})\delta\tau$, if it exists. Next, we consider, with $\tau = \text{id} + \delta\tau$, $\boldsymbol{v} = \boldsymbol{v}(\tau)$, and $p = p(\tau)$ the equation

$$E(\boldsymbol{v}, p, \tau) - E(\bar{\boldsymbol{v}}, \bar{p}, \text{id}) - [A_1(\bar{\boldsymbol{v}}, \text{id})\delta\boldsymbol{v} + A_2(\text{id})\delta p + A_3(\bar{\boldsymbol{v}}, \bar{p}, \text{id})\delta\tau] = 0.$$

This is rearranged to obtain a linear equation for the residual $(\boldsymbol{r}, r_p) = (\boldsymbol{v} - \bar{\boldsymbol{v}} - \delta\boldsymbol{v}, p - \bar{p} - \delta p)$ while the remaining terms are taken to the right-hand side. Again $\boldsymbol{r}$ is not solenoidal and using the operator B the splitting $\boldsymbol{r} = \boldsymbol{r}_0 + \boldsymbol{r}_1$ is obtained such that $\boldsymbol{r}_1$ is solenoidal. Now, again, standard theory for the linearized Navier-Stokes equations can be used, where the estimation of the right-hand side is quite involved. Taking all together, the following can be shown.

Theorem 2.3. *Let the assumption* (T) *hold. Then the operator*

$$\tau \in T_{\text{ad}} \subset W^{2,\infty}(\Omega')^2 \mapsto \boldsymbol{v}(\tau) \in W(I; H_0^1(\Omega_{\text{ref}})^2) + W(I; V(\Omega_{\text{ref}}))$$

is Fréchet differentiable. Note that the range space is continuously imbedded into $L^2(I; H_0^1(\Omega_{\text{ref}})^2) \cap C(I; L^2(\Omega_{\text{ref}})^2)$.

This result implies for example that the time averaged drag on a body B depends differentiable on τ, since it can be rewritten in an appropriate distributed form, see [6].

2.5. Shape gradient calculation with the Navier-Stokes equations

As for every $\tau \in T_{\text{ad}}$ there exists a unique solution $(\boldsymbol{v}, p)$ of the Navier-Stokes equations, it is reasonable to define the following reduced problem on the space of transformations $T(\Omega_{\text{ref}})$:

$$\min j(\tau) := J(y(\tau), \tau) \qquad \text{s.t.} \quad \tau \in T_{\text{ad}},$$

where $y(\tau)$ is given as the solution of the transformed Navier-Stokes eqns. (2.5).

In order to calculate the gradient of j we need to solve the adjoint equations of the transformed Navier-Stokes equations (2.5).

Given a weak solution $(\boldsymbol{v}, p) \in Y(\Omega_{\text{ref}}) := W(I; H_0^1(\Omega_{\text{ref}})^d) \times L^2(I; L_0^2(\Omega_{\text{ref}}))$ we seek $(\boldsymbol{\lambda}, \mu) \in L^2(I; H_0^1(\Omega_{\text{ref}})^d) \times L^2(I; L_0^2(\Omega_{\text{ref}}))$ with

$$\begin{aligned}
&- \int_I \int_{\Omega_{\text{ref}}} \boldsymbol{w}^T \boldsymbol{\lambda}_t \det \tau' \; dt \; dx \\
&+ \int_{\Omega_{\text{ref}}} \boldsymbol{w}(x, T)^T \boldsymbol{\lambda}(x, T) \det \tau' \; dx \\
&+ \int_I \sum_{i=1}^{d} \int_{\Omega_{\text{ref}}} \nu \nabla w_i^T \tau'^{-1} \tau'^{-T} \nabla \lambda_i \det \tau' \; dx \; dt \\
&+ \int_I \int_{\Omega_{\text{ref}}} \left(\boldsymbol{w}^T \tau'^{-T} \nabla \boldsymbol{v} + \boldsymbol{v}^T \tau'^{-T} \nabla \boldsymbol{w} \right) \boldsymbol{\lambda} \det \tau' \; dx \; dt \\
&- \int_I \int_{\Omega_{\text{ref}}} q \; \operatorname{tr}(\tau'^{-T} \nabla \boldsymbol{\lambda}) \det \tau' \; dx \; dt \\
&+ \int_I \int_{\Omega_{\text{ref}}} \mu \; \operatorname{tr}(\tau'^{-T} \nabla \boldsymbol{w}) \det \tau' \; dx \; dt \\
&= - \langle J_{(\boldsymbol{v}, p)}((\boldsymbol{v}, p), \tau), (\boldsymbol{w}, q) \rangle_{Y^*(\Omega_{\text{ref}}), Y(\Omega_{\text{ref}})} \\
&\qquad\qquad \forall (\boldsymbol{w}, q) \in Y(\Omega_{\text{ref}}).
\end{aligned}$$

This is a parabolic equation backwards in time with appropriate initial condition for the adjoint velocity $\boldsymbol{\lambda}$ at time T depending on the objective functional. For details, we refer to [6]. For $\tau = \text{id}$, we arrive at the weak formulation of the usual adjoint system of the Navier-Stokes equations on Ω_{ref}.

For the application of optimization algorithms it is convenient to solve, for a given iterate $\tau_k \in T(\Omega_{\text{ref}})$, an equivalent representation of the optimization problem on the domain $\Omega_k := \tau_k(\Omega_{\text{ref}})$ (for details see [6]). Without loss of generality we assume $\Omega_k = \Omega_{\text{ref}}$ and calculate the reduced gradient $j'(\tau)$ at $\tau = \text{id}$:

1. Find $(\boldsymbol{v}, p)$ by solving the standard Navier-Stokes equations on the domain Ω_{ref}.
2. Find $(\boldsymbol{\lambda}, \mu)$ by solving the standard adjoint Navier-Stokes equations on the domain Ω_{ref}.

3. Calculate the reduced gradient $j'(\mathrm{id})$ via

$$
\begin{aligned}
&\langle j'(\mathrm{id}), V\rangle_{T^*(\Omega_{\mathrm{ref}}),T(\Omega_{\mathrm{ref}})} = \langle J_\tau((\boldsymbol{v},p),\tau), V\rangle_{T^*(\Omega_{\mathrm{ref}}),T(\Omega_{\mathrm{ref}})}\\
&- \int_{\Omega_{\mathrm{ref}}} V^T \nabla \tilde{\boldsymbol{v}}_0(x)\boldsymbol{\lambda}(x,0)\, dx + \int_I\int_{\Omega_{\mathrm{ref}}} \boldsymbol{v}_t^T \boldsymbol{\lambda} \operatorname{div} V\, dx\, dt\\
&+ \int_I\int_{\Omega_{\mathrm{ref}}} \nu\nabla\boldsymbol{v} : \nabla\boldsymbol{\lambda} \operatorname{div} V\, dx\, dt - \sum_{i=1}^{d}\int_I\int_{\Omega_{\mathrm{ref}}} \nu\nabla\boldsymbol{v}_i^T (V' + V'^T)\nabla\boldsymbol{\lambda}_i\, dx\, dt\\
&- \int_I\int_{\Omega_{\mathrm{ref}}} \boldsymbol{v}^T V'^T \nabla\boldsymbol{v}\boldsymbol{\lambda}\, dx\, dt + \int_I\int_{\Omega_{\mathrm{ref}}} \boldsymbol{v}^T\nabla\boldsymbol{v}\boldsymbol{\lambda} \operatorname{div} V\, dx\, dt \qquad (2.7)\\
&+ \int_I\int_{\Omega_{\mathrm{ref}}} p \operatorname{tr}(V'^T\nabla\boldsymbol{\lambda})\, dx\, dt - \int_I\int_{\Omega_{\mathrm{ref}}} p \operatorname{div}\boldsymbol{\lambda} \operatorname{div} V\, dx\, dt\\
&- \int_I\int_{\Omega_{\mathrm{ref}}} \tilde{\boldsymbol{f}}^T\boldsymbol{\lambda} \operatorname{div} V\, dx\, dt - \int_I\int_{\Omega_{\mathrm{ref}}} V^T\nabla\tilde{\boldsymbol{f}}\boldsymbol{\lambda}\, dx\, dt\\
&- \int_I\int_{\Omega_{\mathrm{ref}}} \mu \operatorname{tr}(V'^T\nabla\boldsymbol{v})\, dx\, dt + \int_I\int_{\Omega_{\mathrm{ref}}} \mu \operatorname{div}\boldsymbol{v} \operatorname{div} V\, dx\, dt.
\end{aligned}
$$

Finally, if we assume more regularity for the state and adjoint, we can integrate by parts in the above formula and can represent the shape gradient as a functional on the boundary. However, we prefer to work with the distributed version (2.7), since it is also appropriate for FE-Galerkin approximations, while the integration by parts to obtain the boundary representation is not justified for FE-discretizations with H^1-elements.

2.6. Derivatives with respect to shape parameters

In practical situations, the domain transformations τ are usually not given directly. Instead, one often deals with design parameters $u \in U$ where U is a finite or infinite-dimensional design space defining the boundary Γ_B of the design object. Possible choices for the design parameters u are, e.g., B-spline control points or even the boundary curve itself. In this context, the reduced gradient of the objective function w.r.t u can be computed in a very efficient way.

First of all, u defines a displacement $d(u)$ of the reference object boundary Γ_B. Sensitivities for this mapping are usually easy to calculate, e.g., for B-Spline parametrizations $d(u)$ is a linear function w.r.t. the B-Spline control points u. The boundary displacement $d(u)$ can then be mapped to a domain transformation $\tau(u)$ by solving for example an elliptic PDE (e.g., linear elasticity equation or Poisson equation) with fixed displacement $d(u)$ on the object boundary and homogeneous Dirichlet boundary conditions on $\partial\Omega_{\mathrm{ref}} \setminus \Gamma_B$. In the case of the elasticity equation this leads to an optimization problem of the form

$$\min \quad j(\tau) \quad \text{s.t.} \quad A(\tau, d(u)) = 0, \quad u \in U_{\mathrm{ad}}, \tag{2.8}$$

where A denotes the elasticity equation with boundary conditions. Using adjoint calculus, we get

$$\tilde{j}_d(d(u)) = A_d(\tau, d(u))^* z,$$

where $\tilde{j} := j(\tau(d(u))$ denotes the reduced objective functional w.r.t $d(u)$ and z is the solution of the adjoint elasticity equation

$$\langle A_\tau(\tau, d(u))V, z\rangle_{T^*(\Omega_{\mathrm{ref}}),T(\Omega_{\mathrm{ref}})} = -\langle j'(\tau), V\rangle_{T^*(\Omega_{\mathrm{ref}}),T(\Omega_{\mathrm{ref}})} \qquad \forall V \in T(\Omega_{\mathrm{ref}}).$$

Note that the right-hand side is just (2.7) and that the evaluation of this equation makes up the main part of the work for computing the reduced derivative. Derivatives of $\tilde{j}$ w.r.t. u are then easily obtained by the chain rule and are given by

$$\frac{d}{du}\tilde{j}(d(u)) = d_u(u)^* A_d(\tau, d(u))^* z.$$

3. Discretization

To discretize the instationary Navier-Stokes equations, we use the $cG(1)dG(0)$ space-time finite element method, which uses piecewise constant finite elements in time and piecewise linear finite elements in space and is a variant of the General Galerkin G^2-method developed by Eriksson et al. [11, 12].

Let $\mathcal{I} = \{I_j = (t_{j-1}, t_j] : 1 \leq j \leq N\}$ be a partition of the time interval $(0, T]$ with a sequence of discrete time steps $0 = t_0 < t_1 < \cdots < t_N = T$ and length of the respective time intervals $k_j := |I_j| = t_j - t_{j-1}$. With each time step t_j, we associate a partition $\mathcal{T}_j$ of the spatial domain Ω and the finite element subspaces V_h^j, P_h^j of continuous piecewise linear functions in space.

The $cG(1)dG(0)$ space-time finite element discretization with stabilization can be written as an implicit Euler scheme: $\tilde{\boldsymbol{v}}_h^0 = \tilde{\boldsymbol{v}}_0$ and for $j = 1 \ldots N$, find $(\tilde{\boldsymbol{v}}_h^j, \tilde{p}_h^j) \in V_h^j \times P_h^j$ such that

$$\left(\bar{E}^{h,j}(\tilde{\boldsymbol{v}}_h, \tilde{p}_h), (\tilde{\boldsymbol{w}}_h, \tilde{q}_h)\right) + \left(\bar{SD}_\delta^j(\tilde{\boldsymbol{v}}_h, \tilde{p}_h), (\tilde{\boldsymbol{w}}_h, \tilde{q}_h)\right) = 0$$
$$\forall (\tilde{\boldsymbol{w}}_h, \tilde{q}_h)_{|I_j} \in V_h^j \times P_h^j$$

with the discretized Navier-Stokes equations

$$\left(\bar{E}^{h,j}(\tilde{\boldsymbol{v}}_h, \tilde{p}_h), (\tilde{\boldsymbol{w}}_h, \tilde{q}_h)\right) := \left(\frac{\tilde{\boldsymbol{v}}_h^j - \tilde{\boldsymbol{v}}_h^{j-1}}{k_j}, \tilde{\boldsymbol{w}}_h^j\right) + (\nu \nabla \tilde{\boldsymbol{v}}_h^j, \nabla \tilde{\boldsymbol{w}}_h^j)$$
$$+ (\tilde{\boldsymbol{v}}_h^j \cdot \nabla \tilde{\boldsymbol{v}}_h^j, \tilde{\boldsymbol{w}}_h^j) - (\tilde{p}_h^j, \operatorname{div} \tilde{\boldsymbol{w}}_h^j) + (\operatorname{div} \tilde{\boldsymbol{v}}_h^j, \tilde{q}_h^j) - (\tilde{\boldsymbol{f}}, \tilde{\boldsymbol{w}}_h^j),$$

and with the stabilization

$$\left(\bar{SD}_\delta^j(\tilde{\boldsymbol{v}}_h, \tilde{p}_h), (\tilde{\boldsymbol{w}}_h, \tilde{q}_h)\right) := \left(\tilde{\delta}_1(\tilde{\boldsymbol{v}}_h^j \cdot \nabla \tilde{\boldsymbol{v}}_h^j + \nabla \tilde{p}_h^j - \tilde{\boldsymbol{f}}), \tilde{\boldsymbol{v}}_h^j \cdot \nabla \tilde{\boldsymbol{w}}_h^j + \nabla \tilde{q}_h^j\right)$$
$$+ (\tilde{\delta}_2 \operatorname{div} \tilde{\boldsymbol{v}}_h^j, \operatorname{div} \tilde{\boldsymbol{w}}_h^j).$$

Note that the terms $(\tilde{\boldsymbol{v}}_h^j)_t - \nu \Delta \tilde{\boldsymbol{v}}_h^j$ vanish on each element of $\mathcal{T}_j$ for $cG(1)dG(0)$ elements and could equivalently be included if we understand $\bar{SD}_\delta^j$ as a sum of the contributions on the spacial elements. With this interpretation, the stabilization term vanishes for sufficiently regular solutions of the Navier-Stokes equations (e.g., (2.3) and $\tilde{\boldsymbol{v}} \in L^2(I, H^2(\Omega)^2), \tilde{p} \in L^2(I, H^1(\Omega))$).

The stabilization parameters $\tilde{\delta}_1$ and $\tilde{\delta}_2$ act as a subgrid model in the convection-dominated case by adding a viscosity roughly of the local mesh size, see [19].

In order to obtain gradients which are exact on the discrete level, we consider the discrete Lagrangian functional based on the $cG(1)dG(0)$ finite element method, which is given by

$$\mathcal{L}^h((\boldsymbol{v}_h,p_h),\tau_h,(\boldsymbol{\lambda}_h,\mu_h)) = J^h((\boldsymbol{v}_h,p_h),\tau_h)$$
$$+\underbrace{\sum_{j=1}^{N} k_j\left(E^{h,j}((\boldsymbol{v}_h,p_h),\tau_h),(\boldsymbol{\lambda}_h,\mu_h)\right)}_{=:(E^h((\boldsymbol{v}_h,p_h),\tau_h),(\boldsymbol{\lambda}_h,\mu_h))}+\underbrace{\sum_{j=1}^{N} k_j\left(SD_\delta^j((\boldsymbol{v}_h,p_h),\tau_h),(\boldsymbol{w}_h,q_h)\right)}_{=:(SD_\delta((\boldsymbol{v}_h,p_h),\tau_h),(\boldsymbol{w}_h,q_h))}.$$

To obtain the discrete adjoint equation and the reduced gradient, we take the derivatives of the discrete Lagrangian w.r.t. the state variables and the shape variables, respectively. Similar to the discrete state equation, the discrete adjoint system can be cast in the form of an implicit time-stepping scheme, backward in time, see [6].

For the computation of shape derivatives on the discrete level we use a transformation space $T_h(\Omega_{\text{ref}})$ of piecewise linear continuous functions. Then a discrete version of Lemma 2.2 holds, and we get an analogue of (2.7) on the discrete level. Thus, we obtain the exact shape derivative on the discrete level by using a continuous adjoint approach without the tedious task of computing mesh sensitivities.

4. Goal-oriented error estimation and adaptivity

In this section we investigate the use of goal-oriented a posteriori error estimation and adaptivity in the context of shape optimization. Goal-oriented error estimators based on duality arguments date back to a series of papers by Babuška and Miller, starting with [1], and were systematically investigated in the sequel, see, e.g., [11, 3]. They have already successfully been applied to optimal control and parameter identification problems, both for the elliptic and the parabolic case, see [2, 3, 23]. Goal-oriented estimators measure the error in a quantity of interest depending on the state, rather than the state itself. Thus, in the adaptive process, the computational domain is only resolved more accurately in areas which have an influence on the quantity of interest, which in our case is the objective function value.

4.1. Goal-oriented error estimation in shape optimization

In the context of shape optimization problems governed by the instationary Navier-Stokes equations, certain difficulties arise which keep us from directly using goal-oriented error estimators for optimal control problems as in, e.g., [2]. In particular, we use a nonconforming time discretization, i.e., $Y_h \not\subseteq Y$, which has also been considered in [23], and $T_h \not\subseteq T$. This means that $\mathcal{L}'(y,\tau,\lambda)(\bar{y},\bar{\tau},\bar{\lambda}) = 0$ for $(\bar{y},\bar{\tau},\bar{\lambda}) \in Y_h \times T_h \times Z_h^*$ is not satisfied automatically. Furthermore, we introduce additional stabilization terms which contribute to the error estimator, and E^h and J^h contain additional jump terms compared to their continuous counterparts.

To overcome these difficulties, we assume for the continuous optimal solution (y,τ) that $E^h(y,\tau)' = E(y,\tau)'$, $J^h(y,\tau)' = J(y,\tau)'$, and $SD_\delta(y,\tau) = 0$ (i.e. the stabilization is consistent), and that

$$J'(y,\tau)(\bar{y}_h,\bar{\tau}_h) + \langle \lambda, E'(y,\tau)(\bar{y}_h,\bar{\tau}_h,\bar{\lambda}_h)\rangle = 0 \quad \forall(\bar{y}_h,\bar{\tau}_h,\bar{\lambda}_h) \in Y_h \times T_h \times Z_h^*.$$

We will see that these assumptions are satisfied for the considered $cG(1)dG(0)$ method if the continuous state and adjoint are sufficiently regular.

Assuming further that J and E are three times Gateaux-differentiable and $(y,\tau,\lambda) \in Y \times T \times Z^*$ and $(y_h,\tau_h,\lambda_h) \in Y_h \times T_h \times Z_h^*$ are stationary points of the continuous and discrete Lagrange functionals, resp., the a posteriori error with respect to the objective functional is given by

$$\begin{aligned} J(y,\tau) - J^h(y_h,\tau_h) = {} & \frac{1}{2}\rho^y(y_h,\tau_h,\lambda_h)(\lambda - i_h\lambda) \\ & + \frac{1}{2}\rho^\lambda(y_h,\tau_h,\lambda_h)(y - i_h y) + \frac{1}{2}\rho^\tau(y_h,\tau_h,\lambda_h)(\tau - i_h\tau) + \mathcal{R}, \end{aligned} \tag{4.1}$$

in terms of the residuals of the first-order derivatives of the discrete Lagrangian (including the stabilization terms), where $i_h\lambda$, $i_h y$, and $i_h\tau$ are arbitrary approximations and the remainder $\mathcal{R}$ is of third order. Below we will state appropriate spaces for state and adjoint such that the required differentiability properties hold.

To arrive at a computable error estimator, we drop the residual $\mathcal{R}$. As the exact solution is unknown, we have to approximate the interpolation errors using linear operators: $\Pi_h y_h \approx (y - i_h y)$, $\Pi_h\lambda_h \approx (\lambda - i_h\lambda)$ and $\Pi_h\tau_h \approx (\tau - i_h\tau)$. Thus, we arrive at the computable error estimator

$$\begin{aligned} J(y,\tau) - J^h(y_h,\tau_h) \approx {} & \frac{1}{2}\rho^\lambda(y_h,\tau_h,\lambda_h)(\Pi_h y_h) \\ & + \frac{1}{2}\rho^y(y_h,\tau_h,\lambda_h)(\Pi_h\lambda_h) + \frac{1}{2}\rho^\tau(y_h,\tau_h,\lambda_h)(\Pi_h\tau_h). \end{aligned}$$

The interpolation errors are approximated by interpolating the discrete solution into higher-order finite element spaces. In our computations, we interpolate the computed state and adjoint into the space of functions that are continuous, piecewise linear in time and discontinuous, piecewise quadratic on patches of elements in space.

4.2. Application to the Navier-Stokes equations

To apply the goal-oriented error estimator derived in the previous section to the instationary Navier-Stokes equations, we have to verify the assumptions made in section 4.1. We consider first the case of an inf-sup stable pair of finite elements in space for the state and adjoint. Then no stabilization is needed and we drop the terms SD_δ in the scheme. Under the assumptions of Lemma 2.1 the additional jump terms in J^h and E^h in comparison to J and E vanish. If the objective functional has a structure such that $\tilde{\boldsymbol{\lambda}}(T) \in H(\Omega)$ and that the right-hand side of the adjoint equation is in $L^2(I; H^{-1}(\Omega)^2)$ then the adjoint state satisfies $\tilde{\boldsymbol{\lambda}} \in W(I; V(\Omega))$, see for example [17]. Then we can show that the assumptions of Section 4.1 hold and that we obtain the error representation (4.1) with a remainder term of the form

$$\begin{aligned}\mathcal{R} = O\Big(\big(&\|\boldsymbol{v} - i_h\boldsymbol{v}\|_{L^2(I;H_0^1(\Omega_{\text{ref}})^2)} + \|\boldsymbol{v}_t - (I^h(i_h\boldsymbol{v}))_t\|_{L^1(I;L^2(\Omega_{\text{ref}})^2)} \\ &+ \|\boldsymbol{\lambda} - i_h\boldsymbol{\lambda}\|_{L^\infty(I;L^2(\Omega_{\text{ref}})^2)} + \|\boldsymbol{\lambda} - i_h\boldsymbol{\lambda}\|_{L^2(I;H_0^1(\Omega_{\text{ref}})^2)} \\ &+ \|p - i_h p\|_{L^2(I;L_0^2(\Omega_{\text{ref}}))} + \|\mu - i_h\mu\|_{L^2(I;L_0^2(\Omega_{\text{ref}}))} \\ &+ \|\tau - i_h\tau\|_{W^{1,\infty}(\Omega_{\text{ref}})^2}\big)^3\Big),\end{aligned} \tag{4.2}$$

where $I^h(i_h\boldsymbol{v})$ denotes the continuous piecewise linear interpolation in time of $i_h\boldsymbol{v}$. Here, we assume that J is three times continuously differentiable in the topology given by the norms in (4.2) (otherwise, the necessary norms would also appear). Other choices of the norms are possible. Furthermore, for a continuous, piecewise polynomial approximation of the transformations, $J'(y,\tau)(\bar{y}_h, \bar{\tau}_h) + \langle \lambda, E'(y,\tau)(\bar{y}_h, \bar{\tau}_h, \bar{\lambda}_h)\rangle = 0$ for all $(\bar{y}_h, \bar{\tau}_h, \bar{\lambda}_h) \in Y_h \times T_h \times Z_h^*$ is satisfied under our regularity assumptions on $\tilde{\boldsymbol{v}}, \tilde{p}, \tilde{\boldsymbol{\lambda}}$. For details we refer to [7].

In the presence of a streamline-diffusion type stabilization as considered in section 3, the continuous optimal solution has to be more regular in order to guarantee differentiability of the stabilization terms, for example $\tilde{\boldsymbol{v}}, \tilde{\boldsymbol{\lambda}} \in L^2(I; H^2(\Omega)^2)$ and $\tilde{p}, \tilde{\mu} \in L^2(I; H^1(\Omega))$. This is in particular the case if $\partial\Omega$ is of class C^2, the initial/end data are in $V(\Omega)$, and the source terms of state and adjoint equation are in $L^2(I; H(\Omega))$, see [30, Ch. III, Thm. 3.10] and [17]. The remainder term (4.2) contains then in particular the additional norms $\|\boldsymbol{v} - i_h\boldsymbol{v}\|_{L^\infty(I;H^1(\Omega_{\text{ref}})^2)}$, $\|\nabla\boldsymbol{v} - \nabla i_h\boldsymbol{v}\|_{L^2(I;L^4(\Omega_{\text{ref}})^2)}$, $\|\boldsymbol{\lambda} - i_h\boldsymbol{\lambda}\|_{L^\infty(I;H^1(\Omega_{\text{ref}})^2)}$, $\|\nabla\boldsymbol{\lambda} - \nabla i_h\boldsymbol{\lambda}\|_{L^2(I;L^4(\Omega_{\text{ref}})^2)}$, $\|p - i_h p\|_{L^2(I;H^1(\Omega_{\text{ref}}))}$, $\|\mu - i_h\mu\|_{L^2(I;H^1(\Omega_{\text{ref}}))}$. The proofs are lengthy and will be given in [7].

For example, the contribution of the momentum equation to the error estimator is given by

$$\begin{aligned}&\sum_{j=1}^{N}\int_{\Omega_{\text{ref}}} ([\boldsymbol{v}_h]^{j-1})^T\ \Pi_h\boldsymbol{\lambda}_h^{j-1,+}\ \det\tau_h'\ dx \\ &+ \int_I\int_{\Omega_{\text{ref}}} \sum_{i=1}^{d} \nu\nabla(v_{h,i})^T\tau_h'^{-1}\tau_h'^{-T}\ \nabla(\Pi_h\lambda_{h,i})\ \det\tau_h'\ dx\ dt \\ &+ \int_I\int_{\Omega_{\text{ref}}} \boldsymbol{v}_h^T\tau_h'^{-T}\nabla\boldsymbol{v}_h\ \Pi_h\boldsymbol{\lambda}_h \det\tau_h'\ dx\ dt \\ &- \int_I\int_{\Omega_{\text{ref}}} p_h\ \mathrm{tr}(\tau_h'^{-T}\nabla(\Pi_h\boldsymbol{\lambda}_h))\det\tau_h'\ dx\ dt \\ &- \int_I\int_{\Omega_{\text{ref}}} \tilde{\boldsymbol{f}}(\tau_h(x),t)^T\ \Pi_h\boldsymbol{\lambda}_h\ \det\tau_h'\ dx\ dt,\end{aligned}$$

while the other terms are derived in the same fashion. Here, $[\boldsymbol{v}_h]^{j-1}$ denotes the jump of $\boldsymbol{v}_h$ at timestep t_{j-1} and $\boldsymbol{\lambda}_h^{j-1,+}$ denotes the limit of $\boldsymbol{\lambda}_h$ for $t \searrow t_{j-1}$.

To drive the refinement process, we compute the local contributions to the global error on all pairs of spatial elements and time intervals. The local errors are computed by transforming the residual equations to the discrete physical domain and integrating cellwise by parts. For the momentum equation, the local error on element K and time interval I_j is thus approximated by

$$\left(R^{\tilde{\boldsymbol{v}}}(\tilde{y}_h), \widetilde{\Pi_h \boldsymbol{\lambda}_h}\right)_{I_j,K} + \left(r^{\tilde{\boldsymbol{v}}}(\tilde{y}_h), \widetilde{\Pi_h \boldsymbol{\lambda}_h}\right)_{I_j,\partial K} + \left([\tilde{\boldsymbol{v}}_h]^{j-1}, \widetilde{\Pi_h \boldsymbol{\lambda}_h}^{j-1,+}\right)_K,$$

with

$$R^{\tilde{\boldsymbol{v}}}(\tilde{y}_h)_{|K} := -\nu \Delta \tilde{\boldsymbol{v}}_h + \tilde{\boldsymbol{v}}_h^T \nabla \tilde{\boldsymbol{v}}_h + \nabla \tilde{p}_h - \tilde{\boldsymbol{f}}$$

$$r^{\tilde{\boldsymbol{v}}}(\tilde{y}_h)_{|e} := \begin{cases} \frac{1}{2}[\nu \partial_n \tilde{\boldsymbol{v}}_h - \tilde{p}_h \tilde{n}], & e \not\subset \partial\Omega \\ 0, & e \subset \partial\Omega \end{cases},$$

where e denotes the edges of K.

Remark 4.1. Note that, in this section, we have neglected the presence of constraints on the admissible transformations τ. For elliptic optimization problems, this has been investigated in [31].

5. Numerical results

We now demonstrate our results on numerical model problems. In the previous sections we considered homogeneous Dirichlet boundary conditions for the Navier-Stokes equations. In our numerical tests we will discuss problems with inflow, free outflow and noslip boundaries where we always impose a noslip condition on the boundary Γ_B of the design object. However, we can also derive a formula for the reduced gradient in this setting as well as in the presence of the stabilization terms introduced by the $cG(1)dG(0)$ discretization.

When dealing with inflow and outflow boundaries, i.e., parts of the boundary where conditions of the form $\boldsymbol{v} = \boldsymbol{g} \in L^2(I; H^{1/2})$ and $\nu \partial_n \boldsymbol{v} - pn = 0$ hold, resp., we have to pay attention to some changes. For example in the presence of a free outflow boundary the correct space for the pressure is $L^2(\Omega)$ and not $L_0^2(\Omega)$. Since the admissible transformations equal the identity in a neighborhood of the inflow and outflow boundaries, there is no contribution to the shape derivative from these parts of the boundary. Moreover, for the adjoint equation, we obtain the boundary conditions $\boldsymbol{\lambda} = 0$ on the inflow boundary and, given sufficient regularity, $\nu \partial_n \boldsymbol{\lambda} + \boldsymbol{v}^T n\, \boldsymbol{\lambda} + \mu n = 0$ on the outflow boundary. Because a complete discussion of this setting is too long we will not go into further details. See also [5] and [29] for optimal control problems with the Navier-Stokes equations with inflow and outflow boundary conditions.

5.1. Shape Optimization with goal-oriented adaptivity

The first model problem is based on the DFG benchmark of a 2D instationary flow around a cylinder [27]. We prescribe a steady parabolic inflow profile on the left boundary with $v_{\max} = 1.5\,\mathrm{m/s}$, noslip boundary conditions on the top, bottom and object boundaries, and a free outflow condition on the right. The flow is modeled by the instationary incompressible Navier-Stokes equations with viscosity $\nu = 10^{-4}$, corresponding to a Reynolds number $\mathrm{Re} \approx 400$ for the initial shape. Discretization is done using the $cG(1)dG(0)$ finite element method in Section 3. On the initial level of the adaptive refinement process, we start with a very coarse triangular spatial mesh with 2436 vertices and an initial uniform time step size $k = 10^{-3}$. For the adaptive mesh refinement we use a fixed-fraction strategy, refining 15% of the cells with the largest local errors with a red-green refinement algorithm; in addition, we refine 10% of the time intervals with the largest local error using bisection.

The object boundary Γ_B is parameterized using a cubic B-Spline curve [25] with 7 control points for the upper half, which is reflected at the $y = 0.2$-axis to obtain a y-symmetric closed curve. This parameterization allows for apices at the front and rear of the object, while the remaining boundary is C^2. For the calculation of the reduced gradient we use a linear elasticity equation to extend the boundary displacement to the domain as described in section 2.6. We impose constraints on the volume of the object B as well as bound constraints on the control points. The volume of B can be evaluated analytically as a function of the B-spline control points by expressing the volume as a boundary integral.

We minimize the mean value of the drag on the object boundary Γ_B over the time interval $[0, T]$, given by the formula

$$J((\tilde{\boldsymbol{v}}, \tilde{p}), \Omega) = \frac{1}{T} \int_0^T \int_\Omega \Big((\tilde{\boldsymbol{v}}_t + (\tilde{\boldsymbol{v}} \cdot \nabla)\tilde{\boldsymbol{v}} - \tilde{\boldsymbol{f}})^T \boldsymbol{\Phi} - \tilde{p} \operatorname{div} \boldsymbol{\Phi} + \nu \nabla \tilde{\boldsymbol{v}} : \nabla \boldsymbol{\Phi} \Big)\, dx\, dt.$$

Here, $\boldsymbol{\Phi}$ is a smooth function such that with a unit vector $\boldsymbol{\phi}$ pointing in the mean flow direction holds $\quad \boldsymbol{\Phi}|_{\Gamma_B} \equiv \boldsymbol{\phi}, \quad \boldsymbol{\Phi}|_{\partial\Omega \setminus \Gamma_B} \equiv 0 \quad \forall \Omega \in \mathcal{O}_{\mathrm{ad}}$.

This formula is an alternative formula for the mean value of the drag on Γ_B,

$$c_d := \frac{1}{T} \int_0^T \int_{\Gamma_B} \boldsymbol{n} \cdot \sigma(\tilde{\boldsymbol{v}}, \tilde{p}) \cdot \boldsymbol{\phi}\, dS,$$

with normal vector n and stress tensor $\sigma(\tilde{\boldsymbol{v}}, \tilde{p}) = \nu(\nabla \tilde{\boldsymbol{v}} + (\nabla \tilde{\boldsymbol{v}})^T) - \tilde{p}\, I$, and can be obtained through integration by parts. For a detailed derivation, see [18].

Computation of the state, adjoint and shape derivative equations is done using Dolfin [21]. The optimization is carried out using a SQP solver written in Matlab, with a BFGS-approximation for the reduced Hessian.

Table 1 summarizes the computed results for the adaptive shape optimization problem. The columns contain from left the refinement level, the number of vertices, the number of timesteps N, the optimal objective function value J, the value of the (signed) global error η and the number of SQP-iterations.

level	vertices	N	J	η	SQPit
0	2436	2001	0.0330721648	−0.00542514	12
1	4283	2200	0.0314016876	−0.00273496	5
2	7875	2419	0.0325643722	0.000807385	4
3	13666	2658	0.0336900832	0.00145268	2
4	24396	2923	0.0339111094	0.00132779	2

TABLE 1. Optimization Results

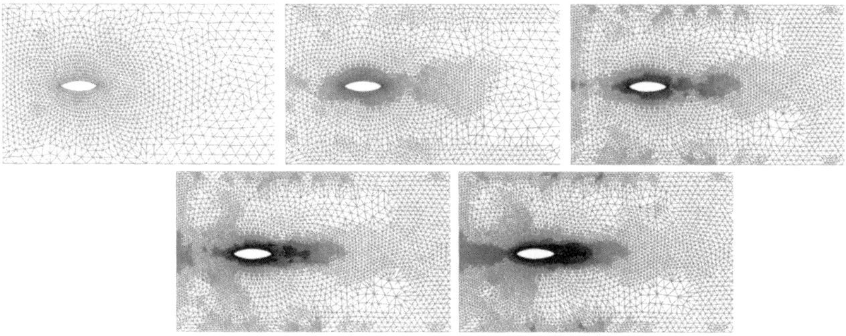

FIGURE 1. Comparison of the meshes for the optimal shape on different refinement levels

On the higher refinement levels, the differences between the objective function values decrease. Most importantly, the number of SQP iterations reduces on the higher levels. Thus, most of the optimization iterations are performed on the cheap coarser grids, significantly reducing the amount of computational work. Furthermore, we obtain meshes that are well adapted to the evaluation of the drag objective function value. Figure 1 shows the computational meshes for the optimal shape on the refinement levels. The adaptive refinement primarily takes place in the vicinity of the object and in its wake. The instationary, time-periodic behavior of the flow starts to be resolved on level 2 and is only fully resolved on levels 3 and 4, which explains the increase in the error estimator from level 2 to 3.

5.2. Shape optimization with two objects

The shape derivative calculus that was introduced in Section 2 can readily be applied to shape optimization problems with more than one object.

Figure 2 shows the optimal solution for a shape optimization problem with two objects in a row, where the drag was minimized for a flow governed by the stationary Navier-Stokes equations with $\nu = 10^{-2}$. The design parameters are the grid points on the boundary curve with 80 boundary points per object and separate volume constraints are imposed on both objects. The front point of the first object is fixed while the front point of the second object can move in x-direction within

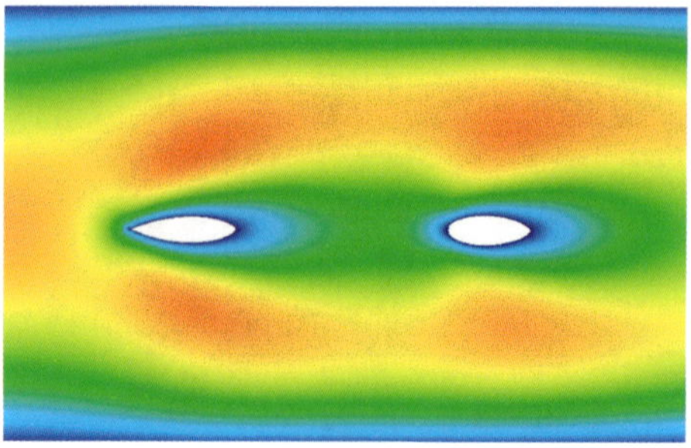

FIGURE 2. Flow around 2 objects

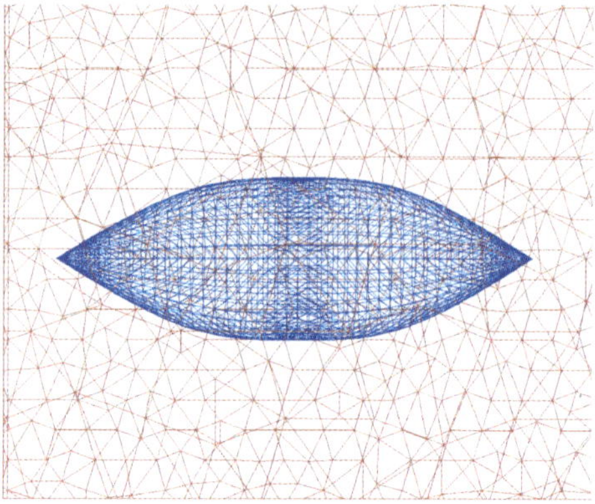

FIGURE 3. 3D object

box constraints. In the optimal solution the second object took the maximum distance to the first object. The optimization was carried out using Ipopt [32], while the computation of the state, adjoint and shape derivative equations was done with Sundance [22].

5.3. Shape optimization in three dimensions

Currently we are extending our approach to three-dimensional problems. While transferring the equations is straightforward, solving the arising systems of equations becomes numerically challenging due to the huge amount of degrees of freedom. Therefore, techniques such as multigrid methods and problem specific preconditioners, in conjunction with adaptive mesh refinement, have to be used.

Solving the state equation includes the successive solve of linear systems of the form $\left(\begin{smallmatrix} F & \tilde{B}^T \\ B & C \end{smallmatrix}\right)\left(\begin{smallmatrix} s_v \\ s_p \end{smallmatrix}\right) = b$. We analyzed different solvers and preconditioners in this context and finally use GMRESR with the SIMPLEC preconditioner [10]. The preconditioner needs to solve two subsystems for F^{-1} and for the approximation of the Schur complement. Numerical tests showed that we need to solve these subsystems only up to precision 10^{-3} which is done by BiCGStab with an algebraic multigrid preconditioner. The implementation is done in Trilinos [16] and uses parallelization on 16 processors.

Figure 3 shows the optimal shape for a 3D object which is parameterized with cubic B-spline surface patches. The mean value of the drag was minimized for a flow governed by the instationary Navier-Stokes equations with $\nu = 10^{-4}$ under a volume constraint.

6. Conclusions

We have presented a continuous approach to shape optimization which is based on the instationary Navier-Stokes equations. In this setting the appropriate function spaces for the transformations and states were characterized and a differentiability result for the design-to-state operator was presented. The approach for calculating the shape gradients allows the solution of the state and adjoint equation on the physical domain, hence existing solvers can be used. Furthermore, the approach is flexible enough to conveniently use arbitrary types of shape parameterizations, for example free form deformation or parameterized boundary displacements, also with multiple objects.

The combination with error estimators and multilevel techniques can reduce the number of optimization iterations on the fine grids and the necessary degrees of freedom significantly. Currently, the developed techniques are extended to 3-dimensional shape optimization problems.

References

[1] I. Babuška and A. Miller, *The post-processing approach in the finite element method. Part 1: Calculation of displacements, stresses and other higher derivatives of the displacements*, Int. J. Numer. Methods Eng., 20, pp. 1085–1109, 1984.

[2] R. Becker and H. Kapp and R. Rannacher, *Adaptive finite element methods for optimal control of partial differential equations: Basic concepts*, SIAM J. Control Optim., 39(1), pp. 113–132, 2000.

[3] R. Becker and R. Rannacher, *An optimal control approach to a posteriori error estimation in finite element methods*, Acta Numerica 10, pp. 1–102, 2001.

[4] J.A. Bello, E. Fernandez-Cara, J. Lemoine, and J. Simon, *The differentiability of the drag with respect to the variations of a Lipschitz domain in Navier-Stokes flow*, SIAM J. Control Optim., 35(2), pp. 626–640, 1997.

[5] M. Berggren, *Numerical solution of a flow-control problem: vorticity reduction by dynamic boundary action*, SIAM J. Sci. Comput. 19, no. 3, pp. 829–860, 1998.

[6] C. Brandenburg, F. Lindemann, M. Ulbrich, and S. Ulbrich, *A continuous adjoint approach to shape optimization for Navier Stokes flow*, in *Optimal control of coupled systems of partial differential equations*, K. Kunisch, G. Leugering, and J. Sprekels, Eds., Int. Ser. Numer. Math. 158, Birkhäuser, Basel, pp. 35–56, 2009.

[7] C. Brandenburg and S. Ulbrich, *Goal-oriented error estimation for shape optimization with the instationary Navier-Stokes equations*, Preprint in preparation, Fachbereich Mathematik, TU Darmstadt, 2010.

[8] K.K. Choi and N.-H. Kim, *Structural sensitivity analysis and optimization 2: Nonlinear systems and applications*, Mechanical Engineering Series, Springer, 2005.

[9] M.C. Delfour and J.-P. Zolésio, *Shapes and geometries: Analysis, differential calculus, and optimization*, SIAM series on Advances in Design and Control, 2001.

[10] J. Doormaal and G.D. Raithby, *Enhancements of the SIMPLE method for predicting incompressible fluid flows*, Num. Heat Transfer, 7, pp. 147–163, 1984.

[11] K. Eriksson, D. Estep, P. Hansbo, and C. Johnson, *Introduction to adaptive methods for differential equations*, Acta Numerica, pp. 105–158, 1995.

[12] K. Eriksson, D. Estep, P. Hansbo, and C. Johnson, *Computational differential equations*, Cambridge University Press, 1996.

[13] M. Geissert, H. Heck, and M. Hieber, *On the equation* $\operatorname{div} u = g$ *and Bogovski's operator in Sobolev spaces of negative order*, in *Partial differential equations and functional analysis*, E. Koelink, J. van Neerven, B. de Pagter, and G. Sweers, eds., Oper. Theory Adv. Appl., 168, Birkhäuser, Basel, pp. 113–121, 2006.

[14] P. Guillaume and M. Masmoudi, *Computation of high-order derivatives in optimal shape design*, Numer. Math., 67, pp. 231–250, 1994.

[15] J. Haslinger and R.A.E. Mäkinen, *Introduction to shape optimization: Theory, approximation, and computation*, SIAM series on Advances in Design and Control, 2003.

[16] M.A. Heroux, J.M. Willenbring, and R. Heaphy, *Trilinos developers guide*, Sandia National Laboratories, SAND2003-1898, 2003.

[17] M. Hinze, *Optimal and instantaneous control of the instationary Navier-Stokes equations*, Habilitation, TU Dresden, 2002.

[18] J. Hoffman and C. Johnson, *Adaptive Finite Element Methods for Incompressible Fluid Flow*, Error estimation and solution adaptive discretization in CFD: Lecture Notes in Computational Science and Engineering, Springer Verlag, 2002.

[19] J. Hoffman and C. Johnson, *A new approach to computational turbulence modeling*, Comput. Meth. Appl. Mech. Engrg., 195, pp. 2865–2880, 2006.

[20] F. Lindemann, M. Ulbrich, and S. Ulbrich, *Fréchet differentiability of time-dependent incompressible Navier-Stokes flow with respect to domain variations*, Preprint in preparation, Fakultät für Mathematik, TU München, 2010.

[21] A. Logg and G.N. Wells, *DOLFIN: Automated finite element computing*, ACM Transactions on Mathematical Software, 37(2), 2010.

[22] K. Long, *Sundance: A rapid prototyping tool for parallel PDE-constrained optimization*, in *Large-scale pde-constrained optimization*, L.T. Biegler, M. Heinkenschloss, O. Ghattas, and B. van Bloemen Wanders, eds., Lecture Notes in Computational Science and Engineering, 30, Springer, pp. 331–341, 2003.

[23] D. Meidner and B. Vexler, *Adaptive space-time finite element methods for parabolic optimization problems*, SIAM J. Control Optim., 46, pp. 116–142, 2007.

[24] B. Mohammadi and O. Pironneau, *Applied shape optimization for fluids*, Oxford University Press, 2001.

[25] M.E. Mortensen, *Geometric modeling*, Wiley, 1985.

[26] F. Murat and S. Simon, *Etudes de problems d'optimal design*, Lectures Notes in Computer Science, 41, pp. 54–62, 1976.

[27] M. Schäfer and S. Turek, *Benchmark computations of laminar flow around a cylinder*, Preprints SFB 359, No. 96-03, Universität Heidelberg, 1996.

[28] J. Sokolowski and J.-P. Zolésio, *Introduction to shape optimization*, Series in Computational Mathematics, Springer, 1992.

[29] T. Slawig, *PDE-constrained control using Femlab – Control of the Navier-Stokes equations*, Numerical Algorithms, 42, pp. 107–126, 2006.

[30] R. Temam, *Navier-Stokes equations: Theory and numerical analysis* 3rd Edition, Elsevier Science Publishers, 1984.

[31] B. Vexler and W. Wollner, *Adaptive finite elements for elliptic optimization problems with control constraints*, SIAM J. Control Optim., 47, pp. 509–534, 2008.

[32] A. Wächter and L.T. Biegler, *On the implementation of a primal-dual interior point filter line search algorithm for large-scale nonlinear programming*, Math. Program., 106, pp. 25–57, 2006.

Christian Brandenburg and Stefan Ulbrich
Fachbereich Mathematik
Technische Universität Darmstadt
Dolivostr. 15
D-64293 Darmstadt, Germany
e-mail: brandenburg@mathematik.tu-darmstadt.de
ulbrich@mathematik.tu-darmstadt.de

Florian Lindemann and Michael Ulbrich
Lehrstuhl für Mathematische Optimierung
Zentrum Mathematik, M1
TU München
Boltzmannstr. 3
D-85747 Garching bei München, Germany
e-mail: lindemann@ma.tum.de
mulbrich@ma.tum.de

International Series of Numerical Mathematics, Vol. 160, 277–288

Shape Optimization for Free Boundary Problems – Analysis and Numerics

Karsten Eppler and Helmut Harbrecht

Abstract. In this paper the solution of a Bernoulli type free boundary problem by means of shape optimization is considered. Four different formulations are compared from an analytical and numerical point of view. By analyzing the shape Hessian in case of matching data it is distinguished between well-posed and ill-posed formulations. A nonlinear Ritz-Galerkin method is applied for the discretization of the shape optimization problem. In case of well-posedness existence and convergence of the approximate shapes is proven. In combination with a fast boundary element method efficient first and second-order shape optimization algorithms are obtained.

Mathematics Subject Classification (2000). 49Q10, 49K20, 35R35, 65N38.

Keywords. Shape optimization, free boundary problems, sufficient optimality conditions, boundary element method.

1. Problem formulation

The present paper is dedicated to the solution of a generalized *Bernoulli exterior free boundary problem* which serves as a prototype of many shape optimization problems. Let $T \subset \mathbb{R}^n$ denote a bounded domain with *free boundary* $\partial T = \Gamma$. Inside the domain T we assume the existence of a simply connected subdomain $S \subset T$ with *fixed* boundary $\partial S = \Sigma$. The resulting annular domain $T \setminus \overline{S}$ is denoted by Ω, see Figure 1 for an illustration.

The exterior free boundary problem under consideration might be formulated as follows: For given data f, g, h, seek the domain Ω and the associated function u such that the overdetermined boundary value problem

$$-\Delta u = f \text{ in } \Omega, \quad -\frac{\partial u}{\partial \mathbf{n}} = h, \; u = 0 \text{ on } \Gamma, \quad u = g \text{ on } \Sigma, \tag{1}$$

is satisfied. Here, $g, h > 0$ and $f \geq 0$ are sufficiently smooth *functions* on $\mathbb{R}^n$ such that u provides enough regularity for a second-order shape calculus. We like to

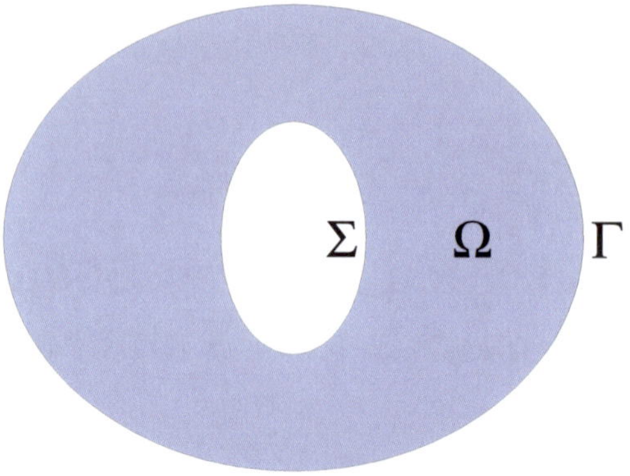

FIGURE 1. The domain Ω and its boundaries Γ and Σ.

stress that the non-negativity of the Dirichlet data implies that u is positive on Ω and thus it holds in fact $\partial u/\partial \mathbf{n} < 0$.

Shape optimization is a well-established tool to solve free boundary value problems like (1), see, e.g., [9, 27, 28, 31, 38, 39] and the references therein. The authors themselves provided analytical results and numerical algorithms in several papers, see in particular [14, 16, 17, 18].

The problem under consideration can be viewed as the prototype of a free boundary problem arising in many applications. The growth of anodes in electrochemical processes might be modeled like above with $f \equiv 0$, $g \equiv 1$, $h =$ const and corresponds to the original Bernoulli free boundary problem. In the exterior magnetic shaping of liquid metals the state equation is an exterior Poisson equation where the uniqueness is ensured by a volume constraint [7, 15, 34, 35]. The maximization of the torsional stiffness of an elastic cylindrical bar under simultaneous constraints on its volume and bending rigidity fits also in the above general setup, see [3, 12] for the details. The detection of voids or inclusions in electrical impedance tomography is slightly different since the roles of Σ and Γ are interchanged [13, 36] which amounts to a severely ill-posed problem, see [4, 5, 6, 30] and the references therein.

We do not consider the interesting question of existence of optimal solutions in this paper. Instead, we will tacitly assume the existence of optimal domains, being sufficiently regular to allow for a second-order-shape calculus. For the existence of solutions to the free boundary problem (1) we refer the reader to, e.g., [2], see also [25] for the related *interior* free boundary problem. Results concerning the geometric form of the solutions can be found in [1] and the references therein.

The outline is as follows: In Section 2 we introduce the shape functionals under consideration. Necessary and sufficient conditions are derived in Sections 3 and 4. Sections 5 and 6 are dedicated to the discretization of the shape. The solution of the state equation is considered in Section 7. Numerical experiments are carried out in Section 8.

Throughout the paper, in order to avoid the repeated use of generic but unspecified constants, by $C \lesssim D$ we mean that C can be bounded by a multiple of D, independently of parameters which C and D may depend on. Obviously, $C \gtrsim D$ is defined as $D \lesssim C$, and $C \sim D$ as $C \lesssim D$ and $C \gtrsim D$.

2. Free boundary problems as shape optimization problems

To numerically solve (1) by means of shape optimization we shall introduce two different state functions, namely

$$\begin{aligned} -\Delta v = f \text{ in } \Omega, \qquad & v = g \text{ on } \Sigma, \qquad && v = 0 \text{ on } \Gamma, \\ -\Delta w = f \text{ in } \Omega, \qquad & -\frac{\partial w}{\partial \mathbf{n}} = h \text{ on } \Sigma, \qquad && w = 0 \text{ on } \Gamma. \end{aligned} \tag{2}$$

Here, the state v solves the pure Dirichlet problem whereas the state w solves a mixed boundary value problem.

We will consider the following four formulations, where the infimum has always to be taken over all sufficiently smooth domains which include the domain S.

(i) An energy variational formulation is derived by using the Dirichlet energy. The solution $\big(\Omega, u(\Omega)\big)$ of (1) is the minimizer (cf. [14]) of the Dirichlet energy functional

$$J_1(\Omega) = \int_\Omega \{\|\nabla v\|^2 - 2fv + h^2\} d\mathbf{x} \to \inf. \tag{3}$$

(ii) A variational least-squares cost function, firstly proposed by Kohn and Vogelius [32] in the context of the inverse conductivity problem, is considered as second formulation:

$$J_2(\Omega) = \int_\Omega \|\nabla(v - w)\|^2 d\mathbf{x} = -\int_\Gamma v \left(h + \frac{\partial w}{\partial \mathbf{n}}\right) d\sigma \to \inf. \tag{4}$$

This functional seems to be very attractive since, due to

$$J_2(\Omega) \sim \|w\|_{H^{1/2}(\Gamma)} \left\| h + \frac{\partial v}{\partial \mathbf{n}} \right\|_{H^{-1/2}(\Gamma)},$$

the Dirichlet and Neumann data are both tracked in their natural trace spaces.

(iii) One can also consider the solution v of the pure Dirchlet problem and track the Neumann data in a least-squares sense relative to $L^2(\Gamma)$, that is

$$J_3(\Omega) = \frac{1}{2} \int_\Gamma \left(h + \frac{\partial v}{\partial \mathbf{n}}\right)^2 d\sigma \to \inf. \tag{5}$$

(iv) Correspondingly, if the Neumann datum h is assumed to be prescribed, the L^2-least square tracking of the Dirichlet boundary condition at Γ reads as

$$J_4(\Omega) = \frac{1}{2} \int_\Gamma w^2 d\sigma \to \inf. \tag{6}$$

3. Necessary conditions

The Hadamard representations of the shape gradients of the shape functionals (3)–(6) have been computed in the papers [14, 16, 17, 18], see also [29, 31]. For a general overview on shape calculus, mainly based on the perturbation of identity (Murat and Simon) or the speed method (Sokolowski and Zolesio), we refer the reader for example to [9, 33, 37, 38] and the references therein.

Existence of the shape gradients of J_1, J_2 and J_4 is ensured if $\Gamma \in C^2$, whereas the fixed boundary Σ needs to be only Lipschitz continuous. In case of J_3 the regularity of the free boundary has to be increased to $\Gamma \in C^3$. Correspondingly, a boundary variation $\mathbf{V}: \Gamma \to \mathbb{R}^n$ needs to be in $C^2(\Gamma)$ or $C^3(\Gamma)$. We remark that the higher regularity for J_3 ensures a sufficiently smooth adjoint state (see (10) below).

The shape gradients of J_1 and J_2 are respectively given by

$$dJ_1(\Omega)[\mathbf{V}] = \int_\Gamma \langle \mathbf{V}, \mathbf{n}\rangle \left\{ h^2 - \left[\frac{\partial v}{\partial \mathbf{n}}\right]^2 \right\} d\sigma, \tag{7}$$

$$dJ_2(\Omega)[\mathbf{V}] = \int_\Gamma \langle \mathbf{V}, \mathbf{n}\rangle \left\{ h^2 - \left[\frac{\partial v}{\partial \mathbf{n}}\right]^2 - \|\nabla_\Gamma w\|^2 + 2w\left[f - \mathcal{H}h - \frac{\partial h}{\partial \mathbf{n}}\right] \right\} d\sigma \tag{8}$$

(see [14, 18] for the derivation). Here, $\mathcal{H}$ denotes the mean curvature (for convenience scaled by the factor $n-1$) and ∇_Γ the surface gradient. Both problems are self-adjoint and thus no adjoint state needs to be defined.

According to [16, 31], the shape gradient of the functional J_3 reads as

$$dJ_3(\Omega)[\mathbf{V}] = \int_\Gamma \langle \mathbf{V}, \mathbf{n}\rangle \left\{ p\left[\frac{\partial h}{\partial \mathbf{n}} - \mathcal{H}\frac{\partial v}{\partial \mathbf{n}} - f\right] + \mathcal{H}\frac{p^2}{2} - \frac{\partial v}{\partial \mathbf{n}}\frac{\partial p}{\partial \mathbf{n}} \right\} d\sigma, \tag{9}$$

where p denotes the adjoint state function defined by

$$\Delta p = 0 \text{ in } \Omega, \quad p = 0 \text{ on } \Sigma, \quad p = \frac{\partial v}{\partial \mathbf{n}} + h \text{ on } \Gamma. \tag{10}$$

In case of the shape functional J_4 (cf. [17, 29]) we obtain

$$\begin{aligned} dJ_4(\Omega)[\mathbf{V}] = \int_\Gamma \langle \mathbf{V}, \mathbf{n}\rangle \bigg\{ &\mathcal{H}\left(\frac{w^2}{2} - qh\right) \\ &+ q\left(f - \frac{\partial h}{\partial \mathbf{n}}\right) - \langle \nabla_\Gamma q, \nabla_\Gamma w\rangle - wh \bigg\} d\sigma \end{aligned} \tag{11}$$

with q fulfilling the adjoint state equation

$$\Delta q = 0 \text{ in } \Omega, \quad \frac{\partial q}{\partial \mathbf{n}} = w \text{ on } \Gamma, \quad q = 0 \text{ on } \Sigma. \tag{12}$$

The following theorem is an obvious consequence of the explicit representation formulae of the shape gradients:

Theorem 1 ([14, 16, 17, 18]). *Let the domain $\Omega^\star$ be such that the overdetermined boundary value problem* (1) *is satisfied. Then, the domain $\Omega^\star$ fulfills for $i = 1, 2, 3, 4$ the necessary optimality condition*

$$dJ_i(\Omega^\star)[\mathbf{V}] = 0 \text{ for all sufficiently smooth } \mathbf{V}. \tag{13}$$

4. Sufficient optimality conditions

Based on a shape calculus via boundary variations, developed in [10, 11], we computed the boundary integral representations of the shape gradients and Hessians of the four formulations in [14, 16, 17, 18]. With the shape Hessian at hand we are able to investigate the stability of the global minimizer $\Omega^\star$.

We shall introduce a fixed smooth reference manifold $\widehat{\Gamma}$ to compute explicit expressions of the second derivatives. A domain is called admissible if it can be represented by a function $r \in X := C^{k,\alpha}(\widehat{\Gamma})$ according to

$$\Gamma = \{\mathbf{x} \in \mathbb{R}^n : \mathbf{x} = \widehat{\mathbf{x}} + r(\widehat{\mathbf{x}})\widehat{\mathbf{n}}(\widehat{\mathbf{x}}),\ \widehat{\mathbf{x}} \in \widehat{\Gamma}\}, \tag{14}$$

where $\widehat{\mathbf{n}}$ denotes the normal relative to the reference manifold. The smoothness requirements are $k = 2$ in case of the functionals J_1, J_2 and J_4 and $k = 3$ in case of the functional J_3. Additionally, the Hölder coefficient α is required to be positive. Via (14), the domain Ω is identified with a scalar-valued function r and this identification is one-to-one. In fact, solving the shape optimization problems (3)–(6) corresponds then just to the solution of a nonlinear pseudodifferential equation for r.

We consider the variations $dr, dr_1, dr_2 \in X$. The perturbed shape $\Gamma_\varepsilon[dr]$ is defined via a particular perturbation of identity, using the normal $\widehat{\mathbf{n}}$ with respect to the reference manifold, that is

$$\Gamma_\varepsilon[dr] = \big\{\mathbf{x} \in \mathbb{R}^n : \mathbf{x} = \widehat{\mathbf{x}} + \big(r(\widehat{\mathbf{x}}) + \varepsilon dr(\widehat{\mathbf{x}})\big)\widehat{\mathbf{n}}(\widehat{\mathbf{x}}),\ \widehat{\mathbf{x}} \in \widehat{\Gamma}\big\}.$$

Likewise, second-order variations are given as

$$\Gamma_{\varepsilon_1,\varepsilon_2}[dr_1, dr_2] = \big\{\mathbf{x} \in \mathbb{R}^n : \mathbf{x} = \widehat{\mathbf{x}} + \big(r(\widehat{\mathbf{x}}) + \varepsilon_1 dr_1(\widehat{\mathbf{x}}) + \varepsilon_2 dr_2(\widehat{\mathbf{x}})\big)\widehat{\mathbf{n}}(\widehat{\mathbf{x}}),\ \widehat{\mathbf{x}} \in \widehat{\Gamma}\big\}.$$

Then, all formulations own a shape Hessian which defines a continuous bilinear form $d^2J_i(\Omega) : H^s(\widehat{\Gamma}) \times H^s(\widehat{\Gamma}) \to \mathbb{R}$ with respect to the *energy space* $H^s(\Gamma)$, that is

$$\big|d^2J_i(\Omega)[dr_1, dr_2]\big| \le c_S \|dr_1\|_{H^s(\widehat{\Gamma})} \|dr_2\|_{H^s(\widehat{\Gamma})}.$$

Precisely, one has $s = 1/2$ for the first and $s = 1$ the other formulations, see [14, 16, 17, 18]. Accordingly, the second-order Taylor remainder $R_2\big(J_i(\Omega), dr\big)$ satisfies

$$\big|R_2\big(J_i(\Omega), dr\big)\big| = o\big(\|dr\|_X\big)\|dr\|^2_{H^s(\widehat{\Gamma})}$$

where $X \subsetneq H^s(\widehat{\Gamma})$ is the space of differentiation. Therefore, a local minimum is *stable* if the shape Hessian $d^2J_i(\Omega^\star)$ is strictly coercive in its energy space $H^s(\widehat{\Gamma})$

$$d^2J_i(\Omega^\star)[dr, dr] \ge c_E \|dr\|^2_{H^s(\widehat{\Gamma})}, \quad c_E > 0.$$

The shape problem under consideration is then *well posed* and a nonlinear Ritz-Galerkin method produces approximate shapes that converge quasi-optimal with respect to the energy norm, see the next section.

At the optimal domain, even though the shape gradients look quite different, the shape Hessians surprisingly consist of the same ingredients. Namely, in [14, 16, 17, 18], the following expressions have been proven for the shape Hessian at the optimal domain:

$$\begin{aligned}
d^2 J_1(\Omega^\star)[dr_1, dr_2] &= \big((\Lambda + \mathcal{A})\mathcal{M}dr_1, \mathcal{M}dr_2\big)_{L^2(\Gamma^\star)},\\
d^2 J_2(\Omega^\star)[dr_1, dr_2] &= \big((\Lambda + \mathcal{A})\mathcal{M}dr_1, \Lambda^{-1}(\Lambda + \mathcal{A})\mathcal{M}dr_2\big)_{L^2(\Gamma^\star)},\\
d^2 J_3(\Omega^\star)[dr_1, dr_2] &= \big((\Lambda + \mathcal{A})\mathcal{M}dr_1, (\Lambda + \mathcal{A})\mathcal{M}dr_2\big)_{L^2(\Gamma^\star)},\\
d^2 J_4(\Omega^\star)[dr_1, dr_2] &= \big(\Lambda^{-1}(\Lambda + \mathcal{A})\mathcal{M}dr_1, \Lambda^{-1}(\Lambda + \mathcal{A})\mathcal{M}dr_2\big)_{L^2(\Gamma^\star)},
\end{aligned}$$

where $\mathcal{M} : L^2(\Gamma^\star) \to L^2(\Gamma^\star)$ is a *bijective* multiplication operator, $\Lambda : H^{1/2}(\Gamma^\star) \to H^{-1/2}(\Gamma^\star)$ is the Dirichlet-to-Neumann map (associated with the pure Dirichlet problem in (2)) and

$$\mathcal{A} := \mathcal{H} + \left[\frac{\partial h}{\partial \mathbf{n}} - f\right] \Big/ g : L^2(\Gamma^\star) \to L^2(\Gamma^\star)$$

is a multiplication operator. Notice that $\Lambda^{-1} : H^{-1/2}(\Gamma^\star) \to H^{1/2}(\Gamma^\star)$ is the inverse of Λ, which has to be understood as the Neumann-to-Dirichlet map in the sense of the mixed boundary value problem in (2).

We shall employ that $H^m(\widehat{\Gamma})$ and $H^m(\Gamma^\star)$ are isomorphic for the whole range of interesting m. Then, since Λ is $H^{1/2}$-coercive, the condition $\mathcal{A} \geq 0$ is *sufficient* to ensure the positivity of the above shape Hessians.

Theorem 2 ([14, 16, 17, 18]). *Let $\Omega^\star$ be a stationary domain of the shape functional J_i. Then, the shape Hessian is $H^t(\Gamma^\star)$-coercive if $\mathcal{A} \geq 0$. Here, we have $t = s$ if $i = 1, 3$, $t = s - 1/2$ if $i = 2$, and $t = s - 1$ if $i = 4$.*

Consequently, in case of the formulations (i) and (iii) the positiveness is given with respect to the energy space $H^s(\widehat{\Gamma})$ which implies the well-posedness of these formulations. Whereas in case of the formulations (ii) and (iv) the positivity holds only in the weaker spaces $H^{1/2}(\widehat{\Gamma})$ and $L^2(\widehat{\Gamma})$, respectively, that is

$$d^2 J_2(\Omega^\star)[dr, dr] \geq c_E \|dr\|^2_{H^{1/2}(\Gamma^\star)}, \quad d^2 J_4(\Omega^\star)[dr, dr] \geq c_E \|dr\|^2_{L^2(\Gamma^\star)}, \quad c_E > 0.$$

This implies the algebraically *ill-posedness* of these formulations. In particular, tracking the Dirichlet data in the L^2-norm is not sufficient. We strongly assume that they have to be tracked relative to H^1. Our results are summarized in Table 1.

functional	energy space	positivity space	posedness
J_1	$H^{1/2}$	$H^{1/2}$	well-posed
J_2	H^1	$H^{1/2}$	algebraically ill-posed
J_3	H^1	H^1	well-posed
J_4	H^1	L^2	algebraically ill-posed

TABLE 1. The energy spaces and the positivity spaces of the shape Hessians.

5. Nonlinear Ritz-Galerkin approximation for the shape

In order to solve the optimization problems (3)–(6), we seek a stationary point $\Omega^\star$ which satisfies the necessary condition (13). In accordance with the previous section, we shall identify the domain $\Omega \subset \mathbb{R}^n$ with a parametrization of its boundary.

Let $\widehat{\Gamma} \subset \mathbb{R}^n$ be a smooth closed reference surface of the same topological type as Γ. Then, Γ can be parametrized by a smooth mapping

$$\boldsymbol{\gamma} = [\gamma_1, \ldots, \gamma_n]^T : \widehat{\Gamma} \to \Gamma, \tag{15}$$

which is one-to-one, preserves orientation, and $\boldsymbol{\gamma}'(\widehat{\mathbf{x}})$ is invertible for all $\widehat{\mathbf{x}} \in \widehat{\Gamma}$.

We focus first on parametrizations of the form

$$\boldsymbol{\gamma}(\widehat{\mathbf{x}}) = \widehat{\mathbf{x}} + r(\widehat{\mathbf{x}})\widehat{\mathbf{n}}(\widehat{\mathbf{x}}), \quad \widehat{\mathbf{x}} \in \widehat{\Gamma}, \tag{16}$$

with $r \in X$. Then, we choose suitable basis functions $\varphi_i \in X$ and consider the approximation space $V_N = \operatorname{span}\{\varphi_1, \varphi_2, \ldots, \varphi_N\} \subset X$. By making the ansatz $r_N = \sum_{i=1}^N a_i\varphi_i \in V_N$, $a_i \in \mathbb{R}$, we arrive at the finite-dimensional parametrization

$$\boldsymbol{\gamma}_N(\widehat{\mathbf{x}}) = \widehat{\mathbf{x}} + r_N(\widehat{\mathbf{x}})\widehat{\mathbf{n}}(\widehat{\mathbf{x}}), \quad \widehat{\mathbf{x}} \in \widehat{\Gamma}. \tag{17}$$

Employing respectively (16) and (17), we are able to discretize the necessary condition (13) along the lines of a nonlinear Ritz-Galerkin scheme:

$$\text{seek } r_N^\star \in V_N \text{ such that } dJ_i(r_N^\star)[dr] = 0 \text{ for all } dr \in V_N. \tag{18}$$

Notice that this is the necessary condition associated with the finite-dimensional optimization problem $J_i(r_N) \to \min_{r_N \in V_N}$.

Remark. A quite canonical choice is to choose $\widehat{\Gamma}$ as the n-dimensional unit sphere $\mathbb{S}^{n-1}$. Then, the parametrization (16) yields the class of starshaped domains and we may even simplify the representation according to $\boldsymbol{\gamma}(\widehat{\mathbf{x}}) = r(\widehat{\mathbf{x}})\widehat{\mathbf{x}}$ since $\widehat{\mathbf{n}}(\widehat{\mathbf{x}}) = \widehat{\mathbf{x}}$. The first N spherical harmonics in $\mathbb{R}^n$ can be used as appropriate basis functions. This was the choice for our numerical experiments in Section 8 and the papers [12, 13, 14, 15, 16, 17, 18, 27].

For the numerical solution of the nonlinear variational equation (18) we apply the quasi-Newton method, updated by the inverse BFGS-rule without damping. A second-order approximation is used for performing the line search update if the descent does not satisfy the Armijo rule. For all the details and a survey on available optimization algorithms we refer to [23, 24, 26] and the references therein.

In case of well-posedness, the existence and convergence of the approximate shapes can be proven.

Theorem 3 ([20]). *Assume that the shape Hessian is strictly $H^s(\Gamma^\star)$-coercive at the stationary domain $r^\star \in X$. Then, there exists a neighbourhood $U(r^\star) \subset X$ such that* (18) *admits a unique solution $r_N^\star \in V_N \cap U(r^\star)$ provided that N is large enough. The approximation error stays in the energy norm proportional to the best approximation in V_N, that is*

$$\|r_N^\star - r^\star\|_{H^s(\widehat{\Gamma})} \lesssim \inf_{r_N \in V_N} \|r_N - r^\star\|_{H^s(\widehat{\Gamma})}.$$

6. Flexible shape representation

For realizing gradient based shape optimization algorithms, we may employ a more general boundary representation than the somehow restrictive approach (16). We can discretize each coordinate of the parametric representation (15) separately by functions of V_N. In this manner, we derive the discretization

$$\boldsymbol{\gamma}_N : \widehat{\Gamma} \to \mathbb{R}^n, \quad \boldsymbol{\gamma}_N = \sum_{i=1}^N \mathbf{a}_i \varphi_i \in V_N^3, \tag{19}$$

where $\mathbf{a}_i \in \mathbb{R}^n$ are *vector-valued* coefficients. The corresponding nonlinear Ritz-Galerkin scheme reads as

$$\text{seek } \boldsymbol{\gamma}_N^\star \in V_N^3 \text{ such that } J(\boldsymbol{\gamma}_N^\star)[\mathbf{V}] = 0 \text{ for all } \mathbf{V} \in V_N^3.$$

On the one hand, the ansatz (19) does not impose any restriction to the topology of the domain except for its genus. On the other hand, the representation (15) of the boundary Γ is not unique. In fact, if $\Xi : \widehat{\Gamma} \to \widehat{\Gamma}$ denotes any smooth diffeomorphism, then the composed function $\boldsymbol{\gamma}_N \circ \Xi$ describes another parametrization of Γ.

For our purpose some parametrizations are preferable to others. Therefore, we introduce a penalty term for finding an appropriate parametrization of the free boundary. In order to discretize functions on the free boundary we like to use the parametrization (15) to map a subdivision of the parameter space $\widehat{\Gamma}$ to the actual boundary Γ. From this point of view it is quite obvious that, for numerical computations, a “nice” parametrization maps a uniform and shape regular mesh of the reference surface to a uniform and shape regular mesh on Γ. To realize this claim, we shall introduce a suitable *mesh functional* $M(\Omega)$, penalizing bad parametrizations, and solve for small $\beta > 0$ the regularized shape problem

$$\text{seek } \boldsymbol{\gamma}_N^\star \in V_N^n \text{ such that } dJ(\boldsymbol{\gamma}_N^\star)[\mathbf{V}] + \beta dM(\boldsymbol{\gamma}_N^\star)[\mathbf{V}] = 0 \text{ for all } \mathbf{V} \in V_N^n$$

instead of the original problem (13). We refer the reader to, e.g., [15, 19] for the choice of appropriate mesh functionals and further details.

7. Numerical method to compute the state

Even though the shape gradients (7)–(11) look quite different, they can numerically be computed by using the same method. We propose to apply a boundary element method since then only the free boundary needs to be discretized which avoids the complicated triangulation of the moving domain Ω. Let the single and double layer operators defined by

$$(\mathcal{V}u)(\mathbf{x}) = \int_{\partial\Omega} E(\mathbf{x},\mathbf{y})u(\mathbf{y})d\sigma_{\mathbf{y}}, \quad (\mathcal{K}u)(\mathbf{x}) = \int_{\partial\Omega} \frac{\partial E(\mathbf{x},\mathbf{y})}{\partial \mathbf{n}(\mathbf{y})} u(\mathbf{y})d\sigma_{\mathbf{y}}, \quad \mathbf{x} \in \partial\Omega,$$

where $E(\mathbf{x},\mathbf{y})$ denotes the fundamental solution of the Laplacian in $\mathbb{R}^n$. Thus, the Neumann data of v from (2) are given by the *Dirichlet-to-Neumann map*

$$\mathcal{V}\frac{\partial v}{\partial n} = \left(\frac{1}{2} + \mathcal{K}\right)(g\chi_\Sigma - N_f) + \mathcal{V}\frac{\partial N_f}{\partial \mathbf{n}} \quad \text{on } \partial\Omega \tag{20}$$

with N_f denoting a Newton potential satisfying $-\Delta N_f = f$. A similar equation can be derived for the adjoint state p from (10). By reordering the unknowns of (20), which is a straightforward task, one obtains equations for the mixed boundary value problems for w and q.

The Newton potential needs to be computed only once in advance on a domain $\widehat{\Omega}$ which contains all iterates $\Omega^{(n)}$. Since $\widehat{\Omega}$ can be chosen arbitrary simple, for example as a cube, very efficient Poisson solvers like FFT or multigrid methods are applicable. The boundary integral equation itself can be solved in essentially linear complexity if fast boundary element methods like multipole [21], panel clustering [22], or wavelet Galerkin schemes [8] are used as we do. We refer the reader to the papers [14, 16, 17, 18] for all the details.

Having the complete Dirichlet and Neumann data of the states and their adjoints at hand, gradient based algorithms are realizable for all shape functionals under consideration. Particularly, to enable the line-search in case of the shape functional J_1, which is of domain integral type, we can apply integration by parts. By involving the Newton potential we find the expression

$$J_1(\Omega) = \int_\Omega \{h^2 - N_f f\}d\mathbf{x} + \int_\Sigma \frac{\partial(N_f + v)}{\partial \mathbf{n}} g \, d\sigma_{\mathbf{x}} - \int_{\partial\Omega} N_f \frac{\partial v}{\partial \mathbf{n}} \, d\sigma_{\mathbf{x}}.$$

The remaining domain integral is computed by applying the domain quadrature algorithm developed in [19].

8. Numerical results

We solve the free boundary problem (1) for $f = 0$, $g = 1$ and various settings for $h = \text{const}$. The boundary $\Sigma = \partial S$ is chosen as the boundary of the T-shape

$$S := \big((-3/8, 3/8) \times (-1/4, 0)\big) \cup \big((-1/8, 1/8) \times [0, 1/4)\big) \subset \mathbb{R}^2.$$

The free boundary is numerically approximated by a Fourier series of 65 coefficients according to (17). The (adjoint) state equation is solved by a fast wavelet

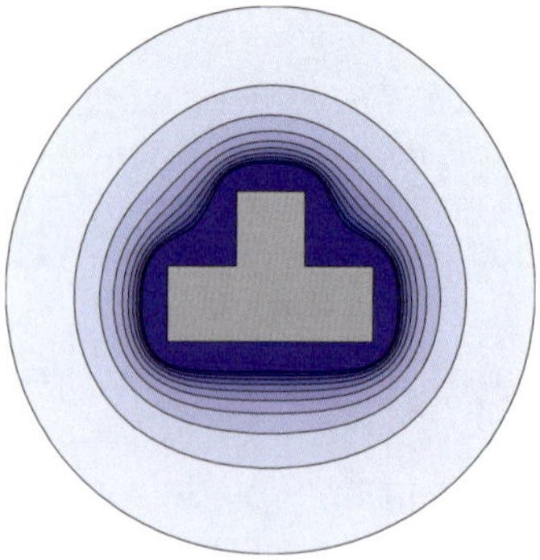

FIGURE 2. The optimal domains $\Omega^\star$.

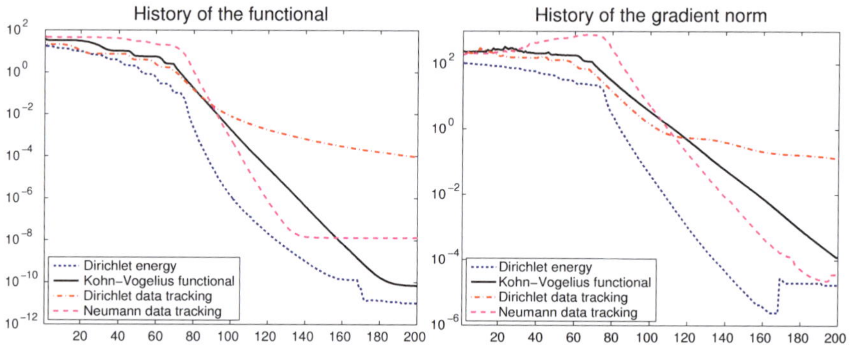

FIGURE 3. Histories of the shape functionals and the gradients.

based boundary element method along the lines of the authors' previous work, see, e.g., [12, 14] for the details. About 2000 piecewise linear boundary elements are spent to discretize the functions on the boundary. The optimization is performed by a quasi-Newton method (see Section 5) based on the information of the last ten gradients. The initial guess is always a circle of radius 0.75. We compute the free boundary for all integers $h = 1, 2, \ldots, 10$. The resulting domains are depicted in Figure 2, where the outermost boundary corresponds to $h = 1$ and the innermost boundary to $h = 10$.

For $h = 5$, we plotted the histories of the shape functionals and their gradient in Figure 3. In fact, it turns out that the Dirichlet energy functional and the Neumann data tracking functional converge with identical rates, referring to their well-posedness. It is clearly seen that the convergence rate of Kohn-Vogelius functional is lower which issues from its algebraically ill-posedness of one order (see Table 1). Whereas, the convergence of the Dirichlet data tracking functional slows down which is an effect of its algebraic ill-posedness of order 2. These results clearly validate the present analysis.

References

[1] A. Acker. On the geometric form of Bernoulli configurations. *Math. Meth. Appl. Sci.* **10** (1988) 1–14.

[2] H.W. Alt and L.A. Caffarelli. Existence and regularity for a minimum problem with free boundary. *J. reine angew. Math.* **325** (1981) 105–144.

[3] N.V. Banichuk and B.L. Karihaloo. Minimum-weight design of multi-purpose cylindrical bars. *International Journal of solids and Structures* **12** (1976) 267–273.

[4] M. Brühl. Explicit characterization of inclusions in electrical impedance tomography. *SIAM J. Math. Anal.* **32** (2001) 1327–1341.

[5] M. Brühl and M. Hanke. Numerical implementation of two noniterative methods for locating inclusions by impedance tomography. *Inverse Problems* **16** (2000) 1029–1042.

[6] R. Chapko and R. Kress. A hybrid method for inverse boundary value problems in potential theory. *J. Inverse Ill-Posed Probl.* **13** (2005) 27–40.

[7] O. Colaud and A. Henrot. Numerical approximation of a free boundary problem arising in electromagnetic shaping. *SIAM J. Numer. Anal.* **31** (1994) 1109–1127.

[8] W. Dahmen, H. Harbrecht and R. Schneider. Compression techniques for boundary integral equations – optimal complexity estimates. *SIAM J. Numer. Anal.* **43** (2006) 2251–2271.

[9] M. Delfour and J.-P. Zolesio. *Shapes and Geometries.* SIAM, Philadelphia, 2001.

[10] K. Eppler. Boundary integral representations of second derivatives in shape optimization. *Discuss. Math. Differ. Incl. Control Optim.* **20** (2000) 63–78.

[11] K. Eppler. Optimal shape design for elliptic equations via BIE-methods. *Appl. Math. Comput. Sci.* **10** (2000) 487–516.

[12] K. Eppler and H. Harbrecht. Numerical solution of elliptic shape optimization problems using wavelet-based BEM. *Optim. Methods Softw.* **18** (2003) 105–123.

[13] K. Eppler and H. Harbrecht. A regularized Newton method in electrical impedance tomography using shape Hessian information. *Control Cybern.* **34** (2005) 203–225.

[14] K. Eppler and H. Harbrecht. Efficient treatment of stationary free boundary problems. *Appl. Numer. Math.* **56** (2006) 1326–1339.

[15] K. Eppler and H. Harbrecht. Wavelet based boundary element methods in exterior electromagnetic shaping. *Eng. Anal. Bound. Elem.* **32** (2008) 645–657.

[16] K. Eppler and H. Harbrecht. Tracking Neumann data for stationary free boundary problems. *SIAM J. Control Optim.* **48** (2009) 2901–2916.

[17] K. Eppler and H. Harbrecht. Tracking the Dirichlet data in L^2 is an ill-posed problem. *J. Optim. Theory Appl.* **145** (2010) 17–35.

[18] K. Eppler and H. Harbrecht. On a Kohn-Vogelius like formulation of free boundary problems. *Comput. Optim. Appl.* (to appear).

[19] K. Eppler, H. Harbrecht, and M.S. Mommer. A new fictitious domain method in shape optimization. *Comput. Optim. Appl.*, **40** (2008) 281–298.

[20] K. Eppler, H. Harbrecht, and R. Schneider. On Convergence in Elliptic Shape Optimization. *SIAM J. Control Optim.* **45** (2007) 61–83.

[21] L. Greengard and V. Rokhlin. A fast algorithm for particle simulation. *J. Comput. Phys.* **73** (1987), 325–348.

[22] W. Hackbusch and Z.P. Nowak. On the fast matrix multiplication in the boundary element method by panel clustering. *Numer. Math.* **54** (1989), 463–491.

[23] A.V. Fiacco and G.P. McCormick. *Nonlinear Programming: Sequential Unconstrained Minimization Techniques*. Wiley, New York, 1968.

[24] R. Fletcher. *Practical Methods for Optimization, volume* 1*,* 2. Wiley, New York, 1980.

[25] M. Flucher and M. Rumpf. Bernoulli's free-boundary problem, qualitative theory and numerical approximation. *J. reine angew. Math.* **486** (1997) 165–204.

[26] C. Grossmann and J. Terno. *Numerik der Optimierung*. B.G. Teubner, Stuttgart, 1993.

[27] H. Harbrecht. A Newton method for Bernoulli's free boundary problem in three dimensions. *Computing* **82** (2008) 11–30.

[28] J. Haslinger, T. Kozubek, K. Kunisch, and G. Peichl. Shape optimization and fictitious domain approach for solving free boundary problems of Bernoulli type. *Comput. Optim. Appl.* **26** (2003) 231–251.

[29] J. Haslinger, K. Ito, T. Kozubek, K. Kunisch, and G. Peichl. On the shape derivative for problems of Bernoulli type. *Interfaces and Free Boundaries* **11** (2009) 317–330.

[30] F. Hettlich and W. Rundell The determination of a discontinuity in a conductivity from a single boundary measurement. *Inverse Problems* **14** (1998) 67–82.

[31] K. Ito, K. Kunisch, and G. Peichl. Variational approach to shape derivatives for a class of Bernoulli problems. *J. Math. Anal. Appl.* **314** (2006) 126–149.

[32] R. Kohn and M. Vogelius. Determining conductivity by boundary measurements. *Comm. Pure Appl. Math.* **37** (1984) 289–298.

[33] F. Murat and J. Simon. Étude de problèmes d'optimal design. in *Optimization Techniques, Modeling and Optimization in the Service of Man*, edited by J. Céa, Lect. Notes Comput. Sci. 41, Springer-Verlag, Berlin, 54–62 (1976).

[34] A. Novruzi and J.R. Roche. Newton's method in shape optimisation: a three-dimensional case. *BIT* **40** (2000) 102–120.

[35] M. Pierre and J.-R. Roche Computation of free surfaces in the electromagnetic shaping of liquid metals by optimization algorithms. *Eur. J. Mech. B/Fluids* **10** (1991) 489–500.

[36] J.-R. Roche and J. Sokolowski Numerical methods for shape identification problems. *Control Cybern.* **25** (1996) 867–894.

[37] J. Simon. Differentiation with respect to the domain in boundary value problems. *Numer. Funct. Anal. Optimization* **2** (1980) 649–687.

[38] J. Sokolowski and J.-P. Zolesio. *Introduction to Shape Optimization*. Springer, Berlin, 1992.

[39] T. Tiihonen. Shape optimization and trial methods for free-boundary problems. *RAIRO Model. Math. Anal. Numér.* **31** (1997) 805–825.

Karsten Eppler
Institut für Numerische Mathematik, Technische Universität Dresden
Zellescher Weg 12–14, D-01062 Dresden, Germany
e-mail: `karsten.eppler@tu-dresden.de`

Helmut Harbrecht
Mathematisches Institut, Universität Basel
Rheinsprung 21, CH-4051 Basel, Schweiz
e-mail: `helmut.harbrecht@unibas.ch`

International Series of Numerical Mathematics, Vol. 160, 289–300

Non-parametric Aerodynamic Shape Optimization

Nicolas Gauger, Caslav Ilic, Stephan Schmidt and Volker Schulz

Abstract. Numerical schemes for large scale shape optimization are considered. Exploiting the structure of shape optimization problems is shown to lead to very efficient optimization methods based on non-parametric surface gradients in Hadamard form. The resulting loss of regularity is treated using higher-order shape Newton methods where the shape Hessians are studied using operator symbols. The application ranges from shape optimization of obstacles in an incompressible Navier–Stokes fluid to super- and transonic airfoil and wing optimizations.

Mathematics Subject Classification (2000). 65K10, 49M25, 76D55, 76N25, 49Q10, 49Q12.

Keywords. Shape optimization, one-shot, shape SQP methods.

1. Introduction

1.1. Paradigms in aerodynamic shape optimization

There are two paradigms to solve aerodynamic shape optimization problems: parametric and non-parametric. The non-parametric approach is traditionally used to derive analytically optimal shapes that can be globally represented as the graph of a function or by a deformation of the submanifold of the surface of the flow obstacle. With this paradigm, optimality of certain rotationally symmetric ogive shaped bodies in supersonic, irrotational, inviscid potential flows can be shown [15]. In the incompressible regime, optimal shapes for a viscous Stokes flow are derived in [20].

Any actual optimization so far follows the parametric paradigm. After choosing a finite-dimensional design vector $q \in \mathbb{R}^{n_q}$, the gradient is computed by a formal Lagrangian approach:

$$\frac{df}{dq}(u(q), q) = \frac{\partial f}{\partial q} - \lambda^T \frac{\partial c}{\partial q},$$

where f is the objective function, u is the solution of the flow state $c(u, q)$, and λ is the adjoint variable. The mesh sensitivity Jacobian $\frac{\partial c}{\partial q}$ is a dense matrix. Often, the computation of the mesh sensitivity involves a mesh deformation procedure, making the computation of this Jacobian very costly. As the computational time and storage requirements increase quickly with the number of design parameters n_q, there is a strong desire to use as few design parameters as possible: Usually, the shape is deformed by a small number of smooth ansatz functions, where the coefficients of these functions are the design parameters. The de facto standard is a parametrization by Hicks-Henne functions [16], but sometimes b-spline parameters are also used.

Overcoming the limited search space of such a global parametrization in the form of coordinates of boundary points as design variables has long been desired, but the computational costs become prohibitive very quickly and the resulting loss of regularity is not well understood. A formulation of the gradient for the highly complex, nonlinear, hyperbolic equations describing compressible flows which can be computed without the design chain has long been sought after by both academia, [7] and [11], and industry [28, 29, 30, 31, 32]. More in line with the theoretical non-parametric approach, the problem is seldom treated from a true shape optimization perspective, except in [1, 2] for pressure tracking or in [4]. Due to the complexity of the shape differentiation techniques, none of the approaches above, which omit the design chain of the formal Lagrangian approach, have so far been successfully applied on a large scale drag reduction problem.

The work presented here follows mainly the non-parametric approach as a means to overcome the difficulties of the standard approach described above. Non-parametric shape gradients for various objectives and flow regimes will be discussed. Furthermore, shape Hessians can be exploited for convergence acceleration in higher-order optimization approaches. The resulting shape-Newton methods define a new efficient algorithmic approach in aerodynamic non-parametric shape optimization. Such optimization problems are usually solved by a two loop approach: The outer loop is given by a gradient based optimization scheme, while solving the flow and adjoint equations creates the inner loop. This nested loop is rarely broken up, as in [14, 19] or in [13, 26]. The latter one-shot optimization relies on the structure of a standard non-linear finite-dimensional optimization problem of the parametric approach [12]. Therefore, the need of a repeated mesh sensitivity computation usually reduces the effective speed-up for a large-scale shape parametrization. Therefore, similar to [23] a shape one-shot method is presented, which works outside the structure of a finite-dimensional nonlinear problem and features a significant speed-up. As the standard approximation of the Hessian by BFGS-updates is questionable in this setting, we present a Hessian approximation based on operator symbols.

1.2. Shape calculus

Shape calculus describes mathematical concepts when the geometry is the variable. Forsaking the shape problem origin by parametrizing, most – if not all – of the standard calculus of finite spaces is applicable. The alternative is to retain concepts of differential calculus, spaces of geometries, evolution equations, etc. to geometric domains. Excellent overviews about these concepts applied to shapes can be found in [6, 8, 27]. As both approaches are usually called "shape analysis", this field of research appears much more unified than it actually is. For example in the area of continuum mechanics and structural mechanics of elastic bodies the thickness of the material can be used to create a distributed parameter set. Alternatively, one can employ direct shape calculus techniques on the moving boundaries and topological derivatives in the interior [3, 5]. Outside of this compliance analysis, non-parametric shape calculus enables very elegant and efficient descriptions of sensitivities of general partial differential equations with respect to changes in the domain. However, there are many open questions when using these analytical objects numerically. For example, the proper discretization of analytic shape Hessians or finding reliable update formulas like BFGS need to be discussed more in order to better establish optimization schemes beyond shape gradient steepest descent. As such, higher-order non-parametric shape optimization schemes are rare. In [9] the shape Hessian is studied via sinusoidal perturbations of the annulus, in [10] the shape Hessian for potential flow pressure tracking in star-shaped domains is considered, and in [17, 18] a non-parametrized image segmentation approach is shown, also employing shape Hessians for the minimization of a shape functional without any additional PDE-constraints.

2. Impulse response approach for characterizing shape Hessians in Stokes and Navier–Stokes flow

The research presented in this section focuses on finding reliable and easily applicable shape Hessian approximations for flow problems governed by the incompressible Navier–Stokes equations. We study shapes that minimize the conversion of kinetic energy into heat, which is physically closely related to a proper drag reduction using the formulation based on surface forces. The model problem for finding shape Hessians is given by:

$$\min_{(u,p,\Omega)} J(u,\Omega) := \int_\Omega \nu \sum_{i,j=1}^{3} \left(\frac{\partial u_i}{\partial x_j}\right)^2 dA \tag{2.1}$$

$$\begin{aligned} \text{s.t.}\ -\nu\Delta u + \rho u \nabla u + \nabla p &= 0 \quad \text{in } \Omega \\ \operatorname{div} u &= 0 \\ u &= 0 \quad \text{on } \Gamma_1 \\ u &= u_\infty \quad \text{on } \Gamma_2 \\ \text{Volume}(\Omega) &= V_0, \end{aligned} \tag{2.2}$$

where $u = (u_1, u_2)^T$ is the speed of the fluid, ν is the kinematic viscosity, p denotes the pressure, and ρ is the density which is constant in an incompressible fluid. Also,

$\Gamma_1 \subset \partial\Omega$, the no-slip surface of the flow obstacle, is the unknown to be found. The shape derivative for this problem is given by:

$$dJ(u,\Omega)[V] = \int_\Gamma \langle V, n\rangle \left[-\nu \sum_{k=1}^{3} \left(\frac{\partial u_k}{\partial n}\right)^2 - \frac{\partial u_k}{\partial n}\frac{\partial \lambda_k}{\partial n}\right] dS,$$

where $\lambda = (\lambda_1, \lambda_2)^T$ and λ_p again satisfy the adjoint equation

$$\begin{aligned} -\nu\Delta\lambda - \rho\lambda\nabla u - \rho\,(\nabla\lambda)^T u + \nabla\lambda_p &= -2\Delta u \quad \text{in } \Omega \\ \operatorname{div} \lambda_p &= 0 \qquad\quad \text{in } \Omega. \end{aligned}$$

For more details see [22, 24, 25]. The idea is to characterize the Hessian of this problem by its symbol, which is defined as the image of a single Fourier mode $\tilde{q}(x) := \hat{q}e^{i\omega x}$ under the Hessian H of the problem.

If we assume $\Omega = \{(x, y) : x \in \mathbb{R}, y > 0\}$ to be the upper half-plane, then we need to track the Fourier mode $\alpha(x) := e^{i\omega_1 x}$ through the perturbation of the local shape derivatives

$$\begin{aligned} u_i'[\alpha] &= \hat{u}_i e^{i\omega_1 x} e^{\omega_2 y} \qquad (2.3) \\ p'[\alpha] &= \hat{p} e^{i\omega_1 x} e^{\omega_2 y}. \end{aligned}$$

In the limit of the Stokes case, $\rho = 0$, the perturbed shape gradient $dG_\Gamma[\alpha]$ is given by

$$dG_\Gamma[\alpha] = -2\nu \sum_{i=1}^{2} \frac{\partial u_i}{\partial y}\frac{\partial u_i'[\alpha]}{\partial y},$$

which means that the Hessian H is defined by the mapping

$$\frac{\partial u_i'[\alpha]}{\partial y} = H\alpha.$$

Thus, if the angular frequency ω_1 can be made explicit in the left-hand side, we have exactly the definition of the operator symbol as stated above. To achieve this, the PDE defining the local shape derivatives is transformed into the Fourier space. There, the solvability requirement of the PDE and its boundary conditions in the frequency space defines an implicit function relating ω_2 to ω_1 in (2.3). Essentially, this implicit function is given by the roots of the characteristic polynomial of the PDE for the local shape derivatives and gives for the Stokes problem:

$$\det dC[\alpha] = \nu(-\omega_1^2 + \omega_2^2)\omega_2^2 - \nu(-\omega_1^2 + \omega_2^2)\omega_1^2. \qquad (2.4)$$

Using this method, one can show that the shape Hessian for the Stokes problem is a pseudo-differential operator of order +1, closely related to the Poincaré-Stecklov operator. The task of identifying the shape Hessian is thus transformed to finding an explicit representation of a certain implicit function in the Fourier space.

Finding explicit representations of implicit functions analytically is still infeasible for more complex problems. For an empirical determination of the symbol of the Navier–Stokes shape Hessian see [22, 24], where the operator is found to be a pseudo-differential operator with the symbol $|\omega|$, again closely related to the

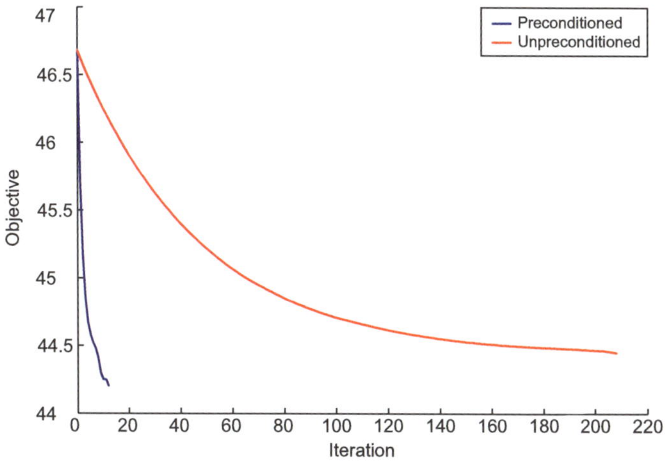

FIGURE 1. Preconditioning accelerates the Stokes problem by 96%.

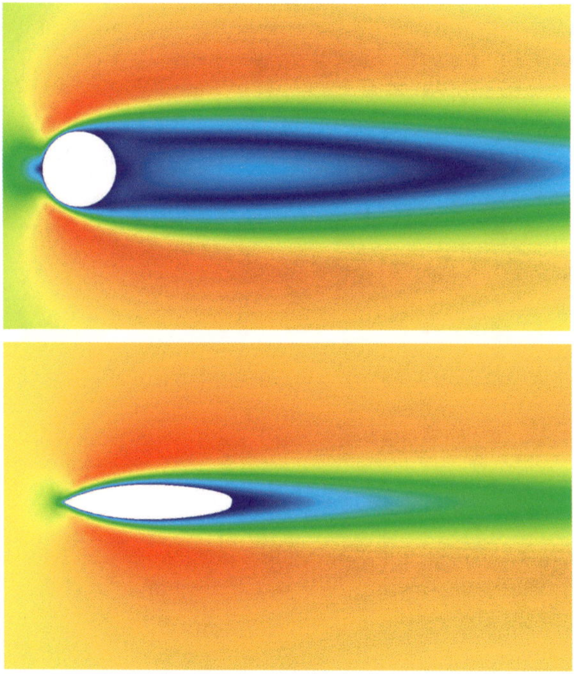

FIGURE 2. Initial and optimized Navier-Stokes shapes. Color denotes speed.

Poincaré-Stecklov operator. We approximate this operator by a damped Laplace-Beltrami operator for preconditioning these problems, where the damping is determined by the frequency spectrum of the standing waves possible on the given mesh resolution of the surface. This leads to a significant speed-up of 96% in the Stokes case and of 80% in the Navier-Stokes case. The speed-up for the Stokes case is shown in Figure 1 and the initial and optimal shapes for the Navier–Stokes problem can be seen in Figure 2: The double vortex behind the initial circle has been completely removed.

3. Exploitation of shape calculus for supersonic and transonic Euler flow

3.1. Introduction

As the usual cruise speed is Mach 0.7 and more, the incompressible Navier–Stokes equations considered above are not a sophisticated enough flow model: viscosity effects become negligible compared to compression effects, which means that we must now at least consider the compressible Euler equations as a model for the flow. At transonic and supersonic flow conditions, shock waves form that dominate the drag of the aircraft. This means not only that the discontinuous solutions must be computed correctly for the forward and adjoint problem, but also that the shape derivative must give correct results under a PDE constraint which has a discontinuous solution. We found that the nodal shape-Newton method works very well given a discontinuous state. The objective function is now a proper force minimization and the whole drag optimization problem is stated as:

$$\begin{aligned}
\min_{(U,\Omega)} F_{\text{drag}}(U,\Omega) &:= \int_\Gamma \langle p_d, n\rangle \, dS \\
\text{s.t.} \sum_{i=1}^{3} A_i(U_p)\frac{\partial U}{\partial x_i} &= 0 \text{ in } \Omega \text{ (Euler equations)} \\
\langle u, n\rangle &= 0 \text{ on } \Gamma \text{ (Euler slip condition)} \\
F_{\text{lift}}(U,\Omega) := \int_\Gamma \langle p_l, n\rangle \, dS &\geq l_0 \text{ (lift force)} \\
\text{Volume}(\Omega) &= V_0.
\end{aligned}$$

For 2D applications there is also the additional constraint that the leading edge is fixed at $(0,0)^T$ and the trailing edge must be fixed at $(1,0)^T$. Otherwise, the optimization changes the reference length of the airfoil, which would lead to a wrong non-dimensionalization of the flow quantities. The volume constraint again prevents the solution from degenerating into a flat line and is always active. From physical considerations it is known that the lift constraint will also be active and as such it is treated as an equality constraint in the lifting case considered below.

Additionally, $U := (\rho, \rho u_1, \rho u_2, \rho u_3, \rho E)^T$ is the vector of the unknown state and denotes the conserved variables, where ρ is the density of the fluid, u_i are the velocity components, and E is the total energy of the fluid. Likewise, $U_p :=$

$(\rho, u_1, u_2, u_3, E)^T$ denotes the primitive variables which enter the Euler flux Jacobians $A_i := \frac{\partial F_i}{\partial U}$ of the inviscid fluxes F_i. Using the non-conservative formulation of the Euler equations in terms of the flux Jacobians simplifies the derivation of the adjoint equations. For a given angle of attack α, we define $p_d := p \cdot (\cos\alpha, \sin\alpha)^T$ and $p_l := p \cdot (-\sin\alpha, \cos\alpha)^T$, where p denotes the pressure, which is related to the conserved variables by the perfect gas law $p = (\gamma - 1)\rho(E - \frac{1}{2}(u_1^2 + u_2^2 + u_3^2))$. Here, γ is the isentropic expansion factor, i.e., the heat capacity ratio, of air.

The numerous mappings of the pressure p to the conserved variables U makes the derivation of the shape derivative non-trivial. In fact, the shape derivative for this problem was long sought after for the benefits stated above. In the engineering literature often called "surface formulation of the gradient", there have been previous attempts to find the shape derivative [4, 7, 11, 28]. The surface measure variation dS_ϵ must be considered, because a boundary integral objective function requires an integration by parts on the surface of submanifolds, which usually introduces additional curvature terms. Considering this, the shape derivative of the Euler drag reduction is given by

$$\begin{aligned} dF_{\text{drag}}(\Omega)[V] &= \int_\Gamma \langle V, n\rangle \left[\langle \nabla p_d\, n, n\rangle + \kappa\langle p_d, n\rangle\right] + (p_d - \lambda U_H u) dn[V]\; dS \qquad (3.1)\\ &= \int_\Gamma \langle V, n\rangle \left[\langle \nabla p_d\, n, n\rangle + \operatorname{div}_\Gamma(p_d - \lambda U_H u)\right]\; dS, \end{aligned}$$

where λ solves the adjoint Euler equations

$$-\frac{\partial}{\partial x_1}\left(A_1^T(U_p)\lambda\right) - \frac{\partial}{\partial x_2}\left(A_2^T(U_p)\lambda\right) - \frac{\partial}{\partial x_3}\left(A_3^T(U_p)\lambda\right) = 0 \quad \text{in } \Omega$$

and the wall boundary condition

$$(\lambda_2, \lambda_3, \lambda_4)\, n + n_d = 0$$

on the wing. Here, $U_H := (\rho, \rho u_1, \rho u_2, \rho u_3, \rho H)^T$ are the conserved variables with the last component replaced by the enthalpy $\rho H = \rho E + p$. Also, $\operatorname{div}_\Gamma$ denotes the divergence in the tangent space of Γ, which is sometimes also called "surface divergence". Also, κ denotes additive mean curvature. Thus, a correct computation of the shape derivative also requires discrete differential geometry, as the curvature and normal variation or surface divergence must be computed correctly on the given CFD mesh.

The following results were achieved with the DLR flow solver TAU, which is an unstructured finite volume code, in Euler mode. It features an implementation of the continuous adjoint and is also the production code of Airbus, making the following computations examples of real world applications. According to [1, 2], the Hessian is a pseudo-differential operator of order +2. For the smoothing procedure we always employed the Laplace-Beltrami operator.

3.2. Supersonic airfoil optimization

The first test was conducted for a fully supersonic flow at no angle of attack and no lift constraint. The fully supersonic case is considered easier, because the initial NACA0012 airfoil produces a detached bow shock due to its blunt nose. Thus,

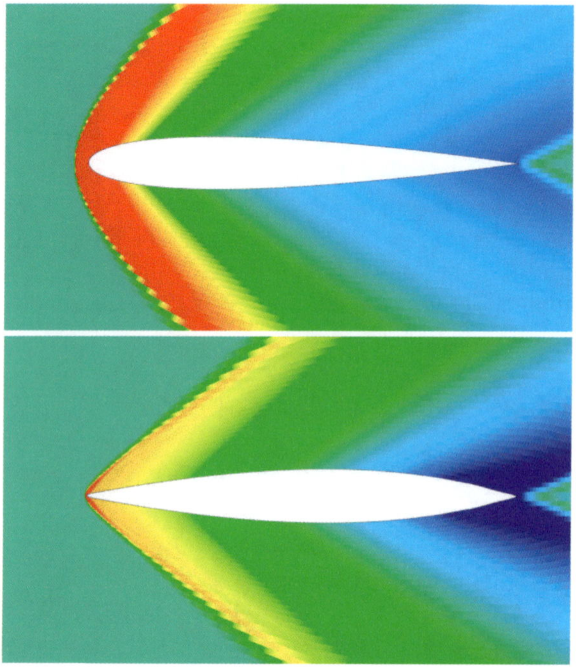

FIGURE 3. Non-lifting optimization, supersonic flow Mach 2.0. NACA-0012 airfoil deforms to a Haack Ogive shape. Color denotes pressure.

the state is continuous where the shape derivative must be evaluated. The shapes can be seen in Figure 3. We have the automatic formation of a sharp leading edge without user intervention and the computed optimal ogive shape matches very well the analytic predictions from [15]. The strong detached bow shock of the blunt nose body is transformed to a weaker attached shock of a body with a sharp leading edge. The optimal shape, a Haack Ogive, was expected from the literature, as it is analytically known from simpler supersonic inviscid flow models that such shapes are optimal, making this an excellent test to gauge the method against. The correct shape was very efficiently found: From an initial $C_D = 9.430 \cdot 10^{-2}$, the optimal $C_D = 4.721 \cdot 10^{-2}$ was found with 2.5 times the cost of the simulation alone. Using the shape Hessian, the shape derivative, and a one-shot approach, we can solve this problem in about 100 seconds. The classical approach of solving a non-linear optimization problem post discretization requires 2.77 hours. The CPU time reduction of one-shot, shape derivative, and shape Hessian are all cumulative,

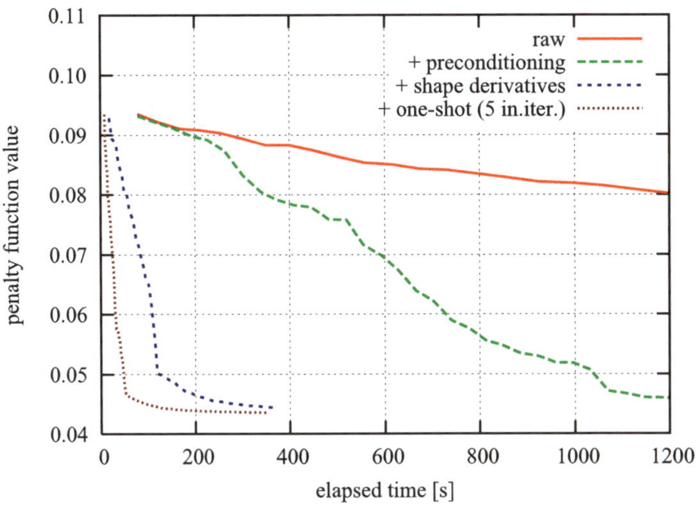

FIGURE 4. Speed-up in CPU time due to shape Hessian preconditioning, shape derivative, and one-shot.

making the nodal one-shot approach 99% faster. The effects of each ingredient can be seen in Figure 4.

3.3. Onera M6 wing optimization

We conclude with the optimization of the Onera M6 wing in three dimensions. During cruise condition of Mach 0.83 and 3.01° angle of attack, the wing features a lift coefficient of $C_L = 2.761 \cdot 10^{-1}$ and a drag coefficient of $C_D = 1.057 \cdot 10^{-2}$. This drag is mainly created due to two interacting shock waves on the upper side of the wing. Thus, in this three-dimensional application, the wing shape must be optimized such that the shock waves vanish while at the same time maintaining lift and internal volume, which adds another constrained compared to the problem considered above. We conduct a multi-level optimization using all CFD mesh surface nodes as design parameter. The coarse mesh features 18, 285 surface nodes and a finer mesh is created adaptively during the optimization with has 36, 806 surface nodes. Surface finite elements in a curved space are used to compute the Laplace–Beltrami operator for the Hessian approximation. Likewise, the curvature is computed by discretely constructing the second fundamental tensor **II** similar to [21]. The resulting optimal shapes are shown in Figure 5. The optimized wing is shock free with a drag coefficient of $C_D = 7.27 \cdot 10^{-3}$ and maintains lift with $C_L = 2.65 \cdot 10^{-1}$.

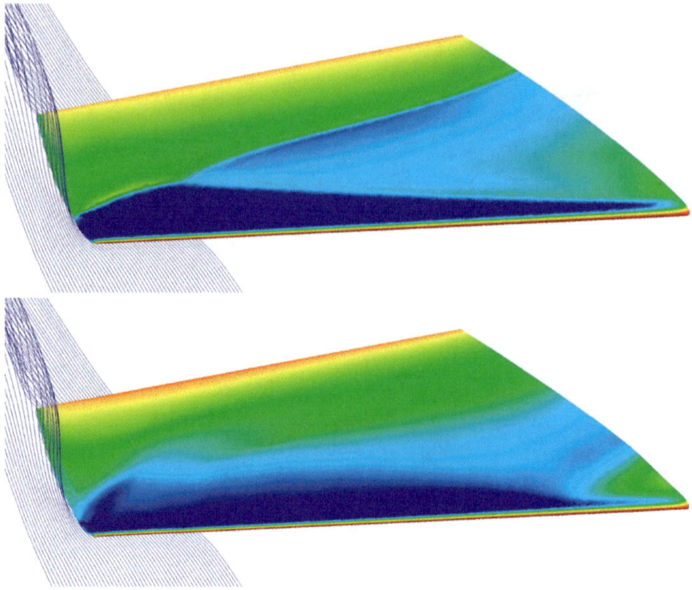

FIGURE 5. Initial and optimized Onera M6 wing. Color denotes pressure. The upper surface shock waves are completely removed.

4. Conclusions

Large scale aerodynamic shape optimization was considered. While usually the actual computation of optimal shapes is based on a parametric approach, here we focus on a non-parametric shape sensitivity analysis in order to very efficiently compute the shape gradients. Paired with a one-shot optimization approach, this creates a highly efficient numerical scheme exploiting the nature of shape optimization problems, e.g., the possibility of computing the gradient using surface quantities alone. The resulting loss of regularity is treated using higher-order optimization methods where the shape Hessian is approximated using operator symbols. Both the incompressible Navier–Stokes equations as well as the compressible Euler equations are considered as a model for the fluid and both the shape optimization of obstacles in a flow channel as well as super- and transonic airfoil and wing optimizations are presented.

Acknowledgment

This research was funded by the German Science Foundation (DFG) as part of the priority program SPP 1253: "Optimization with Partial Differential Equations."

References

[1] E. Arian and S. Ta'Asan. Analysis of the Hessian for aerodynamic optimization: Inviscid flow. *ICASE*, 96-28, 1996.

[2] E. Arian and V.N. Vatsa. A preconditioning method for shape optimization governed by the Euler equations. *ICASE*, 98-14, 1998.

[3] M. Bendsøe. *Methods for Optimization of structural topology, shape and material.* Springer, 1995.

[4] C. Castro, C. Lozano, F. Palacios, and E. Zuazua. Systematic continuous adjoint approach to viscous aerodynamic design on unstructured grids. *AIAA*, 45(9):2125–2139, 2007.

[5] K.-T. Cheng and N. Olhoff. An investigation concerning optimal design of solid elastic plates. *International Journal of Solids and Structures*, 17:305–323, 1981.

[6] M.C. Delfour and J.-P. Zolésio. *Shapes and Geometries: Analysis, Differential Calculus, and Optimization.* Advances in Design and Control. SIAM Philadelphia, 2001.

[7] O. Enoksson. *Shape Optimization in compressible inviscid flow.* Licentiate thesis, Linköpings Universitet, S-581 83 Linköping, Sweden, 2000.

[8] K. Eppler. *Efficient Shape Optimization Algorithms for Elliptic Boundary Value Problems.* Habilitation thesis, Technische Universität Chemnitz, Germany, 2007.

[9] K. Eppler and H. Harbrecht. A regularized Newton method in electrical impedance tomography using shape Hessian information. *Control and Cybernetics*, 34(1):203–225, 2005.

[10] K. Eppler, S. Schmidt, V. Schulz, and C. Ilic. Preconditioning the pressure tracking in fluid dynamics by shape Hessian information. *Journal of Optimization Theory and Applications*, 141(3):513–531, 2009.

[11] N. Gauger. *Das Adjungiertenverfahren in der aerodynamischen Formoptimierung.* PhD thesis, TU Braunschweig, 2003.

[12] I. Gherman. *Approximate Partially Reduced SQP Approaches for Aerodynamic Shape Optimization Problems.* PhD thesis, University of Trier, 2008.

[13] I. Gherman and V. Schulz. Preconditioning of one-shot pseudo-timestepping methods for shape optimization. *PAMM*, 5(1):741–742, 2005.

[14] A. Griewank. Projected Hessians for preconditioning in one-step one-shot design optimization. *Nonconvex Optimization and its application*, 83:151–172, 2006.

[15] W. Haack. Geschoßformen kleinsten Wellenwiderstandes. *Bericht der Lilienthal-Gesellschaft*, 136(1):14–28, 1941.

[16] R.M. Hicks and P.A. Henne. Wing design by numerical optimization. *Journal of Aircraft*, 15:407–412, 1978.

[17] M. Hintermüller and W. Ring. An inexact Newton-CG-type active contour approach for the minimization of the Mumford-Shah functional. *Journal of Mathematical Imaging and Vision*, 20(1–2):19–42, 2004.

[18] M. Hintermüller and W. Ring. A second-order shape optimization approach for image segmentation. *SIAM Journal on Applied Mathematics*, 64(2):442–467, 2004.

[19] A. Iollo, G. Kuruvila, and S. Ta'Asan. Pseudo-time method for optimal shape design using Euler equation. Technical Report 95-59, ICASE, 1995.

[20] O. Pironneau. On optimum profiles in stokes flow. *Journal of Fluid Mechanics*, 59(1):117–128, 1973.

[21] S. Rusinkiewicz. Estimating curvatures and their derivatives on triangle meshes. In *Symposium on 3D Data Processing, Visualization, and Transmission*, 2004.

[22] S. Schmidt. *Efficient Large Scale Aerodynamic Design Based on Shape Calculus*. PhD thesis, University of Trier, Germany, 2010.

[23] S. Schmidt, C. Ilic, N. Gauger, and V. Schulz. Shape gradients and their smoothness for practical aerodynamic design optimization. Technical Report Preprint-Nr.: SPP1253-10-03, DFG-SPP 1253, 2008 (submitted to OPTE).

[24] S. Schmidt and V. Schulz. Impulse response approximations of discrete shape Hessians with application in CFD. *SIAM Journal on Control and Optimization*, 48(4):2562–2580, 2009.

[25] S. Schmidt and V. Schulz. Shape derivatives for general objective functions and the incompressible Navier–Stokes equations. *Control and Cybernetics*, 2010 (to appear issue 3/2010).

[26] V. Schulz and I. Gherman. One-shot methods for aerodynamic shape optimization. In N. Kroll, D. Schwamborn, K. Becker, H. Rieger, and F. Thiele, editors, *MEGADESIGN and MegaOpt – German Initiatives for Aerodynamic Simulation and Optimization in Aircraft Design*, volume 107 of *Notes on Numerical Fluid Mechanics and Multidisciplinary Design*, pages 207–220. Springer, 2009.

[27] J. Sokolowski and J.-P. Zolésio. *Introduction to Shape Optimization: Shape Sensitivity Analysis*. Springer, 1992.

[28] P. Weinerfeld. Aerodynamic optimization using control theory and surface mesh points as control variables. Technical Report FAU-97.044, SAAB Aerospace, Linköping, 1997.

[29] P. Weinerfeld. Alternative gradient formulation for aerodynamic shape optimization based on the Euler equations. Technical Report FF-2001-0042, SAAB Aerospace, Linköping, 2001.

[30] P. Weinerfeld. Gradient formulation for aerodynamic shape optimization based on the Navier-Stokes equations. Technical Report FF-2001-0043, SAAB Aerospace, Linköping, 2001.

[31] P. Weinerfeld. Gradient formulations for aerodynamic shape optimization based on Euler equations. Technical Report FF-2001-0041, SAAB Aerospace, Linköping, 2001.

[32] P. Weinerfeld. Some theorems related to the variation of metric terms and integrals. Technical Report FF-2001-0040, SAAB Aerospace, Linköping, 2001.

Nicolas Gauger
RWTH Aachen University, Department of Mathematics and CCES
Schinkelstr. 2, D-52062 Aachen, Germany
e-mail: gauger@mathcces.rwth-aachen.de

Caslav Ilic
Deutsches Zentrum für Luft- und Raumfahrt, e.V.
Lilienthalplatz 7, D-38108 Braunschweig, Germany
e-mail: Caslav.Ilic@dlr.de

Stephan Schmidt and Volker Schulz
Universitätsring 15, D-54296 Trier, Germany
e-mail: Stephan.Schmidt@uni-trier.de
Volker.Schulz@uni-trier.de

Part III

Model Reduction

Introduction to Part III
Model Reduction

This part summarizes several results of recent research in model reduction techniques by means of goal oriented mesh refinement techniques for both elliptic and parabolic optimization problems possibly subject to additional pointwise control or state constraints. It consists of three independent sections:

Andreas Günther, Michael Hinze and Moulay Hicham Tber consider, in *A Posteriori Error Representations for Elliptic Optimal Control Problems with Control and State Constraints*, adaptive finite element methods for the evaluation of the cost functional of an elliptic optimal control problem with Moreau-Yosida regularized state constraints.

Dominik Meidner and Boris Vexler review, in *Adaptive Space-Time Finite Element Methods for Parabolic Optimization Problems*, the state of the art in goal oriented space-time adaptive finite element methods for parabolic optimization problems.

Rolf Rannacher, Boris Vexler and Winnifried Wollner review, in *A Posteriori Error Estimation in PDE-constrained Optimization with Pointwise Inequality Constraints*, the most recent progress in goal-oriented adaptive finite element methods for elliptic optimization problems with pointwise bounds on control and state variables. The article is concerned both with the direct treatment of the constraints as well as with its indirect treatment through regularization.

Rolf Rannacher

International Series of Numerical Mathematics, Vol. 160, 303–317

A Posteriori Error Representations for Elliptic Optimal Control Problems with Control and State Constraints

Andreas Günther, Michael Hinze and Moulay Hicham Tber

Abstract. In this work we develop an adaptive algorithm for solving elliptic optimal control problems with simultaneously appearing state and control constraints. Building upon the concept proposed in [9] the algorithm applies a Moreau-Yosida regularization technique for handling state constraints. The state and co-state variables are discretized using continuous piecewise linear finite elements while a variational discretization concept is applied for the control. To perform the adaptive mesh refinement cycle we derive local error representations which extend the goal-oriented error approach to our setting. The performance of the overall adaptive solver is demonstrated by a numerical example.

Mathematics Subject Classification (2000). 49J20; 65N30; 65N50.

Keywords. Elliptic optimal control problem, control constraints, state constraints, goal-oriented adaptivity, error estimates.

1. Introduction

Optimal control problems with state constraints have been the topic of an increasing number of theoretical and numerical studies. The challenging character of these problems has its origin in the fact that state constraints feature Lagrange multipliers of low regularity only [4, 7]. In the presence of additional control constraints, the solution may exhibit subsets where both control and state are active simultaneously. In this case, the Lagrange multipliers associated to the control and state constraints may not be unique [22]. To overcome this difficulty several techniques in the literature have been proposed. Very popular are relaxation concepts for state constraints such as Lavrentiev, interior point and Moreau-Yosida regularization. The former one is investigated in [23] and [19]. Barrier methods in

This work was supported by the DFG priority program 1253 grant No. DFG 06/382.

function space [26] applied to state constrained optimal control problems are considered in [20]. Relaxation by Moreau-Yosida regularization is considered for the fully discrete case in [3, 5], and in function space in [15]. Residual-type a posteriori error estimators for mixed control-state constrained problems are derived in [18]. The dual weighted residual method proposed in [1] is investigated in [12, 24] with the presence of control constraints. Goal-oriented adaptive approaches with state constraints is subject matter in the works [2, 9, 13]. Within the framework of goal-oriented adaptive function space algorithms based on an interior point method we mention [21] and [27].

In this note we combine results from [9] and [10] to design an adaptive finite element algorithm for solving elliptic optimal control problems with pointwise control and state constraints. Following [16], our algorithm combines a Moreau-Yosida regularization approach with a semi-smooth Newton solver [14]. We apply variational discretization [8, 17] to the regularized optimal control problem. For a fixed regularization parameter, we develop a goal-oriented a posteriori error estimate by extending the error representation obtained in [9] to the control and state constrained case. Let us note that in this work we are not interested in controlling the error contribution stemming from the regularization. In a further investigation the aim of balancing the errors in the objective functional that arise from discretization and Moreau-Yosida regularization could be carried out combining techniques from [11] and [27].

The rest of this paper is organized as follows: In the next section we present the optimal control problem under consideration and recall its first-order necessary optimality system. Section 3 is devoted to the purely state-constrained case and collects results from [9] which in Section 4 are extended to simultaneously appearing control and state constraints. We introduce a regularized version of the original problem and derive an error representation in terms of the objective functional. Finally, a numerical experiment is reported in Section 5.

2. Optimal control problem

To simplify the presentation we assume Ω to be a bounded polygonal and convex domain in $\mathbb{R}^d$ $(d = 2, 3)$ with boundary $\partial\Omega$. We consider the general elliptic partial differential operator $\mathcal{A} : H^1(\Omega) \longrightarrow H^1(\Omega)^*$ defined by

$$\mathcal{A}y := \sum_{i,j=1}^{d} \partial_{x_j}(a_{ij}y_{x_i}) + \sum_{i=1}^{d} b_i y_{x_i} + cy$$

along with its formal adjoint operator $\mathcal{A}^*$

$$\mathcal{A}^*y = \sum_{i=1}^{d} \partial_{x_i}\Big(\sum_{j=1}^{d} a_{ij}y_{x_j} + b_i y\Big) + cy.$$

We subsequently assume the coefficients a_{ij}, b_i and c $(i, j = 1, \ldots, d)$ to be sufficiently smooth functions on $\bar{\Omega}$. Moreover we suppose that there exists $c_0 > 0$ such that $\sum_{i,j=1}^d a_{ij}(x)\xi_i\xi_j \geq c_0|\xi|^2$ for almost all x in Ω and all ξ in $\mathbb{R}^d$. Corresponding to the operator $\mathcal{A}$ we associate the bilinear form $a(\cdot,\cdot) : H^1(\Omega) \times H^1(\Omega) \longrightarrow \mathbb{R}$ with

$$a(y,v) := \int_\Omega \Big(\sum_{i,j=1}^d a_{ij} y_{x_i} v_{x_j} + \sum_{i=1}^d b_i y_{x_i} v + cyv \Big).$$

Suppose that the form a is coercive on $H^1(\Omega)$, i.e., there exists $c_1 > 0$ such that $a(v,v) \geq c_1 \|v\|^2_{H^1(\Omega)}$ for all v in $H^1(\Omega)$. This follows for instance if

$$\operatorname*{ess\,inf}_{x\in\Omega} \Big(c - \frac{1}{2}\sum_{i=1}^d \partial_{x_i} b_i \Big) > 0 \text{ and } \operatorname*{ess\,inf}_{x\in\partial\Omega} \Big(\sum_{i=1}^d b_i \nu_i \Big) \geq 0$$

holds. Here ν denotes the unit outward normal at $\partial\Omega$.

For given $u \in L^2(\Omega)$ and fixed $f \in L^2(\Omega)$ the homogeneous Neumann boundary value problem

$$\begin{aligned} \mathcal{A}y &= u + f && \text{in } \Omega \\ \partial_{\nu_\mathcal{A}} y := \textstyle\sum_{i,j=1}^d a_{ij} y_{x_i} \nu_j &= 0 && \text{on } \partial\Omega \end{aligned} \tag{2.1}$$

has a unique solution $y =: \mathcal{G}(u) \in H^2(\Omega)$. Moreover, there exists a constant C depending on the domain Ω such that

$$\|\mathcal{G}(u)\|_{H^2(\Omega)} \leq C(\|u\|_{L^2(\Omega)} + \|f\|_{L^2(\Omega)}).$$

The regularity of y is an immediate consequence of the assumptions made on the domain Ω. For more general domains the analysis also applies whenever $y \in W^{1,p}(\Omega)$ for $p > d$. The weak form of (2.1) is given by

$$a(y,v) = (u+f,v) \qquad \forall v \in H^1(\Omega), \tag{2.2}$$

where $(\cdot,\cdot)$ denotes the standard inner product in $L^2(\Omega)$.

For given $u_d, y_d \in L^2(\Omega)$, $\alpha > 0$, u_a, $u_b \in \mathbb{R}$ with $u_a < u_b$, y_a and $y_b \in C(\Omega)$ with $y_a < y_b$ we focus on the optimal control problem

$$\begin{aligned} & J(y,u) := \tfrac{1}{2}\|y - y_d\|^2_{L^2(\Omega)} + \tfrac{\alpha}{2}\|u - u_d\|^2_{L^2(\Omega)} \to \min \\ & \text{s.t.} \quad y = \mathcal{G}(u), \quad u \in U_{\text{ad}}, \quad \text{and} \quad y_a \leq y \leq y_b \quad \text{a.e. in } \bar{\Omega}, \end{aligned} \tag{2.3}$$

where U_{ad} is the set of admissible controls given by

$$U_{\text{ad}} = \big\{ u \in L^2(\Omega) \,:\, u_a \leq u \leq u_b \quad \text{in } \Omega \big\}.$$

We require the Slater condition

$$\exists\, u_s \in U_{\text{ad}} : \qquad y_a < \mathcal{G}(u_s) < y_b \quad \text{in } \Omega,$$

and the space of Radon measures $\mathcal{M}(\bar{\Omega})$ which is defined as the dual space of $C^0(\bar{\Omega})$ such that

$$\langle \mu, y \rangle := \int_{\bar{\Omega}} y \, d\mu \quad \forall \mu \in \mathcal{M}(\bar{\Omega}) \; \forall y \in C^0(\bar{\Omega}).$$

The proof of the following theorem follows from [6, 7].

Theorem 2.1. *The optimal control problem (2.3) has a unique solution* $(y,u) \in H^2(\Omega) \times U_{\mathrm{ad}}$. *Moreover there exist* $p \in W^{1,s}(\Omega)$ *for all* $1 \le s < d/(d-1)$, λ_a, $\lambda_b \in L^2(\Omega)$ *and* μ_a, $\mu_b \in \mathcal{M}(\bar\Omega)$ *satisfying the optimality system*

$$
\begin{aligned}
y &= \mathcal{G}(u),\\
(p, \mathcal{A}v) = (y - y_d, v) + \langle \mu_a + \mu_b, v\rangle \ \forall v \in W^{1,\frac{s}{s-1}}(\Omega) \text{ with } \partial_{\nu_{\mathcal{A}}} v|_{\partial\Omega} &= 0,\\
\alpha(u - u_d) + p + \lambda_a + \lambda_b &= 0,\\
\lambda_a \le 0, \quad u \ge u_a, \quad (\lambda_a, u - u_a) &= 0,\\
\lambda_b \ge 0, \quad u \le u_b, \quad (\lambda_b, u - u_b) &= 0,\\
\mu_a \le 0, \quad y \ge y_a, \quad \langle \mu_a, y - y_a\rangle &= 0,\\
\mu_b \ge 0, \quad y \le y_b, \quad \langle \mu_b, y - y_b\rangle &= 0.
\end{aligned}
\tag{2.4}
$$

3. The purely state-constrained case

First let us consider problem (2.3) without control constraints, i.e., $U_{\mathrm{ad}} = U = L^2(\Omega)$. Then in (2.4) $\lambda_a = \lambda_b = 0$ and p, μ_a, and μ_b are uniquely determined with $p = -\alpha(u - u_d)$, see, e.g.,[6].

3.1. Finite element discretization

In the sequel we consider a shape-regular simplicial triangulation $\mathcal{T}_h$ of Ω. Since Ω is assumed to be a polyhedral, the boundary $\partial\Omega$ is exactly represented by the boundaries of simplices $T \in \mathcal{T}_h$. We refer to $\mathcal{N}_h = \cup_{i=1}^{np}\{x_i\}$ as the set of nodes of $\mathcal{T}_h$. The overall mesh size is defined by $h := \max_{T \in \mathcal{T}_h} \operatorname{diam} T$. Further, we associate with $\mathcal{T}_h$ the continuous piecewise linear finite element space

$$
V_h = \{v \in C_0(\bar\Omega) \,:\, v|_T \in P_1(T),\ \forall T \in \mathcal{T}_h\},
$$

where $P_1(T)$ is the space of first-order polynomials on T. The standard nodal basis of V_h denoted by $\{\phi_i\}_{i=1}^{np}$ satisfies $\phi_i(x_j) = \delta_{ij}$ for all x_j in $\mathcal{N}_h$ and $i, j \in \{1, \dots, np\}$. Furthermore for all $v \in C^0(\bar\Omega)$ we denote by $I_h v := \sum_{i=1}^{np} v(x_i)\phi_i$ the Lagrange interpolation of v. In analogy to (2.2) we define for given $u \in L^2(\Omega)$ the discrete solution operator $\mathcal{G}_h$ by

$$
y_h =: \mathcal{G}_h(u) \iff y_h \in V_h \text{ and } a(y_h, v_h) = (u + f, v_h) \quad \forall v_h \in V_h.
$$

Problem (2.3) is now approximated by the following sequence of variational discrete control problems depending on the mesh parameter h:

$$
\begin{aligned}
&\min_{u \in U} J_h(y_h, u) := \frac{1}{2}\|y_h - y_d\|^2_{L^2(\Omega)} + \frac{\alpha}{2}\|u - u_{d,h}\|^2_U\\
&\text{subject to } \ y_h = \mathcal{G}_h(u) \text{ and } y_a(x_j) \le y_h(x_j) \le y_b(x_j) \text{ for } j = 1, \dots, np.
\end{aligned}
\tag{3.1}
$$

Here, $u_{d,h} \in V_h$ denotes an approximation to u_d.

Problem (3.1) represents a convex infinite-dimensional optimization problem of similar structure as problem (2.3), but with only finitely many equality and inequality constraints for the state, which define a convex set of admissible functions. Since for $h > 0$ small enough we have $y_a < \mathcal{G}_h(u_s) < y_b$ such that [6, 7] can again be applied to obtain

Lemma 3.1. *Problem* (3.1) *has a unique solution* $u_h \in U$. *There exist unique* $\mu_1^a, \ldots, \mu_{np}^a, \mu_1^b, \ldots, \mu_{np}^b \in \mathbb{R}$ *and a unique function* $p_h \in V_h$ *such that with* $y_h = \mathcal{G}_h(u_h)$, $\mu_{a,h} = \sum_{j=1}^{np} \mu_j^a \delta_{x_j}$ *and* $\mu_{b,h} = \sum_{j=1}^{np} \mu_j^b \delta_{x_j}$ *we have*

$$a(v_h, p_h) = \int_\Omega (y_h - y_d) v_h + \int_{\bar{\Omega}} v_h \, d(\mu_{a,h} + \mu_{b,h}) \qquad \forall\, v_h \in V_h, \quad (3.2)$$

$$p_h + \alpha(u_h - u_{d,h}) = 0, \quad (3.3)$$

$$\mu_j^a \le 0, \quad y_h(x_j) \ge y_a(x_j), \quad j = 1, \ldots, np \text{ and } \int_{\bar{\Omega}} \big(y_h - I_h y_a\big)\, d\mu_{a,h} = 0, \quad (3.4)$$

$$\mu_j^b \ge 0, \quad y_h(x_j) \le y_b(x_j), \quad j = 1, \ldots, np \text{ and } \int_{\bar{\Omega}} \big(y_h - I_h y_b\big)\, d\mu_{b,h} = 0. \quad (3.5)$$

Here, δ_x denotes the Dirac measure concentrated at x.

Remark 3.2. Problem (3.1) is still an infinite-dimensional optimization problem, but with finitely many state constraints. By (3.3) it follows that $u_h \in V_h$, i.e., the optimal discrete solution is discretized implicitly through the optimality condition of the discrete problem. Hence in (3.1) U may be replaced by V_h to obtain the same discrete solution u_h, which results in a finite-dimensional discrete optimization problem instead.

3.2. Local error indicators

In this section we extend the dual weighted residual method proposed in [1] to the presence of state constraints (cf. [2, 9, 13]). For goal-oriented adaptivity with control constraints we again refer to [12, 24].

From here onwards we assume $u_d = u_{d,h}$, so that $J = J_h$ holds. This assumption is fulfilled by affine linear functions u_d. Including more general desired controls u_d would lead to additional weighted data oscillation contributions $(u_d - u_{d,h}, \cdot)$ in the error representation (4.5). For their treatment in the context of residual type a posteriori estimators we refer to [18].

Let us abbreviate

$$\mu := \mu_a + \mu_b, \quad \mu_h := \mu_{a,h} + \mu_{b,h}.$$

Following [1] we introduce the dual, control and primal residual functionals determined by the discrete solution y_h, u_h, p_h, μ_h^a and μ_h^b of (3.2)–(3.5) by

$$\begin{aligned}
\rho^p(\cdot) &:= J_y(y_h, u_h)(\cdot) - a(\cdot, p_h) + \langle \mu_h, \cdot \rangle, \\
\rho^u(\cdot) &:= J_u(y_h, u_h)(\cdot) + (\cdot, p_h) \text{ and} \\
\rho^y(\cdot) &:= -a(y_h, \cdot) + (u_h + f, \cdot).
\end{aligned}$$

In addition we set

$$e^\mu(y) := \langle \mu + \mu_h, y_h - y \rangle.$$

It follows from (3.3) that $\rho^u(\cdot) \equiv 0$. This is due to variational discretization, i.e., we do not discretize the control, so that the discrete structure of the solution u_h of problem (3.1) is induced by the optimality condition (3.3).

The proof of the following theorem can be found in [9].

Theorem 3.3. *There hold the error representations*

$$2(J(y,u) - J(y_h,u_h)) = \rho^p(y - i_h y) + \rho^y(p - i_h p) + e^\mu(y), \tag{3.6}$$

and

$$\begin{aligned} 2(J(y,u) - J(y_h,u_h)) \;=\;& J_y(y_h,u_h)(y-y_h) - a(y-y_h,p_h) \\ & -a(y_h, p - i_h p) + (u_h + f, p - i_h p) \\ & +(y - y_d, y - y_h) - a(y - y_h, p) \end{aligned} \tag{3.7}$$

with arbitrary quasi-interpolants $i_h y$ and $i_h p \in V_h$.

Following the lines of Remark 3.5 in [1] we split the above equation into a cellwise representation and integrate by parts. This gives rise to define

$$\begin{aligned} R^{y_h}_{|T} &= u_h + f - \mathcal{A} y_h \\ R^{p_h}_{|T} &= y_h - y_d - \mathcal{A}^* p_h \\ R^{p}_{|T} &= y - y_d - \mathcal{A}^* p \\ r^{y_h}_{|\Gamma} &= \begin{cases} \frac{1}{2}\nu \cdot [\nabla y_h \cdot (a_{ij})] & \text{for } \Gamma \subset \partial T \setminus \partial\Omega \\ \nu \cdot (\nabla y_h \cdot (a_{ij})) & \text{for } \Gamma \subset \partial\Omega \end{cases} \\ r^{p_h}_{|\Gamma} &= \begin{cases} \frac{1}{2}\nu \cdot [(a_{ij})\nabla p_h] & \text{for } \Gamma \subset \partial T \setminus \partial\Omega \\ \nu \cdot ((a_{ij})\nabla p_h + p_h b) & \text{for } \Gamma \subset \partial\Omega \end{cases} \\ r^{p}_{|\Gamma} &= \begin{cases} \frac{1}{2}\nu \cdot [(a_{ij})\nabla p] & \text{for } \Gamma \subset \partial T \setminus \partial\Omega \\ \nu \cdot ((a_{ij})\nabla p + p b) & \text{for } \Gamma \subset \partial\Omega \end{cases}, \end{aligned}$$

where $[\cdot]$ defines the jump across the inter-element edge Γ. Now (3.7) can be rewritten in the form

$$\begin{aligned} 2(J(y,u) - J(y_h,u_h)) = \sum_{T\in\mathcal{T}_h} & (y - y_h, R^{p_h}_{|T})_T - (y - y_h, r^{p_h}_{|\partial T})_{\partial T} \\ & + (R^{y_h}_{|T}, p - i_h p)_T - (r^{y_h}_{|\partial T}, p - i_h p)_{\partial T} \\ & + (y - y_h, R^{p}_{|T})_T - (y - y_h, r^{p}_{|\partial T})_{\partial T}. \end{aligned}$$

Since this localized sum still contains unknown quantities, we make use of local higher-order approximation ([1, Sec. 5.1]) which has shown to be a successful heuristic technique for a posteriori error estimation. More precisely we take the local higher-order quadratic interpolant operator $i^{(2)}_{2h} : V_h \to P_2(T)$ for some $T \in \mathcal{T}_h$. In detail for $d = 2$ the local interpolant $i^{(2)}_{2h} v_h$ for an arbitrary function $v_h \in V_h$ on a triangle T is defined by

$$(i^{(2)}_{2h} v_h)(x_1,x_2) := a + bx_1 + cx_2 + dx_1x_2 + ex_1^2 + fx_2^2, \qquad (x_1,x_2)^T \in \Omega,$$

where the coefficients $a, b, c, d, e, f \in \mathbb{R}$ are obtained by the solution of a linear system demanding the exact interpolation in the sampling nodes shown in Figure 1. The technique for computing $i^{(2)}_{2h} v_h$ for some $v_h \in V_h$ can easily be carried over to three space dimensions. Although this construction requires additional care at the domains boundary in order to obtain enough neighboring information a special

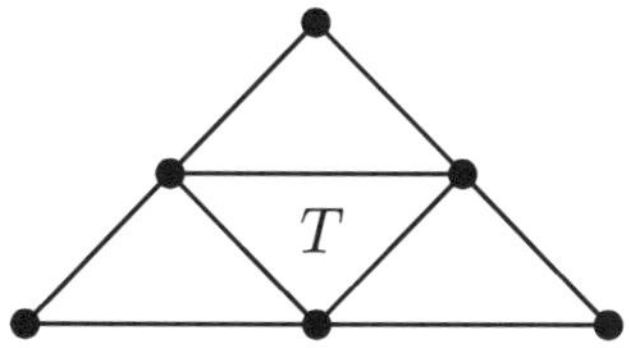

FIGURE 1. Used sampling nodes for $i_{2h}^{(2)}$, $d = 2$.

mesh structure is not required. If that still is the case then, by exploiting this fact, one would benefit from it in terms of less CPU-time too.

We substitute $R^p_{|T}$ and $r^p_{|\Gamma}$ by

$$R^{i_{2h}^{(2)}p_h}_{|T} = i_{2h}^{(2)} y_h - y_d - \mathcal{A}^* i_{2h}^{(2)} p_h$$

$$r^{i_{2h}^{(2)}p_h}_{|\Gamma} = \begin{cases} 0 & \text{for } \Gamma \subset \partial T \setminus \partial\Omega \\ \nu \cdot ((a_{ij})\nabla i_{2h}^{(2)} p_h + i_{2h}^{(2)} p_h b) & \text{for } \Gamma \subset \partial\Omega \end{cases}$$

and define

$$\begin{aligned} \eta := \frac{1}{2} \sum_{T \in \mathcal{T}_h} & (i_{2h}^{(2)} y_h - y_h, R^{p_h}_{|T})_T - (i_{2h}^{(2)} y_h - y_h, r^{p_h}_{|\partial T})_{\partial T} \\ & + (R^{y_h}_{|T}, i_{2h}^{(2)} p_h - p_h)_T - (r^{y_h}_{|\partial T}, i_{2h}^{(2)} p_h - p_h)_{\partial T} \\ & + (i_{2h}^{(2)} y_h - y_h, R^{i_{2h}^{(2)}p_h}_{|T})_T - (i_{2h}^{(2)} y_h - y_h, r^{i_{2h}^{(2)}p_h}_{|\partial T})_{\partial T}. \end{aligned}$$

For numerical experiments we refer to [9] and Section 5.

4. Control- and state-constraints simultaneously

Let us now consider problem (2.3) with control and state constraints appearing simultaneously. The situation then becomes more involved since Lagrange multipliers may not be unique if active sets of control and state intersect [22]. Since the appearance of such situations is not known in advance robust numerical methods are required for the numerical solution of the optimization problem, which do not rely on uniqueness of the multipliers. A numerically example where this situation appears is presented in [10]. If control and state constraints are active simultaneously on the same part of the domain it is required that the bounds satisfy the PDE (2.1) there. This certainly is not the generic case. The case of separated active sets is figured out in the enclosed numerical example.

To overcome difficulties arising from the above discussion we consider

4.1. Moreau-Yosida regularization

In Moreau-Yosida regularization the state constraints $y_a \leq y \leq y_b$ are substituted by appropriate regularization terms which are added to the objective functional J.

The corresponding regularized optimal control problem reads

$$\begin{aligned} J^\gamma(y,u) := J(y,u) + \tfrac{\gamma}{2}\|\min(0, y-y_a)\|^2 + \tfrac{\gamma}{2}\|\max(0,y-y_b)\|^2 \to \min \\ \text{s.t.} \quad y = \mathcal{G}(u) \quad \text{and} \quad u \in U_{\mathrm{ad}}, \end{aligned} \tag{4.1}$$

where $\gamma > 0$ denotes a regularization parameter tending to $+\infty$ later on. The max- and min-expressions in the regularized objective functional J^γ arise from regularizing the indicator function corresponding to the set of admissible states. Notice that (4.1) is a purely control constrained optimal control problem that has a unique solution $(y^\gamma, u^\gamma) \in H^2(\Omega) \times U_{\mathrm{ad}}$. Furthermore, we can prove the existence of Lagrange multipliers $(p^\gamma, \lambda_a^\gamma, \lambda_b^\gamma) \in L^2(\Omega)\times L^2(\Omega)\times L^2(\Omega)$ using standard theory of mathematical programming in Banach spaces such that

$$\begin{aligned} y^\gamma &= \mathcal{G}(u^\gamma) \\ (p^\gamma, \mathcal{A}v) = (y^\gamma - y_d, v) + (\mu_a^\gamma + \mu_b^\gamma, v) \;\forall v \in H^2(\Omega) \text{ with } \partial_{\nu_{\mathcal{A}}} v|_{\partial\Omega} &= 0, \\ \alpha(u^\gamma - u_d) + p^\gamma + \lambda_a^\gamma + \lambda_b^\gamma &= 0, \\ \lambda_a^\gamma \le 0, \quad u^\gamma \ge u_a, \quad (\lambda_a^\gamma, u^\gamma - u_a) &= 0, \\ \lambda_b^\gamma \ge 0, \quad u^\gamma \le u_b, \quad (\lambda_b^\gamma, u^\gamma - u_b) &= 0 \end{aligned} \tag{4.2}$$

holds, where

$$\mu_a^\gamma = \gamma \min(0, y^\gamma - y_a) \quad \text{and} \quad \mu_b^\gamma = \gamma\max(0, y^\gamma - y_b).$$

For the convergence of the solutions of the regularized problems to the solution of the limit problem (2.3) we refer the reader to [10, Thm. 3.1]. To recover the solution of the optimal control problem (2.3) an overall algorithm can be designed by solving (4.1) for a sequence $\gamma \to \infty$. For (4.1) with γ fixed, a locally superlinear semi-smooth Newton method can be applied (see [16]).

4.2. Finite element discretization

For the convenience of the reader we assume y_a and y_b to be real numbers. Again we apply variational discretization [17], but now to problem (4.1). We therefore consider

$$\begin{aligned} J_h^\gamma(y_h,u_h) := J_h(y_h,u_h) + \tfrac{\gamma}{2}\|\min(0, y_h-y_a)\|^2 + \tfrac{\gamma}{2}\|\max(0,y_h-y_b)\|^2 \to \min \\ \text{s.t.} \quad y_h = \mathcal{G}_h(u_h) \quad \text{and} \quad u_h \in U_{\mathrm{ad}}. \end{aligned} \tag{4.3}$$

The existence of a solution of (4.3) as well as of Lagrange multipliers again follows from standard arguments. The corresponding first-order optimality system associated to (4.3) leads to the variationally discretized counterpart of (4.2)

$$\begin{aligned} y_h^\gamma &= \mathcal{G}_h(u_h^\gamma), \\ a(v_h, p_h^\gamma) = (v_h, y_h^\gamma - y_d + \mu_{a,h}^\gamma + \mu_{b,h}^\gamma) &\quad \forall v_h \in V_h, \\ \alpha(u_h^\gamma - u_{d,h}) + p_h^\gamma + \lambda_{a,h}^\gamma + \lambda_{b,h}^\gamma &= 0, \\ \lambda_{a,h}^\gamma \le 0, \quad u_h^\gamma \ge u_a, \quad (\lambda_{a,h}^\gamma, u_h^\gamma - u_a) &= 0, \\ \lambda_{b,h}^\gamma \ge 0, \quad u_h^\gamma \le u_b, \quad (\lambda_{b,h}^\gamma, u_h^\gamma - u_b) &= 0, \end{aligned} \tag{4.4}$$

where y_h^γ, $p_h^\gamma \in V_h$ and u_h^γ, $\lambda_{a,h}^\gamma$, $\lambda_{b,h}^\gamma \in L^2(\Omega)$. The multipliers corresponding to the regularized state constraints $\mu_{a,h}^\gamma$ and $\mu_{b,h}^\gamma$ are given by

$$\mu_{a,h}^\gamma = \gamma \min(0, y_h^\gamma - y_a) \quad \text{and} \quad \mu_{b,h}^\gamma = \gamma \max(0, y_h^\gamma - y_b).$$

We mention here that (4.3) is a function space optimization problem and the optimal control u_h^γ is not necessarily an element of the finite element space. However, regarding (4.4), u_h^γ corresponds to the projection of a finite element function onto the admissible set U_{ad}, namely

$$u_h^\gamma = \Pi_{[u_a,u_b]}\left(-\frac{1}{\alpha}p_h^\gamma + u_{d,h}\right),$$

where $\Pi_{[u_a,u_b]}$ is the orthogonal projection onto U_{ad}.

4.3. Error representation and estimator

For a fixed regularization parameter γ we now derive an error representation in J for the solutions of (4.1) and (4.3) respectively.

Similar as before we define the following residuals

$$\rho^{p^\gamma}(\cdot) := J_y(y_h^\gamma, u_h^\gamma)(\cdot) - a(\cdot, p_h^\gamma) + (\mu_h^\gamma, \cdot),$$

and

$$\rho^{y^\gamma}(\cdot) := \; - a(y_h^\gamma, \cdot) + (u_h^\gamma + f, \cdot)$$

with

$$\mu^\gamma := \gamma \min(0, y^\gamma - y_a) + \gamma \max(0, y^\gamma - y_b),$$

and

$$\mu_h^\gamma := \gamma \min(0, y_h^\gamma - y_a) + \gamma \max(0, y_h^\gamma - y_b).$$

The functions μ^γ and μ_h^γ play the role of the Lagrange multipliers μ, μ_h corresponding to state constraints in the limit problem (2.3) (compare with [9, Thm. 4.1, Rem. 4.1]). Furthermore we abbreviate

$$\lambda^\gamma := \lambda_a^\gamma + \lambda_b^\gamma \quad \text{and} \quad \lambda_h^\gamma := \lambda_{a,h}^\gamma + \lambda_{b,h}^\gamma.$$

Theorem 4.1. *Let (u^γ, y^γ) and (u_h^γ, y_h^γ) be the solutions of the optimal control problems (4.1) and (4.3) with corresponding adjoint states p^γ, p_h^γ and multipliers $\lambda^\gamma, \lambda_h^\gamma, \mu^\gamma, \mu_h^\gamma$ associated to control and state constraints respectively. Then*

$$\begin{aligned} &2(J(y^\gamma, u^\gamma) - J_h(y_h^\gamma, u_h^\gamma)) \\ &\quad = \rho^{p^\gamma}(y^\gamma - i_h y^\gamma) + \rho^{y^\gamma}(p^\gamma - i_h p^\gamma) \\ &\qquad + (\mu^\gamma + \mu_h^\gamma, y_h^\gamma - y^\gamma) + (\lambda^\gamma + \lambda_h^\gamma, u_h^\gamma - u^\gamma). \end{aligned} \tag{4.5}$$

For the proof we refer the reader to [10]. Applying integration by parts the error representation (4.5) can be localized as follows,

$$\begin{aligned}2(J(y^\gamma,u^\gamma)-J_h(y_h^\gamma,u_h^\gamma)) = \sum_{T\in\mathcal{T}_h} &(y^\gamma - y_h^\gamma, R_{|T}^{p_h^\gamma})_T - (y^\gamma - y_h^\gamma, r_{\partial T}^{p_h^\gamma})_{\partial T}\\
&+ (R_{|T}^{y_h^\gamma}, p^\gamma - i_h p^\gamma)_T - (r_{|\partial T}^{y_h^\gamma}, p^\gamma - i_h p^\gamma)_{\partial T}\\
&+ (y^\gamma - y_h^\gamma, R_{|T}^{p^\gamma})_T - (y^\gamma - y_h^\gamma, r_{|\partial T}^{p^\gamma})_{\partial T}\\
&+ (\lambda^\gamma + \lambda_h^\gamma, u_h^\gamma - u^\gamma)_T.\end{aligned}$$

Here the interior and edge residuals $R_{|T}^{\bullet}, r_{|\partial T}^{\bullet}$ are similarly defined as in the purely state constrained case. In order to derive a computable estimator we again replace the unknown functions y^γ and p^γ in (4.5) by $i_{2h}^{(2)} y_h^\gamma$ and $i_{2h}^{(2)} p_h^\gamma$. Since $u^\gamma = \Pi_{[u_a,u_b]}\left(-\frac{1}{\alpha}p^\gamma + u_d\right)$ holds, a reasonable locally computable approximation then is given by

$$\tilde{u}^\gamma = \Pi_{[u_a,u_b]}\left(-\frac{1}{\alpha} i_{2h}^{(2)} p_h^\gamma + u_d\right),$$

as is already suggested in [24]. Similarly for $\lambda^\gamma = -p^\gamma - \alpha(u^\gamma - u_d)$ we locally compute

$$\tilde{\lambda}^\gamma = -i_{2h}^{(2)} p_h^\gamma - \alpha\left(\tilde{u}^\gamma - u_d\right)$$

instead. The estimator η^γ now reads

$$\eta^\gamma = \sum_{T\in\mathcal{T}_h} \eta_T^\gamma,$$

where

$$\begin{aligned}2\eta_T^\gamma =&(i_{2h}^{(2)} y_h^\gamma - y_h^\gamma, R_{|T}^{p_h^\gamma})_T - (i_{2h}^{(2)} y_h^\gamma - y_h^\gamma, r_{|\partial T}^{p_h^\gamma})_{\partial T}\\
&+ (R_{|T}^{y_h^\gamma}, i_{2h}^{(2)} p_h^\gamma - p_h^\gamma)_T - (r_{|\partial T}^{y_h^\gamma}, i_{2h}^{(2)} p_h^\gamma - p_h^\gamma)_{\partial T}\\
&+ (i_{2h}^{(2)} y_h^\gamma - y_h^\gamma, R_{|T}^{i_{2h}^{(2)} p_h^\gamma})_T - (i_{2h}^{(2)} y_h^\gamma - y_h^\gamma, r_{|\partial T}^{i_{2h}^{(2)} p_h^\gamma})_{\partial T}\\
&+ (\tilde{\lambda}^\gamma + \lambda_h^\gamma, u_h^\gamma - \tilde{u}^\gamma)_T.\end{aligned}$$

While for most quantities in η_T^γ quadrature rules of moderate order are well suited, one has to take care for the last term

$$(\tilde{\lambda}^\gamma + \lambda_h^\gamma, u_h^\gamma - \tilde{u}^\gamma)_T = \int_T (\tilde{\lambda}^\gamma + \lambda_h^\gamma)(u_h^\gamma - \tilde{u}^\gamma). \tag{4.6}$$

The integrand has a support within the symmetric difference of the control active set of the variational discrete solution and the locally improved quantities. Such a situation is depicted in Figure 2. One recognizes that $\tilde{u}^\gamma$ captures the activity structure of u_h^γ but smoothes out the control active boundary towards the exact control active boundary. The kidney-shaped green area resolves the true control active set from the example of Section 5 already very well even on a coarse mesh (compare also Figure 3 (right)). Finally for computing (4.6) we just provide the

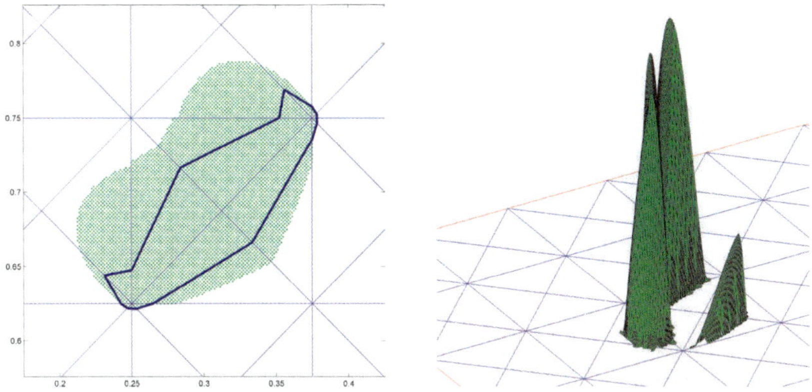

FIGURE 2. u_a active set: blue by u_h^γ, green by $\tilde{u}^\gamma$ (left), integrand $(\tilde{\lambda}^\gamma + \lambda_h^\gamma)(u_h^\gamma - \tilde{u}^\gamma)$ with support on symmetric difference of active sets (right).

integrand and a desired tolerance and apply the adaptive quadrature routine of [25, Algo. 31] for triangles containing the boundary of the control active set.

In order to study the effectivity index of our implemented estimator, we define

$$I_{\text{eff}} := \frac{J(y^\gamma, u^\gamma) - J_h(y_h^\gamma, u_h^\gamma)}{\eta^\gamma}.$$

Remark 4.2. The adjoint variable p admits less regularity at state active sets so that higher-order interpolation with regard to the adjoint variable is not completely satisfying. However this circumstance only leads to local higher weights in the estimator and therefore reasonably suggests to refine the corresponding regions. The effectivity index of the estimator is not affected as we are going to see in the numerical experiment. Another possible technique to derive a computable approximation for $p^\gamma - i_h p^\gamma$ is to substitute p_h^γ for p^γ and compute $p_h^\gamma - p_h^\gamma(x_T)$, where x_T denotes the barycenter of the element T.

5. Numerical experiment

We consider problem (2.3) with data

$$\Omega = (0,\, 1)^2, \quad \mathcal{A} = -\Delta + \text{Id}, \quad y_d = \sin(2\pi x_1)\sin(2\pi x_2), \quad f = u_d = 0,$$
$$u_a = -30, \quad u_b = 30, \quad y_a = -0.55, \quad y_b = 0.55, \quad \alpha = 10^{-4}.$$

Its analytic solution is not known, so for obtaining the effectivity index we compute a reference solution on a uniform grid (level $l = 14$, $np = 525313$, $h = 0.00195$) which delivers an approximation of $J(y^\gamma, u^\gamma)$. The numerical solution in terms of $-\frac{1}{\alpha}p_h^\gamma$ as well as the optimal state y_h^γ is displayed in Figure 3 for $\gamma = 10^{14}$ on the mesh $l = 14$. We solve the arising linear systems directly due to their expected high condition number. The projection of $-\frac{1}{\alpha}p_h^\gamma$ onto $[u_a, u_b]$ corresponds to the

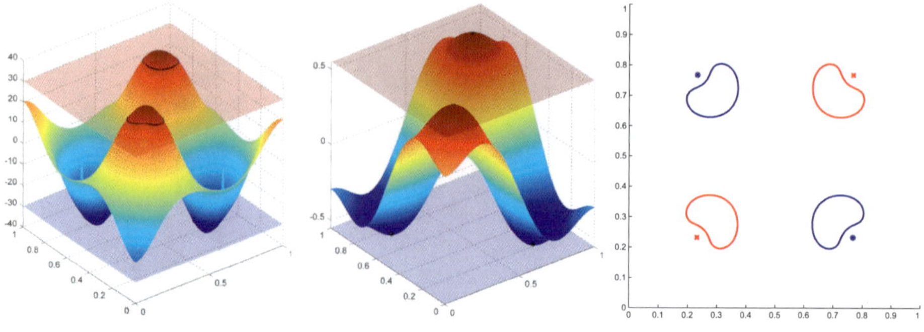

FIGURE 3. $u_a, -\frac{1}{\alpha}p_h^\gamma, u_b$ (left), $y_a \le y_h^\gamma \le y_b$ (middle) and active sets (right) for $l = 14$.

k	np	$J(y^\gamma, u^\gamma) - J_h(y_h^\gamma, Su_h^\gamma)$	I_{eff}
1	81	$4.275 \cdot 10^{-3}$	1.622
10	3123	$7.148 \cdot 10^{-5}$	1.144
18	63389	$3.996 \cdot 10^{-6}$	1.290

TABLE 1. Adaptive refinement.

optimal control u_h^γ which together with y_h^γ represents our best approximation to the solution of (4.1). The boundaries of the control active sets are depicted as solid lines, while the state active sets are displayed as star and cross markers. The color blue corresponds to the lower bound while the color red highlights the upper bound. We numerically approximate $J(y^\gamma, u^\gamma)$ by 0.0375586175. In Table 1 we depict the effectivity coefficient and the convergence history of the goal. Notice that the values

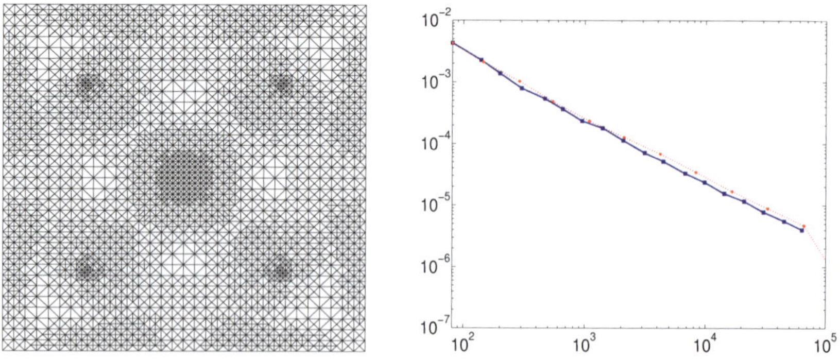

FIGURE 4. Adaptive mesh for $k = 10$ (left), comparison of error decrease in the goal (right).

of I_{eff} are close to 1 which illustrates the good performance of our error estimator. A comparison between our adaptive finite element algorithm and a uniform mesh refinement in terms of number of degrees of freedom $N_{\text{dof}} := np$ is reported in Figure 4. The adaptive refinement process performs well even though the benefit in this example is not big since the characteristic features of the optimal solution already occupy a considerable area of the computational domain, as is illustrated by the adapted grid in Figure 4. Our motivation for including this example is to illustrate the variational discretization effect on the mesh refinement process. If variational discretization for the control would not have been used, also some refinement at the boundary of the control active set would be expected. For the details in terms of our marking strategy, stopping criterion of the adaptive finite element algorithm and one more example we refer the reader to [10].

Acknowledgment

The authors gratefully acknowledge support by the DFG Priority Program 1253 for meetings in Graz and Hamburg through grants DFG06-382, and by the Austrian Ministry of Science and Research and the Austrian Science Fund FWF under START-grant Y305 "Interfaces and Free Boundaries".

References

[1] R. Becker and R. Rannacher, *An optimal control approach to a posteriori error estimation in finite element methods.* Acta Numerica **10** (2001), 1–102.

[2] O. Benedix and B. Vexler, *A posteriori error estimation and adaptivity for elliptic optimal control problems with state constraints.* Comput. Optim. Appl. **44** (2009), 3–25.

[3] M. Bergounioux, M. Haddou, M. Hintermüller, and K. Kunisch, *A comparison of a Moreau-Yosida based active strategy and interior point methods for constrained optimal control problems.* SIAM J. Optim. **11** (2000), 495–521.

[4] M. Bergounioux and K. Kunisch, *On the structure of the Lagrange multiplier for state-constrained optimal control problems.* Syst. Control Lett. **48** (2002), 169–176.

[5] M. Bergounioux and K. Kunisch, *Primal-dual strategy for state-constrained optimal control problems.* Comput. Optim. Appl. **22** (2002), 193–224.

[6] E. Casas, *Boundary control of semilinear elliptic equations with pointwise state constraints.* SIAM J. Control Optim. **31** (1993), 993–1006.

[7] E. Casas, *Control of an elliptic problem with pointwise state constraints.* SIAM J. Control Optim. **24** (1986), 1309–1318.

[8] K. Deckelnick and M. Hinze, *A finite element approximation to elliptic control problems in the presence of control and state constraints.* Hamburger Beiträge zur Angewandten Mathematik, Universität Hamburg, preprint No. HBAM2007-01 (2007).

[9] A. Günther and M. Hinze, *A posteriori error control of a state constrained elliptic control problem.* J. Numer. Math. **16** (2008), 307–322.

[10] A. Günther and M.H. Tber, *A goal-oriented adaptive Moreau-Yosida algorithm for control- and state-constrained elliptic control problems.* DFG Schwerpunktprogramm 1253, preprint No. SPP1253-089 (2009).

[11] Michael Hintermüller and Michael Hinze. Moreau-yosida regularization in state constrained elliptic control problems: error estimates and parameter adjustment. *SIAM J. Numerical Analysis*, 47:1666–1683, 2009.

[12] M. Hintermüller and R.H.W. Hoppe, *Goal-oriented adaptivity in control constrained optimal control of partial differential equations.* SIAM J. Control Optim. **47** (2008), 1721–1743.

[13] M. Hintermüller and R.H.W. Hoppe, *Goal-oriented adaptivity in pointwise state constrained optimal control of partial differential equations.* SIAM J. Control Optim. **48** (2010), 5468–5487.

[14] M. Hintermüller, K. Ito and K. Kunisch, *The primal-dual active set strategy as a semismooth Newton method.* SIAM J. Optim. **13** (2003), 865–888.

[15] M. Hintermüller and K. Kunisch, *Feasible and noninterior path-following in constrained minimization with low multiplier regularity.* SIAM J. Control Optim. **45** (2006), 1198–1221.

[16] M. Hintermüller and K. Kunisch, *Pde-constrained optimization subject to pointwise constraints on the control, the state and its derivative.* SIAM J. Optim. **20** (2009), 1133–1156.

[17] M. Hinze, *A variational discretization concept in control constrained optimization: the linear-quadratic case.* Comput. Optim. Appl. **30** (2005), 45–63.

[18] R.H. Hoppe and M. Kieweg, *Adaptive finite element methods for mixed control-state constrained optimal control problems for elliptic boundary value problems.* Comput. Optim. Appl. **46** (2010), 511–533.

[19] C. Meyer, A. Rösch and F. Tröltzsch, *Optimal control of PDEs with regularized pointwise state constraints.* Comput. Optim. Appl. **33** (2006), 209–228.

[20] A. Schiela, *State constrained optimal control problems with states of low regularity.* SIAM J. Control Optim. **48** (2009), 2407–2432.

[21] A. Schiela and A. Günther, *An interior point algorithm with inexact step computation in function space for state constrained optimal control.* 35 pp. To appear in Numerische Mathematik, doi:10.1007/s00211-011-0381-4, 2011.

[22] A. Shapiro, *On uniqueness of Lagrange multipliers in optimization problems subject to cone constraints.* SIAM J. Optim. **7** (1997), 508–518.

[23] F. Tröltzsch, *Regular Lagrange multipliers for control problems with mixed pointwise control-state constraints.* SIAM J. Optim. **15** (2005), 616–634.

[24] B. Vexler and W. Wollner, *Adaptive finite elements for elliptic optimization problems with control constraints.* SIAM J. Control Optim. **47** (2008), 509–534.

[25] W. Vogt, *Adaptive Verfahren zur numerischen Quadratur und Kubatur.* IfMath TU Ilmenau, preprint No. M 1/06 (2006).

[26] M. Weiser, *Interior point methods in function space.* SIAM J. Control Optim. **44** (2005), 1766–1786.

[27] W. Wollner, *A posteriori error estimates for a finite element discretization of interior point methods for an elliptic optimization problem with state constraints.* Comput. Optim. Appl. **47** (2010), 133–159.

Andreas Günther
Konrad-Zuse-Zentrum für Informationstechnik Berlin (ZIB)
Takustr. 7
D-14195 Berlin-Dahlem, Germany
e-mail: `guenther@zib.de`

Michael Hinze
University of Hamburg
Bundesstr. 55
D-20146 Hamburg, Germany
e-mail: `michael.hinze@uni-hamburg.de`

Moulay Hicham Tber
Department of mathematics
Faculty of Science and Technology
B.P. 523
Beni-Mellal, Morocco
e-mail: `hicham.tber@gmail.com`

International Series of Numerical Mathematics, Vol. 160, 319–348

Adaptive Space-Time Finite Element Methods for Parabolic Optimization Problems

Dominik Meidner and Boris Vexler

Abstract. In this paper we summerize recent results on a posteriori error estimation and adaptivity for space-time finite element discretizations of parabolic optimization problems. The provided error estimates assess the discretization error with respect to a given quantity of interest and separate the influences of different parts of the discretization (time, space, and control discretization). This allows us to set up an efficient adaptive strategy producing economical (locally) refined meshes for each time step and an adapted time discretization. The space and time discretization errors are equilibrated, leading to an efficient method.

Mathematics Subject Classification (2000). 65N30, 49K20, 65M50, 35K55, 65N50.

Keywords. Parabolic equations, optimal control, parameter identification, a posteriori error estimation, mesh refinement, space-time finite elements, dynamic meshes.

1. Introduction

In this paper, we discuss adaptive algorithms for the efficient solution of time-dependent optimization problems governed by parabolic partial differential equations.

The use of adaptive techniques based on a posteriori error estimation is well accepted in the context of finite element discretization of partial differential equations, see, e.g., [4, 12, 39]. The application of these techniques is also investigated for optimization problems governed by elliptic partial differential equations. Energy-type error estimators for the error in the state, control, and adjoint variable are developed, e.g., in [18, 21, 26, 27] – see also [29] and the references therein – in the context of distributed and boundary control problems subject to a elliptic state equation and pointwise control constraints.

For optimal control problems governed by linear parabolic equations a posteriori error estimates with respect to some norms are presented in [25, 28]. In [25], the state equation is discretized by space-time finite elements, and the reliability of error estimates is shown under the H^2-regularity assumption on the domain. In [34], an anisotropic error estimate is derived for the error due to the space discretization of an optimal control problem governed by the linear heat equation. In [37], an adaptive algorithm for optimal control of time-dependent differential algebraic equations is presented. Space-time adaptivity within an SQP method is investigated in [10]. In [23], adaptive techniques are applied for boundary control of time-dependent equations describing radiative heat transfer in glass.

However, in many applications, one is interested in the precise computation of certain physical quantities ("quantities of interest") rather than in reducing the discretization error with respect to global norms. In [2, 4], a general concept for a posteriori estimation of the discretization error with respect to the cost functional in the context of stationary optimal control problems is presented. In the papers [5, 6], this approach was extended to the estimation of the discretization error with respect to an arbitrary functional depending on both, the control and the state variable, i.e., with respect to a quantity of interest. This allows – among other things – an efficient treatment of parameter identification and model calibration problems. Moreover, a posteriori error estimates for optimal control problems with pointwise inequality constraints were derived, see, e.g., [7, 17, 19, 41].

Here, we discuss the extension of this "quantity of interest" driven approach of error estimation to optimization problems governed by parabolic partial differential equations, see [30, 31]. The results and the numerical examples of [31] are extended to the case of dynamically changing meshes, see [36], which is essential for an efficient adaptive algorithm.

We consider optimization problems under constraints of (nonlinear) parabolic differential equations:

$$\begin{aligned} \partial_t u + A(q,u) &= f \\ u(0) &= u_0(q). \end{aligned} \tag{1.1}$$

Here, the state variable is denoted by u and the control variable by q. Both, the differential operator A and the initial condition u_0 may depend on q. This allows a simultaneous treatment of optimal control and parameter identification problems. For optimal control problems, the operator A is typically given by

$$A(q,u) = \bar{A}(u) - B(q),$$

with a (nonlinear) operator $\bar{A}$ and a (usually linear) control operator B. In parameter identification problems, the variable q denotes the unknown parameters to be determined and may enter the operator A in a nonlinear way. The case of initial control is included via the q-dependent initial condition $u_0(q)$. The target of the optimization is to minimize a given cost functional $J(q,u)$ subject to the state equation (1.1).

For the numerical solution of this optimization problem, the state variable has to be discretized in space and in time. Moreover, if the control (parameter) space is infinite dimensional, it is usually discretized too, cf. [17] for the approach without discretizing the control in the context of elliptic problems. Both, time and space discretization of the state equation are based on the finite element method as proposed, e.g., in [13, 14]. In [3] we have shown that this type of discretization allows for a natural translation of the optimality conditions from the continuous to the discrete level. This gives rise to exact computation of the derivatives required in the optimization algorithms on the discrete level.

Our main contribution is the development of a posteriori error estimates which assess the error between the solutions of the continuous and the discrete optimization problem with respect to a given quantity of interest. This quantity of interest may coincide with the cost functional or expresses another goal of the computation. In order to set up an efficient adaptive algorithm, we separate the influences of the time and space discretizations on the error in the quantity of interest. This allows to balance different types of errors and to successively improve the accuracy by constructing locally refined meshes for time and space discretizations.

We introduce σ as a general discretization parameter including the space, time, and control discretization and denote the solution of the discrete problem by (q_σ, u_σ). For this discrete solution we derive an a posteriori error estimate with respect to the cost functional J of the following form:

$$J(q,u) - J(q_\sigma, u_\sigma) \approx \eta_k^J + \eta_h^J + \eta_d^J. \tag{1.2}$$

Here, η_k^J, η_h^J, and η_d^J denote the error estimators which can be evaluated from the computed discrete solution: η_k^J assess the error due to the time discretization, η_h^J due to the space discretization, and η_d^J due to the discretization of the control space. In the case of not discretizing the control space – cf. [17] for elliptic problems – the contribution of η_d^J vanishes. However, in many situations it is reasonable, to use coarser time grids for the discretization of the control variable than for the state variable, see, e.g., [30]. The structure of the error estimate (1.2) allows for equilibration of different discretization errors within an adaptive refinement algorithm to be described in the sequel.

The error estimator for the error with respect to the cost functional does in general not provide a bound for the norm of the error $\|q - q_\sigma\|$. This allows for very economical meshes – in terms of degrees of freedom – for approximating the optimal value $J(q,u)$. If an estimate for $\|q - q_\sigma\|$ is requested, our technique can be adapted using the (local) coercivity of the reduced cost functional, see [1] for this approach in the context of elliptic problems.

For many optimization problems the quantity of physical interest coincides with the cost functional which explains the choice of the error measure (1.2). However, in the case of parameter identification or model calibration problems, the cost functional is only an instrument for the estimation of the unknown parameters. Therefore, the value of the cost functional in the optimum and the corresponding discretization error are of secondary importance. This motivates error estimation with respect to a given functional I depending on the state and the control (parameter) variable. In this paper we discuss the extension of the corresponding results from [5, 6, 40] to parabolic problems, see [31], and present an a posteriori error estimator of the form

$$I(q,u) - I(q_\sigma, u_\sigma) \approx \eta_k^I + \eta_h^I + \eta_d^I,$$

where again η_k^I and η_h^I estimate the temporal and spatial discretization errors and η_d^I estimates the discretization error due to the discretization of the control space.

In Section 5.2 we describe an adaptive algorithm based on these error estimators. Within this algorithm the time, space, and control discretizations are refined separately to achieve an efficient reduction of the total error by equilibrating different types of the error. This local refinement relies on the computable representation of the error estimators as a sum of local contributions (error indicators), see the discussion in Section 5.1.

The outline of the paper is as follows: In the next section we introduce the functional analytic setting of the considered optimization problem and describe necessary optimality conditions for the problem under consideration. In Section 3 we present the space time finite element discretization of the optimization problem. Section 4 is devoted to the derivation of the error estimators in a general setting. In Section 5 we discuss numerical evaluation of these error estimators and the adaptive algorithm in details. In the last section we present two numerical examples illustrating the behavior of the proposed methods. The first example deals with the optimal control of surface hardening of steel, whereas the second one is concerned with the identification of Arrhenius parameters in a simplified gaseous combustion model.

2. Optimization problems

The optimization problems considered in this paper are formulated in the following abstract setting: Let Q be a Hilbert space for the controls (parameters) with inner product $(\cdot,\cdot)_Q$. Moreover, let V and H be Hilbert spaces, which build together with the dual space V^* of V a Gelfand triple $V \hookrightarrow H \hookrightarrow V^*$. The duality pairing between the Hilbert spaces V and its dual V^* is denoted by $\langle \cdot,\cdot \rangle_{V^* \times V}$ and the inner product in H by $(\cdot,\cdot)$. A typical choice for these space could be

$$V = \left\{ v \in H^1(\Omega) \;\middle|\; v\big|_{\partial\Omega_{\mathrm{D}}} = 0 \right\} \text{ and } H = L^2(\Omega), \tag{2.1}$$

where $\partial\Omega_{\mathrm{D}}$ denotes the part of the boundary of the computational domain Ω with prescribed Dirichlet boundary conditions.

For a time interval $(0,T)$ we introduce the Hilbert space $X := W(0,T)$ defined as

$$W(0,T) = \left\{ v \;\middle|\; v \in L^2((0,T),V) \text{ and } \partial_t v \in L^2((0,T),V^*) \right\}. \tag{2.2}$$

It is well known that the space X is continuously embedded in $C([0,T],H)$, see, e.g., [11]. Furthermore, we use the inner product of $L^2((0,T),H)$ given by

$$(u,v)_I := (u,v)_{L^2((0,T),H)} = \int_0^T (u(t),v(t))\,dt \tag{2.3}$$

for setting up the weak formulation of the state equation.

By means of the spatial semi-linear form $\bar{a}\colon Q \times V \times V \to \mathbb{R}$ defined for a differential operator $A\colon Q \times V \to V^*$ by

$$\bar{a}(q,\bar{u})(\bar{\varphi}) := \langle A(q,\bar{u}),\bar{\varphi}\rangle_{V^*\times V},$$

we can define the semi-linear form $a(\cdot,\cdot)(\cdot)$ on $Q \times L^2(I,V) \times L^2(I,V)$ as

$$a(q,u)(\varphi) := \int_0^T \bar{a}(q,u(t))(\varphi(t))\,dt$$

which is assumed to be linear in the third argument and three times directional differentiable with derivatives which are linear in the direction. This would be implied if the semi-linear form is Gâteaux differentiable. However, in our analysis, we do not require the boundedness of the derivatives.

Remark 2.1. If the control variable q depends on time, this has to be incorporated by an obvious modification of the definitions of the semi-linear forms.

After these preliminaries, we pose the *state equation* in a weak form: Find for given control $q \in Q$ the *state* $u \in X$ such that

$$\begin{aligned} (\partial_t u,\varphi)_I + a(q,u)(\varphi) &= (f,\varphi)_I \quad \forall \varphi \in X,\\ u(0) &= u_0(q), \end{aligned} \tag{2.4}$$

where $f \in L^2((0,T),V^*)$ represents the right-hand side of the state equation and $u_0\colon Q \to H$ denotes a three times directional differentiable mapping with derivatives which are linear in the direction describing parameter-dependent initial conditions.

Remark 2.2. There are several sets of assumptions on the nonlinearity in $\bar{a}(\cdot,\cdot)(\cdot)$ and its dependence on the control variable q allowing the state equation (2.4) to be well posed. Typical examples are different semi-linear equations, where the form $\bar{a}(\cdot,\cdot)(\cdot)$ consists of a linear elliptic part and a nonlinear term depending on u and ∇u. Due to the fact that the development of the proposed adaptive algorithm does not depend on the particular structure of the nonlinearity in $\bar{a}$, we do not specify a set of assumptions on it, but assume that the state equation (2.4) possess a unique solution $u \in X$ for each $q \in Q$.

The cost functional $J\colon Q \times \{\, v \in L^2(I,V) \mid v(T) \in H \,\} \to \mathbb{R}$ is defined using two three times direction differentiable functionals $J_1\colon V \to \mathbb{R}$ and $J_2\colon H \to \mathbb{R}$ with derivatives which are linear in the direction by

$$J(q,u) = \int_0^T J_1(u)\,dt + J_2(u(T)) + \frac{\alpha}{2}\|q-\bar{q}\|_Q^2, \tag{2.5}$$

where the regularization (or cost) term is added which involves $\alpha \geq 0$ and a reference control (parameter) $\bar{q} \in Q$. We note, that due to the embedding $X \hookrightarrow C(\bar{I}, H)$ the functional J is well defined for arguments $u \in X$.

The corresponding optimization problem is formulated as follows:

$$\text{Minimize } J(q,u) \text{ subject to the state equation (2.4)}, \ (q,u) \in Q \times X. \tag{2.6}$$

The question of existence and uniqueness of solutions to such optimization problems is discussed, e.g., in [24, 16, 38]. Throughout the paper, we assume problem (2.6) to admit a (locally) unique solution. Moreover, we assume the existence of a neighborhood $W \subset Q \times X$ of the optimal solution, such that the linearized form $\bar{a}'_u(q, u(t))(\cdot,\cdot)$ considered as a linear operator

$$\bar{a}'_u(q, u(t)) \colon V \to V^*$$

is an isomorphism for all $(q,u) \in W$ and almost all $t \in (0,T)$. This assumption will allow all considered linearized and adjoint problems to be well posed.

To set up the optimality system we introduce the Lagrangian $\mathcal{L} \colon Q \times X \times X \to \mathbb{R}$, defined as

$$\mathcal{L}(q,u,z) = J(q,u) + (f - \partial_t u, z)_I - a(q,u)(z) - (u(0) - u_0(q), z(0)). \tag{2.7}$$

By definition, $\mathcal{L}$ inherits the differentiability properties from those of a, u_0, and J.

The optimality system of the considered optimization problem (2.6) is given by the derivatives of the Lagrangian:

$$\begin{aligned} \mathcal{L}'_z(q,u,z)(\varphi) &= 0, \quad \forall \varphi \in X \qquad \text{(State equation)},\\ \mathcal{L}'_u(q,u,z)(\varphi) &= 0, \quad \forall \varphi \in X \qquad \text{(Adjoint state equation)},\\ \mathcal{L}'_q(q,u,z)(\psi) &= 0, \quad \forall \psi \in Q \qquad \text{(Gradient equation)}. \end{aligned} \tag{2.8}$$

For the explicit formulation of the dual equation in this setting see, e.g., [3].

3. Discretization

In this section, we discuss the discretization of the optimization problem (2.6). To this end, we use Galerkin finite element methods in space and time to discretize the state equation. Our systematic approach to a posteriori error estimation relies on using such Galerkin-type discretizations.

The first of the following subsections is devoted to the semi-discretization in time by *discontinuous Galerkin* (dG) methods. Subsection 3.2 deals with the space discretization of the semi-discrete problems arising from time discretization. For the numerical analysis of these schemes we refer to [13]. In the context of optimal control problems, a priori error estimates for this type of discretization are derived in [32, 33].

The discretization of the control space Q is kept rather abstract by choosing a finite-dimensional subspace $Q_d \subset Q$. A possible concretion of this choice is shown in the numerical examples in Section 6.

3.1. Time discretization of the states

To define a semi-discretization in time, let us partition the time interval $[0,T]$ as

$$[0,T] = \{0\} \cup I_1 \cup I_2 \cup \cdots \cup I_M$$

with subintervals $I_m = (t_{m-1}, t_m]$ of size k_m and time points

$$0 = t_0 < t_1 < \cdots < t_{M-1} < t_M = T.$$

We define the discretization parameter k as a piecewise constant function by setting $k\big|_{I_m} = k_m$ for $m = 1, \ldots, M$.

By means of the subintervals I_m, we define for $r \in \mathbb{N}_0$ the semi-discrete space X_k^r by

$$X_k^r = \left\{ v_k \in L^2((0,T),V) \;\middle|\; v_k\big|_{I_m} \in \mathcal{P}^r(I_m,V) \text{ and } v_k(0) \in H \right\}.$$

Here, $\mathcal{P}^r(I_m,V)$ denotes the space of polynomials up to order r defined on I_m with values in V. Thus, the functions in X_k^r may have discontinuities at the edges of the subintervals I_m. This space is used in the sequel as trial and test space in the discontinuous Galerkin method. For functions in X_k^r we use the notation

$$(v,w)_{I_m} := (v,w)_{L^2(I_m,L^2(\Omega))}.$$

To define the dG(r) discretization we employ the following definition for functions $v_k \in X_k^r$:

$$v_{k,m}^+ := \lim_{t\to 0^+} v_k(t_m+t), \quad v_{k,m}^- := \lim_{t\to 0^+} v_k(t_m-t) = v_k(t_m), \quad [v_k]_m := v_{k,m}^+ - v_{k,m}^-.$$

Then, the dG(r) semi-discretization of the state equation (2.4) reads: Find for given control $q_k \in Q$ a state $u_k \in X_k^r$ such that

$$\begin{aligned} \sum_{m=1}^{M} (\partial_t u_k, \varphi)_{I_m}\, dt + a(q_k,u_k)(\varphi) + \sum_{m=0}^{M-1} ([u_k]_m, \varphi_m^+) &= (f,\varphi)_I, \quad \forall \varphi \in X_k^r, \\ u_{k,0}^- &= u_0(q_k). \end{aligned} \tag{3.1}$$

Remark 3.1. This equation is assumed to posses a unique solution for each $q_k \in Q$, cf. Remark 2.2. In special cases, the existence and uniqueness can be shown by separation of variables and using the fact that X_k^r is finite dimensional with respect to time.

The semi-discrete optimization problem for the dG(r) time discretization has the form:

$$\text{Minimize } J(q_k,u_k) \text{ subject to the state equation (3.1)}, \quad (q_k,u_k) \in Q \times X_k^r. \tag{3.2}$$

Then, we pose the Lagrangian $\widetilde{\mathcal{L}}\colon Q \times X_k^r \times X_k^r \to \mathbb{R}$ associated with the dG(r) time discretization for the state equation as

$$\begin{aligned}\widetilde{\mathcal{L}}(q_k, u_k, z_k) = J(q_k, u_k) + (f, z_k)_I - \sum_{m=1}^{M} (\partial_t u_k, z_k)_{I_m}\, dt \\ - a(q_k, u_k)(z_k) - \sum_{m=0}^{M-1} ([u_k]_m, z_{k,m}^+) - (u_{k,0}^- - u_0(q_k), z_{k,0}^-).\end{aligned}$$

As on the continuous level for $\mathcal{L}$, $\widetilde{\mathcal{L}}$ inherits here the differentiability properties from those of a, u_0, and J.

3.2. Space discretization of the states

In this subsection, we first describe the finite element discretization in space. To this end, we consider two- or three-dimensional shape-regular meshes, see, e.g., [9]. A mesh consists of quadrilateral or hexahedral cells K which constitute a non-overlapping cover of the computational domain $\Omega \subset \mathbb{R}^n$, $n \in \{2, 3\}$. The corresponding mesh is denoted by $\mathcal{T}_h = \{K\}$, where we define the discretization parameter h as a cellwise constant function by setting $h\big|_K = h_K$ with the diameter h_K of the cell K.

We allow dynamic mesh changes in time whereas the time steps k_m are kept constant in space. Therefore, we associate with each time point t_m a triangulation $\mathcal{T}_h^m$ and a corresponding finite element space $V_h^{s,m} \subseteq V$ which is used as spatial trial and test space in the adjacent time interval I_m.

The finite element space $V_h^{s,m} \subset V$ is constructed in a standard way by

$$V_h^{s,m} = \left\{\, v \in V \;\middle|\; v\big|_K \in \mathcal{Q}^s(K) \text{ for } K \in \mathcal{T}_h^m \,\right\},$$

where $\mathcal{Q}^s(K)$ consists of shape functions obtained via bi- or tri-linear transformations of polynomials in $\widehat{\mathcal{Q}^s}(\widehat{K})$ defined on the reference cell $\widehat{K} = (0,1)^n$, see, e.g., [9] for details.

To obtain the fully discretized versions of the time discretized state equation (3.1), we utilize the space-time finite element space

$$X_{k,h}^{r,s} = \left\{\, v_{kh} \in L^2((0.T), H) \;\middle|\; v_{kh}\big|_{I_m} \in \mathcal{P}^r(I_m, V_h^{s,m}) \text{ and } v_{kh}(0) \in V_h^{s,0} \,\right\}.$$

Then, the cG(s)dG(r) discretization has the form: Find for given control $q_{kh} \in Q$ a state $u_{kh} \in X_{k,h}^{r,s}$ such that

$$\begin{aligned}\sum_{m=1}^{M} (\partial_t u_{kh}, \varphi)_{I_m}\, dt + a(q_{kh}, u_{kh})(\varphi) + \sum_{m=0}^{M-1} ([u_{kh}]_m, \varphi_m^+) + (u_{kh,0}^-, \varphi_0^-) \\ = (f, \varphi)_I + (u_0(q_{kh}), \varphi_0^-) \quad \forall \varphi \in X_{k,h}^{r,s}. \quad (3.3)\end{aligned}$$

These fully discretized state equations are assumed to posses unique solutions for each $q_{kh} \in Q$, see Remark 3.1.

Thus, the optimization problem with fully discretized state is given by

$$\text{Minimize } J(q_{kh}, u_{kh}) \text{ subject to the state equation (3.3),} \qquad (q_{kh}, u_{kh}) \in Q \times X_{k,h}^{r,s}. \quad (3.4)$$

The definition of the Lagrangian $\widetilde{\mathcal{L}}$ for fully discretized states can directly be transferred from the formulations for semi-discretization in time just by restriction of the state space X_k^r to the subspaces $X_{k,h}^{r,s}$, respectively.

3.3. Discretization of the controls

As proposed in the beginning of the current section, the discretization of the control space Q is kept rather abstract. It is done by choosing a finite-dimensional subspace $Q_d \subset Q$. Then, the formulation of the state equation, the optimization problems and the Lagrangians defined on the fully discretized state space can directly be transferred to the level with fully discretized control and state spaces by replacing Q by Q_d. The full discrete solutions will be indicated by the subscript σ which collects the discretization indices k, h, and d.

4. A posteriori error estimators

In this section, we establish a posteriori error estimators for the error arising due to the discretization of the control and state spaces in terms of the cost functional J and in terms of an arbitrary quantity of interest I.

4.1. Error estimator for the cost functional

In the sequel, we derive error estimators for the discretization error with respect to the cost functional J:

$$J(q, u) - J(q_\sigma, u_\sigma)$$

Here, $(q, u) \in Q \times X$ denotes the continuous optimal solution of (2.6) and $(q_\sigma, u_\sigma) = (q_{khd}, u_{khd}) \in Q_d \times X_{k,h}^{r,s}$ is the optimal solution of the fully discretized problem.

To separate the influences of the different discretizations on the discretization error, we split

$$\begin{aligned} J(q, u) - J(q_\sigma, u_\sigma) = {} & J(q, u) - J(q_k, u_k) \\ & + J(q_k, u_k) - J(q_{kh}, u_{kh}) \\ & + J(q_{kh}, u_{kh}) - J(q_\sigma, u_\sigma), \end{aligned}$$

where $(q_k, u_k) \in Q \times X_k^r$ is the solution of the time discretized problem (3.2) and $(q_{kh}, u_{kh}) \in Q \times X_{k,h}^{r,s}$ is the solution of the time and space discretized problem (3.4) with still undiscretized control space Q.

Theorem 4.1. *Let $\xi = (q, u, z)$, $\xi_k = (q_k, u_k, z_k)$, $\xi_{kh} = (q_{kh}, u_{kh}, z_{kh})$, and $\xi_\sigma = (q_\sigma, u_\sigma, z_\sigma)$ be stationary points of $\mathcal{L}$ resp. $\widetilde{\mathcal{L}}$ on the different levels of discretization, i.e.,*

$$\begin{aligned}
\mathcal{L}'(\xi)(\hat{q}, \hat{u}, \hat{z}) = \widetilde{\mathcal{L}}'(\xi)(\hat{q}, \hat{u}, \hat{z}) &= 0, \quad \forall (\hat{q}, \hat{u}, \hat{z}) \in Q \times X \times X,\\
\widetilde{\mathcal{L}}'(\xi_k)(\hat{q}_k, \hat{u}_k, \hat{z}_k) &= 0, \quad \forall (\hat{q}_k, \hat{u}_k, \hat{z}_k) \in Q \times X_k^r \times X_k^r,\\
\widetilde{\mathcal{L}}'(\xi_{kh})(\hat{q}_{kh}, \hat{u}_{kh}, \hat{z}_{kh}) &= 0, \quad \forall (\hat{q}_{kh}, \hat{u}_{kh}, \hat{z}_{kh}) \in Q \times X_{k,h}^{r,s}, \times X_{k,h}^{r,s}\\
\widetilde{\mathcal{L}}'(\xi_\sigma)(\hat{q}_\sigma, \hat{u}_\sigma, \hat{z}_\sigma) &= 0, \quad \forall (\hat{q}_\sigma, \hat{u}_\sigma, \hat{z}_\sigma) \in Q_d \times X_{k,h}^{r,s} \times X_{k,h}^{r,s}.
\end{aligned}$$

Then, there holds for the errors with respect to the cost functional due to the time, space, and control discretizations:

$$\begin{aligned}
J(q,u) - J(q_k, u_k) &= \frac{1}{2}\widetilde{\mathcal{L}}'(\xi_k)(q - \hat{q}_k, u - \hat{u}_k, z - \hat{z}_k) + \mathcal{R}_k\\
J(q_k, u_k) - J(q_{kh}, u_{kh}) &= \frac{1}{2}\widetilde{\mathcal{L}}'(\xi_{kh})(q_k - \hat{q}_{kh}, u_k - \hat{u}_{kh}, z_k - \hat{z}_{kh}) + \mathcal{R}_h\\
J(q_{kh}, u_{kh}) - J(q_\sigma, u_\sigma) &= \frac{1}{2}\widetilde{\mathcal{L}}'(\xi_\sigma)(q_{kh} - \hat{q}_\sigma, u_{kh} - \hat{u}_\sigma, z_{kh} - \hat{z}_\sigma) + \mathcal{R}_d.
\end{aligned}$$

Here,

$$(\hat{q}_k, \hat{u}_k, \hat{z}_k) \in Q \times X_k^r \times X_k^r, \quad (\hat{q}_{kh}, \hat{u}_{kh}, \hat{z}_{kh}) \in Q \times X_{k,h}^{r,s} \times X_{k,h}^{r,s},$$

and

$$(\hat{q}_\sigma, \hat{u}_\sigma, \hat{z}_\sigma) \in Q_d \times X_{k,h}^{r,s} \times X_{k,h}^{r,s}$$

can be chosen arbitrary. The remainder terms $\mathcal{R}_k$, $\mathcal{R}_h$, and $\mathcal{R}_d$ are of third order in the error, see [31]. $\mathcal{R}_k$, for instance, is given for $e_k := \xi - \xi_k$ by

$$\frac{1}{2}\int_0^1 \widetilde{\mathcal{L}}'''(\xi_k + se)(e, e, e) \cdot s \cdot (1 - s)\, ds.$$

Remark 4.2. Due to the structure of $\widetilde{\mathcal{L}}$, the remainder terms $\mathcal{R}_k$, $\mathcal{R}_h$, and $\mathcal{R}_d$ depend only on third derivatives of the semi-linear form a and the cost functional J. If the cost functional is of tracking type, then J''' vanishes and the remainder term depend only on a'' and a'''. For given applications, one can explicitly calculate the concrete form of the remainder terms.

By means of the residuals of the three equations building the optimality system (2.8)

$$\begin{aligned}
\rho^u(q,u)(\varphi) := \widetilde{\mathcal{L}}'_z(\xi)(\varphi), \quad \rho^z(q,u,z)(\varphi) := \widetilde{\mathcal{L}}'_u(\xi)(\varphi),\\
\rho^q(q,u,z)(\varphi) := \widetilde{\mathcal{L}}'_q(\xi)(\varphi),
\end{aligned}$$

the statement of Theorem 4.1 can be rewritten as

$$J(q,u)-J(q_k,u_k)\approx\frac{1}{2}\Big(\rho^u(q_k,u_k)(z-\hat{z}_k)+\rho^z(q_k,u_k,z_k)(u-\hat{u}_k)\Big) \tag{4.1a}$$

$$J(q_k,u_k)-J(q_{kh},u_{kh})\approx\frac{1}{2}\Big(\rho^u(q_{kh},u_{kh})(z_k-\hat{z}_{kh})+\rho^z(q_{kh},u_{kh},z_{kh})(u_k-\hat{u}_{kh})\Big) \tag{4.1b}$$

$$J(q_{kh},u_{kh})-J(q_\sigma,u_\sigma)\approx\frac{1}{2}\rho^q(q_\sigma,u_\sigma,z_\sigma)(q_{kh}-\hat{q}_\sigma). \tag{4.1c}$$

Here, we employed the fact, that the terms

$$\begin{aligned}&\rho^q(q_k,u_k,z_k)(q-\hat{q}_k), &&\rho^q(q_{kh},u_{kh},z_{kh})(q_k-\hat{q}_{kh}),\\ &\rho^u(q_\sigma,u_\sigma)(z_{kh}-\hat{z}_\sigma), &&\rho^z(q_\sigma,u_\sigma,z_\sigma)(u_{kh}-\hat{u}_\sigma)\end{aligned}$$

are zero for the choice

$$\begin{aligned}&\hat{q}_k=q\in Q, &&\hat{q}_{kh}=q_k\in Q,\\ &\hat{z}_\sigma=z_{kh}\in X^{r,s}_{k,h}, &&\hat{u}_\sigma=u_{kh}\in X^{r,s}_{k,h}.\end{aligned}$$

This is possible since for the errors $J(q,u)-J(q_k,u_k)$ and $J(q_k,u_k)-J(q_{kh},u_{kh})$ only the state space is discretized and for $J(q_{kh},u_{kh})-J(q_\sigma,u_\sigma)$ we keep the discrete state space while discretizing the control space Q.

4.2. Error estimator for an arbitrary functional

We now tend toward an error estimation of the different types of discretization errors in terms of a given functional $I\colon Q\times X\to\mathbb{R}$ describing the quantity of interest. This will be done using solutions of some auxiliary problems. In order to ensure the solvability of these problems, we assume that the semi-discrete and the fully discrete optimal solutions (q_k,u_k), (q_{kh},u_{kh}), and (q_σ,u_σ) are in the neighborhood $W\subset Q\times X$ of the optimal solution (q,u) introduced in Section 2.

We define exterior Lagrangians $\mathcal{M}\colon [Q\times X\times X]^2\to\mathbb{R}$ and $\widetilde{\mathcal{M}}\colon [Q\times X^r_k\times X^r_k]^2\to\mathbb{R}$ as

$$\mathcal{M}(\xi,\chi)=I(q,u)+\mathcal{L}'(\xi)(\chi)$$

with $\xi=(q,u,z),\chi=(p,v,y)$ and

$$\widetilde{\mathcal{M}}(\xi_k,\chi_k)=I(q_k,u_k)+\widetilde{\mathcal{L}}'(\xi_k)(\chi_k)$$

with $\xi_k=(q_k,u_k,z_k),\chi_k=(p_k,v_k,y_k)$.

Now we are in a similar setting as in the subsection before: We split the total discretization error with respect to I as

$$\begin{aligned}I(q,u)-I(q_\sigma,u_\sigma)=\;&I(q,u)-I(q_k,u_k)\\&+I(q_k,u_k)-I(q_{kh},u_{kh})\\&+I(q_{kh},u_{kh})-I(q_\sigma,u_\sigma)\end{aligned}$$

and obtain the following theorem:

Theorem 4.3. *Let (ξ,χ), (ξ_k,χ_k), (ξ_{kh},χ_{kh}), and (ξ_σ,χ_σ) be stationary points of $\mathcal{M}$ resp. $\widetilde{\mathcal{M}}$ on the different levels of discretization, i.e.,*

$$\begin{aligned}
\mathcal{M}'(\xi,\chi)(\hat{\xi},\hat{\chi}) = \widetilde{\mathcal{M}}'(\xi,\chi)(\hat{\xi},\hat{\chi}) &= 0, \quad \forall(\hat{\xi},\hat{\chi}) \in [Q \times X \times X]^2,\\
\widetilde{\mathcal{M}}'(\xi_k,\chi_k)(\hat{\xi}_k,\hat{\chi}_k) &= 0, \quad \forall(\hat{\xi}_k,\hat{\chi}_k) \in [Q \times X_k^r \times X_k^r]^2,\\
\widetilde{\mathcal{M}}'(\xi_{kh},\chi_{kh})(\hat{\xi}_{kh},\hat{\chi}_{kh}) &= 0, \quad \forall(\hat{\xi}_{kh},\hat{\chi}_{kh}) \in [Q \times X_{k,h}^{r,s} \times X_{k,h}^{r,s}]^2,\\
\widetilde{\mathcal{M}}'(\xi_\sigma,\chi_\sigma)(\hat{\xi}_\sigma,\hat{\chi}_\sigma) &= 0, \quad \forall(\hat{\xi}_\sigma,\hat{\chi}_\sigma) \in [Q_d \times X_{k,h}^{r,s} \times X_{k,h}^{r,s}]^2.
\end{aligned}$$

Then, there holds for the errors with respect to the quantity of interest due to the time, space, and control discretizations:

$$\begin{aligned}
I(q,u) - I(q_k,u_k) &= \frac{1}{2}\widetilde{\mathcal{M}}'(\xi_k,\chi_k)(\xi-\hat{\xi}_k,\chi-\hat{\chi}_k) + \mathcal{R}_k,\\
I(q_k,u_k) - I(q_{kh},u_{kh}) &= \frac{1}{2}\widetilde{\mathcal{M}}'(\xi_{kh},\chi_{kh})(\xi_k-\hat{\xi}_{kh},\chi_k-\hat{\chi}_{kh}) + \mathcal{R}_h,\\
I(q_{kh},u_{kh}) - I(q_\sigma,u_\sigma) &= \frac{1}{2}\widetilde{\mathcal{M}}'(\xi_\sigma,\chi_\sigma)(\xi_{kh}-\hat{\xi}_\sigma,\chi_{kh}-\hat{\chi}_\sigma) + \mathcal{R}_d.
\end{aligned}$$

Here,

$$(\hat{\xi}_k,\hat{\chi}_k) \in [Q \times X_k^r \times X_k^r]^2, \quad (\hat{\xi}_{kh},\hat{\chi}_{kh}) \in [Q \times X_{k,h}^{r,s} \times X_{k,h}^{r,s}]^2,$$

and

$$(\hat{\xi}_\sigma,\hat{\chi}_\sigma) \in [Q_d \times X_{k,h}^{r,s} \times X_{k,h}^{r,s}]^2$$

can be chosen arbitrary and the remainder terms $\mathcal{R}_k$, $\mathcal{R}_h$, and $\mathcal{R}_d$ are of third order in the error, see [31].

To apply Theorem 4.3 for instance to $I(q_{kh},u_{kh}) - I(q_\sigma,u_\sigma)$, we have to require that

$$\widetilde{\mathcal{M}}'(\xi_\sigma,\chi_\sigma)(\hat{\xi}_\sigma,\hat{\chi}_\sigma) = 0, \quad \forall(\hat{\xi}_\sigma,\hat{\chi}_\sigma) \in [X_{k,h}^{r,s} \times X_{k,h}^{r,s} \times Q_d]^2.$$

For solving this system, we have to consider the concrete form of $\widetilde{\mathcal{M}}'$:

$$\begin{aligned}
&\widetilde{\mathcal{M}}'(\xi_\sigma,\chi_\sigma)(\delta\xi_\sigma,\delta\chi_\sigma)\\
&\qquad = I_q'(q_\sigma,u_\sigma)(\delta q_\sigma) + I_u'(q_\sigma,u_\sigma)(\delta u_\sigma) + \widetilde{\mathcal{L}}'(\xi_\sigma)(\delta\chi_\sigma) + \widetilde{\mathcal{L}}''(\xi_\sigma)(\chi_\sigma,\delta\xi_\sigma)
\end{aligned}$$

Since (q_σ,u_σ) is the solution of the discrete optimization problem, $\xi_\sigma = (q_\sigma,u_\sigma,z_\sigma)$ is a stationary point of the Lagrangian $\widetilde{\mathcal{L}}$ and consequently, it fulfills $\widetilde{\mathcal{L}}'(\xi_\sigma)(\delta\chi_\sigma) = 0$. Thus, the solution triple $\chi_\sigma = (p_\sigma,v_\sigma,y_\sigma) \in Q_d \times X_{k,h}^{r,s} \times X_{k,h}^{r,s}$ has to fulfill

$$\begin{aligned}
&\widetilde{\mathcal{L}}''(\xi_\sigma)(\chi_\sigma,\delta\xi_\sigma)\\
&\quad = -I_q'(q_\sigma,u_\sigma)(\delta q_\sigma) - I_u'(q_\sigma,u_\sigma)(\delta u_\sigma), \quad \forall\delta\xi_\sigma \in Q_d \times X_{k,h}^{r,s} \times X_{k,h}^{r,s}. \qquad (4.2)
\end{aligned}$$

Solving this system of equations is apart from a different right-hand side equivalent to the execution of one step of a (reduced) SQP-type method, see [31] for details.

By means of the residuals of the equations for p, v and y, i.e.,

$$\rho^v(\xi, p, v)(\varphi) := \widetilde{\mathcal{L}}''_{uz}(\xi)(v, \varphi) + \widetilde{\mathcal{L}}''_{qz}(\xi)(p, \varphi)$$
$$\rho^y(\xi, p, v, y)(\varphi) := \widetilde{\mathcal{L}}''_{zu}(\xi)(y, \varphi) + \widetilde{\mathcal{L}}''_{qu}(\xi)(p, \varphi) + \widetilde{\mathcal{L}}''_{uu}(\xi)(v, \varphi) + I'_u(q, u)(\varphi)$$
$$\rho^p(\xi, p, v, y)(\varphi) := \widetilde{\mathcal{L}}''_{qq}(\xi)(p, \varphi) + \widetilde{\mathcal{L}}''_{uq}(\xi)(v, \varphi) + \widetilde{\mathcal{L}}''_{zq}(\xi)(y, \varphi) + I'_q(q, u)(\varphi),$$

and the already defined residuals ρ^u, ρ^z, and ρ^q the result of Theorem 4.3 can be expressed as

$$I(q, u) - I(q_k, u_k) \approx \frac{1}{2}\Big(\rho^u(q_k, u_k)(y - \hat{y}_k) + \rho^z(q_k, u_k, z_k)(v - \hat{v}_k) + \rho^v(\xi_k, p_k, v_k)(z - \hat{z}_k) + \rho^y(\xi_k, p_k, v_k, y_k)(u - \hat{u}_k)\Big)$$

$$I(q_k, u_k) - I(q_{kh}, u_{kh}) \approx \frac{1}{2}\Big(\rho^u(q_{kh}, u_{kh})(y_k - \hat{y}_{kh}) + \rho^z(q_{kh}, u_{kh}, z_{kh})(v_k - \hat{v}_{kh}) + \rho^v(\xi_{kh}, p_{kh}, v_{kh})(z_k - \hat{z}_{kh}) + \rho^y(\xi_{kh}, p_{kh}, v_{kh}, y_{kh})(u_k - \hat{u}_{kh})\Big)$$

$$I(q_{kh}, u_{kh}) - I(q_\sigma, u_\sigma) \approx \frac{1}{2}\Big(\rho^q(q_\sigma, u_\sigma, z_\sigma)(p_{kh} - \hat{p}_\sigma) + \rho^p(\xi_\sigma, p_\sigma, v_\sigma, y_\sigma)(q_{kh} - \hat{q}_\sigma)\Big).$$

As for the estimator for the error in the cost functional, we employed here the fact, that the terms

$$\begin{aligned} &\rho^q(q_k, u_k, z_k)(p - \hat{p}_k), && \rho^p(\xi_k, p_k, v_k, y_k)(q - \hat{q}_k),\\ &\rho^q(q_{kh}, u_{kh}, z_{kh})(p_k - \hat{p}_{kh}), && \rho^p(\xi_{kh}, p_{kh}, v_{kh}, y_{kh})(q_k - \hat{q}_{kh}),\\ &\rho^u(q_\sigma, u_\sigma)(y_{kh} - \hat{y}_\sigma), && \rho^z(q_\sigma, u_\sigma, z_\sigma)(v_{kh} - \hat{v}_\sigma),\\ &\rho^v(\xi_\sigma, p_\sigma, v_\sigma)(z_{kh} - \hat{z}_\sigma), && \rho^y(\xi_\sigma, p_\sigma, v_\sigma, y_\sigma)(u_{kh} - \hat{u}_\sigma) \end{aligned}$$

vanish if $\hat{p}_k$, $\hat{q}_k$, $\hat{p}_{kh}$, $\hat{q}_{kh}$, $\hat{y}_\sigma$, $\hat{v}_\sigma$, $\hat{z}_\sigma$, $\hat{u}_\sigma$ are chosen appropriately.

Remark 4.4. For the error estimation with respect to the cost functional no additional equations have to be solved. The error estimation with respect to a given quantity of interest requires the computation of the auxiliary variables p_σ, v_σ, y_σ. The additional numerical effort is similar to the execution of one step of the SQP or Newton's method.

5. Numerical realization

5.1. Evaluation of the error estimators

In this subsection, we concretize the a posteriori error estimator developed in the previous section for the cG(1)dG(0) space-time discretization on quadrilateral meshes in two space dimensions. That is, we consider the combination of dG(0) time discretization with piecewise bi-linear finite elements for the space discretization.

The error estimates presented in the previous section involve interpolation errors of the time, space, and the control discretizations. We approximate these

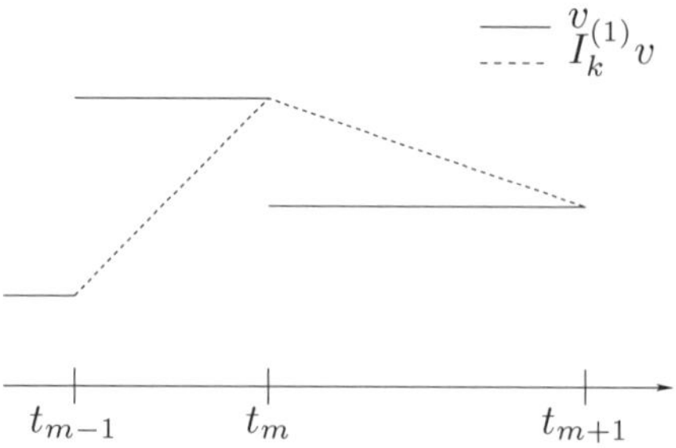

FIGURE 1. Temporal Interpolation

errors using interpolations in higher-order finite element spaces. To this end, we introduce linear operators Π_h, Π_k, and Π_d, which will map the computed solutions to the approximations of the interpolation errors:

$$
\begin{aligned}
z - \hat{z}_k &\approx \Pi_k z_k & \qquad u - \hat{u}_k &\approx \Pi_k u_k \\
z_k - \hat{z}_{kh} &\approx \Pi_h z_{kh} & \qquad u_k - \hat{u}_{kh} &\approx \Pi_h u_{kh} \\
q_{kh} - \hat{q}_\sigma &\approx \Pi_d q_\sigma & &
\end{aligned}
$$

$$
\begin{aligned}
y - \hat{y}_k &\approx \Pi_k y_k & \qquad v - \hat{v}_k &\approx \Pi_k v_k \\
y_k - \hat{y}_{kh} &\approx \Pi_h y_{kh} & \qquad v_k - \hat{v}_{kh} &\approx \Pi_h v_{kh} \\
p_{kh} - \hat{p}_\sigma &\approx \Pi_d p_\sigma & &
\end{aligned}
$$

For the here considered case of cG(1)dG(0) discretization of the state space, the operators are chosen as

$$
\begin{aligned}
\Pi_k &= I_k^{(1)} - \mathrm{id} & \quad \text{with} \quad & I_k^{(1)} : X_k^0 \to \widetilde{X}_k^1, \\
\Pi_h &= I_{2h}^{(2)} - \mathrm{id} & \quad \text{with} \quad & I_{2h}^{(2)} : X_{k,h}^{0,1} \to \widetilde{X}_{k,2h}^{0,2}.
\end{aligned}
$$

Here, $\widetilde{X}_k^1$ denotes the space of continuous piecewise linear polynomials with values in V. The action of the piecewise linear interpolation operator $I_k^{(1)}$ in time is depicted in Figure 1.

The piecewise bi-quadratic spatial interpolation $I_{2h}^{(2)}$ can be easily computed if the underlying mesh provides a patch structure. That is, one can always combine four adjacent cells to a macro cell on which the bi-quadratic interpolation can be defined. An example of such an patched mesh is shown in Figure 2.

The choice of Π_d depends on the discretization of the control space Q. If the finite-dimensional subspaces Q_d are constructed similar to the discrete state spaces, one can directly choose for Π_d a modification of the operators Π_k and Π_h defined above. If for example the controls q only depend on time and the discretization is done with piecewise constant polynomials, we can choose $\Pi_d = I_d^{(1)} - \mathrm{id}$. If the control space Q is already finite-dimensional and Q_d is chosen equal to Q, which is usually the case in the context of parameter estimation, it is possible to choose

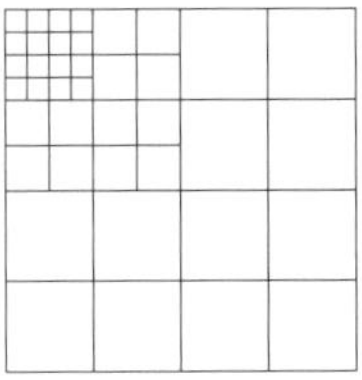

FIGURE 2. Patched Mesh

$\Pi_d = 0$ and thus, the estimator for the error $J(q_{kh}, u_{kh}) - J(q_\sigma, u_\sigma)$ is zero – as well as this discretization error itself.

In order to make the error representations from the previous section computable, we replace the residuals linearized on the solution of semi-discretized problems by the linearization at full discrete solutions.

We finally obtain the following computable a posteriori error estimator for the cost functional J

$$J(q, u) - J(q_\sigma, u_\sigma) \approx \eta_k^J + \eta_h^J + \eta_d^J$$

with

$$\begin{aligned}
\eta_k^J &:= \frac{1}{2}\Big(\rho^u(q_\sigma, u_\sigma)(\Pi_k z_\sigma) + \rho^z(q_\sigma, u_\sigma, z_\sigma)(\Pi_k u_\sigma)\Big)\\
\eta_h^J &:= \frac{1}{2}\Big(\rho^u(q_\sigma, u_\sigma)(\Pi_h z_\sigma) + \rho^z(q_\sigma, u_\sigma, z_\sigma)(\Pi_h u_\sigma)\Big)\\
\eta_d^J &:= \frac{1}{2}\rho^q(q_\sigma, u_\sigma, z_\sigma)(\Pi_d q_\sigma).
\end{aligned}$$

For the quantity of interest I the error estimator is given by

$$I(q, u) - I(q_\sigma, u_\sigma) \approx \eta_k^I + \eta_h^I + \eta_d^I$$

with

$$\begin{aligned}
\eta_k^I := \frac{1}{2}\Big(&\rho^u(q_\sigma, u_\sigma)(\Pi_k y_\sigma) + \rho^z(q_\sigma, u_\sigma, z_\sigma)(\Pi_k v_\sigma)\\
&+ \rho^v(\xi_\sigma, v_\sigma, p_\sigma)(\Pi_k z_\sigma) + \rho^y(\xi_\sigma, v_\sigma, y_\sigma, p_\sigma)(\Pi_k u_\sigma)\Big)\\
\eta_h^I := \frac{1}{2}\Big(&\rho^u(q_\sigma, u_\sigma)(\Pi_h y_\sigma) + \rho^z(q_\sigma, u_\sigma, z_\sigma)(\Pi_h v_\sigma)\\
&+ \rho^v(\xi_\sigma, v_\sigma, p_\sigma)(\Pi_h z_\sigma) + \rho^y(\xi_\sigma, v_\sigma, y_\sigma, p_\sigma)(\Pi_h u_\sigma)\Big)\\
\eta_d^I := \frac{1}{2}\Big(&\rho^q(q_\sigma, u_\sigma, z_\sigma)(\Pi_d p_\sigma) + \rho^p(\xi_\sigma, v_\sigma, y_\sigma, p_\sigma)(\Pi_d q_\sigma)\Big).
\end{aligned}$$

The presented a posteriori error estimators are directed towards two aims: assessment of the discretization error and improvement of the accuracy by local refinement. For the second aim, the information provided by the error estimators have to be localized to cellwise or nodewise contributions (local error indicators).

We concretize this procedure here for the error estimators η_k^J and η_h^J assessing the error with respect to the cost functional. For concrete choices of discretizations for the control space one can proceed with η_d^J in the same manner. Of course, the error indicators η_k^I, η_h^I, and η_d^I can be treated similarly, too.

For localizing the error estimators, we split up the error estimates η_k^J and η_h^J into their contributions on each subinterval I_m by

$$\eta_k^J = \sum_{m=1}^{M} \eta_k^{J,m} \quad \text{and} \quad \eta_h^J = \sum_{m=0}^{M} \eta_h^{J,m},$$

where the contributions $\eta_k^{J,m}$ and $\eta_h^{J,m}$ are given in terms of the time stepping residuals ρ_m^u and ρ_m^z as

$$\begin{aligned}
\eta_k^{J,m} &= \frac{1}{2}\Big\{\rho_m^u(q_\sigma, u_\sigma)(P_k z_\sigma) + \rho_m^z(q_\sigma, u_\sigma, z_\sigma)(P_k u_\sigma)\Big\}, \\
\eta_h^{J,m} &= \frac{1}{2}\Big\{\rho_m^u(q_\sigma, u_\sigma)(P_h z_\sigma) + \rho_m^z(q_\sigma, u_\sigma, z_\sigma)(P_h u_\sigma)\Big\}.
\end{aligned}$$

Thereby, the time stepping residuals ρ_m^u and ρ_m^z are those parts of the global residuals ρ^u and ρ^z belonging to the time interval I_m or to the initial time $t = 0$ for $m = 0$.

Whereas the temporal indicators $\eta_k^{J,m}$ can be used directly for determining the time intervals to be refined, the indicators $\eta_h^{J,m}$ for the spatial discretization error have to be further localized to indicators on each spatial mesh. For details of this further localization procedure we refer, e.g., to [8, 36].

5.2. Adaptive algorithm

Goal of the adaption of the different types of discretizations has to be the equilibrated reduction of the corresponding discretization errors. If a given tolerance TOL has to be reached, this can be done by refining each discretization as long as the value of this part of the error estimator is greater than $1/3$ TOL. We want to present here a strategy which will equilibrate the different discretization errors even if no tolerance is given.

Aim of the equilibration algorithm presented in the sequel is to obtain discretization such that

$$|\eta_k| \approx |\eta_h| \approx |\eta_d|$$

and to keep this property during the further refinement. Here, the estimators η_i denote the estimators η_i^J for the cost functional J or η_i^I for the quantity of interest I.

For doing this equilibration, we choose an "equilibration factor" $e \approx 2 - 5$ and propose the following strategy: We compute a permutation (a, b, c) of the discretization indices (k, h, d) such that

$$|\eta_a| \geq |\eta_b| \geq |\eta_c|,$$

and define the relations

$$\gamma_{ab} := \left|\frac{\eta_a}{\eta_b}\right| \geq 1, \qquad \gamma_{bc} := \left|\frac{\eta_b}{\eta_c}\right| \geq 1.$$

Relation between the estimators	Discretizations to be refined
$\gamma_{ab} \le e$ and $\gamma_{bc} \le e$	a, b, and c
$\gamma_{bc} > e$	a and b
else ($\gamma_{ab} > e$ and $\gamma_{bc} \le e$)	a

TABLE 1. Equilibration Strategy

Then we decide by means of Table 1 in every repetition of the adaptive refinement algorithm which discretization shall be refined. For every discretization to be adapted we select by means of the local error indicators the cells for refinement. For this purpose there are several strategies available, see, e.g., [4].

6. Numerical examples

6.1. Surface hardening of steel

We consider the optimal control of laser surface hardening of steel. In this process, a laser beam moves along the surface of a workpiece. The heating induced by the laser is accompanied by a phase transition, in which austenite, the high temperature phase in steel, is produced. Due to further phase transitions (which are not contained in the considered model) the desired hardening effect develops.

The goal is to control this hardening process such that a desired hardening profile is produced. Since in practical applications, the moving velocity of the laser beam is kept constant, the most important control parameter is the energy of the laser beam. Especially when there are large variations in the thickness of the workpiece or in regions near the boundaries of the workpiece, the proper adjustment of the laser energy is crucial to meet the given hardening profile.

The configuration of the control problem to be investigated in this section is taken from [15, 20]:

$$\begin{aligned}
\partial_t a &= \frac{1}{\tau(\theta)}(a_{\mathrm{eq}}(\theta) - a)\mathcal{H}_\delta(a_{\mathrm{eq}}(\theta) - a) && \text{in } \Omega \times (0,T), \\
\rho c_{\mathrm{p}} \partial_t \theta - \varepsilon \Delta \theta &= -\rho L \partial_t a + q\Lambda && \text{in } \Omega \times (0,T), \\
\partial_n \theta &= 0 && \text{on } \partial\Omega \times (0,T) \\
a &= 0 && \text{on } \Omega \times \{0\}, \\
\theta &= \theta_0 && \text{on } \Omega \times \{0\},
\end{aligned} \tag{6.1}$$

Here a is the volume fraction of austenite, a_{eq} is the equilibrium volume fraction of austenite, and τ is a time constant. Both a_{eq} and τ depend on the temperature θ. Furthermore, as in [20], $\mathcal{H}_\delta$ is a monotone regularization of the Heaviside function and the density ρ, the heat capacity c_{p}, the heat conductivity ε, and the latent heat L are assumed to be positive constants. The term $q(t)\Lambda(x,t)$ describes the volumetric heat source due to laser radiation, where q acts as time-dependent control variable. Thus, the optimal control is searched for in $Q = L^2(0,T)$.

Here, because of the Neumann boundary conditions for θ, the space X is defined using $V = H^1(\Omega)$ and $H = L^2(\Omega)$.

As cost functional to be minimized, we choose

$$J(q,u) = \frac{\beta}{2}\int_0^T \|a(t) - \hat{a}(t)\|^2_{L^2(\Omega)}\,dt + \frac{\alpha}{2}\int_0^T q(t)^2\,dt,$$

where $\hat{a}$ is a given desired volume fraction of austenite. For the computations, we choose the physical parameters for the heat equation accordingly to [20] as

$$\rho c_{\mathrm{p}} = 1.17, \qquad \varepsilon = 0.153, \qquad \text{and} \qquad \rho L = 150.$$

The equilibrium volume fraction a_{eq} and the time constant τ are constructed by cubic spline interpolation from values taken also from [20]. The initial condition for the temperature is set to $\theta_0 = 20$. The laser source Λ is modeled by

$$\Lambda(x,t) = \frac{4\kappa A}{\pi D^2}\exp\Bigl(-\frac{2(x_1 - vt)^2}{D^2}\Bigr)\exp(\kappa x_2), \quad x = (x_1,x_2)^T,$$

where the values of the parameters are taken from [20] as $D = 0.47$, $\kappa = 60$, $A = 0.3$, and $v = 1.15$.

For the numerical computations, we choose the domain Ω to be $(0,5)\times(-1,0)$ and determine the final time T such that the laser, which moves from $(0,0)$ to $(5,0)$, reaches the boundary at $(5,0)$ at time T. Thus, we set $T = 5/v \approx 4.34782$. The desired volume fraction $\hat{a}$ (cf. Figure 3(c)) is chosen as

$$\hat{a}(x) := \begin{cases} 1 & \text{for } 0 \geq x_1 \geq -\frac{1}{8} \\ 0 & \text{for } -\frac{1}{8} > x_1 \geq -1 \end{cases}, \qquad x = (x_1,x_2)^T,$$

and for the parameters α and β from the definition of the objective functional J we take $\alpha = 10^{-4}$ and $\beta = 3500$. The parameter δ for the regularized Heaviside function is chosen as $\delta = 0.15$.

For discretizing the state space, we employ the cG(1)dG(0) discretization Since the control space is given by $Q = L^2((0,T),\mathbb{R})$, we have to discretize the controls only in time. Correspondingly to the state space, we choose a dG(0) discretization based on a possibly coarser step size than the step size used for discretizing the state space.

At first, we investigate the qualitative behavior of the optimization algorithm. Figure 3 presents the distribution of austenite at final time T before (a) and after (b) the optimization on a fine discretization of the state space. To compare with, the desired state is depicted in Figure 3(c).

In Figure 4, we present a comparison of different refinement strategies for the spatial discretization. We depict the development of the error $e_h^J := J(q_{kd},u_{kd}) - J(q_\sigma,u_\sigma)$ caused by the spatial discretization of the state space. Thereby, we consider the following three types of refinement:

- Uniform refinement
- Local refinement based on the error indicator η_h^J with one fixed mesh for all time steps

(a) uncontrolled

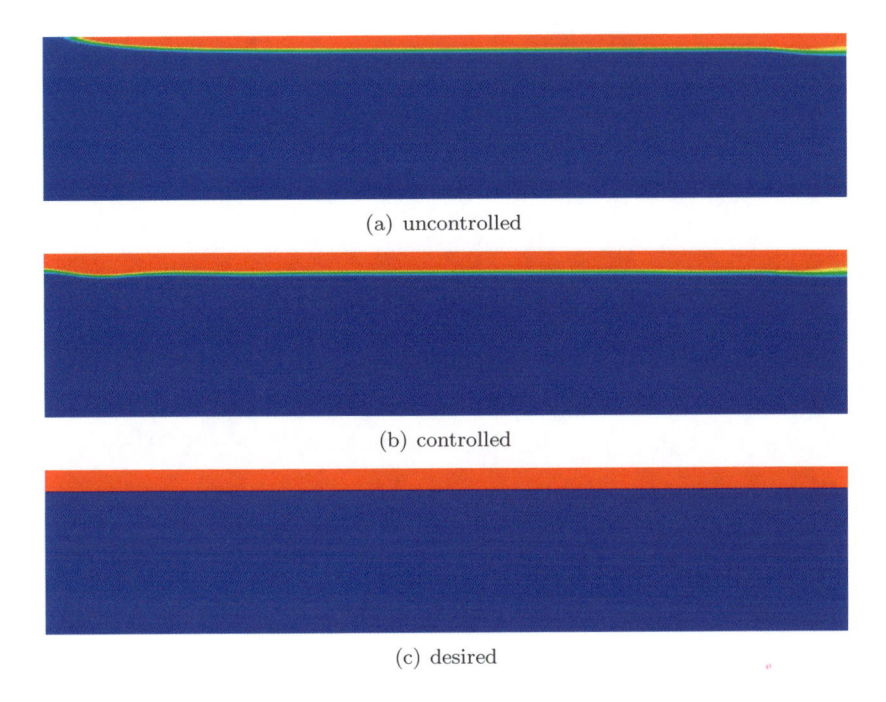

(b) controlled

(c) desired

FIGURE 3. Distribution of austenite at final time T

- Local refinement based on η_h^J but allowing separate spatial meshes for each time interval by using dynamically changing meshes

We observe that by the usage of local refinement the number of grid points can be reduced from $N = 16{,}641$ to $N = 5{,}271$ to achieve an error of 10^{-5}. Moreover, if we allow dynamically changing meshes, we only need $N_{\max} = 3{,}873$ grid points. The total number of degrees of freedom in the space discretization $(\dim X_{k,h}^{0,1})$ is reduced even by a factor of 5.7 when employing local refinement on dynamic meshes.

In Figure 5, a selection from the sequence of locally refined meshes is given. Thereby, we detect a strong refinement at the position where the laser currently acts and at the region around the transition from hardened to not hardened steel. In this region, the optimal distribution of austenite as well as the desired hardening profile exhibits spatial discontinuities.

We now couple the temporal and spatial estimators by the equilibration strategy described in Section 5.2. Since in this example, we do not benefit from local refinement in time, we allow only uniform refinement of the time steps. In space, we allow the adaptation procedure to use dynamically changing meshes. Results of this computation are given in Table 2. Therein, we observe that for the chosen initial levels of discretizations, the contribution from the spatial discretization error to the overall error is much smaller than the contribution from the temporal

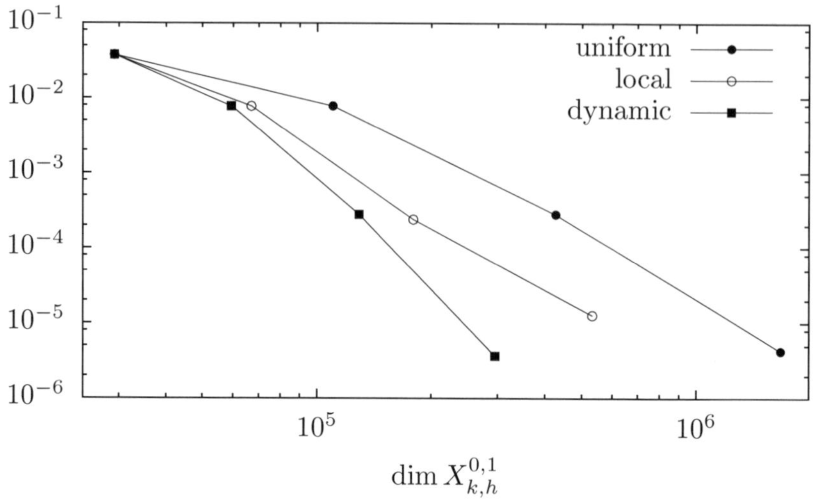

FIGURE 4. Comparison of the relative error $|e_h^J|/J$ for uniform and local refinement of the triangulation using dynamically changing meshes

N_{tot}	$N_{\max}$	M	η_h^J/J	η_k^J/J	$\eta_h^J/J + \eta_k^J/J$	e_{kh}^J/J	I_{eff}
14739	289	50	$-7.4 \cdot 10^{-03}$	$2.3 \cdot 10^{-02}$	$1.622 \cdot 10^{-02}$	$-4.916 \cdot 10^{-03}$	−0.30
59325	675	100	$-2.8 \cdot 10^{-03}$	$1.3 \cdot 10^{-02}$	$1.049 \cdot 10^{-02}$	$7.828 \cdot 10^{-03}$	0.74
257867	1659	200	$-3.9 \cdot 10^{-04}$	$6.3 \cdot 10^{-03}$	$6.040 \cdot 10^{-03}$	$7.445 \cdot 10^{-03}$	1.23
515115	1659	400	$-3.9 \cdot 10^{-04}$	$3.2 \cdot 10^{-03}$	$2.827 \cdot 10^{-03}$	$3.454 \cdot 10^{-03}$	1.22
1029611	1659	800	$-4.3 \cdot 10^{-04}$	$1.6 \cdot 10^{-03}$	$1.193 \cdot 10^{-03}$	$1.424 \cdot 10^{-03}$	1.19
4721397	3911	1600	$9.8 \cdot 10^{-06}$	$7.8 \cdot 10^{-04}$	$8.143 \cdot 10^{-04}$	$9.375 \cdot 10^{-04}$	1.15

TABLE 2. Local refinement on dynamic meshes with equilibration

discretization error. Consequently, the equilibration procedure decides for example to keep the spatial meshes fixed while increasing the number of time steps from 200 over 400 to 800 time steps. The effectivity index given in the last column of this table is defined as usual by

$$I_{\text{eff}} := \frac{e_{kh}^J}{\eta_h^J + \eta_k^J}.$$

This implies that uniform refinement of the time and space discretizations without the knowledge of the size of the different error contributions can not be competitive. For the efficient equilibration – and thus the efficient reduction of the error – estimations of the size of each involved discretization errors are essential. Furthermore, the table demonstrates that the estimator $\eta_h^J/J + \eta_k^J/J$ is in very good agreement with the relative error $|e_{kh}^J|/J$ Thereby, the error e_{kh}^J is defined as $e_{kh}^J := J(q_d, u_d) - J(q_\sigma, u_\sigma)$ with an approximation of $J(q_d, u_d)$ obtained by

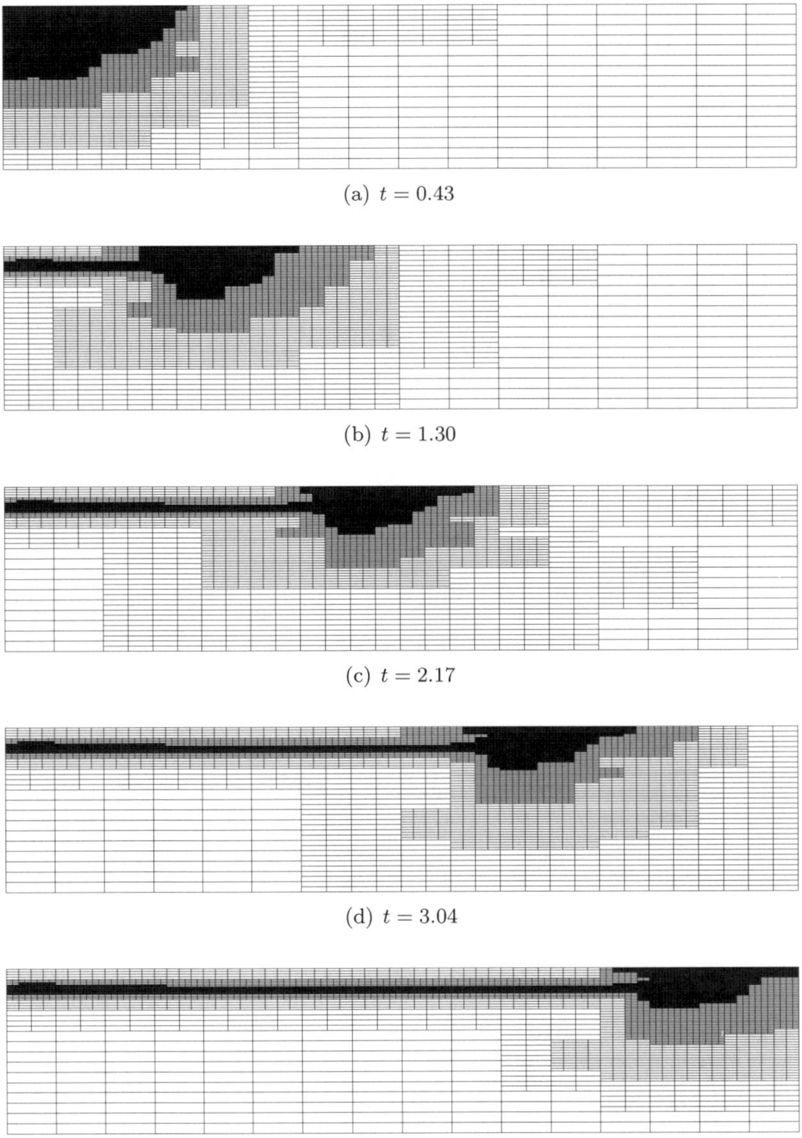

(a) $t = 0.43$

(b) $t = 1.30$

(c) $t = 2.17$

(d) $t = 3.04$

(e) $t = 3.91$

FIGURE 5. Locally refined meshes for $t \in \{ 0.43, 1.30, 2.17, 3.04, 3.91 \}$

extrapolation. For normalizing the errors and the estimators we use J, which denotes an approximation of the exact value of the cost functional $J(q, u)$.

7. Propagation of laminar flames

In this section, we consider a parameter estimation problem arising in chemical combustion processes. We aim at the identification of an unknown parameter in a reaction mechanism governed by an Arrhenius law. This formulation is employed to model the propagation of laminar flames through a channel. The channel is narrowed by two heat absorbing obstacles influencing the travelling of the flame.

The identification of the unknown parameter is done employing measurements of the solution components at four spatial points at final time. Using these values, the cost functional is constructed by means of a least-squares formalism.

The governing equations for the considered problem are taken from an example given in [22]. They describe the major part of gaseous combustion under the low Mach number hypothesis. In this approach, the dependency of the fluid density on the pressure is eliminated while the temperature dependence remains. If additionally the dependence on the temperature is neglected, the motion of the fluid becomes independent on the temperature and the species concentration. Hence, one can solve the temperature and the species equation alone specifying any solenoidal velocity field v. In particular, $v = 0$ is an interesting case.

Introducing the dimensionless temperature θ, denoting by Y the species concentration, and assuming constant diffusion coefficients yields the system of equations

$$\begin{aligned} \partial_t\theta - \Delta\theta &= \omega(Y,\theta) && \text{in } \Omega \times (0,T),\\ \partial_t Y - \frac{1}{\mathrm{Le}}\Delta Y &= -\omega(Y,\theta) && \text{in } \Omega \times (0,T),\\ \theta &= \theta_0 && \text{on } \Omega \times \{\,0\,\},\\ Y &= Y_0 && \text{on } \Omega \times \{\,0\,\}, \end{aligned} \tag{7.1}$$

where the Lewis number Le is the ratio of diffusivity of heat and diffusivity of mass. We use a simple one-species reaction mechanism governed by an Arrhenius law given by

$$\omega(Y,\theta) = \frac{\beta^2}{2\mathrm{Le}} Y \mathrm{e}^{\frac{\beta(\theta-1)}{1+\alpha(\theta-1)}}, \tag{7.2}$$

in which an approximation for large activation energy has been employed.

Here, we consider a freely propagating laminar flame described by (7.1) and its response to a heat absorbing obstacle, a set of cooled parallel rods with rectangular cross section (cf. Figure 6). The computational domain has width 16 and length 60. The obstacle covers half of the width and has length $L/4$. The boundary conditions are chosen as

$$\begin{aligned} \theta &= 1 && \text{on } \Gamma_{\mathrm{D}} \times (0,T), & \partial_n\theta &= 0 && \text{on } \Gamma_{\mathrm{N}} \times (0,T),\\ Y &= 0 && \text{on } \Gamma_{\mathrm{D}} \times (0,T), & \partial_n Y &= 0 && \text{on } \Gamma_{\mathrm{N}} \times (0,T), \end{aligned}$$

$$\begin{aligned} \partial_n\theta &= -\kappa\theta && \text{on } \Gamma_{\mathrm{R}} \times (0,T),\\ \partial_n Y &= 0 && \text{on } \Gamma_{\mathrm{R}} \times (0,T), \end{aligned}$$

where the heat absorption is modeled by boundary conditions of Robin type on Γ_{R}.

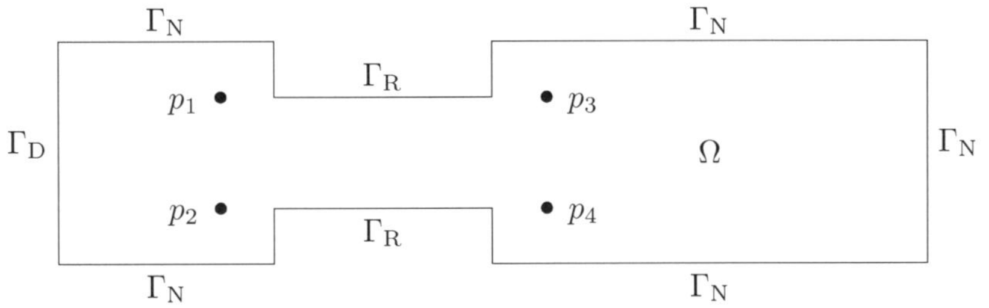

FIGURE 6. Computational domain Ω and measurement points p_i

The initial condition is the analytical solution of a one-dimensional right-travelling flame in the limit $\beta \to \infty$ located left of the obstacle:

$$\theta_0(x) = \begin{cases} 1 & \text{for } x_1 \leq \tilde{x}_1 \\ e^{\tilde{x}_1 - x_1} & \text{for } x_1 > \tilde{x}_1 \end{cases},$$

$$Y_0(x) = \begin{cases} 0 & \text{for } x_1 \leq \tilde{x}_1 \\ 1 - e^{\mathrm{Le}(\tilde{x}_1 - x_1)} & \text{for } x_1 > \tilde{x}_1 \end{cases}.$$

For the computations, the occurring parameters are set as in [22] to

$$\mathrm{Le} = 1, \qquad \beta = 10, \qquad \kappa = 0.1, \qquad \tilde{x}_1 = 9,$$

whereas the temperature ratio α, which determines the gas expansion in non-constant density flows, is the objective of the parameter estimation.

We define the pair of solution components $u := (\theta, Y) \in \tilde{u} + X^2$ and denote the parameter α to be estimated by $q \in Q := \mathbb{R}$. For the definition of the state space X, we use here the spaces V and H given as

$$V = \left\{ v \in H^1(\Omega) \,\middle|\, v\big|_{\Gamma_\mathrm{D}} = 0 \right\} \qquad \text{and} \qquad H = L^2(\Omega).$$

The function $\tilde{u}$ is defined to fulfill the prescribed Dirichlet data as $\tilde{u}\big|_{\Gamma_\mathrm{D}} = (1, 0)$.

The unknown parameter α is estimated here using information from pointwise measurements of θ and Y at four points $p_i \in \Omega$, $i = 1, 2, 3, 4$, at final time $T = 60$. This parameter identification problem can be formulated by means of a cost functional of least-squares type, that is

$$J(q, u) = \frac{1}{2} \sum_{i=1}^{4} \left(\theta(p_i, T) - \hat{\theta}_i\right)^2 + \frac{1}{2} \sum_{i=1}^{4} \left(Y(p_i, T) - \hat{Y}_i\right)^2.$$

The values of the artificial measurements $\hat{\theta}_i$ and $\hat{Y}_i$, $i = 1, 2, 3, 4$, are obtained from a reference solution computed on fine space and time discretizations.

The consideration of point measurements does not fulfill the assumption on the cost functional in (2.5), since the point evaluation is not bounded as a functional on H. Therefore, the point functionals here have to be understood as regularized functionals defined on H. An a priori analysis of parameter estimation

N	M	η_h^I	η_k^I	$\eta_h^I + \eta_k^I$	e^I	I_{eff}
269	512	$4.3 \cdot 10^{-02}$	$-8.4 \cdot 10^{-03}$	$3.551 \cdot 10^{-02}$	$-2.889 \cdot 10^{-02}$	-0.81
635	512	$5.5 \cdot 10^{-03}$	$-9.1 \cdot 10^{-03}$	$-3.533 \cdot 10^{-03}$	$-4.851 \cdot 10^{-02}$	13.72
1847	722	$-1.5 \cdot 10^{-02}$	$-3.6 \cdot 10^{-03}$	$-1.889 \cdot 10^{-02}$	$-3.024 \cdot 10^{-02}$	1.60
5549	1048	$-6.5 \cdot 10^{-03}$	$-2.5 \cdot 10^{-03}$	$-9.074 \cdot 10^{-03}$	$-1.097 \cdot 10^{-02}$	1.20
14419	1088	$-2.4 \cdot 10^{-03}$	$-2.5 \cdot 10^{-03}$	$-5.064 \cdot 10^{-03}$	$-5.571 \cdot 10^{-03}$	1.10
43343	1102	$-8.5 \cdot 10^{-04}$	$-2.5 \cdot 10^{-03}$	$-3.453 \cdot 10^{-03}$	$-3.693 \cdot 10^{-03}$	1.06

TABLE 3. Local refinement on a *fixed mesh* with equilibration

problems governed by elliptic equations and using such types of point functionals can be found in [35].

For the considered type of parameter estimation problems, one is usually not interested in reducing the discretization error measured in terms of the cost functional J. The focus is rather on the error in the parameter q itself. Hence, we define the quantity of interest I as

$$I(q, u) = q,$$

and apply the techniques presented in Section 4 for estimating the discretization error with respect to I. Since the control space in this application is given by $Q = \mathbb{R}$, we do not discretize Q. Thus, there is no discretization error due to the control discretization and the a posteriori error estimator η^I consists only of η_k^I and η_h^I.

For the computations, the state is discretized using the cG(1)dG(0) approach. We define the temporal and spatial discretization errors e_k^I and e_h^I as

$$e_k^I := I(q_h, u_h) - I(q_{kh}, u_{kh}) \qquad \text{and} \qquad e_h^I := I(q_k, u_k) - I(q_{kh}, u_{kh}).$$

The values of $I(q_h, u_h)$ and $I(q_k, u_k)$ are extrapolated from computations on a sequence of fine time and space discretizations, respectively. Since we have $I(q, u) = q = \alpha \approx 0.8$, there is no difference between the consideration of relative or absolute errors.

We apply the adaptive algorithm described above to the considered problem using simultaneous refinement of the space and time discretizations. Thereby, the refinements are coupled by the equilibration strategy introduced in Section 5.2. The Tables 3 and 4 demonstrate the effectivity of the error estimator $\eta_h^I + \eta_k^I$ on locally refined discretizations using fixed and dynamically changing spatial triangulations. The effectivity index given in the last column of these tables is defined as usual by

$$I_{\text{eff}} := \frac{e^I}{\eta_h^I + \eta_k^I}.$$

In Figure 7, we compare uniform refinement of the space and time discretizations with local refinement of both discretizations on a fixed spatial triangulation and on dynamically changing triangulations. We gain an remarkable reduction of

N_{tot}	$N_{\max}$	M	η_h^I	η_k^I	$\eta_h^I + \eta_k^I$	e^I	I_{eff}
137997	269	512	$4.3 \cdot 10^{-02}$	$-8.4 \cdot 10^{-03}$	$3.551 \cdot 10^{-02}$	$-2.889 \cdot 10^{-02}$	-0.81
238187	663	512	$3.5 \cdot 10^{-03}$	$-8.6 \cdot 10^{-03}$	$-5.192 \cdot 10^{-03}$	$-5.109 \cdot 10^{-02}$	9.84
633941	1677	724	$-1.6 \cdot 10^{-02}$	$-3.5 \cdot 10^{-03}$	$-2.015 \cdot 10^{-02}$	$-3.227 \cdot 10^{-02}$	1.60
1741185	2909	1048	$-7.3 \cdot 10^{-03}$	$-2.5 \cdot 10^{-03}$	$-9.869 \cdot 10^{-03}$	$-1.214 \cdot 10^{-02}$	1.23
3875029	4785	1098	$-2.2 \cdot 10^{-03}$	$-2.5 \cdot 10^{-03}$	$-4.792 \cdot 10^{-03}$	$-5.432 \cdot 10^{-03}$	1.13
9382027	10587	1140	$-7.9 \cdot 10^{-04}$	$-2.5 \cdot 10^{-03}$	$-3.301 \cdot 10^{-03}$	$-3.588 \cdot 10^{-03}$	1.08
23702227	25571	1160	$-2.8 \cdot 10^{-04}$	$-2.4 \cdot 10^{-03}$	$-2.756 \cdot 10^{-03}$	$-2.944 \cdot 10^{-03}$	1.06

TABLE 4. Local refinement on *dynamic meshes* with equilibration

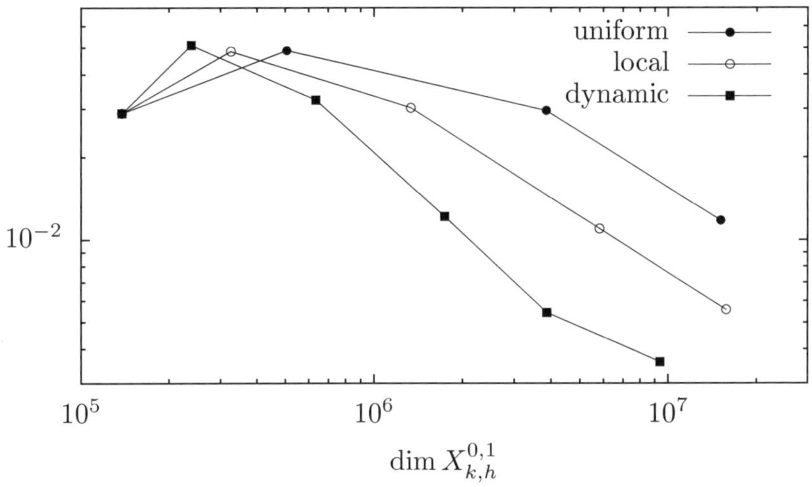

FIGURE 7. Comparison of the error $|e^I|$ for different refinement strategies

the required degrees of freedom for reaching a given tolerance. To meet for instance an error of $|e^I| \approx 10^{-2}$, the uniform refinement requires in total 15,056,225 degrees of freedom, the local refinement needs 5,820,901 degrees of freedom, and the dynamical refinement necessitates only 1,741,185 degrees of freedom. Thus, we gain a reduction of about 8.6.

Figure 8 depicts the distribution of the temporal step size k resulting from a fully adaptive computation on dynamic meshes. We observe a strong refinement of the time steps at the beginning of the time interval, whereas the time steps at the end are determined by the adaptation to be eight times larger.

Before presenting a sequence of dynamically changing meshes, we show in Figure 9 a typical locally refined mesh obtained by computations on a fixed spatial triangulation. We note, that the refinement is especially concentrated at the four reentrant corners and the two measurement points behind the obstacle. The interior of the region with restricted cross section is also strongly refined.

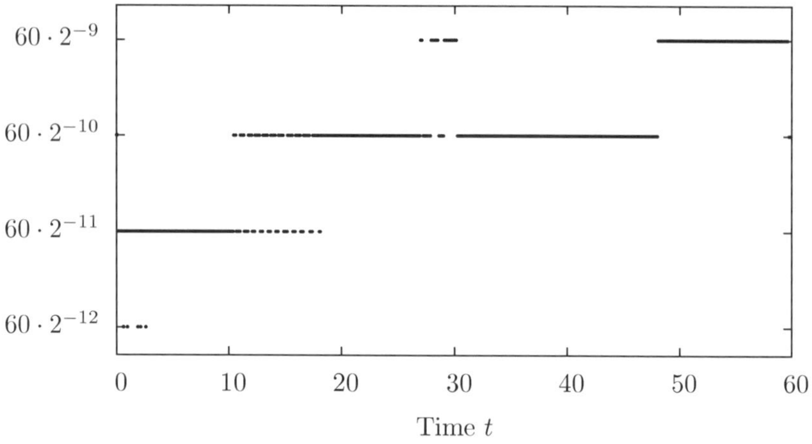

FIGURE 8. Visualization of the adaptively determined time step size k

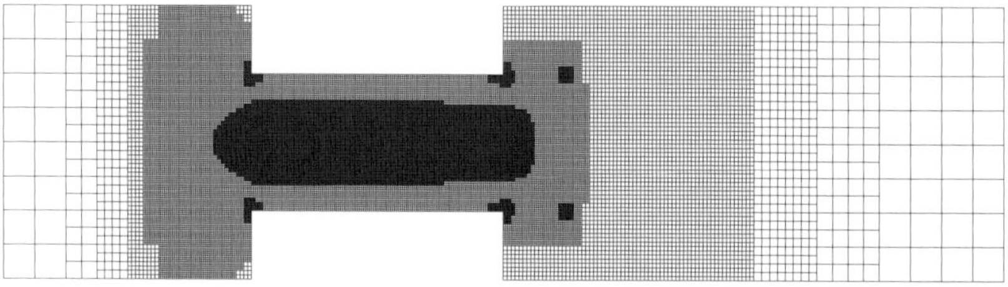

FIGURE 9. Locally refined fixed mesh

Finally, Figure 10 shows the spatial triangulation and the reaction rate ω for certain selected time points. Thereby, ω is computed from the numerical solution by means of formula (7.2). We observe, that the refinement traces the front of the reaction rate ω until $t \approx 56$ (cf. Figure 10(d)). Afterwards, the mesh around the front becomes coarser and the refinement is concentrated at the four measurement points p_i. Compared to the usage of one fixed triangulation, the usage of dynamically changing meshes enables us here to reduce the discretization error in terms of the quantity of interest at lower computational costs; cf. Figure 7.

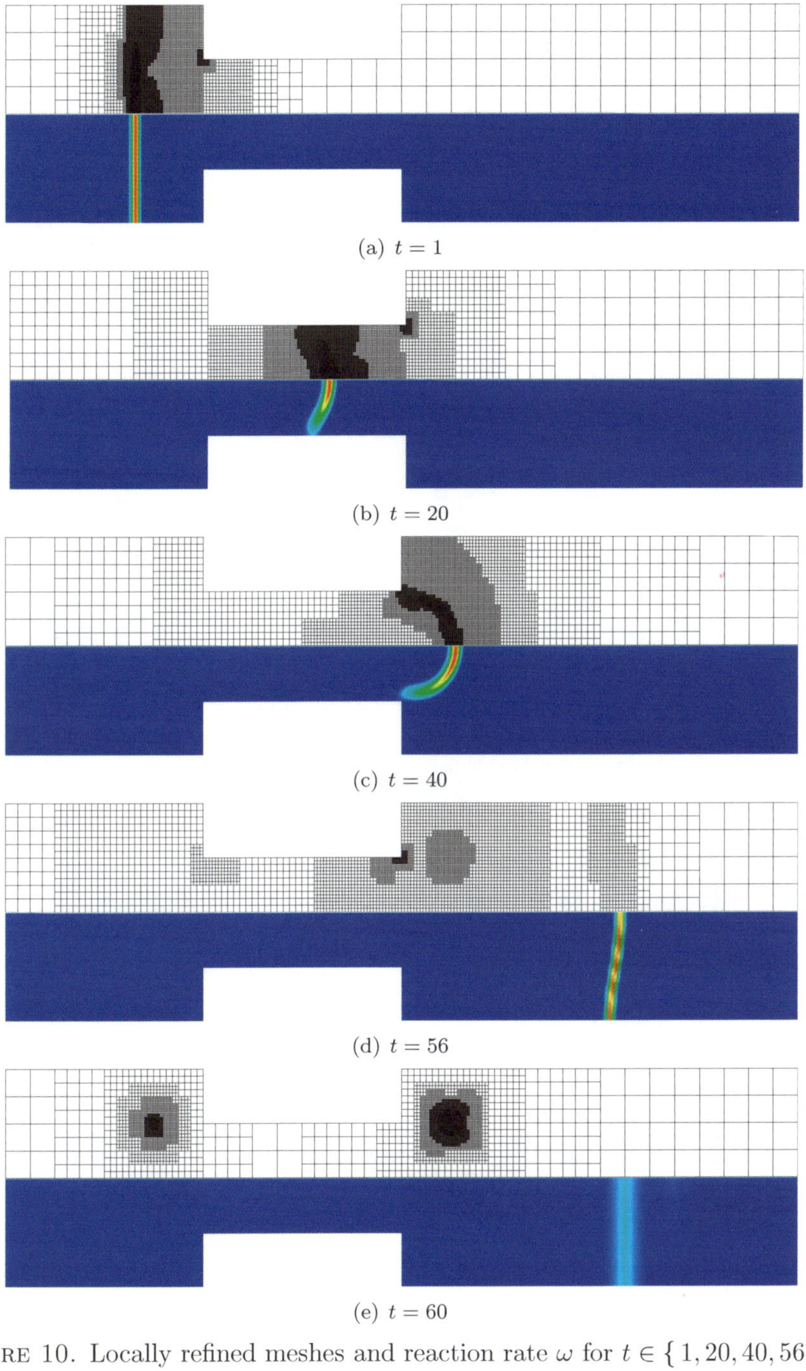

(a) $t = 1$

(b) $t = 20$

(c) $t = 40$

(d) $t = 56$

(e) $t = 60$

FIGURE 10. Locally refined meshes and reaction rate ω for $t \in \{1, 20, 40, 56, 60\}$

References

[1] R. Becker: *Estimating the control error in discretized PDE-constraint optimization.* J. Numer. Math. **14** (2006), 163–185.

[2] R. Becker, H. Kapp, and R. Rannacher: *Adaptive finite element methods for optimal control of partial differential equations: Basic concepts.* SIAM J. Control Optimization **39** (2000), 113–132.

[3] R. Becker, D. Meidner, and B. Vexler: *Efficient numerical solution of parabolic optimization problems by finite element methods.* Optim. Methods Softw. **22** (2007), 818–833.

[4] R. Becker and R. Rannacher: *An optimal control approach to a-posteriori error estimation.* In A. Iserles, editor, *Acta Numerica* 2001, pages 1–102. Cambridge University Press, 2001.

[5] R. Becker and B. Vexler: *A posteriori error estimation for finite element discretizations of parameter identification problems.* Numer. Math. **96** (2004), 435–459.

[6] R. Becker and B. Vexler: *Mesh refinement and numerical sensitivity analysis for parameter calibration of partial differential equations.* J. Comp. Physics **206** (2005), 95–110.

[7] O. Benedix and B. Vexler: *A posteriori error estimation and adaptivity for elliptic optimal control problems with state constraints.* Comput. Optim. Appl. **44** (2009), 3–25.

[8] M. Braack and A. Ern: *A posteriori control of modeling errors and discretization errors.* Multiscale Model. Simul. **1** (2003), 221–238.

[9] P.G. Ciarlet: *The Finite Element Method for Elliptic Problems*, volume 40 of *Classics Appl. Math.* SIAM, Philadelphia, 2002.

[10] D. Clever, J. Lang, S. Ulbrich, and J. C. Ziems: *Combination of an adaptive multilevel SQP method and a space-time adaptive PDAE solver for optimal control problems.* Preprint SPP1253-094, SPP 1253, 2010.

[11] R. Dautray and J.-L. Lions: *Mathematical Analysis and Numerical Methods for Science and Technology: Evolution Problems I*, volume 5. Springer-Verlag, Berlin, 1992.

[12] K. Eriksson, D. Estep, P. Hansbo, and C. Johnson: *Introduction to adaptive methods for differential equations.* In A. Iserles, editor, *Acta Numerica* 1995, pages 105–158. Cambridge University Press, 1995.

[13] K. Eriksson, D. Estep, P. Hansbo, and C. Johnson: *Computational differential equations.* Cambridge University Press, Cambridge, 1996.

[14] K. Eriksson, C. Johnson, and V. Thomée: *Time discretization of parabolic problems by the discontinuous Galerkin method.* RAIRO Modelisation Math. Anal. Numer. **19** (1985), 611–643.

[15] J. Fuhrmann and D. Hömberg: *Numerical simulation of the surface hardening of steel.* Internat. J. Numer. Methods Heat Fluid Flow **9** (1999), 705–724.

[16] A.V. Fursikov: *Optimal Control of Distributed Systems: Theory and Applications*, volume 187 of *Transl. Math. Monogr.* AMS, Providence, 1999.

[17] A. Günther and M. Hinze: *A posteriori error control of a state constrained elliptic control problem.* J. Numer. Math. **16** (2008), 307–322.

[18] M. Hintermüller, R. Hoppe, Y. Iliash, and M. Kieweg: *An a posteriori error analysis of adaptive finite element methods for distributed elliptic control problems with control constraints.* ESIAM Control Optim. Calc. Var. **14** (2008), 540–560.

[19] M. Hintermüller and R.H.W. Hoppe: *Goal-oriented adaptivity in control constrained optimal control of partial differential equations.* SIAM J. Control Optim. **47** (2008), 1721–1743.

[20] D. Hömberg and S. Volkwein: *Suboptimal control of laser surface hardening using proper orthogonal decomposition.* Preprint 639, WIAS Berlin, 2001.

[21] R. Hoppe, Y. Iliash, C. Iyyunni, and N. Sweilam: *A posteriori error estimates for adaptive finite element discretizations of boundary control problems.* J. Numer. Math. **14** (2006), 57–82.

[22] J. Lang: *Adaptive Multilevel Solution of Nonlinear Parabolic PDE Systems. Theory, Algorithm, and Applications*, volume 16 of *Lecture Notes in Earth Sci.* Springer-Verlag, Berlin, 1999.

[23] J. Lang: *Adaptive computation for boundary control of radiative heat transfer in glass.* J. Comput. Appl. Math. **183** (2005), 312–326.

[24] J.-L. Lions: *Optimal Control of Systems Governed by Partial Differential Equations*, volume 170 of *Grundlehren Math. Wiss.* Springer-Verlag, Berlin, 1971.

[25] W. Liu, H. Ma, T. Tang, and N. Yan: *A posteriori error estimates for discontinuous Galerkin time-stepping method for optimal control problems governed by parabolic equations.* SIAM J. Numer. Anal. **42** (2004), 1032–1061.

[26] W. Liu and N. Yan: *A posteriori error estimates for distributed convex optimal control problems.* Adv. Comput. Math **15** (2001), 285–309.

[27] W. Liu and N. Yan: *A posteriori error estimates for control problems governed by nonlinear elliptic equations.* Appl. Num. Math. **47** (2003), 173–187.

[28] W. Liu and N. Yan: *A posteriori error estimates for optimal control problems governed by parabolic equations.* Numer. Math. **93** (2003), 497–521.

[29] W. Liu and N. Yan: *Adaptive finite element methods for optimal control governed by PDEs*, volume 41 of *Series in Information and Computational Science.* Science Press, Beijing, 2008.

[30] D. Meidner: *Adaptive Space-Time Finite Element Methods for Optimization Problems Governed by Nonlinear Parabolic Systems.* PhD Thesis, Institut für Angewandte Mathematik, Universität Heidelberg, 2007.

[31] D. Meidner and B. Vexler: *Adaptive space-time finite element methods for parabolic optimization problems.* SIAM J. Control Optim. **46** (2007), 116–142.

[32] D. Meidner and B. Vexler: *A priori error estimates for space-time finite element approximation of parabolic optimal control problems. Part I: Problems without control constraints.* SIAM J. Control Optim. **47** (2008), 1150–1177.

[33] D. Meidner and B. Vexler: *A priori error estimates for space-time finite element approximation of parabolic optimal control problems. Part II: Problems with control constraints.* SIAM J. Control Optim. **47** (2008), 1301–1329.

[34] M. Picasso: *Anisotropic a posteriori error estimates for an optimal control problem governed by the heat equation.* Int. J. Numer. Methods PDE **22** (2006), 1314–1336.

[35] R. Rannacher and B. Vexler: *A priori error estimates for the finite element discretization of elliptic parameter identification problems with pointwise measurements.* SIAM J. Control Optim. **44** (2005), 1844–1863.

[36] M. Schmich and B. Vexler: *Adaptivity with dynamic meshes for space-time finite element discretizations of parabolic equations.* SIAM J. Sci. Comput. **30** (2008), 369–393.

[37] R. Serban, S. Li, and L.R. Petzold: *Adaptive algorithms for optimal control of time-dependent partial differential-algebraic equation systems.* Int. J. Numer. Meth. Engng. **57** (2003), 1457–1469.

[38] F. Tröltzsch: *Optimale Steuerung partieller Differentialgleichungen.* Vieweg + Teubner, Wiesbaden, 2nd edition 2009.

[39] R. Verfürth: *A Review of A Posteriori Error Estimation and Adaptive Mesh-Refinement Techniques.* Wiley/Teubner, New York-Stuttgart, 1996.

[40] B. Vexler: *Adaptive Finite Elements for Parameter Identification Problems.* PhD Thesis, Institut für Angewandte Mathematik, Universität Heidelberg, 2004.

[41] B. Vexler and W. Wollner: *Adaptive finite elements for elliptic optimization problems with control constraints.* SIAM J. Control Optim. **47** (2008), 509–534.

Dominik Meidner and Boris Vexler
Lehrstuhl für Mathematische Optimierung
Technische Universität München
Fakultät für Mathematik
Boltzmannstraße 3
D-85748 Garching b. München, Germany
e-mail: `meidner@ma.tum.de`
`vexler@ma.tum.de`

International Series of Numerical Mathematics, Vol. 160, 349–373

A Posteriori Error Estimation in PDE-constrained Optimization with Pointwise Inequality Constraints

Rolf Rannacher, Boris Vexler, Winnifried Wollner

Abstract. This article summarizes several recent results on goal-oriented error estimation and mesh adaptation for the solution of elliptic PDE-constrained optimization problems with additional inequality constraints. The first part is devoted to the control constrained case. Then some emphasis is given to pointwise inequality constraints on the state variable and on its gradient. In the last part of the article regularization techniques for state constraints are considered and the question is addressed, how the regularization parameter can adaptively be linked to the discretization error.

Mathematics Subject Classification (2000). 65N30, 65K10; 90C59.

Keywords. Goal oriented error estimation, PDE-constrained optimization, control and state constraints, regularization, adaptivity, adaptive finite elements.

1. Introduction

In this paper, we summarize our results on a posteriori error estimation and adaptivity for the solution of PDE-constrained optimization problems with pointwise inequality constraints.

The use of adaptive techniques based on a posteriori error estimation is well accepted in the context of finite element discretization of partial differential equations. There are mainly two approaches in this context: error estimation with respect to natural norms, see, e.g., [Ver96, AO00, BS01] for surveys, and a "quantity of interest" driven (goal-oriented) approach going back to [EEHJ96, BR96, BR01].

In the last years both approaches were extended to optimal control problems governed by partial differential equations. In articles [GHIK07, HHIK08, HIIS06, LLMT02, LY01] the authors provide a posteriori error estimates with respect to natural norms for elliptic optimal control problems with distributed or Neumann control subject to box constraints on the control variable. In [GHIK07] convergence of an adaptive algorithm for a control constrained optimal control problem is shown. For an optimal control with pointwise state constraints a posteriori error estimates were derived in [HK09b].

The motivation of the goal-oriented adaptivity is the fact, that in many applications one is interested in the precise computation of certain physical quantities, we will call them quantities of interest, rather than in reducing the discretization error with respect to global norms. In particular, although convergence in a sufficiently strong global norm will give convergence in the quantity of interest, the order of convergence in this quantity with respect to the unknowns may not be optimal. In [BKR00, BR01] a general concept is presented for a posteriori estimation of the discretization error with respect to the cost functional in the context of optimal control problems. In papers [BV04, BV05] this approach was extended to the estimation of the discretization error with respect to an arbitrary functional – a so-called quantity of interest – depending on both the control and the state variable. This allows, among other things, an efficient treatment of parameter identification and model calibration problems. An extension to optimization problems governed by parabolic PDEs was done in [MV07, Mei07], where separate error estimators for temporal and spatial discretization errors are derived. These error contributions are balanced in the corresponding adaptive algorithm.

Only recently goal-oriented error estimation was considered for optimal control problems subject to inequality constraints. The case of pointwise control constraints is treated in [VW08, HH08b]. In [Bec06] similar techniques are used explicitly to estimate the error in the control with respect to its natural norm, as the error in the natural norm can be bounded by the error in the cost functional, provided the cost functional is sufficiently coercive. For problems with pointwise state constraints recent work has been done simultaneously by [GH08, BV09, Wol10a].

The outline of the paper is as follows: We begin with a brief statement of the problem class we are concerned with and the corresponding optimality systems in Section 2. In Section 3 the finite element discretization on a fixed mesh is discussed. Section 4 is devoted to a posteriori error estimation for optimal control problems with inequality constraints and is structured as follows: In Section 4.1 we summarize the results from [VW08] on error estimation with respect to the cost functional for control constrained problems. The extension to an arbitrary quantity of interest is not considered here but can be found in [VW08]. The case of pointwise state constraints is discussed in 4.2 following the presentation from [BV09], see also [GH08] for similar results. Finally in Section 4.3 we consider the case, where the state constraints are eliminated using a barrier or penalty method. For the barrier approach a discussion of error estimates – for both regularization as well as discretization errors – can be found in [Wol10a]. A posteriori error estimates for the discretization error in a Moreau-Yosida regularization of state-constrained problems are derived in [GT09]. Here, we will present error estimates for the regularization error in the context of Moreau-Yosida regularization. The approach is similar to the one from [Wol10a]. This allows the balancing of the different error contributions in an adaptive algorithm.

2. Problem formulation & regularization

Let $\Omega \subset \mathbb{R}^d$ be a bounded domain. To avoid technicalities, we assume that Ω is the union of quadrangular domains. Let $\Gamma = \partial\Omega$ be its boundary and $\Gamma_D \subset \Gamma$ a relatively open subset. Let V be a Hilbert space, e.g., $V = H^1_D(\Omega) = \{v \in H^1(\Omega) \,|\, v|_{\Gamma_D}= 0\}$ with its dual space denoted by V^*. Further let W and Q be reflexive Banach spaces where $W \subset V$, the embedding being dense and continuous, and $Q_{\text{ad}} \subset Q$ being a closed, convex, and non empty subset. For simplicity, we choose $Q = L^2(\Omega)$ and $Q_{\text{ad}} = \{q \in Q \,|\, q_a \le q \le q_b \;\text{ a.e. in } \Omega\}$ with some $q_a, q_b \in \mathbb{R} \cup \{\pm\infty\}$, $q_a < q_b$. The choice of the space W depends on the type of constraints. In the case of control constraints only it is sufficient to consider $W = V$. In the case of state constraints W is typically chosen such that the constraint maps W into $\subset C(\overline{\Omega})$. Hence for pointwise state constraints one may choose $W = W^{1,p}(\Omega) \cap V$ with $p > d$. For pointwise constraints on the derivatives $W = W^{2,p}(\Omega) \cap V$ with $p > d$ can be chosen.

Finally, let $a\colon Q \times V \times V \to \mathbb{R}$ be a semi-linear "energy form" which is three times continuously differentiable.

With this notation, we can now state our state equation in abstract form. For given $f \in V^*$ and $q \in Q_{\text{ad}}$ find $u \in V$ such that

$$a(q, u)(\phi) = f(\phi) \quad \forall \phi \in V. \tag{2.1}$$

In the following, we assume that for given $(q, f) \in Q_{\text{ad}} \times V^*$ problem (2.1) has a unique solution u_q, such that the mapping $(q, f) \mapsto u_q$ is at least once continuously differentiable. We denote the directional derivatives of $a(q, u)(z)$ with respect to the variable u in direction δu by $a'_u(q, u)(\delta u, z)$. All other derivatives are denoted analogously.

As it is not suitable in general to consider (2.1) under minimal regularity assumptions, we will pose the following conditions on the differential operator. Let there be a space $Z \supset V$ with dense embedding, and Z^* be its dual space. We assume that for $q \in Q_{\text{ad}}$ and $f \in Z^*$ any solution u_q of (2.1) possesses the additional regularity $u_q \in W$, where again the mapping $(q, f) \mapsto u_q$ is assumed to be continuously differentiable.

Remark 2.1. We remark that the additional regularity $u_q \in W$ might seem rather restrictive. However in the case of state constraints on may chose $W = W^{1,p}(\Omega) \cap V$ with some $p > d$ which can be achieved in many situations. In the case of constraints on the derivative of the state the choice $W = W^{2,p}(\Omega) \cap V$ imposes a rather strong condition on the angles in the corners or along the edges of the domain even for the simple equation $-\Delta u = q$. However, the assumption $u_q \in W$ for all $q \in Q_{\text{ad}}$ can be avoided, in this case, by a more careful analysis, see [Wol10b] for details.

We continue by introducing the pointwise state constraints. For this purpose, we define the compact set $\Omega^C \subset \overline{\Omega}$ with non empty interior. We assume that the mapping $g\colon u \mapsto g(u)$ is three times continuously differentiable from W into $G := C(\Omega^C)$. We employ the usual ordering to state the constraint $g(u) \le 0$. In addition, we require that $g'(u)(\phi) \in L^1(\Omega^C)$ for $\phi \in V$ and $g(u) \in L^2(\Omega^C)$ for $u \in V$. Consider, e.g., $g(u) = u - u_b$ for the pointwise constraint $u \le u_b$ on Ω^C. Finally, we introduce the cost functional $J\colon Q \times V \to \mathbb{R}$, which is also assumed to be three times continuously differentiable.

We can now formulate our abstract optimization problem as follows:

$$\text{Minimize } J(q,u), \tag{2.2a}$$

$$\text{subject to } \begin{cases} a(q,u)(\phi) = f(\phi) \quad \forall\, \phi \in V, \\ (q,u) \in Q_{\text{ad}} \times V, \\ g(u) \le 0. \end{cases} \tag{2.2b}$$

In order to avoid technicalities, we assume that there exists a solution to (2.2), which satisfies the usual necessary optimality condition. Let $(\overline{q},\overline{u}) \in Q_{\text{ad}} \times V$ be a solution to (2.2), then $\overline{u} \in W$ and there exists $\overline{z} \in Z$, $\overline{\mu} \in M(\Omega^C) = C(\Omega^C)^* = G^*$ such that

$$\begin{aligned}
a(\overline{q},\overline{u})(\psi) &= f(\psi) && \forall\ \psi \in V, && (2.3a)\\
a'_u(\overline{q},\overline{u})(\phi,\overline{z}) &= J'_u(\overline{q},\overline{u})(\phi) + \langle \overline{\mu}\,, g'(\overline{u})(\phi)\,\rangle_{G^*\times G} && \forall\ \phi \in W, && (2.3b)\\
J'_q(\overline{q},\overline{u})(\delta q - \overline{q}) &\ge a'_q(\overline{q},\overline{u})(\delta q - \overline{q},\overline{z}) && \forall\, \delta q \in Q_{\text{ad}}, && (2.3c)\\
\langle \overline{\mu}\,, \phi \rangle_{G^*\times G} &\le 0 && \forall\ \phi \in G;\ \phi \le 0, && (2.3d)\\
\langle \overline{\mu}\,, g(\overline{u})\,\rangle_{G^*\times G} &= 0. && && (2.3e)
\end{aligned}$$

That this assumption is actually satisfied has been established for various types of problems, see, e.g., [Cas86, CB88, Ber92] for state constraints with distributed control, or [Ber93] for boundary control. For the case of first-order state constraints the existence of the Lagrangian multipliers has been shown in [CF93].

For our presentation it is convenient to assume that the cost functional is subject to a Tikhonov regularization, e.g.,

$$J(q,u) = J_1(u) + \frac{\alpha}{2}\|q\|^2, \tag{2.4}$$

with some $\alpha > 0$. Then, we introduce a projection operator $\mathcal{P}_{Q_{\text{ad}}}\colon Q \to Q_{\text{ad}}$ by

$$\mathcal{P}_{Q_{\text{ad}}}(p) = \max\big(q_a, \min(p,q_b)\big),$$

which holds pointwise almost everywhere. Following [Trö05]), this allows us to rewrite the variational inequality (2.3c) as

$$\overline{q} = \mathcal{P}_{Q_{\text{ad}}}\left(\frac{1}{\alpha}a'_q(\overline{q},\overline{u})(\cdot,\overline{z})\right), \tag{2.5}$$

where $a'_q(\overline{q},\overline{u})(\cdot,\overline{z})$ is understood as a Riesz representative of a linear functional on Q.

For a solution $(\overline{q}, \overline{u})$ of (2.2), we introduce active sets ω_- and ω_+ of the control as follows:

$$\omega_- = \{ x \in \Omega \,|\, \overline{q}(x) = q_a \}, \tag{2.6}$$

$$\omega_+ = \{ x \in \Omega \,|\, \overline{q}(x) = q_b \}. \tag{2.7}$$

Let $(\overline{q}, \overline{u}, \overline{z}, \overline{\mu}) \in Q \times V \times Z \times G^*$ be a solution to (2.3), then we introduce an additional Lagrange multiplier $\overline{\mu}_Q \in Q$ by the following identification:

$$(\overline{\mu}_Q, \delta q) = -\alpha(\overline{q}, \delta q) + a'_q(\overline{q}, \overline{u})(\delta q, \overline{z}), \quad \delta q \in Q. \tag{2.8}$$

The variational inequality (2.3c) or the projection formula (2.5) are known to be equivalent to the following conditions:

$$\overline{\mu}_Q(x) \leq 0 \quad \text{a.e. on } \omega_-, \tag{2.9a}$$

$$\overline{\mu}_Q(x) \geq 0 \quad \text{a.e. on } \omega_+, \tag{2.9b}$$

$$\overline{\mu}_Q(x) = 0 \quad \text{a.e. on } \Omega \setminus (\omega_- \cup \omega_+). \tag{2.9c}$$

There are two possible ways to deal with the constraint $g(u) \leq 0$. Either one discretizes problem (2.2) directly, or one removes the state constraint using for instance barrier or penalty methods and then discretizes the regularized problem. For barrier methods a posteriori error estimates were already derived in [Wol10a]. Here, we will only consider penalty methods.

We begin by defining the penalty term $P_\gamma \colon W \to \mathbb{R}$ for $\gamma \in \mathbb{R}$:

$$P_\gamma(u) := \frac{\gamma}{2} \|(g(u))^+\|^2_{L^2(\Omega^C)}, \tag{2.10}$$

where, as usual, the superscript $^+$ denotes the positive part of the function. With this setting, we define the following abstract Moreau-Yosida regularized problem

$$\text{Minimize } J_\gamma(q,u) := J(q,u) + P_\gamma(u), \tag{2.11a}$$

$$\text{subject to } \begin{cases} a(q,u)(\phi) = f(\phi) \quad \forall\, \phi \in V, \\ \qquad (q,u) \in Q_{\text{ad}} \times V. \end{cases} \tag{2.11b}$$

Then, we can formally state the corresponding first-order necessary conditions. Let $(\overline{q}_\gamma, \overline{u}_\gamma) \in Q_{\text{ad}} \times V$ be a solution of (2.11). Then, $\overline{u} \in W$ and there exist $\overline{\mu}_\gamma = \gamma g(\overline{u}_\gamma)^+$ and $\overline{z}_\gamma \in V$, such that

$$a(\overline{q}_\gamma, \overline{u}_\gamma)(\phi) = f(\phi) \qquad \forall\, \phi \in V, \tag{2.12a}$$

$$a'_u(\overline{q}_\gamma, \overline{u}_\gamma)(\phi, \overline{z}_\gamma) = J'_u(\overline{q}_\gamma, \overline{u}_\gamma)(\phi) + (g'(\overline{u}_\gamma)(\phi), \overline{\mu}_\gamma)_{\Omega^C} \qquad \forall\, \phi \in V, \tag{2.12b}$$

$$J'_q(\overline{q}_\gamma, \overline{u}_\gamma)(\delta q - \overline{q}_\gamma) \geq a_q(\overline{q}_\gamma, \overline{u}_\gamma)(\delta q - \overline{q}_\gamma, \overline{z}_\gamma) \qquad \forall\, \delta q \in Q_{\text{ad}}. \tag{2.12c}$$

This has been shown for linear elliptic equations and zero- as well as first-order constraints in [HK09a] together with the convergence towards the solution of (2.2) as $\gamma \to \infty$.

Remark 2.2. The above setting can be easily adapted to various types of control by considering $Q = L^2(\omega)$ for some set ω, e.g., $\omega = \Omega$ for distributed control, $\omega = \partial\Omega$ for boundary control, or $\omega = \{1, \ldots, n\}$ for finite-dimensional controls.

Analogously, the case of finitely many state constraints can be handled in this framework by setting $\Omega^C \subset \{1,\dots,n\}$ for some n instead of $\Omega^C \subset \overline{\Omega}$. In this case $G := C(\Omega^C) = \mathbb{R}^n$ where its dual G^* can be identified with G. The condition on Ω^C to have non empty interior is fulfilled, as every point in $\{1,\dots,n\}$ is open in the trivial metric.

3. Discretization

We consider a shape-regular mesh $\mathcal{T}_h$ consisting of quadrilateral cells K. The mesh parameter h is defined as a cellwise constant function by setting $h\big|_K = h_K$ where h_K is the diameter of K. On the mesh $\mathcal{T}_h$, we consider a conforming finite element space $V_h \subset V$ consisting of cellwise bilinear shape functions, see, e.g., [Bra07, BS08] for details.

In order to ease the mesh refinement, we allow the cells to have nodes which lie on midpoints of edges of neighboring cells. But at most one of such *hanging nodes* is permitted per edge. Consideration of meshes with hanging nodes requires additional care. These irregular nodes do not carry degrees of freedom as the corresponding values of the finite element function are determined by pointwise interpolation. We refer, e.g., to [CO84] for implementation details.

Throughout, we will denote the set of nodes of $\mathcal{T}_h$ by $\mathcal{N}_h = \{x_i\}$ and the corresponding nodewise basis functions of V_h by $\{\phi_i\}$. The space V_h will be used for the discretization of the state variable.

The finite-dimensional space $Q_h \subset Q$ can be constructed as an analog of V_h consisting of cellwise bilinear shape functions or as a space consisting of cellwise constant functions. Another possibility is to choose $Q_h = Q$ also in the case of an infinite-dimensional control space following the approach proposed in [Hin05]. Assume that Q_h is discretized. Then we introduce a basis of Q_h by

$$\mathcal{B} = \{\psi_i\}, \text{ with } \psi_i \geq 0, \sum_i \psi_i = 1, \max_{x\in\Omega} \psi_i(x) = 1.$$

Remark 3.1. As pointed out, e.g., in [Liu05, MV07], it might be desirable to use different meshes for the control and the state variables. The error estimator presented below can provide information for separate assessment of the errors due to the control and the state discretization. The refinement then follows an equilibration strategy for both estimators, cf. [MV07].

For a given discrete control $q_h \in Q_h$ the solution $u_h \in V_h$ of the discrete state equation is determined by

$$a(q_h, u_h)(\varphi_h) = f(\phi_h) \quad \forall \varphi_h \in V_h. \tag{3.1}$$

As on the continuous level, we assume the existence of a unique solution $u_h \in V_h$ of (3.1) for each $q_h \in Q_h$. This allows the definition of the discrete solution operator $S_h \colon Q_h \to V_h$.

For clarity of the presentation, here we restrict ourselves to the case $g(u) = u - u_b$ with $u_b \in C(\bar{\Omega})$. Then, the state constraint considered has the form

$$u(x) \le u_b(x) \quad \forall x \in \bar{\Omega}.$$

For the discretization of this state constraint, we use a nodal interpolation operator $i_h : C(\bar{\Omega}) \to V_h$ and define the discrete admissible set as

$$V_{\mathrm{ad},h} = \{\, u_h \in V_h \mid u_h(x) \le u_b(x) \ \ \forall x \in \mathcal{N}_h \,\}.$$

The discrete admissible set for the controls is defined by

$$Q_{\mathrm{ad},h} = Q_{\mathrm{ad}} \cap Q_h.$$

Altogether, the discretized optimal control problem is formulated as follows:

$$\text{Minimize } J(q_h, u_h), \tag{3.2a}$$

$$\text{subject to } \begin{cases} a(q_h, u_h)(\varphi_h) = 0 \qquad \forall \varphi_h \in V_h, \\ (q_h, u_h) \in Q_{\mathrm{ad},h} \times V_h, \\ u_h(x) \le u_b(x) \ \forall x \in \mathcal{N}_h. \end{cases} \tag{3.2b}$$

In various situations, see, e.g., [BV09], one can derive the necessary optimality conditions for (3.2) assuming a discrete Slater condition. In order to write the discrete optimality system similarly to the continuous case, we introduce a space of discrete Lagrange multipliers,

$$\mathcal{M}_h = \left\{ \mu_h = \sum_i \mu_i \delta_{x_i} \;\middle|\; x_i \in \mathcal{N}_h \right\} \subset \mathcal{M}(\Omega),$$

where δ_{x_i} denotes the Dirac measure concentrated at the point x_i. Then, for a solution $(\overline{q}_h, \overline{u}_h) \in Q_{\mathrm{ad},h} \times V_h$ of (3.2) there exists $(\overline{z}_h, \overline{\mu}_h) \in V_h \times \mathcal{M}_h$ such that

$$a(\overline{q}_h, \overline{u}_h)(\phi_h) = f(\phi_h) \qquad \forall\ \phi_h \in V_h, \tag{3.3a}$$

$$a'_u(\overline{q}_h, \overline{u}_h)(\phi_h, \overline{z}_h) = J'_u(\overline{q}_h, \overline{u}_h)(\phi_h) + \langle \overline{\mu}_h , \phi_h \rangle_{G^* \times G} \qquad \forall\ \phi_h \in V_h, \tag{3.3b}$$

$$J'_q(\overline{q}_h, \overline{u}_h)(\delta q_h - \overline{q}_h) \ge a'_q(\overline{q}_h, \overline{u}_h)(\delta q_h - \overline{q}_h), \overline{z}_h) \qquad \forall\, \delta q_h \in Q_{\mathrm{ad},h}, \tag{3.3c}$$

$$\langle \overline{\mu}_h, u_b - \overline{u}_h \rangle = 0, \quad \overline{\mu}_h \ge 0. \tag{3.3d}$$

We are now in the position to formulate the discretized version of the regularized problem (2.11):

$$\text{Minimize } J_\gamma(q_h, u_h), \tag{3.4a}$$

$$\text{subject to } \begin{cases} a(q_h, u_h)(v_h) = f(v_h) \quad \forall v_h \in V_h, \\ (q_h, u_h) \in Q_{\mathrm{ad},h} \times V_h. \end{cases} \tag{3.4b}$$

As for the continuous problem, we assume that this discrete problem admits a solution $(\overline{q}^h_\gamma, \overline{u}^h_\gamma)$ satisfying the following first-order necessary condition. Let

$(\overline{q}_\gamma^h, \overline{u}_\gamma^h) \in Q_{\mathrm{ad},h} \times V_h$ be a solution to (3.4), then there exist $\overline{z}_\gamma^h \in V_h$ and $\overline{\mu}_\gamma^h = \gamma g(\overline{u}_\gamma^h)^+$, such that

$$a(\overline{q}_\gamma^h, \overline{u}_\gamma^h)(\phi) = (f, \phi^h) \qquad \forall\ \phi^h \in V_h, \tag{3.5a}$$

$$a'_u(\overline{q}_\gamma^h, \overline{u}_\gamma^h)(\phi^h, \overline{z}_\gamma^h) = J'_u(\overline{q}_\gamma^h, \overline{u}_\gamma^h)(\phi^h) + (g'(\overline{u}_\gamma^h)(\phi^h), \overline{\mu}_\gamma^h)_{\Omega^C} \ \forall\ \phi^h \in V_h, \tag{3.5b}$$

$$J'_q(\overline{q}_\gamma^h, \overline{u}_\gamma^h)(\delta q^h - \overline{q}_\gamma^h) \geq a_q(\overline{q}_\gamma^h, \overline{u}_\gamma^h)(\delta q^h - \overline{q}_\gamma^h, \overline{z}_\gamma^h) \qquad \forall\, \delta q^h \in Q_{\mathrm{ad},h}. \tag{3.5c}$$

It seems unreasonable to directly solve this problem because of its increasing ill-conditioning for $\gamma \to \infty$, which particularly affects the solution of the linear systems for computing descend directions, see, e.g., [Loo69, Mur71]. A way to overcome this difficulty is by balancing mesh size h and regularization parameter γ such that the error contributions are equilibrated. To do so, one can either use a priori information [HH08a, HS11] or more efficiently use an a posteriori error estimate, see [Wol10a] for a barrier method or Section 4.3 for the penalty method case. An application in the context of inexact Newton methods can be found in [GS11].

Analogously to (2.8) a discrete multiplier for the control constraints is given by

$$(\overline{\mu}_{Q,h}, \delta q_h) = -\alpha(\overline{q}_h, \delta q) + a'_q(\overline{q}_h, \overline{u}_h)(\delta q, \overline{z}_h) \quad \forall\, \delta q_h \in Q_h. \tag{3.6}$$

Moreover, we introduce $\overline{\mu}_{Q,h}^- \in Q_h$ and $\overline{\mu}_{Q,h}^+ \in Q_h$ by

$$\overline{\mu}_{Q,h}^+ - \overline{\mu}_{Q,h}^- = \overline{\mu}_{Q,h}, \ (\overline{\mu}_{Q,h}^-, \psi_i)_Q \geq 0, \ (\overline{\mu}_{Q,h}^+, \psi_i)_Q \geq 0 \quad \forall \psi_i \in \mathcal{B}, \tag{3.7}$$

$$(\overline{\mu}_{Q,h}^-, \overline{q}_h - q_a)_Q = (\overline{\mu}_{Q,h}^+, q_b - \overline{q}_h)_Q = 0, \tag{3.8}$$

which uniquely determines $\overline{\mu}_{Q,h}^\pm$.

4. A posteriori error estimates

In this section, we will discuss the derivation of a posteriori error estimates for the discretization described above. We begin with a discussion of the control constrained case in Subsection 4.1, i.e., there are no state constraints present either by assuming $g(u) \equiv 0$, resulting in $\overline{\mu} = 0$ in (2.3), or by considering the regularized version (2.11). These results are collected from [VW08], and are an extension of the results for the unconstrained case derived in [BKR00, BR01, BV04, BV05]. We will then continue in 4.2 with a look at the purely state constrained problem where the results are reported from [BV09]. Similar results were obtained by [GH08]. Finally, we will consider the estimation of the error in the cost functional introduced by the regularization (2.11). The results described here can also be derived for the case of a barrier formulation, see [Wol10a].

4.1. Control constraints

The section is structured as follows. First, we will derive two a posteriori estimates for the error with respect to the cost functional. The first one is based on the first-order necessary condition (2.3), which involves a variational inequality.

For a comparison with analogous "weighted" error estimators for variational inequalities see [BS00a, BS00b]. The second estimate uses the information obtained from the Lagrange multipliers for the inequality constraints. Both estimates can be evaluated in terms of the solution to the discretized optimality condition to (2.3). For the case of an arbitrary quantity of interest, we refer to [VW08]. Here, we are considering only control constrained problems and therefore set $g(u) \equiv 0$ implying $\overline{\mu} = \overline{\mu}_h = 0$.

Throughout this section, we shall denote a solution to the optimization problem (2.2) by $(\overline{q}, \overline{u})$, the corresponding solution to the optimality system (2.3) by $\overline{\xi} = (\overline{q}, \overline{u}, \overline{z}) \in \mathcal{X}_{\mathrm{ad}} = Q_{\mathrm{ad}} \times V \times W$ and its discrete counterpart (3.2) by $\overline{\xi}_h = (\overline{q}_h, \overline{u}_h, \overline{z}_h) \in \mathcal{X}_{\mathrm{ad},h} = Q_{\mathrm{ad},h} \times V_h \times V_h$. The corresponding solution including the Lagrange multipliers (2.8) and its discrete counterpart will be abbreviated as $\overline{\chi} = (\overline{q}, \overline{u}, \overline{z}, \overline{\mu}_Q^-, \overline{\mu}_Q^+)$ and $\overline{\chi}_h = (\overline{q}_h, \overline{u}_h, \overline{z}_h, \overline{\mu}_{Q,h}^-, \overline{\mu}_{Q,h}^+)$, respectively, where $\overline{\mu}_{Q,h}^{\pm}$ are the discrete multipliers for the control constraints as given by (3.6)–(3.8).

4.1.1. Error estimates without multipliers for control constraints. For the derivation of the error estimate with respect to the cost functional, we introduce the residual functionals $\rho_u(\xi_h)(\cdot)$, $\rho_z(\xi_h)(\cdot) \in V'$ and $\rho_q(\xi_h)(\cdot) \in Q'$ by

$$\rho_u(\xi_h)(\cdot) := f(\cdot) - a(q_h, u_h)(\cdot), \tag{4.1}$$

$$\rho_z(\xi_h)(\cdot) := J_1'(u_h)(\cdot) - a_u'(q_h, u_h)(\cdot, z_h), \tag{4.2}$$

$$\rho_q(\xi_h)(\cdot) := \alpha(q_h, \cdot) - a_q'(u_h, q_h)(\cdot, z_h). \tag{4.3}$$

The following theorem is an extension of the result from [BR01] for the control-constrained case.

Theorem 4.1. *Let $\overline{\xi} \in \mathcal{X}_{\mathrm{ad}}$ be a solution to the first-order necessary system* (2.3) *and $\overline{\xi}_h \in \mathcal{X}_{\mathrm{ad},h}$ its Galerkin approximation* (3.3)*. Then, the following estimate holds:*

$$\begin{aligned} J(\overline{q}, \overline{u}) - J(\overline{q}_h, \overline{u}_h) \le{}& \frac{1}{2}\rho_u(\overline{\xi}_h)(\overline{z} - \tilde{z}_h) + \frac{1}{2}\rho_z(\overline{\xi}_h)(\overline{u} - \tilde{u}_h) \\ &+ \frac{1}{2}\rho_q(\overline{\xi}_h)(\overline{q} - q_h) + R_1, \end{aligned} \tag{4.4}$$

where $\tilde{u}_h, \tilde{z}_h \in V_h$ are arbitrarily chosen and R_1 is a remainder term, which is cubic in the error, see [VW08] *for details.*

We note that, in contrast to the terms involving the residuals of the state and the adjoint equations, the error $\overline{q} - \overline{q}_h$ in the term $\rho_q(\overline{\xi}_h)(\overline{q} - \overline{q}_h)$ in (4.4) cannot be replaced by $\overline{q} - \tilde{q}_h$ with an arbitrary $\tilde{q}_h \in Q_{\mathrm{ad},h}$. This is caused by the control constraints. However, we may replace $\rho_q(\overline{\xi}_h)(\overline{q} - \overline{q}_h)$ by $\rho_q(\overline{\xi}_h)(\overline{q} - \overline{q}_h + \tilde{q}_h)$ with arbitrary $\tilde{q}_h$ satisfying $\operatorname{supp}(\tilde{q}_h) \subset \omega \setminus (\omega_{-,h} \cup \omega_{+,h})$ due to the structure of $\rho_q(\overline{\xi}_h)(\cdot)$. A similar situation occurs in the case of error estimation in the context of variational inequalities, see, e.g., [BS00b].

In order to use the estimate from the theorem above for computable error estimation, we proceed as follows. First, we choose $\tilde{u}_h = i_h \overline{u}$ and $\tilde{z}_h = i_h \overline{z}$, with an interpolation operator $i_h \colon V \to V_h$. Then, we have to approximate the corresponding interpolation errors $\overline{u}-i_h\overline{u}$ and $\overline{z}-i_h\overline{z}$. There are several heuristic techniques to do this, see for instance [BR01]. We assume that there is an operator $\pi\colon V_h \to \tilde{V}_h$, with $\tilde{V}_h \neq V_h$, such that $\overline{u} - \pi\overline{u}_h$ has a better local asymptotical behavior as $\overline{u} - i_h\overline{u}$. Then, we approximate as follows:

$$\rho_u(\xi_h)(\overline{z} - i_h\overline{z}) \approx \rho_u(\xi_h)(\pi\overline{z}_h - \overline{z}_h), \quad \rho_z(\xi_h)(\overline{u} - i_h\overline{u}) \approx \rho_z(\xi_h)(\pi\overline{u}_h - \overline{u}_h).$$

Such an operator can be constructed for example by the interpolation of the computed bilinear finite element solution in the space of biquadratic finite elements on patches of cells. For this operator the improved approximation property relies on local smoothness of $\overline{u}$ and super-convergence properties of the approximation $\overline{u}_h$. The use of such "local higher-order approximation" is observed to work very successfully in the context of "goal-oriented" a posteriori error estimation, see, e.g., [BR01].

The approximation of the term $\rho_q(\overline{\xi}_h)(\overline{q}-\overline{q}_h)$ requires more care. In contrast to the state $\overline{u}$ and the adjoint state $\overline{z}$, the control variable $\overline{q}$ can generally not be approximated by "local higher-order approximation" for the following reasons:

- In the case of finite-dimensional control space Q, there is no "patch-like" structure allowing for "local higher-order approximation".
- If $\overline{q}$ is a distributed control, it typically does not possess sufficient smoothness (due to the inequality constraints) for the improved approximation property.

We therefore suggest another approximation of $\rho_q(\overline{\xi}_h)(\overline{q} - \overline{q}_h)$ based on the projection formula (2.5). To this end, we introduce $\pi^q\colon Q_h \to Q_{\mathrm{ad}}$ by

$$\pi^q\overline{q}_h = \mathcal{P}_{Q_{\mathrm{ad}}}\Big(\frac{1}{\alpha}a_q'(\overline{q}_h, \pi\overline{u}_h)(\cdot, \pi\overline{z}_h)\Big). \tag{4.5}$$

In some cases, one can show better approximation behavior of $\overline{q}-\pi^q\overline{q}_h$ in comparison with $\overline{q}-\overline{q}_h$, see [MR04] and [Hin05] for similar considerations in the context of a priori error analysis. This construction results in the following computable a posteriori error estimator:

$$\eta_h^{(1)} := \frac{1}{2}\big(\rho_u(\overline{\xi}_h)(\pi\overline{z}_h - \overline{z}_h) + \rho_z(\overline{\xi}_h)(\pi\overline{u}_h - \overline{u}_h) + \rho_q(\overline{\xi}_h)(\pi^q\overline{q}_h - \overline{q}_h)\big).$$

In order to use this error estimator for guiding mesh refinement, we have to localize it to cell-wise or node-wise contributions. A direct localization of the terms like $\rho_u(\overline{\xi}_h)(\pi\overline{z}_h - \overline{z}_h)$ leads, in general, to local contributions of wrong order (overestimation) due to oscillatory behavior of the residual terms [CV99]. To overcome this problem, one may integrate the residual terms by parts, see, e.g., [BR01], or use a filtering operator, see [Vex04, SV08] for details.

We should note, that (4.4) does not provide an estimate for the absolute value of $J(\overline{q},\overline{u}) - J(\overline{q}_h,\overline{u}_h)$, which is due to the inequality sign in (4.4). In the next section, we will overcome this difficulty utilizing the Lagrange multiplier (2.8).

4.1.2. Error estimates with multipliers for control constraints. In order to derive an error estimate for $J(\overline{q},\overline{u}) - J(\overline{q}_h,\overline{u}_h)$, we introduce the following additional residual functionals $\tilde{\rho}_q(\chi_h)(\cdot)$, $\tilde{\rho}_{\mu^-}(\chi_h)(\cdot)$, $\tilde{\rho}_{\mu^+}(\chi_h)(\cdot) \in Q'$ by

$$\tilde{\rho}_q(\chi_h)(\cdot) = \alpha(q_h,\cdot)_Q - a'_q(q_h,u_h)(\cdot,z_h) + (\mu^+_{Q,h} - \mu^-_{Q,h},\cdot)_Q, \tag{4.6}$$

$$\tilde{\rho}_{\mu^-}(\chi_h)(\cdot) = (\cdot,q_a - q_h)_Q, \tag{4.7}$$

$$\tilde{\rho}_{\mu^+}(\chi_h)(\cdot) = (\cdot,q_h - q_b)_Q. \tag{4.8}$$

In the following the last two residual functionals will also be evaluated at the point χ where they read as follows:

$$\tilde{\rho}_{\mu^-}(\chi)(\cdot) = (\cdot,q_a - q)_Q, \quad \tilde{\rho}_{\mu^+}(\chi)(\cdot) = (\cdot,q - q_b)_Q.$$

Then, analogously to Theorem 4.1 we obtain the following result.

Theorem 4.2. *Let $\overline{\chi} = (\overline{q},\overline{u},\overline{z},\overline{\mu}^-_Q,\overline{\mu}^-_Q)$ be a solution to the system (2.3) and the corresponding multiplier (2.8) and $\overline{\chi}_h$ be its Galerkin approximation defined by (3.3) and (3.6)–(3.8). Then, we have the error representation*

$$\begin{aligned} J(\overline{q},\overline{u}) - J(\overline{q}_h,\overline{u}_h) = {} & \frac{1}{2}\rho_u(\overline{\chi}_h)(\overline{z}-\tilde{z}_h) + \frac{1}{2}\rho_z(\overline{\chi}_h)(\overline{u}-\tilde{u}_h) + \frac{1}{2}\tilde{\rho}_q(\overline{\chi}_h)(\overline{q}-\tilde{q}_h) \\ & + \frac{1}{2}\tilde{\rho}_{\mu^-}(\overline{\chi}_h)(\overline{\mu}^-_Q-\tilde{\mu}^-_h) + \frac{1}{2}\tilde{\rho}_{\mu^+}(\overline{\chi}_h)(\overline{\mu}^+_Q-\tilde{\mu}^+_h) \\ & + \frac{1}{2}\tilde{\rho}_{\mu^-}(\overline{\chi})(\tilde{\mu}^- - \overline{\mu}^-_{Q,h}) + \frac{1}{2}\tilde{\rho}_{\mu^+}(\overline{\chi})(\tilde{\mu}^+ - \overline{\mu}^+_{Q,h}) + R_2, \end{aligned} \tag{4.9}$$

where $\tilde{u}_h,\tilde{z}_h \in V_h$, $\tilde{q}_h \in Q_h$, $\tilde{\mu}^-_h \in Q_{-,h}$, $\tilde{\mu}^+_h \in Q_{+,h}$, $\tilde{\mu}^- \in Q_-$, $\tilde{\mu}^+ \in Q_+$ are arbitrarily chosen and R_2 is a cubic remainder term, see [VW08] *for details.*

To derive a computable error estimator from (4.9), we proceed analogously as in the previous section. In order to deal with the new residual functionals, we utilize (2.8) and construct an approximation for $\overline{\mu}_Q$ by

$$\tilde{\mu}_Q = -\alpha\pi^q\overline{q}_h + a'_q(\pi^q\overline{q}_h,\pi\overline{u}_h)(\cdot,\pi\overline{z}_h), \tag{4.10}$$

where $\pi^q\overline{q}_h$ is given by (4.5). This leads us to the computable a posteriori error estimator,

$$\begin{aligned} \eta_h^{(2)} := {} & \frac{1}{2}\big(\rho_u(\overline{\chi}_h)(\pi\overline{z}_h - \overline{z}_h) + \rho_z(\overline{\chi}_h)(\pi u_h - \overline{u}_h) + \tilde{\rho}_q(\overline{\chi}_h)(\pi^q\overline{q}_h - \overline{q}_h) \\ & + \tilde{\rho}_{\mu^-}(\overline{\chi}_h)(\tilde{\mu}^-_Q - \overline{\mu}^-_{Q,h}) + \tilde{\rho}_{\mu^+}(\overline{\chi}_h)(\tilde{\mu}^+_Q - \overline{\mu}^+_{Q,h}) \\ & + \tilde{\rho}_{\mu^-}(\tilde{\chi})(\tilde{\mu}^-_Q - \overline{\mu}^-_{Q,h}) + \tilde{\rho}_{\mu^+}(\tilde{\chi})(\tilde{\mu}^+_Q - \overline{\mu}^+_{Q,h})\big), \end{aligned}$$

where $\tilde{\chi} = (\pi^q\overline{q}_h,\pi\overline{u}_h,\pi\overline{u}_h,\tilde{\mu}^-,\tilde{\mu}^+)$.

Remark 4.3. We note that the a posteriori error estimates derived in Theorem 4.1 and Theorem 4.2 coincide if the control constraints are inactive, e.g., if $Q_{\text{ad}} = Q$.

4.1.3. Numerical results. In this section, we discuss a numerical example illustrating the behavior of mesh adaptation based on the "weighted" a posteriori error estimates of Theorems 4.1 and 4.2. We use bilinear (H^1-conforming) finite elements for the discretization of the state variable and cell-wise constant discretization of the control space. The computation is done using the optimization library RoDoBo [RoD] and the finite element toolkit Gascoigne [Gas].

Example on a domain with a hole. We consider the following nonlinear optimization problem:

$$\text{Minimize} \quad \frac{1}{2}\|u - u^d\|^2_{L^2(\Omega)} + \frac{\alpha}{2}\|q\|^2_{L^2(\Omega)}, \quad u \in V,\ q \in Q_{\text{ad}}, \tag{4.11}$$

subject to

$$\begin{aligned} -\Delta u + 30u^3 + u &= f + q \qquad && \text{in } \Omega, \\ u &= 0 && \text{on } \partial\Omega, \end{aligned} \tag{4.12}$$

where $\Omega = (0,1)^2 \setminus [0.4, 0.6]^2$, $V = H^1_0(\Omega)$, $Q = L^2(\Omega)$, and the admissible set Q_{ad} is give by

$$Q_{\text{ad}} = \{\, q \in Q \mid -7 \le q(x) \le 20 \text{ a.e. on } \Omega \,\}.$$

The target state u^d and the right-hand side f are defined as

$$u^d(x) = x_1 \cdot x_2, \quad f(x) = (x_1 - 0.5)^{-2}(x_2 - 0.5)^{-2},$$

and the regularization parameter is chosen as $\alpha = 10^{-4}$. We note that the state equation (4.12) involves a monotone semi-linear operator and therefore possesses a unique solution $u \in V$ for each $q \in Q$. The proof of the existence of a global solution as well as the derivation of necessary and sufficient optimality conditions for the corresponding optimization problem (4.11)–(4.12) can be found, e.g., in [Trö05].

Above, we presented two different error estimators for the error with respect to the cost functional. In order to compare the quality of these estimators, we use their "effectivity indices", which are defined as follows:

$$I_{\text{eff}}(\eta^{(1)}) = \frac{J(\overline{q}, \overline{u}) - J(\overline{q}_h, \overline{u}_h)}{\eta_h^{(1)}}, \quad I_{\text{eff}}(\eta^{(2)}) = \frac{J(\overline{q}, \overline{u}) - J(\overline{q}_h, \overline{u}_h)}{\eta_h^{(2)}}. \tag{4.13}$$

In Table 1 these effectivity indices are listed for a sequence of randomly refined meshes as well as for a sequence of globally refined meshes. We see that both error estimators yield quantitatively correct information about the discretization error. We note that the results for $\eta^{(1)}$ and $\eta^{(2)}$ are very close to each other in this example. In addition, it can be clearly seen that by local mesh refinement based on these error estimators a substantial saving is achieved in the number of degrees of freedom for reaching a prescribed level of accuracy.

In Figure 1 the dependence of the error in the cost functional on the number of degrees of freedom used is shown for different refinement criteria: global (uniform) refinement, refinement based on the error estimator $\eta^{(2)}$ for the cost functional, and refinement based on a standard "energy-norm" error estimator η_{energy} for the

(A) Random refinement

N	$I_{\text{eff}}(\eta^{(1)})$	$I_{\text{eff}}(\eta^{(2)})$
432	1.1	1.1
906	1.1	1.1
2328	1.3	1.2
5752	1.2	1.2
13872	1.3	1.3
33964	1.3	1.3

(B) Global refinement

N	$I_{\text{eff}}(\eta^{(1)})$	$I_{\text{eff}}(\eta^{(2)})$
120	1.2	1.2
432	1.1	1.1
1631	1.1	1.1
6336	1.1	1.1
24960	1.1	1.1
99072	1.2	1.2

TABLE 1. Effectivity indices

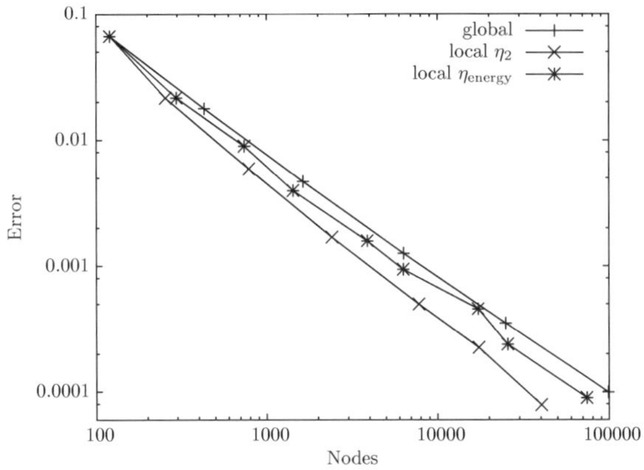

FIGURE 1. Discretization error in J for different refinement criteria

state and the adjoint variable. We observe the best behavior of error in the cost functional if the mesh is refined based on $\eta^{(2)}$. Especially it becomes clear that a refinement according to the energy-norm error estimator does not yield the best results in terms of needed degrees of freedom for a given error in J.

Figure 2 shows a series of meshes generated according to the information obtained from the error estimators together with the optimal control $\overline{q}$ and the corresponding state $\overline{u}$.

4.2. State constraints

Next, we consider the case $g(u) = u - u_b$, $Q_{\text{ad}} = Q$ and hence $\overline{\mu}_Q = \overline{\mu}_{Q,h} = 0$. The results presented here are collected from [BV09].

Theorem 4.4. *Let* $(\overline{q}, \overline{u}, \overline{z}, \overline{\mu})$ *be a solution of the optimality system* (2.3) *and* $(\overline{q}_h, \overline{u}_h, \overline{z}_h, \overline{\mu}_h)$ *a solution of the corresponding discrete optimality system* (3.3).

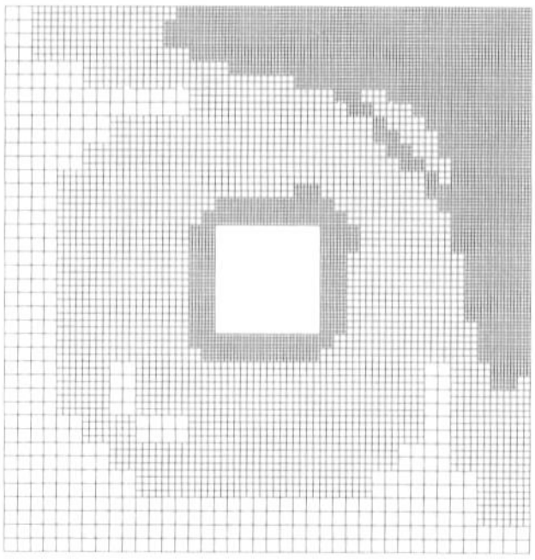

(A) Mesh 3 from $\eta^{(2)}$

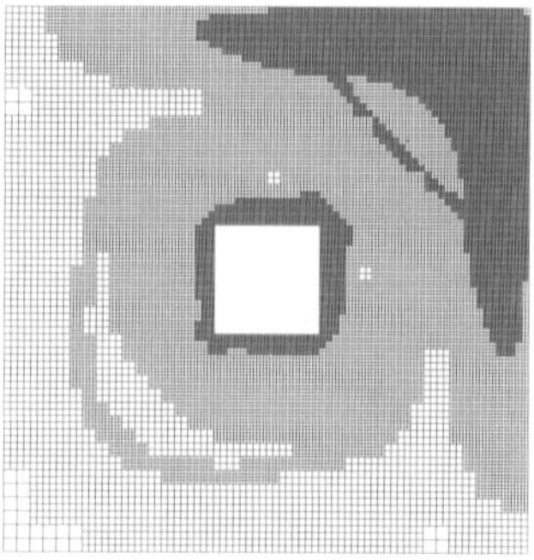

(B) Mesh 4 from $\eta^{(2)}$

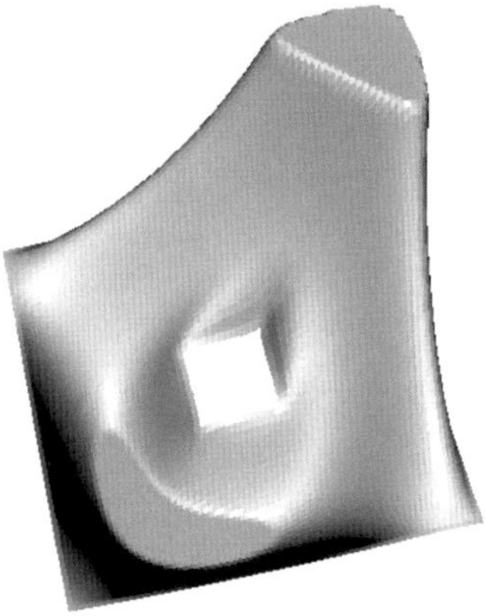

(C) Optimal control

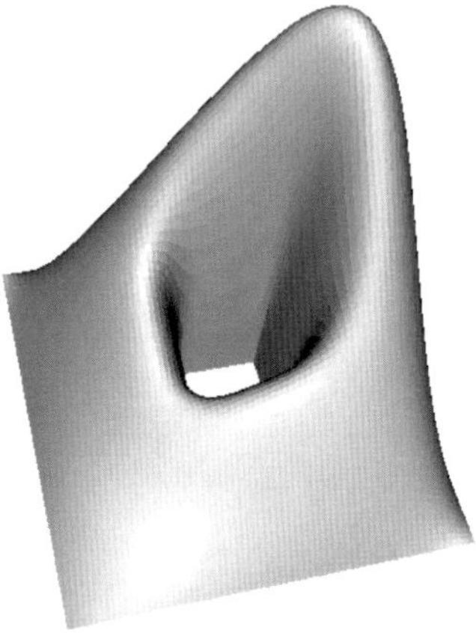

(D) Optimal state

FIGURE 2. Locally refined meshes and solution variables

Then, there holds

$$\begin{aligned} J(\overline{q},\overline{u}) - J(\overline{q}_h,\overline{u}_h) = \frac{1}{2}\Big(& J'_u(\overline{q}_h,\overline{u}_h)(\overline{u}-\overline{u}_h) - a'_u(\overline{q}_h,\overline{u}_h)(\overline{u}-\overline{u}_h,\overline{z}_h) \\ & + J'_q(\overline{q}_h,\overline{u}_h)(\overline{q}-\tilde{q}_h) - a'_q(\overline{q}_h,\overline{u}_h)(\overline{q}-\tilde{q}_h,z_h) \\ & + f(\overline{z}-\tilde{z}_h) - a(\overline{q}_h,\overline{u}_h)(\overline{z}-\tilde{z}_h) - \langle\overline{\mu},\overline{u}-\overline{u}_h\rangle\Big) + \mathcal{R}, \end{aligned} \tag{4.14}$$

where $\tilde{q}_h \in Q_h$ and $\tilde{z}_h \in V_h$ can be arbitrarily chosen and $\mathcal{R}$ is a remainder term that is cubic in the error, see [BV09] *for details.*

Remark 4.5. Note, that the errors $\overline{q}-\overline{q}_h$ and $\overline{z}-\overline{z}_h$ are replaced by $\overline{q}-\tilde{q}_h$ and $\overline{z}-\tilde{z}_h$, respectively, which can be viewed as interpolation errors. However, the error in the state variable $\overline{u}-\overline{u}_h$ can not be directly replaced in this way because of the structure of the optimality system.

Remark 4.6. The residual of the gradient equation

$$J'_q(\overline{q}_h, \overline{u}_h)(\overline{q} - \tilde{q}_h) - a'_q(\overline{q}_h, \overline{u}_h)(\overline{q} - \tilde{q}_h, \overline{z}_h)$$

in the above error representation can be shown to be zero in some cases, see, e.g., [BV09].

Remark 4.7. One can directly show that the error in the Lagrange multiplier $\overline{\mu}$ does not appear explicitly in the remainder term $\mathcal{R}$. Therefore the remainder term can usually be neglected as a higher-order term, see, e.g., [BV09].

The error representation formula (4.14) still contains continuous solutions $(\overline{q}, \overline{u}, \overline{z}, \overline{\mu})$, which have to be approximated in order to obtain a computable error estimator. For details see the discussion in Section 4.1.1. In the notation of Section 4.1.1, we approximate as follows:

$$\begin{aligned} J'_u(\overline{q}_h, \overline{u}_h)(\overline{u} - \overline{u}_h) &- a'_u(\overline{q}_h, \overline{u}_h)(\overline{u} - \overline{u}_h, \overline{z}_h) \\ &\approx J'_u(\overline{q}_h, \overline{u}_h)(\pi\overline{u}_h - \overline{u}_h) - a'_u(\overline{q}_h, \overline{u}_h)(\pi\overline{u}_h - \overline{u}_h, \overline{z}_h), \end{aligned}$$

$$\begin{aligned} J'_q(\overline{q}_h, \overline{u}_h)(\overline{q} - \tilde{q}_h) &- a'_q(\overline{q}_h, \overline{u}_h)(\overline{q} - \tilde{q}_h, \overline{z}_h) \\ &\approx J'_q(\overline{q}_h, \overline{u}_h)(\pi^q\overline{q}_h - \overline{q}_h) - a'_q(\overline{q}_h, \overline{u}_h)(\pi^q\overline{q}_h - \overline{q}_h, \overline{z}_h), \end{aligned}$$

and

$$f(\overline{z} - \tilde{z}_h) - a(\overline{q}_h, \overline{u}_h)(\overline{z} - \tilde{z}_h) \approx f(\pi\overline{z}_h - \overline{z}_h) - a(\overline{q}_h, \overline{u}_h)(\pi\overline{z}_h - \overline{z}_h).$$

The construction of an approximation for the term $\langle \overline{\mu}, \overline{u} - \overline{u}_h \rangle$ is more involved, since no direct analog of higher-order interpolation can be used for the multipliers $\overline{\mu}$ being Borel measures. A simple approximation $\overline{\mu} \approx \overline{\mu}_h$ is not directly useful, because $\langle \overline{\mu}_h, \pi\overline{u}_h - \overline{u}_h \rangle$ is identical to zero if the biquadratic interpolation is taken for π. In [GH08] the authors therefore use another construction for π. Instead, we express the term $\langle \overline{\mu}, \overline{u} - \overline{u}_h \rangle$ using the adjoint equation (2.3b) leading us to

$$\langle \overline{\mu}, \overline{u} - \overline{u}_h \rangle = -J'_u(\overline{q}, \overline{u})(\overline{u} - \overline{u}_h) + a'_u(\overline{q}, \overline{u})(\overline{u} - \overline{u}_h, \overline{z}).$$

This expression does not involve any Lagrange multiplier and can be approximated like above,

$$\langle \overline{\mu}, \overline{u} - \overline{u}_h \rangle \approx -J'_u(\pi^q\overline{q}_h, \pi\overline{u}_h)(\pi\overline{u}_h - \overline{u}_h) + a'_u(\pi^q\overline{q}_h, \pi\overline{u}_h)(\pi\overline{u}_h - \overline{u}_h, \pi\overline{z}_h).$$

Taking all these considerations into account, the error representation formula (4.14) motivates the definition of the following error estimator:

$$\begin{aligned} \eta := \frac{1}{2}\Big(&J'_q(\overline{q}_h, \overline{u}_h)(\pi^q\overline{q}_h - \overline{q}_h) - a'_q(\overline{q}_h, \overline{u}_h)(\pi^q\overline{q}_h - \overline{q}_h, \overline{z}_h) \\ &+ J'_u(\overline{q}_h, \overline{u}_h)(\pi\overline{u}_h - \overline{u}_h) - a'_u(\overline{q}_h, \overline{u}_h)(\pi\overline{u}_h - \overline{u}_h, \overline{z}_h) \\ &+ J'_u(\pi^q\overline{q}_h, \pi\overline{u}_h)(\pi_h\overline{u}_h - \overline{u}_h) - a'_u(\pi^q\overline{q}_h, \pi\overline{u}_h)(\pi\overline{u}_h - \overline{u}_h, \pi\overline{z}_h) \\ &+ f(\pi\overline{z}_h - \overline{z}_h) - a(\overline{q}_h, \overline{u}_h)(\pi\overline{z}_h - \overline{z}_h)\Big). \end{aligned} \tag{4.15}$$

Remark 4.8. The use of the operators π and π^q in the estimation of local approximation errors can be rigorously justified only for smooth solutions $\overline{q}, \overline{u}, \overline{z}$ employing super-convergence effects. However, the adjoint solution $\overline{z}$ and consequently the control variable $\overline{q}$ in general possess only reduced regularity, i.e., $\overline{z} \in W^{1,p}(\Omega)$ for some $p < 2$. Nevertheless, we expect a good behavior of the proposed error estimator, since the operators π and π^q are defined locally and the regions, where the adjoint state $\overline{z}$ is not smooth, are usually strongly localized.

4.2.1. Numerical example. In this section, we present a numerical result illustrating the behavior of the error estimator and the adaptive algorithm developed in this paper. Further examples can be found in [BV09]. The optimal control problem is solved on sequences of meshes produced either by uniform refinement or by the refinement according to the proposed error estimator. This allows us to demonstrate the saving in the number of degrees of freedom required to reach a prescribed error tolerance if local refinement is employed. Moreover, we investigate the quality of the error estimation by calculating the effectivity index of the error estimator defined by

$$I_{\text{eff}} = \frac{J(\overline{q}, \overline{u}) - J(\overline{q}_h, \overline{u}_h)}{\eta}. \tag{4.16}$$

On each mesh the discrete optimal control problem (3.2) is solved using a primal-dual-active-set strategy implemented in the software packages RoDoBo [RoD] and Gascoigne [Gas]. For visualization, we use the visualization tool VisuSimple [Vis].

We consider the following optimal control problem governed by a semilinear elliptic equation on the unit square $\Omega = (0,1)^2$:

$$(P) \begin{cases} \text{Minimize } J(u,q) = \dfrac{1}{2}\|u - u^d\|^2_{L^2(\Omega)} + \dfrac{\alpha}{2}\|q\|^2_{L^2(\Omega)}, \\ \text{subject to } \begin{cases} -\Delta u + u^3 = q + f & \text{in } \Omega, \\ u = 0 & \text{on } \partial\Omega, \\ u_a \le u \le u_b & \text{in } \overline{\Omega}, \end{cases} \end{cases}$$

with $\alpha = 0.001$, $f = 0$, $u_b = 0$, and

$$u^d = 16x(1-x)^2(x-y) + \frac{3}{5}, \quad u_a = -0.08 - 4\left(x - \frac{1}{4}\right)^2 - 4\left(y - \frac{27}{32}\right)^2.$$

For this example problem no exact solution is available. Therefore, we use the value $J(\overline{q}_{h^*}, \overline{u}_{h^*})$ computed on a very fine mesh $\mathcal{T}_{h^*}$ as an approximation of the exact value $J(\overline{q}, \overline{u})$. The plots of the optimal control and state for this example are shown in Figure 3.

We observe that the active set A^+ corresponding to the upper bound is a two-dimensional set with nonempty interior and the active set A^- contains apparently only one point. A typical locally refined mesh, which captures the main features of the problem under consideration is shown in Figure 4.

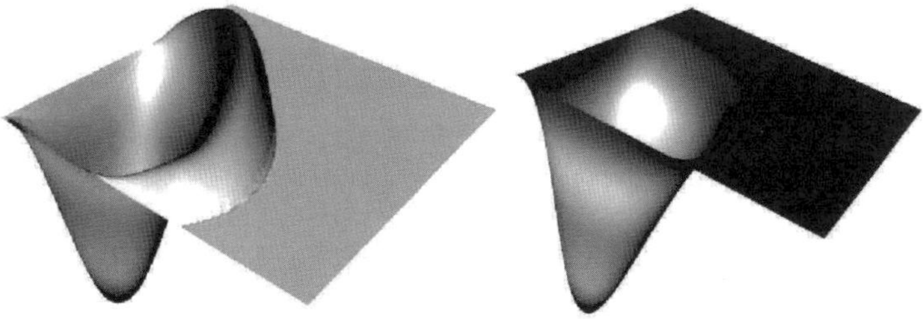

(A) control variable (B) state variable

FIGURE 3. Plots of the optimal control and the optimal state for (P)

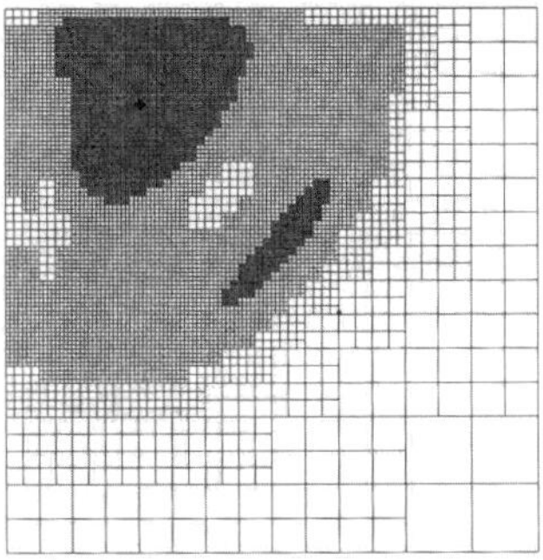

FIGURE 4. Example of a locally refined mesh for (P)

Table 2 shows the behavior of the discretization errors and the effectivity indices as defined in (4.16) for both, uniform and local, mesh refinements. Figure 5 gives the comparison of both refinement strategies with respect to the required number of degrees of freedom to reach a prescribed error tolerance.

4.3. Regularized state constraints

Finally, we consider the case of a penalty method (2.11). As in the previous section, we concentrate on the case without control constraints, e.g., $Q_{\mathrm{ad}} = Q$.

In order to estimate the error introduced by the regularization parameter γ, we define the Lagrangian functional $\mathcal{M}\colon Q \times W \times Z \times M(\Omega_C) \to \mathbb{R}$ by

$$\mathcal{M}(q, u, z, \mu) = J(q, u) + (f, \phi) - a(q, u)(z) + \int_{\Omega_C} g(u)\, d\mu.$$

Then, we obtain the following error representation.

Theorem 4.9. *Let $(\overline{q}, \overline{u}, \overline{z}, \overline{\mu})$ be a solution to the system (2.3) and let $(\overline{q}_\gamma, \overline{u}_\gamma, \overline{z}_\gamma, \overline{\mu}_\gamma)$ be a solution to the corresponding system (2.12) of the penalty problem. Then, the*

(A) local refinement

N	$J(q,u)-J(q_h,u_h)$	I_{eff}
25	5.38e-04	-1.41
41	-1.16e-04	0.43
99	-4.48e-05	0.33
245	-2.68e-05	0.60
541	-1.04e-05	0.56
1459	-6.04e-06	0.89
4429	-1.54e-06	0.83

(B) uniform refinement

N	$J(q,u)-J(q_h,u_h)$	I_{eff}
25	5.38e-04	-1.41
81	-1.58e-04	0.62
289	-6.18e-05	0.87
1089	-1.58e-05	0.87
4225	-3.99e-06	0.89
16641	-7.45e-07	0.66

TABLE 2. Discretization errors and efficiency indices under mesh refinement for (P)

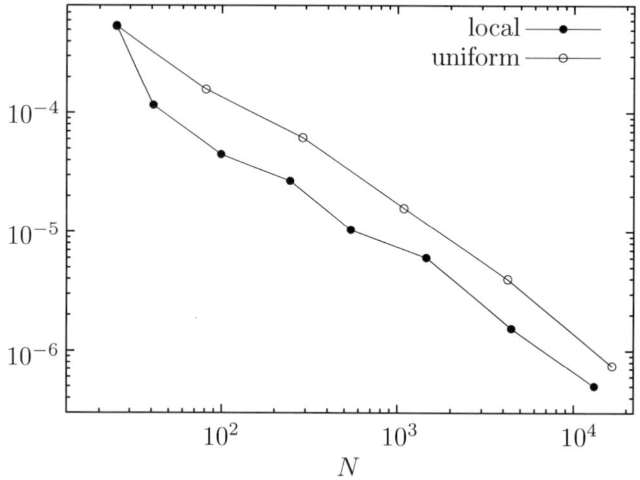

FIGURE 5. Discretization error vs. degrees of freedom for the two refinement strategies for (P)

following error identity holds:

$$J(\overline{q},\overline{u}) - J_\gamma(\overline{q}_\gamma,\overline{u}_\gamma) = \frac{1}{2}\int_{\Omega_C} g(\overline{u}_\gamma)\,d\overline{\mu} - \frac{1}{2}\int_{\Omega_C} g(\overline{u})\,d\overline{\mu}_\gamma + \mathcal{R}_{\text{reg}} \tag{4.17}$$

with a remainder term $\mathcal{R}_{\text{reg}}$ *given by:*

$$\mathcal{R}_{\text{reg}} = \frac{1}{2}\int_0^1 \mathcal{M}'''(\overline{\xi}_\gamma + s(\overline{\xi}-\overline{\xi}_\gamma))(\overline{\xi}-\overline{\xi}_\gamma,\overline{\xi}-\overline{\xi}_\gamma,\overline{\xi}-\overline{\xi}_\gamma)s(s-1)\,ds. \tag{4.18}$$

For details see [Wol10b].

Observing $0 \le J(\overline{q},\overline{u}) - J_\gamma(\overline{q}_\gamma,\overline{u}_\gamma)$, we obtain from the first-order necessary conditions that

$$0 \le J(\overline{q},\overline{u}) - J_\gamma(\overline{q}_\gamma,\overline{u}_\gamma) \le \frac{1}{2}\int_{\Omega_C} g(\overline{u}_\gamma) - g(\overline{u})\, d(\overline{\mu}+\overline{\mu}_\gamma).$$

We proceed heuristically by assuming that $g(\overline{u}) \approx 0$ on the support of $\overline{\mu}+\overline{\mu}_\gamma$, to obtain

$$\frac{1}{2}\int_{\Omega_C} g(\overline{u}_\gamma) - g(\overline{u})\, d(\overline{\mu}+\overline{\mu}_\gamma) \approx \frac{1}{2}\int_{\Omega_C} g(\overline{u}_\gamma)\, d(\overline{\mu}+\overline{\mu}_\gamma) \approx \int_{\Omega_C} g(\overline{u}_\gamma)\, d\overline{\mu}_\gamma.$$

This motivates the use of the following estimator for the regularization error:

$$\eta_{\text{reg}} = \int_{\Omega_C} g(\overline{u}_\gamma^h)\, d\overline{\mu}_\gamma^h,$$

where $\overline{u}_\gamma^h$, $\overline{\mu}_\gamma^h$ are solutions to the Galerkin discretization (3.5).

The derivation of the error estimator for the discretization error is rather straightforward and is therefore omitted. The main difficulty in its proof lies in the fact, that $g(\,\cdot\,)^+$ is only directionally differentiable. In the case of zero-order constraints it can be found in [GT09].

4.3.1. Numerical example. Once again the computations are done using the software packages RoDoBo [RoD] and Gascoigne [Gas]. In order to demonstrate the behavior of our estimator we consider the following test problem from [DGH08]:

$$\text{Minimize } J(q,u) = \frac{1}{2}\|u-u^d\|^2_{L^2(\Omega)} + \frac{1}{2}\|q\|^2_{L^2(\Omega)}$$

$$\text{subject to } \begin{cases} (\nabla u, \nabla \phi) = (f+q,\phi) & \forall \phi \in H^1_0(\Omega),\\ \frac{1}{4} - |\nabla u(x)|^2 \ge 0 & \forall x \in \overline{\Omega},\\ q \in Q_{\text{ad}}, \end{cases}$$

with given functions u^d and f such that the solution is known and the optimal value of J is $\frac{\pi}{2}$. The domain Ω is given as $\{x \in \mathbb{R}^2 \,|\, |x| \le 2\}$.

In order to conveniently embed this problem into our framework, we consider the following reformulation where we introduce an additional state variable $w = |\nabla u|^2$, i.e., we consider the problem

$$\text{Minimize } J(q,u) = \frac{1}{2}\|u-u^d\|^2_{L^2(\Omega)} + \frac{1}{2}\|q\|^2_{L^2(\Omega)}$$

$$\text{subject to } \begin{cases} (\nabla u, \nabla \phi) = (f+q,\phi) & \forall \phi \in H^1_0(\Omega),\\ (w,\phi) = (|\nabla u|^2,\phi) & \forall \phi \in L^2(\Omega),\\ \frac{1}{4} - w \ge 0 & \forall x \in \overline{\Omega},\\ q \in Q_{\text{ad}}. \end{cases}$$

We begin by a consideration of the convergence behavior of the cost functional. Figure 6 shows the convergence of the quantity $J(\overline{q},\overline{u}) - J_\gamma(\overline{q}_\gamma^h,\overline{u}_\gamma^h)$ which behaves

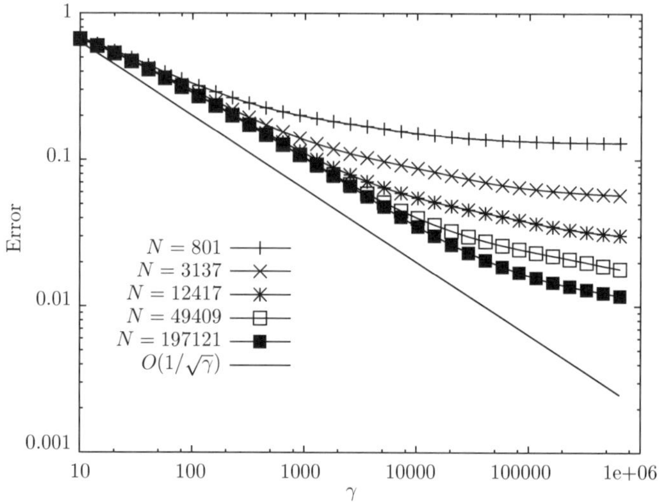

FIGURE 6. Convergence behavior of $J(\overline{q}, \overline{u}) - J_\gamma(\overline{q}_\gamma^h, \overline{u}_\gamma^h)$

	γ						
N	$4 \cdot 10^1$	$8 \cdot 10^1$	$2 \cdot 10^2$	$3 \cdot 10^2$	$6 \cdot 10^2$	$1 \cdot 10^3$	$3 \cdot 10^3$
801	1.4	1.4	1.5	1.7	2.0	2.3	2.5
3137	1.3	1.3	1.2	1.3	1.4	1.7	1.9
12417	1.3	1.2	1.2	1.1	1.1	1.2	1.3
49409	1.3	1.2	1.1	1.1	1.1	1.1	1.1
197121	1.3	1.2	1.1	1.1	1.1	1.1	1.1

TABLE 3. Efficiency of $\eta_{\text{disc}} + \eta_{\text{reg}}$ on various meshes. Values with $|\eta_{\text{reg}}|$ below the discretization error are on gray background.

like $O(\gamma^{-1/2})$. By transferring the results of [HH08a] to the case of first-order state constraints a convergence order between $O(\gamma^{-1/2})$ and $O(\gamma^{-1})$ has to be expected. We remark that this directly shows that an a priori choice of the relation between h and γ is difficult, as the convergence behavior is not a priori known, but has to be determined in the course of the computation.

We now investigate whether the proposed method of estimating the regularization error is sufficiently accurate. To do so, we consider a range of parameters for γ such that we will be able to see the behavior of the regularization estimate in the vicinity of the equilibrium of regularization and discretization error, see Figure 6.

In Table 3 we have depicted the effectivity index

$$I_{\text{eff}} = \frac{|J(\overline{q}, \overline{u}) - J_\gamma(\overline{q}_\gamma^h, \overline{u}_\gamma^h)|}{|\eta_{\text{disc}}| + |\eta_{\text{reg}}|}$$

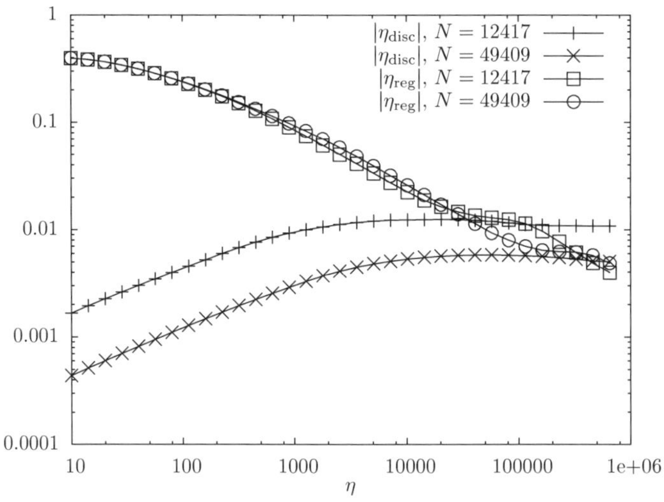

FIGURE 7. Convergence behavior of the error indicators

on different meshes for various choices of γ. The sequence of γ was obtained by starting from $\gamma_0 = 10$ and then successively increasing γ by a factor of two. The results clearly show that the estimator η_{reg} provides a good estimate for the influence of the regularization error, although it slightly underestimates the true error.

Finally we will have a short look on the interplay between the discretization error estimate and the regularization error estimate. For this, we consider the behavior for both estimators separately on globally refined meshes with 12 417 and 49 409 vertices. The results are depicted in Figure 7. Here we can see that the discretization error estimators grow towards a limit for $\gamma \to \infty$. This is exactly what we expect, as a solution to the limiting problem exists and should be harder to approximate by a discretization due to the measure in the right-hand side of the adjoint equation. Next, we observe that for $|\eta_{\text{reg}}| \gg |\eta_{\text{disc}}|$ the estimator for the regularization remains almost unchanged under mesh refinement, hence it makes sense to call $|\eta_{\text{reg}}|$ an estimate for the regularization error. However, when $|\eta_{\text{reg}}| \approx |\eta_{\text{disc}}|$ the estimator remains stagnant for a short range of γ values, before they become again almost constant. This behavior indicates that when balancing the contributions of both estimators one should not aim for $|\eta_{\text{reg}}| \ll |\eta_{\text{disc}}|$ in order to obtain an efficient algorithm. From Figure 8, we immediately obtain that this is indeed the case, as we can clearly see, and that both strategies of balancing, either balancing the error contributions or letting $\gamma \to \infty$, lead to comparable results concerning the error. But the computational costs for the balancing strategy are far less.

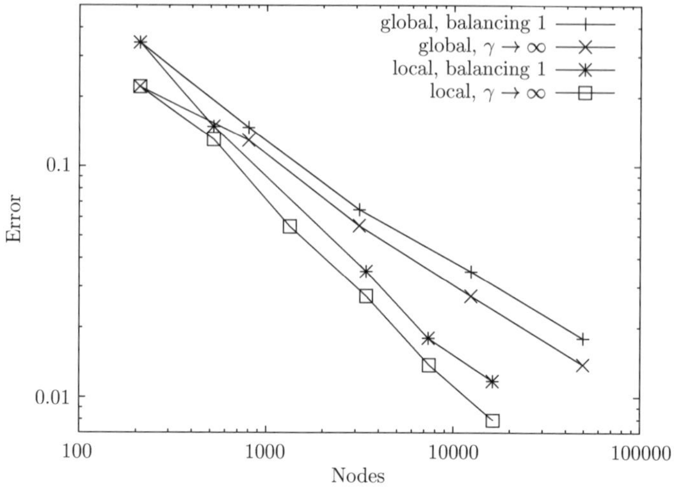

FIGURE 8. Convergence behavior of the error indicators

References

[AO00] Mark Ainsworth and John Tinsley Oden. *A Posteriori Error Estimation in Finite Element Analysis.* Pure and Applied Mathematics (New York). Wiley-Interscience [John Wiley & Sons], New York, 2000.

[Bec06] Roland Becker. Estimating the control error in discretized PDE-constraint optimization. *J. Numer. Math.*, 14(3):163–185, 2006.

[Ber92] Maïtine Bergounioux. A penalization method for optimal control of elliptic problems with state constraints. *SIAM J. Control Optim.*, 30(2):305–323, 1992.

[Ber93] Maïtine Bergounioux. On boundary state constrained control problems. *Numer. Funct. Anal. Optim.*, 14(5&6):515–543, 1993.

[BKR00] Roland Becker, Hartmut Kapp, and Rolf Rannacher. Adaptive finite element methods for optimal control of partial differential equations: Basic concept. *SIAM J. Control Optim.*, 39(1):113–132, 2000.

[BR96] Roland Becker and Rolf Rannacher. A feed-back approach to error control in finite element methods: Basic analysis and examples. *East-West J. Numer. Math.* 4, 4:237–264, 1996.

[BR01] Roland Becker and Rolf Rannacher. An optimal control approach to a posteriori error estimation. In A. Iserles, editor, *Acta Numerica* 2001, pages 1–102. Cambridge University Press, 2001.

[Bra07] Dietrich Braess. *Finite Elements: Theory, Fast Solvers and Applications in Solid Mechanics.* Cambridge University Press, Cambridge, 2007.

[BS00a] Heribert Blum and Franz-Theo Suttmeier. An adaptive finite element discretization for a simplified signorini problem. *Calcolo*, 37(2):65–77, 2000.

[BS00b] Heribert Blum and Franz-Theo Suttmeier. Weighted error estimates for finite element solutions of variational inequalities. *Computing*, 65(2):119–134, 2000.

[BS01] Ivo Babuška and Theofanis Strouboulis. *The Finite Element Method and its Reliability.* Numerical Mathematics and Scientific Computation. The Clarendon Press Oxford University Press, New York, 2001.

[BS08] Susanne C. Brenner and Larkin Ridgway Scott. *The Mathematical Theory of Finite Element Methods.* Springer Verlag, New York, 3. edition, 2008.

[BV04] Roland Becker and Boris Vexler. A posteriori error estimation for finite element discretization of parameter identification problems. *Numer. Math.*, 96:435–459, 2004.

[BV05] Roland Becker and Boris Vexler. Mesh refinement and numerical sensitivity analysis for parameter calibration of partial differential equations. *J. Comp. Physics*, 206(1):95–110, 2005.

[BV09] Olaf Benedix and Boris Vexler. A posteriori error estimation and adaptivity for elliptic optimal control problems with state constraints. *Comput. Optim. Appl.*, 44(1):3–25, 2009.

[Cas86] Eduardo Casas. Control of an elliptic problem with pointwise state constraints. *SIAM J. Control Optim.*, 24(6):1309–1318, 1986.

[CB88] Eduardo Casas and Joseph Frédéric Bonnans. Contrôle de systèmes elliptiques semilinéares comportant des contraintes sur l'état. In *Nonlinear Partial Differential Equations and their Applications. Collège de France seminar, Vol. VIII* (*Paris,* 1984–1985), volume 166 of *Pitman Res. Notes Math. Ser.*, pages 69–86. Longman Sci. Tech., Harlow, 1988.

[CF93] Eduardo Casas and Luis Alberto Fernández. Optimal control of semilinear elliptic equations with pointwise constraints on the gradient of the state. *Appl. Math. Optim.*, 27:35–56, 1993.

[CO84] Graham F. Carey and J. Tinsley Oden. *Finite Elements. Computational Aspects*, volume 3. Prentice-Hall, 1984.

[CV99] Carsten Carstensen and Rüdiger Verfürth. Edge residuals dominate a posteriori error estimates for low-order finite element methods. *SIAM J. Numer. Anal.*, 36(5):1571–1587, 1999.

[DGH08] Klaus Deckelnick, Andreas Günther, and Michael Hinze. Finite element approximation of elliptic control problems with constraints on the gradient. *Numer. Math.*, 111:335–350, 2008.

[EEHJ96] Kenneth Eriksson, Don Estep, Peter Hansbo, and Claes Johnson. *Computational Differential Equations.* Cambridge University Press, Cambridge, 1996.

[Gas] The finite element toolkit Gascoigne. `http://www.gascoigne.uni-hd.de`.

[GH08] Andreas Günther and Michael Hinze. A posteriori error control of a state constrained elliptic control problem. *J. Numer. Math.*, 16:307–322, 2008.

[GHIK07] Alexandra Gaevskaya, Roland H.W. Hoppe, Yuri Iliash, and Michael Kieweg. Convergence analysis of an adaptive finite element method for distributed control problems with control constraints. In *Control of Coupled Partial Differential Equations*, International Series of Numerical Mathematics. Birkhäuser, 2007.

[GS11] Andreas Günther and Anton Schiela. An interior point algorithm with inexact step computation in function space for state constrained optimal control. 35 pp. To appear in *Numerische Mathematik*, doi:10.1007/s00211-011-0381-4, 2011.

[GT09] Andreas Günther and Moulay Hicham Tber. A goal-oriented adaptive moreau-yosida algorithm for control- and state-constrained elliptic control problems. Preprint SPP1253-089, DFG Priority Program 1253, 2009.

[HH08a] Michael Hintermüller and Michael Hinze. Moreau-Yosida regularization in state constrained elliptic control problems: Error estimates and parameter adjustment. *SIAM J. Numer. Anal.*, 47:1666–1683, 2008.

[HH08b] Michael Hintermüller and Ronald H.W. Hoppe. Goal-oriented adaptivity in control constrained optimal control of partial differential equations. *SIAM J. Control Optim.*, 47(4):1721–1743, 2008.

[HHIK08] Michael Hintermüller, Ronald H.W. Hoppe, Yuri Iliash, and Michael Kieweg. An a posteriori error analysis of adaptive finite element methods for distributed elliptic control problems with control constraints. *ESIAM Control Optim. Calc. Var.*, 14:540–560, 2008.

[HIIS06] Ronald H.W. Hoppe, Yuri Iliash, Chakradhar Iyyunni, and Nasser H. Sweilam. A posteriori error estimates for adaptive finite element discretizations of boundary control problems. *J. Numer. Math.*, 14(1):57–82, 2006.

[Hin05] Michael Hinze. A variational discretization concept in control constrained optimization: The linear-quadratic case. *Comp. Optim. Appl.*, 30(1):45–61, 2005.

[HK09a] Michael Hintermüller and Karl Kunisch. PDE-constrained optimization subject to pointwise constraints on the control, the state, and its derivative. *SIAM J. Optim.*, 20(3):1133–1156, 2009.

[HK09b] Ronald H.W. Hoppe and Michael Kieweg. A posteriori error estimation of finite element approximations of pointwise state constrained distributed control problems. *J. Numer. Math.*, 17(3):219–244, 2009.

[HS11] Michael Hinze and Anton Schiela. Discretization of interior point methods for state constrained elliptic optimal control problems: Optimal error estimates and parameter adjustment. *Comput. Optim. Appl.*, 48(3):581–600, 2011.

[Liu05] Wenbin Liu. Adaptive multi-meshes in finite element approximation of optimal control. *Contemporary Mathematics*, (383):113–132, 2005.

[LLMT02] Ruo Li, Wenbin Liu, Heping Ma, and Tao Tang. Adaptive finite element approximation for distributed elliptic optimal control problems. *SIAM J. Control Optim.*, 41(5):1321–1349, 2002.

[Loo69] Freerk Auke Lootsma. Hessian matrices of penalty functions for solving constrained-optimization problems. *Philips Res. Repts.*, 24:322–330, 1969.

[LY01] Wenbin Liu and Ningning Yan. A posteriori error estimates for distributed convex optimal control problems. *Adv. Comput. Math.*, 15(1):285–309, 2001.

[Mei07] Dominik Meidner. *Adaptive Space-Time Finite Element Methods for Optimization Problems Governed by Nonlinear Parabolic Systems.* PhD thesis, Mathematisch-Naturwissenschaftliche Gesamtfakultät, Universität Heidelberg, 2007.

[MR04] Christian Meyer and Arnd Rösch. Superconvergence properties of optimal control problems. *SIAM J. Control Optim.*, 43(3):970–985, 2004.

[Mur71] Walter Murray. Analytical expressions for the eigenvalues and eigenvectors of the Hessian matrices of barrier and penalty functions. *J. Optim. Theory Appl.*, 7(3):189–196, 1971.

[MV07] Dominik Meidner and Boris Vexler. Adaptive space-time finite element methods for parabolic optimization problems. *SIAM J. Control Optim.*, 46(1):116–142, 2007.

[RoD] RoDoBo: A C++ library for optimization with stationary and nonstationary PDEs. http://www.rodobo.uni-hd.de.

[SV08] Michael Schmich and Boris Vexler. Adaptivity with dynamic meshes for space-time finite element discretizations of parabolic equations. *SIAM J. Sci. Comput.*, 30(1):369–393, 2008.

[Trö05] Fredi Tröltzsch. *Optimale Steuerung partieller Differentialgleichungen.* Vieweg, 1. edition, 2005.

[Ver96] Rüdiger Verfürth. *A Review of A Posteriori Error Estimation and Adaptive Mesh-Refinement Techniques.* Wiley/Teubner, New York-Stuttgart, 1996.

[Vex04] Boris Vexler. *Adaptive Finite Element Methods for Parameter Identification Problems.* PhD thesis, Mathematisch-Naturwissenschaftliche Gesamtfakultät, Universität Heidelberg, 2004.

[Vis] VisuSimple: An interactive VTK-based visualization and graphics/mpeg-generation program. http://www.visusimple.uni-hd.de.

[VW08] Boris Vexler and Winnifried Wollner. Adaptive finite elements for elliptic optimization problems with control constraints. *SIAM J. Control Optim.*, 47(1):509–534, 2008.

[Wol10a] Winnifried Wollner. A posteriori error estimates for a finite element discretization of interior point methods for an elliptic optimization problem with state constraints. *Comput. Optim. Appl.*, 47(3):133–159, 2010.

[Wol10b] Winnifried Wollner. *Adaptive Methods for PDE based Optimal Control with Pointwise Inequality Constraints.* PhD thesis, Mathematisch-Naturwissenschaftliche Gesamtfakultät, Universität Heidelberg, 2010.

Rolf Rannacher
Universität Heidelberg, Institut für angewandte Mathematik
Im Neuenheimer Feld 293/294, D-69120 Heidelberg, Germany
e-mail: rannacher@iwr.uni-heidelberg.de

Boris Vexler
Technische Universität München, Lehrstuhl für Mathematische Optimierung
Zentrum Mathematik, M1, Boltzmannstr. 3, D-85748 Garching bei München, Germany
e-mail: vexler@ma.tum.de

Winnifried Wollner
Universität Hamburg, Fachbereich Mathematik
Bundesstrasse 55, D-20146 Hamburg, Germany
e-mail: winnifried.wollner@math.uni-hamburg.de

Part IV

Discretization: Concepts and Analysis

Introduction to Part IV
Discretization: Concepts and Analysis

This part summarizes recent trends and addresses future research directions in the field of tailored discrete concepts for PDE constrained optimization with elliptic and parabolic PDEs in the presence of pointwise constraints. It covers the range from the treatment of mesh grading over relaxation techniques, and adaptive a posteriori finite element concepts, to the modern treatment of optimal control problems with parabolic PDEs.

Thomas Apel and Dieter Sirch deal, in *A Priori Mesh Grading for Distributed Optimal Control Problems*, with L^2-error estimates for finite element approximations of control constrained distributed optimal control problems governed by linear partial differential equations whose solutions develop singularities. In order to avoid a reduced convergence order, graded finite element meshes are used.

Michael Hinze and Arnd Rösch review, with *Discretization of Optimal Control Problems*, the state of the art in designing discrete concepts for optimization problems with PDE constraints with emphasis on structure conservation of solutions on the discrete level, and on error analysis for the discrete variables involved.

Kristina Kohls, Arnd Rösch and Kunibert G. Siebert contribute, with *A Posteriori Error Estimators for Control Constrained Optimal Control Problems*, a framework for the a posteriori error analysis of control constrained optimal control problems with linear PDE constraints, which is solely based on reliable and efficient error estimators for the corresponding linear state and adjoint equations.

Ira Neitzel and Fredi Tröltzsch review, in *Numerical Analysis of State-constrained Optimal Control Problems for PDEs*, Lavrentiev-type regularization of both distributed and boundary control problems, as well as a priori error estimates for elliptic control problems with finite dimensional control space and finitely many state-constraints.

Dominik Meidner and Boris Vexler finally summarize, in *A Priori Error Estimates for Space-Time Finite Element Discretization of Parabolic Optimal Control Problems*, their results on a priori error analysis for space-time finite element discretizations of optimization problems governed by parabolic equations.

The editor of this part wishes to thank all authors who contributed to this volume, and also all involved referees, whose notes and comments were very valuable in preparing this part.

Michael Hinze

International Series of Numerical Mathematics, Vol. 160, 377–389

A Priori Mesh Grading for Distributed Optimal Control Problems

Thomas Apel and Dieter Sirch

Abstract. This paper deals with L^2-error estimates for finite element approximations of control constrained distributed optimal control problems governed by linear partial differential equations. First, general assumptions are stated that allow to prove second-order convergence in control, state and adjoint state. Afterwards these assumptions are verified for problems where the solution of the state equation has singularities due to corners or edges in the domain or nonsmooth coefficients in the equation. In order to avoid a reduced convergence order, graded finite element meshes are used.

Mathematics Subject Classification (2000). 65N30, 65N15.

Keywords. Linear-quadratic optimal control problems, control constraints, finite element method, error estimates, graded meshes, anisotropic elements.

1. Introduction

In this paper we consider the optimal control problem

$$\begin{aligned}\min_{(y,u)\in Y\times U} J(y,u) &:= \frac{1}{2}\|y-y_d\|_Z^2 + \frac{\nu}{2}\|u\|_U^2, \\ \text{subject to} \quad y &= Su, \quad u \in U^{\text{ad}},\end{aligned} \tag{1.1}$$

where Z, $U = U^*$ are Hilbert spaces and Y is a Banach space with $Y \hookrightarrow Z \hookrightarrow Y^*$. We introduce a Banach space $X \hookrightarrow Z$ and demand $y_d \in X$. The operator $S : U \to Y \hookrightarrow U$ is the solution operator of a linear elliptic partial differential equation. We assume ν to be a fixed positive number and $U^{\text{ad}} \subset U$ to be non-empty, convex and closed.

A general review of results is given by Hinze and Rösch in this volume [8]; they shall not be repeated here. We focus on results where the solution of the state equation has singularities due to corners or edges in the domain or nonsmooth coefficients in the equation [1, 3, 4, 16]. In Section 2, general assumptions are stated that allow to prove second-order convergence in control, state and adjoint state. Afterwards, in Section 3, these assumptions are verified for a scalar elliptic state equation with discontinuous coefficients in a polygonal domain and isotropic graded meshes, for the Poisson equation in a nonconvex prismatic domain and anisotropic graded meshes and for the Stokes equations as state equation in a nonconvex prismatic domain and a nonconforming discretization on anisotropic meshes.

For further use we recall now part of the theory of control constrained optimal control problems.

Remark 1.1. *Problem* (1.1) *is equivalent to the reduced problem*

$$\min_{u\in U^{\mathrm{ad}}} \hat{J}(u) \tag{1.2}$$

with

$$\hat{J}(u) := J(Su,u) = \frac{1}{2}\|Su - y_d\|_Z^2 + \frac{\nu}{2}\|u\|_U^2.$$

The following theorem can be proved with well-known arguments, see, e.g., [10].

Theorem 1.1. *The optimal control problem* (1.1) *has a unique optimal solution* $(\bar{y},\bar{u})$. *Furthermore, for* S^* *being the adjoint of* S, *the optimality conditions*

$$\bar{y} = S\bar{u}, \tag{1.3}$$

$$\bar{p} = S^*(S\bar{u} - y_d) \tag{1.4}$$

$$\bar{u} \in U^{\mathrm{ad}}, \quad (\nu\bar{u} + \bar{p}, u - \bar{u})_U \geq 0 \quad \forall u \in U^{\mathrm{ad}} \tag{1.5}$$

are necessary and sufficient.

Lemma 1.1. *Let* $\Pi_{U^{\mathrm{ad}}} : U \to U^{\mathrm{ad}}$ *be the projection on* U^{ad}, *i.e.,*

$$\Pi_{U^{\mathrm{ad}}}(u) \in U^{\mathrm{ad}}, \quad \|\Pi_{U^{\mathrm{ad}}}(u) - u\|_U = \min_{v\in U^{\mathrm{ad}}} \|v - u\|_U \quad \forall u \in U.$$

Then the projection formula

$$\bar{u} = \Pi_{U^{\mathrm{ad}}}\left(-\frac{1}{\nu}\bar{p}\right) \tag{1.6}$$

is equivalent to the variational inequality (1.5).

Proof. The assertion is motivated in [11]. A detailed proof is given, e.g., in [7]. The assertion follows from Lemma 1.11 in that book by setting $\gamma = 1/\nu$. □

2. Discretization and error estimates

The results of this subsection are also published in [17]. Detailed proofs can be found there. We consider a family of triangulations $\mathcal{T}_h = \{T\}$ of Ω, that is admissible in Ciarlet's sense, see [5, Assumptions $(\mathcal{T}_h 1)$–$(\mathcal{T}_h 5)$]. The operators $S_h : U \to Y_h$ and $S_h^* : Y^* \to Y_h$ are finite element approximations of S and S^*, respectively where Y_h is a suitable finite element space.

2.1. Variational discrete approach

In [6] Hinze introduced a discretization concept for the optimal control problem (1.2) which is based on the discretization of the state space only. The control space is not discretized. Instead, the discrete optimal control $\bar{u}_h^s$ is defined via the variational inequality

$$(\nu\bar{u}_h^s + S_h^*(S_h\bar{u}_h^s - y_d), u - \bar{u}_h^s)_U \geq 0 \quad \forall u \in U^{\mathrm{ad}}. \tag{2.1}$$

In the following we formulate two assumptions that are are sufficient to prove optimal error estimates.

Assumption VAR1. *The operators S_h and S_h^* are bounded, i.e., the inequalities*

$$\|S_h\|_{U\to Y} \leq c \qquad \text{and} \qquad \|S_h^*\|_{Y^*\to U^*} \leq c$$

are valid.

As usual in numerical analysis the generic constant c used here and in the sequel does not depend on h.

Assumption VAR2. *The estimates*

$$\|(S - S_h)u\|_U \leq ch^2\|u\|_U \quad \forall u \in U,$$
$$\|(S^* - S_h^*)z\|_U \leq ch^2\|z\|_Z \quad \forall z \in Z$$

hold true.

We introduce the optimal discrete state $\bar{y}_h^s := S_h\bar{u}_h^s$ and optimal adjoint state $\bar{p}_h^s := S_h^*(S_h\bar{u}_h^s - y_d)$ and formulate the error estimates in the following theorem.

Theorem 2.1. *Let Assumptions* VAR1 *and* VAR2 *hold. Then the estimates*

$$\|\bar{u} - \bar{u}_h^s\|_U \leq ch^2 \left(\|\bar{u}\|_U + \|y_d\|_Z\right), \tag{2.2}$$
$$\|\bar{y} - \bar{y}_h^s\|_U \leq ch^2 \left(\|\bar{u}\|_U + \|y_d\|_Z\right), \tag{2.3}$$
$$\|\bar{p} - \bar{p}_h^s\|_U \leq ch^2 \left(\|\bar{u}\|_U + \|y_d\|_Z\right) \tag{2.4}$$

hold.

Proof. The first estimate is proved in [6]. For the proof of the second assertion we write

$$\begin{aligned}\|\bar{y} - \bar{y}_h^s\|_U &= \|S\bar{u} - S_h\bar{u}_h^s\|_U \\ &\leq \|(S - S_h)\bar{u}\| + \|S_h(\bar{u} - \bar{u}_h^s)\|_U.\end{aligned}$$

Inequality (2.3) follows then from Assumptions VAR1 and VAR2 and (2.2). For the third assertion we can conclude similarly to above

$$\begin{aligned}\|\bar{p} - \bar{p}_h^s\|_U &= \|S^*(Su - y_d) - S_h^*(S_hu_h^s - y_d)\|_U \\ &= \|S^*(S - S_h)\bar{u} + (S^* - S_h^*)S_h\bar{u} + S_h^*S_h(\bar{u} - \bar{u}_h^s) - (S^* - S_h^*)y_d\|_U.\end{aligned}$$

With the triangle inequality the assertion (2.3) follows from the boundedness of S^*, Assumptions VAR1 and VAR2 and (2.2). □

2.2. Postprocessing approach

We consider the reduced problem (1.2) and choose

$$U = Z = L^2(\Omega)^d, \quad Y = H_0^1(\Omega)^d \text{ or } Y = H^1(\Omega)^d,$$

where $d \in \{1, 2, 3\}$ depending on the problem under consideration. As space of admissible controls we use

$$U^{\mathrm{ad}} := \{u \in U : u_a \leq u \leq u_b \text{ a.e.}\},$$

where $u_a \leq u_b$ are constant vectors from $\mathbb{R}^d$. Then the projection in the admissible set reads for a continuous function f as

$$(\Pi_{U^{\mathrm{ad}}} f)(x) := \max(u_a, \min(u_b, f(x))).$$

This formula is to be understood componentwise for vector-valued functions f. We introduce the discrete control space U_h,

$$U_h = \{u_h \in U : u_h|_T \in (\mathcal{P}_0)^d \text{ for all } T \in \mathcal{T}_h\} \quad \text{and} \quad U_h^{\mathrm{ad}} = U_h \cap U^{\mathrm{ad}}.$$

Then the discretized optimal control problem can be written as

$$\begin{aligned} J_h(\bar{u}_h) &= \min_{u_h \in U_h^{\mathrm{ad}}} J_h(u_h), \\ J_h(u_h) &:= \frac{1}{2}\|S_h u_h - y_d\|_{L^2(\Omega)}^2 + \frac{\nu}{2}\|u_h\|_{L^2(\Omega)}^2. \end{aligned} \tag{2.5}$$

As in the continuous case, this is a strictly convex and radially unbounded optimal control problem. Consequently, (2.5) admits a unique solution $\bar{u}_h$, that satisfies the necessary and sufficient optimality conditions

$$\begin{aligned} \bar{y}_h &= S_h \bar{u}_h, \\ \bar{p}_h &= S_h^*(\bar{y}_h - y_d), \\ (\nu \bar{u}_h + \bar{p}_h, u_h - \bar{u}_h)_U &\geq 0 \quad \forall u_h \in U_h^{\mathrm{ad}}. \end{aligned} \tag{2.6}$$

For later use, we introduce the affine operators $Pu = S^*(Su - y_d)$ and $P_h u = S_h^*(S_h u - y_d)$, that maps a given control u to the adjoint state $p = Pu$ and the approximate adjoint state $p_h = P_h u$, respectively.

The approximate control $\tilde{u}_h$ is constructed as projection of the approximate adjoint state in the set of admissible controls,

$$\tilde{u}_h = \Pi_{U^{\mathrm{ad}}}\Big(-\frac{1}{\nu}\bar{p}_h\Big). \tag{2.7}$$

This postprocessing technique was originally introduced by Meyer and Rösch [14]. In the following we state four rather general assumptions, that allow to prove optimal discretization error estimates. To this end, we first define two projection operators.

Definition 2.2. *For continuous functions f we define the projection R_h in the space $\mathcal{P}_0$ of piecewise constant functions by*

$$(R_h f)(x) := f(S_T) \text{ if } x \in T \tag{2.8}$$

where S_T denotes the centroid of the element T.

The operator Q_h projects L^2-functions g in the space $\mathcal{P}_0$ of piecewise constant functions,

$$(Q_h g)(x) := \frac{1}{|T|} \int_T g(x)\,\mathrm{d}x \text{ for } x \in T. \tag{2.9}$$

Both operators are defined componentwise for vector-valued functions.

Assumption PP1. *The discrete solution operators S_h and S_h^* are bounded,*

$$\|S_h\|_{U\to H_h^1(\Omega)^d} \le c, \qquad \|S_h^*\|_{U\to H_h^1(\Omega)^d} \le c,$$
$$\|S_h\|_{U\to L^\infty(\Omega)^d} \le c, \qquad \|S_h^*\|_{U\to L^\infty(\Omega)^d} \le c,$$

with the space

$$H_h^1(\Omega)^d := \left\{ v : \Omega \to \mathbb{R}^d : \sum_{T\in\mathcal{T}h} \|v\|^2_{H^1(T)^d} < \infty \right\}.$$

Notice that Assumption PP1 implies

$$\|S_h\|_{U\to U} \le c \quad \text{and} \quad \|S_h^*\|_{U\to U} \le c$$

by the embedding $H_h^1(\Omega) \hookrightarrow U$.

Assumption PP2. *The finite element error estimates*

$$\|(S - S_h)u\|_U \le ch^2\|u\|_U \quad \forall u \in U,$$
$$\|(S^* - S_h^*)z\|_U \le ch^2\|z\|_Z \quad \forall z \in Z$$

hold.

Assumption PP3. *The optimal control $\bar u$ is contained in X and the corresponding adjoint state $\bar p$ satisfy the inequality*

$$\|Q_h\bar p - R_h\bar p\|_U \le ch^2 \left(\|\bar u\|_X + \|y_d\|_X\right).$$

for a space $X \hookrightarrow U$. In particular, $\bar p$ is continuous, such that $R_h\bar p$ is well defined.

Assumption PP4. *The optimal control $\bar u$ is contained in X and for all functions $\varphi_h \in Y_h$ the inequality*

$$(Q_h\bar u - R_h\bar u, \varphi_h)_U \le ch^2\|\varphi_h\|_{L^\infty(\Omega)^d} \left(\|\bar u\|_X + \|y_d\|_X\right)$$

holds. In particular, $\bar u$ is continuous, such that $R_h\bar u$ is well defined.

First, we recall a property of Q_h that is proved in [4].

Lemma 2.1. *For $f,\ g \in H^1(T)$ the inequality*

$$(f - Q_h f, g)_{L^2(T)} \le ch_T^2|f|_{H^1(T)}|g|_{H^1(T)}$$

is valid where h_T denotes the diameter of the element T.

Now we can prove the following properties of the operator R_h.

Lemma 2.2. *Assume that the Assumptions* PP1 *and* PP4 *hold. Then the estimates*

$$\|S_h\bar{u} - S_hR_h\bar{u}\|_U \leq ch^2\left(\|\bar{u}\|_X + \|y_d\|_X\right) \tag{2.10}$$

$$\|P_h\bar{u} - P_hR_h\bar{u}\|_U \leq ch^2\left(\|\bar{u}\|_X + \|y_d\|_X\right) \tag{2.11}$$

are valid.

Proof. The proof of this lemma is similar to the one given by Apel and Winkler in [4] in the special case of optimal control of the Poisson equation and a discretization with linear finite elements. A proof under assumptions like PP1 and PP4 for the optimal control of the Stokes equation is given in [16]. A detailed proof for the general case can be found in [17]. □

Lemma 2.3. *The inequality*

$$\nu\|R_h\bar{u} - \bar{u}_h\|_U^2 \leq (R_h\bar{p} - \bar{p}_h, \bar{u}_h - R_h\bar{u})_U \tag{2.12}$$

holds.

Proof. This lemma was originally proved in [14] and is based on a combination of the variational inequalities (1.5) and (2.6). □

Now we are able to prove the following supercloseness result.

Theorem 2.3. *Assume that Assumptions* PP1–PP4 *hold. Then the inequality*

$$\|\bar{u}_h - R_h\bar{u}\|_U \leq ch^2\left(\|\bar{u}\|_X + \|y_d\|_X\right)$$

is valid.

Proof. The following proof is similar to the one of Theorem 4.21 of [18] which is given in the context of optimal control of the Poisson equation. We give the details here to illustrate the validity under the Assumptions PP1–PP4. From Lemma 2.3 we have

$$\begin{aligned}\nu\|\bar{u}_h - R_h\bar{u}\|_U^2 &\leq (R_h\bar{p} - \bar{p}_h, \bar{u}_h - R_h\bar{u})_U \\ &= (R_h\bar{p} - \bar{p}, \bar{u}_h - R_h\bar{u})_U + (\bar{p} - P_hR_h\bar{u}, \bar{u}_h - R_h\bar{u})_U \\ &\quad + (P_hR_h\bar{u} - \bar{p}_h, \bar{u}_h - R_h\bar{u})_U.\end{aligned} \tag{2.13}$$

We estimate these three terms separately. For the first term, we use that Q_h is an L^2-projection and get

$$\begin{aligned}(R_h\bar{p} - \bar{p}, \bar{u}_h - R_h\bar{u})_U &= (R_h\bar{p} - Q_h\bar{p}, \bar{u}_h - R_h\bar{u})_U + (Q_h\bar{p} - \bar{p}, \bar{u}_h - R_h\bar{u})_U \\ &= (R_h\bar{p} - Q_h\bar{p}, \bar{u}_h - R_h\bar{u})_U.\end{aligned}$$

The Cauchy-Schwarz inequality yields together with Assumption PP3

$$\begin{aligned}(R_h\bar{p} - \bar{p}, \bar{u}_h - R_h\bar{u})_U &\leq \|R_h\bar{p} - Q_h\bar{p}\|_U\|\bar{u}_h - R_h\bar{u}\|_U \\ &\leq ch^2\left(\|\bar{u}\|_X + \|y_d\|_X\right)\|\bar{u}_h - R_h\bar{u}\|_U.\end{aligned} \tag{2.14}$$

For the second term we apply again the Cauchy-Schwarz inequality and use $\bar{p} = P\bar{u}$, so that we arrive at

$$(\bar{p} - P_hR_h\bar{u}, \bar{u}_h - R_h\bar{u})_U \leq \|P\bar{u} - P_hR_h\bar{u}\|_U\|\bar{u}_h - R_h\bar{u}\|_U.$$

With Assumptions PP1 and PP2, Lemma 2.2 and the embedding $X \hookrightarrow U$, one can conclude

$$\begin{aligned}\|P\bar{u} - P_h R_h \bar{u}\|_U &\le \|S_h^*\|_{U\to U}\|S\bar{u} - S_h\bar{u}\|_U + \|S^* y_d - S_h^* y_d\|_U \\ &\quad + \|P_h\bar{u} - P_h R_h\bar{u}\|_U \\ &\le ch^2\left(\|\bar{u}\|_X + \|y_d\|_X\right),\end{aligned}$$

and therefore

$$(\bar{p} - P_h R_h \bar{u}, u_h - R_h\bar{u})_U \le ch^2\left(\|\bar{u}\|_X + \|y_d\|_X\right)\|\bar{u}_h - R_h\bar{u}\|_U. \tag{2.15}$$

The third term can simply be omitted since

$$\begin{aligned}(P_h R_h\bar{u} - \bar{p}_h, \bar{u}_h - R_h\bar{u})_U &= (P_h R_h \bar{u} - P_h\bar{u}_h, \bar{u}_h - R_h\bar{u})_U \\ &= (S_h(R_h\bar{u} - \bar{u}_h), S_h(\bar{u}_h - R_h\bar{u}))_U \\ &\le 0.\end{aligned} \tag{2.16}$$

Thus, one can conclude from the estimates (2.13)–(2.16)

$$\nu\|\bar{u}_h - R_h\bar{u}\|_U^2 \le ch^2\left(\|\bar{u}\|_X + \|y_d\|_X\right)\|u_h - R_h\bar{u}\|_U$$

what yields the assertion. □

Now we are able to formulate the main result of this section.

Theorem 2.4. *Assume that the Assumptions* PP1–PP4 *hold. Then the estimates*

$$\|\bar{y} - \bar{y}_h\|_U \le ch^2\left(\|\bar{u}\|_X + \|y_d\|_X\right), \tag{2.17}$$

$$\|\bar{p} - \bar{p}_h\|_U \le ch^2\left(\|\bar{u}\|_X + \|y_d\|_X\right), \tag{2.18}$$

$$\|\bar{u} - \tilde{u}_h\|_U \le ch^2\left(\|\bar{u}\|_X + \|y_d\|_X\right) \tag{2.19}$$

are valid.

Proof. In order to prove the first assertion we apply the triangle inequality and get

$$\begin{aligned}\|\bar{y} - \bar{y}_h\|_U &= \|S\bar{u} - S_h\bar{u}_h\|_U \\ &\le \|Su - S_h u\|_U + \|S_h\bar{u} - S_h R_h\bar{u}\|_U + \|S_h(R_h\bar{u} - \bar{u}_h)\|_U.\end{aligned}$$

The first term is a finite element error and is estimated in the first inequality of Assumption PP2. For the second term an upper bound is given in Lemma 2.2. For the third term we use the supercloseness result of Theorem 2.3 and the boundedness of S_h given in Assumption PP1. These three estimates yield assertion (2.17). In a similar way one can prove inequality (2.18). By using the Lipschitz continuity of the projection operator, we get

$$\|\bar{u} - \tilde{u}_h\|_U = \left\|\Pi_{U^{\mathrm{ad}}}\left(-\frac{1}{\nu}\bar{p}\right) - \Pi_{U^{\mathrm{ad}}}\left(-\frac{1}{\nu}\bar{p}_h\right)\right\|_U \le \frac{1}{\nu}\|\bar{p} - \bar{p}_h\|_U$$

and inequality (2.19) is a direct consequence of estimate (2.18). □

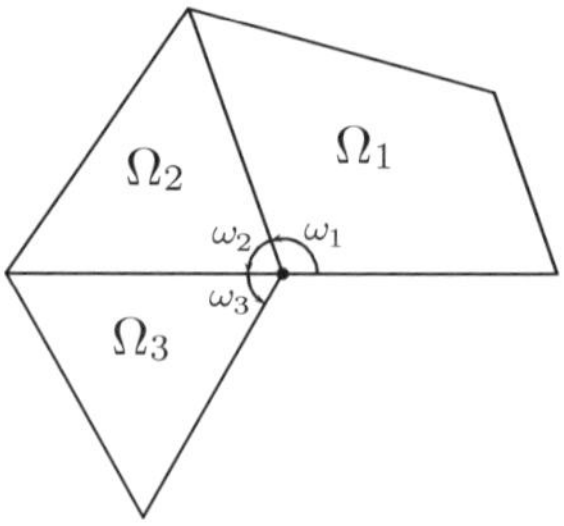

FIGURE 1. Example for subdomains Ω_i in interface problem

3. Examples

3.1. Scalar elliptic state equation with discontinuous coefficients in polygonal domain

We consider the optimal control problem (1.2) with the interface problem for the Laplacian as state equation. We assume that the domain Ω can be partitioned into disjoint, open, polygonal Lipschitz subdomains Ω_i, $i = 1, \ldots, n$, on which the diffusion coefficient k has the constant value k_i. Since the singular behaviour is a local phenomenon we restrict our considerations to one corner located at the origin and assume that no singularities occur at the other corners. The interior angle of the subdomains Ω_i at this corner is denoted by ω_i, see Figure 1 for an example. Notice, that for $n = 1$ the state equation reduces to the Poisson equation. This case is treated in [1].

The variational formulation of the state equation reads as

$$\text{Find } y \in V_0 : \qquad a_I(y, v) = (u, v)_{L^2(\Omega)} \qquad \forall v \in V_0 \tag{3.1}$$

with bilinear form $a_I : H^1(\Omega) \times H^1(\Omega) \to \mathbb{R}$,

$$a_I(y, v) := \int_\Omega k \nabla y \cdot \nabla v$$

and $V_0 := \{v \in H^1(\Omega) : v|_{\partial\Omega} = 0\}$. We triangulate the domain Ω by an isotropic graded mesh with element size $h_T := \operatorname{diam} T$, that satisfy

$$\begin{aligned} c_1 h^{1/\mu} &\le h_T \le c_2 h^{1/\mu} && \text{for } r_T = 0, \\ c_1 h r_T^{1-\mu} &\le h_T \le c_2 h r_T^{1-\mu} && \text{for } r_T > 0, \end{aligned}$$

where μ is the grading parameter and r_T the distance of the triangle T to the corner. We assume that the triangulation $\mathcal{T}_h$ of Ω is aligned with the partition of Ω, i.e., the boundary $\partial\Omega_i$ is made up of edges of triangles in $\mathcal{T}_h$. The discrete state $y_h = S_h u$ is given as solution of

$$\text{Find } y_h \in V_{0h} : \qquad a_I(y_h, v_h) = (f, v_h)_{L^2(\Omega)} \qquad \forall v_h \in V_{0h}.$$

with

$$V_{0h} = \{v_h \in C(\bar{\Omega}) : v_h|_T \in \mathcal{P}_1 \ \forall T \in \mathcal{T}_h \text{ and } v_h = 0 \text{ on } \partial\Omega\} \tag{3.2}$$

We introduce the singular exponent λ_I as smallest positive solution of the *Sturm-Liouville* eigenvalue problem

$$-\Phi_i''(\varphi) = \lambda_I^2 \Phi_i(\varphi), \quad \varphi \in (\omega_{i-1}, \omega_i), \quad i = 1, \dots, n \tag{3.3}$$

with the boundary and interface conditions

$$\begin{aligned} \Phi_1(0) &= \Phi_n(\omega) = 0, \\ \Phi_i(\omega_i) &= \Phi_{i+1}(\omega_i) \quad i = 1, \dots, n-1, \\ k_i \Phi_i'(\omega_i) &= k_{i+1} \Phi_{i+1}'(\omega_i) \quad i = 1, \dots, n-1. \end{aligned}$$

This is derived, e.g., in [15, Example 2.29]. In the following we assume that the mesh grading parameter satisfies $\mu < \lambda_I$. For the Poisson equation Assumptions VAR1 and PP1 are proved in [1] under the condition $\mu > 1/2$, which is a reasonable assumption since $\lambda_I > 1/2$ in that case. For $n > 1$ the singular exponent $\lambda_I > 0$ can become arbitrary small such that a more involved proof is necessary. This is given in [17, Lemma 5.16] for smooth coefficients. The proof can be easily adapted for the interface problem (3.1). The finite element error estimates of Assumptions VAR2 and PP2 can be verified using interpolation error estimates in weighted Sobolev spaces and the fact that the triangulation is aligned with the partition (Ω_i) of Ω, see [17, Theorem 4.29]. Assumptions PP3 and PP4 follow with similar arguments to the case of the Poisson equation in a prismatic domain (see Subsection 3.2). The proofs are even less complicated since one has not to exploit an anisotropic behaviour of the solution. Detailed proofs are given in [17, Lemma 5.46 and 5.47].

Altogether this means that the results of Theorems 2.1 and 2.4 hold for this example. In [17, Subsection 5.2.3.3] one can find numerical tests for the postprocessing approach that confirm the theoretical findings.

3.2. Examples in prismatic domains

In this section we consider a scalar elliptic equation and the Stokes equation as state equation in a nonconvex prismatic domain and show that Assumptions VAR1–VAR2 and PP1–PP4 hold on anisotropic graded meshes. Let $\Omega = G \times Z$ be a domain with boundary $\partial\Omega$, where $G \subset \mathbb{R}^2$ is a bounded polygonal domain and $Z := (0, z_0) \subset \mathbb{R}$ is an interval. Since situations with more than one singular edge can be reduced to the case of only one reentrant edge by a localization argument, see, e.g., [9], we assume that the cross-section G has only one corner with interior angle $\omega > \pi$ located at the origin.

To construct such an anisotropic graded mesh we first triangulate the two-dimensional domain G by an isotropic graded mesh, then extrude this mesh in x_3-direction with uniform mesh size h and finally divide each of the resulting pentahedra into tetrahedra. If one denotes by $h_{T,i}$ the length of the projection of an element T on the x_i-axis, these element sizes satisfy

$$\begin{aligned} c_1 h^{1/\mu} &\le h_{T,i} \le c_2 h^{1/\mu} && \text{for } r_T = 0, \\ c_1 h r_T^{1-\mu} &\le h_{T,i} \le c_2 h r_T^{1-\mu} && \text{for } r_T > 0, \\ c_1 h &\le h_{T,3} \le c_2 h, \end{aligned} \tag{3.4}$$

for $i = 1, 2$, where r_T is the distance of the element T to the edge,

$$r_T := \inf_{x \in T} \sqrt{x_1^2 + x_2^2},$$

and μ is the grading parameter. Note that these meshes are anisotropic, i.e., their elements are not shape regular. Anisotropic refinement near edges is more effective than grading with isotropic (shape-regular) elements. Nevertheless, the latter is also investigated, see [4].

3.2.1. Scalar elliptic state equation. We consider the optimal control problem (1.2) with operator S that associates the state $y = Su$ to the control u as the solution of the Dirichlet problem

$$\text{Find } y \in V_0 := \{v \in H^1(\Omega) : v = 0 \text{ on } \partial\Omega\} : \quad a_D(y, v) = (f, v)_{L^2(\Omega)} \quad \forall v \in V_0, \tag{3.5}$$

where the bilinear form $a_D : H^1(\Omega) \times H^1(\Omega) \to \mathbb{R}$ is defined as

$$a_D(y, v) = \int_\Omega \nabla y \cdot \nabla v,$$

or as the solution of the Neumann problem

$$\text{Find } y \in H^1(\Omega) : \qquad a_N(y, v) = (f, v)_{L^2(\Omega)} \qquad \forall v \in H^1(\Omega) \tag{3.6}$$

where the bilinear form $a_N : H^1(\Omega) \times H^1(\Omega) \to \mathbb{R}$ is defined as

$$a_N(y, v) = \int_\Omega \nabla y \cdot \nabla v + \int_\Omega y \cdot v.$$

The corresponding finite element approximation $y_h = S_h u$ is given as the unique solution of

$$a_D(y_h, v_h) = (u, v_h)_{L^2(\Omega)} \quad \forall v_h \in V_{0h}$$

or

$$a_N(y_h, v_h) = (u, v_h)_{L^2(\Omega)} \quad \forall v_h \in V_h$$

respectively, where the spaces V_{0h} and V_h are defined as

$$\begin{aligned} V_{0h} &= \left\{ v \in C(\bar\Omega) : v|_T \in \mathcal{P}_1 \ \forall T \in T_h \text{ and } v_h = 0 \text{ on } \partial\Omega \right\}, \\ V_h &= \left\{ v \in C(\bar\Omega) : v|_T \in \mathcal{P}_1 \ \forall T \in T_h \right\}. \end{aligned}$$

In the following we assume that the domain Ω is discretized according to (3.4) with grading parameter $\mu < \pi/\omega$. The boundedness of the operators S_h and S_h^* is proved in [18, Subsection 3.6] by using Green function techniques. This means Assumptions VAR1 and PP1 are true. For the proof of the finite element error estimates of Assumptions VAR2 and PP2 one had to deal with quasi-interpolation operators and exploit the regularity properties of the solution. For details we refer to [2]. Assumption PP3 is proved in [17, Lemma 5.38]. The main idea is to split Ω into the sets $K_s = \bigcup_{\{T\in\mathcal{T}_h : r_T=0\}} T$ and $K_r = \Omega\backslash\bar{K}_s$ and to utilize the different regularity properties of the optimal adjoint state $\bar{p}$ in these subdomains. Additionally, one has to take account of the fact that the number of elements in K_s is bounded by ch^{-1}. To prove Assumption PP4 we have to assume that the boundary of the active set has finite two-dimensional measure. Furthermore we utilize the boundedness of $r^\beta\nabla\bar{p}$ for $\beta > 1-\lambda$, which can be proved with the help of a regularity result by Maz'ya and Rossmann [13]. For a detailed proof we refer to [17].

In summary this means that Theorems 2.1 and 2.4 hold true in this setting. Numerical tests for the postprocessing approach that confirm the theoretical results are given in [17, Subsection 5.2.2.3].

3.2.2. Stokes equation as state equation. For the Stokes equation the state has actually two components, namely velocity and pressure. Therefore we slightly change the notation. We denote by v the velocity field and by q the pressure. The velocity field v plays the role of the state y in Section 2. Consequently, we substitute y_d by v_d, such that the optimal control problem (1.1) reads as

$$\begin{aligned} J(\bar{u}) &= \min_{u\in U^{\mathrm{ad}}} J(u) \\ J(u) &:= \frac{1}{2}\|Su - v_d\|^2_{L^2(\Omega)^3} + \frac{\nu}{2}\|u\|^2_{L^2(\Omega)^3}. \end{aligned} \tag{3.7}$$

The operator S maps the control u to the velocity v as solution of the Stokes equations,

$$\text{Find } (v,q)\in X\times M : \begin{cases} a(v,\varphi) + b(\varphi,q) = (u,\varphi) & \forall\varphi\in X \\ b(v,\psi) = 0 & \forall\psi\in M \end{cases}$$

with the bilinear forms $a : X\times X\to\mathbb{R}$ and $b : X\times M\to\mathbb{R}$ defined as

$$a(v,\varphi) := \sum_{i=1}^{3}\int_\Omega \nabla v_i\cdot\nabla\varphi_i \quad\text{and}\quad b(\varphi,q) := -\int_\Omega q\nabla\cdot\varphi,$$

and the spaces

$$X = \left\{v\in(H^1(\Omega))^3 : v|_{\partial\Omega} = 0\right\} \quad\text{and}\quad M = \left\{q\in L^2(\Omega) : \int_\Omega q = 0\right\}.$$

The finite element solution $v_h = S_h u$ is given as the unique solution of

$$\begin{aligned} &\text{Find } (v_h,q_h)\in X_h\times M_h \text{ such that} \\ &\qquad a_h(v_h,\varphi_h) + b_h(\varphi_h,q_h) = (u,\varphi_h) \qquad \forall\varphi_h\in X_h \\ &\qquad b_h(v_h,\psi_h) = 0 \qquad \forall\psi_h\in M_h. \end{aligned}$$

with the weaker bilinear forms $a_h : X_h\times X_h\to\mathbb{R}$ and $b_h : X_h\times M_h\to\mathbb{R}$,

$$a_h(v_h,\varphi_h) := \sum_{T\in\mathcal{T}_h}\sum_{i=1}^{3}\int_T \nabla v_{h,i}\cdot\nabla\varphi_{h,i} \quad\text{and}\quad b_h(\varphi_h,p_h) := -\sum_{T\in\mathcal{T}_h}\int_T p_h\nabla\cdot\varphi_h.$$

Here, the ith component of the vectors v_h and φ_h is denoted by $v_{h,i}$ and $\varphi_{h,i}$, respectively. We approximate the velocity by Crouzeix-Raviart elements,

$$X_h := \left\{ v_h \in L^2(\Omega)^3 : v_h|_T \in (\mathcal{P}_1)^3\ \forall T, \int_F [v_h]_F = 0\ \forall F \right\}$$

where F denotes a face of an element and $[v_h]_F$ means the jump of v_h on the face F,

$$[v_h(x)]_F := \begin{cases} \lim\limits_{\alpha\to 0} (v_h(x + \alpha n_F) - v_h(x - \alpha n_F)) & \text{for an interior face F,} \\ v_h(x) & \text{for a boundary face F.} \end{cases}$$

Here n_F is a fixed normal of F. For the approximation of the pressure we use piecewise constant functions, this means

$$M_h := \left\{ q_h \in L^2(\Omega) : q_h|_T \in \mathcal{P}_0\ \forall T, \int_\Omega q_h = 0 \right\}.$$

For our further considerations we assume that the mesh is graded according to (3.4) with parameter $\mu < \lambda_s$, where λ_s is the smallest positive solution of

$$\sin(\lambda_s \omega) = -\lambda_s \sin \omega.$$

This eigenvalue equation is given in [12, Theorem 6.1]. With the help of a discrete Poincaré inequality in X_h one can prove the boundedness of the operators S_h and S_h^*, comp. [16], such that Assumptions VAR1 and PP1 hold true. The finite element error estimates of Assumptions VAR2 and PP2 are shown in [17, Lemma 4.38]. For the proof of Assumption PP3 one cannot apply the arguments of the scalar elliptic case componentwise since there are only regularity results for the derivatives of the solution in edge direction in $L^2(\Omega)$, but not in $L^p(\Omega)$ for general p. For the proof of this assumption we refer to [17, Lemma 5.57]. The proof of Assumption PP4 is similar to the one for the scalar elliptic case. The weaker regularity result in edge direction, however, results in the condition, that the number of elements in the set K_1 is bounded by ch^{-2}, which is slightly stronger than the condition in Subsection 3.2.1, where the active set is assumed to have bounded two-dimensional measure.

Altogether one can conclude that Theorems 2.1 and 2.4 are valid. Corresponding numerical tests for the postprocessing approach that confirm the theoretical findings can be found in [16, 17].

References

[1] Th. Apel and A. Rösch and G. Winkler, *Optimal control in non-convex domains: a priori discretization error estimates.* Calcolo, **44** (2007), 137–158.

[2] Th. Apel and D. Sirch, *L^2-error estimates for the Dirichlet and Neumann problem on anisotropic finite element meshes.* Appl. Math. **56** (2011), 177–206.

[3] Th. Apel, D. Sirch and G. Winkler, *Error estimates for control constrained optimal control problems: Discretization with anisotropic finite element meshes.* DFG Priority Program 1253, 2008, preprint, SPP1253-02-06, Erlangen.

[4] Th. Apel and G. Winkler, *Optimal control under reduced regularity.* Appl. Numer. Math., **59** (2009), 2050–2064.

[5] P.G. Ciarlet, *The finite element method for elliptic problems.* North-Holland, 1978.

[6] M. Hinze, *A variational discretization concept in control constrained optimization: The linear-quadratic case.* Comput. Optim. Appl., **30** (2005), 45–61.

[7] M. Hinze, R. Pinnau, M. Ulbrich and S. Ulbrich, *Optimization with PDE Constraints.* Springer, 2008.

[8] M. Hinze and A. Rösch, *Discretization of optimal control problems.* International Series of Numerical Mathematics, Vol. 160 (this volume), Springer, Basel, 2011, 391–430.

[9] A. Kufner and A.-M. Sändig, *Some Applications of Weighted Sobolev Spaces.* Teubner, Leipzig, 1987.

[10] J.L. Lions, *Optimal Control of Systems Governed by Partial Differential Equations.* Springer, 1971. Translation of the French edition "Contrôle optimal de systèmes gouvernés par de équations aux dérivées partielles", Dunod and Gauthier-Villars, 1968.

[11] K. Malanowski, *Convergence of approximations vs. regularity of solutions for convex, control-constrained optimal-control problems.* Appl. Math. Optim., **8** (1981), 69–95.

[12] V.G. Maz'ya and B.A. Plamenevsky, *The first boundary value problem for classical equations of mathematical physics in domains with piecewise smooth boundaries, part I, II.* Z. Anal. Anwend., **2** (1983), 335–359, 523–551. In Russian.

[13] V.G. Maz'ya and J. Rossmann, *Schauder estimates for solutions to boundary value problems for second-order elliptic systems in polyhedral domains.* Applicable Analysis, **83** (2004), 271–308.

[14] C. Meyer and A. Rösch, *Superconvergence properties of optimal control problems.* SIAM J. Control Optim., **43** (2004), 970–985.

[15] S. Nicaise, *Polygonal Interface Problems.* Peter Lang GmbH, Europäischer Verlag der Wissenschaften, volume 39 of Methoden und Verfahren der mathematischen Physik, Frankfurt/M., 1993.

[16] S. Nicaise and D. Sirch, *Optimal control of the Stokes equations: Conforming and non-conforming finite element methods under reduced regularity.* Comput. Optim. Appl. **49** (2011), 567–600.

[17] D. Sirch, *Finite element error analysis for PDE-constrained optimal control problems: The control constrained case under reduced regularity.* TU München, 2010, `http://mediatum2.ub.tum.de/node?id=977779`.

[18] G. Winkler, *Control constrained optimal control problems in non-convex three-dimensional polyhedral domains.* PhD thesis, TU Chemnitz, 2008. `http://archiv.tu-chemnitz.de/pub/2008/0062`.

Thomas Apel and Dieter Sirch
Universität der Bundeswehr München
Institut für Mathematik und Bauinformatik
D-85579 Neubiberg, Germany
e-mail: thomas.apel@unibw.de
dieter.sirch@gmx.de

International Series of Numerical Mathematics, Vol. 160, 391–430

Discretization of Optimal Control Problems

Michael Hinze and Arnd Rösch

Abstract. Solutions to optimization problems with pde constraints inherit special properties; the associated state solves the pde which in the optimization problem takes the role of a equality constraint, and this state together with the associated control solves an optimization problem, i.e., together with multipliers satisfies first- and second-order necessary optimality conditions. In this note we review the state of the art in designing discrete concepts for optimization problems with pde constraints with emphasis on structure conservation of solutions on the discrete level, and on error analysis for the discrete variables involved. As model problem for the state we consider an elliptic pde which is well understood from the analytical point of view. This allows to focus on structural aspects in discretization. We discuss the approaches *First discretize, then optimize* and *First optimize, then discretize*, and consider in detail two variants of the *First discretize, then optimize* approach, namely variational discretization, a discrete concept which avoids explicit discretization of the controls, and piecewise constant control approximations. We consider general constraints on the control, and also consider pointwise bounds on the state. We outline the basic ideas for providing optimal error analysis and accomplish our analytical findings with numerical examples which confirm our analytical results. Furthermore we present a brief review on recent literature which appeared in the field of discrete techniques for optimization problems with pde constraints.

Mathematics Subject Classification (2000). 49J20, 49K20, 35B37.

Keywords. Elliptic optimal control problem, state & control constraints, error analysis.

1. Introduction

In PDE-constrained optimization, we have usually a pde as state equation and constraints on control and/or state. Let us write the pde for the state $y \in Y$ with the control $u \in U$ in the form $e(y, u) = 0$ in Z. Assuming smoothness, we are then lead to optimization problems of the form

$$\min_{(y,u)\in Y\times U} J(y, u) \text{ s.t. } e(y, u) = 0, \quad c(y) \in \mathcal{K}, \quad u \in U_{\text{ad}}, \tag{1}$$

where $e : Y \times U \to Z$ and $c : Y \to \mathcal{R}$ are continuously Fréchet differentiable, $\mathcal{K} \subset \mathcal{R}$ is a closed convex cone representing the state constraints, and $U_{\rm ad} \subset U$ is a closed convex set representing the control constraints.

Let us give two examples.

Example 1.1. Consider the distributed optimal control of a semilinear elliptic PDE:

$$\begin{array}{l} \min J(y,u) := \frac{1}{2}\|y-y_d\|^2_{L^2(\Omega)} + \frac{\alpha}{2}\|u\|^2_{L^2(\Omega)} \\ \text{subject to} \\ -\Delta y + y^3 = \gamma u \quad \text{on } \Omega, \\ y = 0 \quad \text{on } \partial\Omega, \\ a \le u \le b \text{ on } \Omega, \text{ and } y \le c \text{ on } D, \end{array} \tag{2}$$

where $\gamma \in L^\infty(\Omega) \setminus \{0\}$, $a, b \in L^\infty(\Omega)$, and $a \le b$. We require $D \subset\subset \Omega$ to avoid restrictions on the bound c which would have to be imposed in the case $D \equiv \Omega$ due to homogeneous boundary conditions required for y. Let $n \le 3$. By the theory of monotone operators one can show that there exists a unique bounded solution operator of the state equation

$$u \in U_{\rm ad} := \{v \in L^2(\Omega); a \le u \le b \text{ a.e. }\} \to y \in Y := H^1_0(\Omega).$$

Here $c(y) = y$ with $\mathcal{K} = \{y \in Y; y \le c \text{ on } D \subset\subset \Omega\} \subset \mathcal{R} := C^0(\bar{\Omega})$. Let $A : H^1_0(\Omega) \to H^1_0(\Omega)^*$ be the operator associated with the bilinear form $a(y,v) = \int_\Omega \nabla y \cdot \nabla v\, dx$ for the Laplace operator $-\Delta y$ and let $N : y \to y^3$. Then the weak formulation of the state equation can be written in the form

$$e(y,u) := Ay + N(y) - \gamma u = 0.$$

Example 1.2. We consider optimal control of the time-dependent incompressible Navier-Stokes system for the velocity field $y \in \mathbb{R}^d$ and the pressure p. Let $\Omega \subset \mathbb{R}^d$ denote the flow domain, let $f : [0,T] \times \Omega \to \mathbb{R}^d$ be the force per unit mass acting on the fluid and denote by $y_0 : \Omega \to \mathbb{R}^d$ the initial velocity of the fluid at $t = 0$. Then the Navier-Stokes equations can be written in the form

$$\begin{aligned} y_t - \nu\Delta y + (y\cdot\nabla)y + \nabla p = f \quad & \text{on } \Omega_T := (0,T)\times\Omega, \\ \nabla \cdot y = 0 \quad & \text{on } \Omega_T, \\ y(0,\cdot) = y_0 \quad & \text{on } \Omega, \end{aligned} \tag{3}$$

and have to be accomplished by appropriate boundary conditions. For the functional analytic setting let us define the Hilbert spaces

$$V := \mathrm{cl}_{H^1_0(\Omega)^2}\{y \in C_c^\infty(\Omega)^2; \nabla\cdot y = 0\},\ H := \mathrm{cl}_{L^2(\Omega)^2}\{y \in C_c^\infty(\Omega)^2; \nabla\cdot y = 0\},$$

and the associated parabolic solution space

$$Y := W(I) \equiv W(I;H,V) = \{y \in L^2(I;V); y_t \in L^2(I;V^*)\}.$$

We say that

$$y_t + (y\cdot\nabla)y - \nu\Delta y = f \text{ in } L^2(I;V^*) =: Z :\Longleftrightarrow$$
$$\langle y_t, v\rangle_{V^*,V} + \nu(\nabla y, \nabla v)_{L^2(\Omega)^{2\times 2}} + \langle y\cdot\nabla)y, v\rangle_{V^*,V} = \langle f, v\rangle_{V^*,V}\ \forall\, v \in V.$$

Let $U_{\text{ad}} \subset U$ be nonempty, convex and closed with U denoting a Hilbert space, and $B : U \to L^2(I, V^*)$ a linear, bounded control operator. Furthermore we set $\mathcal{K} := Y$, i.e., we omit state constraints. Let finally $J : Y \times U \to \mathbb{R}$ be a Fréchet differentiable functional. Then we may consider the following optimal control problem

$$\min_{u \in U_{\text{ad}}, y \in Y} J(y,u) \text{ s.t. } e(y,u) = 0 \text{ in } Z,$$

where the state equation $e(y,u) = 0$ is the weak Navier-Stokes equation, i.e.,

$$e : Y \times U \to Z \times H,\ e(y,u) = \begin{pmatrix} y_t + (y \cdot \nabla)y - \nu \Delta y - Bu \\ y(0,\cdot) - y_0 \end{pmatrix}.$$

We are interested in illuminating discrete approaches to problem (1), where we place particular emphasis on structure preservation on the discrete level, and also on analysing the contributions to the total error of the discretization errors in the variables and multipliers involved. To approach an optimal control problem of the form (1) numerically one may either discretize this problem by substituting all appearing function spaces by finite-dimensional spaces, and all appearing operators by suitable approximate counterparts which allow their numerical evaluation on a computer, say. Denoting by h the discretization parameter, one ends up with the problem

$$\min_{(y_h,u_h) \in Y_h \times U_h} J_h(y_h,u_h) \text{ s.t } e_h(y_h,u_h) = 0 \text{ and } c_h(y_h) \in \mathcal{K}_h,\ u_h \in U^h_{\text{ad}}, \tag{4}$$

where $J_h : Y_h \times U_h \to \mathbb{R}$, $e_h : Y_h \times U_h \to Z$, and $c_h : Y_h \to R$ with $\mathcal{K}_h \subset R$. For the finite-dimensional subspaces one may require $Y_h \subset Y, U_h \subset U$, say, and $\mathcal{K}_h \subseteq R$ a closed and convex cone, $U^h_{\text{ad}} \subseteq U_h$ closed and convex. This approach in general is referred to as first discretize, then optimize. On the other hand one may switch to the Karush-Kuhn-Tucker system associated to (1)

$$e(\bar{y},\bar{u}) = 0, \quad c(\bar{y}) \in \mathcal{K}, \tag{5}$$

$$\bar{\lambda} \in \mathcal{K}^\circ, \quad \langle \bar{\lambda}, c(\bar{y}) \rangle_{\mathcal{R}^*,\mathcal{R}} = 0, \tag{6}$$

$$L_y(\bar{y},\bar{u},\bar{p}) + c'(\bar{y})^* \bar{\lambda} = 0, \tag{7}$$

$$\bar{u} \in U_{\text{ad}}, \quad \langle L_u(\bar{y},\bar{u},\bar{p}), u - \bar{u} \rangle_{U^*,U} \geq 0 \quad \forall\, u \in U_{\text{ad}}. \tag{8}$$

and substitute all appearing function spaces and operators accordingly, where $L(y,u,p) := J(y,u) - \langle p, e(y,u) \rangle_{Z^*,Z}$ denotes the Lagrangian associated to (1). This leads to solving

$$e_h(y_h,u_h) = 0, \quad c_h(y_h) \in \mathcal{K}_h, \tag{9}$$

$$\lambda_h \in {\mathcal{K}_h}^\circ, \quad \langle \lambda_h, c_h(y_h) \rangle_{\mathcal{R}^*,\mathcal{R}} = 0, \tag{10}$$

$$L_{h_y}(y_h,u_h,p_h) + c_h'(y_h)^* \lambda_h = 0, \tag{11}$$

$$\bar{u}_h \in U^h_{\text{ad}}, \quad \langle L_{h_u}(y_h,u_h,p_h), u - u_h \rangle_{U^*,U} \geq 0 \quad \forall\, u \in U^h_{\text{ad}} \tag{12}$$

for $\bar{y}_h, \bar{u}_h, \bar{p}_h, \bar{\lambda}_h$, where L_h denotes a discretized version of L. Of course, $L \equiv L_h$ is possible. This approach in general is referred to as first optimize, then discretize, since it builds the discretization upon the first-order necessary optimality conditions.

Instead of applying discrete concepts to problem (1) or (5)–(8) directly we may first apply an SQP approach on the continuous level and then apply first discretize, then optimize to the related linear quadratic constrained subproblems, or first optimize, then discretize to the SQP systems appearing in each iteration of the Newton method on the infinite-dimensional level. This motivates us to illustrate all discrete concepts at hand of linear model pdes which are well understood w.r.t. analysis and discretization concepts and to focus the presentation on structural aspects inherent to optimal control problems with pde constraints. However, error analysis for optimization problems with nonlinear state equations in the presence of constraints on controls and/or state is not straightforward and requires special techniques such as extensions of Newton-Kontorovich-type theorems, and second-order sufficient optimality conditions. This complex of questions also will be discussed.

The outline of this work is as follows. In Section 2, we consider an elliptic model optimal control problem containing many relevant features which need to be resolved by a numerical approach. In Section 3 we preview the case of nonlinear state equations and highlight the solution approaches taken so far, as well as the analytical difficulties one is faced with in this situation. Section 4 is devoted to the finite element element method for the discretization of the state equation in our model problem. We propose two different approximation approaches of the *First discretize, then optimize*-type to the optimal control problem, including detailed numerical analysis. In Section 5 we present an introduction to relaxation approaches used in presence of state constraints. For Lavrentiev regularization applied to the model problem we present numerical analysis which allows to adapt the finite element discretization error to the regularization error. Finally, we present in Section 6 a brief review on recent literature which appeared in the field of discrete techniques for optimization problems with pde constraints.

2. A model problem

As model problem with pointwise bounds on the state we take the Neumann problem

$$(\mathbb{S}) \quad \begin{cases} \min_{(y,u)\in Y\times U_{\mathrm{ad}}} J(y,u) := \frac{1}{2}\int_\Omega |y-y_0|^2 + \frac{\alpha}{2}\|u\|_U^2 \\ \text{s.t.} \\ \left.\begin{array}{rcll} Ay & = & Bu & \text{in } \Omega, \\ \partial_\eta y & = & 0 & \text{on } \Gamma, \end{array}\right\} :\Longleftrightarrow y = \mathcal{G}(Bu) \\ \text{and} \\ y \in Y_{\mathrm{ad}} := \{y\in Y, y(x) \le b(x) \text{ a.e. in } \Omega\}. \end{cases} \tag{13}$$

Here, $Y := H^1(\Omega)$, A denotes an uniformly elliptic operator, for example $Ay = -\Delta y + y$, and $\Omega \subset \mathbb{R}^d$ ($d = 2,3$) denotes an open, bounded sufficiently smooth (or polyhedral) domain. Furthermore, we suppose that $\alpha > 0$ and that $y_0 \in L^2(\Omega)$, and $b \in W^{2,\infty}(\Omega)$ are given. $(U, (\cdot,\cdot)_U)$ denotes a Hilbert space and $B : U \to L^2(\Omega) \subset H^1(\Omega)^*$ the linear, continuous control operator. By $R : U^* \to U$ we denote the inverse of the Riesz isomorphism. Furthermore, we associate to A the continuous, coercive bilinear form $a(\cdot,\cdot)$.

Example 2.1. There are several examples for the choice of B and U.

(i) Distributed control: $U = L^2(\Omega)$, $B = Id : L^2(\Omega) \to Y'$.

(ii) Boundary control: $U = L^2(\partial\Omega)$, $Bu(\cdot) = \int_{\partial\Omega} u\gamma_0(\cdot)\, dx \in Y'$, where γ_0 is the boundary trace operator defined on Y.

(iii) Linear combinations of input fields: $U = \mathbb{R}^n$, $Bu = \sum_{i=1}^n u_i f_i$, $f_i \in Y'$.

If not stated otherwise we from here onwards consider the situation (i) of the previous example. In view of $\alpha > 0$, it is standard to prove that problem (13) admits a unique solution $(y,u) \in Y_{\text{ad}} \times U_{\text{ad}}$. In pde constrained optimization, the pde for given data frequently is uniquely solvable. In equation (13) this is also the case, so that for every control $u \in U_{\text{ad}}$ we have a unique state $y = \mathcal{G}(Bu) \in H^1(\Omega) \cap C^0(\bar\Omega)$. We need $y \in C^0(\bar\Omega)$ to satisfy the Slater condition required below. Problem (13) therefore is equivalent to the so-called reduced optimization problem

$$\min_{v \in U_{\text{ad}}} \hat{J}(v) := J(\mathcal{G}(Bv), v) \text{ s.t. } \mathcal{G}(Bv) \in Y_{\text{ad}}. \tag{14}$$

The key to the proper numerical treatment of problems (13) and (14) can be found in the first-order necessary optimality conditions associated to these control problems. To formulate them properly we require the following constraint qualification, often referred to as *Slater condition.* It requires the existence of a state in the interior of the set Y_{ad} considered as a subset of $C^0(\bar\Omega)$ and ensures the existence of a Lagrange multiplier in the associated dual space. Moreover, it is useful for deriving error estimates.

Assumption 2.2. $\exists \tilde{u} \in U_{\text{ad}} \quad \mathcal{G}(B\tilde{u})(x) < b(x)$ for all $x \in \bar\Omega$.

Following Casas [9, Theorem 5.2] for the problem under consideration we now have the following theorem, which specifies the KKT system (5)–(8) for the setting of problem (13).

Theorem 2.3. *Let $u \in U_{\text{ad}}$ denote the unique solution to* (13). *Then there exist $\mu \in \mathcal{M}(\bar\Omega)$ and $p \in L^2(\Omega)$ such that with $y = \mathcal{G}(Bu)$ there holds*

$$\int_\Omega pAv = \int_\Omega (y - y_0)v + \int_{\bar\Omega} v d\mu \qquad \forall v \in H^2(\Omega) \text{ with } \partial_\eta v = 0 \text{ on } \partial\Omega, \tag{15}$$

$$(RB^*p + \alpha u, v - u)_U \geq 0 \qquad \forall v \in U_{\text{ad}}, \tag{16}$$

$$\mu \geq 0,\ y(x) \leq b(x) \text{ in } \Omega \text{ and } \int_{\bar\Omega} (b - y) d\mu = 0. \tag{17}$$

Here, $\mathcal{M}(\bar\Omega)$ denotes the space of Radon measures which is defined as the dual space of $C^0(\bar\Omega)$ and endowed with the norm

$$\|\mu\|_{\mathcal{M}(\bar\Omega)} = \sup_{f \in C^0(\bar\Omega), |f| \leq 1} \int_{\bar\Omega} f d\mu.$$

Since $\hat{J}'(v) = B^*p + \alpha(\cdot, u)_U$ a short calculation shows that the variational inequality (16) is equivalent to

$$u = P_{U_{\text{ad}}}(u - \sigma R\hat{J}'(u)) \ (\sigma > 0),$$

where $P_{U_{\text{ad}}}$ denotes the orthogonal projection in U onto U_{ad}. This nonsmooth operator equation constitutes a relation between the optimal control u and its associated adjoint state p. In the present situation, when we consider the special case $U \equiv L^2(\Omega)$ with B denoting the injection from $L^2(\Omega)$ into $H^1(\Omega)^*$, and without control constraints, i.e., $U_{\text{ad}} \equiv L^2(\Omega)$, this relation boils down to

$$\alpha u + p = 0 \text{ in } L^2(\Omega),$$

since $\sigma > 0$. This relation already gives a hint to the discretization of the state y and the control u in problem (13), if one wishes to conserve the structure of this algebraic relation also on the discrete level.

3. Nonlinear state equations

Practical applications are usually characterized by nonlinear partial differential equations, see Example 1.1 and Example 1.2. We will here only focus on optimization and discretization aspects. Let us assume that there is a solution operator S:

$$e(y, u) = 0 \iff y = S(u) \tag{18}$$

which maps U in Y. The nonlinearity of S results in a nonconvex optimization problem (1). Therefore, we have to replace the Slater condition (Assumption 2.2) by a Mangasarian-Fromovitz constraint qualification to get the necessary optimality condition. Of course, one needs differentiability properties of the solution operator S. Let us assume that the operator S is two times Fréchet differentiable. All these assumptions are satisfied for both examples.

Moreover, let us assume there is a discrete solution operator $S_h : U \to Y_h$ with

$$e_h(y_h, u) = 0 \iff y_h = S_h(u). \tag{19}$$

Consequently, the discretization error of the PDE is described by $\|S(u) - S_h(u)\|_Z$. Usually, such a priori error estimates are known for a lot of discretizations and for different spaces Z. However, this is only a first small step in estimating the discretization error for the optimization problem.

Since the optimal control problem is not convex, we have to work with local minima and local convexity properties. We can only expect that a numerical optimization method generates a sequence of locally optimal (discrete) solutions converging to a local optimal solution of the undiscretized problem. To get error estimates one has to deal with local convexity properties. These local convexity properties are described by second-order sufficient optimality conditions. However, these properties lead only to a priori error estimates if one already knows that the discretized solution is sufficiently close to the undiscretized one. This complicate situation requires innovative techniques to obtain the desired results.

Until now such techniques are known only for control constrained optimal control problems. Let us first sketch a technique which was presented in [16]. Here, an auxiliary problem is introduced with the additional constraint

$$\|u - \bar{u}\|_{U_1} \leq r. \tag{20}$$

The choice of the space U_1 is connected to the differentiability properties of the operator S. In general the U_1-norm is a stronger norm than the U-norm. For instance, $U = L^2(\Omega)$ and $U_1 = L^\infty(\Omega)$ is a typical choice. Let us mention that the two-norm discrepancy can be avoided for certain elliptic optimal control problems. The radius r is chosen in such a way that the auxiliary problem is now a strictly convex problem. Consequently, the solution $\bar{u}$ is the unique solution of the auxiliary optimal control problem.

Using the second-order sufficient optimality condition one shows in a next step that

$$\|\bar{u}_h^r - \bar{u}\|_U + \|\bar{y}_h^r - \bar{y}\|_Y \leq ch^\kappa \tag{21}$$

where $\bar{u}_h^r$ is the solution of a discretized version of the auxiliary problem and y_h^r the corresponding state. Note, that the convergence order does not depend on r. A similar estimate is obtained for the adjoint state. A projection formula is used to derive an error estimate

$$\|\bar{u}_h^r - \bar{u}\|_{U_1} \leq ch^{\hat{\kappa}}.$$

If h is sufficiently small, then the additional inequality (20) cannot be active. Consequently $\bar{u}_h^r$ is also a local minimizer of the discretized problem without this inequality and the error estimate (21) is valid. Let us mention the practical drawback of that result. Since the solution $\bar{u}$ is unknown, we have no information about what h *is small enough* means.

A second approach was used in [68, 67]. Only information on the numerical solution are used in that approach. The main idea is to construct a ball around the numerical solution where the objective value of the undiscretized problem on the surface of that ball is greater than the objective value of the discretized control.

Let us mention that the available techniques cannot be applied to state constrained problems. Low regularity properties, instability of dual variables, and missing smoothing properties are some of the reasons that a priori error estimates for nonlinear state constrained problems are challenging.

Let us sketch a third approach which is based on the first-order necessary optimality conditions (5)–(8) and (9)–(12), respectively, and which does not use second-order sufficient optimality conditions. To begin with we consider

$$\min_{u \in U_{\mathrm{ad}}} \hat{J}(u) \equiv J(S(u), u), \tag{22}$$

with J as in problem (13). In this situation (7) reduces to

$$\langle \alpha(u, \cdot) + B^* p, v - u \rangle_{U^*U} \geq 0 \text{ for all } v \in U_{\mathrm{ad}}. \tag{23}$$

Then this variational inequality is equivalent to the semi-smooth operator equation

$$G(u) := u - P_{U_{\text{ad}}}\left(-\frac{1}{\alpha}B^*p\right) = 0 \text{ in } U, \tag{24}$$

where $P_{U_{\text{ad}}}$ denotes the orthogonal projection onto U_{ad} in U, and where we assume that the Riesz isomorphism is the identity map. Analogously, for the variational discrete approach and its numerical solutions $u_h \in U_{\text{ad}}$ (see the next section),

$$G_h(u_h) := u_h - P_{U_{\text{ad}}}\left(-\frac{1}{\alpha}B^*p_h\right) = 0 \text{ in } U. \tag{25}$$

We now pose the following two question; 1. *Given a solution $u \in U_{\text{ad}}$ to* (22), *does there exist a solution $u_h \in U_{\text{ad}}$ of* (25) *in a neighborhood of u?* 2. *If yes, is this solution unique?* It is clear that solutions to (22) might not be local solutions to the optimization problem, and that every local solution is a solution to (22). In this respect the following exposition generalizes the classical Newton-Kantorovich concept.

To provide positive answers to these questions we have to pose appropriate assumptions on a solution $u \in U_{\text{ad}}$ of (22).

Definition 3.1. A solution $u \in U_{\text{ad}}$ of (22) is called regular, if $M \in \partial G(u)$ exists with $\|G(v) - G(u) - M(v-u)\|_U = o(\|v-u\|)$ for $v \to u$, and M is invertible with bounded inverse M^{-1}.

We note that in the case of box constraints with $U := L^2(\Omega)$ and $U_{\text{ad}} = \{v \in U; a \leq u \leq b\}$ this regularity requirement is satisfied if the gradient of the adjoint state associated to u admits a non-vanishing gradient on the boarder of the active set, see [28].

In the following we write $G'(u) := M$. Now let $u \in U_{\text{ad}}$ denote a regular solution to (22) and consider the operator

$$\Phi(v) := v - G'(u)^{-1}G_h(v). \tag{26}$$

We now, under certain assumptions, show that Φ has a fixed point u_h in a neighborhood of u which we then consider as discrete approximation to the solution $u \in U_{\text{ad}}$ of (22). A positive answer to question 1 is given in

Theorem 3.2. *Let $u \in U_{\text{ad}}$ denote a regular solution to* (22). *Furthermore let for $v \in U_{\text{ad}}$ the error estimate*

$$\|G_h(v) - G(v)\| \leq ch^\kappa \textit{ for } h \to 0$$

be satisfied and let Φ be compact. Then a neighborhood $B_r(u) \subset U$ exists such that Φ admits a least one fixed point $u_h \in U_{\text{ad}} \cap B_r(u)$. For u_h the error estimate

$$\|u - u_h\|_U \leq Ch^\kappa \textit{ for all } 0 < h \leq h_0$$

holds.

If we strengthen the regularity requirement on u by requiring strict differentiability of G at u, also uniqueness can be argued. Details are given in [28], where also error estimates for approximation schemes related to Example (1.2) are presented.

4. Finite element discretization

For the convenience of the reader we recall the finite element setting. To begin with let $\mathcal{T}_h$ be a triangulation of Ω with maximum mesh size $h := \max_{T\in\mathcal{T}_h} \operatorname{diam}(T)$ and vertices $x_1, \ldots, x_m$. We suppose that $\bar{\Omega}$ is the union of the elements of $\mathcal{T}_h$ so that element edges lying on the boundary are curved. In addition, we assume that the triangulation is quasi-uniform in the sense that there exists a constant $\kappa > 0$ (independent of h) such that each $T \in \mathcal{T}_h$ is contained in a ball of radius $\kappa^{-1}h$ and contains a ball of radius κh. Let us define the space of linear finite elements,

$$X_h := \{v_h \in C^0(\bar{\Omega}) \mid v_h \text{ is a linear polynomial on each } T \in \mathcal{T}_h\}$$

with the appropriate modification for boundary elements. In what follows it is convenient to introduce a discrete approximation of the operator $\mathcal{G}$. For a given function $v \in L^2(\Omega)$ we denote by $z_h = \mathcal{G}_h(v) \in X_h$ the solution of the discrete Neumann problem

$$a(z_h, v_h) \quad = \quad \int_\Omega v v_h \qquad \text{for all } v_h \in X_h.$$

It is well known that for all $v \in L^2(\Omega)$

$$\|\mathcal{G}(v) - \mathcal{G}_h(v)\| \quad \leq \quad Ch^2\|v\|, \tag{27}$$

$$\|\mathcal{G}(v) - \mathcal{G}_h(v)\|_{L^\infty} \quad \leq \quad Ch^{2-\frac{d}{2}}\|v\|. \tag{28}$$

The estimate (28) can be improved provided one strengthens the assumption on v.

Lemma 4.1.

(a) *Suppose that* $v \in W^{1,s}(\Omega)$ *for some* $1 < s < \frac{d}{d-1}$. *Then*

$$\|\mathcal{G}(v) - \mathcal{G}_h(v)\|_{L^\infty} \leq Ch^{3-\frac{d}{s}}|\log h|\ \|v\|_{W^{1,s}}.$$

(b) *Suppose that* $v \in L^\infty(\Omega)$. *Then*

$$\|\mathcal{G}(v) - \mathcal{G}_h(v)\|_{L^\infty} \leq Ch^2|\log h|^2\ \|v\|_{L^\infty}.$$

Proof. (a): Let $z = \mathcal{G}(v)$, $z_h = \mathcal{G}_h(v)$. Elliptic regularity theory implies that $z \in W^{3,s}(\Omega)$ from which we infer that $z \in W^{2,q}(\Omega)$ with $q = \frac{ds}{d-s}$ using a well-known embedding theorem. Furthermore, we have

$$\|z\|_{W^{2,q}} \leq c\|z\|_{W^{3,s}} \leq c\|v\|_{W^{1,s}}. \tag{29}$$

Using Theorem 2.2 and the following estimate from [69] we have

$$\|z - z_h\|_{L^\infty} \leq c|\log h| \inf_{\chi\in X_h} \|z - \chi\|_{L^\infty}, \tag{30}$$

which, combined with a well-known interpolation estimate, yields

$$\|z - z_h\|_{L^\infty} \le ch^{2-\frac{d}{q}} |\log h| \|z\|_{W^{2,q}} \le ch^{3-\frac{d}{s}} |\log h| \|v\|_{W^{1,s}}$$

in view (29) and the relation between s and q.

(b): Elliptic regularity theory in the present case implies that $z \in W^{2,q}(\Omega)$ for all $1 \le q < \infty$ with

$$\|z\|_{W^{2,q}} \le Cq \|v\|_{L^q}$$

where the constant C is independent of q. For the dependence on q in this estimate we refer to the work of Agmon, Douglis and Nirenberg [1], see also [31] and [33, Chapter 9]. Proceeding as in (a) we have

$$\|z - z_h\|_{L^\infty} \le Ch^{2-\frac{d}{q}} |\log h| \|z\|_{W^{2,q}} \le Cqh^{2-\frac{d}{q}} |\log h| \|v\|_{L^q} \le Cqh^{2-\frac{d}{q}} |\log h| \|v\|_{L^\infty},$$

so that choosing $q = |\log h|$ gives the result. □

An important ingredient in our analysis is an error bound for a solution of a Neumann problem with a measure-valued right-hand side. Let A be as above and consider

$$\begin{aligned} A^* q &= \tilde{\mu}_\Omega \quad \text{in } \Omega \\ \sum_{i=1}^d \big(\sum_{j=1}^d a_{ij} q_{x_j} + b_i q\big)\nu_i &= \tilde{\mu}_{\partial\Omega} \quad \text{on } \partial\Omega. \end{aligned} \tag{31}$$

Theorem 4.2. Let $\tilde{\mu} \in \mathcal{M}(\bar{\Omega})$. Then there exists a unique weak solution $q \in L^2(\Omega)$ of (31), i.e.,

$$\int_\Omega qAv = \int_{\bar{\Omega}} v d\tilde{\mu} \qquad \forall v \in H^2(\Omega) \text{ with } \sum_{i,j=1}^d a_{ij} v_{x_i} \nu_j = 0 \text{ on } \partial\Omega.$$

Furthermore, q belongs to $W^{1,s}(\Omega)$ for all $s \in (1, \frac{d}{d-1})$. For the finite element approximation $q_h \in X_h$ of q defined by

$$a(v_h, q_h) = \int_{\bar{\Omega}} v_h d\tilde{\mu} \qquad \text{for all } v_h \in X_h$$

the following error estimate holds:

$$\|q - q_h\| \le Ch^{2-\frac{d}{2}} \|\tilde{\mu}\|_{\mathcal{M}(\bar{\Omega})}. \tag{32}$$

Proof. A corresponding result is proved in [7] for the case of an operator A without transport term subject to Dirichlet conditions, but the arguments can be adapted to our situation. We omit the details. □

4.1. Variational discretization

The discretization of the partial differential equations induces a natural discretization of the control via the optimality condition. Every a priori discretization of the control introduces a significant additional error which may reduce the approximation rates. Therefore, only the partial differential equations are discretized in the variational discretization concept. Problem (13) is now approximated by the following sequence of so-called *variational discrete* control problems [44] depending on the mesh parameter h:

$$\begin{aligned} &\min_{u \in U_{\mathrm{ad}}} \hat{J}_h(u) := \frac{1}{2}\int_\Omega |y_h - y_0|^2 + \frac{\alpha}{2}\|u\|_U^2 \\ &\text{subject to } y_h = \mathcal{G}_h(Bu) \text{ and } y_h(x_j) \le b(x_j) \text{ for } j = 1, \ldots, m. \end{aligned} \tag{33}$$

Notice that the integer m is not fixed and tends to infinity as $h \to 0$, so that the number of state constraints in this optimal control problem increases with decreasing mesh size of underlying finite element approximation of the state space. This discretization approach can be understood as a generalization of the *First discretize, then optimize* approach in that it avoids discretization of the control space U. Problem (33) represents a convex infinite-dimensional optimization problem of similar structure as problem (13), but with only finitely many equality and inequality constraints for the state, which form a convex admissible set. So we are again in the setting of (1) with Y replaced by the finite element space X_h (compare also the analysis of Casas presented in [10])

Lemma 4.3. *Problem* (33) *has a unique solution* $u_h \in U_{\mathrm{ad}}$. *There exist* $\mu_1, \ldots, \mu_m \in \mathbb{R}$ *and* $p_h \in X_h$ *such that with* $y_h = \mathcal{G}_h(Bu_h)$ *and* $\mu_h = \sum_{j=1}^m \mu_j \delta_{x_j}$ *we have*

$$a(v_h, p_h) = \int_\Omega (y_h - y_0)v_h + \int_{\bar{\Omega}} v_h d\mu_h \qquad \forall v_h \in X_h, \tag{34}$$

$$(RB^* p_h + \alpha u_h, v - u_h)_U \ge 0 \qquad \forall v \in U_{\mathrm{ad}}, \tag{35}$$

$$\mu_j \ge 0,\ y_h(x_j) \le b(x_j), j = 1, \ldots, m \text{ and } \int_{\bar{\Omega}} (I_h b - y_h) d\mu_h = 0. \tag{36}$$

Here, δ_x denotes the Dirac measure concentrated at x and I_h is the usual Lagrange interpolation operator. We have $\hat{J}_h'(v) = B^* p_h + \alpha(\cdot, u_h)_U$, so that the considerations after Theorem 2.3 also apply in the present setting, but with p replaced by the discrete function p_h, i.e., there holds

$$u_h = P_{U_{\mathrm{ad}}}(u_h - \sigma R \hat{J}_h'(u_h)) \ (\sigma > 0).$$

For $\sigma = \frac{1}{\alpha}$ we obtain

$$u = P_{U_{\mathrm{ad}}}\left(-\frac{1}{\alpha}RB^* p\right) \text{ and } u_h = P_{U_{\mathrm{ad}}}\left(-\frac{1}{\alpha}RB^* p_h\right). \tag{37}$$

Due to the presence of $P_{U\mathrm{ad}}$ in variational discretization the function $u_h \in U_{\mathrm{ad}}$ will in general not belong to X_h even in the case $U = L^2(\Omega), B = Id$. This is different for the purely state constrained problem, for which $P_{U_{\mathrm{ad}}} \equiv Id$, so that in this specific setting $u_h = -\frac{1}{\alpha}p_h \in X_h$ by (37). In that case the space $U = L^2(\Omega)$ in (33) may be replaced by X_h to obtain the same discrete solution u_h, which results in a finite-dimensional discrete optimization problem instead. However, we emphasize, that the infinite-dimensional formulation of (33) is very useful in numerical analysis [46, Chap. 3].

As a first result for (33) it is proved in, e.g., [46, Chap. 3] that the sequence of optimal controls, states and the measures μ_h are uniformly bounded.

Lemma 4.4. *Let $u_h \in U_{\mathrm{ad}}$ be the optimal solution of (33) with corresponding state $y_h \in X_h$ and adjoint variables $p_h \in X_h$ and $\mu_h \in \mathcal{M}(\bar\Omega)$. Then there exists $\bar h > 0$ so that*

$$\|y_h\|,\ \|u_h\|_U,\ \|\mu_h\|_{\mathcal{M}(\bar\Omega)} \le C \qquad \text{for all } 0 < h \le \bar h.$$

Proof. Since $\mathcal{G}(B\tilde u)$ is continuous, Assumption 2.2 implies that there exists $\delta > 0$ such that

$$\mathcal{G}(B\tilde u) \le b - \delta \qquad \text{in } \bar\Omega. \tag{38}$$

It follows from (28) that there is $h_0 > 0$ with

$$\mathcal{G}_h(B\tilde u) \le b \quad \text{in } \bar\Omega \text{ for all } 0 < h \le h_0$$

so that $J_h(u_h) \le J_h(\tilde u) \le C$ uniformly in h giving

$$\|u_h\|_U,\ \|y_h\| \le C \qquad \text{for all } h \le h_0. \tag{39}$$

Next, let u denote the unique solution to problem (13). We infer from (38) and (28) that $v := \frac{1}{2}u + \frac{1}{2}\tilde u$ satisfies

$$\begin{aligned}
\mathcal{G}_h(Bv) \ &\le\ \frac{1}{2}\mathcal{G}(Bu) + \frac{1}{2}\mathcal{G}(B\tilde u) + Ch^{2-\frac{d}{2}}(\|Bu\| + \|B\tilde u\|) \\
&\le\ b - \frac{\delta}{2} + Ch^{2-\frac{d}{2}}(\|u\|_U + \|\tilde u\|_U) \le b - \frac{\delta}{4} \qquad \text{in } \bar\Omega
\end{aligned} \tag{40}$$

provided that $h \le \bar h, \bar h \le h_0$. Since $v \in U_{\mathrm{ad}}$, (35), (34), (39) and (40) imply

$$\begin{aligned}
0 &\le (RB^* p_h + \alpha(u_h - u_{0,h}), v - u_h)_U = \int_\Omega B(v - u_h)p_h + \alpha(u_h - u_{0,h}, v - u_h)_U \\
&= a(\mathcal{G}_h(Bv) - y_h, p_h) + \alpha(u_h - u_{0,h}, v - u_h)_U \\
&= \int_\Omega (\mathcal{G}_h(Bv) - y_h)(y_h - y_0) + \int_{\bar\Omega} (\mathcal{G}_h(Bv) - y_h)d\mu_h + \alpha(u_h - u_{0,h}, v - u_h)_U \\
&\le C + \sum_{j=1}^m \mu_j\big(b(x_j) - \frac{\delta}{4} - y_h(x_j)\big) = C - \frac{\delta}{4}\sum_{j=1}^m \mu_j
\end{aligned}$$

where the last equality is a consequence of (36). It follows that

$$\|\mu_h\|_{\mathcal{M}(\bar\Omega)} \le C$$

and the lemma is proved. □

4.2. Piecewise constant controls

We consider the special case $U = L^2(\Omega)$, so that B denotes the injection of $L^2(\Omega)$ into $H^1(\Omega)^*$ with box constraints $a_l \le u \le a_r$ on the control. Controls are approximated by element-wise piecewise constant functions. For details we refer to [23]. We define the space of piecewise constant functions

$$Y_h := \{v_h \in L^2(\Omega) \,|\, v_h \text{ is constant on each } T \in \mathcal{T}_h\}.$$

and denote by $Q_h : L^2(\Omega) \to Y_h$ the orthogonal projection onto Y_h so that

$$(Q_h v)(x) := \fint_T v, \quad x \in T, T \in \mathcal{T}_h,$$

where $\fint_T v$ denotes the average of v over T. In order to approximate (13) we introduce a discrete counterpart of U_{ad},

$$U_{\text{ad}}^h := \{v_h \in Y_h \,|\, a_l \le v_h \le a_u \text{ in } \Omega\}.$$

Note that $U_{\text{ad}}^h \subset U_{\text{ad}}$ and that $Q_h v \in U_{\text{ad}}^h$ for $v \in U_{\text{ad}}$. Since $Q_h v \to v$ in $L^2(\Omega)$ as $h \to 0$ we infer from the continuous embedding $H^2(\Omega) \hookrightarrow C^0(\bar{\Omega})$ and Lemma 4.1 that

$$\mathcal{G}_h(Q_h v) \to \mathcal{G}(v) \text{ in } L^\infty(\Omega) \text{ for all } v \in U_{\text{ad}}. \tag{41}$$

Problem (13) here now is approximated by the following sequence of control problems depending on the mesh parameter h:

$$\begin{aligned} &\min_{u \in U_{\text{ad}}^h} J_h(u) := \frac{1}{2}\int_\Omega |y_h - y_0|^2 + \frac{\alpha}{2}\int_\Omega |u|^2 \\ &\text{subject to } \; y_h = \mathcal{G}_h(u) \text{ and } y_h(x_j) \le b(x_j) \text{ for } j = 1, \dots, m. \end{aligned} \tag{42}$$

Problem (42), as problem (33), represents a convex finite-dimensional optimization problem of similar structure as problem (13), but with only finitely many equality and inequality constraints for state and control, which form a convex admissible set. The following optimality conditions can be argued as those given in (4.3) for problem (33).

Lemma 4.5. *Problem* (42) *has a unique solution $u_h \in U_{\text{ad}}^h$. There exist $\mu_1, \dots, \mu_m \in \mathbb{R}$ and $p_h \in X_h$ such that with $y_h = \mathcal{G}_h(u_h)$ and $\mu_h = \sum_{j=1}^m \mu_j \delta_{x_j}$ we have*

$$a(v_h, p_h) = \int_\Omega (y_h - y_0) v_h + \int_{\bar{\Omega}} v_h \, d\mu_h \qquad \forall v_h \in X_h, \tag{43}$$

$$\int_\Omega (p_h + \alpha u_h)(v_h - u_h) \ge 0 \qquad \forall v_h \in U_{\text{ad}}^h, \tag{44}$$

$$\mu_j \ge 0,\; y_h(x_j) \le b(x_j), j = 1, \dots, m \text{ and } \int_{\bar{\Omega}} (I_h b - y_h) \, d\mu_h = 0. \tag{45}$$

Here, δ_x denotes the Dirac measure concentrated at x and I_h is the usual Lagrange interpolation operator.

For (42) we now prove bounds on the discrete states and the discrete multipliers. Similar to Lemma 4.4 we have

Lemma 4.6. *Let $u_h \in U_{\text{ad}}^h$ be the optimal solution of* (42) *with corresponding state $y_h \in X_h$ and adjoint variables $p_h \in X_h$ and $\mu_h \in \mathcal{M}(\bar{\Omega})$. Then there exists $\bar{h} > 0$ such that*

$$\|y_h\|, \|\mu_h\|_{\mathcal{M}(\bar{\Omega})} \le C, \quad \|p_h\|_{H^1} \le C\gamma(d, h) \qquad \text{for all } 0 < h \le \bar{h},$$

where $\gamma(2, h) = \sqrt{|\log h|}$ and $\gamma(3, h) = h^{-\frac{1}{2}}$.

Proof. Since $\mathcal{G}(\tilde{u}) \in C^0(\bar{\Omega})$, Assumption 2.2 implies that there exists $\delta > 0$ such that

$$\mathcal{G}(\tilde{u}) \le b - \delta \qquad \text{in } \bar{\Omega}. \tag{46}$$

It follows from (41) that there is $\bar{h} > 0$ with

$$\mathcal{G}_h(Q_h\tilde{u}) \le b - \frac{\delta}{2} \quad \text{in } \bar{\Omega} \text{ for all } 0 < h \le \bar{h}. \tag{47}$$

Since $Q_h\tilde{u} \in U_{\rm ad}^h$, (45), (44) and (47) imply

$$\begin{aligned}
0 \;&\le\; \int_\Omega (p_h + \alpha u_h)(Q_h\tilde{u} - u_h) = \int_\Omega p_h(Q_h\tilde{u} - u_h) + \alpha \int_\Omega u_h(Q_h\tilde{u} - u_h) \\
&=\; a(\mathcal{G}_h(Q_h\tilde{u}) - y_h, p_h) + \alpha \int_\Omega u_h(Q_h\tilde{u} - u_h) \\
&=\; \int_\Omega (\mathcal{G}_h(Q_h\tilde{u}) - y_h)(y_h - y_0) + \int_{\bar{\Omega}} (\mathcal{G}_h(Q_h\tilde{u}) - y_h) d\mu_h + \alpha \int_\Omega u_h(Q_h\tilde{u} - u_h) \\
&\le\; C - \frac{1}{2}\|y_h\|^2 + \sum_{j=1}^m \mu_j \big(b(x_j) - \frac{\delta}{2} - y_h(x_j)\big) = C - \frac{1}{2}\|y_h\|^2 - \frac{\delta}{2}\sum_{j=1}^m \mu_j
\end{aligned}$$

where the last equality is a consequence of (45). It follows that $\|y_h\|, \|\mu_h\|_{\mathcal{M}(\bar{\Omega})} \le C$. In order to bound $\|p_h\|_{H^1}$ we insert $v_h = p_h$ into (44) and deduce with the help of the coercivity of A, a well-known inverse estimate and the bounds we have already obtained that

$$\begin{aligned}
c_1\|p_h\|_{H^1}^2 \;&\le\; a(p_h, p_h) = \int_\Omega (y_h - y_0)p_h + \int_{\bar{\Omega}} p_h d\mu_h \\
&\le\; \|y_h - y_0\|\,\|p_h\| + \|p_h\|_{L^\infty}\|\mu_h\|_{\mathcal{M}(\bar{\Omega})} \le C\|p_h\| + C\gamma(d,h)\|p_h\|_{H^1}.
\end{aligned}$$

Hence $\|p_h\|_{H^1} \le C\gamma(d,h)$ and the lemma is proved. □

Similar considerations hold for control approximations by continuous, piecewise polynomial functions. Discrete approaches to problem (13) relying on control approximations directly lead to fully discrete optimization problems like (42). These approaches lead to large-scale finite-dimensional optimization problems, since the discretization of the pde in general introduces a large number of degrees of freedom. Numerical implementation then is easy, which certainly is an important advantage of control approximations over variational discretization, whose numerical implementation is more involved. The use of classical NLP solvers for the numerical solution of the underlying discretized problems only is feasible, if the solver allows to exploit the underlying problem structure, e.g., by providing user interfaces for first- and second-order derivatives.

On the other hand, the numerical implementation of variational discretization is not straightforward. The great advantage of variational discretization however is its property of optimal approximation accuracy, which is completely determined by that of the related state and adjoint state. Fig. 1 compares active sets obtained by variational discretization and piecewise linear control approximations in the presence of box constraints. One clearly observes that the active sets are

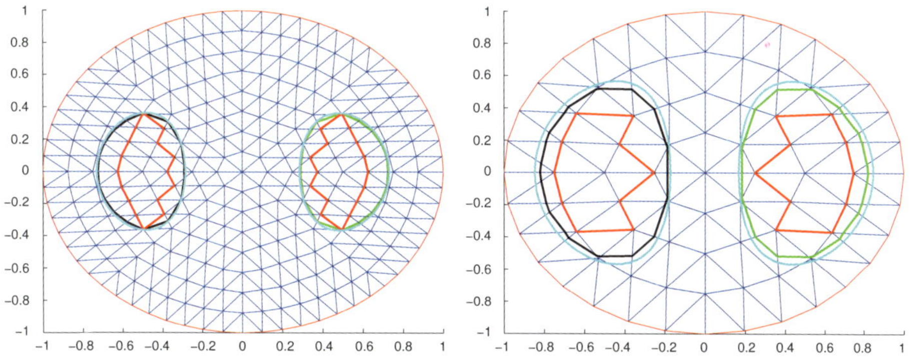

FIGURE 1. Numerical comparison of active sets obtained by variational discretization, and those obtained by a conventional approach with piecewise linear, continuous controls: $h = \frac{1}{8}$ and $\alpha = 0.1$ (left), $h = \frac{1}{4}$ and $\alpha = 0.01$ (right). The red line depicts the boarder of the active set in the conventional approach, the cyan line the exact boarder, the black and green lines, respectively the boarders of the active set in variational discretization.

resolved much more accurate when using variational discretization. In particular, the boundary of the active set is in general different from finite element edges.

The error analysis for problem (13) relies on the regularity of the involved variables, which is reflected by the optimality system presented in (15)–(17). If only control constraints are present, neither the multiplier μ in (15) nor the complementarity condition (17) appear. Then the variational inequality (16) restricts the regularity of the control u, and thus also that of the state y. If the desired state y_0 is regular enough, the adjoint variable p admits the highest regularity properties among all variables involved in the optimality system. Error analysis in this case then should involve the adjoint variable p and exploit its regularity properties.

If pointwise state constraints, are present, the situation is completely different. Now the adjoint variable only admits low regularity due to the presence of the multiplier μ, which in general is only a measure. The state now admits the highest regularity in the optimality system. This fact should be exploited in the error analysis. However, the presence of the complementarity system (17) requires L^∞-error estimates for the state.

4.3. Error bounds

For the approximation error of variational discretization we have the following theorem, whose proof can be found in [46, Chap. 3].

Theorem 4.7. *Let u and u_h be the solutions of* (13) *and* (33) *respectively. Then*

$$\|u - u_h\|_U + \|y - y_h\|_{H^1} \leq Ch^{1-\frac{d}{4}}.$$

If in addition $Bu \in W^{1,s}(\Omega)$ *for some* $s \in (1, \frac{d}{d-1})$ *then*

$$\|u - u_h\|_U + \|y - y_h\|_{H^1} \leq Ch^{\frac{3}{2}-\frac{d}{2s}}\sqrt{|\log h|}.$$

If $Bu, Bu_h \in L^\infty(\Omega)$ *with* $(Bu_h)_h$ *uniformly bounded in* $L^\infty(\Omega)$ *also*

$$\|u - u_h\|_U,\ \|y - y_h\|_{H^1} \leq Ch|\log h|,$$

where the latter estimate is valid for $d = 2, 3$.

Proof. We test (16) with u_h, (35) with u and add the resulting inequalities. This gives

$$(RB^*(p - p_h) - \alpha(u_0 - u_{0,h}) + \alpha(u - u_h), u_h - u)_U \geq 0,$$

which in turn yields

$$\alpha\|u - u_h\|_U^2 \leq \int_\Omega B(u_h - u)(p - p_h) - \alpha\,(u_0 - u_{0,h}, u_h - u)_U\,. \tag{48}$$

Let $y^h := \mathcal{G}_h(Bu) \in X_h$ and denote by $p^h \in X_h$ the unique solution of

$$a(w_h, p^h) = \int_\Omega (y - y_0)w_h + \int_{\bar\Omega} w_h d\mu \qquad \text{for all } w_h \in X_h.$$

Applying Theorem 4.2 with $\tilde\mu = (y - y_0)dx + \mu$ we infer

$$\|p - p^h\| \leq Ch^{2-\frac{d}{2}}\big(\|y - y_0\| + \|\mu\|_{\mathcal{M}(\bar\Omega)}\big). \tag{49}$$

Recalling that $y_h = \mathcal{G}_h(Bu_h), y^h = \mathcal{G}_h(Bu)$ and observing (34) as well as the definition of p^h we can rewrite the first term in (48)

$$\begin{aligned}
\int_\Omega B(u_h - u)(p - p_h) &= \int_\Omega B(u_h - u)(p - p^h) + \int_\Omega B(u_h - u)(p^h - p_h)\\
&= \int_\Omega B(u_h - u)(p - p^h) + a(y_h - y^h, p^h - p_h) \qquad (50)\\
&= \int_\Omega B(u_h - u)(p - p^h) + \int_\Omega (y - y_h)(y_h - y^h) + \int_{\bar\Omega}(y_h - y^h)d\mu - \int_{\bar\Omega}(y_h - y^h)d\mu_h\\
&= \int_\Omega B(u_h - u)(p - p^h) - \|y - y_h\|^2 + \int_\Omega (y - y_h)(y - y^h)\\
&\quad + \int_{\bar\Omega}(y_h - y^h)d\mu + \int_{\bar\Omega}(y^h - y_h)d\mu_h.
\end{aligned}$$

After inserting (50) into (48) and using Young's inequality we obtain in view of (49), (27) and the properties of the L^2-projection

$$\begin{aligned}
&\frac{\alpha}{2}\|u - u_h\|_U^2 + \frac{1}{2}\|y - y_h\|^2 \qquad (51)\\
&\quad\leq C\big(\|p - p^h\|^2 + \|y - y^h\|^2 + \|u_0 - u_{0,h}\|^2\big) + \int_{\bar\Omega}(y_h - y^h)d\mu + \int_{\bar\Omega}(y^h - y_h)d\mu_h\\
&\quad\leq Ch^{4-d} + \int_{\bar\Omega}(y_h - y^h)d\mu + \int_{\bar\Omega}(y^h - y_h)d\mu_h.
\end{aligned}$$

It remains to estimate the integrals involving the measures μ and μ_h. Since

$$y_h - y^h \leq (I_h b - b) + (b - y) + (y - y^h) \qquad \text{in } \bar{\Omega}$$

we deduce with the help of (17)

$$\int_{\bar{\Omega}} (y_h - y^h) d\mu \leq \|\mu\|_{\mathcal{M}(\bar{\Omega})} \left(\|I_h b - b\|_\infty + \|y - y^h\|_\infty \right).$$

Similarly, (36) implies

$$\int\limits_{\bar{\Omega}} (y^h - y_h) d\mu_h \leq \|\mu_h\|_{\mathcal{M}(\bar{\Omega})} \left(\|b - I_h b\|_\infty + \|y - y^h\|_\infty \right).$$

Inserting the above estimates into (51) and using Lemma 4.4 as well as an interpolation estimate we infer

$$\|u - u_h\|_U^2 + \|y - y_h\|^2 \leq Ch^{4-d} + C\|y - y^h\|_{L^\infty}. \tag{52}$$

The estimates on $\|u - u_h\|_U$ now follow from (28) and Lemma 4.1 respectively. Finally, in order to bound $\|y - y_h\|_{H^1}$ we note that

$$a(y - y_h, v_h) = \int_\Omega B(u - u_h) v_h$$

for all $v_h \in X_h$, from which one derives the desired estimates using standard finite element techniques and the bounds on $\|u - u_h\|_U$. In order to avoid the dependence on the dimension we should avoid finite element approximations of the adjoint variable p, which due to its low regularity only allows error estimates in the L^2 norm. We therefore provide a proof technique which completely avoids the use of finite element approximations of the adjoint variable. To begin with we start with the basic estimate (48)

$$\alpha\|u - u_h\|_U^2 \leq \int_\Omega B(u_h - u)(p - p_h) - \alpha\left(u_0 - u_{0,h}, u_h - u\right)_U$$

and write

$$\begin{aligned} \int_\Omega B(u_h - u)(p - p_h) &= \int\limits_\Omega pA(\tilde{y} - y) - a(y_h - y^h, p_h) \\ &= \int\limits_\Omega (y - y_0)(\tilde{y} - y) + \int\limits_{\bar{\Omega}} \tilde{y} - y d\mu - \int\limits_\Omega (y_h - y_0)(y_h - y^h) + \int\limits_{\bar{\Omega}} y_h - y^h d\mu_h, \end{aligned}$$

where $\tilde{y} := \mathcal{G}(u_h)$. Proceeding similar as in the proof of the previous theorem we obtain

$$\begin{aligned} &\int\limits_\Omega (y - y_0)(\tilde{y} - y) + \int\limits_{\bar{\Omega}} \tilde{y} - y d\mu - \int\limits_\Omega (y_h - y_0)(y_h - y^h) + \int\limits_{\bar{\Omega}} y_h - y^h d\mu_h \\ &\quad \leq C\{\|\mu\|_{\mathcal{M}(\bar{\Omega})} + \|\mu_h\|_{\mathcal{M}(\bar{\Omega})}\}\{\|b - I_h b\|_\infty + \|y - y^h\|_\infty + \|\tilde{y} - y_h\|_\infty\} \\ &\qquad - \|y - y_h\|^2 + C\{\|y - y^h\| + \|\tilde{y} - y_h\|\}. \end{aligned}$$

Using Lemma 4.4 together with Lemma 4.1 then yields

$$\alpha\|u-u_h\|_U^2+\|y-y_h\|^2\le C\{h^2+h^2|\log h|^2\},$$

so that the claim follows as in the proof of the previous theorem. □

Remark 4.8. Let us note that the approximation order of the controls and states in the presence of control and state constraints is the same as in the purely state constrained case, if $Bu\in W^{1,s}(\Omega)$. This assumption holds for the important example $U=L^2(\Omega)$, $B=Id$ and $u_{0_h}=P_hu_0$, with $u_0\in H^1(\Omega)$ and $P_h:L^2(\Omega)\to X_h$ denoting the L^2-projection, and subsets of the form

$$U_{\mathrm{ad}}=\{v\in L^2(\Omega), a_l\le v\le a_u \text{ a.e. in } \Omega\},$$

with bounds $a_l,a_u\in W^{1,s}(\Omega)$, since $u_0\in H^1(\Omega)$, and $p\in W^{1,s}(\Omega)$. Moreover, $u,u_h\in L^\infty(\Omega)$ with $\|u_h\|_\infty\le C$ uniformly in h holds if for example $a_l,a_u\in L^\infty(\Omega)$.

For piecewise constant control approximations and the setting of Section 4.2 the following theorem is proved in [23].

Theorem 4.9. *Let u and u_h be the solutions of* (13) *and* (42) *respectively with $(u_h)_h\subset L^\infty(\Omega)$ uniformly bounded. Then we have for $0<h\le\bar h$*

$$\|u-u_h\|+\|y-y_h\|_{H^1}\le\begin{cases}Ch|\log h|, & \text{if } d=2\\ C\sqrt{h}, & \text{if } d=3.\end{cases}$$

Proof. We test (16) with u_h, (45) with Q_hu and add the resulting inequalities. Keeping in mind that $u-Q_hu\perp Y_h$ we obtain

$$\int_\Omega\big(p-p_h+\alpha(u-u_h)\big)(u_h-u)$$
$$\ge\int_\Omega\big(p_h+\alpha u_h\big)(u-Q_hu)=\int_\Omega(p_h-Q_hp_h)(u-Q_hu).$$

As a consequence,

$$\alpha\|u-u_h\|^2\le\int_\Omega(u_h-u)(p-p_h)-\int_\Omega(p_h-Q_hp_h)(u-Q_hu)\equiv I+II. \quad (53)$$

Let $y^h:=\mathcal{G}_h(u)\in X_h$ and denote by $p^h\in X_h$ the unique solution of

$$a(w_h,p^h)=\int_\Omega(y-y_0)w_h+\int_{\bar\Omega}w_h\,d\mu \qquad \text{for all } w_h\in X_h.$$

Applying Theorem 4.2 with $\tilde{\mu} = (y - y_0)dx + \mu$ we infer

$$\|p - p^h\| \le Ch^{2-\frac{d}{2}}\big(\|y - y_0\| + \|\mu\|_{\mathcal{M}(\bar{\Omega})}\big). \tag{54}$$

Recalling that $y_h = \mathcal{G}_h(u_h), y^h = \mathcal{G}_h(u)$ and observing (44) as well as the definition of p^h we can rewrite the first term in (53)

$$\begin{aligned}
I &= \int_\Omega (u_h - u)(p - p^h) + \int_\Omega (u_h - u)(p^h - p_h) \\
&= \int_\Omega (u_h - u)(p - p^h) + a(y_h - y^h, p^h - p_h) \qquad (55)\\
&= \int_\Omega (u_h - u)(p - p^h) + \int_\Omega (y - y_h)(y_h - y^h) + \int_{\bar{\Omega}} (y_h - y^h)d\mu - \int_{\bar{\Omega}} (y_h - y^h)d\mu_h \\
&= \int_\Omega (u_h - u)(p - p^h) - \|y - y_h\|^2 + \int_\Omega (y - y_h)(y - y^h) \\
&\quad + \int_{\bar{\Omega}} (y_h - y^h)d\mu + \int_{\bar{\Omega}} (y^h - y_h)d\mu_h.
\end{aligned}$$

Applying Young's inequality we deduce

$$\begin{aligned}
|I| \;&\le\; \frac{\alpha}{4}\|u - u_h\|^2 - \frac{1}{2}\|y - y_h\|^2 + C\big(\|p - p^h\|^2 + \|y - y^h\|^2\big) \\
&\quad + \int_{\bar{\Omega}} (y_h - y^h)d\mu + \int_{\bar{\Omega}} (y^h - y_h)d\mu_h. \qquad (56)
\end{aligned}$$

Let us estimate the integrals involving the measures μ and μ_h. Since $y_h - y^h \le (I_h b - b) + (b - y) + (y - y^h)$ in $\bar{\Omega}$ we deduce with the help of (17), Lemma 4.1 and an interpolation estimate

$$\int_{\bar{\Omega}} (y_h - y^h)d\mu \le \|\mu\|_{\mathcal{M}(\bar{\Omega})}\big(\|I_h b - b\|_\infty + \|y - y^h\|_\infty\big) \le Ch^2|\log h|^2.$$

On the other hand $y^h - y_h \le (y^h - y) + (b - I_h b) + (I_h b - y_h)$, so that (45), Lemma 4.1 and Lemma 4.6 yield

$$\int_{\bar{\Omega}} (y^h - y_h)d\mu_h \le \|\mu_h\|_{\mathcal{M}(\bar{\Omega})}\big(\|b - I_h b\|_\infty + \|y - y^h\|_\infty\big) \le Ch^2|\log h|^2.$$

Inserting these estimates into (56) and recalling (27) as well as (32) we obtain

$$|I| \le \frac{\alpha}{4}\|u - u_h\|^2 - \frac{1}{2}\|y - y_h\|^2 + Ch^{4-d} + Ch^2|\log h|^2. \tag{57}$$

Let us next examine the second term in (53). Since $u_h = Q_h u_h$ and Q_h is stable in $L^2(\Omega)$ we have

$$\begin{aligned}
|II| \;&\le\; 2\|u - u_h\|\,\|p_h - Q_h p_h\| \le \frac{\alpha}{4}\|u - u_h\|^2 + Ch^2\|p_h\|_{H^1}^2 \\
&\le\; \frac{\alpha}{4}\|u - u_h\|^2 + Ch^2\gamma(d,h)^2
\end{aligned}$$

using an interpolation estimate for Q_h and Lemma 4.6. Combining this estimate with (57) and (53) we finally obtain

$$\|u - u_h\|^2 + \|y - y_h\|^2 \leq Ch^{4-d} + Ch^2|\log h|^2 + Ch^2\gamma(d,h)^2$$

which implies the estimate on $\|u - u_h\|$. In order to bound $\|y - y_h\|_{H^1}$ we note that

$$a(y - y_h, v_h) = \int_\Omega (u - u_h)v_h$$

for all $v_h \in X_h$, from which one derives the desired estimate using standard finite element techniques and the bound on $\|u - u_h\|$. □

Remark 4.10. An inspection of the proof of Theorem 4.7 shows that we also could avoid to use error estimates for the auxiliary function p^h if we would use a technique for the term I similar to that used in the proof of the third part of Theorem 4.7. However, our approach to estimate II is based on inverse estimates which finally lead to the dimension-dependent error estimate presented in Theorem 4.7.

The theorems above have in common that a control error estimate is only available for $\alpha > 0$. However, the appearance of α in these estimates indicates that in the *bang-bang*-case $\alpha = 0$ an error estimate for $\|y - y_h\|_{L^2}$ still is available, whereas no information for the control error $\|u - u_h\|_U$ seems to remain. In [25] a refined analysis of bang-bang controls without state constraints also provides estimates for the control error on inactive regions in the L^1-norm. We further observe that piecewise constant control approximations in 2 space dimensions deliver the same approximation quality as variational discrete controls. Only in 3 space dimensions variational discretization provides a better error estimate. This is caused by the fact that state constraints limit the regularity of the adjoint state, so that optimal error estimates can be expected by techniques which avoid its use. Currently the analysis for piecewise constant control approximations involves an inverse estimate for $\|p_h\|_{H^1}$, which explains the lower approximation order in the case $d = 3$.

Let us mention that the bottleneck in the analysis here is not formed by control constraints, but by the state constraints. In fact, if one uses $U_{\text{ad}} = U$, then variational discretization (33) delivers the same numerical solution as the approach (42) with piecewise linear, continuous control approximations. Variational discretization really pays off if only control constraints are present and the adjoint variable is smooth, compare [44], [46, Chap. 3].

For the numerical solution of problem (33), (42) several approaches exist in the literature. Common are so-called regularization methods which relax the state constraints in (13) by either substituting it by a mixed control-state constraint (Lavrentiev relaxation [62]), or by adding suitable penalty terms to the cost functional instead requiring the state constraints (barrier methods [47, 70], penalty methods [38, 40]. These approaches will be discussed in the following section.

5. Regularization and discretization

5.1. Motivation

In this section we will focus on optimal control problems with pointwise state constraints. Let us consider

$$\min F(y,u) = \frac{1}{2}\|y - y_d\|^2_{L^2(\Omega)} + \frac{\nu}{2}\|u\|^2_{L^2(\Omega)} \tag{58}$$

subject to the state equation

$$y = S(u), \tag{59}$$

and pointwise state constraints

$$y \geq y_c \quad \text{a.e. on } D, \tag{60}$$

and $u \in U_{\text{ad}}$ which may be the whole space $U = L^2(\Omega)$ or contain additional control constraints. Moreover, $D \subset \Omega$ denotes the set where the state constraints are given. The operator S plays the role of a solution operator of a linear or nonlinear partial differential equation. Example 1.1 represents the nonlinear case.

Let us now focus on linear partial differential equations. For the numerical solution of problem (33), (42) several approaches exist in the literature. Common are so-called regularization methods which relax the state constraints in (13) by either substituting it by a mixed control-state constraint (Lavrentiev relaxation [62]), or by adding suitable penalty terms to the cost functional instead requiring the state constraints (barrier methods [47, 70], penalty methods [38, 40].

The analysis of unregularized and regularized optimal control problems is quite similar. Regularization leads often to more smooth solution. However, this effect disappears when the regularization parameter tends to zero. Consequently solving methods and numerical aspects are the main reason for regularization.

Numerical approaches for discretized problems

Discretized optimal control problems with pointwise state constraints can be attacked by different techniques. Projected or conditional gradient methods are very robust, but slow.

Active set strategies become very popular in the recent years, see [5, 6, 52]. Active and inactive sets are fixed in every iteration. In contrast to the most classical techniques in nonlinear optimization, whole sets can change from one iteration to the next. In each iteration a problem has to be solved without inequality constraints. In many cases such methods can be interpreted as semismooth Newton methods, see [39]. Thus, active set strategies are fast convergent and mesh independent solving methods [42]. However, a direct application of active set strategies to the discretized is often impossible: If one starts completely inactive, then all violated inequalities leads to elements of active sets. It is easy to construct situations where the number of free optimization variables is smaller then the number of new active constraints. Moreover, subproblems in the active set algorithm may be ill posed or ill conditioned. Therefore, a direct application of active set methods to the discretized problem cannot be recommended.

Interior point methods and barrier techniques are an alternative approach. The objective functional is modified by a penalty term in such a way that the feasible iterates stay away from the bounds in the inequalities. In contrast to the active set strategies the new problem is smooth, but nonlinear. The penalty term modifies the problem. The obtained solution is not the solution of the original problem. In general the solution of the original problem is obtained by tending the penalty parameter to zero or infinity.

Regularization techniques

We have introduced numerical techniques for solving optimal control problems with inequality constraints. The most techniques introduce some parameters and modify the (discretized) optimal control problem. One can have different views on the whole solution process. Our first approach was to discretize the optimal control problem first and then to find a solving method. Another view is to fix a discretization and a regularization or penalty parameter and to look for over all error.

Exactly this is the issue of this section. The tuning of discretization and regularization can significantly reduce the computational effort without loss on accuracy. Before we start to find error estimates for this combined approach, we will shortly explain the different techniques. For simplicity we chose a linear operator S.

1. Moreau-Yosida-regularization

The Moreau-Yosida-regularization (see [40]) uses a quadratic penalty term

$$\min F_\gamma(y,u) = \frac{1}{2}\|y - y_d\|_Y^2 + \frac{\nu}{2}\|u\|_U^2 + \frac{\gamma}{2}\|(y_c - y)_+\|_{L^2(D)}^2. \tag{61}$$

Moreover the inequality constraint (60) is dropped. The quantity γ plays the role of a regularization parameter. The original problem is obtained in the limit $\gamma \to \infty$. Combined error estimates should help to find a reasonable size of γ for a given discretization. Let us mention that the nonsmooth penalty term can be treated by a semismooth Newton approach which can be reinterpreted as an active set approach.

2. Lavrentiev regularization

The Lavrentiev regularization (see [59, 62]) modifies only the inequality constraint

$$y + \lambda u \geq y_c \quad \text{a.e. on } D. \tag{62}$$

This approach is only possible if the control u acts on the whole set D. This modification overcomes the ill-posedness effect in active set strategies, since

$$y_c = Su + \lambda u \text{ on } D, \tag{63}$$

requires no longer the inversion of a compact operator. In contrast to the original problem, the Lagrange multipliers associated with the mixed constraints are regular functions. This technique works well for problems without additional control constraints. However, there are difficulties for problems with additional control constraints if the control constraints and the mixed constraints are active simultaneously. Then the dual variables are not uniquely determined. Consequently, the corresponding active set strategy is not well defined. In this approach, the Lavrentiev parameter λ has to tend to zero to obtain the solution of the original problem. A generalization of the Lavrentiev regularization is used in the source representation method, see [72, 73].

3. Virtual control approach

The virtual control approach (see [50, 51]) modifies the complete problem. A new control v is introduced on the domain D:

$$\min F(y,u) = \frac{1}{2}\|y - y_d\|_Y^2 + \frac{\nu}{2}\|u\|_U^2 + \frac{f(\varepsilon)}{2}\|v\|_{L^2(D)}^2 \tag{64}$$

subject to

$$y = Su + g(\varepsilon)Tv, \tag{65}$$

and

$$y + h(\varepsilon)v \geq y_c \quad \text{a.e. on } D, \tag{66}$$

with suitable chosen functions f, g, and h. The operator T represents a solution operator of a partial differential equation for the source term $g(\varepsilon)v$. Let us mention that this technique can be interpreted as Moreau-Yosida approach for the choice $g \equiv 0$. The original problem is obtained for $\varepsilon \to 0$. In contrast to the Lavrentiev regularization, the dual variables are unique. Moreover, this approach guarantees well defined subproblems in active set strategies in contrast to the Lavrentiev regularization.

4. Barrier methods – Interior point methods

Interior point methods or barrier methods deliver regularized solution which are feasible for the original problem. The objective is modified to

$$\min F_\mu(y,u) = \frac{1}{2}\|y - y_d\|_Y^2 + \frac{\nu}{2}\|u\|_U^2 + \mu\varphi(y - y_c) \tag{67}$$

where φ denotes a suitable smooth barrier function, see [70]. A typical choice would be a logarithmic function, i.e.,

$$\varphi(y - y_c) = -\int_D \log(y - y_c)\,dx.$$

The inequality constraints are dropped. The regularization parameter μ tends to zero or infinity to obtain the original problem. Interior point methods have the advantage that no nonsmooth terms occurs. However, the barrier function φ generates a new nonlinearity.

Before we start with the presentation of the main ideas, we give an overview on combined regularization and discretization error estimates. Error estimates for the Lavrentiev regularization can be found in [18, 45]. The Moreau-Yosida approach is analyzed in [38]. Discretization error estimates for virtual control concept are derived in [48]. Results for the interior point approach are published in [47].

5.2. Error estimates for variational discretization

In this subsection we will demonstrate the technique to obtain error estimates for the variational discretization concept. We will focus on linear solution operators S here. The case of a nonlinear solution operator is discussed later.

Regularization error

The regularization techniques presented above modify the original problem in the objective, the state equation or in the inequality constraints. To deal with all these concepts in a general framework would lead to a very technical presentation. Therefore we pick a specific approach.

Let us explain the main issues for the Lavrentiev regularization Here, the inequality constraint was changed to

$$y + \lambda u \geq y_c \quad \text{a.e. on } D. \tag{68}$$

Remember, that this regularization is only possible, if the set D is a subset of the set where the control acts. The inequality (68) changes the set of admissible controls. Therefore, one has to ensure that the admissible set of the regularized problem is nonempty. This is one motivation to require a Slater type condition:

Assumption 5.1. There exists a control $\hat{u} \in U_{\text{ad}}$ and a real number $\tau > 0$ with $\hat{y} = S\hat{u} \geq y_c + \tau$ and $\|\hat{u}\|_{L^\infty(D)} \leq c$.

Assumption 2.2 ensure the existence of at least one feasible point if $\lambda \leq \tau/c$. Now one has at least three possibilities to derive regularization error estimates:

1. Work with the complete optimality system including Lagrange multipliers and use the uniform boundedness of the Lagrange multipliers. That technique was used in Section 4.
2. Use the optimality conditions for a multiplier free formulation. Test the corresponding variational inequalities with suitable functions. We will demonstrate this technique in this section.
3. Work again multiplier free. The definition of admissible sets is the same as in approach 2. Now, the aim is to construct estimates of the form

$$|J(\bar{y}, \bar{u}) - J(\bar{y}_\lambda, \bar{u}_\lambda)| \leq \psi(\lambda).$$

This can be used to obtain the desired estimates for the regularization error. All three approaches are used in literature to derive estimates. We will focus on the second approach. Let us define the sets:

$$\begin{aligned} U^0 &= \{u \in U_{\text{ad}} : \ Su \geq y_c\}, \\ U^\lambda &= \{u \in U_{\text{ad}} : \ \lambda u + Su \geq y_c\}. \end{aligned}$$

Then the first-order optimal conditions for the solution $\bar{u}$ of the original problem and the solution $\bar{u}_\lambda$ for the regularized problem read

$$(S^*(S\bar{u} - y_d) + \nu\bar{u}, u - \bar{u}) \ \geq \ 0 \text{ for all } u \in U^0 \tag{69}$$

$$(S^*(S\bar{u}_\lambda - y_d) + \nu\bar{u}_\lambda, u - \bar{u}_\lambda) \ \geq \ 0 \text{ for all } u \in U^\lambda. \tag{70}$$

Now one has to look for suitable test functions u. Suitable test functions should be feasible for one of these problems and close to the solutions of the other problem. We assume $u_0 \in U^0$ and $u_\lambda \in U^\lambda$ and obtain

$$(S^*(S\bar{u} - y_d) + \nu\bar{u}, u_0 - \bar{u}) \geq 0 \tag{71}$$

$$(S^*(S\bar{u}_\lambda - y_d) + \nu\bar{u}_\lambda, u_\lambda - \bar{u}_\lambda) \geq 0. \tag{72}$$

Adding these inequalities yields

$$\begin{aligned}(S^*(S\bar{u} - y_d) + \nu\bar{u}, u_0 - \bar{u}_\lambda) + (S^*(S\bar{u} - y_d) + \nu\bar{u}, \bar{u}_\lambda - \bar{u}) +\\ (S^*(S\bar{u}_\lambda - y_d) + \nu\bar{u}_\lambda, u_\lambda - \bar{u}) + (S^*(S\bar{u}_\lambda - y_d) + \nu\bar{u}_\lambda, \bar{u} - \bar{u}_\lambda) \geq 0.\end{aligned} \tag{73}$$

We obtain for the first and the third term

$$(S^*(S\bar{u} - y_d) + \nu\bar{u}, u_0 - \bar{u}_\lambda) \leq c\|u_0 - \bar{u}_\lambda\|_U \tag{74}$$

$$(S^*(S\bar{u}_\lambda - y_d) + \nu\bar{u}_\lambda, u_\lambda - \bar{u}) \leq c\|u_\lambda - \bar{u}\|_U. \tag{75}$$

For the sum of the second and the fourth term we find

$$\begin{aligned}(S^*(S\bar{u} - y_d) + \nu\bar{u}, \bar{u}_\lambda - \bar{u}) +\\ (S^*(S\bar{u}_\lambda - y_d) + \nu\bar{u}_\lambda, u_\lambda - \bar{u}) = -\nu\|\bar{u} - \bar{u}_\lambda\|_U^2 - \|S(\bar{u} - \bar{u}_\lambda)\|_Y^2.\end{aligned} \tag{76}$$

Consequently, we end up with

$$\nu\|\bar{u} - \bar{u}_\lambda\|_U^2 + \|S(\bar{u} - \bar{u}_\lambda)\|_Y^2 \leq c\|u_0 - \bar{u}_\lambda\|_U + c\|u_\lambda - \bar{u}\|_U. \tag{77}$$

Thus, the choice of the test functions u_0, u_λ is an important issue. We will investigate two cases.

First case

In the first case we assume that

$$u \in U_{\text{ad}} \quad \Rightarrow \quad \|u\|_{L^\infty(D)} \leq K.$$

Let us start with the construction of the function u_0. A reasonable choice is

$$u_0 = (1 - \delta)\bar{u}_\lambda + \delta\hat{u}.$$

Of course we have $u_0 \in U_{\text{ad}}$. Therefore, we have only to check the state constraints

$$\begin{aligned} Su_0 &= (1 - \delta)S\bar{u}_\lambda + \delta S\hat{u} \\ &\geq (1 - \delta)(y_c - \lambda\bar{u}) + \delta(y_c + \tau) \\ &\geq y_c + \delta\tau - (1 - \delta)\lambda K. \end{aligned}$$

Consequently, we have to choose $\delta \sim \lambda$ to satisfy the state constraints. In the same way we get for

$$u_\lambda = (1 - \sigma)\bar{u} + \sigma\hat{u}$$

the relation

$$Su_\lambda + \lambda u_\lambda \geq y_c + \sigma(\tau - \lambda\|\hat{u}\|_{L^\infty(D)}) - (1 - \sigma)\lambda K$$

and we need for feasibility $\sigma \sim \lambda$. This leads to the final regularization error estimate

$$\nu\|\bar{u} - \bar{u}_\lambda\|_U^2 + \|S(\bar{u} - \bar{u}_\lambda)\|_Y^2 \leq c\lambda. \tag{78}$$

Second case

In the second case we assume $U_{\text{ad}} = U$. Here we have no bounds for the supremum norm of the control u on the set D. Therefore, the way to derive the two inequalities for u_0, u_λ is not longer possible.

To get error estimates one has to require that the operator S is sufficiently smoothing, self-adjoint and $S + \lambda I$ is continuously invertible. Let us assume that we have

$$\|Su\|_{H^s(\Omega)} \leq c\|u\|_U$$

with $s > d/2$.

Since the control $\bar{u}$ may be unbounded in a point where the state constraint is active, we have to find a new construction of u_λ. We choose

$$u_\lambda = (\lambda I + S)^{-1} S\bar{u} \tag{79}$$

and a simple computation shows that this function satisfies the regularized state constraints. Moreover, we get

$$u_\lambda - \bar{u} = -\lambda(\lambda I + S)^{-1}\bar{u}.$$

Note that the operator $(\lambda I + S)^{-1}$ becomes unbounded for $\lambda \to 0$. The optimality condition with Lagrange multipliers yields a representation

$$\bar{u} = S^* w = Sw$$

with some $w \in (L^\infty(D))^*$. This can be written as

$$\bar{u} = S^k(S^{1-k}w)$$

and $S^{1-k}w$ is an L^2-function. An easy computation yields $0 < k < 1 - \frac{d}{2s}$. Such a property is called source representation in the theory of inverse problems. Applying standard spectral methods from that theory, we obtain

$$\|u_\lambda - \bar{u}\|_U \leq c\lambda^k \tag{80}$$

with $0 < k < 1 - \frac{d}{2s}$, see [17]. It remains to construct u_0. Here we can choose again the construction

$$u_0 = (1 - \delta)\bar{u}_\lambda + \delta\hat{u},$$

but we have to change the estimation technique since the constant K appears in the estimates in the first case. We have to avoid the term $\|\bar{u}_\lambda\|_{L^\infty(D)}$ to get an error estimate. Our aim is now to replace the $L^\infty(D)$-norm by the $L^2(D)$-norm.

Due to the form of the objective, the controls u_λ are uniformly bounded in U. Next we use

$$\|Su\|_{H^s(\Omega)} \leq c\|u\|_U$$

and obtain a uniform bound of Su_0 in $H^s(\Omega)$. This space is continuously embedded in $C^{0,s-d/2}(\bar{\Omega})$ if $s - d/2 < 1$ and we get

$$\|Su_0\|_{C^{0,\gamma}(\bar{\Omega})} \leq c$$

with $\gamma = s - d/2$. Moreover, one needs the estimate

$$\|f\|_{L^\infty(D)} \leq c\|f\|_{L^2(D)}^{\frac{\gamma}{\gamma + d/2}}$$

for the specific function $f := (y_c - Su_0)_+$, see [50], Lemma 3.2. A short computation yields

$$\frac{\gamma}{\gamma + d/2} = 1 - \frac{d}{2s}.$$

Combining these inequalities, we find

$$\|(y_c - Su_0)_+\|_{L^\infty(D)} \le c\|(y_c - Su_0)_+\|_{L^2(D)}^{1-\frac{d}{2s}}.$$

This is essential ingredient for the error estimate. Now one can proceed like in the first case.

The final estimate is given by

$$\|u_0 - \bar{u}_\lambda\|_U \le c\lambda^{1-\frac{d}{2s}}. \tag{81}$$

Together with (77) and (80) we end up by

$$\nu\|\bar{u} - \bar{u}_\lambda\|_U^2 + \|S(\bar{u} - \bar{u}_\lambda)\|_Y^2 \le c\lambda^k \tag{82}$$

with $0 < k < 1 - \frac{d}{2s}$.

Discretization error

Only the partial differential equations are discretized in the variational discretization concept. This is reflected by the discretized state equation

$$y_h = S_h u_h. \tag{83}$$

Let us define the set of admissible controls for the discretized and regularized problem

$$U_h^\lambda = \{u_h \in U_{\text{ad}} : \; \lambda u_h^\lambda + S_h u_h^\lambda \ge y_c\}.$$

We obtain for the optimal solution $\bar{u}_h^\lambda$ the following necessary and sufficient optimality condition

$$(S_h^*(S_h\bar{u}_h^\lambda - y_d) + \nu\bar{u}_h^\lambda, u_h - \bar{u}_h^\lambda) \ge 0 \text{ for all } u_h \in U_h^\lambda. \tag{84}$$

We estimate the total regularization and discretization error by means of the triangle inequality

$$\|\bar{u} - \bar{u}_h^\lambda\|_U \le \|\bar{u} - \bar{u}_\lambda\|_U + \|\bar{u}_\lambda - \bar{u}_h^\lambda\|_U. \tag{85}$$

The first term was already estimated in (78) and (82). Let us mention that a direct estimate of the total error will lead to the same result, see [45]. The estimation of the second term can be done in a similar manner as for the regularization error. Let us define

$$u_0^\lambda = (1-\delta)\bar{u}_h^\lambda + \delta\hat{u}.$$

Then we obtain

$$\begin{aligned} \lambda u_0^\lambda + S u_0^\lambda &= (1-\delta)(\lambda\bar{u}_h^\lambda + S u_h^\lambda) + \delta(\lambda\hat{u} + S\hat{u}) \\ &\ge y_c + (1-\delta)(S\bar{u}_h^\lambda - S_h\bar{u}_h^\lambda) + \delta\tau - \delta\lambda\|\hat{u}\|_{L^\infty(D)}. \end{aligned}$$

For $\lambda < \frac{\tau}{\|\hat{u}\|_{L^\infty(D)}}$ we obtain

$$\lambda u_0^\lambda + S u_0^\lambda \geq y_c + \frac{\delta\tau}{2} - (1-\delta)\|S\bar{u}_h^\lambda - S_h\bar{u}_h^\lambda\|_{L^\infty(D)}$$

which allows a choice $\delta \sim \|S\bar{u}_h^\lambda - S_h\bar{u}_h^\lambda\|_{L^\infty(D)}$. The same technique can be applied to

$$u_h^\lambda = (1-\sigma)\bar{u}_\lambda + \sigma\hat{u}.$$

Again, we find for $\lambda < \frac{\tau}{\|\hat{u}\|_{L^\infty(D)}}$

$$\lambda u_h^\lambda + S_h u_h^\lambda \geq y_c + \frac{\sigma\tau}{2} - (1-\sigma)\|S\bar{u}_\lambda - S_h\bar{u}_\lambda\|_{L^\infty(D)} - \sigma\|S\hat{u} - S_h\hat{u}\|_{L^\infty(D)}.$$

Consequently, we can choose $\sigma \sim \|S\bar{u}_\lambda - S_h\bar{u}_\lambda\|_{L^\infty(D)} + \|S\hat{u} - S_h\hat{u}\|_{L^\infty(D)}$.

The derivation of the error estimate is similar to that one of the regularization error. We start with

$$\begin{aligned}
(S^*(S\bar{u}_\lambda - y_d) + \nu\bar{u}_\lambda, u_0^\lambda - \bar{u}_\lambda) &\geq 0\\
(S_h^*(S_h\bar{u}_h^\lambda - y_d) + \nu\bar{u}_h^\lambda, u_h^\lambda - \bar{u}_h^\lambda) &\geq 0
\end{aligned}$$

and add these two inequalities. However, the different operators S and S_h leads to modifications. Let us estimate the term

$$\begin{aligned}
&(S^*(S\bar{u}_\lambda - y_d) + \nu\bar{u}_\lambda, \bar{u}_h^\lambda - \bar{u}_\lambda) + (S_h^*(S_h\bar{u}_h^\lambda - y_d) + \nu\bar{u}_h^\lambda, \bar{u}^\lambda - \bar{u}_h^\lambda)\\
&= -\nu\|\bar{u}^\lambda - \bar{u}_h^\lambda\|_U^2 + (S^*(S\bar{u}_\lambda - y_d) - S_h^*(S_h\bar{u}_h^\lambda - y_d), \bar{u}_h^\lambda - \bar{u}_\lambda)\\
&\leq -\nu\|\bar{u}^\lambda - \bar{u}_h^\lambda\|_U^2 + \|S^*y_d - S_h^*y_d\|_U\|\bar{u}^\lambda - \bar{u}_h^\lambda\|_U - \|S_h(\bar{u}^\lambda - \bar{u}_h^\lambda)\|_Y^2\\
&\quad + \|S^*S\bar{u}^\lambda - S_h^*S_h\bar{u}^\lambda\|_U\|\bar{u}^\lambda - \bar{u}_h^\lambda\|_U.
\end{aligned}$$

The final estimate is obtained by means of Young's inequality

$$\begin{aligned}
\frac{\nu}{2}\|\bar{u}^\lambda - \bar{u}_h^\lambda\|_U^2 + \|S_h(\bar{u}^\lambda - \bar{u}_h^\lambda)\|_Y^2 \leq c\big(&\|S^*S\bar{u}^\lambda - S_h^*S_h\bar{u}^\lambda\|_U^2 + \|S^*y_d - S_h^*y_d\|_U^2\\
&+ \|S\bar{u}_h^\lambda - S_h\bar{u}_h^\lambda\|_{L^\infty(D)} + \|S\bar{u}_\lambda - S_h\bar{u}_\lambda\|_{L^\infty(D)}\\
&+ \|S\hat{u} - S_h\hat{u}\|_{L^\infty(D)}\big). \qquad (86)
\end{aligned}$$

Let us specify the quantities for the problems (13), (33) for the elliptic equation

$$-\Delta y = u \text{ in } \Omega, \quad y = 0 \text{ on } \Gamma$$

where the domain Ω is a polygonal (polyhedral) domain or has smooth boundary. Because of the Dirichlet boundary condition we require $D \subset\subset \Omega$. Moreover, we assume a standard quasiuniform finite element discretization. The first two terms of (86) are of higher order

$$\begin{aligned}
\|S^*S\bar{u}^\lambda - S_h^*S_h\bar{u}^\lambda\|_U &\leq ch^2\|u^\lambda\|_{L^2(\Omega)},\\
\|S^*y_d - S_h^*y_d\|_U &\leq ch^2\|y_d\|_{L^2(\Omega)}.
\end{aligned}$$

The three remaining terms of (86) are responsible for the approximation rate

$$\begin{aligned}\|S\bar{u}_h^\lambda - S_h\bar{u}_h^\lambda\|_{L^\infty(D)} &\leq c|\log h|^2 h^2 \|\bar{u}_h^\lambda\|_{L^\infty(\Omega)}\\ \|S\bar{u}_\lambda - S_h\bar{u}_\lambda\|_{L^\infty(D)} &\leq c|\log h|^2 h^2 \|\bar{u}^\lambda\|_{L^\infty(\Omega)}\\ \|S\hat{u} - S_h\hat{u}\|_{L^\infty(D)} &\leq c|\log h|^2 h^2 \|\hat{u}\|_{L^\infty(\Omega)}.\end{aligned}$$

Combining all results we end up with

$$\|\bar{u} - \bar{u}_h^\lambda\|_U \leq c(\sqrt{\lambda} + |\log h| h)$$

for the case of additional control constraints. Consequently a choice $\lambda \sim |\log h|^2 h^2$ leads to a balanced error contribution in this case. The problem without additional control constraints is more difficult. Then, norm like $\|\bar{u}^\lambda\|_{L^\infty(\Omega)}$ are not uniformly bounded with respect to λ. In that case one has to deal with weaker error estimates where the corresponding norms of $\bar{u}^\lambda$ are uniformly bounded with respect to λ.

5.3. Full discretization

Now, we will discuss a full discretization. In an abstract setting, we replace the control space U by an arbitrary finite-dimensional control subspace U_h. The admissible discrete control set is defined by $U_{\mathrm{ad}}^h = U_{\mathrm{ad}} \cap U_h$.

Let us directly estimate the norm $\|\bar{u} - \bar{u}_h^\lambda\|_U$. We will only emphasize the key points in the estimation process. Again, one test function can be constructed by

$$u_0 = (1-\delta)\bar{u}_h^\lambda + \delta\hat{u}. \tag{87}$$

This term can be analyzed similar to the variational discretization concept. The construction of the other test function depends again on the presence of additional control constraints. If additional control constraints are given, then a choice

$$u_h^\lambda = (1-\sigma)P_h\bar{u} + \sigma P_h\hat{u} \tag{88}$$

is reasonable. The test function u_h^λ has to belong to U_h and has to satisfy the control constraints. This is reflected by the choice of a suitable projection or interpolation operator P_h. For piecewise constant controls one can choose P_h as the L^2-projection operator to U_h. A quasi-interpolation operator can be used as P_h for piecewise linear controls. Similar to the variational discretization, a choice of λ in the size of $L^\infty(D)$-error $\lambda \sim \|S\bar{u} - S_h\bar{u}\|_{L^\infty(D)}$ leads again to a balanced error contribution. For problems without control constraints we need a construction similar to (79).

Next, we will point out a specific feature of the derivation process. Let us recall the inequality (75)

$$(S^*(S\bar{u}_\lambda - y_d) + \nu\bar{u}_\lambda, u_\lambda - \bar{u}) \leq c\|u_\lambda - \bar{u}\|_U$$

for the variational discretization. Proceeding the same way for the full discretization we would get

$$(S_h^*(S_h\bar{u}_h^\lambda - y_d) + \nu\bar{u}_h^\lambda, u_h^\lambda - \bar{u}) \leq c\|u_h^\lambda - \bar{u}\|_U.$$

Using (88), we find

$$\|u_h^\lambda - \bar{u}\|_U \leq \sigma\|P_h\hat{u} - \bar{u}\|_U + (1-\sigma)\|P_h\bar{u} - \bar{u}\|_U.$$

The first term becomes small because of the factor σ. The second term has bad approximation properties because of the low regularity properties of the control.

A modification of the estimation process yields

$$(S_h^*(S_h\bar{u}_h^\lambda - y_d) + \nu\bar{u}_h^\lambda, P_h\bar{u} - \bar{u}) = (S_h^*(S_h\bar{u}_h^\lambda - y_d), P_h\bar{u} - \bar{u}) + \nu(\bar{u}_h^\lambda, P_h\bar{u} - \bar{u}).$$

We find for the first term

$$\begin{aligned}&(S_h^*(S_h\bar{u}_h^\lambda - y_d), P_h\bar{u} - \bar{u})\\ &\qquad \le c(\|SP_h\bar{u} - S\bar{u}\|_Y + \|S^*S\bar{u}_h^\lambda - S_h^*S_h\bar{u}_h^\lambda\|_U + \|S^*y_d - S_h^*y_d\|_U)\end{aligned}$$

and all terms have good approximation properties. Let us assume that P_h is the L^2-projection to U_h. By orthogonality we get

$$\nu(\bar{u}_h^\lambda, P_h\bar{u} - \bar{u}) = 0.$$

and the all problems are solved. However, this choice is possible only for spaces of piecewise constant functions if control constraints are given. For piecewise linear functions a quasiinterpolation operator yields the desired results, see [29]. In the final result the $L^\infty(D)$-error dominates again the approximation behavior. For the complete derivation of the results we refer to [18].

5.4. A short note to nonlinear state equations

We already addressed the main difficulties in Section 3. Let us mention that the approach of the last subsections cannot be used, since the admissible sets U^0, U^λ, and U_h^λ are not convex.

There are two techniques available to tackle this problem. A first approach works mainly with objective values. Again feasible points were constructed. Then, the difference of objective values is estimated. The desired error estimate can be derived by means of a local quadratic growth condition. This technique was used in [49].

Local quadratic growth is usually shown by a second-order sufficient optimality condition. If the dual variables of the unregularized problems are unique, then the second-order sufficient optimality conditions are also satisfied for regularized problems. This is true for the Moreau-Yosida approach and for the virtual control concept. For both techniques one has uniqueness of dual variables for the regularized problems by construction. This is not the case for the Lavrentiev regularization. Dual variables are not unique for the Lavrentiev regularization if control constraints and mixed constraints are active simultaneously. Thus, separation of strongly active sets is needed to get the corresponding local uniqueness result, see [63]. These papers contain regularization error results of the form (78) for nonlinear problems.

Another possible technique would be to work with the complete optimality system with Lagrange multipliers. However, this approach was used only for state constrained linear-quadratic problems until now.

Our motivation for the regularization of state constrained problems was that the resulting problems can be solved efficiently. This statement is also correct for nonlinear problems. Main issues for a good performance are local convexity properties of the regularized problems and local uniqueness of stationary points (including dual variables). This can be guaranteed by second-order sufficient optimality conditions.

6. A brief discussion of further literature

6.1. Literature related to control constraints

There are many contributions to finite element analysis for elliptic control problems with constraints on the controls. For an introduction to the basic techniques we refer to the book [71] of Tröltzsch. Falk [30], and Geveci [32] present finite element analysis for piecewise constant approximations of the controls. For semilinear state equations Arada, Casas, and Tröltzsch in [4] present a finite element analysis for piecewise constant discrete controls. Among other things they prove that the sequence $(u_h)_h$ of discrete controls contains a subsequence converging to a solution u of the continuous optimal control problem. Assuming certain second-order sufficient conditions for u they are also able to prove optimal error estimates of the form

$$\|u - u_h\| = \mathcal{O}(h) \text{ and } \|u - u_h\|_\infty = \mathcal{O}(h^\lambda),$$

with $\lambda = 1$ for triangulations of non-negative type, and $\lambda = 1/2$ in the general case. In [15] these results are extended in that Casas and Tröltzsch prove that every nonsingular local solution u (i.e., a solution satisfying a second-order sufficient condition) locally can be approximated by a sequence $(u_h)_h$ of discrete controls, also satisfying these error estimates. There are only few results considering uniform estimates. For piecewise linear controls in the presence of control constraints Meyer and Rösch in [61] for two-dimensional bounded domains with $C^{1,1}$-boundary prove the estimate

$$\|u - u_h\|_\infty = \mathcal{O}(h),$$

which seems to be optimal with regard to numerical results reported in [46, Chap. 3], and which is one order less than the approximation order obtained with variational discretization. The same authors in [60] propose post processing for elliptic optimal control problems which in a preliminary step computes a piecewise constant optimal control $\bar{u}$ and with its help a projected control u^P through $u^P = P_{U_{\text{ad}}}(-\frac{1}{\alpha}B^* p_h(\bar{u}))$ which then satisfies

$$\|u - u^P\| = \mathcal{O}(h^2).$$

Casas, Mateos and Tröltzsch in [13] present numerical analysis for Neumann boundary control of semilinear elliptic equations and prove the estimate

$$\|u - u_h\|_{L^2(\Gamma)} = \mathcal{O}(h)$$

for piecewise constant control approximations. In [12] Casas and Mateos extend these investigations to piecewise linear, continuous control approximations, and also to variational discrete controls. Requiring a second-order sufficient conditions at the continuous solution u they are able to prove the estimates

$$\|u - u_h\|_{L^2(\Gamma)} = o(h), \quad \text{and} \quad \|u - \bar{u}_h\|_{L^\infty(\Gamma)} = o(h^{1/2}),$$

for a general class of control problems, where u_h denotes the piecewise linear, continuous approximation to u. For variational discrete controls u_h^v they show the better estimate

$$\|u - u_h^v\|_{L^2(\Gamma)} = \mathcal{O}(h^{\frac{3}{2}-\epsilon}) \quad (\epsilon > 0).$$

Furthermore, they improve their results for objectives which are quadratic w.r.t. the control and obtain

$$\|u - u_h\|_{L^2(\Gamma)} = \mathcal{O}(h^{\frac{3}{2}}), \text{ and } \|u - u_h\|_{L^\infty(\Gamma)} = \mathcal{O}(h).$$

The dependence of the approximation with respect to the largest angle ω of a polygonal domain is studied in Mateos and Rösch [54]. This allows to obtain error estimates of the form

$$\|u - u_h\|_{L^2(\Gamma)} = \mathcal{O}(h^\kappa)$$

with $\kappa > 3/2$ for convex domains ($\omega < \pi$) and $\kappa > 1$ for concave domains ($\omega > \pi$).

Let us finally recall the contribution [14] of Casas and Raymond to numerical analysis of Dirichlet boundary control, who for two-dimensional convex polygonal domains prove the optimal estimate

$$\|u - u_h\|_{L^2(\Gamma)} \le Ch^{1-1/q},$$

where u_h denotes the optimal discrete boundary control which they sought in the space of piecewise linear, continuous finite elements on Γ. Here $q \ge 2$ depends on the smallest angle of the boundary polygon. May, Rannacher and Vexler study Dirichlet boundary control without control constraints in [55]. They also consider two-dimensional convex polygonal domains and among other things provide optimal error estimates in weaker norms. In particular they address

$$\|u - u_h\|_{H^{-1}(\Gamma)} + \|y - y_h\|_{H^{-1/2}(\Omega)} \sim h^{2-2/q}.$$

Vexler in [74] for $U_{\text{ad}} = \{u \in \mathbb{R}^n; a \le u \le b\}$ and $Bu := \sum_{i=1}^n u_i f_i$ with $f_i \in H^{5/2}(\Gamma)$ provides finite element analysis for Dirichlet boundary control in bounded, two-dimensional polygonal domains. Among other things he in [74, Theorem 3.4] shows that

$$|u - u_h| \le Ch^2.$$

Error analysis for general two- and three-dimensional curved domains is presented by Deckelnick, Günther and Hinze in [27]. They prove the error bound

$$\|u - \tilde{u}_h\|_{0,\Gamma} + \|y - \tilde{y}_h\|_{0,\Omega} \le Ch\sqrt{|\log h|},$$

and for piecewise $O(h^2)$ regular triangulations of two-dimensional domains the superconvergence result

$$\|u - \tilde{u}_h\|_{0,\Gamma} + \|y - \tilde{y}_h\|_{0,\Omega} \le Ch^{3/2}.$$

Let us shortly comment on a priori error estimates for non-uniform grids. Mesh grading for reentrant corners was investigated by Apel, Rösch, and Winkler

[3] and Apel Rösch, and Sirch [2] for optimal approximation error in the L^2-norm and in the L^∞-norm, respectively. A detailed overview on results with non-uniform grids can be found in the paper of Apel and Sirch inside this book.

6.2. Literature for (control and) state constraints

To the authors knowledge only few attempts have been made to develop a finite element analysis for elliptic control problems in the presence of control and state constraints. In [10] Casas proves convergence of finite element approximations to optimal control problems for semi-linear elliptic equations with finitely many state constraints. Casas and Mateos extend these results in [11] to a less regular setting for the states and prove convergence of finite element approximations to semi-linear distributed and boundary control problems. In [58] Meyer considers a fully discrete strategy to approximate an elliptic control problem with pointwise state and control constraints. He obtains the approximation order

$$\|\bar{u} - \bar{u}_h\| + \|\bar{y} - \bar{y}_h\|_{H^1} = \mathcal{O}(h^{2-d/2-\epsilon}) \quad (\epsilon > 0),$$

where d denotes the spatial dimension. His results confirm those obtained by the Deckelnick and Hinze in [21] for the purely state constrained case, and are in accordance with Theorem 4.7. Meyer also considers variational discretization and in the presence of L^∞ bounds on the controls shows

$$\|\bar{u} - \bar{u}_h\| + \|\bar{y} - \bar{y}_h\|_{H^1} = \mathcal{O}(h^{1-\epsilon}|\log h|) \quad (\epsilon > 0),$$

which is a result of a similar quality as that given in the third part of Theorem 4.7.

Let us comment also on further approaches that tackle optimization problems for pdes with control and state constraints. A *Lavrentiev-type regularization* of problem (13) is investigated by Meyer, Rösch and Tröltzsch in [62]. In this approach the state constraint $y \le b$ in (13) is replaced by the mixed constraint $\epsilon u + y \le b$, with $\epsilon > 0$ denoting a regularization parameter. It turns out that the associated Lagrange multiplier μ_ϵ belongs to $L^2(\Omega)$. Numerical analysis for this approach with emphasis on the coupling of gridsize and regularization parameter ϵ is presented by Hinze and Meyer in [45]. The resulting optimization problems are solved either by interior-point methods or primal-dual active set strategies, compare the work [59] by Meyer, Prüfert and Tröltzsch.

Hintermüller and Kunisch in [40, 41] consider the Moreau-Yosida relaxation approach to problem classes containing (13). In this approach the state constraint is relaxed in that it is dropped and a L^2 regularization term of the form $\frac{1}{2\gamma}\int_\Omega |\max(0, \gamma\mathcal{G}(Bu))|^2$ is added to the cost functional instead, where γ denotes the relaxation parameter. Numerical analysis for this approach with emphasis on the coupling of gridsize and relaxation parameter γ is presented by Hintermüller and Hinze in [38].

Schiela in [70] chooses a different way to relax state constraints in considering barrier functionals of the form $-\mu\int_\Omega \log(-\mathcal{G}(Bu))dx$ which penalize the state constraints. In [47] he together with Hinze presents numerical analysis for this approach with emphasis on the coupling of gridsize and barrier parameter μ.

6.3. Gradient constraints

In many practical applications pointwise constraints on the gradient of the state are required, for example if one aims on avoiding large van Mises stresses, see [26] for a discussion. For elliptic optimal control problems with these kind of constraints Deckelnick, Günter and Hinze in [26] propose a mixed finite element approximation for the state combined with variational discretization and prove the error estimate

$$\|u - u_h\| + \|y - y_h\| \le Ch^{\frac{1}{2}}|\log h|^{\frac{1}{2}},$$

which is valid for two- and three-dimensional spatial domains. The classical finite element approach using piecewise linear, continuous approximations for the states is investigated by Günter and Hinze in [35]. They are able to show the estimates

$$\|y - y_h\| \le Ch^{\frac{1}{2}(1-\frac{d}{r})}, \text{ and } \|u - u_h\|_{L^r} \le Ch^{\frac{1}{r}(1-\frac{d}{r})},$$

which are valid for variational discretization as well as for piecewise constant control approximations. Here, $d=2,3$ denotes the space dimension, and $r>d$ the integration order of the L^r-control penalization term in the cost functional. Ortner and Wollner in [65] for the same discretization approach obtain similar results adapting the proof technique of [21] to investigate the numerical approximation of elliptic optimal control problems with pointwise bounds on the gradient of the state.

6.4. Literature on control of time-dependent problems

In the literature only few contributions to numerical analysis for control problems with time-dependent pdes can be found. For unconstrained linear quadratic control problems with the time-dependent Stokes equation in two- and three-dimensional domains Deckelnick and Hinze in [19] prove the error bound

$$\|u - u_{h,\sigma}\|_{L^2((0,T)\times\Omega)} = \mathcal{O}(\sigma + h^2).$$

Here and below σ denotes the discretization parameter for the controls. They use a fully implicit variant of Eulers method for the time discretization which is equivalent to the $dG(0)$ approximation. In space the use Taylor-Hood finite elements. Using [19, (3.1),(3.6)] this estimate directly extends also to the control constrained case.

Boundary control for the heat equation in one spatial dimension is considered by Malanowski in [53] with piecewise constant, and by Rösch in [66] with piecewise linear, continuous control approximations. Requiring strict complementarity for the continuous solution Rösch is able to prove the estimate

$$\|u - u_\sigma\| = \mathcal{O}(\sigma^{3/2}).$$

Malanowski proves the estimate

$$\|u - u_{h,\sigma}\|_{L^2((0,T)\times\Omega)} = \mathcal{O}(\sigma + h),$$

where h denotes the discretization parameter for the space discretizations.

In a recent work [56, 57] Meidner and Vexler present extensive research on control problems governed by parabolic equations and their discrete approximation based on $dG(0)$ in time and finite element in space, where they consider the heat equation as mathematical model on a two- or three-dimensional convex polygonal domain. For variational discretization of [44] they prove the estimate

$$\|u - u_{h,\sigma}\|_{L^2((0,T)\times\Omega)} = \mathcal{O}(\sigma + h^2),$$

which under the assumption of strict complementarity of the continuous solution also holds for post-processing [60].

For control problems with nonlinear time-dependent equations one only finds few contributions in the literature. In [36, 37] Gunzburger and Manservisi present a numerical approach to control of the instationary Navier-Stokes equations (3) using the first discretize then optimize approach. The first optimize then discretize approach applied to the same problem class is discussed by Hinze in [43]. Deckelnick and Hinze provide numerical analysis for a general class of control problems with the instationary Navier-Stokes system (3) in [20]. Among other things they prove existence and local uniqueness of variational discrete controls in neighborhoods of nonsingular continuous solutions, and for semi-discretization in space with Taylor-Hood finite elements provide the error estimate

$$\int_0^T \|u - u_h\|_U^2 \, dt \ \le Ch^4.$$

Here, u, u_h denote the continuous and variational discrete optimal control, respectively. This result also carries over to the case of control constraints under the assumptions made in Section 3.

For problems with state constraints only a few contributions are known. Deckelnick and Hinze in [24] investigate variational discretization for parabolic control problems in the presence of state constraints. Among other things they prove an error bound

$$\alpha\|u - u_h\|^2 + \|y - y_h\|^2 \le C \begin{cases} h\sqrt{|\log h|}, & (d = 2) \\ \sqrt{h}, & (d = 3). \end{cases}$$

under the natural regularity assumption $y = \mathcal{G}(Bu) \in W = \{v \in C^0([0,T]; H^2),$ $v_t \in L^2(H^1)\}$ with time stepping $\delta t \sim h^2$. Exploiting results of Nochetto and Verdi [64] in the case $d = 2$ and $Bu \in L^\infty(\Omega_T)$ it seems possible to us that an error bound of the form

$$\alpha\|u - u_h\|^2 + \|y - y_h\|^2 \lesssim C\big(h^2 + \tau\big)$$

can be proved.

Very recently Giles and S. Ulbrich [34] considered the numerical approximation of optimal control problems for scalar conservation laws and provided a detailed numerical analysis for the discrete treatment of control problem.

References

[1] Agmon, S., Douglis, A., Nirenberg, L.: Estimates near the boundary for solutions of elliptic partial differential equations satisfying general boundary conditions. Comm. Pure Appl. Math., **12**, 623–727 (1959).

[2] Apel, T., Rösch, A., Sirch, D.: L^∞-Error Estimates on Graded Meshes with Application to Optimal Control. SIAM Journal Control and Optimization, 48(3):1771–1796, 2009.

[3] Apel, T., Rösch, A., Winkler, G: Optimal control in non-convex domains: a priori discretization error estimates. Calcolo, 44(3), 137–158 (2007).

[4] Arada, N., Casas, E., Tröltzsch, F.: Error estimates for the numerical approximation of a semilinear elliptic control problem. Computational Optimization and Applications **23**, 201–229 (2002).

[5] Bergounioux, M., Ito, K., and Kunisch, K.: Primal-dual strategy for constrained optimal control problems. *SIAM J. Control and Optimization*, 37:1176–1194, 1999.

[6] Bergounioux, M. and Kunisch, K.: Primal-dual strategy for state-constrained optimal control problems. *Computational Optimization and Applications*, 22:193–224, 2002.

[7] Casas, E.: L^2 estimates for the finite element method for the Dirichlet problem with singular data. Numer. Math. **47**, 627–632 (1985).

[8] Casas, E.: Control of an elliptic problem with pointwise state constraints. SIAM J. Cont. Optim. **4**, 1309–1322 (1986).

[9] Casas, E.: Boundary control of semilinear elliptic equations with pointwise state constraints. SIAM J. Cont. Optim. **31**, 993–1006 (1993).

[10] Casas, E.: Error Estimates for the Numerical Approximation of Semilinear Elliptic Control Problems with Finitely Many State Constraints. ESAIM, Control Optim. Calc. Var. **8**, 345–374 (2002).

[11] Casas, E., Mateos, M.. Uniform convergence of the FEM. Applications to state constrained control problems. Comp. Appl. Math. **21**, (2002).

[12] Casas, E., Mateos, M.: Error Estimates for the Numerical Approximation of Neumann Control Problems. Comp. Appl. Math. 39(3):265–295, 2008.

[13] Casas, E., Mateos, M., Tröltzsch, F.: Error estimates for the numerical approximation of boundary semilinear elliptic control problems. Comput. Optim. Appl. **31**, 193–219 (2005).

[14] Casas, E., Raymond, J.P.: Error estimates for the numerical approximation of Dirichlet Boundary control for semilinear elliptic equations. SIAM J. Cont. Optim. **45**, 1586–1611 (2006).

[15] Casas, E., Tröltzsch, F.: Error estimates for the finite element approximation of a semilinear elliptic control problems. Contr. Cybern. **31**, 695–712 (2005).

[16] Casas, E., Mateos, M., and Tröltzsch, F.: Error estimates for the numerical approximation of boundary semilinear elliptic control problems. *Comput. Optim. Appl.*, 31(2):193–219, 2005.

[17] Cherednichenko, S., Krumbiegel, K., and Rösch, A.: Error estimates for the Lavrentiev regularization of elliptic optimal control problems. *Inverse Problems*, 24(5):055003, 2008.

[18] Cherednichenko, S. and Rösch, A.: Error estimates for the discretization of elliptic control problems with pointwise control and state constraints. *Computational Optimization and Applications*, 44:27–55, 2009.

[19] Deckelnick, K., Hinze, M.: Error estimates in space and time for tracking-type control of the instationary Stokes system. ISNM **143**, 87–103 (2002)

[20] Deckelnick, K., Hinze, M.: Semidiscretization And Error Estimates For Distributed Control Of The Instationary Navier-Stokes Equations. Numer. Math. **97**, 297–320 (2004).

[21] Deckelnick, K., Hinze, M.: Convergence of a finite element approximation to a state constrained elliptic control problem. SIAM J. Numer. Anal. **45**, 1937–1953 (2007).

[22] Deckelnick, K., Hinze, M.: A finite element approximation to elliptic control problems in the presence of control and state constraints. Hamburger Beiträge zur Angewandten Mathematik HBAM2007-01 (2007).

[23] Deckelnick, K., Hinze, M.: Numerical analysis of a control and state constrained elliptic control problem with piecewise constant control approximations. Proceedings of the ENUMATH (2007).

[24] Deckelnick, K., Hinze,M: Variational Discretization of Parabolic Control Problems in the Presence of Pointwise State Constraints. Journal of Computational Mathematics 29:1–16, 2011.

[25] Deckelnick, K., Hinze, M.: A note on the approximation of elliptic control problems with bang-bang controls. Comput. Optim. Appl. DOI: 10.1007/s10589-010-9365-z (2010).

[26] Deckelnick, K., Günther, A., Hinze, M.: Finite element approximations of elliptic control problems with constraints on the gradient. Numer. Math. 111:335-350 (2009).

[27] Deckelnick, K., Günther, A, Hinze, M.: Finite Element Approximation of Dirichlet Boundary Control for Elliptic PDEs on Two- and Three-Dimensional Curved Domains. SIAM J. Cont. Optim. 48:2798–2819 (2009).

[28] Deckelnick, K., Hinze, M., Matthes, U., Schiela, A.: Approximation schemes for constrained optimal control with nonlinear pdes. In preparation (2010).

[29] de los Reyes, J.C., Meyer, C., and Vexler, B.: Finite element error analysis for state-constrained optimal control of the Stokes equations. *Control and Cybernetics*, 37(2):251–284, 2008.

[30] Falk, R.: Approximation of a class of optimal control problems with order of convergence estimates. J. Math. Anal. Appl. **44**, 28–47 (19973).

[31] Gastaldi, L., Nochetto, R.H.: On L^∞-accuracy of mixed finite element methods for second-order elliptic problems. Mat. Apl. Comput. **7**, 13–39 (1988).

[32] Geveci, T.: On the approximation of the solution of an optimal control problem governed by an elliptic equation. Math. Model. Numer. Anal. **13**, 313–328 (1979).

[33] Gilbarg, D., Trudinger, N.S.: Elliptic partial differential equations of second order (2nd ed.). Springer (1983).

[34] Giles, M., Ulbrich, S.: Convergence of linearised and adjoint approximations for discontinuous solutions of conservation laws. Part 1: Siam J. Numer. Anal. 48:882–904, Part 2: Siam J. Numer. Anal. 48:905–921 (2010).

[35] Günther, A., Hinze, M.: Elliptic Control Problems with Gradient Constraints – Variational Discrete Versus Piecewise Constant Controls, Comput. Optim. Appl. 49:549–566, 2011.

[36] Gunzburger, M.D, Manservisi, S.: Analysis and approximation of the velocity tracking problem for Navier-Stokes flows with distributed control. Siam J. Numer. Anal. **37**, 1481–1512, 2000.

[37] Gunzburger, M.D., Manservisi, S.: The velocity tracking problem for Navier-Stokes flows with boundary controls. Siam J. Control and Optimization **39**, 594–634 (2000).

[38] Hintermüller, M. and Hinze, M.: Moreau-yosida regularization in state constrained elliptic control problems: error estimates and parameter adjustment. *SIAM J. Numerical Analysis*, 47:1666–1683, 2009.

[39] Hintermüller, M., Ito, K., and Kunisch, K.: The primal-dual active set strategy as a semismooth Newton method. *SIAM J. Optimization*, 13:865–888, 2003.

[40] Hintermüller, M. and Kunisch, K.: Path-following methods for a class of constrained minimization problems in function space. *SIAM J. Optimization*, 17:159–187, 2006.

[41] Hintermüller, M., Kunisch, K.: Feasible and non-interior path following in constrained minimization with low multiplier regularity. Report, Universität Graz (2005).

[42] Hintermüller, M., Ulbrich, M.: A mesh independence result for semismooth Newton methods. *Mathematical Programming* 101:151–184, 2004

[43] Hinze, M.: Optimal and instantaneous control of the instationary Navier-Stokes equations. Habilitationsschrift, Fachbereich Mathematik, Technische Universität Berlin (2000).

[44] Hinze, M.: A variational discretization concept in control constrained optimization: the linear-quadratic case. Comp. Optim. Appl. 30:45–63, 2005.

[45] Hinze, M. and Meyer, C.: Variational discretization of Lavrentiev-regularized state constrained elliptic optimal control problems. Comput. Optim. Appl. 46:487–510, 2010.

[46] Hinze, M., Pinnau, R., Ulbrich, M., and Ulbrich, S.: *Optimization with PDE constraints* MMTA 23, Springer 2009.

[47] Hinze, M. and Schiela, A.: Discretization of interior point methods for state constrained elliptic optimal control problems: optimal error estimates and parameter adjustment. *Computational Optimization and Applications*, DOI: 10.1007/s10589-009-9278-x, 2010.

[48] Krumbiegel, K., Meyer, C., and Rösch. A.: A priori error analysis for linear quadratic elliptic Neumann boundary control problems with control and state constraints. SIAM Journal Control and Optimization 48(8):5108–5142, 2010.

[49] Krumbiegel, K., Neitzel, I., and Rösch, A.: Regularization for semilinear elliptic optimal control problems with pointwise state and control constraints. Computational Optimization and Applications, DOI 10.1007/s10589-010-9357-z.

[50] Krumbiegel, K. and Rösch, A.: On the regularization error of state constrained Neumann control problems. *Control and Cybernetics*, 37:369–392, 2008.

[51] Krumbiegel, K. and Rösch, A.: A virtual control concept for state constrained optimal control problems. *Computational Optimization and Applications*, 43:213–233, 2009.

[52] Kunisch, K. and Rösch, A.: Primal-dual active set strategy for a general class of constrained optimal control problems. *SIAM J. Optimization*, 13(2):321–334, 2002.

[53] Malanowski, K.: Convergence of approximations vs. regularity of solutions for convex, control constrained optimal-control problems. Appl. Math. Optim. **8** 69–95 (1981).

[54] Mateos, M. and Rösch, A. On saturation effects in the Neumann boundary control of elliptic optimal control problems. Computational Optimization and Applications, 49:359–378, 2011.

[55] May, S., Rannacher, R., Vexler, B.: A priori error analysis for the finite element approximation of elliptic Dirichlet boundary control problems. Proceedings of ENUMATH 2007, Graz (2008).

[56] Meidner, D., Vexler, B.: A Priori Error Estimates for Space-Time Finite Element Discretization of Parabolic Optimal Control Problems. Part I: Problems without Control Constraints. SIAM Journal on Control and Optimization **47**, 1150–1177 (2008).

[57] Meidner, D., Vexler, B.: A Priori Error Estimates for Space-Time Finite Element Discretization of Parabolic Optimal Control Problems. Part II: Problems with Control Constraints. SIAM Journal on Control and Optimization **47**, 1301–1329 (2008).

[58] Meyer, C.: Error estimates for the finite element approximation of an elliptic control problem with pointwise constraints on the state and the control, WIAS Preprint 1159 (2006).

[59] Meyer, C., Prüfert, U., and Tröltzsch, F.: On two numerical methods for state-constrained elliptic control problems. *Optimization Methods and Software*, 22:871–899, 2007.

[60] Meyer, C., Rösch, A.: Superconvergence properties of optimal control problems. SIAM J. Control Optim. **43**, 970–985 (2004).

[61] Meyer, C., Rösch, A.: L^∞-estimates for approximated optimal control problems. SIAM J. Control Optim. **44**, 1636–1649 (2005).

[62] Meyer, C., Rösch, A., and Tröltzsch, F.: Optimal control of PDEs with regularized pointwise state constraints. *Computational Optimization and Applications*, 33(2-3):209–228, 2006.

[63] Neitzel, I. and Tröltzsch, F.: On regularization methods for the numerical solution of parabolic control problems with pointwise state constraints. *Control and Cybernetics*, 37:1013–1043, 2008.

[64] Nochetto, R.H., Verdi, C.: *Convergence past singularities for a fully discrete approximation of curvature-driven interfaces*, SIAM J. Numer. Anal. **34**, 490–512 (1997).

[65] Ortner, C., Wollner, W.: *A priori error estimates for optimal control problems with pointwise constraints on the gradient of the state.* DFG Schwerpunktprogramm 1253, preprint No. SPP1253-071, (2009).

[66] Rösch, A.: Error estimates for parabolic optimal control problems with control constraints, Zeitschrift für Analysis und ihre Anwendungen ZAA **23**, 353–376 (2004).

[67] Rösch, A. and Wachsmuth, D.: *How to check numerically the sufficient optimality conditions for infinite-dimensional optimization problems*, in Optimal control of coupled systems of partial differential equations, eds. Kunisch, K., Leugering, G., Sprekels, J., and Tröltzsch, F., pages 297–318, Birkhäuser 2009.

[68] Rösch, A. and Wachsmuth, D.: *Numerical verification of optimality conditions*, SIAM Journal Control and Optimization, 47, 5 (2008), 2557–2581.

[69] Schatz, A.H.: Pointwise error estimates and asymptotic error expansion inequalities for the finite element method on irregular grids. I: Global estimates. Math. Comput. **67**(223), 877–899 (1998).

[70] Schiela, A.: Barrier methods for optimal control problems with state constraints. *SIAM J. Optimization*, 20:1002–1031, 2009.

[71] Tröltzsch, F.: Optimale Steuerung mit partiellen Differentialgleichungen, (2005).

[72] Tröltzsch, F. and Yousept, I.: A regularization method for the numerical solution of elliptic boundary control problems with pointwise state constraints. *Computational Optimization and Applications*, 42:43–66, 2009.

[73] Tröltzsch, F., and Yousept, I.: Source representation strategy for optimal boundary control problems with state constraints. *Journal of Analysis and its Applications*, 28:189–203, 2009.

[74] Vexler, B.: Finite element approximation of elliptic Dirichlet optimal control problems. Numer. Funct. Anal. Optim. **28**, 957–975 (2007).

Michael Hinze
Department of Mathematics
University of Hamburg
Bundesstrasse 55
D-20146 Hamburg, Germany
e-mail: `michael.hinze@uni-hamburg.de`

Arnd Rösch
Fakultät für Mathematik
Universität Duisburg-Essen
Forsthausweg 2
D-47057 Duisburg, Germany
e-mail: `arnd.roesch@uni-due.de`

International Series of Numerical Mathematics, Vol. 160, 431–443

A Posteriori Error Estimators for Control Constrained Optimal Control Problems

Kristina Kohls, Arnd Rösch and Kunibert G. Siebert

Abstract. In this note we present a framework for the a posteriori error analysis of control constrained optimal control problems with linear PDE constraints. It is solely based on reliable and efficient error estimators for the corresponding *linear* state and adjoint *equations*. We show that the sum of these estimators gives a reliable and efficient estimator for the optimal control problem.

Mathematics Subject Classification (2000). 49J20, 65N30, 65N15.

Keywords. Finite elements, a posteriori error estimators, optimal control, control constraints.

1. Introduction

The a priori error analysis of optimal control problems is a well-established mathematical field since the 1970s starting with the publications of Falk [2] and Geveci [4]. More references are listed in the survey by Hinze and Rösch in this book. In contrast to this, a posteriori error analysis is quite young and was initiated by Liu and Yan at the beginning of this century [10]. Compared to the vast amount of literature about a posteriori error estimators for linear problems the existing results for optimal control problems are rather limited. For sake of brevity we only refer to [5, 11] among others. We would also like to mention the dual weighted residual method as an alternative approach; compare with [1].

Compared to linear problems the a posteriori error analysis for control constrained optimal control problems gets inevitably more technical and complicated due to its intrinsic nonlinear character. Looking into existing papers on error estimation for control problems one gets the following impression. On the one hand, many arguments are similar or sometimes even exactly the same as in the linear case. On the other hand, any new PDE constraint seems to require a new analysis without directly employing existing results from the linear theory.

In this note we derive a framework that allows for a unified a posteriori error analysis of control constrained optimal control problems for a general class of PDE constraints, error estimators and objective functionals. We use the variational discretization by Hinze [7] and we only assume to have access to reliable and efficient error estimators for the corresponding *linear* state and adjoint *equations*. We show that the error of the optimal control problem is bounded by the sum of the errors in the state and the adjoint equations. Reliability of the estimators for the linear problems then in turn implies that their sum gives an upper bound for the error of the control problem. The efficiency of the two estimators yields a lower bound of their sum for the error of the control problem up to oscillation.

Summarizing, the a posteriori analysis for a specific optimal control problem reduces to deriving reliable and efficient estimators for the state and the adjoint equations.

2. The Continuous and the discretized problem

We first state the abstract control constrained optimal control problem together with assumptions on the structure of the objective functional, spaces, and bilinear form for the constraint. We then turn to existence and uniqueness and to the discretization by conforming finite elements. We conclude the section with two important applications of the abstract framework.

2.1. The optimal control problem

Let $\mathbb{Y}$ be an L^2-based Hilbert space with norm $\|\cdot\|_{\mathbb{Y}}$ for the state over a bounded, polygonal domain $\Omega \subset \mathbb{R}^d$ and denote by $\mathbb{Y}^*$ its (topological) dual space. Let $\mathbb{U}$ be the control space that is also an L^2-based Hilbert space with scalar product $\langle \cdot, \cdot \rangle_{\mathbb{U}}$. Suppose $\mathbb{Y} \hookrightarrow \mathbb{U} \hookrightarrow \mathbb{Y}^*$ in the sense that $v \in \mathbb{Y}$ implies $v \in \mathbb{U}$ with $\|v\|_{\mathbb{U}} \le C\|v\|_{\mathbb{Y}}$ and $u \in \mathbb{U}$ implies $u \in \mathbb{Y}^*$ by

$$\langle u, v \rangle = \langle u, v \rangle_{\mathbb{Y}^* \times \mathbb{Y}} := \langle u, v \rangle_{\mathbb{U}} \qquad \forall v \in \mathbb{Y}.$$

We set

$$\mathcal{J}[u, y] := \psi(y) + \tfrac{\alpha}{2}\|u\|_{\mathbb{U}}^2,$$

where $\psi \colon \mathbb{Y} \to \mathbb{R}$ is a quadratic and convex objective functional. We suppose that for any $z \in \mathbb{Y}$ the Fréchet derivative $\psi'(z)$ belongs to $\mathbb{Y}^*$. Let $\mathcal{B} \colon \mathbb{Y} \times \mathbb{Y} \to \mathbb{R}$ be a continuous bilinear form satisfying the inf-sup condition

$$\inf_{\substack{v \in \mathbb{Y} \\ \|v\|_{\mathbb{Y}}=1}} \sup_{\substack{w \in \mathbb{Y} \\ \|w\|_{\mathbb{Y}}=1}} \mathcal{B}[v, w] = \inf_{\substack{w \in \mathbb{Y} \\ \|w\|_{\mathbb{Y}}=1}} \sup_{\substack{v \in \mathbb{Y} \\ \|v\|_{\mathbb{Y}}=1}} \mathcal{B}[v, w] = \gamma > 0. \tag{2.1}$$

Finally, let $\emptyset \neq \mathbb{U}^{\mathrm{ad}} \subset \mathbb{U}$ be a convex set of admissible controls and $\alpha > 0$ be a cost parameter. We then consider optimal control problems of the form

$$\min_{u \in \mathbb{U}^{\mathrm{ad}}} \mathcal{J}[u, y] \qquad \text{subject to} \qquad \mathcal{B}[y, w] = \langle u, w \rangle_{\mathbb{U}} \quad \forall w \in \mathbb{Y}. \tag{2.2}$$

We have a particular interest in bilinear forms that arise from the variational formulation of partial differential equations. In §2.4 below we give examples of PDEs with corresponding Hilbert spaces $\mathbb{Y}$ and bilinear forms $\mathcal{B}$, control spaces $\mathbb{U}$ and sets of admissible controls $\mathbb{U}^{\text{ad}}$, as well as objective functionals ψ for the desired state.

2.2. Existence and uniqueness

We show that (2.2) admits a unique solution. The inf-sup condition (2.1) is equivalent to solvability of the state equation

$$y \in \mathbb{Y}: \qquad \mathcal{B}[y,\, w] = \langle f,\, w\rangle \qquad \forall\, w \in \mathbb{Y} \tag{2.3a}$$

for any $f \in \mathbb{Y}^*$, as well as solvability of the adjoint equation

$$p \in \mathbb{Y}: \qquad \mathcal{B}[v,\, p] = \langle g,\, v\rangle \qquad \forall\, v \in \mathbb{Y} \tag{2.3b}$$

for any $g \in \mathbb{Y}^*$ [13, Theorem 5.3]; compare also [14, §2.3] for a more detailed discussion of the inf-sup theory. In particular this implies the existence of solution operators $S, S^*\colon \mathbb{Y}^* \to \mathbb{Y}$ of the state and adjoint equation. This means for given $f, g \in \mathbb{Y}^*$ the unique solutions of (2.3a) and (2.3b) are $y = S(f)$, respectively $p = S^*(g)$. Moreover, there holds

$$\|S\|_{L(\mathbb{Y}^*,\mathbb{Y})} \leq \gamma^{-1} \qquad \text{and} \qquad \|S^*\|_{L(\mathbb{Y}^*,\mathbb{Y})} \leq \gamma^{-1}.$$

Utilizing the solution operator S, we introduce the reduced cost functional $\hat{\mathcal{J}}[u] := \mathcal{J}[u,\, Su]$. The assumptions on ψ, $\alpha > 0$, and the inf-sup condition (2.1) imply that $\hat{\mathcal{J}}[u]$ is bounded from below and strictly convex. Moreover, if $\mathbb{U}^{\text{ad}}$ is not bounded $\hat{\mathcal{J}}[u]$ is radially unbounded. In combination with the fact that $\mathbb{U}^{\text{ad}}$ is non-empty and convex the following theorem emerges; compare with [9, 17].

Theorem 2.1 (Existence and uniqueness). *The constrained optimal control problem* (2.2) *has a unique solution* $(\hat{u}, \hat{y}) = (\hat{u}, S\hat{u}) \in \mathbb{U}^{\text{ad}} \times \mathbb{Y}$.

Introducing the *adjoint state* $\hat{p} \in \mathbb{Y}$, the solution $(\hat{y}, \hat{p}, \hat{u}) \in \mathbb{Y} \times \mathbb{Y} \times \mathbb{U}^{\text{ad}}$ is characterized by the *first-order optimality system*

$$\begin{aligned} &\text{state equation:} & \mathcal{B}[\hat{y},\, w] &= \langle \hat{u},\, w\rangle_{\mathbb{U}} & \forall\, w &\in \mathbb{Y}, & (2.4a)\\ &\text{adjoint equation:} & \mathcal{B}[v,\, \hat{p}] &= \langle \psi'(\hat{y}),\, v\rangle & \forall\, v &\in \mathbb{Y}, & (2.4b)\\ &\text{gradient equation:} & \langle \alpha\hat{u} + \hat{p},\, \hat{u} - u\rangle_{\mathbb{U}} &\leq 0 & \forall\, u &\in \mathbb{U}^{\text{ad}}. & (2.4c) \end{aligned}$$

Note, that (2.4c) characterizes $\hat{u}$ as the best approximation of $-\frac{1}{\alpha}\hat{p}$ in the convex set $\mathbb{U}^{\text{ad}}$. Hereafter, we denote by $\Pi\colon \mathbb{U} \to \mathbb{U}^{\text{ad}}$ the nonlinear projection operator such that $\Pi(u)$ is the best approximation of $-\frac{1}{\alpha}u$ in $\mathbb{U}^{\text{ad}}$.

2.3. The discretized optimal control problem

We next introduce the discretized problem. Hereby we only discretize the state space $\mathbb{Y}$. To be more precise: Let $\mathcal{T}$ be a conforming, exact and shape-regular partition of Ω into simplices. Suppose that $\mathbb{Y}(\mathcal{T}) \subset \mathbb{Y}$ is a conforming finite element space over $\mathcal{T}$ such that $\mathcal{B}$ satisfies the discrete inf-sup condition

$$\inf_{\substack{V\in\mathbb{Y}(\mathcal{T})\\ \|V\|_{\mathbb{Y}}=1}} \sup_{\substack{W\in\mathbb{Y}(\mathcal{T})\\ \|W\|_{\mathbb{Y}}=1}} \mathcal{B}[V,\, W] = \gamma(\mathcal{T}) > 0. \tag{2.5}$$

We then consider the discretized problem

$$\min_{u\in\mathbb{U}^{\mathrm{ad}}} \mathcal{J}[u, Y] \qquad \text{subject to} \qquad \mathcal{B}[Y, W] = \langle u, W\rangle_{\mathbb{U}} \qquad \forall W \in \mathbb{Y}(\mathcal{T}). \tag{2.6}$$

Note, that the single inf-sup condition (2.5) implies the second one

$$\inf_{\substack{W\in\mathbb{Y}(\mathcal{T})\\ \|W\|_{\mathbb{Y}}=1}} \sup_{\substack{V\in\mathbb{Y}(\mathcal{T})\\ \|V\|_{\mathbb{Y}}=1}} \mathcal{B}[V, W] = \gamma(\mathcal{T})$$

thanks to $\mathbb{Y}(\mathcal{T})$ being finite dimensional. Consequently, the inf-sup theory implies that for any $f, g \in \mathbb{Y}(\mathcal{T})^*$ there exist unique solutions to the discrete state and adjoint equation

$$Y \in \mathbb{Y}(\mathcal{T}): \qquad \mathcal{B}[Y, W] = \langle f, W\rangle \qquad \forall W \in \mathbb{Y}(\mathcal{T}), \tag{2.7a}$$

$$P \in \mathbb{Y}(\mathcal{T}): \qquad \mathcal{B}[V, P] = \langle g, V\rangle \qquad \forall V \in \mathbb{Y}(\mathcal{T}). \tag{2.7b}$$

Using the same arguments as for the continuous problem we deduce existence and uniqueness of a solution $(\hat{U}, \hat{Y}) \in \mathbb{U}^{\mathrm{ad}} \times \mathbb{Y}(\mathcal{T})$ to the discretized optimal control problem (2.6). Introducing the discrete adjoint state $\hat{P} \in \mathbb{Y}(\mathcal{T})$ the discrete solution $(\hat{Y}, \hat{P}, \hat{U}) \in \mathbb{Y}(\mathcal{T}) \times \mathbb{Y}(\mathcal{T}) \times \mathbb{U}^{\mathrm{ad}}$ is characterized by the discrete first-order optimality system

$$\mathcal{B}[\hat{Y}, W] = \langle \hat{U}, W\rangle_{\mathbb{U}} \qquad \forall W \in \mathbb{Y}(\mathcal{T}), \tag{2.8a}$$

$$\mathcal{B}[V, \hat{P}] = \langle \psi'(\hat{Y}), V\rangle \qquad \forall V \in \mathbb{Y}(\mathcal{T}) \tag{2.8b}$$

$$\langle \alpha\hat{U} + \hat{P}, \hat{U} - u\rangle_{\mathbb{U}} \le 0 \qquad \forall u \in \mathbb{U}^{\mathrm{ad}}. \tag{2.8c}$$

Again the optimal control $\hat{U}$ is characterized by (2.8c) as $\Pi(\hat{P})$. In general it is not a discrete function. Efficient solution techniques for solving (2.8) are described in [7, 8].

2.4. Applications

We next give two examples of spaces $\mathbb{Y}, \mathbb{U}$, bilinear forms $\mathcal{B}$ and sets of admissible controls $\mathbb{U}^{\mathrm{ad}}$.

Poisson problem with boundary control. The PDE constraint is given by the Poisson problem

$$-\Delta y = 0 \text{ in } \Omega, \qquad y = 0 \text{ on } \Gamma_D, \qquad \partial_{\boldsymbol{n}} y = u \text{ on } \Gamma_N, \tag{2.9}$$

i.e., the control u equals the normal derivative $\partial_{\boldsymbol{n}} y$ of the state on the Neumann boundary Γ_N. We suppose that the Dirichlet boundary $\Gamma_D = \partial\Omega \setminus \Gamma_N$ has positive Hausdorff measure. Furthermore, we ask the triangulation $\mathcal{T}$ to be consistent with the splitting of boundary, i.e., Γ_D and Γ_N are both unions of boundary sides of mesh elements.

We let $\mathbb{U} = L^2(\Gamma_N)$ and for given $a, b \in L^2(\Gamma)$ with $a \le b$ we use box-constraints, i.e., $\mathbb{U}^{\mathrm{ad}} = \{u \in \mathbb{U} \mid a \le u \le b\}$. Obviously, $\mathbb{U}^{\mathrm{ad}}$ is a non-empty, convex, and bounded subset of the Hilbert space $\mathbb{U}$. The variational formulation of (2.9) utilizes on the Hilbert space $\mathbb{Y} = \{v \in H^1(\Omega) \mid v = 0 \text{ on } \Gamma_D\}$ the bilinear form $\mathcal{B}[v, w] = \int_\Omega \nabla v \cdot \nabla w \, d\Omega$, and reads for given $u \in \mathbb{U}$

$$y \in \mathbb{Y}: \qquad \mathcal{B}[y, w] = \int_{\Gamma_N} u\, w \, d\Gamma =: \langle u, w \rangle_{\mathbb{U}} \qquad \forall w \in \mathbb{Y}.$$

The bilinear form $\mathcal{B}$ is continuous and coercive on $\mathbb{Y}$ since Γ_D has positive Hausdorff measure. Coercivity of $\mathcal{B}$ implies the inf-sup condition (2.1). The required embeddings $\mathbb{Y} \hookrightarrow \mathbb{U} \hookrightarrow \mathbb{Y}^*$ follow from the trace theorem.

Let $\emptyset \neq \Omega_d \subset \Omega$ be meshed by a subset of grid elements in $\mathcal{T}$. For given $y_d \in L^2(\Omega_d)$ we consider the objective functional

$$\psi(z) := \tfrac{1}{2} \|z - y_d\|^2_{L^2(\Omega_d)}.$$

Consequently, for given $z \in \mathbb{Y}$ the adjoint problem reads

$$p \in \mathbb{Y}: \qquad \mathcal{B}[v, p] = \int_{\Omega_d} (z - y_d)\, v \, d\Omega \qquad \forall v \in \mathbb{Y}.$$

The discrete space $\mathbb{Y}(\mathcal{T})$ is constructed from Courant finite elements, i.e., continuous, piecewise affine functions over $\mathcal{T}$, that vanish on Γ_D. This is the most simple discretization of (2.9). Coercivity is inherited to any subspace, which implies (2.5) with a constant $\gamma(\mathcal{T})$ that is bounded from below by the continuous inf-sup constant γ of (2.1).

Stokes problem with distributed control. We consider the Stokes problem

$$-\Delta \boldsymbol{y} + \nabla \pi = \boldsymbol{u} \quad \text{in } \Omega, \qquad \nabla \cdot \boldsymbol{y} = 0 \quad \text{in } \Omega, \qquad \boldsymbol{y} = \boldsymbol{0} \quad \text{on } \partial\Omega,$$

where the control $\boldsymbol{u}$ acts as an external force on the fluid with velocity $\boldsymbol{y}$ and pressure π. We consider distributed control $\mathbb{U} = L^2(\Omega; \mathbb{R}^d)$ with the constraint $\mathbb{U}^{\text{ad}} = \boldsymbol{f} + \{\boldsymbol{u} \in \mathbb{U} \mid |\boldsymbol{u}|_2 \leq R\}$ for some $R > 0$ and fixed $\boldsymbol{f} \in \mathbb{U}$.

For the weak form we define on $\mathbb{Y} = H^1_0(\Omega; \mathbb{R}^d) \times L^2_0(\Omega)$ the continuous bilinear form

$$\mathcal{B}[(\boldsymbol{v}, r), (\boldsymbol{w}, q)] := \int_\Omega \nabla \boldsymbol{w} : \nabla \boldsymbol{v} \, d\Omega - \int_\Omega \nabla \cdot \boldsymbol{w}\, r \, d\Omega - \int_\Omega q \nabla \cdot \boldsymbol{v} \, d\Omega,$$

which satisfies an inf-sup condition; compare for instance with [3, Theorem III.3.1]. Obviously $\mathbb{Y} \hookrightarrow \mathbb{U} \hookrightarrow \mathbb{Y}^*$ holds. Given a desired velocity profile $\boldsymbol{y_d} \in L^2(\Omega)$ we consider with $z = (\boldsymbol{z}, q)$ the objective functional

$$\psi(z) := \tfrac{1}{2} \|\boldsymbol{z} - \boldsymbol{y}_d\|^2_{L^2(\Omega)}.$$

Therefore, for given $\boldsymbol{u} \in \mathbb{U}$ and $\boldsymbol{y} \in \mathbb{Y}$ the primal and adjoint problem reads

$$y = (\boldsymbol{y}, \pi) \in \mathbb{Y}: \qquad \mathcal{B}[y, w] = \int_\Omega \boldsymbol{u} \cdot \boldsymbol{w} \, d\Omega =: \langle \boldsymbol{u}, \boldsymbol{w} \rangle_{\mathbb{U}} \qquad \forall w = (\boldsymbol{w}, q) \in \mathbb{Y},$$

$$p = (\boldsymbol{p}, r) \in \mathbb{Y}: \qquad \mathcal{B}[v, p] = \int_\Omega (\boldsymbol{y} - \boldsymbol{y}_d) \cdot \boldsymbol{v} \, d\Omega \qquad \forall v = (\boldsymbol{v}, q) \in \mathbb{Y}.$$

We discretize the Stokes problem with the Taylor-Hood element of order $\ell \geq 2$, which is a stable discretization, i.e., the bilinear form $\mathcal{B}$ satisfies an inf-sup condition on $\mathbb{Y}(\mathcal{T})$ with a constant $\gamma(\mathcal{T})$ that only depends on the shape-regularity of $\mathcal{T}$.

3. A posteriori error analysis

In this section we state and prove the main result of this note. Basically we show that the sum of reliable and efficient estimators for the primal and the dual problems (2.3a) and (2.3b) gives a reliable and efficient estimator for the optimal control problem (2.2). Note, that there exists a well-established theory of different kinds of estimators for a huge class of linear problems (2.3a) and (2.3b). We start with the assumptions on the estimators for the linear problems, then prove the main result, and finally show how the applications from §2.4 fit into the framework.

3.1. Assumptions on the estimators for the linear problems

Given $u \in \mathbb{U}$ we consider the primal problem

$$y \in \mathbb{Y}: \qquad \mathcal{B}[y,\, w] = \langle u,\, w\rangle_{\mathbb{U}} \qquad \forall\, w \in \mathbb{Y}, \tag{3.1a}$$

and for given $z \in \mathbb{Y}$ we consider the adjoint problem

$$p \in \mathbb{Y}: \qquad \mathcal{B}[v,\, p] = \langle \psi'(z),\, v\rangle \qquad \forall\, v \in \mathbb{Y}. \tag{3.1b}$$

Utilizing the solution operators S, S^* we may write $y = S(u)$ and $p = S^*(\psi'(z))$.

Let $Y \in \mathbb{Y}(\mathcal{T})$ be the Galerkin approximation to y, i.e.,

$$Y \in \mathbb{Y}(\mathcal{T}): \qquad \mathcal{B}[Y,\, W] = \langle u,\, W\rangle_{\mathbb{U}} \qquad \forall\, W \in \mathbb{Y}(\mathcal{T}), \tag{3.2a}$$

and denote by $P \in \mathbb{Y}(\mathcal{T})$ the Galerkin approximation to p, i.e.,

$$P \in \mathbb{Y}(\mathcal{T}): \qquad \mathcal{B}[V,\, P] = \langle \psi'(z),\, V\rangle \qquad \forall\, V \in \mathbb{Y}(\mathcal{T}). \tag{3.2b}$$

We are now in the position to pose the requirements on the estimators for these linear problems.

In this process we use $a \lesssim b$ for $a \le C\,b$, where C or any other explicitly stated constant C only depends on data of the bilinear form $\mathcal{B}$, like the inf-sup constant γ or the continuity constant $\|\mathcal{B}\|$, data α and ψ of the objective $\mathcal{J}$, the shape regularity of the grid $\mathcal{T}$, the domain Ω, and the dimension d. Any such constant *must not* depend on the input u or z of the source terms u or $\psi'(z)$ of (3.1).

Assumption 3.1 (Estimators for the linear problems). There exist a posteriori error estimators $\mathcal{E}_{\mathcal{T}}$ and $\mathcal{E}^*_{\mathcal{T}}$ for (3.1) and (3.2) with the following properties for any source $u \in \mathbb{U}$ and $\psi'(z)$ with $z \in \mathbb{Y}$ and corresponding solutions $y = S(u)$ respectively $p = S^*(\psi'(z))$, and Galerkin approximations Y respectively P:

(1) **Reliability:** The estimators $\mathcal{E}_{\mathcal{T}}(Y, u; \mathcal{T})$ and $\mathcal{E}^*_{\mathcal{T}}(P, \psi'(z); \mathcal{T})$ provide an upper bound for the true error, i.e.,

$$\|Y - y\|_{\mathbb{Y}} \lesssim \mathcal{E}_{\mathcal{T}}(Y, u; \mathcal{T}), \tag{3.3a}$$
$$\|P - p\|_{\mathbb{Y}} \lesssim \mathcal{E}^*_{\mathcal{T}}(P, \psi'(z); \mathcal{T}). \tag{3.3b}$$

(2) **Efficiency:** The estimators $\mathcal{E}_{\mathcal{T}}(Y,u;\mathcal{T})$ and $\mathcal{E}^*_{\mathcal{T}}(P,\psi'(z);\mathcal{T})$ are lower bounds for the true error up to oscillation, i.e.,

$$\mathcal{E}_{\mathcal{T}}(Y,u;\mathcal{T}) \lesssim \|Y-y\|_{\mathbb{Y}} + \operatorname{osc}_{\mathcal{T}}(Y,u;\mathcal{T}), \tag{3.4a}$$

$$\mathcal{E}^*_{\mathcal{T}}(P,\psi'(z);\mathcal{T}) \lesssim \|P-p\|_{\mathbb{Y}} + \operatorname{osc}^*_{\mathcal{T}}(P,\psi'(z);\mathcal{T}). \tag{3.4b}$$

Hereafter, $\operatorname{osc}_{\mathcal{T}}$ and $\operatorname{osc}^*_{\mathcal{T}}$ are the typical oscillation terms that have to be present in the lower bound.

(3) **Lipschitz continuity of oscillation:** The oscillation terms are Lipschitz continuous with respect to their second arguments, i.e., for fixed $Y,P \in \mathbb{Y}(\mathcal{T})$ and any $u_1,u_2 \in \mathbb{U}$, $z_1,z_2 \in \mathbb{Y}$ there holds

$$|\operatorname{osc}_{\mathcal{T}}(Y,u_1;\mathcal{T}) - \operatorname{osc}_{\mathcal{T}}(Y,u_2;\mathcal{T})| \lesssim \|u_1-u_2\|_{\mathbb{U}}, \tag{3.5a}$$

$$|\operatorname{osc}^*_{\mathcal{T}}(P,\psi'(z_1);\mathcal{T}) - \operatorname{osc}^*_{\mathcal{T}}(P,\psi'(z_2);\mathcal{T})| \lesssim \|\psi'_1\|_{L(\mathbb{Y};\mathbb{Y}^*)}\|z_1-z_2\|_{\mathbb{Y}}, \tag{3.5b}$$

where ψ'_1 is the linear part of the Fréchet derivative ψ'.

The assumptions posed for the estimators $\mathcal{E}_{\mathcal{T}}$ and $\mathcal{E}^*_{\mathcal{T}}$ of the linear problems are standard; compare with the discussion of the applications in §3.3 below.

3.2. Error analysis for the optimal control problem

We next show that the sum of these estimators give a reliable and efficient estimator for the optimal control problem (2.2). In doing so we use the notation

$$\|y,p,u\|_{\mathbb{Y}\times\mathbb{Y}\times\mathbb{U}} := \left(\|y\|^2_{\mathbb{Y}} + \|p\|^2_{\mathbb{Y}} + \|u\|^2_{\mathbb{U}}\right)^{1/2} \qquad \forall\,(y,p,u) \in \mathbb{Y}\times\mathbb{Y}\times\mathbb{U}.$$

Theorem 3.2 (Upper bound). *Let* $(\hat{y},\hat{p},\hat{u}) \in \mathbb{Y}\times\mathbb{Y}\times\mathbb{U}^{\mathrm{ad}}$ *be the solution of the optimal control problem* (2.2) *and let* $(\hat{Y},\hat{P},\hat{U}) \in \mathbb{Y}(\mathcal{T})\times\mathbb{Y}(\mathcal{T})\times\mathbb{U}^{\mathrm{ad}}$ *be the discrete solution of* (2.6)*. Assume that the estimators* $\mathcal{E}_{\mathcal{T}}$, $\mathcal{E}^*_{\mathcal{T}}$ *for the linear problems* (3.1) *satisfy Assumption* 3.1(1)*.*

Then their sum is an upper bound for the error of the optimal control problem:

$$\|\hat{Y}-\hat{y},\, \hat{P}-\hat{p},\, \hat{U}-\hat{u}\|_{\mathbb{Y}\times\mathbb{Y}\times\mathbb{U}} \lesssim \mathcal{E}_{\mathcal{T}}(\hat{Y},\hat{U};\mathcal{T}) + \mathcal{E}^*_{\mathcal{T}}(\hat{P},\psi'(\hat{Y});\mathcal{T}).$$

Proof. [1] We realize that the discrete state $\hat{Y}$ is the Galerkin approximation to the solution $\bar{y}$ of (3.1a) with source $\hat{U}$, i.e., $\bar{y} = S(\hat{U})$. Consequently, (3.3a) implies

$$\|\hat{Y}-\bar{y}\|_{\mathbb{Y}} \lesssim \mathcal{E}_{\mathcal{T}}(\hat{Y},\hat{U};\mathcal{T}). \tag{3.6}$$

In the same vein, $\hat{P}$ is the Galerkin approximation to the solution $\bar{p}$ of (3.1b) with source $\psi'(\hat{Y})$, in short, $\bar{p} = S^*(\psi'(\hat{Y}))$. Therefore, (3.3b) implies

$$\|\hat{P}-\bar{p}\|_{\mathbb{Y}} \lesssim \mathcal{E}^*_{\mathcal{T}}(\hat{P},\psi'(\hat{Y});\mathcal{T}). \tag{3.7}$$

We show below

$$\|\hat{Y}-\hat{y},\, \hat{P}-\hat{p},\, \hat{U}-\hat{u}\|_{\mathbb{Y}\times\mathbb{Y}\times\mathbb{U}} \lesssim \|\hat{Y}-\bar{y}\|_{\mathbb{Y}}, + \|\hat{P}-\bar{p}\|_{\mathbb{Y}}, \tag{3.8}$$

which immediately implies the claim by (3.6) and (3.7).

[2] We first estimate the error in the state. The triangle inequality in conjunction with

$$\|\bar{y}-\hat{y}\|_{\mathbb{Y}} = \|S(\hat{U}-\hat{u})\|_{\mathbb{Y}} \le \gamma^{-1}\|\hat{U}-\hat{u}\|_{\mathbb{Y}^*} \le \gamma^{-1}C\|\hat{U}-\hat{u}\|_{\mathbb{U}}$$

yields

$$\|\hat{Y} - \hat{y}\|_{\mathbb{Y}} \le \|\hat{Y} - \bar{y}\|_{\mathbb{Y}} + \|\bar{y} - \hat{y}\|_{Y} \le \|\hat{Y} - \bar{y}\|_{\mathbb{Y}} + \gamma^{-1} C \|\hat{U} - \hat{u}\|_{\mathbb{U}}.$$

[3] We next address the error for the adjoint state and observe first

$$\|\bar{p} - \hat{p}\|_{\mathbb{Y}} = \|S^*(\psi'(\hat{Y}) - \psi'(\hat{y}))\|_{\mathbb{Y}} \le \gamma^{-1} \|\psi_1'\|_{L(\mathbb{Y};\mathbb{Y}^*)} \|\hat{Y} - \hat{y}\|_{\mathbb{Y}}.$$

In combination with [2] this yields

$$\begin{aligned} \|\hat{P} - \hat{p}\|_{\mathbb{Y}} &\le \|\hat{P} - \bar{p}\|_{\mathbb{Y}} + \|\bar{p} - \hat{p}\|_{\mathbb{Y}} \\ &\le \|\hat{P} - \bar{p}\|_{\mathbb{Y}} + \gamma^{-1} \|\psi_1'\|_{L(\mathbb{Y};\mathbb{Y}^*)} \|\hat{Y} - \bar{y}\|_{\mathbb{Y}} + \gamma^{-2} C \|\psi_1'\|_{L(\mathbb{Y};\mathbb{Y}^*)} \|\hat{U} - \hat{u}\|_{\mathbb{U}}. \end{aligned}$$

[4] We next turn to the error in the control and start with

$$\begin{aligned} \alpha \|\hat{U} - \hat{u}\|_{\mathbb{U}}^2 &= \langle \alpha \hat{U} + \hat{P},\, \hat{U} - \hat{u} \rangle_{\mathbb{U}} + \langle \alpha \hat{u} + \hat{p},\, \hat{u} - \hat{U} \rangle_{\mathbb{U}} + \langle \hat{p} - \hat{P},\, \hat{U} - \hat{u} \rangle_{\mathbb{U}} \\ &\le \langle \hat{p} - \hat{P},\, \hat{U} - \hat{u} \rangle_{\mathbb{U}}, \end{aligned}$$

as a consequence of the discrete and continuous gradient equation (2.8c) and (2.4c) implying $\langle \alpha \hat{U} + \hat{P},\, \hat{U} - \hat{u} \rangle_{\mathbb{U}} \le 0$ and $\langle \alpha \hat{u} + \hat{p},\, \hat{u} - \hat{U} \rangle_{\mathbb{U}} \le 0$ since $\hat{U}, \hat{u} \in \mathbb{U}^{\mathrm{ad}}$.

Recalling $\bar{y} = S(\hat{U})$ and $\hat{Y} = S(\hat{u})$, we proceed by

$$\begin{aligned} \alpha \|\hat{U} - \hat{u}\|_{\mathbb{U}}^2 &\le \langle \hat{U} - \hat{u},\, \hat{p} - \hat{P} \rangle_{\mathbb{U}} = \mathcal{B}[\bar{y} - \hat{y},\, \hat{p} - \hat{P}] \\ &= \mathcal{B}[\bar{y} - \hat{y},\, \bar{p} - \hat{P}] + \mathcal{B}[\bar{y} - \hat{y},\, p - \bar{p}] + \mathcal{B}[\bar{y} - \hat{y},\, \hat{p} - p], \end{aligned}$$

where we use $p = S^*(\psi'(\bar{y}))$. Since $\hat{p} = S^*(\psi'(\hat{y}))$ we can estimate the last term on the right-hand side

$$\mathcal{B}[\bar{y} - \hat{y},\, \hat{p} - p] = \langle \psi'(\hat{y}) - \psi'(\bar{y}),\, \bar{y} - \hat{y} \rangle \le 0$$

by employing convexity of ψ. The identity $\mathcal{B}[\bar{y} - \hat{y},\, p - \bar{p}] = \langle \psi'(\bar{y}) - \psi'(\hat{Y}),\, \bar{y} - \hat{y} \rangle$ is a consequence of $\bar{p} = S^*(\psi'(\hat{Y}))$ and in combination with the bound $\|\bar{y} - \hat{y}\|_{\mathbb{Y}} \le \gamma^{-1} C \|\hat{U} - \hat{u}\|_{\mathbb{U}}$ it allows us to estimate

$$\begin{aligned} \alpha \|\hat{U} - \hat{u}\|_{\mathbb{U}}^2 &\le \mathcal{B}[\bar{y} - \hat{y},\, \bar{p} - \hat{P}] + \langle \psi'(\bar{y}) - \psi'(\hat{Y}),\, \bar{y} - \hat{y} \rangle \\ &\le \big(\|\mathcal{B}\| \|\bar{p} - \hat{P}\|_{\mathbb{Y}} + \|\psi_1'\|_{L(\mathbb{Y};\mathbb{Y}^*)} \|\bar{y} - \hat{Y}\|_{\mathbb{Y}} \big) \gamma^{-1} C \|\hat{U} - \hat{u}\|_{\mathbb{U}}. \end{aligned}$$

Thus we have shown

$$\|\hat{U} - \hat{u}\| \le (\alpha\gamma)^{-1} C \big(\|B\| \|\hat{P} - \bar{p}\|_{\mathbb{Y}} + \|\psi_1'\|_{L(\mathbb{Y};\mathbb{Y}^*)} \|\hat{Y} - \bar{y}\|_{\mathbb{Y}} \big).$$

[5] Collecting the estimates of steps [1]–[4] we have shown (3.8), which finishes the proof. □

We next prove efficiency of the estimator.

Theorem 3.3 (Lower bound). *Let $(\hat{y}, \hat{p}, \hat{u}) \in \mathbb{Y} \times \mathbb{Y} \times \mathbb{U}^{\mathrm{ad}}$ be the solution of the optimal control problem* (2.2) *and let $(\hat{Y}, \hat{P}, \hat{U}) \in \mathbb{Y}(\mathcal{T}) \times \mathbb{Y}(\mathcal{T}) \times \mathbb{U}^{\mathrm{ad}}$ be the discrete solution of* (2.6). *Suppose that the estimators $\mathcal{E}_{\mathcal{T}}$, $\mathcal{E}_{\mathcal{T}}^*$ for the linear problems* (3.1) *satisfy Assumption* 3.1(2).

Then their sum is a lower bound for the error of the optimal control problem up to oscillation, i.e.,

$$\mathcal{E}_\mathcal{T}(\hat{Y},\hat{U};\mathcal{T}) + \mathcal{E}^*_\mathcal{T}(\hat{P},\psi'(\hat{Y});\mathcal{T}) \\ \lesssim \|\hat{Y}-\hat{y},\ \hat{P}-\hat{p},\ \hat{U}-\hat{u}\|_{\mathbb{Y}\times\mathbb{Y}\times\mathbb{U}} + \operatorname{osc}_\mathcal{T}(\hat{Y},\hat{U};\mathcal{T}) + \operatorname{osc}^*_\mathcal{T}(\hat{P},\psi'(\hat{Y});\mathcal{T}).$$

If, in addition, oscillation is Lipschitz-continuous with respect to its second argument, i.e., Assumption 3.1(3) *holds, then we have the lower bound*

$$\mathcal{E}_\mathcal{T}(\hat{Y},\hat{U};\mathcal{T}) + \mathcal{E}^*_\mathcal{T}(\hat{P},\psi'(\hat{Y});\mathcal{T}) \\ \lesssim \|\hat{Y}-\hat{y},\ \hat{P}-\hat{p},\ \hat{U}-\hat{u}\|_{\mathbb{Y}\times\mathbb{Y}\times\mathbb{U}} + \operatorname{osc}_\mathcal{T}(\hat{Y},\hat{u};\mathcal{T}) + \operatorname{osc}^*_\mathcal{T}(\hat{P},\psi'(\hat{y});\mathcal{T}).$$

Proof. We recall that $\hat{Y}$ is the Galerkin approximation to the solution $\bar{y} = S(\hat{U})$ of (3.1a). Consequently, (3.4a) in combination with the triangle inequality implies

$$\begin{aligned} \mathcal{E}_\mathcal{T}(\hat{Y},\hat{U};\mathcal{T}) &\lesssim \|\hat{Y}-\bar{y}\|_\mathbb{Y} + \operatorname{osc}_\mathcal{T}(\hat{Y},\hat{U};\mathcal{T}) \\ &\le \|\hat{Y}-\hat{y}\|_\mathbb{Y} + \|\bar{y}-\hat{y}\|_\mathbb{Y} + \operatorname{osc}_\mathcal{T}(\hat{Y},\hat{U};\mathcal{T}) \\ &\le \|\hat{Y}-\hat{y}\|_\mathbb{Y} + \gamma^{-1}\|\hat{U}-\hat{u}\|_\mathbb{Y} + \operatorname{osc}_\mathcal{T}(\hat{Y},\hat{U};\mathcal{T}). \end{aligned}$$

This proves that $\mathcal{E}_\mathcal{T}(\hat{Y},\hat{U};\mathcal{T})$ is a lower bound for the true error up to oscillation $\operatorname{osc}_\mathcal{T}(\hat{Y},\hat{U};\mathcal{T})$. The lower bound up to oscillation $\operatorname{osc}_\mathcal{T}(\hat{Y},\hat{u};\mathcal{T})$ is a direct consequence of the Lipschitz-continuity (3.5a) of $\operatorname{osc}_\mathcal{T}$:

$$\operatorname{osc}_\mathcal{T}(\hat{Y},\hat{U};\mathcal{T}) \lesssim \operatorname{osc}_\mathcal{T}(\hat{Y},\hat{u};\mathcal{T}) + \|\hat{U}-\hat{u}\|_\mathbb{U}.$$

Efficiency of $\mathcal{E}^*_\mathcal{T}(\hat{P},\psi'(\hat{Y});\mathcal{T})$ is proved in the same way by utilizing that $\hat{P}$ is the Galerkin approximation to $\bar{p} = S^*(\psi'(\hat{Y}))$ of (3.1b). □

Since we do not discretize the control, the upper bound stated in Theorem 3.2 is insensitive to the approximability of the continuous optimal control $\hat{u}$ by discrete functions. In contrast to this, the oscillation term $\operatorname{osc}_\mathcal{T}(\hat{Y},\hat{u};\mathcal{T})$ in the lower bound of Theorem 3.3 encodes such an approximation of $\hat{u}$ by a suitable finite element function over $\mathcal{T}$; compare with §3.3 below. This means, if $\hat{u}$ is not resolved well over the grid $\mathcal{T}$, the estimator $\mathcal{E}_\mathcal{T}(\hat{Y},\hat{u};\mathcal{T})$ may overestimate the true error in the state equation.

3.3. Estimators for the applications

We return to the applications from §2.4. We give examples of estimators for the linear sub-problems and show the genuineness of Assumption 3.1.

Poisson problem with boundary control. We are going to use a hierarchical estimator for both primal and dual problem. The idea of hierarchical estimators is based upon evaluating the residual with sufficiently many discrete functions that do not belong to $\mathbb{Y}(\mathcal{T})$. Given $u \in \mathbb{U}$ and $z \in \mathbb{Y}$ the residuals $\mathcal{R}(Y,u), \mathcal{R}^*(P, z-y_d) \in \mathbb{Y}^*$

in Y and P of the primal and dual problem are

$$\langle \mathcal{R}(Y,u),\, w\rangle := \int_\Omega \nabla w \cdot \nabla Y \, d\Omega - \int_{\Gamma_N} u\, w\, d\Gamma \qquad \forall w \in \mathbb{Y},$$

$$\langle \mathcal{R}^*(P, z-y_d),\, v\rangle := \int_\Omega \nabla P \cdot \nabla v \, d\Omega - \int_{\Omega_d} (z-y_d)\, v\, d\Omega \qquad \forall v \in \mathbb{Y}.$$

Let $\mathcal{S}_\Omega$ be the union of all interior sides and $\mathcal{S}_N$ be the union of Neumann boundary sides. To each side $\sigma \in \mathcal{S} := \mathcal{S}_\Omega \cup \mathcal{S}_N$ associate a bubble function ϕ_σ with $\phi_\sigma \geq 0$, $\operatorname{supp}(\phi_\sigma) \subset \omega_\sigma$, and $\|\nabla \phi_\sigma\|_{L^2(\Omega)} = 1$, where ω_σ is the patch consisting of elements in $\mathcal{T}$ having σ as a side. Moreover, to each $T \in \mathcal{T}$ associate a bubble function ϕ_T with $\phi_T \geq 0$, $\operatorname{supp}(\phi_T) \subset \omega_T$, and $\|\nabla \phi_T\|_{L^2(\Omega)} = 1$.

We first define oscillation and then the estimators. For the primal problem and $\sigma \in \mathcal{S}_N$ define $\operatorname{osc}_\mathcal{T}(Y,u;\sigma)^2 := h_\sigma \|u - P_\mathcal{S} u\|^2_{L^2(\sigma)}$, where $P_\mathcal{S}$ is the L^2 projection onto the piecewise constant functions over $\mathcal{S}_N$ and h_σ the diameter of σ. For $\sigma \in \mathcal{S}_\Omega$ we set $\operatorname{osc}_\mathcal{T}(Y,u;\sigma) := 0$. For the dual problem and $T \in \mathcal{T}$ with $T \subset \Omega_d$ set $\operatorname{osc}^*_\mathcal{T}(P,\psi'(z);T)^2 := h_T^2 \|(z-y_d) - P_\mathcal{T}(z-y_d)\|^2_{L^2(T)}$, where $P_\mathcal{T}$ is the L^2 projection onto the piecewise constant functions over $\mathcal{T}$ and h_T the diameter of T. For $T \not\subset \Omega_d$ we let $\operatorname{osc}^*_\mathcal{T}(P,\psi'(z);T) := 0$. It is easy to check that oscillation is Lipschitz with respect to the second argument, i.e., Assumption 3.1(3) holds.

The estimators are then given as

$$\mathcal{E}_\mathcal{T}(Y,u;\mathcal{T})^2 := \sum_{T\in\mathcal{T}} \langle \mathcal{R}(Y,u),\, \phi_T\rangle^2 + \sum_{\sigma\in\mathcal{S}} \langle \mathcal{R}(Y,u),\, \phi_\sigma\rangle^2 + \operatorname{osc}_\mathcal{T}(Y,u;\sigma)^2,$$

$$\begin{aligned} \mathcal{E}^*_\mathcal{T}(P,\psi'(z);\mathcal{T})^2 := {} & \sum_{T\in\mathcal{T}} \langle \mathcal{R}^*(P,\psi'(z)),\, \phi_T\rangle^2 + \operatorname{osc}^*_\mathcal{T}(P,\psi'(z);T)^2 \\ & + \sum_{\sigma\in\mathcal{S}} \langle \mathcal{R}^*(P,\psi'(z)),\, \phi_\sigma\rangle^2. \end{aligned}$$

These estimators are reliable and efficient, which means that Assumption 3.1(1) and (2) hold; compare for instance with [20, Proposition 1.14] or [16, 18].

Stokes problem with distributed control. We are going to use the standard residual estimator as derived by Verfürth [19]. For given $\boldsymbol{u} \in \mathbb{U}$ the estimator for the primal problem is given for $Y = (\boldsymbol{Y}, \Pi)$ as

$$\mathcal{E}_\mathcal{T}(Y,\boldsymbol{u};\mathcal{T})^2 := \sum_{T\in\mathcal{T}} h_T^2 \|{-\Delta \boldsymbol{Y}} + \nabla \Pi - \boldsymbol{u}\|^2_{L^2(T)} + h_T \|\,[\![\nabla \boldsymbol{Y}]\!]\,\|^2_{L^2(\partial T\cap\Omega)} + \|\nabla\cdot \boldsymbol{Y}\|^2_{L^2(T)},$$

where $[\![\boldsymbol{Y}]\!]$ is the normal flux of the discrete velocity across inter-element sides. Oscillation is given as

$$\operatorname{osc}_\mathcal{T}(Y,\boldsymbol{u};\mathcal{T})^2 := \sum_{T\in\mathcal{T}} h_T^2 \|\boldsymbol{u} - P_\mathcal{T}\boldsymbol{u}\|^2_{L^2(T)},$$

where $P_\mathcal{T}$ is the L^2 projection onto the space of piecewise polynomials of degree $\ell - 1$. Since $\boldsymbol{\mathcal{B}}$ is symmetric and due to the structure of ψ the estimator and oscillation of the dual problem simply read

$$\mathcal{E}^*_{\mathcal{T}}(P, \psi'(z); \mathcal{T}) = \mathcal{E}_{\mathcal{T}}(P, \psi'(z); \mathcal{T}) \quad \text{and} \quad \operatorname{osc}^*_{\mathcal{T}}(P, \psi'(z); \mathcal{T}) = \operatorname{osc}_{\mathcal{T}}(P, \psi'(z); \mathcal{T}).$$

The estimators and oscillation satisfy all the assumptions posed in Assumption 3.1.

4. Concluding remarks

In this note we have shown how to derive from standard results of the a posteriori error analysis for linear problems an a posteriori error estimator for control constrained optimal control problems. This enlarges the class of problems that can be treated and the class of estimators that can be used drastically. We want to conclude with the following remarks.

(1) Usually, the lower bound of Theorem 3.3 is shown in a local variant, i.e., the local error indicator on a single element is a lower bound for the local error on a patch around the element; compare for instance with [19, 20]. It is possible to derive such a local lower bound also for the optimal control problem. In this process one has to involve the residual and cannot work directly with norms.

(2) During the last decade the convergence analysis of adaptive finite elements has made a substantial progress. Under quite moderate assumptions on the modules of the adaptive algorithm it can be shown that the sequence of adaptively generated discrete solutions converges to the exact one. The most general result for a quite general class of linear problems can be found in [12, 15]. We have shown that these ideas transfer to control constrained optimal control problems and this result will be published soon.

(3) Another common approach is to also discretize the control and to solve the discretized optimality system by standard semi-smooth Newton methods; compare for instance with [6]. The presented analysis can be generalized such that the discretization of the control is included. Naturally, the analysis gets more complicated and more terms have to be treated. These results will be published elsewhere.

References

[1] R. Becker, M. Braack, D. Meidner, R. Rannacher, and B. Vexler, *Adaptive finite element methods for PDE-constrained optimal control problems*, in Reactive flows, diffusion and transport, Springer, Berlin, 2007, pp. 177–205.

[2] R. Falk, *Approximation of a class of optimal control problems with order of convergence estimates*, J. Math. Anal. Appl., 44 (1973), pp. 28–47.

[3] G.P. Galdi, *An introduction to the mathematical theory of the Navier-Stokes equations. Vol. 1: Linearized steady problems*, Springer Tracts in Natural Philosophy, 38, New York, NY, 1994.

[4] T. Geveci, *On the approximation of the solution of an optimal control problem governed by an elliptic equation*, Math. Model. Numer. Anal. 13 (1979), pp. 313–328.

[5] M. Hintermüller, R.H. Hoppe, Y. Iliash, and M. Kieweg, *An a posteriori error analysis of adaptive finite element methods for distributed elliptic control problems with control constraints*, ESAIM, Control Optim. Calc. Var., 14 (2008), pp. 540–560.

[6] M. Hintermüller, K. Ito, and K. Kunisch, *The primal-dual active set strategy as a semismooth Newton method*, SIAM J. Optim., 13 (2002), pp. 865–888 (electronic) (2003).

[7] M. Hinze, *A variational discretization concept in control constrained optimization: the linear-quadratic case*, Comput. Optim. Appl., 30 (2005), pp. 45–61.

[8] M. Hinze and M. Vierling, *Variational discretization and semi-smooth Newton methods; implementation, convergence and globalization in PDE constrained optimization with control constraints.* Hamburger Beiträge zur Angewandten Mathematik No. 2009-13, 2009.

[9] J. L. Lions, *Optimal Control of Systems Governed by Partial Differential Equations*, vol. 170 of Die Grundlehren der mathematischen Wissenschaften, Springer-Verlag, 1971.

[10] W. Liu and N. Yan, *A posteriori error estimates for distributed convex optimal control problems*, Adv. Comput. Math, 15 (2001), pp. 285–309.

[11] W. Liu and N. Yan, *A posteriori error estimates for control problems governed by Stokes equations*, SIAM J. Numer. Anal., 40 (2002), pp. 1850–1869.

[12] P. Morin, K.G. Siebert, and A. Veeser, *A basic convergence result for conforming adaptive finite elements*, Math. Models Methods Appl. 18 (2008), pp. 707–737.

[13] J. Nečas, *Sur une méthode pour resoudre les équations aux dérivées partielles du type elliptique, voisine de la variationnelle*, Ann. Sc. Norm. Super. Pisa, Sci. Fis. Mat., III. Ser., 16 (1962), pp. 305–326.

[14] R.H. Nochetto, K.G. Siebert, and A. Veeser, *Theory of adaptive finite element methods: An introduction*, in Multiscale, Nonlinear and Adaptive Approximation, R.A. DeVore and A. Kunoth, eds., Springer, 2009, pp. 409–542.

[15] K.G. Siebert, *A convergence proof for adaptive finite elements without lower bound*, 2009. Preprint Universität Duisburg-Essen No. 694, to appear in IMA J. of Numer. Anal.

[16] K.G. Siebert and A. Veeser, *A unilaterally constrained quadratic minimization with adaptive finite elements*, SIAM J. Optim., 18 (2007), pp. 260–289.

[17] F. Tröltzsch, *Optimal Control of Partial Differential Equations. Theory, Methods and Applications*, vol. 112 of Graduate Studies Series in Mathematics, AMS, Providence, 2010.

[18] A. Veeser, *Convergent adaptive finite elements for the nonlinear Laplacian*, Numer. Math., 92 (2002), pp. 743–770.

[19] R. Verfürth, *A posteriori error estimators for the Stokes equations*, Numer. Math., 55 (1989), pp. 309–325.

[20] ——, *A Review of A Posteriori Error Estimation and Adaptive Mesh-Refinement Techniques*, Adv. Numer. Math., John Wiley, Chichester, UK, 1996.

Kristina Kohls and Kunibert G. Siebert
Institut für Angewandte Analysis und Numerische Simulation
Fachbereich Mathematik
Universität Stuttgart
Pfaffenwaldring 57
D-70569 Stuttgart, Germany
e-mail: kristina.kohls@ians.uni-stuttgart.de
kg.siebert@ians.uni-stuttgart.de

URL: http://www.ians.uni-stuttgart.de/nmh/

Arnd Rösch
Fakultät für Mathematik
Universität Duisburg-Essen
Forsthausweg 2
D-47057 Duisburg, Germany
e-mail: arnd.roesch@uni-due.de

URL: http://www.uni-due.de/mathematik/agroesch/

International Series of Numerical Mathematics, Vol. 160, 445–466

A Priori Error Estimates for Space-Time Finite Element Discretization of Parabolic Optimal Control Problems

Dominik Meidner and Boris Vexler

Abstract. In this article we summarize recent results on a priori error estimates for space-time finite element discretizations of linear-quadratic parabolic optimal control problems. We consider the following three cases: problems without inequality constraints, problems with pointwise control constraints, and problems with state constraints pointwise in time. For all cases, error estimates with respect to the temporal and to the spatial discretization parameters are derived. The results are illustrated by numerical examples.

Mathematics Subject Classification (2000). 35K20, 49J20, 49M05, 49M15, 49M25, 49M29, 49N10, 65M12, 65M15, 65M60.

Keywords. Optimal control, parabolic equations, error estimates, finite elements, control constraints, state constraints, discretization error.

1. Introduction

In this paper we summarize our results mainly taken from [21, 23, 24] on a priori error analysis for space-time finite element discretizations of optimization problems governed by parabolic equations. We consider the state equation of the form

$$\begin{aligned} \partial_t u - \Delta u &= f + q \qquad \text{in } (0,T) \times \Omega, \\ u(0) &= u_0 \qquad \text{in } \Omega, \end{aligned} \tag{1.1}$$

combined with either homogeneous Dirichlet or homogeneous Neumann boundary conditions on $(0,T) \times \partial\Omega$. The control variable q is searched for in the space $Q = L^2((0,T), L^2(\Omega))$, whereas the state variable u is from $X = W(0,T)$, see Section 2 for a precise functional analytic formulation. The optimization problem is then given as

$$\text{Minimize } J(q,u) \text{ subject to (1.1) and } q \in Q_{\text{ad}}, u \in X_{\text{ad}}, \tag{1.2}$$

where $J\colon Q \times X \to \mathbb{R}$ is a quadratic cost functional (see Section 2) and the admissible sets Q_{ad} and X_{ad} are given through control and state constrains, respectively. In this paper we consider the following three situations:

(I) problems without constraints, i.e., $Q_{\text{ad}} = Q$, $X_{\text{ad}} = X$.

(II) problems with control constraints, i.e., $X_{\text{ad}} = X$ and

$$Q_{\text{ad}} := \{\, q \in Q \mid q_a \leq q(t,x) \leq q_b \text{ a.e. in } I \times \Omega \,\}, \tag{1.3}$$

with some $q_a, q_b \in \mathbb{R}$, $q_a < q_b$,

(III) problems with state and control constraints, i.e., Q_{ad} as in (II) and

$$X_{\text{ad}} := \left\{ u \in X \;\middle|\; \int_\Omega u(t,x)\omega(x)\,dx \leq b \text{ for all } t \in [0,T] \right\} \tag{1.4}$$

with $\omega \in L^2(\Omega)$ and $b \in \mathbb{R}$.

While the a priori error analysis for finite element discretizations of optimal control problems governed by elliptic equations is discussed in many publications, see, e.g., [1, 5, 11, 13, 14, 25], there are only few published results on this topic for parabolic problems.

Here, we use discontinuous finite element methods for time discretization of the state equation (1.1), as proposed, e.g., in [9, 10]. The spatial discretization is based on usual H^1-conforming finite elements. In [2] it was shown that this type of discretization allows for a natural translation of the optimality conditions from the continuous to the discrete level. This gives rise to exact computation of the derivatives required in the optimization algorithms on the discrete level. In [22] a posteriori error estimates for this type of discretization are derived and an adaptive algorithm is developed.

Throughout, we use a general discretization parameter σ consisting of three discretization parameters $\sigma = (k, h, d)$, where k corresponds to the time discretization of the state variable, h to the space discretization of the state variable, and d to the discretization of the control variable q, respectively. Although the space and time discretization of the control variable may in general differ from the discretization of the state, cf. [23, 24], we consider here the same temporal and spatial meshes for both the state and the control discretization, i.e., $d = (k_d, h_d) = (k, h)$. The aim of the a priori error analysis is to derive error estimates for the error between the optimal solution $\bar{q}$ of (1.2) and the optimal solution $\bar{q}_\sigma$ of its discrete counterpart in terms of discretization parameters k and h.

For some results on the problem without inequality constraints (I) we refer to [20, 31] and a recent paper [6]. Our a priori error estimates for this case from [23] are summarized in Section 5.1. In this case, the regularity of the solution is limited only by the regularity of the domain and the data. Therefore higher-order error estimates can be shown, if using higher-order discretization schemes, see [23]. Here, we present estimates of order $\mathcal{O}(k+h^2)$ for piecewise constant temporal and cellwise (bi-/tri-)linear spatial discretization, see the discussion after Theorem 5.1.

For the problem in the case (II), the presence of control constraints leads to some restrictions of the regularity of the optimal control $\bar{q}$, which are often reflected in a reduction of the order of convergence of finite element discretizations, see [17, 19, 28] for some known results in this case. In [24] we extended the state-of-the-art techniques known for elliptic problems to the problem under consideration. As in [24], we discuss here the following four approaches for the discretization of the control variable and the corresponding error estimates:

1. Discretization using cellwise constant ansatz functions with respect to space and time: In this case we obtain similar to [17, 19] the order of convergence $\mathcal{O}(h+k)$. The result is obtained under weaker regularity assumptions than in [17, 19]. Moreover, we separate the influences of the spatial and temporal regularity on the discretization error, see Theorem 5.2.
2. Discretization using cellwise (bi-/tri-)linear, H^1-conforming finite elements in space and piecewise constant functions in time: For this type of discretization we obtain the improved order of convergence $\mathcal{O}(k+h^{\frac{3}{2}-\frac{1}{p}})$, see Theorem 5.5. Here, p depends on the regularity of the adjoint solution. In two space dimensions we show the assertion for any $p<\infty$, whereas in three space dimensions the result is proved for $p \le 6$. Under an additional regularity assumption, one can choose $p=\infty$ leading to $\mathcal{O}(k+h^{\frac{3}{2}})$. Again the influences of spatial and temporal regularity as well as of the spatial and temporal discretizations are clearly separated.
3. The discretization following the variational approach from [14], where no explicit discretization of the control variable is used: In this case, we obtain an optimal result $\mathcal{O}(k+h^2)$, see Theorem 5.6. The usage of this approach requires a non-standard implementation and more involved stopping criteria for optimization algorithms (see a recent preprint [15] for details), since the control variable does not lie in any finite element space associated with the given mesh. However, there are no additional difficulties caused by the time discretization.
4. The post-processing strategy extending the technique from [25] to parabolic problems: In this case, we use the cellwise constant ansatz functions with respect to space and time. For the discrete solution $(\bar{q}_\sigma, \bar{u}_\sigma)$, a post-processing step based on a projection formula is proposed leading to an approximation $\tilde{q}_\sigma$ with order of convergence $\|\bar{q}-\tilde{q}_\sigma\|_{L^2((0,T),L^2(\Omega))} = \mathcal{O}(k+h^{2-\frac{1}{p}})$, see Theorem 5.7. Here, p can be chosen as discussed for the cellwise linear discretization. Under an additional regularity assumption, one can also choose $p=\infty$ leading to $\mathcal{O}(k+h^2)$.

The main difficulty in the numerical analysis of optimal control problems of the kind (III) with state constraints is the lack of regularity caused by the fact that the Lagrange multiplier corresponding to the state constraint (1.4) is a Borel measure $\mu \in C([0,T])^*$. This fact affects the regularity of the adjoint state and of the optimal control $\bar{q}$. Especially the lack of temporal regularity complicates the derivation of a priori error estimates for finite element discretizations of the optimal control problem under consideration.

The main result of [21] summarized in Section 5.3 is the estimate

$$\|\bar{q} - \bar{q}_\sigma\|_{L^2((0,T),L^2(\Omega))} \leq C\Big(\ln \frac{T}{k}\Big)^{\frac{1}{2}} \{k^{\frac{1}{2}} + h\}, \tag{1.5}$$

see Theorem 5.8. One of the essential tools for the proof of this result are error estimates with respect to the $L^\infty((0,T), L^2(\Omega))$ norm for the state equation with low regularity of the data, see Theorem 4.3 and Theorem 4.4. The derivation of these estimates in [21] is based on the techniques from [18, 26].

To the authors knowledge the result (1.5) from [21] is the first published error estimate for the optimal control problem in the case (III). For an optimal control problem with pointwise state constraints in space and time of the form

$$u(t,x) \leq b \quad \text{for all } (t,x) \in [0,T] \times \bar{\Omega} \tag{1.6}$$

the estimates of order $|\ln h|^{\frac{1}{4}}(h^{\frac{1}{2}} + k^{\frac{1}{4}})$ in 2d and $h^{\frac{1}{4}} + h^{-\frac{1}{4}}k^{\frac{1}{4}}$ in 3d are derived in [8].

The paper is organized as follows: In the next section we introduce the functional analytic setting of the considered optimization problems and describe necessary and sufficient optimality conditions for the problems under consideration. In Section 3 we present the space time finite element discretization of the optimization problems. Section 4 is devoted to the analysis of the discretization error of the state equation. There, we present a priori error estimates for the solution of the uncontrolled state equation in the norms of $L^2((0,T), L^2(\Omega))$ and $L^\infty((0,T), L^2(\Omega))$. In Section 5, we present the main results of this article, which are estimates for the error between the continuous optimal control and its numerical approximation for the three considered problem classes (I), (II), and (III). In the last section we present numerical experiments for the three considered problem classes substantiating the theoretical results.

2. Optimization problems

In this section we briefly discuss the precise formulation of the optimization problems under consideration. Furthermore, we recall theoretical results on existence, uniqueness, and regularity of optimal solutions as well as optimality conditions.

To set up a weak formulation of the state equation (1.1), we introduce the following notation: For a convex polygonal domain $\Omega \subset \mathbb{R}^n$, $n \in \{2, 3\}$, we denote V to be either $H^1(\Omega)$ or $H_0^1(\Omega)$ depending on the prescribed type of boundary conditions (homogeneous Neumann or homogeneous Dirichlet). Together with $H = L^2(\Omega)$, the Hilbert space V and its dual V^* build a Gelfand triple $V \hookrightarrow H \hookrightarrow V^*$. Here and in what follows, we employ the usual notion for Lebesgue and Sobolev spaces.

For a time interval $I = (0,T)$ we introduce the state space

$$X := \{\, v \mid v \in L^2(I,V) \text{ and } \partial_t v \in L^2(I,V^*) \,\}$$

and the control space

$$Q = L^2(I, L^2(\Omega)).$$

In addition, we use the following notations for the inner products and norms on $L^2(\Omega)$ and $L^2(I, L^2(\Omega))$:

$$(v, w) := (v, w)_{L^2(\Omega)}, \qquad (v, w)_I := (v, w)_{L^2(I,L^2(\Omega))},$$
$$\|v\| := \|v\|_{L^2(\Omega)}, \qquad \|v\|_I := \|v\|_{L^2(I,L^2(\Omega))}.$$

In this setting, a standard weak formulation of the state equation (1.1) for given control $q \in Q$, $f \in L^2(I, H)$, and $u_0 \in V$ reads: Find a state $u \in X$ satisfying

$$\begin{aligned} (\partial_t u, \varphi)_I + (\nabla u, \nabla \varphi)_I &= (f + q, \varphi)_I \qquad \forall \varphi \in X, \\ u(0) &= u_0. \end{aligned} \tag{2.1}$$

For simplicity of notation, we skip here and throughout the paper the dependence of the solution variable on x and t.

It is well known that for fixed control $q \in Q$, $f \in L^2(I, H)$, and $u_0 \in V$ there exists a unique solution $u \in X$ of problem (2.1). Moreover the solution exhibits the improved regularity

$$u \in L^2(I, H^2(\Omega) \cap V) \cap H^1(I, L^2(\Omega)) \hookrightarrow C(\bar{I}, V).$$

To formulate the optimal control problem we introduce the admissible set Q_{ad} collecting the inequality constraints (1.3) as

$$Q_{\mathrm{ad}} := \{\, q \in Q \mid q_a \le q(t, x) \le q_b \text{ a.e. in } I \times \Omega \,\},$$

where the bounds $q_a, q_b \in \mathbb{R} \cup \{\, \pm\infty \,\}$ fulfill $q_a < q_b$.

Furthermore, we define for $\omega \in H$ the functional $G\colon H \to \mathbb{R}$ by

$$G(v) := (v, \omega).$$

The application of G to time-dependent functions $u\colon I \to H$ is defined by the setting $G(u)(t) := G(u(t))$. The state constraint (1.4) can then be formulated as

$$G(u) \le b \quad \text{in } \bar{I} \tag{2.2}$$

for $b \in \mathbb{R} \cup \{\, \infty \,\}$.

Introducing the cost functional $J\colon Q \times L^2(I, H) \to \mathbb{R}$ defined as

$$J(q, u) := \frac{1}{2}\|u - \hat{u}\|_I^2 + \frac{\alpha}{2}\|q\|_I^2,$$

the weak formulation of the optimal control problem (1.2) is given as

$$\text{Minimize } J(q, u) \text{ subject to (2.1), (2.2), and } (q, u) \in Q_{\mathrm{ad}} \times X, \tag{2.3}$$

where $\hat{u} \in L^2(I, H)$ is a given desired state and $\alpha > 0$ is the regularization parameter.

Throughout, we assume the following Slater condition:

$$\exists \tilde{q} \in Q_{\mathrm{ad}} : G(u(\tilde{q})) < b \quad \text{in } \bar{I} \tag{2.4}$$

where $u(\tilde{q})$ is the solution of (2.1) for the particular control $\tilde{q}$.

Remark 2.1. Because of the prescribed initial condition $u_0 \in H$, the condition $G(u_0) < b$ is necessary for the assumed Slater condition.

By standard arguments, the existence of the Slater point $\tilde{q}$ ensures the existence and uniqueness of optimal solutions to problem (2.3).

As already mentioned in the introduction, we consider three different variants of the considered optimization problem (1.2):

(I) without constraints: $q_a = -\infty$, $q_b = b = \infty$,
(II) with control constraints: $b = \infty$,
(III) with state and control constraints: $-\infty < q_a, q_b, b < \infty$.

As discussed in [21, 23, 24], the optimal solutions of these problems exhibit the following regularities:

(I) $\bar{q} \in L^2(I, H^2(\Omega)) \cap H^1(I, L^2(\Omega))$,
(II) $\bar{q} \in L^2(I, W^{1,p}(\Omega)) \cap H^1(I, L^2(\Omega))$ with $p < \infty$ for $n = 2$ and $p \leq 6$ for $p = 3$,
(III) $\bar{q} \in L^2(I, H^1(\Omega)) \cap L^\infty(I \times \Omega)$.

To formulate optimality conditions, we employ the dual space of $C(\bar{I})$ denoted by $C(\bar{I})^*$ with the operator norm $\|\mu\|_{C(\bar{I})^*}$ and the duality product $\langle \cdot, \cdot \rangle$ between $C(\bar{I})$ and $C(\bar{I})^*$ given by

$$\langle v, \mu \rangle := \int_{\bar{I}} v \, d\mu.$$

Proposition 2.2. *A control $\bar{q} \in Q_{\mathrm{ad}}$ with associated state $\bar{u}$ is optimal solution of problem (2.3) if and only if $G(\bar{u}) \leq b$ and there exists an adjoint state $\bar{z} \in L^2(I, V)$ and a Lagrange multiplier $\mu \in C(\bar{I})^*$ with $\mu \geq 0$ such that*

$$(\partial_t \varphi, \bar{z})_I + (\nabla \varphi, \nabla \bar{z})_I = (\varphi, \bar{u} - \hat{u})_I + \langle G(\varphi), \mu \rangle \qquad \forall \varphi \in X, \ \varphi(0) = 0 \tag{2.5}$$

$$(\alpha \bar{q} + \bar{z}, q - \bar{q})_I \geq 0 \qquad \forall q \in Q_{\mathrm{ad}} \tag{2.6}$$

$$\langle b - G(\bar{u}), \mu \rangle = 0. \tag{2.7}$$

The variational inequality (2.6) can be equivalently rewritten using the pointwise projection $P_{Q_{\mathrm{ad}}}$ on the set of admissible controls Q_{ad}:

$$\bar{q} = P_{Q_{\mathrm{ad}}}\left(-\frac{1}{\alpha} \bar{z} \right). \tag{2.8}$$

Remark 2.3. This proposition reflects the general situation. In the absence of the state constraint (case (II)), the Lagrange multiplier μ vanishes and the adjoint state possesses the improved regularity $\bar{z} \in L^2(I, H^2(\Omega)) \cap H^1(I, L^2(\Omega))$. If additionally the control constraints are not present (case (I)), the variational inequality (2.6) becomes an equality.

3. Discretization

In this section we describe the space-time finite element discretization of the optimal control problem (2.3).

3.1. Semidiscretization in time

At first, we present the semidiscretization in time of the state equation by discontinuous Galerkin methods. We consider a partitioning of the time interval $\bar{I} = [0, T]$ as

$$\bar{I} = \{0\} \cup I_1 \cup I_2 \cup \cdots \cup I_M \tag{3.1}$$

with subintervals $I_m = (t_{m-1}, t_m]$ of size k_m and time points

$$0 = t_0 < t_1 < \cdots < t_{M-1} < t_M = T.$$

We define the discretization parameter k as a piecewise constant function by setting $k\big|_{I_m} = k_m$ for $m = 1, 2, \ldots, M$. Moreover, we denote by k the maximal size of the time steps, i.e., $k = \max k_m$. Moreover, we assume the following two conditions on the size of the time steps:

(i) There are constants $c, \gamma > 0$ independent of k such that

$$\min_{m=1,2,\ldots,M} k_m \geq ck^\gamma.$$

(ii) There is a constant $\kappa > 0$ independent of k such that for all $m = 1, 2, \ldots, M-1$

$$\frac{1}{\kappa} \leq \frac{k_m}{k_{m+1}} \leq \kappa.$$

(iii) It holds $k \leq \frac{T}{4}$.

The semidiscrete trial and test space is given as

$$X_k^r = \left\{ v_k \in L^2(I, V) \;\middle|\; v_k\big|_{I_m} \in \mathcal{P}_r(I_m, V),\ m = 1, 2, \ldots, M \right\}.$$

Here, $\mathcal{P}_r(I_m, V)$ denotes the space of polynomials up to order r defined on I_m with values in V. On X_k^r we use the notations

$$(v, w)_{I_m} := (v, w)_{L^2(I_m, L^2(\Omega))} \quad \text{and} \quad \|v\|_{I_m} := \|v\|_{L^2(I_m, L^2(\Omega))}.$$

To define the discontinuous Galerkin (dG(r)) approximation using the space X_k^r we employ the following definitions for functions $v_k \in X_k^r$:

$$v_{k,m}^+ := \lim_{t \to 0^+} v_k(t_m + t), \quad v_{k,m}^- := \lim_{t \to 0^+} v_k(t_m - t) = v_k(t_m), \quad [v_k]_m := v_{k,m}^+ - v_{k,m}^-$$

and define the bilinear form $B(\cdot, \cdot)$ for $u_k, \varphi \in X_k^r$ by

$$B(u_k, \varphi) := \sum_{m=1}^{M} (\partial_t u_k, \varphi)_{I_m} + (\nabla u_k, \nabla \varphi)_I + \sum_{m=2}^{M} ([u_k]_{m-1}, \varphi_{m-1}^+) + (u_{k,0}^+, \varphi_0^+). \tag{3.2}$$

Then, the dG(r) semidiscretization of the state equation (2.1) for a given control $q \in Q$ reads: Find a state $u_k = u_k(q) \in X_k^r$ such that

$$B(u_k, \varphi) = (f + q, \varphi)_I + (u_0, \varphi_0^+) \quad \forall \varphi \in X_k^r. \tag{3.3}$$

The existence and uniqueness of solutions to (3.3) can be shown by using Fourier analysis, see [30] for details.

Remark 3.1. Using a density argument, it is possible to show that the exact solution $u = u(q) \in X$ also satisfies the identity

$$B(u, \varphi) = (f + q, \varphi)_I + (u_0, \varphi_0^+) \quad \forall \varphi \in X_k^r.$$

Thus, we have here the property of Galerkin orthogonality

$$B(u - u_k, \varphi) = 0 \quad \forall \varphi \in X_k^r,$$

although the dG(r) semidiscretization is a nonconforming Galerkin method ($X_k^r \not\subset X$).

Throughout the paper we restrict ourselves to the case $r = 0$. The resulting dG(0) scheme is a variant of the implicit Euler method. Because of this, the notation for the discontinuous piecewise constant functions $v_k \in X_k^0$ can be simplified. We set $v_{k,m} := v_{k,m}^-$.Then, this implies $v_{k,m}^+ = v_{k,m+1}$ and $[v_k]_m = v_{k,m+1} - v_{k,m}$.

Since $u_k \in X_k^0$ is piecewise constant in time, the state constraint $G(u_k) \leq b$ can be written as finitely many constraints:

$$G(u_k)\big|_{I_m} \leq b \quad \text{for } m = 1, 2, \dots, M. \tag{3.4}$$

The semi-discrete optimization problem for the dG(0) time discretization has the form:

$$\text{Minimize } J(q_k, u_k) \text{ subject to (3.3), (3.4), and } (q_k, u_k) \in Q_{\mathrm{ad}} \times X_k^0. \tag{3.5}$$

Remark 3.2. Note, that the optimal control $\bar{q}_k$ is searched for in the subset Q_{ad} of the continuous space Q and the subscript k indicates the usage of the semidiscretized state equation.

Similar to the continuous setting, we can formulate the following optimality condition:

Proposition 3.3. *A control $\bar{q}_k \in Q_{\mathrm{ad}}$ with associated state $\bar{u}_k$ is optimal solution of problem* (3.5) *if and only if $G(\bar{u}_k)\big|_{I_m} \leq b$ for $m = 1, 2, \dots, M$ and there exists an adjoint state $\bar{z}_k \in X_k^0$ and a Lagrange multiplier $\mu_k \in C(\bar{I})^*$ given for any $v \in C(\bar{I})$ by*

$$\langle v, \mu_k \rangle = \sum_{l=1}^{M} \frac{\mu_{k,l}}{k_l} \int_{I_l} v(t)\, dt \quad \textit{with} \quad \mu_{k,l} \in \mathbb{R}_+ \ (l = 1, 2, \dots, M) \tag{3.6}$$

such that

$$B(\varphi, \bar{z}_k) = (\varphi, \bar{u}_k - \hat{u})_I + \langle G(\varphi), \mu_k \rangle \qquad \forall \varphi \in X_k^0 \tag{3.7}$$

$$(\alpha \bar{q}_k + \bar{z}_k, q - \bar{q}_k)_I \geq 0 \qquad \forall q \in Q_{\mathrm{ad}} \tag{3.8}$$

$$\langle b - G(\bar{u}_k), \mu_k \rangle = 0. \tag{3.9}$$

Similar to the discussion in Remark 2.3, these optimality conditions can be simplified in the absence of constraints.

3.2. Discretization in space

To define the finite element discretization in space, we consider two- or three-dimensional shape-regular meshes, see, e.g., [7]. A mesh $\mathcal{T}_h = \{K\}$ consists of quadrilateral or hexahedral cells K, which constitute a non-overlapping cover of the computational domain Ω. Here, we define the discretization parameter h as a cellwise constant function by setting $h\big|_K = h_K$ with the diameter h_K of the cell K. We use the symbol h also for the maximal cell size, i.e., $h = \max h_K$.

On the mesh $\mathcal{T}_h$ we construct a conform finite element space $V_h \subset V$ in a standard way:

$$V_h^s = \left\{ v \in V \;\middle|\; v\big|_K \in \mathcal{Q}_s(K) \text{ for } K \in \mathcal{T}_h \right\}.$$

Here, $\mathcal{Q}_s(K)$ consists of shape functions obtained via (bi-/tri-)linear transformations of polynomials in $\widehat{\mathcal{Q}_s}(\widehat{K})$ defined on the reference cell $\widehat{K} = (0,1)^n$, where

$$\widehat{\mathcal{Q}_s}(\widehat{K}) = \operatorname{span}\left\{ \prod_{j=1}^{n} x_j^{\alpha_j} \;\middle|\; \alpha_j \in \mathbb{N}_0,\ \alpha_j \le s \right\}.$$

To obtain the fully discretized versions of the time discretized state equation (3.3), we utilize the space-time finite element space

$$X_{k,h}^{r,s} = \left\{ v_{kh} \in L^2(I, V_h^s) \;\middle|\; v_{kh}\big|_{I_m} \in \mathcal{P}_r(I_m, V_h^s) \right\} \subset X_k^r.$$

The so-called cG(s)dG(r) discretization of the state equation for given control $q \in Q$ has the form: Find a state $u_{kh} = u_{kh}(q) \in X_{k,h}^{r,s}$ such that

$$B(u_{kh}, \varphi) = (f + q, \varphi)_I + (u_0, \varphi_0^+) \quad \forall \varphi \in X_{k,h}^{r,s}. \tag{3.10}$$

Throughout this paper we will restrict ourselves to the consideration of (bi-/tri-)linear elements, i.e., we set $s = 1$ and consider the cG(1)dG(0) scheme. The state constraint on this level of discretization is given as in Section 3.1 by

$$G(u_{kh})\big|_{I_m} \le b \quad \text{for } m = 1, 2, \dots, M. \tag{3.11}$$

Then, the corresponding optimal control problem is given as

$$\text{Minimize } J(q_{kh}, u_{kh}) \text{ subject to (3.10), (3.11), and } (q_{kh}, u_{kh}) \in Q_{\text{ad}} \times X_{k,h}^{0,1}. \tag{3.12}$$

The optimality conditions for this problem can directly be translated from the time-discrete level by replacing X_k^0 by $X_{k,h}^{0,1}$ and $\bar{z}_k$ by $\bar{z}_{kh}$ fulfilling

$$B(\varphi, \bar{z}_{kh}) = (\varphi, \bar{u}_{kh} - \hat{u})_I + \langle G(\varphi), \mu_{kh} \rangle \quad \forall \varphi \in X_{k,h}^{0,1} \tag{3.13}$$

3.3. Discretization of the controls

In this section, we describe four different approaches for the discretization of the control variable. Choosing a subspace $Q_d \subset Q$, we introduce the corresponding admissible set

$$Q_{d,\text{ad}} = Q_d \cap Q_{\text{ad}}.$$

Note, that in what follows the space Q_d will be either finite dimensional or the whole space Q. The optimal control problem on this level of discretization is given as

$$\text{Minimize } J(q_\sigma, u_\sigma) \text{ subject to } (3.10), (3.11) \text{ and } (q_\sigma, u_\sigma) \in Q_{d,\text{ad}} \times X_{k,h}^{0,1}. \quad (3.14)$$

The optimality conditions on this level of discretization are formulated similar as above, cf. [21].

3.3.1. Cellwise constant discretization. The first possibility for the control discretization is to use cellwise constant functions. Employing the same time partitioning and the same spatial mesh as for the discretization of the state variable we set

$$Q_d = \left\{ q \in Q \;\middle|\; q\big|_{I_m \times K} \in \mathcal{P}_0(I_m \times K),\ m = 1, 2, \ldots, M,\ K \in \mathcal{T}_h \right\}.$$

The discretization error for this type of discretization is analyzed for the problem classes (I) and (II), and (III) in the Sections 5.1.1, 5.2.1, and 5.3, respectively.

3.3.2. Cellwise linear discretization. Another possibility for the discretization of the control variable is to choose the control discretization as for the state variable, i.e., piecewise constant in time and cellwise (bi-/tri-)linear in space. Using a spatial space

$$Q_h = \left\{ v \in C(\bar{\Omega}) \mid v\big|_K \in \mathcal{Q}_1(K) \text{ for } K \in \mathcal{T}_h \right\}$$

we set

$$Q_d = \left\{ q \in Q \;\middle|\; q\big|_{I_m} \in \mathcal{P}_0(I_m, Q_h) \right\}.$$

The state space $X_{k,h}^{0,1}$ coincides with the control space Q_d in case of homogeneous Neumann boundary conditions and is a subspace of it, i.e., $Q_d \supset X_{k,h}^{0,1}$ in the presence of homogeneous Dirichlet boundary conditions.

The discretization error for this type of discretization is analyzed for the problem classes (I) and (II) in the Sections 5.1.2 and 5.2.2, respectively.

3.3.3. Variational approach. Extending the discretization approach presented in [14], we can choose $Q_d = Q$. In this case the optimization problems (3.12) and (3.14) coincide and therefore, $\bar{q}_\sigma = \bar{q}_{kh} \in Q_{\text{ad}}$.

We use the fact that the optimality condition can be rewritten employing the projection (2.8) as

$$\bar{q}_{kh} = P_{Q_{\text{ad}}}\left(-\frac{1}{\alpha} z_{kh}(\bar{q}_{kh}) \right),$$

and obtain that $\bar{q}_{kh}$ is piecewise constant function in time. However, $\bar{q}_{kh}$ is in general not a finite element function corresponding to the spatial mesh $\mathcal{T}_h$. This fact requires more care for the construction of algorithms for computation of $\bar{q}_{kh}$, see [14, 15] for details.

The discretization error for this type of discretization will be analyzed for the problem classes (II) in Section 5.2.3.

3.3.4. Post-processing strategy. The strategy described in this section extends the approach from [25] to parabolic problems. For the discretization of the control space we employ the same choice as in Section 3.3.1, i.e., cellwise constant discretization. After the computation of the corresponding solution $\bar{q}_\sigma$, a better approximation $\tilde{q}_\sigma$ is constructed by a post-processing step making use of the projection operator (2.8):

$$\tilde{q}_\sigma = P_{Q_{\mathrm{ad}}}\left(-\frac{1}{\alpha} z_{kh}(\bar{q}_\sigma)\right). \tag{3.15}$$

Note, that similar to the solution obtained by variational approach in Section 3.3.3, the solution $\tilde{q}_\sigma$ is piecewise constant in time and is general not a finite element function in space with respect to the spatial mesh $\mathcal{T}_h$. This solution can be simply evaluated pointwise, however, the corresponding error analysis requires an additional assumption on the structure of active sets, see the discussion for the problem classes (II) in Section 5.2.4.

4. Analysis of the discretization error for the state equation

The goal of this section is to provide a priori error estimates for the discretization error of the (uncontrolled) state equation.

Let $u \in X$ be the solution of the state equation (2.1) for $q = 0$, $u_k \in X_k^r$ be the solution of the corresponding semidiscretized equation (3.3), and $u_{kh} \in X_{k,h}^{r,s}$ be the solution of the fully discretized state equation (3.10). To separate the influences of the space and time discretization, we split the total discretization error $e := u - u_{kh}$ in its temporal part $e_k := u - u_k$ and its spatial part $e_h := u_k - u_{kh}$.

The proof of the following two theorems can be found in [23]:

Theorem 4.1. *For the error $e_k := u - u_k$ between the continuous solution $u \in X$ of (2.1) and the* dG(0) *semidiscretized solution $u_k \in X_k^0$ of (3.3) with $q = 0$, we have the error estimate*

$$\|e_k\|_I \leq Ck\|\partial_t u\|_I \leq Ck\{\|f\|_I + \|\nabla u_0\|\},$$

where the constant C is independent of the size of the time steps k.

Theorem 4.2. *For the error $e_h := u_k - u_{kh}$ between the* dG(0) *semidiscretized solution $u_k \in X_k^0$ of (3.3) and the fully* cG(1)dG(0) *discretized solution $u_{kh} \in X_{k,h}^{0,1}$ of (3.10) with $q = 0$, we have the error estimate*

$$\|e_h\|_I \leq Ch^2\|\nabla^2 u_k\|_I \leq Ch^2\{\|f\|_I + \|\nabla u_0\|\},$$

where the constant C is independent of the mesh size h and the size of the time steps k.

The following two theorems providing error estimates in the $L^\infty(I, L^2(\Omega))$ norm are proved in [21]:

Theorem 4.3. *For the error $e_k := u - u_k$ between the continuous solution $u \in X$ of (2.1) and the* dG(0) *semidiscretized solution $u_k \in X_k^0$ of (3.3) with $q = 0$, we have the error estimate*

$$\|e_k\|_{L^\infty(I,L^2(\Omega))} \leq Ck\Big(\ln \frac{T}{k}\Big)^{\frac{1}{2}} \big\{\|f\|_{L^\infty(I,L^2(\Omega))} + \|\Delta u_0\|\big\}.$$

Theorem 4.4. *For the error $e_h := u_k - u_{kh}$ between the* dG(0) *semidiscretized solution $u_k \in X_k^0$ of (3.3) and the fully* cG(1)dG(0) *discretized solution $u_{kh} \in X_{k,h}^{0,1}$ of (3.10) with $q = 0$, we have the error estimate*

$$\|e_h\|_{L^\infty(I,L^2(\Omega))} \leq Ch^2 \ln \frac{T}{k} \big\{\|f\|_{L^\infty(I,L^2(\Omega))} + \|\Delta u_0\|\big\}.$$

5. Error analysis for the optimal control problem

In this section, we present the main results of this article, namely the estimates of the error between the solution $\bar{q}$ of the continuous optimal control problem (2.3) and the solution $\bar{q}_\sigma$ of the discretized problem (3.14) for the three considered problem classes (I), (II), and (III).

Throughout this section, we will indicate the dependence of the state and the adjoint state on the specific control $q \in Q$ by the notations introduced in Section 2 and Section 3 like $u(q)$, $z(q)$ on the continuous level, $u_k(q)$, $z_k(q)$ on the semidiscrete and $u_{kh}(q)$, $z_{kh}(q)$ on the discrete level.

5.1. Problems without constraints

In the case (I) with no constraints, the following error estimate was proved in [23]:

Theorem 5.1. *The error between the solution $\bar{q} \in Q$ of the continuous optimization problem (2.3) and the solution $\bar{q}_\sigma \in Q_d$ of the discrete optimization problem (3.14) can be estimated as*

$$\begin{aligned}\|\bar{q} - \bar{q}_\sigma\|_I \leq \frac{C}{\alpha} k\big\{\|\partial_t u(\bar{q})\|_I + \|\partial_t z(\bar{q})\|_I\big\}\\ + \frac{C}{\alpha} h^2\big\{\|\nabla^2 u_k(\bar{q})\|_I + \|\nabla^2 z_k(\bar{q})\|_I\big\} + \Big(2 + \frac{C}{\alpha}\Big) \inf_{p_d \in Q_d} \|\hat{q} - p_d\|_I,\end{aligned}$$

where $\hat{q} \in Q$ can be chosen either as the continuous solution $\bar{q}$ or as the solution $\bar{q}_{kh}$ of the purely state discretized problem (3.12). The constants C are independent of the mesh size h, the size of the time steps k and the choice of the discrete control space $Q_d \subset Q$.

To concertize the result of Theorem 5.1, we discuss two possibilities for the control discretization, cf. Section 3.3:

5.1.1. Cellwise constant discretization. In the case of discretizing the controls by cellwise constant polynomials in space and time (cf. Section 3.3.1), the infimum term of the error estimation from Theorem 5.1 has to be taken into account leading to the discretization error of order

$$\|\bar{q} - \bar{q}_\sigma\|_I = \mathcal{O}(k + h).$$

5.1.2. Cellwise linear discretization. Here, the control is discretized using piecewise constants in time and cellwise (bi-/tri-)linear functions in space, cf. Section 3.3.2. In this case, the infimum term in the estimate of Theorem 5.1 vanishes since $Q_d \supset X_{k,h}^{0,1}$. Thus, Theorem 5.1 implies for the discretization error to be of order

$$\|\bar{q} - \bar{q}_\sigma\|_I = \mathcal{O}(k + h^2).$$

5.2. Problems with control constraints

In this section we provide a priori error estimates for the case (II) of purely control constraint problems when discretizing the control by one of the different discretization approaches described in Section 3. The proofs of the presented theorems can be found in [24].

5.2.1. Cellwise constant discretization. In this subsection we provide an estimate for the error $\|\bar{q} - \bar{q}_\sigma\|_I$ when the control is discretized by cellwise constant polynomials in space and time, see Section 3.3.1.

Theorem 5.2. *Let $\bar{q} \in Q_{\text{ad}}$ be the solution of the optimal control problem* (2.3), *$\bar{q}_\sigma \in Q_{d,\text{ad}}$ be the solution of the discretized problem* (3.14), *where the cellwise constant discretization for the control variable is employed. Then the following estimate holds:*

$$\begin{aligned}\|\bar{q} - \bar{q}_\sigma\|_I &\le \frac{C}{\alpha} k \left\{ \|\partial_t \bar{q}\|_I + \|\partial_t u(\bar{q})\|_I + \|\partial_t z(\bar{q})\|_I \right\} \\ &\quad + \frac{C}{\alpha} h \left\{ \|\nabla \bar{q}\|_I + \|\nabla z(\bar{q})\|_I + h \left(\|\nabla^2 u_k(\bar{q})\|_I + \|\nabla^2 z_k(\bar{q})\|_I \right) \right\} = \mathcal{O}(k + h).\end{aligned}$$

5.2.2. Cellwise linear discretization. This subsection is devoted to the error analysis for the discretization of the control variable by piecewise constants in time and cellwise (bi-/tri-)linear functions in space as described in Section 3.3.2.

The analysis is based on an assumption on the structure of the active sets. For each time interval I_m we group the cells K of the mesh $\mathcal{T}_h$ depending on the value of $\bar{q}_k$ on K into three sets $\mathcal{T}_h = \mathcal{T}_{h,m}^1 \cup \mathcal{T}_{h,m}^2 \cup \mathcal{T}_{h,m}^3$ with $\mathcal{T}_{h,m}^i \cap \mathcal{T}_{h,m}^j = \emptyset$ for $i \neq j$. The sets are chosen as follows:

$$\begin{aligned}\mathcal{T}_{h,m}^1 &:= \{\, K \in \mathcal{T}_h \mid \bar{q}_k(t_m, x) = q_a \text{ or } \bar{q}_k(t_m, x) = q_b \text{ for all } x \in K \,\} \\ \mathcal{T}_{h,m}^2 &:= \{\, K \in \mathcal{T}_h \mid q_a < \bar{q}_k(t_m, x) < q_b \text{ for all } x \in K \,\} \\ \mathcal{T}_{h,m}^3 &:= \mathcal{T}_h \setminus (\mathcal{T}_{h,m}^1 \cup \mathcal{T}_{h,m}^2).\end{aligned}$$

Hence, the set $\mathcal{T}_{h,m}^3$ consists of the cells which contain the free boundary between the active and the inactive sets for the time interval I_m.

Assumption 5.3. We assume that there exists a positive constant C independent of k, h, and m such that

$$\sum_{K\in\mathcal{T}^3_{h,m}} |K| \le Ch$$

separately for all $m = 1, 2, \ldots, M$.

Remark 5.4. A similar assumption is used in [3, 25, 29]. This assumption is valid if the boundary of the level sets

$$\{\, x\in\Omega \mid \bar{q}_k(t_m,x) = q_a \,\} \quad \text{and} \quad \{\, x\in\Omega \mid \bar{q}_k(t_m,x) = q_b \,\}$$

consists of a finite number of rectifiable curves.

Theorem 5.5. *Let $\bar{q}\in Q_{\text{ad}}$ be the solution of the optimal control problem (2.3), $\bar{q}_k \in Q_{\text{ad}}$ be the solution of the semidiscretized optimal control problem (3.5) and $\bar{q}_\sigma \in Q_{d,\text{ad}}$ be the solution of the discrete problem (3.14), where the cellwise (bi-/tri-) linear discretization for the control variable is employed. Then, if Assumption 5.3 is fulfilled, the following estimate holds provided $z_k(\bar{q}_k)\in L^2(I, W^{1,p}(\Omega))$ for $n < p \le \infty$:*

$$\|\bar{q}-\bar{q}_\sigma\|_I \le \frac{C}{\alpha} k\bigl\{\|\partial_t u(\bar{q})\|_I + \|\partial_t z(\bar{q})\|_I\bigr\} + \frac{C}{\alpha}\Bigl(1+\frac{1}{\alpha}\Bigr)\bigl\{h^2\|\nabla^2 u_k(\bar{q}_k)\|_I + h^2\|\nabla^2 z_k(\bar{q}_k)\|_I + h^{\frac{3}{2}-\frac{1}{p}}\|\nabla z_k(\bar{q}_k)\|_{L^2(I,L^p(\Omega))}\bigr\} = \mathcal{O}(k + h^{\frac{3}{2}-\frac{1}{p}}).$$

In the sequel we discuss the result from Theorem 5.5 in more details. This result holds under the assumption that $z_k(\bar{q}_k)\in L^2(I, W^{1,p}(\Omega))$. From the stability result in Proposition 4.1 from [24] and the fact that $\bar{q}_k\in Q_{\text{ad}}$, we know that

$$\|z_k(\bar{q}_k)\|_{L^2(I,H^2(\Omega))} \le C.$$

By a Sobolev embedding theorem we have $H^2(\Omega)\hookrightarrow W^{1,p}(\Omega)$ for all $p<\infty$ in two space dimensions and for $p\le 6$ in three dimensions. This implies the order of convergence $\mathcal{O}\bigl(k + h^{\frac{3}{2}-\frac{1}{p}}\bigr)$ for all $2 < p < \infty$ in 2d and $\mathcal{O}\bigl(k + h^{\frac{4}{3}}\bigr)$ in 3d, respectively. If in addition $\|z_k(\bar{q}_k)\|_{L^2(I,W^{1,\infty}(\Omega))}$ is bounded, then we have in both cases the order of convergence $\mathcal{O}\bigl(k + h^{\frac{3}{2}}\bigr)$.

5.2.3. Variational approach. In this subsection we provide an estimate for the error $\|\bar{q}-\bar{q}_\sigma\|_I$ in the case of no control discretization, see Section 3.3.3. In this case we choose $Q_d = Q$ and thus, $Q_{d,\text{ad}} = Q_{\text{ad}}$. This implies $\bar{q}_\sigma = \bar{q}_{kh}$.

Theorem 5.6. *Let $\bar{q}\in Q_{\text{ad}}$ be the solution of optimization problem (2.3) and $\bar{q}_{kh} \in Q_{\text{ad}}$ be the solution of the discretized problem (3.12). Then the following estimate holds:*

$$\|\bar{q}-\bar{q}_{kh}\|_I \le \frac{C}{\alpha} k\bigl\{\|\partial_t u(\bar{q})\|_I + \|\partial_t z(\bar{q})\|_I\bigr\} + \frac{C}{\alpha} h^2\bigl\{\|\nabla^2 u_k(\bar{q})\|_I + \|\nabla^2 z_k(\bar{q})\|_I\bigr\} = \mathcal{O}(k+h^2).$$

5.2.4. Post-processing strategy. In this subsection, we extend the post-processing techniques initially proposed in [25] to the parabolic case. As described in Section 3.3.4 we discretize the control by piecewise constants in time and space. To improve the quality of the approximation, we additionally employ the post-processing step (3.15).

Theorem 5.7. *Let $\bar{q} \in Q_{\mathrm{ad}}$ be the solution of the optimal control problem* (2.3)*, $\bar{q}_k \in Q_{\mathrm{ad}}$ be the solution of the semidiscretized optimization problem* (3.5)*, and $\tilde{q}_\sigma \in Q_{d,\mathrm{ad}}$ be the solution of the discrete problem* (3.14)*, where the cellwise constant discretization for the control variable is employed. Let, moreover, Assumption* 5.3 *be fulfilled and $z_k(\bar{q}_k) \in L^2(I, W^{1,p}(\Omega))$ for $n < p \leq \infty$. Then, it holds*

$$\begin{aligned}\|\bar{q} - \tilde{q}_\sigma\|_I \leq \frac{C}{\alpha}\Big(1 + \frac{1}{\alpha}\Big) k\big\{\|\partial_t u(\bar{q})\|_I + \|\partial_t z(\bar{q})\|_I\big\}\\ + \frac{C}{\alpha}\Big(1 + \frac{1}{\alpha}\Big) h^2\big\{\|\nabla^2 u_k(\bar{q}_k)\|_I + \frac{1}{\alpha}\|\nabla z_k(\bar{q}_k)\|_I + \Big(1 + \frac{1}{\alpha}\Big)\|\nabla^2 z_k(\bar{q}_k)\|_I\big\}\\ + \frac{C}{\alpha^2}\Big(1 + \frac{1}{\alpha}\Big) h^{2-\frac{1}{p}}\|\nabla z_k(\bar{q}_k)\|_{L^2(I,L^p(\Omega))} = \mathcal{O}\big(k + h^{2-\frac{1}{p}}\big).\end{aligned}$$

The choice of p in Theorem 5.7 follows the description in Section 5.2.2 requiring $z_k(\bar{q}_k) \in L^2(I, W^{1,p}(\Omega))$. Due to the fact that $\|z_k(\bar{q}_k)\|_{L^2(I,H^2(\Omega))}$ is bounded independently of k, the result of Theorem 5.7 holds for any $n < p < \infty$ in the two-dimensional case, leading to the order of convergence $\mathcal{O}\big(k + h^{2-\frac{1}{p}}\big)$. In three space dimensions, we obtain $p = 6$ and therefore $\mathcal{O}\big(k + h^{\frac{11}{6}}\big)$. If in addition $\|z_k(\bar{q}_k)\|_{L^2(I,W^{1,\infty}(\Omega))}$ is bounded, then we have in both cases the order of convergence $\mathcal{O}(k + h^2)$.

5.3. Problems with state and control constraints

In this section, we consider the full problem (2.3) with state and control constraints, i.e., case (III). Compared to the cases (I) and (II) which were analyzed in the sections before, the order of convergence of the discretization error is reduced here due to the lack of regularity of the adjoint state. This is caused by the Lagrange multiplier corresponding to the state constraint which vanishes in the cases (I) and (II). Employing the estimates in the $L^\infty(I, L^2(\Omega))$ norm for the state equation from Section 4, we obtain the following result which is proved in [21]:

Theorem 5.8. *Let $\bar{q} \in Q_{\mathrm{ad}}$ be the solution of the optimal control problem* (2.3) *with optimal state $\bar{u} \in X$ and $\bar{q}_\sigma \in Q_{d,\mathrm{ad}}$ be the solution of the discrete optimal control problem* (3.14) *with discrete optimal state $\bar{u}_\sigma \in X^{0,1}_{k,h}$. Then the following estimate holds:*

$$\sqrt{\alpha}\|\bar{q} - \bar{q}_\sigma\|_I \leq C\Big\{k^{\frac{1}{2}}\Big(\ln\frac{T}{k}\Big)^{\frac{1}{4}} + h\Big(\ln\frac{T}{k}\Big)^{\frac{1}{2}} + \frac{1}{\sqrt{\alpha}}h\Big\}.$$

6. Numerical results

6.1. Problems without constraints

In this section, we validate the a priori error estimates for the error in the control variable numerically. To this end, we consider the following concretion of the model problem (2.3) with known analytical exact solution on $I \times \Omega = (0, 0.1) \times (0, 1)^2$ and homogeneous Dirichlet boundary conditions. The right-hand side f, the desired state $\hat{u}$, and the initial condition u_0 are given for $x = (x_1, x_2)^T \in \Omega$ in terms of the eigenfunctions

$$w_a(t, x) := \exp(a\pi^2 t) \sin(\pi x_1) \sin(\pi x_2), \quad a \in \mathbb{R}$$

of the operator $\pm\partial_t - \Delta$ as

$$\begin{aligned} f(t, x) &:= -\pi^4 w_a(T, x), \\ \hat{u}(t, x) &:= \frac{a^2 - 5}{2 + a}\pi^2 w_a(t, x) + 2\pi^2 w_a(T, x), \\ u_0(x) &:= \frac{-1}{2 + a}\pi^2 w_a(0, x). \end{aligned}$$

For this choice of data and with the regularization parameter α chosen as $\alpha = \pi^{-4}$, the optimal solution triple $(\bar{q}, \bar{u}, \bar{z})$ of the optimal control problem (2.3) is given by

$$\begin{aligned} \bar{q}(t, x) &:= -\pi^4 \{w_a(t, x) - w_a(T, x)\}, \\ \bar{u}(t, x) &:= \frac{-1}{2 + a}\pi^2 w_a(t, x), \\ \bar{z}(t, x) &:= w_a(t, x) - w_a(T, x). \end{aligned}$$

We validate the estimates developed in the previous section by separating the discretization errors. That is, we consider at first the behavior of the error for a sequence of discretizations with decreasing size of the time steps and a fixed spatial triangulation with $N = 1089$ nodes. Secondly, we examine the behavior of the error under refinement of the spatial triangulation for $M = 2048$ time steps.

As throughout this article, the state discretization is chosen as cG(1)dG(0), i.e., $r = 0$, $s = 1$. For the control discretization we use the same temporal and spatial meshes as for the state variable and present the result for two choices of the discrete control space Q_d: cG(1)dG(0) and dG(0)dG(0). For the following computations, we choose the free parameter a to be $-\sqrt{5}$. For this choice the right-hand side f and the desired state $\hat{u}$ do not depend on time what avoids side effects introduced by numerical quadrature.

The optimal control problems are solved by the optimization library RoDoBo [27] using a conjugate gradient method applied to the reduced version of the discrete problem (3.14).

Figure 1(a) depicts the development of the error under refinement of the temporal step size k. Up to the spatial discretization error it exhibits the proved convergence order $\mathcal{O}(k)$ for both kinds of spatial discretization of the control space.

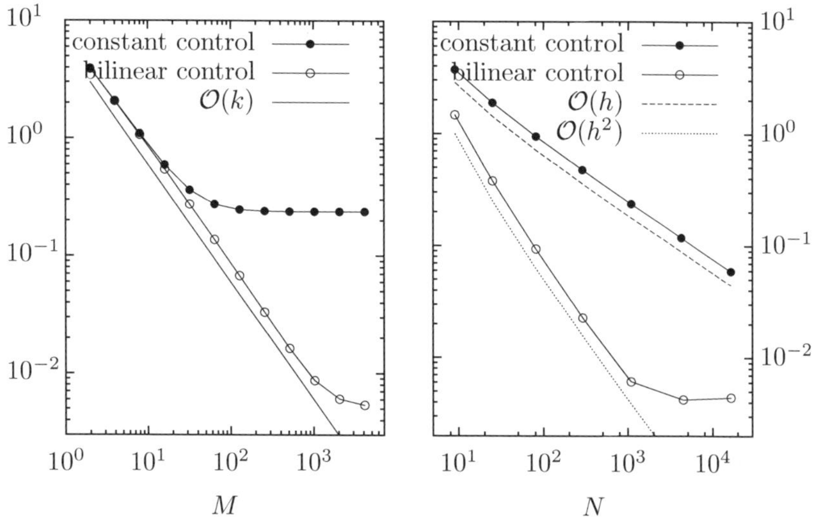

(a) Refinement of the time steps for $N = 1089$ spatial nodes

(b) Refinement of the spatial triangulation for $M = 2048$ time steps

FIGURE 1. Discretization error $\|\bar{q} - \bar{q}_\sigma\|_I$

For piecewise constant control (dG(0)dG(0) discretization), the spatial discretization error is already reached at 128 time steps, whereas in the case of bilinear control (cG(1)dG(0) discretization), the number of time steps could be increased up to $M = 4096$ until reaching the spatial accuracy.

In Figure 1(b) the development of the error in the control variable under spatial refinement is shown. The expected order $\mathcal{O}(h)$ for piecewise constant control (dG(0)dG(0) discretization) and $\mathcal{O}(h^2)$ for bilinear control (cG(1)dG(0) discretization) is observed.

6.2. Problems with control constraints

As in the previous section, we validate the a priori error estimates for the error in the control variable using a concrete optimal control problem with known exact solution on $I \times \Omega = (0, 0.1) \times (0,1)^2$ with homogeneous Dirichlet boundary conditions. The right-hand side f, the desired state $\hat{u}$, and the initial condition u_0 are given for $x = (x_1, x_2)^T \in \Omega$ in terms of w_a defined above as

$$\begin{aligned} f(t,x) &:= -\pi^4 w_a(t,x) - P_{Q_{\mathrm{ad}}}\big(-\pi^4\{w_a(t,x) - w_a(T,x)\}\big), \\ \hat{u}(t,x) &:= \frac{a^2-5}{2+a}\pi^2 w_a(t,x) + 2\pi^2 w_a(T,x), \\ u_0(x) &:= \frac{-1}{2+a}\pi^2 w_a(0,x), \end{aligned}$$

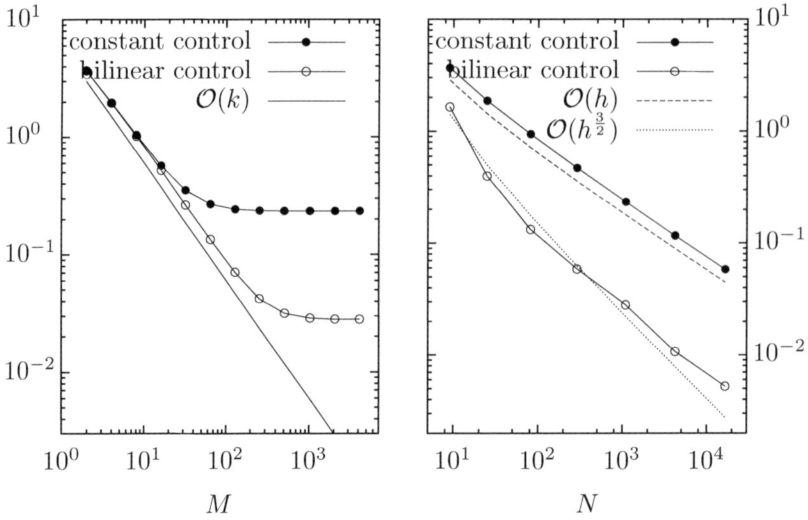

(a) Refinement of the time steps for $N = 1089$ spatial nodes

(b) Refinement of the spatial triangulation for $M = 2048$ time steps

FIGURE 2. Discretization error $\|\bar{q} - \bar{q}_\sigma\|_I$

with $P_{Q_{\text{ad}}}$ given by (2.8) with $q_a = -70$ and $q_b = -1$. For this choice of data and with the regularization parameter α chosen as $\alpha = \pi^{-4}$, the optimal solution triple $(\bar{q}, \bar{u}, \bar{z})$ of the optimal control problem (2.3) is given by

$$\begin{aligned}
\bar{q}(t,x) &:= P_{Q_{\text{ad}}}\big(-\pi^4\{w_a(t,x) - w_a(T,x)\}\big),\\
\bar{u}(t,x) &:= \frac{-1}{2+a}\pi^2 w_a(t,x),\\
\bar{z}(t,x) &:= w_a(t,x) - w_a(T,x).
\end{aligned}$$

Again, we validate the developed estimates by separating the discretization errors using the cG(1)dG(0) discretization for the state variable. For the control discretization we use the same temporal and spatial meshes as for the state variable and present results for cG(1)dG(0) and dG(0)dG(0) control discretization. The free parameter a is here as before chosen as $-\sqrt{5}$.

The optimal control problems are solved by the optimization library RoDoBo [27] using a primal-dual active set strategy (cf. [4, 16]) in combination with a conjugate gradient method applied to the reduced version of the discrete problem (3.14).

Figure 2(a) depicts the development of the error under refinement of the temporal step size k. Up to the spatial discretization error it exhibits the proved convergence order $\mathcal{O}(k)$ for both kinds of spatial discretization of the control space.

For piecewise constant control (dG(0)dG(0) discretization), the spatial discretization error is already reached at 128 time steps, whereas in the case of bilinear control (cG(1)dG(0) discretization), the number of time steps could be increased up to $M = 1024$ until reaching the spatial accuracy. This illustrates the convergence results from Sections 5.2.1 and 5.2.2 with respect to the temporal discretization.

In Figure 2(b) the development of the error in the control variable under spatial refinement is shown. The expected order $\mathcal{O}(h)$ for piecewise constant control (dG(0)dG(0) discretization) and $\mathcal{O}(h^{\frac{3}{2}})$ for bilinear control (cG(1)dG(0) discretization) is observed. This illustrates the convergence results from the Sections 5.2.1 and 5.2.2 with respect to the spatial discretization.

6.3. Problems with state and control constraints

In this section, we validate the a priori error estimates for the error in the control variable for the considered optimization problem with state and control constraints (case (III)). To this end, we consider the following concretion of the optimal control problem (2.3) with known exact solution on $I \times \Omega = (0,1) \times (0,1)^2$ and homogeneous Dirichlet boundary conditions. For the parameters $\gamma \in (0,1)$ and $\lambda := 2\pi^2$, the right-hand side f, the desired state $\hat{u}$, and the initial condition u_0 are given for $x = (x_1, x_2)^T \in \Omega$ in terms of the functions

$$\varepsilon(t,x) := \left(e^{-\frac{\lambda}{2}} - e^{-\lambda t}\right) \sin(\pi x_1) \sin(\pi x_2),$$

$$a(t,x) := \frac{e^{\frac{\lambda}{2} - \lambda t}}{2} \varepsilon(1,x)^{1-\gamma},$$

and

$$c(t,x) := \frac{e^{\frac{\lambda}{2} - \lambda t}}{2\lambda} \varepsilon(1,x)^{1-\gamma}$$

as

$$\begin{aligned} f(t,x) &:= \frac{1}{\lambda}\left(e^{\lambda(t-1)} - 1\right) \sin(\pi x_1) \sin(\pi x_2) \\ &\quad + \frac{e^{\lambda t}}{\lambda(1-\gamma)} \begin{cases} \lambda\left(a(t,x)t + c(t,x) - \frac{a(t,x)}{2}\right) + a(t,x) & t \le \frac{1}{2} \\ \varepsilon(1,x)^{1-\gamma} - \varepsilon(t,x)^{1-\gamma} & t > \frac{1}{2} \end{cases}, \\ \hat{u}(t,x) &:= \sin(\pi x_1)\sin(\pi x_2) \\ &\quad - \frac{e^{\lambda t}}{\lambda(1-\gamma)} \begin{cases} \frac{1}{2\lambda}\varepsilon(1,x)^{1-\gamma} - a(t,x)t - c(t,x) + \frac{a(t,x)}{2} & t \le \frac{1}{2} \\ 0 & t > \frac{1}{2} \end{cases}, \\ u_0(x) &:= -\frac{1}{\lambda(1-\gamma)}\left(\frac{1}{2\lambda}\varepsilon(1,x)^{1-\gamma} - c(0,x) + \frac{a(0,x)}{2}\right). \end{aligned}$$

Furthermore, we choose the regularization parameter α as $\alpha = 1$, the weight ω in the definition of the constraint G as $\omega(x) := \sin(\pi x_1)\sin(\pi x_2)$, and the upper bound $b = 0$.

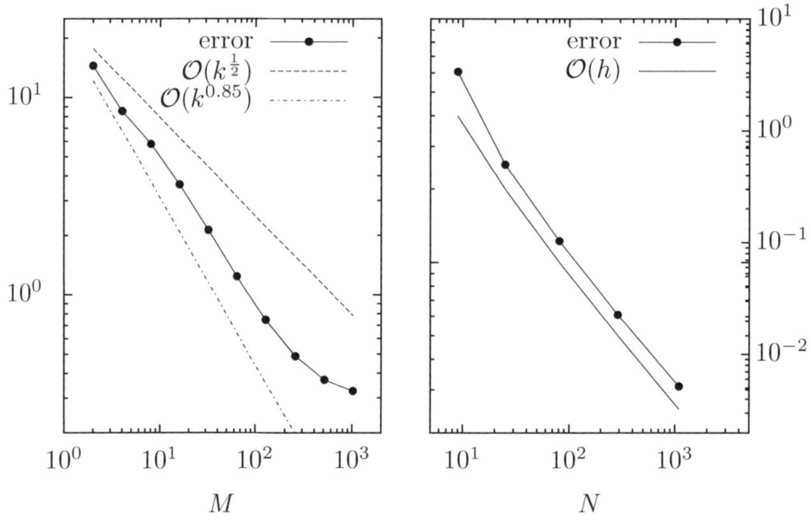

(a) Refinement of the time steps for $N = 1089$ spatial nodes

(b) Refinement of the spatial triangulation for $M = 2048$ time steps

FIGURE 3. Discretization error $\|\bar{q} - \bar{q}_\sigma\|_I$

For this choice of data, the optimal solution triple $(\bar{q}, \bar{u}, \bar{z})$ of control problem (2.3) is given by

$$\begin{aligned} \bar{q}(t,x) &:= -\frac{1}{\lambda}\big(e^{\lambda(t-1)} - 1\big)\sin(\pi x_1)\sin(\pi x_2) \\ &\quad - \frac{e^{\lambda t}}{\lambda(1-\gamma)} \begin{cases} \varepsilon(1,x)^{1-\gamma} & t \le \frac{1}{2} \\ \varepsilon(1,x)^{1-\gamma} - \varepsilon(t,x)^{1-\gamma} & t > \frac{1}{2} \end{cases}, \\ \bar{u}(t,x) &:= \hat{u}(t,x) - \sin(\pi x_1)\sin(\pi x_2), \\ \bar{z}(t,x) &:= -\bar{q}(t,x), \end{aligned}$$

and the Lagrange multiplier μ associated with the state constraint is

$$\mu(t) := \begin{cases} 0 & t \le \frac{1}{2} \\ \big(e^{-\frac{\lambda}{2}} - e^{-\lambda t}\big)^{-\gamma} & t > \frac{1}{2} \end{cases}.$$

The state discretization is again chosen as cG(1)dG(0), whereas the control variable is discretized by piecewise constants on the same temporal and spatial meshes as used for the state variable. For the following computations, we choose the parameter γ to be 0.6.

The optimal control problems are solved by the optimization library RoDoBo [27] using an interior point regularization for the state constrained problem (3.14) and Newton's method combined with an inner conjugate gradient method to solve the regularized problem.

Figure 3(a) depicts the development of the error in the control variable under refinement of the temporal step size k. Up to the spatial discretization error it exhibits at least the proved convergence order $\mathcal{O}(k^{\frac{1}{2}})$. The observed better convergence behavior of approximately $\mathcal{O}(k^{0.85})$ may be due to the constructed problem data which do not exhibit the full irregularity covered by our analysis derived in Section 5.3.

In Figure 3(b) the development of the error in the control variable under spatial refinement is shown. The expected order $\mathcal{O}(h)$ is observed.

References

[1] N. Arada, E. Casas, and F. Tröltzsch: *Error estimates for a semilinear elliptic optimal control problem.* Comput. Optim. Appl. **23** (2002), 201–229.

[2] R. Becker, D. Meidner, and B. Vexler: *Efficient numerical solution of parabolic optimization problems by finite element methods.* Optim. Methods Softw. **22** (2007), 813–833.

[3] R. Becker and B. Vexler: *Optimal control of the convection-diffusion equation using stabilized finite element methods.* Numer. Math. **106** (2007), 349–367.

[4] M. Bergounioux, K. Ito, and K. Kunisch: *Primal-dual strategy for constrained optimal control problems.* SIAM J. Control Optim. **37** (1999), 1176–1194.

[5] E. Casas, M. Mateos, and F. Tröltzsch: *Error estimates for the numerical approximation of boundary semilinear elliptic control problems.* Comput. Optim. Appl. **31** (2005), 193–220.

[6] K. Chrysafinos: *Discontinuous Galerkin approximations for distributed optimal control problems constrained by parabolic PDEs.* Int. J. Numer. Anal. Model. **4** (2007), 690–712.

[7] P.G. Ciarlet: *The Finite Element Method for Elliptic Problems*, volume 40 of *Classics Appl. Math.* SIAM, Philadelphia, 2002.

[8] K. Deckelnick and M. Hinze: *Variational discretization of parabolic control problems in the presence of pointwise state constraints.* Preprint SPP1253–08–08, DFG priority program 1253 "Optimization with PDEs", 2009.

[9] K. Eriksson, D. Estep, P. Hansbo, and C. Johnson: *Computational Differential Equations.* Cambridge University Press, Cambridge, 1996.

[10] K. Eriksson, C. Johnson, and V. Thomée: *Time discretization of parabolic problems by the discontinuous Galerkin method.* M2AN Math. Model. Numer. Anal. **19** (1985), 611–643.

[11] R. Falk: *Approximation of a class of optimal control problems with order of convergence estimates.* J. Math. Anal. Appl. **44** (1973), 28–47.

[12] *The finite element toolkit* GASCOIGNE. `http://www.gascoigne.uni-hd.de`.

[13] T. Geveci: *On the approximation of the solution of an optimal control problem governed by an elliptic equation.* M2AN Math. Model. Numer. Anal. **13** (1979), 313–328.

[14] M. Hinze: *A variational discretization concept in control constrained optimization: The linear-quadratic case.* Comput. Optim. Appl. **30** (2005), 45–61.

[15] M. Hinze and M. Vierling: *Variational discretization and semi-smooth Newton methods; implementation, convergence and globalization in PDE constrained optimization*

with control constraints (2010). Submitted. Preprint 2009-15, Hamburger Beiträge zur Angewandten Mathematik (2009).

[16] K. Kunisch and A. Rösch: *Primal-dual active set strategy for a general class of constrained optimal control problems.* SIAM J. Optim. **13** (2002), 321–334.

[17] I. Lasiecka and K. Malanowski: *On discrete-time Ritz-Galerkin approximation of control constrained optimal control problems for parabolic systems.* Control Cybern. **7** (1978), 21–36.

[18] M. Luskin and R. Rannacher: *On the smoothing property of the Galerkin method for parabolic equations.* SIAM J. Numer. Anal. **19** (1982), 93–113.

[19] K. Malanowski: *Convergence of approximations vs. regularity of solutions for convex, control-constrained optimal-control problems.* Appl. Math. Optim. **8** (1981), 69–95.

[20] R.S. McNight and W.E. Bosarge, jr.: *The Ritz-Galerkin procedure for parabolic control problems.* SIAM J. Control Optim. **11** (1973), 510–524.

[21] D. Meidner, R. Rannacher, and B. Vexler: *A priori error estimates for finite element discretizations of parabolic optimization problems with pointwise state constraints in time.* SIAM J. Control Optim. (2011). Accepted.

[22] D. Meidner and B. Vexler: *Adaptive space-time finite element methods for parabolic optimization problems.* SIAM J. Control Optim. **46** (2007), 116–142.

[23] D. Meidner and B. Vexler: *A priori error estimates for space-time finite element approximation of parabolic optimal control problems. Part I: Problems without control constraints.* SIAM J. Control Optim. **47** (2008), 1150–1177.

[24] D. Meidner and B. Vexler: *A priori error estimates for space-time finite element approximation of parabolic optimal control problems. Part II: Problems with control constraints.* SIAM J. Control Optim. **47** (2008), 1301–1329.

[25] C. Meyer and A. Rösch: *Superconvergence properties of optimal control problems.* SIAM J. Control Optim. **43** (2004), 970–985.

[26] R. Rannacher: *L^∞-Stability estimates and asymptotic error expansion for parabolic finite element equations.* Bonner Math. Schriften **228** (1991), 74–94.

[27] RoDoBo. *A C++ library for optimization with stationary and nonstationary PDEs with interface to* Gascoigne [12]. `http://www.rodobo.uni-hd.de`.

[28] A. Rösch: *Error estimates for parabolic optimal control problems with control constraints.* Z. Anal. Anwend. **23** (2004), 353–376.

[29] A. Rösch and B. Vexler: *Optimal control of the Stokes equations: A priori error analysis for finite element discretization with postprocessing.* SIAM J. Numer. Anal. **44** (2006), 1903–1920.

[30] V. Thomée: *Galerkin Finite Element Methods for Parabolic Problems*, volume 25 of *Spinger Ser. Comput. Math.* Springer, Berlin, 1997.

[31] R. Winther: *Error estimates for a Galerkin approximation of a parabolic control problem.* Ann. Math. Pura Appl. (4) **117** (1978), 173–206.

Dominik Meidner and Boris Vexler
Lehrstuhl für Mathematische Optimierung
Technische Universität München
Fakultät für Mathematik
Boltzmannstraße 3
D-85748 Garching b. München, Germany
e-mail: `meidner@ma.tum.de, vexler@ma.tum.de`

International Series of Numerical Mathematics, Vol. 160, 467–482

Numerical Analysis of State-constrained Optimal Control Problems for PDEs

Ira Neitzel and Fredi Tröltzsch

Abstract. We survey the results of SPP 1253 project "Numerical Analysis of State-constrained Optimal Control Problems for PDEs". In the first part, we consider Lavrentiev-type regularization of both distributed and boundary control. In the second part, we present a priori error estimates for elliptic control problems with finite-dimensional control space and state-constraints both in finitely many points and in all points of a subdomain with nonempty interior.

Mathematics Subject Classification (2000). 49K20, 49M05, 90C06, 90C34, 90C30.

Keywords. Optimal control, state constraints, regularization, finite element discretization.

1. Introduction

Pointwise state constraints play an important role in many real world applications of PDE optimization. For instance, in optimizing the process of hot steel profiles by spraying water on their surface, the temperature differences in the steel must be bounded in order to avoid cracks. Details may be found for example in [12]. Similar restrictions apply to the production process of bulk single crystals, where the temperature in the growth apparatus must be kept between given bounds, see, e.g., [27]. Even in medical applications, pointwise state constraints can be important, as for example in local hyperthermia in cancer treatment. There, the generated temperature in the patient's body must not exceed a certain limit, cf. [11].

All these problems share the mathematical difficulties associated with the presence of pointwise state constraints. One of the related challenges lies in the question of existence and regularity of Lagrange multipliers. For these reasons, we are interested in regularization methods for state constrained problems, where we focus here on time-dependent parabolic problems. In particular, we address a

This work was supported by DFG priority program 1253.

Lavrentiev type regularization method. The low regularity of Lagrange multipliers also presents a challenge in the numerical analysis when, e.g., trying to derive a priori discretization error estimates. We discuss this for two classes of elliptic state-constrained optimal control problems with finitely many real numbers as control variables that we discuss without regularization.

Let us survey some difficulties and questions we have been interested in with the help of two model problems. We will consider optimal control problems, respectively their regularization, of parabolic type with control u and state y, in the spatial domain $\Omega \subset \mathbb{R}^n$ and a time interval $(0, T)$. For convenience, we introduce the time-space cylinder $Q := (0, T) \times \Omega$ and its boundary $\Sigma = (0, T) \times \partial\Omega$. We consider the distributed control problem

$$(P_D) \ \min J(y, u) := \frac{1}{2} \iint\limits_Q (y - y_d)^2 \, dxdt + \frac{\nu}{2} \iint\limits_Q u^2 \, dxdt$$

subject to the semilinear heat equation

$$\begin{aligned} \partial_t y - \Delta y + d(x, t, y) &= u && \text{in } Q \\ y(\cdot, 0) &= y_0 && \text{in } \Omega \\ \partial_n y + \alpha y &= g && \text{on } \Sigma, \end{aligned}$$

the pointwise state constraints

$$y_a \leq y \leq y_b \quad \text{in } Q,$$

and optional control constraints

$$u_a \leq u \leq u_b \quad \text{in } Q.$$

Note that $d(x, t, y) \equiv 0$ can be considered to analyze a linear quadratic case. Moreover, it is possible to consider a more general elliptic differential operator A, as well as a more general objective function under certain conditions.

In addition, we are interested in boundary control problems of the form

$$(P_B) \ \min J(y, u) := \frac{1}{2} \iint\limits_Q (y - y_d)^2 \, dxdt + \frac{\nu}{2} \iint\limits_\Sigma u^2 \, dsdt$$

subject to

$$\begin{aligned} \partial_t y - \Delta y &= f && \text{in } Q \\ y(\cdot, 0) &= y_0 && \text{in } \Omega \\ \partial_n y + \alpha y &= u && \text{on } \Sigma, \end{aligned}$$

and the pointwise state constraints

$$y_a \leq y \leq y_b \quad \text{in } Q,$$

without control constraints. All appearing data is supposed to fulfill typical regularity assumption, and the boundary $\partial\Omega$ is as smooth as desired. The nonlinearity d appearing in the state equation governing (P_D) is assumed to fulfill standard Carathéodory type conditions as well as monotonicity and smoothness, so that for given control u in either $L^\infty(Q)$ or $L^\infty(\Sigma)$ the existence of a unique corresponding state $y(u) \in W(0,T)$ is guaranteed. For a precise formulation of the given setting, we refer to [32], where linear-quadratic problems without control constraints of distributed and boundary control type have been considered, as well as to [31], which is concerned with semilinear distributed control problems with state- and control-constraints.

As mentioned above, the presence of pointwise state constraints leads to difficulties in the analysis and the numerical solution of the problems. One issue is the existence of Lagrange multipliers in order to formulate first-order necessary optimality conditions of Karush-Kuhn-Tucker type. The most common approach is to assume Slater-type conditions. To apply them to pointwise state constraints, the cone of nonnegative functions must have a nonempty interior. This requires continuity of the state functions, because in L^p-spaces with $1 \leq p < \infty$ the cone of nonnegative functions has empty interior, while for $p = \infty$ the dual space is not useful. Depending on the type of problem, however, continuity is not always guaranteed. If, for example, no bounds on the control are given and the control u belongs only to $L^2(Q)$, then the continuity of the associated state y in (P_D) is only granted for spatially one-dimensional domains, cf. for example [4] or the exposition in [39] for associated regularity results. For u in $L^2(\Sigma)$, the parabolic equation in the boundary control problem (P_B) does not generally admit a continuous state y, not even if $\Omega \subset \mathbb{R}$. Therefore, these problems are not well posed a priorily in the sense that first-order necessary optimality conditions of KKT type can be formulated in useful spaces. Of course, a problem may admit a bounded control in L^∞ and an associated continuous state, but this is not clear in advance. Even if Lagrange multipliers do exist, due to the Slater point arguments they are generally only obtained in the space of regular Borel measures. It turns out that regularization concepts are useful to obtain an optimal control problem with more regular Lagrange multipliers in L^p-spaces.

Another difficulty is hidden in the formulation of second-order-sufficient conditions (SSC), which are of interest for nonlinear optimal control problems. While they can be expected for regularized problem formulations, the purely state-constrained parabolic case remains challenging even in cases where Lagrange multipliers exist due to L^∞-bounds as control constraints. For spatio-temporal control functions and pointwise state constraints given in the whole domain Q, a satisfactory theory of SSC is so far only available for one-dimensional distributed control problems, cf. [35], [7]. As part of the research in the SPP 1253, SSC for unregularized problems have been established for higher dimensions in the special setting with finitely many time-dependent controls that are found in practice more often than controls that can vary freely in space and time, cf. [8].

For all these reasons, regularization techniques have been a wide field of active research in the recent past and remain to be of interest. We mention for example a Moreau-Yosida regularization approach by Ito and Kunisch, [20], a Lavrentiev-regularization technique by Meyer, Rösch, and Tröltzsch, [28], or the virtual control concept by Krumbiegel and Rösch, [22], originally developed for elliptic boundary control problems during the first funding period of SPP 1253. Moreover, barrier methods, cf. [38], can be interpreted as regularization methods. We also point out comparisons between different approaches as in, e.g., [1], the analysis of solution algorithms as in, e.g., [15], [17], [18], or discretization error estimates from [14] or [16]. In addition, a combination of Lavrentiev regularization and interior point methods as for example in [33] has been considered. Here, Lavrentiev regularization is used to prove that the barrier method is indeed an interior point method. We lay out in Section 2 how Lavrentiev regularization techniques can be transferred to parabolic control problems, and describe an extension to parabolic boundary control problems. In addition, we comment on additional helpful properties of regularized problems that for example allow to prove a local uniqueness result of local solutions of nonlinear optimal control problems, which is an important property in the context of solving optimal control problems numerically.

A further leading question in the SPP 1253 were error estimates for the numerical approximation of state constrained control problems. Only few results on elliptic problems were known for pointwise state constraints in the whole domain. In [5], [6] convergence of finite element approximations to optimal control problems for semilinear elliptic equations with finitely many state constraints was shown for piecewise constant approximations of the control. Error estimates for elliptic state constrained distributed control functions have been derived in [9] and, with additional control constraints, in [26], [10]. Since in many applications the controls are given by finitely many real parameters, another goal of our SPP project was to investigate the error for associated state-constrained elliptic control problems. Unlike in problems with control functions, the treatment of the finitely many control parameters does not require special attention, and an error estimate without a contribution to the error due to control discretization is automatically obtained. For control *functions*, the same property is exploited by the so-called variational discretization, cf., e.g., [19]. Moreover, in problems with only finite-dimensional controls, it is not exceptional that a state constraint is only active in finitely many points. If the location of these points is known approximately, it is reasonable to prescribe the constraints in these approximate points. As part of the project and a first step towards discussing state constraints in the whole domain, such problems with finitely many state constraints have therefore been considered in [25]. The resulting optimal control problems are equivalent to finite-dimensional mathematical programming problems. Yet, the associated error analysis is not trivial. Maximum norm estimates for the finite element approximation of the semilinear state equation had to be derived. Then results of the stability analysis of nonlinear optimization were applicable to obtain also estimates for the Lagrange multipliers, which are a linear combination of Dirac measures. We survey the results in Section 3.1. If it is necessary to consider the constraints in a subset of Ω with nonempty interior, then the elliptic control problem with finite-dimensional controls is of semi-infinite type. We completed the discussion by considering such elliptic problems and report on this in Section 3.2.

2. Regularization of parabolic state-constrained problems

Let us mention here first that Lavrentiev regularized problems with additional control constraints require a more involved analysis than problems with pure regularized state constraints, and additional assumptions have to be imposed. Therefore, we consider these situations separately, beginning with the purely state-constrained case.

2.1. Problems without control constraints

2.1.1. Distributed control problems. We consider here a linear-quadratic distributed control problem, i.e., consider (P_D) with $d(x,t,y) \equiv 0$. The idea of Lavrentiev regularization for distributed control problems is to replace the pure state constraints by mixed control-state constraints of the form

$$y_a \le \lambda u + y \le y_b \quad \text{a.e. in } Q,$$

where $\lambda \in \mathbb{R}$ is a small regularization parameter. Following [29], the existence of regular Lagrange multipliers in $L^2(Q)$ for arbitrary dimension of Ω is easily shown using a simple substitution technique. The idea is to introduce the new control $w := \lambda u + y$, which yields a purely control-constrained optimal control problem. More precisely, from $w = \lambda u + y$ we obtain $u = (w - y)/\lambda$ so that the state equation can be rewritten as

$$y_t - \Delta y + d(x,t,y) + \lambda^{-1} y = \lambda^{-1} w \quad \text{in } Q,$$

and the objective function can be transformed into

$$\tilde{J}(y,w) := \frac{1}{2}\|y - y_d\|^2_{L^2(Q)} + \frac{\nu}{2\lambda^2}\|w - y - \lambda u_d\|^2_{L^2(Q)}.$$

Then, to prove the existence of regular multipliers of the transformed control constrained problem with

$$y_a \le w \le y_b \quad \text{in } Q$$

and hence of the Lavrentiev regularized problem is standard technique. They are also multipliers for the original state constraints, and are obtained without any Slater condition.

Theorem 2.1. *For each $\lambda > 0$, the linear-quadratic Lavrentiev regularized version of (P_D) admits a unique optimal control $\bar{u}_\lambda$ with associated state $\bar{y}_\lambda$. For arbitrary spatial dimension n, there exist Lagrange multipliers $\mu_a^\lambda, \mu_b^\lambda \in L^2(Q)$ and an adjoint state $p_\lambda \in W(0,T)$ such that:*

$$\begin{aligned} \partial_t \bar{y}_\lambda - \Delta \bar{y}_\lambda &= \bar{u}_\lambda & \qquad -\partial_t p_\lambda - \Delta p_\lambda &= \bar{y}^\lambda - y_d + \mu_b^\lambda - \mu_a^\lambda \\ \bar{y}_\lambda(\cdot,0) &= y_0 & p_\lambda(\cdot,T) &= 0 \\ \partial_n \bar{y}_\lambda + \alpha \bar{y}_\lambda &= g & \partial_n p_\lambda + \alpha p_\lambda &= 0, \end{aligned}$$

$$\begin{aligned} \nu \bar{u}_\lambda + p_\lambda + \lambda(\mu_b^\lambda - \mu_a^\lambda) = 0 & \qquad \textit{a.e. in } Q, \\ y_a \le \lambda \bar{u}_\lambda + \bar{y}_\lambda \le y_b & \qquad \textit{a.e. in } Q, \end{aligned}$$

$$\begin{aligned}
(\mu_a^\lambda, y_a - \lambda \bar{u}_\lambda - \bar{y}_\lambda)_{L^2(Q)} &= 0 \qquad \mu_a^\lambda \geq 0 \qquad a.e.\ in\ Q \\
(\mu_b^\lambda, \lambda \bar{u}_\lambda + \bar{y}_\lambda - y_b)_{L^2(Q)} &= 0 \qquad \mu_b^\lambda \geq 0 \qquad a.e.\ in\ Q.
\end{aligned}$$

For details, we refer to [32]. This simple substitution technique cannot be adapted to the case of additional control constraints. It is, however, possible to show a multiplier rule with L^2-Lagrange multipliers for such problems under a certain separability assumption, cf. [36]. We will consider this type of problem later in Section 2.2. Let us also state a convergence result for the regularized solutions, again for the linear-quadratic version of (P_D).

Theorem 2.2. *Let $\{\lambda_n\}$ be a sequence of positive real numbers converging to zero and denote by $\{u_n\}$ the sequence of associated optimal solutions of the regularized control problem. For $N = 1$, the sequence $\{u_n\}$ converges strongly in $L^2(Q)$ towards $\bar{u}$, where $\bar{u}$ is the unique optimal solution of the unregularized problem. If the optimal control of the unregularized problem is a function in $L^\infty(Q)$ this holds also for dimension $N > 1$.*

The proof has been carried out in detail [32]. From a practical point of view, the boundedness assumption seems reasonable. Indeed, knowing that $\bar{u}$ is essentially bounded, artificial inactive bounds on the control u can be introduced in advance, such that the convergence result from [31] holds. If in a practical application the optimal control is unbounded, then most likely additional bounds on u must be posed.

2.1.2. Boundary control problems. While the distributed control problem (P_D) without control constraints is at least well formulated in one-dimensional cases, the boundary control problem (P_B) may lack the existence of Lagrange multipliers in a suitable space as long as the optimal control is possibly unbounded. It is also quite obvious that the Lavrentiev regularization approach explained above cannot directly be applied to boundary control problems, since the control u and the state y are defined on different sets. In [32], we developed a Lavrentiev-type method for parabolic boundary control problems. Our motivation to extend the Lavrentiev regularization from the distributed case to treat such problems came from [2], where a well-known benchmark problem was introduced.

To treat the state equation in a concise way, we consider the control-to-state mapping $S\colon u \mapsto y$, $S\colon L^2(\Sigma) \to L^2(Q)$. The adjoint operator S^* maps $L^2(Q)$ into $L^2(\Sigma)$. We consider only controls u in the range of S^*, i.e., we introduce an auxiliary control $v \in L^2(Q)$ and set $u = S^*v$. Clearly, this is some smoothing of u, which is motivated by the optimality conditions for the unregularized problems, where we expect $\bar{u} = G^*\mu$ with some measure μ, if it exists. Then the state $y = y(v)$ is given by $y = SS^*v$ and the state constraints can be written as $y_a \leq SS^*v \leq y_b$. Now, we can apply our Lavrentiev regularization to these constraints, i.e., we consider

$$y_a \leq \lambda v + y(v) \leq y_b.$$

The regularizing effect comes from the restriction of $(C(\bar{Q}))^*$ to $L^2(Q)$ by the ansatz $u = S^*v$. This idea also turned out useful in the elliptic case, cf. [40].

Considering a reduced formulation of the optimal control problem in the control v, it is possible to prove first-order optimality conditions with regular Lagrange multipliers, cf. [32].

Theorem 2.3. *Let $\bar{v}_\lambda \in L^2(Q)$ be the optimal control for the Lavrentiev-regularized version of (P_B) with associated boundary control $\bar{u}_\lambda$. Then there exist Lagrange multipliers $\mu_a^\lambda, \mu_b^\lambda \in L^2(Q)$ and adjoint states $p_\lambda, q_\lambda \in W(0,T)$ such that:*

$$\begin{aligned} \partial_t \bar{y}_\lambda - \Delta \bar{y}_\lambda &= f & \qquad -\partial_t z_\lambda - \Delta z_\lambda &= \bar{v}_\lambda \\ \bar{y}_\lambda(\cdot,0) &= y_0 & z_\lambda(\cdot,T) &= 0 \\ \partial_n \bar{y}_\lambda + \alpha \bar{y}_\lambda &= u & \partial_n z_\lambda + \alpha z_\lambda &= 0 \end{aligned}$$

$$\begin{aligned} -\partial_t p_\lambda - \Delta p_\lambda &= \bar{y}_\lambda - y_d + \mu_b^\lambda - \mu_a^\lambda & \qquad \partial_t q_\lambda - \Delta q_\lambda &= 0 \\ p_\lambda(\cdot,T) &= 0 & q_\lambda(\cdot,0) &= 0 \\ \partial_n p_\lambda + \alpha p_\lambda &= 0 & \partial_n q_\lambda + \alpha q_\lambda &= \nu z_\lambda + p_\lambda \end{aligned}$$

$$\begin{aligned} (\mu_a^\lambda, y_a - \lambda \bar{v}_\lambda - \bar{y}_\lambda)_{L^2(Q)} &= 0, \qquad \mu_a^\lambda \geq 0 \\ (\mu_b^\lambda, \lambda \bar{v}_\lambda + \bar{y}_\lambda - y_b)_{L^2(Q)} &= 0, \qquad \mu_b^\lambda \geq 0 \end{aligned}$$

$$\varepsilon \bar{v}_\lambda + q_\lambda + \lambda(\mu_b^\lambda - \mu_a^\lambda) = 0.$$

Here, z_λ is the solution of the adjoint equation for the ansatz $u = S^* v$, i.e., $u = z_{|\Sigma}$. The optimality system also shows the drawback of this approach, since there are twice as many PDEs to be solved as in the unregularized case. Nevertheless, the numerical results are quite satisfying, as we will see in the example of the Betts and Campbell heat transfer problem. Under the reasonable assumption that the optimal control $\bar{u}$ of the unregularized problem is bounded, or at least regular enough to guarantee continuity of the state, we obtain a convergence result for the regularized solution:

Theorem 2.4 ([32]). *Let $\bar{u}$ belong to $L^s(\Sigma)$, $s > N+1$, and let there exist a Slater point $v_0 \in C(\bar{Q})$, such that*

$$y_a + \delta \leq G(\bar{u} + S^* v_0) \leq y_b - \delta,$$

with a given $\delta > 0$, and select the regularization parameter ε by

$$\varepsilon = c_0 \lambda^{1+c_1}, \qquad c_0 > 0, \quad 0 \leq c_1 < 1.$$

Moreover, let $\lambda_n \to 0$ and $\{v_n\}_{n=1}^\infty$ be the sequence of optimal controls of the regularized version of (P_B). Then the sequence $\{S^* v_n\}$ converges strongly in $L^2(\Sigma)$ towards the solution $\bar{u}$ of the unregularized problem.

Using a primal-dual active set strategy, we tested this regularization technique numerically in `Matlab` for the following Robin-boundary control problem that is motivated by the Betts and Campbell heat transfer problem:

$$\min \frac{1}{2} \int_0^5 \int_0^\pi y^2 \, dx dt + \frac{10^{-3}}{2} \int_0^T (u_1^2 + u_\pi^2) dt$$

subject to

$$\begin{aligned} \partial y_t - \Delta y &= 0 && \text{in } (0,\pi)\times(0,5) \\ y(x,0) &= 0 && \text{in } (0,\pi) \\ -\partial_x y(0,t) + \alpha y(0,t) &= \alpha u_1(t) && \text{in } (0,5) \\ \partial_x y(\pi,t) + \alpha y(\pi,t) &= \alpha u_\pi(t) && \text{in } (0,5) \end{aligned}$$

$$\text{as well as} \quad y(x,t) \geq \sin(x)\sin(\frac{\pi t}{5}) - 0.7 \quad \text{in } (0,\pi)\times(0,5).$$

We obtained the optimal control shown in Figure 1(b), compared to the reference solution obtained by `Matlab`'s optimization routine `quadprog`, indicating that the regularization method works quite well. The associated state is shown in Figure 1(a). Notice that the numerical results indicate that the optimal control is indeed bounded, and therefore this model problem is an example for problems that admit Lagrange multipliers in the unregularized case, even though this is not a priorily clear. We have also conducted experiments for the penalization technique by Ito and Kunisch, cf. [20], which yields similar results while solving only two PDEs in each iteration. We also point out the experiments in [30], were a modelling and simulation environment specialized for solving PDEs has been used. In contrast to unregularized state constraints, Lavrentiev regularization permits to make use of projection formulas for the Lagrange multipliers that are equivalent to the complementary slackness conditions and the non-negativity condition of the Lagrange multiplier, e.g.,

$$\mu_a^\lambda = \max\left(0, \frac{\varepsilon}{\lambda^2}(y_a - \bar{y}_\lambda) + \frac{1}{\lambda} q_\lambda\right).$$

Then, the optimality system associated with the regularized version of (P_B), and in a similar way of (P_D), can be supplied in a symbolic way as a coupled system of PDEs to specialized PDE software. If all appearing functions can be handled by the software and a converged solution is returned, this is a very time-efficient way to solve optimal control problems without specialized optimization routines and without much implementational effort. For the above example problem, we obtained satisfying results.

2.2. Lavrentiev regularized distributed control problems with additional control constraints

Let us now consider the semilinear version of (P_D) with additional control constraints. We have already mentioned that then Lagrange multipliers exist as regular Borel measures for the unregularized problem. However, the same dimensional limits as before are needed for a second-order analysis. After Lavrentiev regularization, a generalization of SSC to arbitrary dimensions should be possible in the spirit of [37]. Interestingly, there are some problems associated with the existence of regular Lagrange multipliers. It has been shown in [36] that regular multipliers exist under the assumption that the active sets associated with the different constraints are well separated. If this assumption does not hold, Lagrange multipliers are only known to exist in the space $L^\infty(Q)^*$, which is even less regular than the

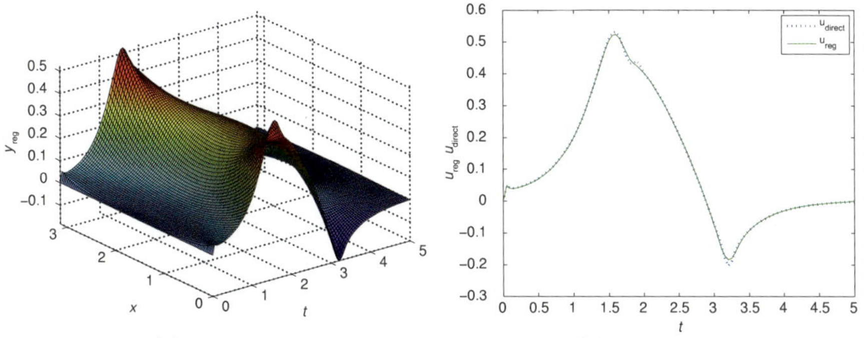

(a) Optimal state (b) Optimal controls

FIGURE 1

space of regular Borel measures. Convergence of local solutions of the Lavrentiev regularized problem has been addressed in detail in [31], along with a global analysis of Moreau-Yosida regularized problems. There we also showed the following helpful result:

Theorem 2.5. *If a locally optimal control $\bar{u}_\lambda$ of the Lavrentiev regularized version of the distributed semilinear control problem (P_D) satisfies additionally a second-order sufficient condition and if, for fixed λ, the active sets of the different constraints are strictly separated, then it is locally unique.*

Strict separation means that at most one constraint can be active or almost active in a given pair $(x,t) \in Q$. Local uniqueness of local solutions is important, since it excludes situations where a local minimum is an accumulation point of a sequence of other local optima. The proof is based on the verification of strong regularity. This property is also helpful for the analysis of SQP methods. We refer to [13] for an associated analysis of elliptic problems. Strong regularity implies local uniqueness of local optima. If one is only interested in local uniqueness, this can be directly deduced from a second-order sufficient condition, if the state constraints are regularized. We refer to [21], where regularization by virtual controls is considered.

3. Finite-element error analysis for state constrained elliptic control problems with finite-dimensional control space

In this section we sketch error estimates for control problems with control vector $u \in \mathbb{R}^m$ in a two-dimensional polygonal convex spatial (open) domain Ω. We consider first the fully finite-dimensional optimal control problem

$$(P_F) \quad \min_{u \in U_{\text{ad}}} J(y,u) := \int_\Omega L(x, y(x), u)\, dx$$

subject to the nonlinear state equation

$$\begin{aligned} -\Delta\, y(x) + d(x, y(x), u) &= 0 \quad \text{in} \quad \Omega \\ y(x) &= 0 \quad \text{on} \quad \Gamma = \partial\Omega, \end{aligned}$$

as well as the finitely many state constraints

$$\begin{aligned} g_i(y(x_i)) &= 0, && \text{for all } i = 1, \ldots, k, \\ g_i(y(x_i)) &\le 0, && \text{for all } i = k+1, \ldots, \ell \end{aligned}$$

given in points $x_i \in \Omega$, and bounds on the control,

$$u \in U_{\mathrm{ad}} = \{u \in \mathbb{R}^m : u_a \le u \le u_b\}$$

with given vectors $u_a \le u_b$ of $\mathbb{R}^m$ that has been analysed in [25]. We assume $l \ge 1$ and set $k = 0$, if only inequality constraints are given and $k = l$, if only equality constraints are given. The precise assumptions on the appearing functions L, d, and g_i are laid out in [25]. In particular, $L, d : \Omega \times \mathbb{R} \times \mathbb{R}^m \to \mathbb{R}$ are supposed to be Hölder continuous with respect to x and d is assumed to be monotone non-decreasing with respect to y. A typical tracking type functional would fit into the given setting. The possibly nonlinear appearance of u does not cause problems with existence of an optimal solution, since u has finite dimension.

Moreover, we consider a model problem of semi-infinite type, given by

$$(P_S) \quad \min_{u \in U_{\mathrm{ad}}} J(y, u) := \frac{1}{2} \int_\Omega (y - y_d)^2 \, dx + \frac{\nu}{2} |u|^2$$

subject to a linear state equation

$$\begin{aligned} -\Delta y(x) &= \sum_{i=1}^{M} u_i e_i(x) && \text{in } \Omega \\ y(x) &= 0 && \text{on } \Gamma, \end{aligned}$$

as well as a pointwise bound $b \in \mathbb{R}$ on the state in a compact interior subdomain of Ω denoted by K,

$$y(x) \le b, \quad \forall x \in K.$$

For the precise assumptions on the given data, we refer to [24], let us just mention that the basis functions e_i, $i = 1, \ldots, M$, are given in $C^{0,\beta}(\bar{\Omega})$, for some $0 < \beta < 1$. The set U_{ad} is defined as in (P_F) with given bounds $u_a \in (\mathbb{R} \cup \{-\infty\})^m$ and $u_b \in (\mathbb{R} \cup \{\infty\})^m$, where $u_a < u_b$.

3.1. The finite-dimensional control problem

We now consider the finite-dimensional problem (P_F) in the equivalent reduced formulation

$$\min_{u \in U_{\mathrm{ad}}} f(u) := J(y(u), u)$$

subject to the constraints

$$G_i(u) = 0, \quad i = 1, \ldots k, \quad G_i(u) \le 0, \quad i = k+1, \ldots \ell,$$

where G is defined as $G(u) = (g_1(y(u)(x_1)), \ldots, g_\ell(y(u)(x_\ell)))$. Using the finite element discretization of the state equation and denoting a corresponding discrete state by $y_h(u)$, let us define

$$f_h(u) = J(y_h(u), u), \qquad G_h(u) = (g_1(y_h(u)(x_1)), \ldots, g_\ell(y_h(u)(x_\ell))).$$

By these terms, we obtain an approximate problem formulation

$$(P_{F,h}) \ \min_{u \in U_{\text{ad}}} f_h(u)$$

$$\text{subject to} \quad G_{h,i}(u) = 0, \quad i = 1, \ldots k, \quad G_{h,i}(u) \le 0, \quad i = k+1, \ldots \ell.$$

Based on [6] and [34], in [25] the following result has been derived for the semilinear state equation: For all $u \in U_{\text{ad}}$, the discretized state equation has a unique discrete solution $y_h(u)$. There exists a constant c independent of h and $u \in U_{\text{ad}}$ such that, for all $u \in U_{\text{ad}}$, there holds

$$\|y(u) - y_h(u)\|_{L^2(\Omega)} + \|y(u) - y_h(u)\|_{C(K)} \le c\, h^2 |\log h|.$$

Due to the finite-dimensional character of this problem, techniques from the perturbation analysis of parametric nonlinear programming problems can be applied. Therefore, also an error of the Lagrange multipliers can be quantified.

Theorem 3.1 ([25]). *Let, under our assumptions, $\bar{u}$ be a locally optimal control of Problem (P_F) satisfying the condition of linear independence of active constraints and the standard strong second-order condition. Then $\bar{u}$ is locally unique and there exists a sequence $\bar{u}_h$ of locally optimal controls of the corresponding finite element approximated problem $(P_{F,h})$ and a constant $C > 0$ independent of h such that the following estimate is satisfied for all sufficiently small h:*

$$|\bar{u} - \bar{u}_h| \le C\, h^2 |\log h|.$$

3.2. A problem of semi-infinite type

Let us finally consider the problem (P_S), which we have discussed in detail in [24]. This problem combines the advantages of a finite-dimensional control space with the difficulties of pointwise state constraints in a domain rather than finitely many points. In contrast to Problem (P_F), the Lagrange multipliers associated with the state constraints will be regular Borel measures rather than vectors of real numbers. In view of this, an estimate not better than $h\sqrt{|\log h|}$ would not surprise.

However, in several computations we observed a much better order of convergence. The associated analytical confirmation turned out to be interesting and surprisingly difficult. Let us briefly outline the main steps.

Due to linearity of the underlying state equation, we can apply the superposition principle to obtain a semi-infinite formulation of Problem (P_S),

$$(P_S) \ \min_{u \in U_{\text{ad}}} f(u) := \frac{1}{2} \left\| \sum_{i=1}^{M} u_i y_i - y_d \right\|^2 + \frac{\nu}{2} |u|^2$$

$$\text{subject to} \quad \sum_{i=1}^{M} u_i y_i(x) \le b, \quad \forall x \in K,$$

where y_i denotes the solution of the state equation associated with $u_i = 1$ and all other components of u taken as zero.

Our results on error estimates are based on a standard Slater condition:

Assumption 3.2. There exist a $u^\gamma \in U_{\text{ad}}$ and a real number $\gamma > 0$ such that

$$y^\gamma(x) := y_{u^\gamma}(x) \leq b - \gamma \quad \forall\, x \in K.$$

Then there exists a non-negative Lagrange multiplier $\bar{\mu}$ in the space of regular Borel measures such that the standard Karush-Kuhn-Tucker conditions are satisfied by $\bar{u}$. However, collecting the state constraints in a feasible set U_{feas} given by

$$U_{\text{feas}} := \{u \in U_{\text{ad}} \colon y(u) \leq b\, \forall x \in K\},$$

the optimality conditions can be formulated with the help of a standard variational inequality due to linearity of the state equation. Let now y_i^h, $i = 1, \ldots, M$, be the discrete states associated with y_i. We obtain the discretized problem formulation

$$(P_{S,h}) \ \min_{u \in U_{\text{ad}}} f^h(u) := \frac{1}{2}\left\|\sum_{i=1}^{M} u_i y_i^h - y_d\right\|^2 + \frac{\nu}{2}|u|^2$$

$$\text{subject to} \quad \sum_{i=1}^{M} u_i y_i^h(x) \leq b, \quad \forall x \in K,$$

where the pointwise state constraints are still prescribed in the whole subdomain K rather than in finitely many discrete points. Under our assumptions, we have

$$\|y_i^h - y_i\|_{L^2(\Omega)} + \|y_i^h - y_i\|_{L^\infty(K)} \leq ch^2|\log h|$$

thanks to an L^∞-error estimate from [34]. Clearly, this error estimate extends to any linear combination $y_u = \sum_{i=1}^{M} u_i y_i$ and $y_u^h = \sum_{i=1}^{M} u_i y_i^h$ for any fixed $u \in \mathbb{R}^M$. As a consequence of the Slater assumption and the error estimate for the state equation, the feasible set of (P^h) is not empty for all sufficiently small $h > 0$. Therefore, there exists a unique optimal control $\bar{u}^h$ of Problem (P^h), with associated optimal state $\bar{y}^h$. Associated with $\bar{y}^h$, there exists a non-negative Lagrange multiplier $\mu_h \in M(\bar{\Omega})$ such that the standard KKT-conditions are satisfied. Again, by introducing a feasible set U^h_{feas} analogously to U_{feas}, the first-order optimality conditions can expressed as a variational inequality. Then, invoking only the Slater condition, the error estimate

$$|\bar{u} - \bar{u}^h| \leq ch\sqrt{|\log h|} \tag{3.1}$$

can be shown in a standard way. While this seems rather obvious, it is to the authors' knowledge the first time that an a priori error estimate for this problem class has been shown. As a consequence we obtain that $\bar{y}^h$ converges uniformly to $\bar{y}$ in K as h tends to zero.

We are able to improve (3.1) under additional assumptions on the structure of the active set. First of all, one obtains under quite natural assumptions that the active set of $\bar{y}$ cannot contain any open subset of K. Still, the set of active points might be fairly irregular, but it is reasonable to assume the following:

Assumption 3.3. The optimal state $\bar{y}$ is active in exactly N points $\bar{x}_1, \ldots, \bar{x}_N \in \operatorname{int} K$, i.e., $\bar{y}(\bar{x}_i) = b$. Moreover, there exists $\sigma > 0$ such that

$$-\langle \xi, \nabla^2 \bar{y}(\bar{x}_j)\xi \rangle \geq \sigma |\xi|^2 \qquad \forall \xi \in \mathbb{R}^n,\ \forall j = 1, \ldots, N.$$

Notice that, in contrast to the fully finite-dimensional case, the location of these active points is not known in advance. To guarantee the existence of sequences $\bar{x}_j^h$ of active points of $\bar{y}^h$ such that $x_j^h \to \bar{x}_j$ as $h \to 0$, we assume strong activity:

Assumption 3.4. All active control and state constraints are strongly active, i.e., the associated Lagrange multipliers are strictly positive.

For simplicity, we do not consider the case of additional weakly active state constraints here. Under this assumption, we are able to show that to any active $\bar{x}_j$ there exists a sequence $\bar{x}_j^h$ of active points for $\bar{y}^h$ such that

$$|\bar{x}_j - \bar{x}_j^h| \leq ch\sqrt{|\log h|}. \tag{3.2}$$

In view of the piecewise linear form of $\bar{y}^h$, we can even assume that all $\bar{x}_j^h$ are node points, and consider a problem formulation where the constraints are only prescribed in the nodes. The proof of this inequality, which is the key estimate in deriving the improved error estimate, is quite elaborate. We refer to [24] for the key ideas.

Assumption 3.5. The number N of active state constraints is equal to the number of inactive control constraints.

Define the $N \times N$-matrix Y with entries $Y_{i_k, j_k} = y_{i_k}(\bar{x}_{j_k})$, $i_k \in \mathcal{I}_{\bar{u}}$, $j_k \in \mathcal{A}_{\bar{y}}$, where $\mathcal{I}_{\bar{u}}$ and $\mathcal{A}_{\bar{y}}$ denote the index sets of inactive control and active state constraints, respectively.

Theorem 3.6. *Let $\bar{u}$ be the optimal solution of Problem (P_S), let $\bar{u}^h$ be optimal for $(P_{S,h})$, and let Assumptions 3.2–3.5 be satisfied. Moreover, let the matrix Y be regular. Then, there exists $h_0 > 0$ such that the following estimate is true for a $c > 0$ independent of h:*

$$|\bar{u} - \bar{u}^h| \leq ch^2 |\log h| \quad \forall h \leq h_0.$$

Assumption 3.5 seems quite restrictive at first glance, and the question is interesting, whether it is indeed necessary for the optimal error estimate of the last theorem. In [23], we constructed simple analytical and numerical examples with more (inactive) controls than active constraints, where the lower-order estimate (3.1) is sharp. On the other hand, the theory of semi-infinite optimization problems says that there can be at most as many strongly active constraints as there are control parameters, cf. [3]. Therefore, the analysis of Problem (P_S) is complete and the estimates are sharp.

References

[1] M. Bergounioux, M. Haddou, M. Hintermüller, and K. Kunisch. A comparison of a Moreau-Yosida-based active set strategy and interior point methods for constrained optimal control problems. *SIAM J. Optimization*, 11:495–521, 2000.

[2] J.T. Betts and S.L. Campbell. Discretize then Optimize. In D.R. Ferguson and T.J. Peters, editors, *Mathematics in Industry: Challenges and Frontiers A Process View: Practice and Theory*. SIAM Publications, Philadelphia, 2005.

[3] F. Bonnans and A. Shapiro. *Perturbation analysis of optimization problems*. Springer, New York, 2000.

[4] E. Casas. Pontryagin's principle for state-constrained boundary control problems of semilinear parabolic equations. *SIAM J. Control and Optimization*, 35:1297–1327, 1997.

[5] E. Casas. Error estimates for the numerical approximation of semilinear elliptic control problems with finitely many state constraints. *ESAIM: Control, Optimization and Calculus of Variations*, 31:345–374, 2002.

[6] E. Casas and M. Mateos. Uniform convergence of the FEM. Applications to state constrained control problems. *J. of Computational and Applied Mathematics*, 21:67–100, 2002.

[7] E. Casas, J. de los Reyes, and F. Tröltzsch. Sufficient second-order optimality conditions for semilinear control problems with pointwise state constraints. *SIAM J. Optimization*, 19(2):616–643, 2008.

[8] J.C. de los Reyes, P. Merino, J. Rehberg, and F. Tröltzsch. Optimality conditions for state-constrained PDE control problems with time-dependent controls. *Control Cybern.*, 37:5–38, 2008.

[9] K. Deckelnick and M. Hinze. Convergence of a finite element approximation to a state constrained elliptic control problem. *SIAM J. Numer. Anal.*, 45:1937–1953, 2007.

[10] K. Deckelnick and M. Hinze. Numerical analysis of a control and state constrained elliptic control problem with piecewise constant control approximations. In K. Kunisch, G. Of, and O. Steinbach, editors, *Proceedings of ENUMATH* 2007*, the* 7*th European Conference on Numerical Mathematics and Advanced Applications*, Heidelberg, September 2007 2008. Springer.

[11] P. Deuflhard, M. Seebass, D. Stalling, R. Beck, and H.-C. Hege. Hyperthermia treatment planning in clinical cancer therapy: Modelling, simulation, and visualization. In A. Sydow, editor, *Computational Physics, Chemistry and Biology*, pages 9–17. Wissenschaft und Technik-Verlag, 1997.

[12] K. Eppler and F. Tröltzsch. Fast optimization methods in the selective cooling of steel. In M. Grötschel, S.O. Krumke, and J. Rambau, editors, *Online Optimization of Large Scale Systems*, pages 185–204. Springer, 2001.

[13] R. Griesse, N. Metla, and A. Rösch. Convergence analysis of the SQP method for nonlinear mixed-constrained elliptic optimal control problems. *ZAMM*, 88(10):776–792, 2008.

[14] M. Hintermüller and M. Hinze. Moreau-Yosida regularization in state constrained elliptic control problems: Error estimates and parameter adjustment. *SIAM J. on Numerical Analysis*, 47:1666–1683, 2009.

[15] M. Hintermüller, K. Ito, and K. Kunisch. The primal-dual active set strategy as a semismooth Newton method. *SIAM J. Optim.*, 13:865–888, 2003.

[16] M. Hintermüller and A. Schiela. Discretization of interior point methods for state constrained elliptic optimal control problems: Optimal error estimates and parameter adjustment. *COAP, published online*, 2009.

[17] M. Hintermüller and K. Kunisch. Path-following methods for a class of constrained minimization problems in function space. *SIAM J. on Optimization*, 17:159–187, 2006.

[18] M. Hintermüller, F. Tröltzsch, and I. Yousept. Mesh-independence of semismooth Newton methods for Lavrentiev-regularized state constrained nonlinear optimal control problems. *Numerische Mathematik*, 108(4):571–603, 2008.

[19] M. Hinze. A variational discretization concept in control constrained optimization: the linear-quadratic case. *J. Computational Optimization and Applications*, 30:45–63, 2005.

[20] K. Ito and K. Kunisch. Semi-smooth Newton methods for state-constrained optimal control problems. *Systems and Control Letters*, 50:221–228, 2003.

[21] K. Krumbiegel, I. Neitzel, and A. Rösch, Regularization for semilinear elliptic optimal control problems with pointwise state and control constraints, Computational optimization and applications, online first. DOI 10.1007/s10589-010-9357-z

[22] K. Krumbiegel and A. Rösch. A virtual control concept for state constrained optimal control problems. *COAP*, 43(2):213–233, 2009.

[23] P. Merino, I. Neitzel, and F. Tröltzsch. Error estimates for the finite element discretization of semi-infinite elliptic optimal control problems. *Discussiones Mathematicae*, Differential Inclusions, Control and Optimization, 30(2):221–236, 2010.

[24] P. Merino, I. Neitzel, and F. Tröltzsch. On linear-quadratic elliptic optimal control problems of semi-infinite type. *Applicable Analysis*, 90(6):1047–1074, 2011.

[25] P. Merino, F. Tröltzsch, and B. Vexler. Error Estimates for the Finite Element Approximation of a Semilinear Elliptic Control Problem with State Constraints and Finite Dimensional Control Space. *ESAIM:Mathematical Modelling and Numerical Analysis*, 44:167–188, 2010.

[26] C. Meyer. Error estimates for the finite-element approximation of an elliptic control problem with pointwise state and control constraints. *Control Cybern.*, 37:51–85, 2008.

[27] C. Meyer and P. Philip. Optimizing the temperature profile during sublimation growth of SiC single crystals: Control of heating power, frequency, and coil position. *Crystal Growth & Design*, 5:1145–1156, 2005.

[28] C. Meyer, A. Rösch, and F. Tröltzsch. Optimal control of PDEs with regularized pointwise state constraints. *Computational Optimization and Applications*, 33(2003-14):209–228, 2006.

[29] C. Meyer and F. Tröltzsch. On an elliptic optimal control problem with pointwise mixed control-state constraints. In A. Seeger, editor, *Recent Advances in Optimization. Proceedings of the 12th French-German-Spanish Conference on Optimization held in Avignon, September 20–24, 2004*, Lectures Notes in Economics and Mathematical Systems. Springer-Verlag, 2005.

[30] I. Neitzel, U. Prüfert, and T. Slawig. Strategies for time-dependent PDE control with inequality constraints using an integrated modeling and simulation environment. *Numerical Algorithms*, 50:241–269, 2009.

[31] I. Neitzel and F. Tröltzsch. On convergence of regularization methods for nonlinear parabolic optimal control problems with control and state constraints. *Control and Cybernetics*, 37(4):1013–1043, 2008.

[32] I. Neitzel and F. Tröltzsch. On regularization methods for the numerical solution of parabolic control problems with pointwise state constraints. *ESAIM Control, Optimisation and Calculus of Variations*, 15(2):426–453, 2009.

[33] U. Prüfert, F. Tröltzsch, and M. Weiser. The convergence of an interior point method for an elliptic control problem with mixed control-state constraints. *Comput. Optim. Appl.*, 39(2):183–218, March 2008.

[34] R. Rannacher and B. Vexler. A priori error estimates for the finite element discretization of elliptic parameter identification problems with pointwise measurements. *SIAM Control Optim.*, 44:1844–1863, 2005.

[35] J. Raymond and F. Tröltzsch. Second-order sufficient optimality conditions for nonlinear parabolic control problems with state constraints. *Discrete Contin. Dyn. Syst.*, 6:431–450, 2000.

[36] A. Rösch and F. Tröltzsch. Existence of regular Lagrange multipliers for a nonlinear elliptic optimal control problem with pointwise control-state constraints. *SIAM J. Control and Optimization*, 45:548–564, 2006.

[37] A. Rösch and F. Tröltzsch. Sufficient second-order optimality conditions for an elliptic optimal control problem with pointwise control-state constraints. *SIAM J. on Optimization*, 17(3):776–794, 2006.

[38] A. Schiela. Barrier methods for optimal control problems with state constraints. *SIAM J. on Optimization*, 20:1002–1031, 2009.

[39] F. Tröltzsch. *Optimal Control of Partial Differential Equations: Theory, Methods and Applications*. AMS, Providence, 2010.

[40] F. Tröltzsch and I. Yousept. A regularization method for the numerical solution of elliptic boundary control problems with pointwise state constraints. *COAP*, 42(1):43–66, 2009.

Ira Neitzel and Fredi Tröltzsch
Technische Universität Berlin
Institut für Mathematik
Str. des 17. Juni 136
D-10623 Berlin, Germany
e-mail: neitzel@math.tu-berlin.de
troeltz@math.tu-berlin.de

Part V

Applications

Introduction to Part V
Applications

In this part we focus on a variety of highly challenging applications. These range from the nano-scale to macroscopic problems, and from electrical and chemical engineering to biology and medicine.

Inga Altrogge, Christof Büskens, Tim Kröger, Heinz-Otto Peitgen, Tobias Preusser and Hanne Tiesler summarize, in *Modeling, Simulation and Optimization of Radio Frequency Ablation*, their results on modeling the patient specific treatment of hepatic lesions, the multi-level gradient descent optimization of the probe placement under the corresponding PDE system and the identification of material parameters from temperature measurements. The focus lies on the uncertainties in the patient specific tissue properties that are represented by a stochastic PDE model. The applicability of the models and algorithms is validated against data from real patients' CT scans.

Eberhard Bänsch, Manfred Kaltenbacher, Günter Leugering, Fabian Schury and Fabian Wein present, in *Optimization of Electro-mechanical Smart Structures*, the most recent progress in topology optimization of piezoelectric loudspeakers using the SIMP method and topology gradient based methods along with analytical and numerical results.

Nikolai Botkin, Karl-Heinz Hoffmann and Varvara Turova present, in *Freezing of Living Cells: Mathematical Models and Design of Optimal Cooling Protocols*, the state of the art in optimal cryopreservation of living cells. Two main injuring effects are modeled and controlled using stable algorithms. Additionally, an optimal shape of cells under shrinkage and swelling during freezing and thawing is computed as a level set of a function that satisfies a Hamilton-Jacobi equation resulting from a Stefan-type condition for the normal velocity of the cell boundary. Examples of the shape evolution computed in two and three dimensions are presented.

Michael Gröschel, Günter Leugering and Wolfgang Peukert consider *Model Reduction, Structure-property Relations and Optimization Techniques for the Production of Nanoscale Particles* within a joint project of the faculties of mathematics and engineering. Parameter identification and optimal control techniques are presented for a precipitation process and an innovative aerosol forming process allowing for a precise control of residence time and temperature.

Frank Haußer, Sandra Janssen and Axel Voigt, in *Control of Nanostructures through Electric Fields and Related Free Boundary Problems*, consider geometric evolution equations to control the evolution of the surface or of the interface using the bulk contribution as a distributed control. The applicability of a phase-field approximation and the corresponding control problem are demonstrated in various numerical examples.

Achim Küpper and Sebastian Engell consider, in *Optimization of Simulated Moving Bed Processes*, periodic chromatographic simulated moving bed SMB processes incorporating rigorous models of the chromatographic columns and the discrete shifts of the inlet and outlet ports. The potential of optimization using a rigorous model and multiple shooting is demonstrated by the ModiCon process where 40% to 50% savings in solvent consumption can be achieved, outperforming other well-established methods.

René Pinnau and Norbert Siedow, in *Optimization and Inverse Problems in Radiative Heat Transfer*, summarize their results on the derivation and investigation of efficient mathematical methods for the solution of optimization and identification problems for radiation dominanted processes, which are described by a nonlinear integro-differential system and diffusive type approximations.

Markus Probst, Michael Lülfesmann, Mike Nicolai, H. Martin Bücker, Marek Behr and Christian H. Bischof report *On the Influence of Constitutive Models on Shape Optimization for Artificial Blood Pumps*, presenting a shape optimization framework that couples a highly parallel finite element solver with a geometric kernel and different optimization algorithms. The entire optimization framework is transformed using automatic differentiation techniques and successfully applied in designing bypass geometries in 2D/3D and ventricular assist devices.

Sebastian Engell and Günter Leugering

International Series of Numerical Mathematics, Vol. 160, 487–499

Modeling, Simulation and Optimization of Radio Frequency Ablation

Inga Altrogge, Christof Büskens, Tim Kröger,
Heinz-Otto Peitgen, Tobias Preusser and Hanne Tiesler

Abstract. The treatment of hepatic lesions with radio-frequency (RF) ablation has become a promising minimally invasive alternative to surgical resection during the last decade. In order to achieve treatment qualities similar to surgical R0 resections, patient specific mathematical modeling and simulation of the biophysical processes during RF ablation are valuable tools. They allow for an a priori estimation of the success of the therapy as well as an optimization of the therapy parameters. In this report we discuss our recent efforts in this area: a model of partial differential equations (PDEs) for the patient specific numerical simulation of RF ablation, the optimization of the probe placement under the constraining PDE system and the identification of material parameters from temperature measurements. A particular focus lies on the uncertainties in the patient specific tissue properties. We discuss a stochastic PDE model, allowing for a sensitivity analysis of the optimal probe location under variations in the material properties. Moreover, we optimize the probe location under uncertainty, by considering an objective function, which is based on the expectation of the stochastic distribution of the temperature distribution. The application of our models and algorithms to data from real patient's CT scans underline their applicability.

Mathematics Subject Classification (2000). 35Q93 and 49J20.

Keywords. Radio frequency ablation, parameter uncertainties, optimal probe placement, stochastic PDE, parameter identification.

1. Introduction

During the last decade the treatment of hepatic lesions with RF ablation has become a promising alternative to surgical resections, in particular for cases where surgical resection is not possible due to the constitution of the patient or the location of the lesion. RF ablation is a minimally invasive technique which destroys tumorous tissue by targeted heating. Thereby, an applicator with electrodes at the tip is placed inside the lesion. An electric current causes heating and therewith a coagulation of the proteins and consequently the destruction of the cells.

For classical surgical resection, the success of the treatment can be controlled via a pathological workup. In so-called R0 resections in which the tumor is cut out with a sufficiently large safety margin, it is checked whether the surface of the cut-out tissue is free of tumor cells. In this case the tumor is considered to be completely removed. However, in the case of the minimally invasive RF ablation the conduction and control of the therapy can solely rely on non-invasive imaging techniques like computed tomography (CT) or magnetic resonance imaging (MRI). In practice this means that to date the success of the ablation depends highly on the experience of the radiologist and the quality of the imaging techniques.

Indeed, complete tumor ablation is difficult to achieve. The main complication is the presence of blood vessels in the local vicinity of the RF probe. The local blood flow leads to a cooling of the tissue and thus makes it hard to achieve the temperatures needed for protein denaturation. Moreover, recent investigations hypothesize that in incomplete tumor destruction, the remaining tumorous cells are more aggressive and may be more resistant towards further heating [12], because of the expression of heat shock proteins during RF ablation [13].

Consequently, in the interest of the patient the outcome of an RF ablation must be available in advance and iterative applications of the treatment in multiple sessions must be avoided. However, the monitoring of RF ablation during the procedure in order to control or adjust it is possible with very specialized MR thermometry imaging only, which is not available in the clinical daily routine. This motivates our investigations on modeling and numerical simulation of RF ablation for prediction and optimization of the outcome of the therapy.

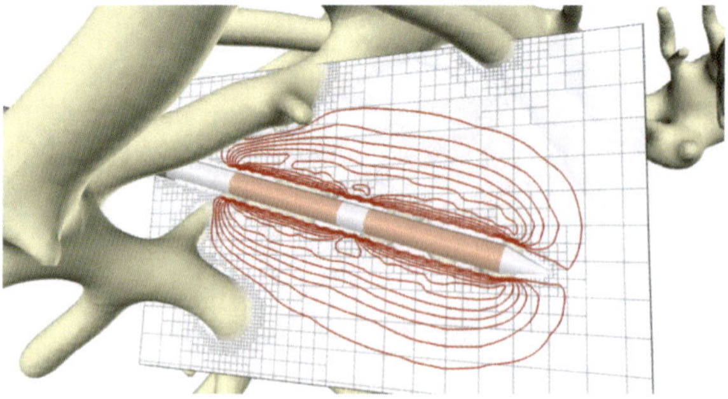

FIGURE 1. A cross section through the adaptive grid for the numerical computation of the tissue temperature is shown. The computational domain contains a bipolar applicator and a blood vessel tree. Isolines indicate the temperature distribution.

This report is organized as follows: Section 2 will give a short overview on the modeling and simulation of RF ablation with a system of partial differential equations (PDEs). In Section 3 we describe the optimization of the probe placement to achieve a successful outcome of the ablation. Moreover, we describe our way of dealing with parameter uncertainties by using stochastic PDEs. Section 4 discusses our investigations to the identification of material parameters from temperature measurements. Finally, in Section 5 we draw conclusions.

2. Modeling and simulation of radio frequency ablation

For the mathematical modeling of the RF ablation we consider a cuboid computational domain $\Omega \subset \mathbb{R}^3$. It contains the tumor $\Omega_{\text{tu}} \subset \Omega$ and adjacent vascular structures $\Omega_{\text{v}} \subset \Omega$. In practical applications these domains are determined in advance from segmented CT image data for every specific patient. Further, we assume that a needle shaped RF probe, denoted by $\Omega_{\text{pr}} \subset \Omega$ is applied. The domains of the positive and negative electrodes are $\Omega_+ \subset \Omega_{\text{pr}}$ and $\Omega_- \subset \Omega_{\text{pr}}$, respectively.

Our model consists of a description of the bio-physical processes taking place during RF ablation. The first component is the electrostatic equation, which describes the electric potential ϕ as the solution of the PDE

$$-\text{div}(\sigma(T,x)\nabla\phi(t,x)) = 0 \qquad \text{in} \quad \mathbb{R}^+ \times \Omega \setminus (\overline{\Omega_+ \cup \Omega_-}), \tag{2.1}$$

$$\phi(t,x) = \phi_0 \qquad \text{in} \quad \mathbb{R}^+ \times \overline{\Omega_+ \cup \Omega_-}, \tag{2.2}$$

with appropriate boundary conditions on $\partial\Omega$ and $\partial\Omega_\pm$.

The second component of the model is the bio-heat transfer equation describing the temperature distribution T inside the tissue under investigation by

$$\partial_t(\rho c T(t,x)) - \text{div}(\lambda(T,x)\nabla T(t,x)) = Q_{\text{perf}}(t,x) + Q_{\text{rf}}(t,x) \quad \text{in } \mathbb{R}^+ \times \Omega \tag{2.3}$$

with source term Q_{rf} and sink term Q_{perf}

$$Q_{\text{rf}}(t,x) = \sigma(T,x) P_{\text{eff}} \left(\int_\Omega \sigma(T,x)|\nabla\phi|^2 \right) |\nabla\phi|^2,$$

$$Q_{\text{perf}}(t,x) = \nu(x)\left(T(t,x) - T_{\text{body}}\right),$$

and boundary conditions on $\partial\Omega$, and $\partial\Omega_{\text{pr}}$ and initial conditions for $t = 0$. The sink Q_{perf} accounts for the cooling effect of the blood perfusion [1]. The source Q_{rf} describes the energy, entering the domain through the electric field of the probe. Here, P_{eff} is a nonlinear function, which models the characteristics of the electric generator and its response to the varying electric impedance of the tissue.

The above PDEs depend on a variety of parameters describing the physical tissue properties involved in the process: electric conductivity σ, tissue density ρ, heat capacity c, thermal conductivity λ and perfusion coefficient ν. These tissue parameters have a dependence on the states of the system, i.e., they change (in parts nonlinearly) with temperature, water content and coagulation status of the tissue [11, 15, 18]. These facts lead to a nonlinear and complicated coupling of the above PDEs through the material parameters. Moreover, since temperatures also reach 100 °C during RF ablation, phase-changes resulting from the evaporation of tissue water must be taken into account [14]. Water evaporation furthermore influences the electric conductivity, which drops rapidly if dry-out of the tissue takes place, making the system stiff and time-stepping schemes complicated. The result of a simulation on an adaptive grid and real patient data is shown in Fig. 1

For practical applications it must furthermore be taken into account that the tissue properties used as parameters in the PDEs vary between different patients,

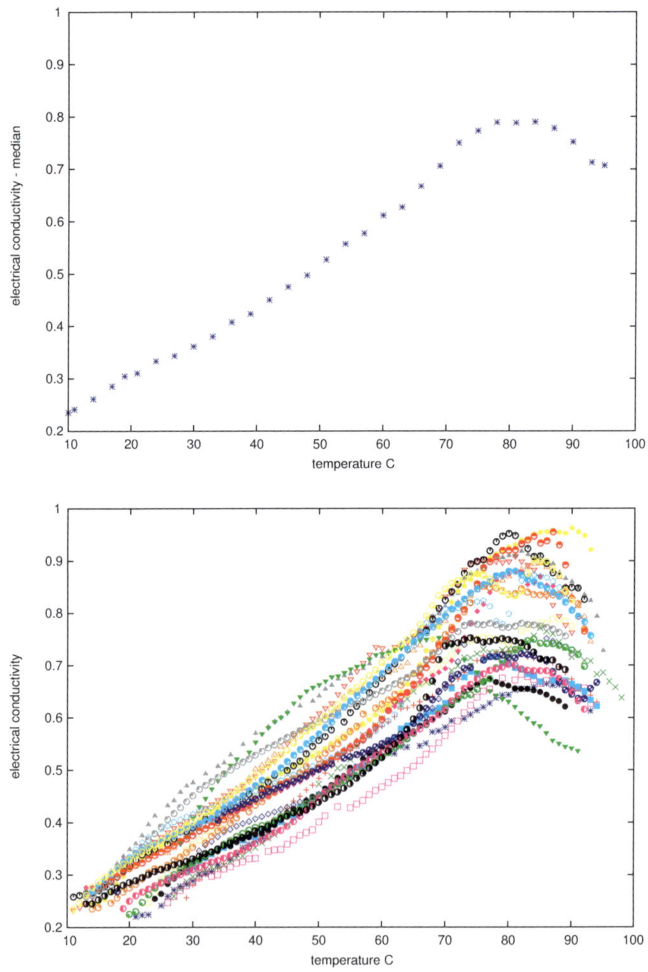

FIGURE 2. The electric conductivity in [S/m] of porcine liver is shown in dependence on the tissue temperature [°C] for one animal (top) and for a group of 25 pigs (bottom). (Data courtesy of U. Zurbuchen, et al., Charite University Clinic, Berlin, [18])

but also for every specific patient over time, depending on the actual constitution (cf. Fig. 2). However, exact patient specific measurements are not possible and currently simulations rely on literature data. Therefore, our investigations also deal with the uncertainty involved in the material properties. In Section 3.2 we discuss the optimization of the probe location in the presence of parameter uncertainty. In Section 4 we discuss our approach to the identification of individual material parameters from temperature measurements during RF ablation.

Investigations by our clinical partners Zurbuchen et al. [18] have shown a dependence of the electric conductivity on the temperature as shown in Fig. 2. Furthermore, Hämmerich et al. [7] show an additional dependency of the electric conductivity on the tissue's type, i.e., tumorous tissue or native tissue. We model this behavior of the electric conductivity by a fourth-order polynomial in the temperature T, which incorporates a spatial dependency of the coefficients, i.e.,

$$\sigma(T,x) = \sigma_0(x)(1+\sigma_1(x)T(1+\sigma_2(x)T(1+\sigma_3(x)T(1+\sigma_4(x)T)))), \tag{2.4}$$

where $\sigma_0,\dots,\sigma_4 : \Omega \to \mathbb{R}$. For the thermal conductivity we use a linear model, based on the investigations of Stein [15], thus

$$\lambda(T,x) = \lambda_{\mathrm{ref}}(x)\left(1+\alpha_\lambda(x)\left(T-T_{\mathrm{ref}}\right)\right). \tag{2.5}$$

3. Optimization of the probe placement

The aim of the therapy is the complete destruction of the tumor with minimum amount of affected native tissue. For the numerical optimization of the probe placement we use a simplified steady state forward simulation with constant material parameters [1, 2, 3]. In practical applications the success of the therapy can be robustly estimated by consideration of the steady state of the temperature. Therefore we do not use the full model as it was sketched in the previous Section 2, which has the good effect that our optimizations run much faster, since we are dealing with elliptic PDEs and no nonlinear coupling of the tissue parameters, only [1, 2, 3].

Moreover, we focus on a temperature based objective function, thereby considering the tumorous tissue to be destroyed if it is heated above a critical high temperature of 60 °C [2, 3]. Thus, our objective function aims at maximizing the minimum temperature inside the lesion. We approximate the negative minimum of the temperature over Ω_{tu} by

$$f(T) = \log\left(\int_{\Omega_{\mathrm{tu}}} \mathrm{e}^{-\alpha T(x)}\,dx\right), \tag{3.1}$$

a function which penalizes lowest temperatures inside the tumor most. Thus, configurations with only small sufficiently hot volume inside the tumor lead to a higher value of the objective function than a uniform tumor heating. The factor α specifies the grade of penalization of a non-uniform heating inside the tumor. Note, that by optimizing the probe placement, not an optimization of the heat intensity is achieved, but only an optimization of the heat distribution. Moreover, our numerical computations have shown that the requirement of a high temperature inside the tumor is sufficient to force the probe's location to be close to the tumor. Moreover, since from the viewpoint of applications it is desirable to rather destroy some more tissue surrounding the lesion (safety margin), we do not incorporate an additional penalization of high temperatures within the native tissue. However, a penalization of high temperatures in other parts of the organ can be used to model the prevention of certain structures from thermal damage (e.g., colon, diaphram).

Our goal is to optimize the probe location $(\bar{x}, \bar{a}) =: u \in U := \Omega \times S^2$ such that the minimal temperature inside the tumor becomes maximal. Here, $\bar{x}$ denotes the position and $\bar{a}$ the orientation of the probe. Thus, we seek for $(\bar{x}, \bar{a})$ such that $F : U \to \mathbb{R}$, $u \mapsto F(u) := f \circ \mathcal{T} \circ \mathcal{Q}(u)$ becomes minimal. Here, the objective function f is considered as a function of the temperature distribution $T = \mathcal{T}(Q_{\mathrm{rf}})$, which in turn is a function of the heat source Q_{rf}, and $Q_{\mathrm{rf}} = \mathcal{Q}(u)$ is a function of the optimization parameter [2, 3].

For the minimization of F, we use a gradient descent method. We do not consider higher-order methods like Lagrange-Newton methods since the optimization with the gradient descent needs only few iteration steps. In the descent the gradient $\nabla_u F(u)$ is calculated with the chain rule, i.e., f' is calculated analytically, $\mathcal{T}'$ is calculated via the adjoint equation [2, 3] and $\partial_{u_j} \mathcal{Q}(u)$ is approximated numerically by central differences. The advantage of using the chain rule instead of a numerical approximation is the reduction of the numerical effort. In our approach we only need few evaluations of the electrostatic equation (2.1).

We repeat the iteration until the norm of the difference $|u^{n+1} - u^n|$ between the new and the old iterate falls below a suitably chosen threshold θ. Furthermore, we split the stopping criterion θ into several components. This allows us to prescribe different accuracies for the location $\bar{x}$ and the orientation $\bar{a}$.

To accelerate the optimization significantly, we take advantage of the hierarchical structure of the grid. We start the optimization of the probe location on a coarse grid and use the results as initial guess on the following finer grid. This approach is related to the multiscale methodology discussed in [6, 5]. It allows us to descend much faster in the energy profile of the objective function. The computational time is reduced by minimizing the number of iteration steps on the finest grid level. The progression of the value of the objective function during the optimization is shown in Fig. 3. The transitions between the grid levels are marked by vertical dotted lines in the energy plot. The steps of the energy graph at the transition points is caused by the re-interpretation of the probe placement on the finer grid.

A further speedup of the multi-level algorithm is achieved by an additional thresholding. Through the hierarchical coarsening we obtain grid cells whose values lie in between the characteristic tissue values. With the thresholding we get back a strict disjunction between the different tissue types. The progression of the corresponding objective function value is shown in the right graph of Fig. 3. Since here the mass of the tumor and the vessels is not conserved on the coarse levels, the values of the objective function increase at the transition stages between grid resolutions.

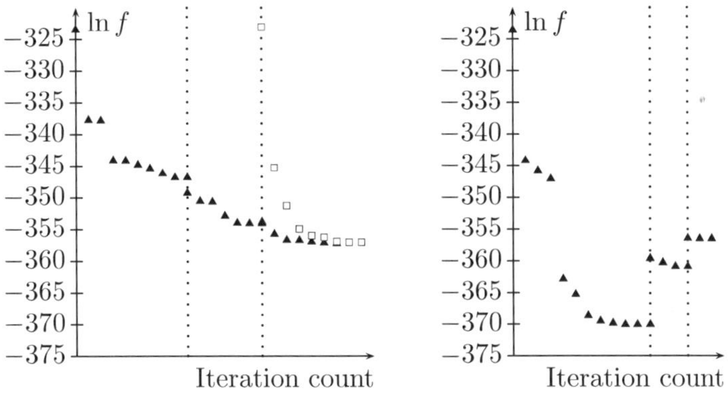

FIGURE 3. Left: The progression of the objective function value is shown for the multi-scale optimization algorithm (▲) and the standard algorithm (□). Here, the multi-scale optimization finds a slightly better minimum with a smaller number of iterations on the finest-grid. Right: The progression of the objective function value is shown for the definition of a coarse-grid tumor and vascular domains which involve a thresholding.

In practice, surgeons and radiologists use several probes either separately or fixed in a cluster for the successful treatment of larger tumors. Therefore we extend the presented optimization method for one probe to the case of a cluster of probes, i. e., we have a fixed geometrical configuration with the same orientation $\bar{a}$ for all probes. The set of optimization parameters is enlarged by a rotation vector $\bar{r} \in S^2$ with $\bar{r} \perp \bar{a}$. Finally, we remark that the placement of umbrella probes, another type of probes that is frequently used by radiologists, can be treated by this extension as well. However, in this case more effort must be spend on the handling of the locally refined grid in order to resolve the fine tines of the umbrella probe.

In Fig. 4 we show the optimization of the probe placement for a real patient case. Thus, the geometries used are obtained from real CT scans and show a segmented tumor and the surrounding vascular system. In this example we use a fixed cluster of three parallel monopolar probes. A stopping criterion of 10^{-4}, 10^{-3} and 10^{-2} for the position, orientation and rotation is used. This stopping criterion is adapted to the accuracy of the probe placement which can be achieved in practice. The optimization is performed on three different levels and the number of iteration steps could be reduced from 8 steps in the case of no multi-scale-optimization to 2 steps on the finest level with multi-scale optimization. Thereby we save about 64 % of the computational time. Indeed one iteration step on the finest level needs about 1 hour, on the next coarser level it takes about 5 minutes and on the next coarser level the time for one iteration is reduced to seconds. In the experiments shown here the finest grid consists of 155 × 103 × 103 nodes and the coarsest grid has 39 × 26 × 26 nodes. The probe's optimal direction is aligned

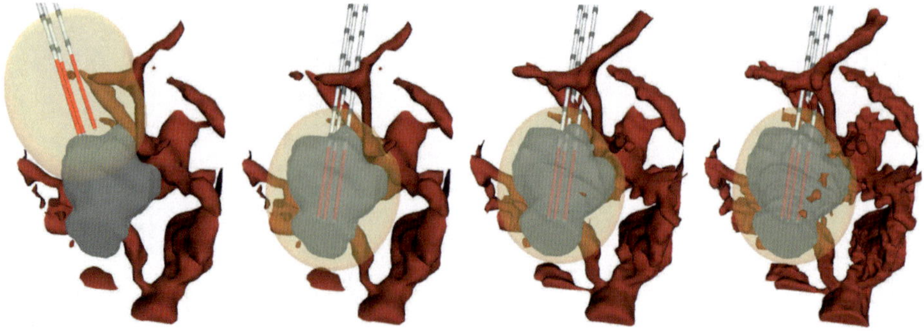

FIGURE 4. Optimization of a cluster of three probes on real patient data with segmented tumor (grey) and vascular system (red) for which the 60 ° isosurface of the temperature is shown (yellow). From left to the right: the initial position at the coarsest grid, the final position at the coarsest grid, the final position at the intermediate grid and the final position at the finest grid.

to the shape of the tumor as well as to the influence of the blood vessels. Beside this, the cooling effect of the blood perfusion is visible from the 60 °-isosurface of the temperature.

3.1. Fast estimation of the heat sink effect of large blood vessels

Solving the forward problem and the optimization of the probe placement for patient specific data is computationally expensive, time-consuming and consequently does not fit into the clinical workflow. A remedy is to separate patient-specific computations from general patient independent computations. This is achieved through tabulating the shape of the thermal coagulation for various reference blood vessel configurations. The reference configurations are parameterized with respect to the thickness of the blood vessel and the distance of the RF probe to the vessel. The contents of the look-up-table are calculated in advance, and they are used for the synthesis of the coagulation necrosis for a patient specific vascular tree. This synthesis is achieved in real-time allowing for an interactive planning of the treatment through a medical doctor [3, 4, 8].

3.2. Optimization considering uncertain material parameters

For the patient specific application of the optimization of the probe placement it must be taken into account that the model presented in section 2 involves uncertain material parameters. To this end we assume that the parameters lie within a range of possible values and that they have a certain probability distribution, which can be obtained from experimental investigations. Thus, we reformulate the coupled system of PDEs, (2.1) and (2.3), as a system of stochastic PDEs. Thereby all parameters and states are replaced by random fields depending on a vector of real-valued random variables $\boldsymbol{\xi} = (\xi_1, \ldots, \xi_N) \in \Gamma, N \in \mathbb{N}$.

Using the joint probability function $\rho(\boldsymbol{\xi}) := \prod_{i=1}^{N} \rho_i(\boldsymbol{\xi}), \forall \boldsymbol{\xi} \in \Gamma$ of these random variables allows us to define an inner product on the Hilbert space $L^2(\Gamma)$

$$\langle X, Y \rangle := \int_\Gamma XY \, \rho(\boldsymbol{\xi}) d\boldsymbol{\xi}. \tag{3.2}$$

Since the expectation of a random variable X is $\mathbb{E}[X] := \int_\Gamma X \rho(\boldsymbol{\xi}) d\boldsymbol{\xi}$, we have $\langle X, Y \rangle = \mathbb{E}[XY]$. This Hilbert-space structure allows to use a Galerkin method in the stochastic space to solve the SPDEs.

In our studies we use a stochastic collocation method, which is based on the stochastic Galerkin method, but has the sampling character like the classical Monte Carlo method. Thereby, we use polynomials to approximate the stochastic processes and random fields in the weak formulation of the stochastic PDE. Using Lagrange interpolation polynomials we can express the approximation of the function $y(\boldsymbol{\xi}, x)$ as

$$\mathcal{I}(y)(\boldsymbol{\xi}, x) = \sum\nolimits_{k=1}^{M} y(\boldsymbol{\xi}_k, x) L_k(\boldsymbol{\xi}), \tag{3.3}$$

where the Lagrange polynomials fulfil $L_i(\boldsymbol{\xi}_j) = \delta_{ij}$ for $i, j = 1, \ldots, M$. Here $\boldsymbol{\xi}_k, k = 1, \ldots, M$ denotes vectors of independent and identically distributed random variables $(\xi_k^i)_{i=1}^N, k = 1, \ldots, M$. The sample points $\boldsymbol{\xi}_k$ at which the deterministic problem has to be evaluated are located on a sparse grid, which is generated by Smolyak's algorithm in the stochastic space. Thus, the sampling character of the stochastic collocation approach means that we evaluate the deterministic problems at locations $\boldsymbol{\xi}_k$ in the stochastic space.

The formulation of the states of the system as random variables allows for the convenient analysis of the sensitivity of the system to perturbations in the material parameters [10]. Our numerical experiments indicate a stronger dependency on the electric conductivity σ than on the thermal conductivity λ. Also the sensitivity of the probe placement is aligned with the structure of the vascular system, whereas the alignment for the thermal conductivity is stronger than for the electric conductivity. For a more detailed report we refer to [3].

Constrained by the SPDE system we can optimize the probe placement in such a way that a maximum of destroyed tumor tissue can be expected. To this end, we reformulate our objective function (3.1) as

$$\begin{aligned} \tilde{f}(T) := \mathbb{E}[f(T, x, \boldsymbol{\xi})] &= \int_\Gamma \log \left(\int_{\Omega_{\text{tu}}} \exp\left(-\alpha T(x, \boldsymbol{\xi})\right) dx \right) \rho(\boldsymbol{\xi}) d\boldsymbol{\xi} \\ &\approx \sum\nolimits_{k=1}^{M} \log\left(f(T, x, \boldsymbol{\xi}_k)\right) \int_\Gamma L_k(\boldsymbol{\xi}) \rho(\boldsymbol{\xi}) d\boldsymbol{\xi}, \end{aligned} \tag{3.4}$$

where the approximation results from the use of the polynomial interpolation for the integration of the expectation. Because of the sampling character, for the computations, the deterministic algorithm can be re-used. Thus, to obtain the expected descent direction we combine all deterministic descent directions in an intermediate step. The effort of the algorithm can be estimated by the product of

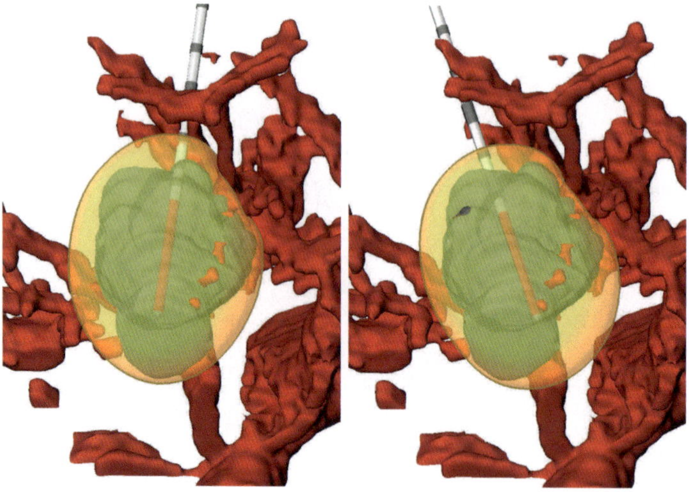

FIGURE 5. We compare the result of the deterministic optimization of the probe location (right) to the optimization of the expected destruction (left).

the deterministic problem's effort and the number of collocation points used. In Fig. 5 we show the result of the stochastic optimization for a real patient's case.

Let us finally remark, that a generalization of this approach to tracking type objective functions involving any combination of stochastic moments of the random fields is possible. For such objectives we have developed a sequential quadratic programming (SQP) method for the optimization under stochastic PDE constraints [16]. Moreover, we investigate the determination of the optimal probe location under parameter uncertainty as a fuzzy optimization problem in [9].

4. Parameter identification

Patient specific information about the parameters and their dynamic behavior can be gained through their identification from temperature measurements during RF ablation. These measurements can be obtained by magnetic resonance imaging (MRI), which however, to date is not available in the clinical standard routine.

We formulate an inverse problem in which we aim at fitting the calculated temperature T at a certain terminal time t_{fin} to the measured temperature T_{meas} by

$$\min_{p \in P} F(T,p) = \tfrac{1}{2}\|T(t_{\text{fin}},x) - T_{\text{meas}}(t_{\text{fin}},x)\|^2, \tag{4.1}$$

subject to box constraints for the material parameters $0 < \lambda(T,x) < b_\lambda$ and $0 < \sigma(T,x) < b_\sigma$, with $b_\lambda, b_\sigma \in \mathbb{R}^+$. The set of optimization variables P consists of all coefficients of the material parameters λ and σ, cf. (2.4) and (2.5).

We approach the full complexity of this problem step by step by first working with spatially varying parameters without temperature dependency. Thereby we split the optimization by successively optimizing for the different parameters one after the other. In the next step we extend the complexity of the parameters by modeling them temperature dependent and spatially dependent like in (2.5) and (2.4). Here we consider the thermal conductivity λ first. This problem is already very complex because of the nonlinear coupling of the temperature and the material parameters. The situation becomes even more complex if we add the temperature-dependent electric conductivity σ. Now the parameters interact via the right-hand side of the bio-heat equation and their temperature dependency.

Another difficulty arises from the varying spatial influence of the electric potential. This means that the quality of the optimization of the electric conductivity depends highly on the probe's position. In fact, it is easier to identify the electric conductivity in the vicinity of the probe than at a larger distance. On the other hand, there is a global influence of the electric conductivity on the whole system and thus on the temperature distribution. We have to account for this different influence on the optimization by scaling the objective appropriately.

To solve the optimal control problem we first discretize with finite elements in space as for the forward simulation and solve the resulting discrete minimization problem with WORHP (www.worhp.de), a combined SQP (Sequential quadratic programming) and primal-dual IP (Interior-Point) method, which aims at solving sparse large-scale NLP problems with more than 1,000,000 variables and constraints. WORHP is based on a reverse communication architecture that offers unique flexibility and control over the optimization process.

An efficient possibility would be to calculate the derivatives via the adjoint equation. However, we are aiming at using the functionality of our solver WORHP to calculate the derivatives by a finite differences method. WORHP provides several alternatives for the calculation of the derivatives, e.g., a method based on group strategies for finite differences or new sparse BFGS update techniques to obtain an approximation of the Hessian. The robustness of WORHP was proved by the CUTEr test set, which consists of 920 sparse large-scale and small dense problems and where WORHP solves 99.5% of all test cases.

5. Conclusions

Radio frequency ablation is a very promising alternative to surgical resection. We have discussed our efforts on modeling, simulation and optimization of the probe location for RF ablation, which aims at giving numerical support for the planning and conduction of RF ablation in order to achieve a treatment quality similar to classical R0 resections. Based on a system of PDEs we have developed a multilevel gradient descent optimization. Since the constraining PDE system depends on a variety of patient-specific material parameters particular emphasis lies on the incorporation of this parameter uncertainty into the optimization. To this end we have considered the stochastic analogs of the PDE system and worked with objective functions involving expectations of the stochastically distributed states. A further approach to the uncertain material parameters is their identification from measured temperature data. Finally, for the prediction of the thermal necrosis we have developed an interactive real-time simulation by synthesizing patient-specific thermal necroses from reference configurations. We have applied our models and algorithms to geometric data resulting from real patient cases.

Acknowledgment

The authors acknowledge T. Stein, A. Roggan (Celon AG), U. Zurbuchen, B. Frericks, K. Lehmann, J.-P. Ritz (Charite Berlin), R.M. Kirby (University of Utah), D. Schmidt (University Clinic Tübingen), P.-L. Pereira (SLK Clinics Heilbronn) for valuable hints, fruitful discussions and collaborations, and for the CT data shown in the figures. Moreover, we thank S. Zentis and C. Hilck from MeVis Medical Solutions AG, Bremen for preprocessing the CT scans.

References

[1] I. Altrogge, T. Kröger, T. Preusser, C. Büskens, P.L. Pereira, D. Schmidt, A. Weihusen, H.-O. Peitgen. *Towards Optimization of Probe Placement for Radio-Frequency Ablation.* Proc. MICCAI, Lect. Not. Comp. Sci. 4190, 486–493, 2006.

[2] I. Altrogge, T. Preusser, T. Kröger, et al., *Multi-Scale Optimization of the Probe Placement for Radio-Frequency Ablation.* Acad. Radiol. **14**(11), 1310–1324, 2007.

[3] I. Altrogge, *Optimization of the Probe Placement for Radiofrequency Ablation.* PhD thesis, University of Bremen, Germany, 2009.

[4] I. Altrogge, T. Pätz, T. Kröger, T. Preusser. *Optimization and Fast Estimation of Vessel Cooling for RF Ablation.* Proc. IFMBE. 25/IV, 1202–1205 (2009).

[5] S.J. Benson, L.C. McInnes, J.J. Moré, J. Sarich. *Scalable Algorithms in Optimization: Computational Experiments.* Preprint ANL/MCS-P1175-0604, Mathematics and Computer Science, Argonne National Laboratory, Argonne, IL, 2004; in Proceedings of the 10th AIAA/ISSMO Multidisciplinary Analysis and Optimization (MA&O) Conference, Albany, NY, 2004.

[6] S. Gratton, A. Sartenaer, Ph.L. Toint. *Recursive Trust-Region Methods for Multiscale Nonlinear Optimization.* SIAM J Optim., 19(1), 414–444 (2008).

[7] D. Hämmerich, S.T. Staelin, J.Z. Tsai, S. Tungjikusolmun, D.M. Mahvi, J.G. Webster. *In Vivo Electrical Conductivity of Hepatic Tumours.*Physiol. Meas. 24, 251–260 (2003).

[8] T. Kröger, T. Pätz, I. Altrogge, A. Schenk, K.S. Lehmann, B.B. Frericks, J.-P. Ritz, H.-O. Peitgen, T. Preusser, *Fast Estimation of the Vascular Cooling in RFA Based on Numerical Simulation*, The Open Biomed. Eng. J. 4, 16–26 (2010).

[9] T. Kröger, S. Pannier, M. Kaliske, I. Altrogge, W. Graf, T. Preusser, *Optimal Applicator Placement in Hepatic Radiofrequency Ablation on the Basis of Rare Data.* Computer Methods in Biomechanics and Biomedical Engineering, to appear, 2009.

[10] T. Kröger, I. Altrogge, O. Konrad, R.M. Kirby, T. Preusser. *Visualization of Parameter Sensitivity Analyses resulting from Stochastic Collocation Methods.* Proc. Simulation and Visualization (SimVis), Magdeburg, 2008.

[11] T. Kröger, I. Altrogge, T. Preusser, P.L. Pereira, D. Schmidt, A. Weihusen, H.-O. Peitgen. *Numerical Simulation of Radio Frequency Ablation with State-Dependent Material Parameters in Three Space Dimensions.* Proc. MICCAI, Lect. Not. Comp. Sci. 4191, 380–388, 2006.

[12] D.D. Mosser, A.W. Caron, L. Bourget, et al., *Role of human heat shock protein* hsp70 *in protection against stress-induced apoptosis.* Mol. Cel. Biol. **17**, 9, 1997.

[13] Mulier, S.; Ni, Y.; Jamart, J.; Michel, L. et al., *Radiofrequency ablation versus resection for resectable colorectal liver metastases: Time for a randomized trial?* Ann. Sur. Oncology **15** 1, 144–157, 2007.

[14] T. Pätz, T. Kröger, T. Preusser. *Simulation of Radiofrequency Ablation including Water Evaporation.* Proc. IFMBE. 25/IV, 1287–1290 (2009).

[15] T. Stein, *Untersuchungen zur Dosimetrie der hochfrequenzstrominduzierten interstitiellen Thermotherapie in bipolarer Technik.* Fortschritte in der Lasermedizin, vol. 22, Müller und Berlien, 2000.

[16] H. Tiesler, R.M. Kirby, D. Xiu, T. Preusser, *Stochastic collocation for optimal control problems with stochastic PDE constraints.* In preparation.

[17] H. Tiesler, C. Büskens, D. Hämmerich, T. Preusser, *On the Optimization of Large Scale nonlinear PDE Systems with Application to Parameter Identification for RF Ablation.* In preparation.

[18] Zurbuchen, U.; Holmer, C.; Lehmann, K.S.; et al., *Determination of the temperature-dependent electric conductivity of liver tissue ex vivo and in vivo: Importance for therapy planning for the radiofrequency ablation of liver tumours.* Int. J. Hyperthermia, **26**, 1, 26–33, 2010.

Inga Altrogge, Tim Kröger, Heinz-Otto Peitgen, Tobias Preusser and Hanne Tiesler
Universitätsallee 29
D-28359 Bremen, Germany
e-mail: `Tobias.Preusser@mevis.fraunhofer.de`
`Hanne.Tiesler@cevis.uni-bremen.de`

Christof Büskens
Bibliothekstraße 1
D-28359 Bremen, Germany
e-mail: `bueskens@math.uni-bremen.de`

International Series of Numerical Mathematics, Vol. 160, 501–519

Optimization of Electro-mechanical Smart Structures

Eberhard Bänsch, Manfred Kaltenbacher, Günter Leugering, Fabian Schury and Fabian Wein

Abstract. We present topology optimization of piezoelectric loudspeakers using the SIMP method and topology gradient based methods along with analytical and numerical results.

Mathematics Subject Classification (2000). 90C30, 74P15.

Keywords. Topology optimization, piezoelectricity, loudspeakers.

1. Project overview

The goal of the project has been to devise mathematical methods capable of making an impact to the solution of the concrete engineering problem of optimizing piezoelectric loudspeakers. Thus, we focused on the SIMP method, which, in the engineering community, is an established method known for its efficiency, see e. g. [3]. Although there is a fast growing literature on SIMP for diverse problems, there are only a few number of publications on the topology optimization of piezoelectric material embedded in multiphysics coupling. In view of all this, it was necessary to develop and test SIMP for our problem setting.

As the analysis of the piezoelectric topology gradient was performed concurrently, expending efforts in the extension of the SIMP method now give the unique situation of having both kinds of topology optimization simultaneously implemented in the multiphysics simulation package of PI M. Kaltenbacher. Thus, comparison and combination of the methods based on realistic large scale 2D and 3D models can be performed on an elaborated numerical base.

1.1. Modeling of electrostatic-mechanical-acoustic transducers

The piezoelectric transducing mechanism is based on the interaction between the electric quantities electric field intensity $\mathbf{E}$ and electric induction $\mathbf{D}$, with the mechanical quantities mechanical stress tensor $[\boldsymbol{\sigma}]$ and strain tensor $[\mathbf{S}]$. By applying a mechanical load (force) to a piezoelectric transducer (e.g., piezoelectric material with top and bottom electrode), one can measure an electric voltage between the two electrodes (sensor effect). This mechanism is called the *direct piezoelectric effect*, and is due to a change in the electric polarization of the material. The so-called *inverse piezoelectric effect* is obtained by loading a piezoelectric transducer with an electric voltage. The transducer will thereby show mechanical deformations (actuator effect) and the setup can be used, e.g., in a positioning system.

The material law describing the piezoelectric effect is given by [24]

$$\boldsymbol{\sigma} = [\mathbf{c}^E]\mathbf{S} - [\mathbf{e}]^T\mathbf{E}, \tag{1.1}$$

$$\mathbf{D} = [\mathbf{e}]\mathbf{S} + [\boldsymbol{\varepsilon}^{\,S}]\mathbf{E}\,. \tag{1.2}$$

Since the stress tensor $[\boldsymbol{\sigma}]$ as well as strain tensor $[\mathbf{S}]$ are symmetric, it is convenient to write them as vectors of six components (the three normal and the three shear components) using *Voigt notation* and to denote them by $\boldsymbol{\sigma}$ and $\mathbf{S}$ [24]. The material tensors $[\mathbf{c}^E]$, $[\boldsymbol{\varepsilon}^{\,S}]$, and $[\mathbf{e}]$ appearing in (1.1) and (1.2) are the tensor of elastic modulus, of dielectric constants, and of piezoelectric moduli, respectively. The superscripts E and S indicate that the corresponding material parameters have to be determined at constant electric field intensity $\mathbf{E}$ and at constant mechanical strain $\mathbf{S}$, respectively. $[\mathbf{e}]^T$ denotes the transposed of $[\mathbf{e}]$.

In order to derive the coupled PDEs for piezoelectricity, we start with Navier's equation

$$\mathbf{f}_\mathrm{V} + \mathcal{B}^T\boldsymbol{\sigma} = \rho_\mathrm{m}\ddot{u}\,, \tag{1.3}$$

describing the mechanical field. In (1.3) $\mathbf{f}_\mathrm{V}$ denotes any mechanical volume force, ρ_m the density, $\mathbf{u}$ the mechanical displacement and $\mathcal{B}$ a differential operator, which computes as follows

$$\mathcal{B} = \begin{pmatrix} \frac{\partial}{\partial x} & 0 & 0 & 0 & \frac{\partial}{\partial z} & \frac{\partial}{\partial y} \\ 0 & \frac{\partial}{\partial y} & 0 & \frac{\partial}{\partial z} & 0 & \frac{\partial}{\partial x} \\ 0 & 0 & \frac{\partial}{\partial z} & \frac{\partial}{\partial y} & \frac{\partial}{\partial x} & 0 \end{pmatrix}^T . \tag{1.4}$$

Expressing $\boldsymbol{\sigma}$ as (1.1) and incorporating the strain–displacement relation $\mathbf{S} = \mathcal{B}\mathbf{u}$ results in

$$\rho_\mathrm{m}\ddot{\mathbf{u}} - \mathcal{B}^T\left([\mathbf{c}^E]\mathcal{B}\mathbf{u} - [\mathbf{e}]^T\mathbf{E}\right) = \mathbf{f}_\mathrm{V}\,. \tag{1.5}$$

Since piezoelectric materials are insulating, i.e., do not contain free-volume charges, and we do not have to consider any magnetic field, the electric field is determined by

$$\boldsymbol{\nabla}\cdot\mathbf{D} = 0, \qquad \boldsymbol{\nabla}\times\mathbf{E} = 0\,. \tag{1.6}$$

According to (1.6) we can express the electric field intensity $\mathbf{E}$ by the gradient of the scalar electric potential V_e

$$\mathbf{E} = -\boldsymbol{\nabla}V_\mathrm{e} = -\tilde{\mathcal{B}}V_\mathrm{e}, \qquad \tilde{\mathcal{B}} = (\partial/\partial x, \partial/\partial y, \partial/\partial z)^T\,. \tag{1.7}$$

By combining these results with (1.2), we obtain

$$\mathcal{B}^T\left([\mathbf{e}]\mathcal{B}\mathbf{u} - [\boldsymbol{\varepsilon}^S]\tilde{\mathcal{B}}V_\mathrm{e}\right) = 0 \tag{1.8}$$

The describing partial differential equations for linear piezoelectricity therefore read as

$$\rho_\mathrm{m}\ddot{\mathbf{u}} - \mathcal{B}^T\left([\mathbf{c}^E]\mathcal{B}\mathbf{u} + [\mathbf{e}]^T\tilde{\mathcal{B}}V_\mathrm{e}\right) = \mathbf{f}_\mathrm{V}, \tag{1.9}$$

$$\mathcal{B}^T\left([\mathbf{e}]\mathcal{B}\mathbf{u} - [\boldsymbol{\varepsilon}^{\ S}]\tilde{\mathcal{B}}V_\mathrm{e}\right) = 0\,. \tag{1.10}$$

For computing the acoustic wave propagation, we choose the scalar acoustic potential ψ, which is related to the acoustic velocity $\mathbf{v}_\mathrm{a}$ by $\mathbf{v}_\mathrm{a} = -\boldsymbol{\nabla}\psi$ and to the acoustic pressure p_a by

$$p_\mathrm{a} = \rho_\mathrm{f}\frac{\partial\psi}{\partial t}, \tag{1.11}$$

with ρ_f the mean density of the fluid (in our case air). The describing PDE is given by the linear wave equation

$$\frac{1}{c^2}\ddot{\psi} - \Delta\psi = 0, \tag{1.12}$$

with c the speed of sound. At the solid/fluid interface, continuity requires that the normal component of the mechanical surface velocity of the solid must coincide with the normal component of the acoustic velocity of the fluid. Thus, the following relation between the velocity $\mathbf{v}$ of the solid (expressed by the mechanical displacement $\mathbf{u}$) and the acoustic particle velocity $\mathbf{v}_a$ (expressed by the acoustic scalar potential ψ) arises

$$\mathbf{v} = \frac{\partial\mathbf{u}}{\partial t} \qquad \mathbf{v}_\mathrm{a} = -\boldsymbol{\nabla}\psi \qquad \mathbf{n}\cdot(\mathbf{v}-\mathbf{v}_\mathrm{a}) = 0$$

$$\mathbf{n}\cdot\frac{\partial\mathbf{u}}{\partial t} = -\mathbf{n}\cdot\boldsymbol{\nabla}\psi = -\frac{\partial\psi}{\partial\mathbf{n}}\,. \tag{1.13}$$

In addition, one has to consider the fact that the ambient fluid causes a pressure load (normal stress $\boldsymbol{\sigma}_\mathrm{n}$) on the solid

$$\boldsymbol{\sigma}_\mathrm{n} = -\mathbf{n}p_\mathrm{a} = -\mathbf{n}\rho_\mathrm{f}\frac{\partial\psi}{\partial t}\,. \tag{1.14}$$

Therefore, the overall problem of the piezoelectric-acoustic-system as displayed in Figure 1, can be stated as follows (for simplicity we neglect any initial conditions and possible mechanical volume forces $\mathbf{f}_\mathrm{V}$)

Given:

$V_\mathrm{e} \quad : \Gamma_\mathrm{hot} \to \mathbb{R}$

Find:

$\mathbf{u}(t) \quad : (\bar{\Omega}_\mathrm{piezo} \cup \bar{\Omega}_\mathrm{plate}) \times [0,T] \to \mathbb{R}^3$

$V_\mathrm{e}(t) \quad : \bar{\Omega}_\mathrm{piezo} \times [0,T] \to \mathbb{R}$

$\psi(t) \quad : \bar{\Omega}_\mathrm{air} \cup \bar{\Omega}_\mathrm{PML} \times [0,T] \to \mathbb{R}$

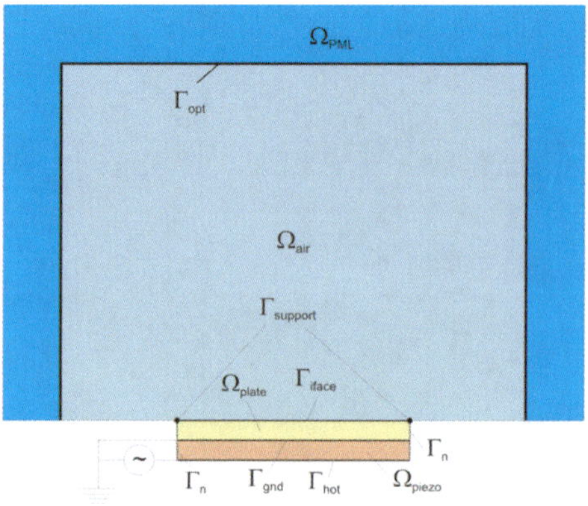

FIGURE 1. Setup of the piezoelectric-mechanical-acoustic system.

$$\begin{aligned}
\rho_{\mathrm{m}}\ddot{\mathbf{u}} - \mathcal{B}^T\left([\mathbf{c}^E]\mathcal{B}\mathbf{u} + [\mathbf{e}]^T\tilde{\mathcal{B}}V_{\mathrm{e}}\right) &= 0 \quad \text{in } \Omega_{\mathrm{piezo}} && (1.15)\\
\mathcal{B}^T\left([\mathbf{e}]\mathcal{B}\mathbf{u} - [\varepsilon^{\,S}]\tilde{\mathcal{B}}V_{\mathrm{e}}\right) &= 0 \quad \text{in } \Omega_{\mathrm{piezo}} && (1.16)\\
\rho_{\mathrm{m}}\ddot{\mathbf{u}} - \mathcal{B}^T[\mathbf{c}]\mathcal{B}\mathbf{u} &= 0 \quad \text{in } \Omega_{\mathrm{plate}} && (1.17)\\
\frac{1}{c^2}\ddot{\psi} - \tilde{\mathcal{B}}^T\tilde{\mathcal{B}}\psi &= 0 \quad \text{in } \Omega_{\mathrm{air}} && (1.18)\\
\frac{1}{c^2}\ddot{\psi} - \mathcal{A}^2\psi &= 0 \quad \text{in } \Omega_{\mathrm{PML}} && (1.19)
\end{aligned}$$

Boundary conditions

$$\begin{aligned}
\mathbf{u} &= \mathbf{0} \;\text{ on } \Gamma_{\mathrm{support}} \times (0,T) && (1.20)\\
\mathbf{n}\cdot[\boldsymbol{\sigma}] &= \mathbf{0} \;\text{ on } (\Gamma_{\mathrm{n}} \cup \Gamma_{\mathrm{hot}}) \times (0,T) && (1.21)\\
V_{\mathrm{e}} &= 0 \;\text{ on } \Gamma_{\mathrm{gnd}} \times (0,T) && (1.22)\\
V_{\mathrm{e}} &= V_{\mathrm{e}}^{\mathrm{p}}(t) \;\text{ on } \Gamma_{\mathrm{hot}} \times (0,T) && (1.23)
\end{aligned}$$

Interface conditions

$$\begin{aligned}
\mathbf{n}\cdot\frac{\partial\mathbf{u}}{\partial t} &= -\frac{\partial\psi}{\partial\mathbf{n}} \;\text{ on } \Gamma_{\mathrm{iface}} \times (0,T) && (1.24)\\
\boldsymbol{\sigma}_{\mathrm{n}} &= -\mathbf{n}\rho_{\mathrm{f}}\frac{\partial\psi}{\partial t} \;\text{ on } \Gamma_{\mathrm{iface}} \times (0,T) && (1.25)
\end{aligned}$$

It should be noted that the electrodes of the piezoelectric actuators (Γ_{gnd} and Γ_{hot} in Figure 2) have been treated as infinitely thin (for most practical cases allowed).

The application of advanced (numerical) methods is essential to the handling of such complex systems in the context of (multi-frequent) optimization. Far field simulation is achieved by the adaptation of the perfectly matched layers (PML)

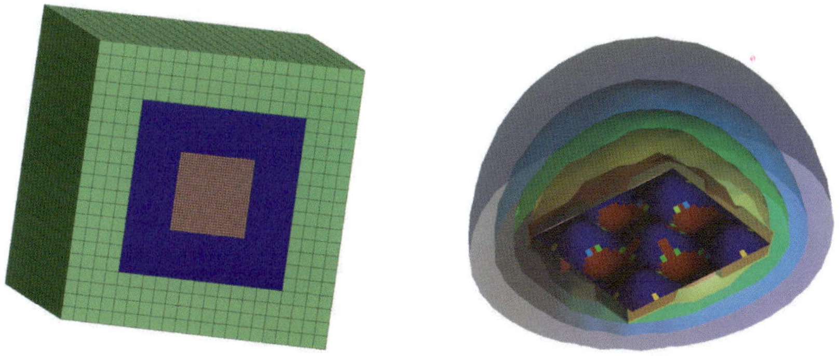

FIGURE 2. Left image: Discretization of the piezoelectric-mechanical-acoustic system using non-matching grids. Right image: Simulation results for an optimized piezoelectric layer (blue: no material, brown: material) excited by a voltage with 2000 Hz and isosurfaces of the pressure within Ω_{air}. No reflections occur due to PML damping.

method (for the original work see J.P. Berenger [4]) to the acoustic wave equation, as given with the quite complex operator $\mathcal{A}$ in (1.19) (see, e.g., [12]). Since currently we solve our coupled problems in the frequency domain, we can adapt the PML in such a way, that we obtain a modified Helmholtz equation (for details see [24]).

Furthermore, the application of non-matching grids for the mechanical-acoustic coupling at Γ_{iface} allows to choose an optimal discretization for the mechanical domain Ω_{plate} as well as for the acoustic domain $\Omega_{\mathrm{air}} \cup \Omega_{\mathrm{PML}}$. This prevents interference as in the case of conforming meshes [19].

The numerical implementation is based on state of the art solver packages, namely the PARDISO (parallel direct solver) package by O. Schenk and K. Gärtner [34] and M. Bollhöfer's ILUPACK [5] (iterative solver package based on an algebraic multilevel preconditioner). All the numerics were done using CFS++ [25].

1.2. SIMP – Introduction

Topology optimization is one of the most challenging research topics in structural optimization. Since the pioneering work of M.P. Bendsøe and N. Kikuchi [2], various approaches have been developed within the fast growing literature. See M.P. Bendsøe and O. Sigmund [3] as the standard textbook in this area or P.W. Christensen and A. Klarbring [11] for a recent textbook with an explicit focus on the mathematical backgrounds.

For the variable density approach (often denoted by the SIMP[1]-method), an artificial density function $0 < \rho \leq 1$ is introduced in order to represent the material distribution within the optimization domain. Via an interpolation function $\mu(\rho)$ the material is locally modified by piecewise constant ρ such that full material

[1]Solid Isotropic Material with Penalization

is represented for ρ close to one and void material for ρ close to a lower bound. Meanwhile, this approach has been adapted to several physical models, mainly elasticity but also piezoelectricity, acoustics (e.g., M.B. Döhring, J.S. Jensen and O. Sigmund [13]), magnetics (J. Lieneman et al.[29]) and many more.

The design variable is not only used as a density parameter but also as a polarization parameter in piezoelectricity (M. Kögl & E.C. Silva [27]), bi-material interpolation (e.g., J. Du & N. Olhoff [18]) or switching between physical models (G.H. Yoon, J.S. Jensen and O. Sigmund [42]).

Generally, additional methods are required to drive ρ towards its bounding values. That is necessary for a physically meaningful interpretation (full or void material). This is usually done by applying a volume constraint limiting the total amount of distributable material to a fraction of complete full material combined with a penalizing interpolation function. The most common for SIMP in the static case is

$$\mu(\rho) = \rho^p$$

with p commonly chosen to three. For the time harmonic case the RAMP [3] interpolation scheme $\mu(\rho) = \frac{\rho}{1+q(1-\rho)}$, with q in the order of 5, is usually chosen. A common phenomenon are mesh-dependent checkerboards which can be avoided by local averaging the gradients [35].

The choice of the volume constraint can be motivated by practical device design considerations (weight and cost) or to achieve the desired 0/1 pattern. In many practical cases this actually gives a multi-criteria optimization problem where the choice of the volume constraint needs to be explicitly motivated.

Significant contributions to SIMP research have been made by M.P. Bendsøe, O. Sigmund, J.S. Jensen and N. Olhoff. E.g., Jensen concentrates on dynamic methods (Padé approximants in topology optimization [23]) and their applications.

The topology optimization of piezoelectric materials is performed by relatively few scientists compared to pure elastic materials. The majority of publications are in cooperation with E.C. Silva (e.g., with N. Kikuchi [36], S. Kishiwaki and N. Kikuchi [37], M. Kögl [27], R.C. Carbonari and S. Nishiwaki [7], R.C. Carbonari and G.H. Paulino [8], P.H. Nakasone [30] or P.H. Nakasone and C.Y. Kiyono [31]). However, all these works are based on the mean transduction which, for the loudspeaker model, results in a vanishing piezoelectric layer when no volume constraint is used (we described this effect in [40]).

Further recent work (2008) has been performed by A. Donoso (with O. Sigmund [16] and J.C. Bellido [15] and [14]) but he uses no full finite element model for the piezoelectric layer and his simplifying assumptions are not valid for a model as used within this project.

B. Zhen, C.-J. Chang and H.C. Gea [43] also published the topology optimization of an energy harvester in 2008, but they considered only the static case and no coupling to further layers. With regard to the less frequent case of piezoelectric-mechanical coupled systems, often only the mechanical part is optimized (e.g., H. Du, G.K. Lau, M.K. Lim and J. Qui [17]).

1.3. Topology gradient – general remarks

While the SIMP-method provided us with a well-established method in order to achieve topology optimization for the relaxed problem as described in Section 1.2, the proposed method involving topological gradients and shape sensitivities guarantees a 0-1-design. However, as was established during the project, the analytical framework for establishing the topological gradient of piezoelectric material is far more involved than the one known for 3D elasticity. Moreover, the existing results on shape sensitivities for 3D elasticity did not directly generalize to the system under consideration.

In order to proceed along the two lines indicated before, we decided to first pursue, in addition to the SIMP-approach, the topology-shape-optimization approach first for 2D and 3D elasticity in order to gain numerical expertise for full piezoelectric-mechanical-acoustic coupling. Even though 2D and even 3D topology optimization based on topology gradients and shape-sensitivities is available in the literature, the implementations known to us were performed in MATLAB (2D, e.g., A. Laurain [20]) or were not available to us. Certainly MATLAB coding would not be feasible for 3D problems of the size that we have to take into account. On the other hand, trying to mimic existing codes for topology optimization for 3D elasticity using FEM would have left us with the problem of properly extending the FEM to piezoelectric-mechanical-acoustic coupling envisaged in this project. As the software package CFS (M. Kaltenbacher) focuses exactly on the latter system and is well established and validated, it appeared mandatory to the team that a solid 3D topology optimization based on topological gradients and shape sensitivities was the avenue along which we should proceed.

The proposed research in this project rests on the same algorithmic paradigm due to Michael Hintermüller [21] and Alliare et al. [1], namely an alternation between topology steps based on topological gradient calculations and shape steps based on shape sensitivities followed by a levelset-step. A principal algorithm would be given by

1. Setup problem
2. Compute topology gradient and create holes
3. Compute shape gradient and deform working domain
4. repeat until an appropriate termination condition is fulfilled

It was reasonable to implement and test recent algorithms due to Céa, Garreau, Guillaume and Masmoudi [10] that are based on a fixed-point iteration for the solution of an optimality condition for the topology optimization problem with and without volume constraints. This new approach has been established for 2D and 3D elasticity problems in the software package CFS of PI M. Kaltenbacher.

A similar approach has also been suggested by Bendsøe et al. [32] where a levelset-function in combination with the topological gradient in a fictitious domain setting is used. This approach has also been implemented in the software CFS.

2. Project achievements

2.1. Own publications

The following publications came out of the project phase:

- *Topology Optimization of Piezoelectric Layers Using the SIMP Method* [40] in cooperation with B. Kaltenbacher, submitted to *Structural and Multidisciplinary Optimization*. It contains the first analysis (to our knowledge) of "void material" in piezoelectricity and results for static optimization using the mean transduction (explaining why it fails for the given setup) and maximization of the displacement. Furthermore, the detailed strong and weak formulations for the forward and adjoint systems are given.
- *Topology optimization of a piezoelectric-mechanical actuator with single- and multiple-frequency excitation* [41], submitted to *Journal for Computation and Mathematics in Electrical and Electronic Engineering* (*COMPEL*). The results are discussed below. Among others, it gives a more efficient computation of the approximated sound pressure optimization presented by J. Du and N. Olhoff [18].
- *Efficient 3D-Finite-Element-Formulation for Thin Mechanical and Piezoelectric Structures* [6] by D. Braess and M. Kaltenbacher, appeared in *International Journal for Numerical Methods in Engineering.* With the developed enhanced formulation it is now possible to avoid locking in thin mechanical as well piezoelectric structures.

The following peer reviewed proceedings paper appeared:

- *Topology Optimization of Piezoelectric Actuators Using the SIMP Method* [38] in *Proceedings of the 10th Workshop on Optimization and Inverse Problems in Electromagnetism – OIPE* 2008.

For SCPIP [44], an extended implementation of the MMA optimization algorithm by Ch. Zillober, we provide an open source C++ interface [39] with the usability of the common IPOPT package. It has already been downloaded 38 times.

2.2. SIMP – results

During the course of this project, an important phenomenon was observed which has not, to our knowledge, been published before: For our models and objective function, we were able to achieve topology optimization of piezoelectric materials without a volume constraint (no appearance of unphysical intermediate material for most cases). Additionally, no penalizing interpolation functions (which change the actual problem) are necessary and no filtering of the objective gradients has to be performed. Without a volume constraint it is furthermore possible to analyze the optimization for different target frequencies (see Figure 5).

We now introduce the notation $\widetilde{(\cdot)}$ for elements modified by the design parameter. The design vector $\boldsymbol{\rho} = (\rho_1 \dots \rho_{N_e})^T$ with $\rho_e \in [\rho_{\min} : 1]$ shall be piecewise constant within the N_e finite elements of the design domain. Limiting the optimization domain to the piezoelectric layer Ω_{piezo}, the following material parameters are formed by applying the design variable

$$[\widetilde{\boldsymbol{c}}_e^E] = \rho_e[\boldsymbol{c}_0^E], \quad \widetilde{\rho}_e^{\mathrm{m}} \;\; = \rho_e\rho^{\mathrm{m}}, \quad [\widetilde{\boldsymbol{e}}_e] = \rho_e[\boldsymbol{e}_0], \quad [\widetilde{\boldsymbol{\epsilon}}_e^S] \;\; = \rho_e[\boldsymbol{\epsilon}_0^S].$$

Applying the finite element method with the new ersatz material properties to the strong formulation given in Section 1.1, the piecewise constant ρ_e can be written in front of the element matrices. Analogously to the example $\widetilde{\boldsymbol{K}}_{uu}^e$ given by

$$\widetilde{k}_{pq}^{uu} = \int_{\Omega_e} (\boldsymbol{\mathcal{B}}_p^u)^T\, [\widetilde{\boldsymbol{c}}_e^E]\boldsymbol{\mathcal{B}}_q^u \, d\Omega = \mu_c(\rho_e) \int_{\Omega_e} (\boldsymbol{\mathcal{B}}_p^u)^T\, [\boldsymbol{c}_0^E]\boldsymbol{\mathcal{B}}_q^u \, d\Omega$$

we get

$$\widetilde{\boldsymbol{K}}_{uu}^e = \mu_c(\rho_e)\boldsymbol{K}_{uu}^e, \quad \widetilde{\boldsymbol{M}}_{uu}^e = \mu_m(\rho_e)\boldsymbol{M}_{uu}^e,$$
$$\widetilde{\boldsymbol{K}}_{u\phi}^e = \mu_e(\rho_e)\boldsymbol{K}_{u\phi}^e, \quad \widetilde{\boldsymbol{K}}_{\phi\phi}^e = \mu_\epsilon(\rho_e)\boldsymbol{K}_{\phi\phi}^e.$$

Assuming a sinusoidal single-frequency excitation, a Fourier transformation of the system solves the steady state solution in the complex domain[2]. Damping is introduced by the Rayleigh model with

$$\boldsymbol{S}(\omega) = \boldsymbol{K} + j\omega\boldsymbol{C} - \omega^2\boldsymbol{M} \quad \text{and} \quad \boldsymbol{C} = \alpha_K\boldsymbol{K} + \alpha_M\boldsymbol{M}.$$

The global system is now written as

$$\begin{pmatrix} \boldsymbol{S}_{\psi\,\psi}(\omega) & \boldsymbol{C}_{\psi\,u_m}(\omega) & \boldsymbol{0} & \boldsymbol{0} \\ \boldsymbol{C}_{\psi\,u_m}^T(\omega) & \boldsymbol{S}_{u_m u_m}(\omega) & \widetilde{\boldsymbol{S}}_{u_m u_p}(\omega,\boldsymbol{\rho}) & \boldsymbol{0} \\ \boldsymbol{0} & \widetilde{\boldsymbol{S}}_{u_m u_p}^T(\omega,\boldsymbol{\rho}) & \widetilde{\boldsymbol{S}}_{u_p u_p}(\omega,\boldsymbol{\rho}) & \widetilde{\boldsymbol{K}}_{u_p\phi}(\boldsymbol{\rho}) \\ \boldsymbol{0} & \boldsymbol{0} & \widetilde{\boldsymbol{K}}_{u_p\phi}^T(\boldsymbol{\rho}) & \widetilde{\boldsymbol{K}}_{\phi\,\phi}(\boldsymbol{\rho}) \end{pmatrix} \begin{pmatrix} \boldsymbol{\psi}(\omega,\boldsymbol{\rho}) \\ \boldsymbol{u}_m(\omega,\boldsymbol{\rho}) \\ \boldsymbol{u}_p(\omega,\boldsymbol{\rho}) \\ \boldsymbol{\phi}(\omega,\boldsymbol{\rho}) \end{pmatrix} = \begin{pmatrix} \boldsymbol{0} \\ \boldsymbol{0} \\ \boldsymbol{0} \\ \bar{\boldsymbol{q}}_\phi(\omega) \end{pmatrix},$$

with $\boldsymbol{C}_{\psi\,u_m}$ the coupling matrix between mechanics and acoustics. In short form we may write

$$\widehat{\widetilde{\mathbf{S}}}(\omega,\boldsymbol{\rho})\,\widehat{\mathbf{u}}(\omega,\boldsymbol{\rho}) = \widehat{\mathbf{f}}(\omega) \quad \text{or} \quad \widehat{\widetilde{\mathbf{S}}}\,\widehat{\mathbf{u}} = \widehat{\mathbf{f}}.$$

Note that with the presented form any topology of the piezoelectric layer can be expressed on the same finite element mesh.

The objective gradient $\partial J/\partial\rho_e$ is found by the adjoint method. The complex solution vector $\widehat{\mathbf{u}}$ is split into a real and imaginary part, enabling the objective function of the form (2.2) or (2.3) to be written as follows

$$J\omega^{-2} = \widehat{\mathbf{u}}_R^T L\,\widehat{\mathbf{u}}_R + \widehat{\mathbf{u}}_I^T L\,\widehat{\mathbf{u}}_I + \boldsymbol{\lambda}_1^T(\widehat{\widetilde{\mathbf{S}}}\,\widehat{\mathbf{u}} - \widehat{\mathbf{f}}) + \boldsymbol{\lambda}_2^T(\widehat{\widetilde{\mathbf{S}}}^*\,\widehat{\mathbf{u}}^* - \widehat{\mathbf{f}}^*),$$

where $\boldsymbol{\lambda}_1$ and $\boldsymbol{\lambda}_2$ are Lagrangian multipliers. Deriving J and eliminating the derivatives of the solution, the sensitivity is given as

$$\frac{\partial J}{\partial\rho_e} = 2\omega^2\mathrm{Re}\{\boldsymbol{\lambda}^T\frac{\partial\widehat{\widetilde{\mathbf{S}}}}{\partial\rho_e}\widehat{\mathbf{u}}\},$$

where $\boldsymbol{\lambda}$ solves the adjoint equation (for details we refer to [22]) $\widehat{\widetilde{\mathbf{S}}}\,\boldsymbol{\lambda} = -2\omega^2\boldsymbol{L}\,\widehat{\mathbf{u}}^*$.

[2]With imaginary unit j as common in electrical engineering.

Concentrating on the application, we consider the complete piezoelectric-mechanical-acoustic system and a corresponding piezoelectric-mechanical setup separately. The smaller system has its own significance and can be solved with less computational effort. This is an important property when elaborating broad frequency optimization.

The acoustic sound power P_{a} is given as

$$P_{\mathrm{a}} = \int_{\Gamma_{\mathrm{target}}} \frac{1}{2} \mathrm{Re}\{p_{\mathrm{a}}\, v_{\mathrm{n}}^*\}\, d\Gamma, \tag{2.1}$$

with the sound pressure p_{a} and v_{n}^* the complex conjugate of the normal particle velocity. The (complex) acoustic impedance Z connects velocity and pressure by $Z(\boldsymbol{x}) = \frac{p_{\mathrm{a}}(\boldsymbol{x})}{v_{\mathrm{n}}(\boldsymbol{x})}$.

Assuming p_{a} and v_{n}^* in phase (Z being constant) and using $v_{\mathrm{n}} = j\omega \mathbf{n}^{\mathbf{T}} \bar{\mathbf{u}}$ where $\mathbf{n}$ is the unit normal in z-direction of target, Du and Olhoff [18] optimize for the sound radiation of a mechanical layer without calculating an acoustic field. In [41] we show the limitations of this approximation but also that the objective function

$$J = \omega^2 \boldsymbol{u}^T \boldsymbol{L} \boldsymbol{u}^*, \tag{2.2}$$

with $\boldsymbol{L}$ a diagonal matrix selecting the z displacements of Γ_{iface} gives a more efficient calculation than the form presented by Du and Olhoff. Selected optimization results are shown in Figure 3.

Considering also the acoustic domain, again a constant acoustic impedance has to be assumed at the optimization region Γ_{opt}. This holds when one optimizes in the acoustic far field and hence the following objective function (by using (1.11)) is the desired optimizer for the sound pressure (2.1):

$$J = \boldsymbol{p}_{\mathrm{a}}^T \boldsymbol{L} \boldsymbol{p}_{\mathrm{a}}^*. \tag{2.3}$$

There are still numerical issues within the optimization process for some frequency ranges to be resolved (see Figure 5).

As we explain in detail in [41], only some of the eigenmodes of the laminated piezoceramic/aluminum structure can be excited electrically by a piezoelectric layer covering the full aluminum plate. The vibrational pattern is then a combination of the lower and upper resonance modes. For some frequency ranges, the optimizer is able to find topologies where resonance actually occurs (see Figure 3c and Figure 4c).

Finding a flat frequency response for a broad frequency range in combination with exploiting resonance effects poses a multi-criteria optimization problem where one has to balance between flat response and sound pressure level. We do this for k base frequencies ω_k weighted by β_k, forming the following scalarized problem

$$\max_{\boldsymbol{\rho},\boldsymbol{\beta}} \sum_{k=1}^{n} \beta_k J(\widehat{\mathbf{u}}_k(\boldsymbol{\rho}_k, \omega_k)). \tag{2.4}$$

The strategy to find the weights is heuristic. In each iteration, the forward and adjoint problems are solved for all base frequencies and β_k is found such that

$$\beta_k^q J_k = \text{const} \quad \text{and} \quad \sum_{k=1}^{n} \beta_k = 1.$$

The damping parameter q controls the switching between the criteria (for more details see [41]).

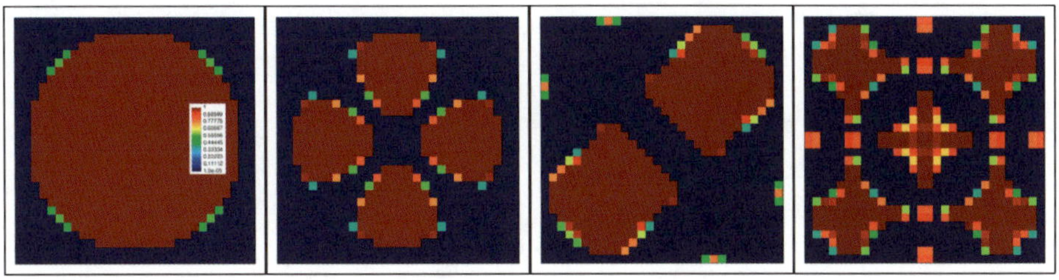

(A) 20 Hz (B) 255 Hz (C) 950 Hz (D) 1995 Hz

FIGURE 3. Selected topologies for the piezoelectric-mechanical optimization for $J = \omega^2 \boldsymbol{u}^T \boldsymbol{L} \boldsymbol{u}^*$ (2.2). Brown represents piezoelectric material, blue expresses void material.

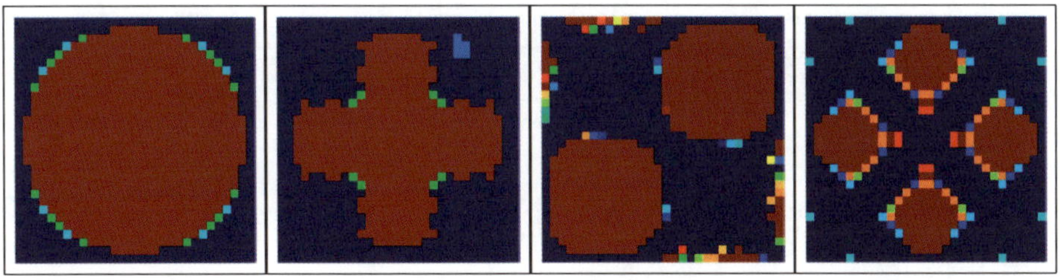

(A) 20 Hz (B) 260 Hz (C) 980 Hz (D) 1990 Hz

FIGURE 4. Selected topologies for the piezoelectric-mechanical-acoustic optimization for $J = \boldsymbol{p}^T \boldsymbol{L} \boldsymbol{p}^*$ (2.3). Colors as in Figure 3.

2.3. Topology gradient and shape gradient

Topology gradient for piezoelectric-mechanical material. In order to establish a substantial joint-work with respect to topology optimization for elastic networks [28], G. Leugering discussed with his colleagues in Nancy. J. Sokołowski, who is founder of the mathematical theory for topological gradient-based optimization, showed strong interest in the subject of analyzing piezoelectric-mechanical materials with respect to topological derivatives. Together with S. Nazarov from St. Petersburg he wrote a very substantial paper on the subject [9]. S. Nazarov subsequently visited Erlangen in December 2009 in order to work with G. Leugering on the particular piezoelectric-mechanical structure relevant for this project. What has been established so far can be understood from the subsequent outline. To

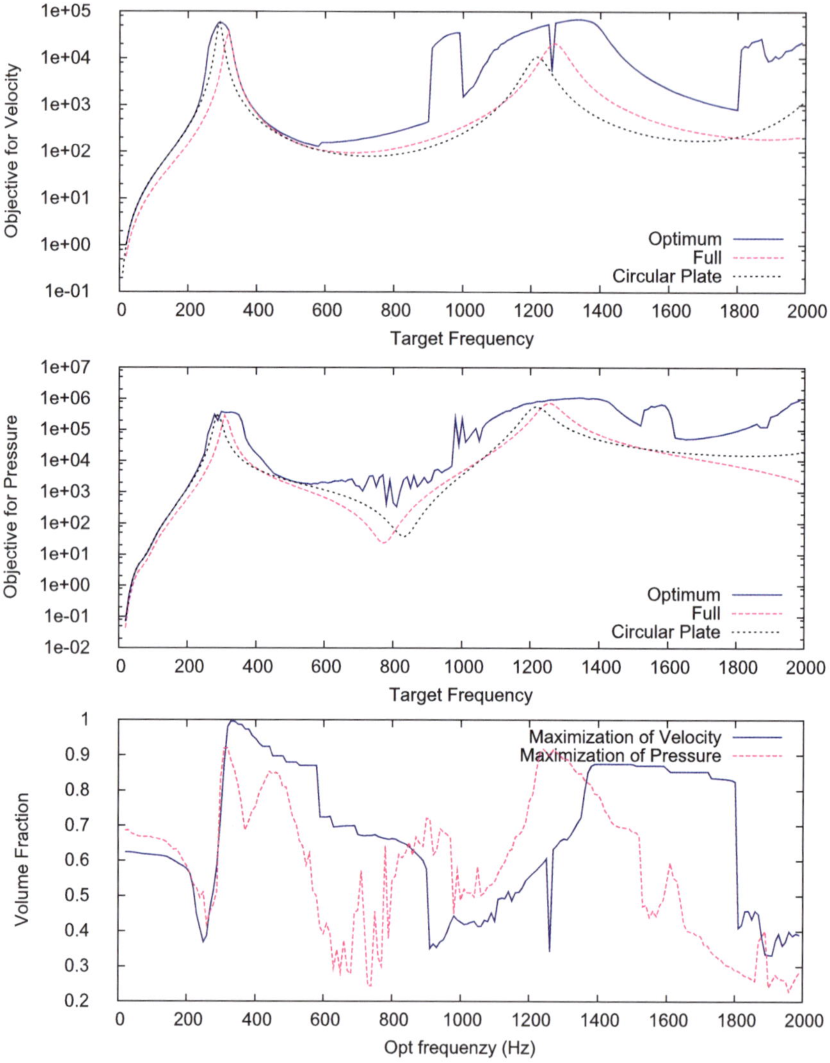

FIGURE 5. Within the target frequency range $\omega_{\text{target}} = 20 \ldots 2000$ Hz we optimize for each single frequency. As reference, the frequency response of a full and a circular setup (optimal static topology) are shown. In the first picture, the plate surface velocity on Γ_{iface} is optimized, in the center figure the acoustic pressure at Γ_{opt}, and the final figure depicts the volume fraction of the optimal topologies.

compute the topology gradient for the time harmonic system, we introduce the mechanical strain ϵ as introduced in Section 1 and define the total strain $\epsilon(u)$ by[3]

$$\epsilon(u) = (\epsilon^M(u^M)^T, \epsilon^E(u^E)^T)^T \quad \text{and} \quad \epsilon(u) = D(\nabla_x)u,$$

with (for the differential operator $\mathcal{B}$ see (1.4))

$$D(\nabla_x)^T = \begin{pmatrix} \mathcal{B} & & 0 \\ \mathbf{O} & \mathbf{O} & \nabla_x^T \end{pmatrix}, \quad \mathbf{O} = (\mathbf{0}, \mathbf{0}, \mathbf{0}).$$

After equivalently reformulating the stresses, we get the stress-strain-relation

$$\sigma(u) = A\epsilon(u).$$

We dig a hole ω_h of radius h into Ω centered at the origin and denote the new domain $\Omega(h) = \Omega \setminus \bar{\omega}_h$. The system then reads

$$\begin{aligned} D(-\nabla_x)^T A(x) D(\nabla_x) u^h(x) &= f(x), & x \in \Omega(h) \\ D(n(x)^h)^T A(x) D(\nabla_x) u^h(x) &= g(x), & x \in \Gamma_N \\ D(n(x)^h)^T A(x) D(\nabla_x) u^h(x) &= 0, & x \in \partial\omega_h \\ u^h(x) &= 0, & x \in \Gamma_D \end{aligned}$$

One can derive a variational form for this set of equations. However, the sesquilinear form is not coercive and one therefore has to deal with an indefinite problem.

After introducing the appropriate material expansions for the sensitivity analysis we can look at the class of shape functionals

$$\mathcal{J}(u;\Omega) = \int_\Omega J(u(x);x)dx,$$

with some canonical regularity conditions and growth bounds.

Finally, we obtain the following formula for the topological gradient

$$\begin{aligned} \mathcal{J}(u^h;\Omega(h)) = {} & \mathcal{J}(u;\Omega) \\ & + h^3 \left(J(u(0);0) - P(0)^T f(0)\mathrm{mes}(\omega) - D(\nabla_x)P(0)^T M\epsilon^0 \right) \\ & + O(h^{3+\cdots}), \end{aligned}$$

where P solves the adjoint problem

$$\begin{aligned} D(-\nabla_x)^T A(x) D(\nabla_x) P(x) &= J'(u(x),x), & x \in \Omega \\ D(n(x))^T A(x) D(\nabla_x) P(x) &= 0, & x \in \Gamma_N \\ P(x) &= 0, & x \in \Gamma_D \end{aligned}$$

and M is the so-called polarization matrix, which has to be computed numerically as no analytical formula is yet available. During the visit of S. Nazarov in Erlangen in December 2009 some important improvements to the theory could be obtained and the cooperation could be further fostered.

---

[3]To be consistent with most mathematical publications we use ϵ for the mechanical strain. Furthermore, from now on vectors are no more written in bold and we use capital letters for tensors.

Using this formula one can now again establish an iterative algorithm for a topology optimization process. In the next step, we wish to obtain an analytical representation of the corresponding shape gradient for the piezoelectric PDEs.

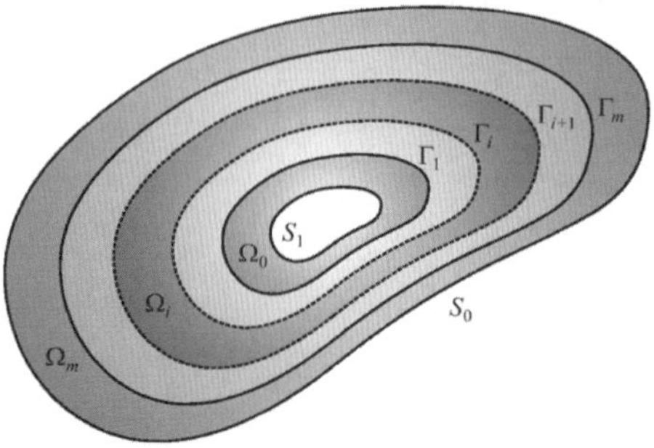

FIGURE 6. Domain Ω with boundary $\partial\Omega = S_0 \cup S_1$.

Shape gradient for the piezoelectric-mechanical problem in the dynamic case. In a second cooperative research attempt we were successful in deriving shape-sensitivities for layered piezoelectric materials. Layered material is conceptually what we need for our final goal. To this end, studying the layered situation was a more attractive option than the coupling between piezoelectric and pure mechanical material. This was the subject of joint work with A. Novotny, G. Perla-Menzala (Rio de Janeiro) and J. Sokołowski (Nancy). The paper is in preparation. The model considered there is as follows:

Let us consider an open bounded domain Ω of $\mathbb{R}^3$ with smooth boundary $\partial\Omega = S$. We assume that Ω has the form $\Omega = \mathcal{B}_0 \setminus \overline{\mathcal{B}_1}$, where $\mathcal{B}_0$ and $\mathcal{B}_1$ are open bounded domains with $\overline{\mathcal{B}_1} \subset \mathcal{B}_0$, with $\overline{(\cdot)}$ used to denote the closure of $(\cdot)$. In addition, $\partial\mathcal{B}_0 = S_0$ and $\partial\mathcal{B}_1 = S_1$, thus $S = S_0 \cup S_1$.

Let $m > 1$ be a given integer. For each i with $1 \le i \le m$ let $\mathcal{D}_i$ be an open subset with smooth boundary Γ_i and such that $\overline{\mathcal{B}_1} \subset \mathcal{D}_i \subset \mathcal{B}_0$, $\overline{\mathcal{D}_i} \subset \mathcal{D}_{i+1}$.

We set $\Omega_0 = \mathcal{D}_1 \setminus \overline{\mathcal{B}_1}$, $\Omega_i = \mathcal{D}_{i+1} \setminus \overline{\mathcal{D}_i}$ for $1 \le i \le m-1$ and $\Omega_m = \mathcal{B}_0 \setminus \overline{\mathcal{D}_m}$. In summary, as shown in Figure 6 we have $\Omega = \cup_{i=0}^m \Omega_i$, such that $\Omega_i \cap \Omega_j = \emptyset$ for $i \neq j$, with boundaries $\partial\Omega = S_0 \cup S_1$, $\partial\Omega_0 = S_1 \cup \Gamma_1$, $\partial\Omega_i = \Gamma_i \cup \Gamma_{i+1}$ for $i = 1, \ldots, m-1$, and $\partial\Omega_m = \Gamma_m \cup S_0$.

The electromechanical interaction phenomenon is modeled by the following coupled system

$$\left\{ \begin{array}{rcl} u_{tt} - \operatorname{div}\sigma & = & 0 \\ \operatorname{div}\varphi & = & 0 \end{array} \right. \quad \text{in} \quad \Omega \times (0,T) \tag{2.5}$$

where σ is again the mechanical stress tensor and φ the electrical displacement field. The material law describing the piezoelectric effect in the linearized case of small mechanical deformations and electric fields essentially reads as before

$$\left\{ \begin{array}{rcl} \sigma(u,q) & = & C\varepsilon(u) - Pe(q) \\ \varphi(u,q) & = & P^T\varepsilon(u) + De(q), \end{array} \right.$$

where C is the elasticity fourth-order tensor, P the piezoelectric coupling third-order tensor and D the dielectric second-order tensor. In addition, the mechanical strain tensor ε and the electric vector field e are given by

$$\varepsilon(u) = \nabla^s u \quad \text{and} \quad e(q) = -\nabla q,$$

where $u = u(x,t)$ is the mechanical displacement and $q = q(x,t)$ the electric potential. We also associate with system (2.5) the following given initial conditions

$$u(x,0) = f(x) \quad \text{and} \quad u_t(x,0) = g(x)$$

and boundary conditions of the form

$$\left\{ \begin{array}{rcl} \sigma n & = & 0 \\ q & = & 0 \end{array} \right. \quad \text{on} \quad S_0 \times (0,T) \quad \text{and} \quad \left\{ \begin{array}{rcl} \varphi \cdot n & = & 0 \\ u & = & 0 \end{array} \right. \quad \text{on} \quad S_1 \times (0,T),$$

where n is the outward unit normal vector pointing toward the exterior of Ω. Finally, we consider the following transmission conditions

$$\left\{ \begin{array}{rcl} [\![\sigma(u,q)]\!] n & = & 0 \\ {[\![u]\!]} & = & 0 \end{array} \right. \quad \text{and} \quad \left\{ \begin{array}{rcl} [\![\varphi(u,q)]\!] \cdot n & = & 0 \\ {[\![q]\!]} & = & 0 \end{array} \right.$$

where for any $(x,t) \in \Gamma_i \times (0,T)$, $i = 1,2,\ldots,m$, the symbol $[\![(\cdot)]\!]$ is used to denote the jump between quantities evaluated on the boundary Γ_i of each pair Ω_{i-1} and Ω_i. That is $[\![(\cdot)]\!] = (\cdot)^{(i)} - (\cdot)^{(i-1)}$ and $n = n^{(i)} = -n^{(i-1)}$ is the unit normal vector pointing toward the exterior of Ω_i.

We now consider the shape functional of the form

$$\mathcal{J}_\Omega(u,q) = \int_0^T J_\Omega(u,q).$$

In order to simplify the further calculation, we introduce the adjoint states v and p, which are solutions of the following variational system: given the final conditions $v(x,T) = 0$ and $v_t(x,T) = 0$, find for each $t \in (0,T)$ the adjoint displacement $v \in \mathcal{W}_M(\Omega)$ and the adjoint electrical potential $p \in \mathcal{W}_E(\Omega)$, such that

$$\left\{ \begin{array}{rcll} \langle v_{tt}, \eta \rangle_\Omega + a_\Omega^{MM}(v,\eta) - a_\Omega^{EM}(p,\eta) & = & -\langle D_u(J_\Omega(u,q)), \eta \rangle & \forall \eta \in \mathcal{W}_M(\Omega) \\ a_\Omega^{EE}(p,\xi) + a_\Omega^{ME}(v,\xi) & = & -\langle D_q(J_\Omega(u,q)), \xi \rangle & \forall \xi \in \mathcal{W}_E(\Omega) \end{array} \right. .$$

We perform the shape sensitivity analysis of the functional $\mathcal{J}_{\Omega_\tau}(u_\tau, q_\tau)$. Thus we need to calculate its derivative with respect to the parameter τ at $\tau = 0$, that is

$$\int_0^T \dot{J}_\Omega(u,q) = \dot{\mathcal{J}}_\Omega(u,q) := \left. \frac{d}{d\tau} \mathcal{J}_{\Omega_\tau}(u_\tau, q_\tau) \right|_{\tau=0}$$

We give an important example for a tracking type functional:

$$J_\Omega(u,q) = \frac{1}{2} \int_\Omega (u - u_\Omega^*)^2.$$

The shape derivative of the above functional can be written as boundary integrals, namely

$$\begin{aligned} \dot{J}_\Omega(u,q) &= \tfrac{1}{2}\int_{S_0}(u-u^*_\Omega)^2 V\cdot n + \int_{\partial\Omega}(u_{tt}\cdot v + Sn\cdot n)V\cdot n \\ &\quad + \sum_{i=1}^m \int_{\Gamma_i}([\![S]\!]n\cdot n)V\cdot n \end{aligned}$$

In the above formula, S is the so-called Ehelby tensor given by

$$S = (\sigma\cdot\nabla^s v - \varphi\cdot\nabla p)I - (\nabla u^T\sigma_a + \nabla v^T\sigma - \nabla q\otimes\varphi_a - \nabla p\otimes\varphi)$$

where v and p are solutions to the adjoint system

$$\left\{\begin{array}{lcl} v_{tt} - \mathrm{div}\sigma_a &=& -(u-u^*_\Omega) \\ -\mathrm{div}\varphi_a &=& 0 \end{array}\right. \quad \text{in } \Omega\times(0,T),$$

with the adjoint stress tensor σ_a and the adjoint electrical displacement φ_a defined by

$$\left\{\begin{array}{lcl} \sigma_a(v,p) &=& C\nabla^s v + Pe(p), \\ \varphi_a &=& -P^T\nabla^s v + De(p). \end{array}\right.$$

In addition, we associate with the final system the final conditions $v(x,T) = v_t(x,T) = 0$, the boundary conditions

$$\left\{\begin{array}{rcl} \sigma_a n &=& 0 \\ p &=& 0 \end{array}\right. \quad \text{on} \quad S_0\times(0,T) \quad \text{and} \quad \left\{\begin{array}{rcl} \varphi_a\cdot n &=& 0 \\ v &=& 0 \end{array}\right. \quad \text{on} \quad S_1\times(0,T),$$

and, for any $(x,t)\in\Gamma_i\times(0,T), i = 1,2,\dots,m$ the transmission conditions of the form

$$\left\{\begin{array}{rcl} [\![\sigma_a]\!]n &=& 0 \\ [\![v]\!] &=& 0 \end{array}\right. \quad \text{and} \quad \left\{\begin{array}{rcl} [\![\varphi_a]\!]\cdot n &=& 0 \\ [\![p]\!] &=& 0 \end{array}\right. .$$

The fully coupled time-dependent problem: Wellposedness. In a third cooperative research initiative, A. Novotny, G. Perla-Menzala, J. Sokołowski and G. Leugering concentrated on the wellposedness of the fully coupled linear system described in detail in the introduction. The team focused on a paper by G. Perla-Menzala, B. Miara and A. Khapitonov [26], where the piezoelectric model described in the previous subsection was analyzed with respect to wellposedness and asymptotic stability as well as controllability. For the fully coupled system, taking also the acoustic wave-equation into account, we succeeded to prove wellposedness. The result is subject to a publication and its description here would be inordinate. The paper is in preparation [33].

Acknowledgment

We would like to thank the reviewers for their helpful comments. We also acknowledge the funding by the DFG Priority Programme 1253 and the Cluster of Excellence in Erlangen for continued funding of ongoing research.

References

[1] G. Allaire, F. Jouve, and A.-M. Toader. Structural optimization using sensitivity analysis and a level-set method. *J. Comp. Physics*, 194:363–393, 2004.

[2] Martin P. Bendsøe and N. Kikuchi. Generating optimal topologies in optimal design using a homogenization method. *Comp. Meth. Appl. Mech. Engn.*, 71:197–224, 1988.

[3] Martin P. Bendsøe and Ole Sigmund. *Topology Optimization: Theory, Methods and Applications.* Springer-Verlag, 2002.

[4] J.P. Berenger. A Perfectly Matched Layer for the Absorption of Electromagnetic Waves. *Journal of Computational Physics*, 1994.

[5] Matthias Bollhöfer and Yousef Saad. Multilevel preconditioners constructed from inverse-based ilus. *SIAM Journal on Scientific Computing*, 27(5):1627–1650, 2006.

[6] D. Braess and Manfred Kaltenbacher. Efficient 3D-Finite-Element-Formulation for Thin Mechanical and Piezoelectric Structures. *Int. J. Numer. Meth. Engng.*, 73:147–161, 2008.

[7] R.C. Carbonari, E.C.N. Silva, and S. Nishiwaki. Design of piezoelectric multi-actuated microtools using topology optimization. *Smart Materials and Structures*, 14(6):1431, 2005.

[8] R.C. Carbonari, E.C.N. Silva, and G.H. Paulino. Topology optimization design of functionally graded bimorph-type piezoelectric actuators. *Smart Materials and Structures*, 16(6):2605, 2007.

[9] G. Cardone, S.A. Nazarov, and Jan Sokołowski. Topological derivatives in piezoelectricity. Preprint, 2008.

[10] Jean Céa, Stéphane Garreau, Philippe Guillaume, and Mohamed Masmoudi. The shape and topological optimizations' connection. *Comput. Methods Appl. Mech. Eng.*, 188(4):713–726, 2000.

[11] P.W. Christensen and A. Klarbring. *An Introduction to Structural Optimization.* Springer Verlag, 2008.

[12] J. Diaz and J. Joly. A time domain analysis of pml models in acoustics. *Comput. Meth. Appl. Mech. Engrg.*, 2005.

[13] Maria B. Döhring, Jakob S. Jensen, and Ole Sigmund. Acoustic design by topology optimization. submitted, 2008.

[14] A. Donoso and J.C. Bellido. Distributed piezoelectric modal sensors for circular plates. *Journal of Sound and Vibration*, 2008.

[15] A. Donoso and J.C. Bellido. Systematic design of distributed piezoelectric modal sensors/actuators for rectangular plates by optimizing the polarization profile. *Structural and Multidisciplinary Optimization*, 2008.

[16] A. Donoso and Ole Sigmund. Optimization of piezoelectric bimorph actuators with active damping for static and dynamic loads. *Structural and Multidisciplinary Optimization*, 2008.

[17] H. Du, G.K. Lau, M.K. Lim, and J. Qiu. Topological optimization of mechanical amplifiers for piezoelectric actuators under dynamic motion. *Smart Materials and Structures*, 9(6):788–800, 2000.

[18] J. Du and N. Olhoff. Minimization of sound radiation from vibrating bi-material structures using topology optimization. *Structural and Multidisciplinary Optimization*, 33(4):305–321, 2007.

[19] B. Flemisch, M. Kaltenbacher, and B.I. Wohlmuth. Elasto-Acoustic and Acoustic-Acoustic Coupling on Nonmatching Grids. *Int. J. Numer. Meth. Engng.*, 67(13): 1791–1810, 2006.

[20] Piotr Fulmanski, Antoine Laurain, Jean-François Scheid, and Jan Sokołowski. A level set method in shape and topology optimization for variational inequalities. *Int. J. Appl. Math. Comput. Sci.*, 17(3):413–430, 2007.

[21] M. Hintermüller. A combined shape-Newton topology optimization technique in real-time image segmentation. Technical report, University of Graz, 2004. Preprint.

[22] Jakob S. Jensen. A note on sensitivity analysis of linear dynamic systems with harmonic excitation. Handout at DCAMM advanced school June 20–26, 2007 at DTU in Lyngby, Denmark, June 2007.

[23] Jakob S. Jensen. Topology optimization of dynamics problems with pade approximants. *Int. J. Numer. Meth. Engng.*, 72:1605–1630, 2007.

[24] M. Kaltenbacher. *Numerical Simulation of Mechatronic Sensors and Actuators.* Springer Berlin-Heidelberg-New York, 2nd edition, 2007. ISBN: 978-3-540-71359-3.

[25] Manfred Kaltenbacher. Advanced Simulation Tool for the Design of Sensors and Actuators. In *Proc. Eurosensors XXIV, Linz, Austria*, September 2010.

[26] B. Kapitonov, B. Miara, and G. Perla-Menzala. Boundary observation and exact control of a quasi-electrostatic piezoelectric system in multilayered media. *SIAM J. CONTROL OPTIM.*, 46(3):1080–1097, 2007.

[27] M. Kögl and E.C.N. Silva. Toplogy optimization of smart structures: design of piezoelectric plate and shell actuators. *Smart Mater. Struct.*, 14:387–399, 2005.

[28] Günter Leugering and Jan Sokołowski. Topological derivatives for elliptic problems on graphs. *Variational Formulations in Mechanics: Theory and Application*, 2006.

[29] J. Lieneman, A. Greiner, J.G. Korvink, and Ole Sigmund. Optimization of integrated magnetic field sensors. *ORBIT*, 2007:2006, 2008.

[30] P.H. Nakasone and C.N.S. Silva. Design of dynamic laminate piezoelectric sensors and actuators using topology optimization. In *Proceedings of the 6th International Conference on Computation of Shell and Spatial Structures.* IASS-IACM 2008, May 2008.

[31] P.H. Nakasone, C.Y. Kiyono, and E.C.N. Silva. Design of piezoelectric sensors, actuators, and energy harvesting devices using topology optimization. In *Proceedings of SPIE*, volume 6932, page 69322W. SPIE, 2008.

[32] Julian A. Norato, Martin P. Bendsøe, Robert B. Haber, and Daniel A. Tortorelli. A topological derivative method for topology optimization. *Struct. Multidisc. Optim.*, 33:375–386, 2007.

[33] G. Perla-Menzala, A. Feij'oo Novotny, Günter Leugering, and Jan Sokołowski. Well-posedness of a dynamic wave-propagation problem in a coupled piezoelectric-elastic-acoustic structure. Preprint, 2008.

[34] O. Schenk and K. Gärtner. Solving unsymmetric sparse systems of linear equations with PARDISO. *Journal of Future Generation Computer Systems*, 20(3):475–487, 2004.

[35] Ole Sigmund. A 99 Line topology optimization code written in MATLAB. *J. Struct. Multidiscip. Optim.*, 21:120–127, 2001.

[36] E.C.N. Silva and N. Kikuchi. Design of piezoelectric transducers using topology optimization. *Smart Mater. Struct.*, 8:350–364, 1999.

[37] E.C. N.Silva, N. Kikuchi, and S. Nishiwaki. Topology optimization design of flextensional actuators. *IEEE Transactions on Ultrasonics, Ferroelectrics and Frequency Control*, 47(3):657–671, 2000.

[38] F. Wein, M. Kaltenbacher, F. Schury, G. Leugering, and E. Bänsch. Topology Optimization of Piezoelectric Actuators using the SIMP Method. In *Proceedings of the 10th Workshop on Optimization and Inverse Problems in Electromagnetism – OIPE* 2008, pages 46–47, September 2008. TU Ilmenau, 14.–17.09.2008.

[39] Fabian Wein. C++SCPIP, a C++ wrapper for SCPIP. Online, August 2007. http://cppmath.sourceforge.net.

[40] Fabian Wein, Manfred Kaltenbacher, Barbara Kaltenbacher, Günter Leugering, Eberhard Bänsch, and Fabian Schury. Topology optimization of piezoelectric layers using the simp method. Submitted for review, July 2008.

[41] Fabian Wein, Manfred Kaltenbacher, Günter Leugering, Eberhard Bänsch, and Fabian Schury. Topology optimization of a piezoelectric-mechanical actuator with single- and multiple-frequency excitation. Submitted for review, October 2008.

[42] Gil Ho Yoon, Jakob S. Jensen, and Ole Sigmund. Topology optimization of acoustic-structure interaction problems using a mixed finite element formulation. *International Journal for Numerical Methods in Engineering*, 70:1049–1075, 2006.

[43] Bin Zheng, Ching-Jui Chang, and Hae Chang Gea. Topology optimization of energy harvesting devices using piezoelectric materials. *Structural and Multidisciplinary Optimization*, 2008.

[44] Ch. Zillober. SCPIP – an efficient software tool for the solution of structural optimization problems. *Structural and Multidisciplinary Optimization*, 24(5):362–371, 2002.

Eberhard Bänsch
Haberstr. 2, D-91052 Erlangen, Germany
e-mail: `baensch@am.uni-erlangen.de`

Manfred Kaltenbacher
Universitätsstr. 65-67, A-9020 Klagenfurt, Austria
e-mail: `manfred.kaltenbacher@uni-klu.ac.at`

Günter Leugering
Martensstr. 3, D-91052 Erlangen, Germany
e-mail: `leugering@am.uni-erlangen.de`

Fabian Schury and Fabian Wein
Nägelsbachstr. 49b, D-91052 Erlangen, Germany
e-mail: `schury@am.uni-erlangen.de`
`fabian.wein@am.uni-erlangen.de`

International Series of Numerical Mathematics, Vol. 160, 521–540

Freezing of Living Cells: Mathematical Models and Design of Optimal Cooling Protocols

Nikolai D. Botkin, Karl-Heinz Hoffmann and Varvara L. Turova

Abstract. Two injuring effects of cryopreservation of living cells are under study. First, stresses arising due to non-simultaneous freezing of water inside and outside of cells are modeled and controlled. Second, dehydration of cells caused by earlier ice building in the extracellular liquid compared to the intracellular one is simulated.

A low-dimensional mathematical model of competitive ice formation inside and outside of living cells during freezing is derived by applying an appropriate averaging technique to partial differential equations describing the dynamics of water-to-ice phase change. This reduces spatially distributed relations to a few ordinary differential equations with control parameters and uncertainties. Such equations together with an objective functional that expresses the difference between the amount of ice inside and outside of a cell are considered as a differential game. The aim of the control is to minimize the objective functional, and the aim of the disturbance is opposite. A stable finite-difference scheme for computing the value function is applied to the problem. On the base of the computed value function, optimal cooling protocols ensuring simultaneous freezing of water inside and outside of living cells are designed. Thus, balancing the inner and outer pressures prevents cells from injuring.

Another mathematical model describes shrinkage and swelling of cells caused by their osmotic dehydration and rehydration during freezing and thawing. The model is based on the theory of ice formation in porous media and Stefan-type conditions describing the osmotic inflow/outflow related to the change of the salt concentration in the extracellular liquid. The cell shape is searched as a level set of a function which satisfies a Hamilton-Jacobi equation resulting from a Stefan-type condition for the normal velocity of the cell boundary. Hamilton-Jacobi equations are numerically solved using finite-difference schemes for finding viscosity solutions as well as by computing reachable sets of an associated conflict control problem. Examples of the shape evolution computed in two and three dimensions are presented.

Mathematics Subject Classification (2000). 49N90, 49L20, 35F21, 65M06.

Keywords. Optimal cooling rate, differential game, Hamilton-Jacobi equations, finite-difference scheme, reachable sets.

Introduction

Our intention "Optimal control in cryopreservation of cells and tissues" is devoted to the application of optimal control theory to the minimization of damaging effects of cooling and thawing to increase the survival rate of frozen and subsequently thawed cells. The main objective of the intention is the development of coupled hierarchical mathematical models that describe the most injuring effects of cryopreservation of living cells. The work has been started with the development of optimized cooling protocols that minimize damaging effects of freezing related to the release of the latent heat and irregular ice formation (see [10] and [11] for results obtained). The present paper continues the work and considers two most damaging factors of cryopreservation: large stresses exerted on cell membranes and rehydration/dehydration effects.

Large stresses exerted on cell membranes occur at slow cooling because of non-simultaneous freezing of extracellular and intracellular fluids. The use of rapid cooling rates is not a perfect solution because water inside cells forms small, irregularly-shaped ice crystals (dendrites) that are relatively unstable. If frozen cells are subsequently thawed, dendrites will aggregate to form larger, more stable crystals that may cause damage. Therefore, it is reasonable to apply control theory to provide simultaneous freezing of extracellular and intracellular fluids even for slow cooling rates. Therefore, a mathematical model describing freezing and thawing is to be formulated with control variables and optimization criteria. Note that models of phase transitions basically utilize partial differential equations that describe the dynamics of phases in each spatial point (see, e.g., [4], [8], and [9]). Nevertheless, the spatial distribution is not very important, if small objects such as living cells and pores of the extracellular matrix are investigated. Our experience shows that appropriate averaging techniques accurately reduce spatially distributed models to a few ordinary differential equations with control parameters and uncertainties. However, such equations contain, as a rule, nonlinear dependencies given by tabular data. This complicates the application of traditional control design methods based on Pontryagin's maximum principle. Nevertheless, dynamic programming principles related to Hamilton-Jacobi-Bellman-Isaacs (HJBI) equations are suitable, if stable grid methods for solving HJBI problems are available. The present paper considers a stable grid procedure that allows us to design optimized controls (cooling protocols) for an ODE system describing competitive ice formation inside and outside of living cells.

The second injuring factor studied in this paper is excessive shrinkage and swelling of cells due to the osmotic dehydration and rehydration occurring in freezing and warming phases, respectively. The paper considers mathematical models of these effects caused by the osmotic pressure arising due to different salt concentrations in the extracellular and intracellular liquids. Conventional models of cell dehydration during freezing (see, e.g., [1] and [14]) describe the change of the cell volume depending on the mass diffusion rate, heat transfer conditions, and the evolution of the freezing front. The cell shape is supposed to be spherical or cylindrical. However, as it is reported by biologists (see, e.g., [5]), keeping both the cell size and shape is very important for cell survival. Mathematical models proposed in the present paper are concerned with the evolution of the cell shape depending on the temperature distribution and the amount of the frozen liquid outside the cell. The models are based on the theory of ice formation in porous media (see [8]) and Stefan-type models (see [4]) describing the osmotic outflow (inflow) caused by the increase (decrease) of salt concentration in the extracellular liquid during its freezing (thawing). Equations of the Hamilton-Jacobi type are solved numerically using both a finite-difference scheme for finding viscosity solutions and the computation of reachable sets of an associated conflict control problem formalized according to [15, 17]. Examples of the shape evolution computed are presented.

1. Mathematical model of ice formation

Remember that cells of a tissue are located in pores of an extracellular matrix (see a sketch in the left part of Figure 1). Each pore is filled by a solution called extracellular liquid and contains a cell. The extracellular fluid freezes earlier than the intracellular one, and the volumetric increase of ice produces a great pressure exerted on cell membranes. The magnitude of the effect can be approximately estimated as $p \approx C^{\text{ice}} \cdot \alpha \cdot (1 - \beta_\ell)$, where p is the pressure, C^{ice} is the elasticity of ice, α is the expansion coefficient of ice, β_ℓ is the unfrozen water fraction. A rough estimate yields: $p \approx 1$Bar, which may be dangerous for cell membranes. The conventional method to avoid that is lowering the freezing point of the extracellular liquid to give a chance the extracellular and intracellular fluids to be frozen simultaneously. Obviously, a proper cooling protocol is necessary to utilize this

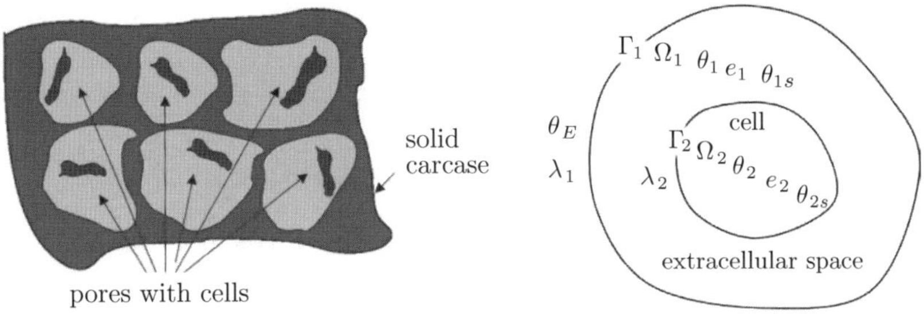

FIGURE 1. Two-dimensional sketch of an extracellular matrix with pores filled with extracellular fluid and living cells containing intracellular liquid (to the left). Schematic notation of variables and three-dimensional regions related to a single pore (to the right). Note that Γ_1 and Γ_2 are the boundaries of the pore and the cell, respectively, Ω_1 is the region lying between Γ_1 and Γ_2 (extracellular space), and Ω_2 is the region of the cell.

chance. Our objective is to develop numerical techniques for the design of such protocols.

In the case were a fluid fills a micro volume, say $\Omega \in R^3$, bounded by a solid wall, the unfrozen water fraction β_ℓ can be computed using the following phase field model (see [8]):

$$\frac{\partial e(\theta)}{\partial t} - \mathcal{K}\Delta\theta = 0, \qquad -\mathcal{K}\frac{\partial \theta}{\partial \nu}\Big|_{\partial\Omega} = \lambda(\theta - \theta_E)|_{\partial\Omega},$$
$$e(\theta) = \rho\mathcal{C}\theta + \rho L\beta_\ell(\theta), \qquad \beta_\ell(\theta) = \phi\left(\frac{L(\theta-\theta_s)}{(T_0+\theta_s)(T_0+\theta)}\right), \tag{1}$$

where θ is the Celsius temperature, $\mathcal{K}$ the heat conductivity coefficient, λ the overall (film) heat transfer coefficient, θ_E the external temperature (outside of Ω), $\mathcal{C}$ the specific heat capacity, ρ the density, L the latent heat, θ_s the freezing (solidification) point, T_0 the Celsius zero point (273°K). The function $e(\theta)$ has the sense of the internal energy. The function ϕ is recovered from experimental data.

Consider now a sketch of a pore shown in Figure 1 (to the right). The notations are self-explanatory. It is only to note that λ_1 and λ_2 denote the film heat transfer coefficients of the pore boundary and the cell membrane, respectively, θ_{1s} and θ_{2s} stand for the freezing points of the extracellular and intracellular liquids, respectively. Integrating the energy balance equation of (1) over Ω_1 and Ω_2, using notations

$$\hat{e}_i = \frac{1}{|\Omega_i|}\int_{\Omega_i} e_i dV, \qquad \hat{\theta}_i = \frac{1}{|\Gamma_i|}\int_{\Gamma_i}\theta_i dS,$$
$$\hat{\theta}_E = \frac{1}{|\Gamma_1|}\int_{\Gamma_1}\theta_E dS, \quad \hat{\alpha}_i = \frac{|\Gamma_2|}{|\Omega_i|}\lambda_2, \qquad \hat{\lambda} = \frac{|\Gamma_1|}{|\Omega_1|}\lambda_1,$$

for mean values, and assuming that $|\Gamma_1|^{-1}\int_{\Gamma_1}\theta_1 dS \approx |\Gamma_2|^{-1}\int_{\Gamma_2}\theta_1 dS$ because θ_1 is almost constant in the small region Ω_1, yield the following coupled system of ordinary differential equations:

$$\frac{d}{dt}\hat{e}_1 = -\hat{\alpha}_1\big[\hat{\theta}_1 - \hat{\theta}_2\big] - \hat{\lambda}\big[\hat{\theta}_1 - \hat{\theta}_E\big], \qquad \frac{d}{dt}\hat{e}_2 = -\hat{\alpha}_2\big[\hat{\theta}_2 - \hat{\theta}_1\big]. \tag{2}$$

The relation between $\hat{e}_i$ and $\hat{\theta}_i$ (see (1)) is given by the formula

$$\hat{e}_i = \rho\mathcal{C}\hat{\theta}_i + \rho L\beta_\ell^i(\hat{\theta}_i), \tag{3}$$

where β_ℓ^1 and β_ℓ^2 are defined by the replacement of θ_s with θ_{1s} and θ_{2s}, respectively, in (1).

Note that the derivatives $\partial\hat{e}_i/\partial\hat{\theta}_i$, $i = 1, 2$, are discontinuous (see equation (3) and the left graph of Figure 2). Therefore, the direct form of equations (2) is not appropriate. It is convenient to express the temperatures $\hat{\theta}_1$ and $\hat{\theta}_2$ through the energies $\hat{e}_1$ and $\hat{e}_2$, respectively, using the relation $\hat{\theta}_i = \Theta_i(\hat{e}_i)$ which is the inverse of relation (3) (see the right graph of Figure 2). By doing that, we obtain

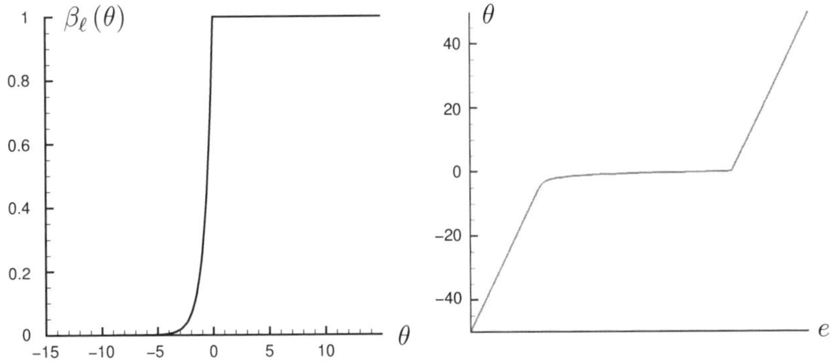

FIGURE 2. Dependence of the unfrozen water fraction on the temperature (to the left). Graph of the inverse function to $e(\theta)$ (to the right).

the following system of ordinary differential equations:

$$\dot{\hat{e}}_1 = -\hat{\alpha}_1\big[\Theta_1(\hat{e}_1) - \Theta_2(\hat{e}_2)\big] - \hat{\lambda}\big[\Theta_1(\hat{e}_1) - \hat{\theta}_E\big], \qquad \dot{\hat{e}}_2 = -\hat{\alpha}_2\big[\Theta_2(\hat{e}_2) - \Theta_1(\hat{e}_1)\big].$$

For simplicity, denote $x = \hat{e}_1,\ y = \hat{e}_2,\ z = \hat{\theta}_E,\ \alpha_i = \hat{\alpha}_i,\ \lambda = \hat{\lambda}$, and consider the following controlled system:

$$\begin{aligned} \dot{x} &= -\alpha_1[\Theta_1(x) - \Theta_2(y)] - \lambda[\Theta_1(x) - z] - d + v_1 && (=: f_1),\\ \dot{y} &= -\alpha_2[\Theta_2(y) - \Theta_1(x)] + v_2 && (=: f_2),\\ \dot{z} &= u && (=: f_3). \end{aligned} \tag{4}$$

Here, z is the temperature outside of the pore (chamber temperature), u is the cooling rate, v_1, v_2 are disturbances interpreted as data errors. The control variable u is restricted by $-2\mu \le u \le 0$, the disturbances v_1, v_2 are bounded by $|v_1| \le \nu$, $|v_2| \le \nu$. Since the zero initial value for z is always assumed, the constant d is introduced to emulate nonzero initial values of z (the initial chamber temperatures).

The definition of the function β_ℓ says that exact simultaneous freezing of the extracellular and intracellular liquids can be expressed as the vanishing of the functional

$$J = \max_{t\in[0,t_f]} \gamma(x(t), y(t)), \tag{5}$$

where

$$\gamma(x,y) = |\beta_\ell^1(\Theta_1(x)) - \beta_\ell^2(\Theta_2(y))|$$

estimates the difference of the ice fractions in the extra- and intracellular regions.

Consider differential game (4), (5) assuming that the objective of the control u is to minimize the functional J, whereas the objective of the disturbance is opposite. It is known, that the value function of a differential game completely defines its solution. The value function will be computed as a viscosity solution to an appropriate HJBI equation, and the optimal feedback control will be designed by applying the procedure of extremal aiming (see [16] and [17]).

2. Hamilton-Jacobi-Bellman-Isaacs equations

The game (4), (5) is formalized as in [16], [17], and [20], which ensures the existence of the value function V. The next important result proved in [21] says that V coincides with a unique viscosity solution (see [6], [7], and [2]) of the HJBI equation

$$V_t + H(x, y, z, V_x, V_y, V_z) = 0, \tag{6}$$

where the Hamiltonian H, the majorization, and terminal conditions are defined as

$$H(x, y, z, p_1, p_2, p_3) = \max_{|v_1|,|v_2|\le\nu} \min_{-2\mu\le u\le 0} \sum_{i=1}^{3} p_i f_i,$$
$$V(t, x, y, z) \ge \gamma(x, y), \quad V(t_f, x, y, z) = \gamma(x, y).$$

In the next section, an approximation scheme for solving this HJBI equation is discussed and a convergence result is given.

3. Approximation scheme and convergence result

Let $\tau, \Delta_x, \Delta_y, \Delta_z$ be time and space discretization step sizes. Introduce the notation

$$V^n(x_i, y_j, z_k) = V(n\tau, i\Delta_x, j\Delta_y, k\Delta_z), \; n = 1, \dots, N, \; N = \frac{t_f}{\tau},$$

and consider a difference scheme

$$V^{n-1}(x_i, y_j, z_k) = V^n(x_i, y_j, z_k) + \tau H(x_i, y_j, z_k, V_x^n, V_y^n, V_z^n),$$
$$V^N(x_i, y_j, z_k) = \gamma(x_i, y_j).$$

Here, the symbols V_x^n, V_y^n, V_z^n denote finite difference approximations (left, right, central, etc.) of the corresponding partial derivatives. The above scheme can be considered as the application of a time step operator Π to the grid function V^n to obtain V^{n-1}, i.e.,

$$V^{n-1} = \Pi(V^n; \tau, \Delta_x, \Delta_y, \Delta_z).$$

It is clear that this operator can be naturally extended to continuum functions.

Definition 1. The operator Π is monotone, if the following implication (the point-wise order is assumed) holds:

$$V \le W \Rightarrow \Pi(V; \tau, \Delta_x, \Delta_y, \Delta_z) \le \Pi(W; \tau, \Delta_x, \Delta_y, \Delta_z).$$

Definition 2. The operator Π has the generator property, if the estimate

$$\left| \frac{\Pi(\phi; \tau, a\tau, b\tau, c\tau)(\vec{r}) - \phi(\vec{r})}{\tau} - H(\vec{r}, D\phi(\vec{r})) \right| \le C\big(1 + \|D\phi\| + \|D^2\phi\|\big)\tau \tag{7}$$

holds for every $\phi \in C_b^2(\mathcal{D})$, $\vec{r} = (x, y, z) \in \mathcal{D}$, and fixed bounded a, b, $c > 0$. Here $\mathcal{D} \subset R^3$ is a large domain containing all states of the game, $C_b^2(\mathcal{D})$ is the space of twice continuously differentiable functions defined on $\mathcal{D}$ and bounded together with their two derivatives, $\|\cdot\|$ denotes the point-wise maximum norm, $D\phi$ and $D^2\phi$ denote the gradient and the Hessian matrix of ϕ.

Theorem 1. (Convergence, [2], [19]). *Assume that the operator* $\Pi(\cdot;\tau,a\tau,b\tau,c\tau)$ *is monotone for any* $\tau>0$ *and satisfies the generator property, then the grid function obtained by*

$$V^{n-1}=\max\{\Pi(V^n;\tau,a\tau,b\tau,c\tau),\ \gamma\},\quad V^N=\gamma, \tag{8}$$

converges point-wise to a unique viscosity solution of (6) *and, therefore, to the value function of the differential game* (4), (5) *as* $\tau\to 0$. *The convergence rate is* $\sqrt{\tau}$.

Remark 1. Theorem 1 refers only to the monotonicity and generator properties of the operator Π. Really, some secondary properties must hold to provide the convergence (see [2] and [19]). We omit here the discussion of these because they obviously hold for an upwind operator Π that will be presented in the next section.

4. Upwind time step operator

We will consider an upwind time step operator proposed in [13] and prove that it is monotone and possesses the generator property, which proves convergence claims. Unfortunately, the convergence arguments given in [13] are very sketchy and not strong. They are solely based on topological considerations and do not take into account the nature of viscosity solutions so that they have little force. Nevertheless, the idea of the operator proposed there is brilliant.

Assume that the right-hand sides $f_i, i=1,2,3$, are now arbitrary functions of the state (x,y,z), control $u\in P\subset R^p$, and disturbance $v\in Q\subset R^q$. Denote $a^+=\max(a,0)$, $a^-=\min(a,0)$. The operator introduced in [13] assumes the following approximations of the spatial derivatives:

$$V_x^n\cdot f_1=p_1^R\cdot f_1^+ + p_1^L\cdot f_1^-,\quad V_y^n\cdot f_2=p_2^R\cdot f_2^+ + p_2^L\cdot f_2^-,\quad V_z^n\cdot f_3=p_3^R\cdot f_3^+ + p_3^L\cdot f_3^-,$$

where f_1, f_2, f_3 are computed at (x_i,y_j,z_k); p_1^R, p_2^R, p_3^R and p_1^L, p_2^L, p_3^L the right and left divided differences, respectively, defined as

$$\begin{aligned}
p_1^R&=[V^n(x_{i+1},y_j,z_k)-V^n(x_i,y_j,z_k]/\Delta_x,\\
p_1^L&=[V^n(x_i,y_j,z_k)-V^n(x_{i-1},y_j,z_k]/\Delta_x,\\
p_2^R&=[V^n(x_i,y_{j+1},z_k)-V^n(x_i,y_j,z_k]/\Delta_y,\\
p_2^L&=[V^n(x_i,y_j,z_k)-V^n(x_i,y_{j-1},z_k]/\Delta_y,\\
p_3^R&=[V^n(x_i,y_j,z_{k+1})-V^n(x_i,y_j,z_k]/\Delta_z,\\
p_3^L&=[V^n(x_i,y_j,z_k)-V^n(x_i,y_j,z_{k-1}]/\Delta_z.
\end{aligned}$$

Finally, the operator is given by

$$\begin{aligned}
&\Pi(V^n;\tau,\Delta_x,\Delta_y,\Delta_z)(x_i,y_j,z_k)=V^n(x_i,y_j,z_k)+\\
&+\tau\max_{v\in Q}\min_{u\in P}\left(p_1^R\cdot f_1^+ + p_1^L\cdot f_1^- + p_2^R\cdot f_2^+ + p_2^L\cdot f_2^- + p_3^R\cdot f_3^+ + p_3^L\cdot f_3^-\right).
\end{aligned}\tag{9}$$

Lemma 1. (Monotonicity and generator property, [3]). *Let M be the bound of the right-hand side of the controlled system. If $a, b, c \geq M\sqrt{3}$, then the operator $\Pi(\cdot; \tau, a\tau, b\tau, c\tau)$ given by (9) is monotone. The generator property (7) holds for any fixed a, b, c.*

The proof is the same as in Lemmas 2 and 3 (see Section 8). Thus, the operator (9) satisfies the conditions of Theorem 1.

Remark 2. If the functions $f_i, i = 1, 2, 3$, are linear in u and v at each fixed state (x, y, z), then the operation $\max_{v \in Q} \min_{u \in P}$ appearing in the definitions of the operator (9) can be replaced by $\max_{v \in \text{ext}\, Q} \min_{u \in \text{ext}\, P}$, where "ext" returns the set of the extremal points. In particular, "max min" can be computed over the set of vertices, if P and Q are polyhedrons. The monotonicity holds independently of the structure of the sets P and Q. To prove the generator property, it is sufficient to observe that the Hamiltonian is equal to that computed using the sets ext P and ext Q whenever the assumption of linearity in u and v at each fixed state point holds. Note that this remark is very important for numerical implementations of the operator (9) because the operation "max min" is applied to a function which is nonlinear and neither convex nor concave in u and v.

5. Control procedure

In this section, the computation of optimal controls for system (4) in accordance with the procedure of extremal aiming (see [16] and [17]) is described.

Let ε be a small positive number, t_n the current time instant. Consider the neighborhood

$$\mathcal{U}_\varepsilon = \{(x, y, z) \in R^3 : |x - x(t_n)| \leq \varepsilon,\ |y - y(t_n)| \leq \varepsilon,\ |z - z(t_n)| \leq \varepsilon\}$$

of the current state $\big(x(t_n), y(t_n), z(t_n)\big)$ of system (4). By searching through all grid points $(x_i, y_j, z_k) \in \mathcal{U}_\varepsilon$, find a point $(x_{i_*}, y_{j_*}, z_{k_*})$ such that

$$V^n(x_{i_*}, y_{j_*}, z_{k_*}) = \min_{(x_i, y_j, z_k) \in \mathcal{U}_\varepsilon} V^n(x_i, y_j, z_k).$$

The current control $u(t_n)$ which is supposed to be applied on the next time interval $[t_n, t_n + \tau]$ is computed from the condition of maximal projection of the system velocity (f_1, f_2, f_3) onto the direction of the vector $\big(x_{i_*} - x(t_n),\ y_{i_*} - y(t_n),\ z_{i_*} - z(t_n)\big)$, i.e.,

$$u(t_n) = \arg \max_{-2\mu \leq u \leq 0} \Big(\big(x_{i_*} - x(t_n)\big) f_1 + \big(y_{i_*} - y(t_n)\big) f_2 + \big(z_{i_*} - z(t_n)\big) f_3\Big).$$

It is clear that the value of the control will be either 0 or -2μ.

6. Simulation results

Let us first consider the following two-dimensional variant of the controlled system:

$$\begin{aligned} \dot{x} &= -\alpha_1[\Theta_1(x) - \Theta_2(y)] - \lambda[\Theta_1(x) - u] - d + v_1 \\ \dot{y} &= -\alpha_2[\Theta_2(y) - \Theta_1(x)] + v_2, \end{aligned} \tag{10}$$

which corresponds to the assumption $z \equiv u$ (infinite cooling rate). The values of the coefficients and bounds on the control and disturbances used for all simulations are: $\alpha_1 = \alpha_2 = 0.1$, $\lambda = 2$, $d = 2$, $\mu = 4$, $\nu = 0.2$. The notation $\beta_\iota^i := 1 - \beta_\ell^i$, $i = 1, 2$, is used for the ice fractions.

We start with the case where the intra- and extracellular liquids have the same freezing points, i.e., $\theta_{1s} = \theta_{2s}$. The plots of the temperatures in the cell and pore regions versus time are given in Figure 3a. Figure 3b shows that the control fails to balance the ice fractions in the pore and in the cell.

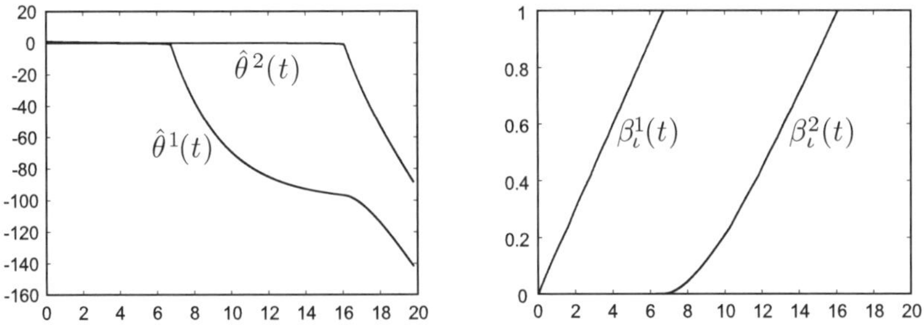

(a) Temperatures in the extracellular space and in the cell versus time; latent heat plateaus are present

(b) Ice fraction in the extracellular space and in the cell versus time

FIGURE 3. The case of infinite cooling rate, $\theta_{1s} = \theta_{2s}$.

The next simulation (see Figure 4) shows the case of different freezing points for the pore and the cell: $\theta_{1s} - \theta_{2s} = -13°\text{C}$. Thus, the freezing point of the extracellular fluid is lowered, e.g., by adding a cryoprotector. Now, we can freeze the intracellular fluid using temperatures laying above the freezing point of the extracellular liquid, which makes possible simultaneous freezing. Figure 5 stands for the same setting as Figure 4 but in the case of finite cooling rate (see equations (4)).

Remember that the central point of the control design is the computation of the value function using the upwind grid method given by (8) and (9). Numerical experiments show a very nice property of this method: the noise usually coming from the boundary of the grid region is absent. The examples are calculated on a Linux computer admitting 64 GB memory and 32 threads. The coefficient of the parallelization is equal to 0.7 per thread (23 times speedup totally). The grid size in three dimensions is 300^3, the number of time steps is 30000 (see the restrictive relation between the space and time step sizes given by Lemma 1). The run time is approximately 60 min. In the case of two dimensions, the run time is several minutes.

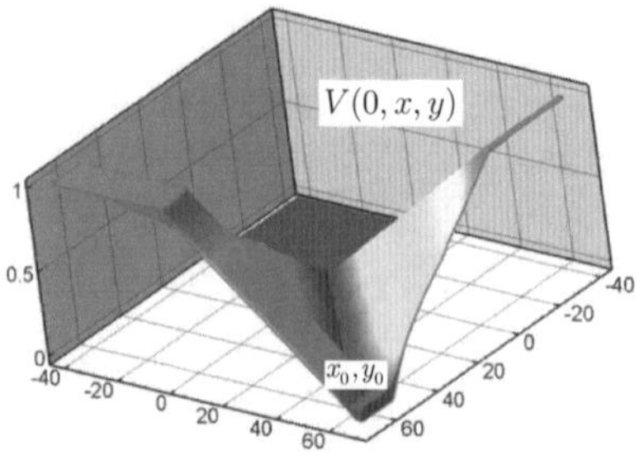

(a) Graph of the value function at $t = 0$

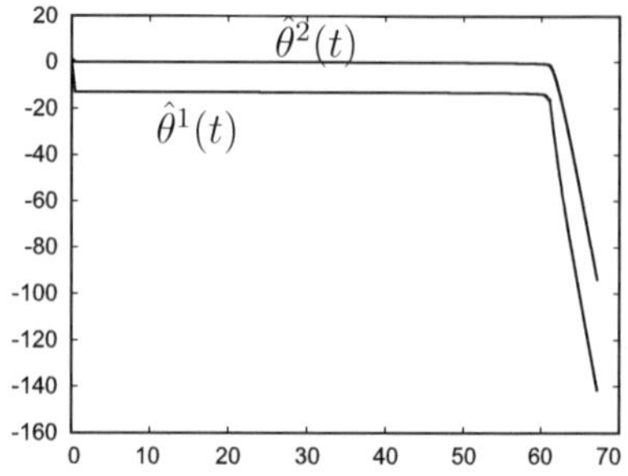

(b) Temperatures in the extracellular space and in the cell versus time; latent heat plateaus are present

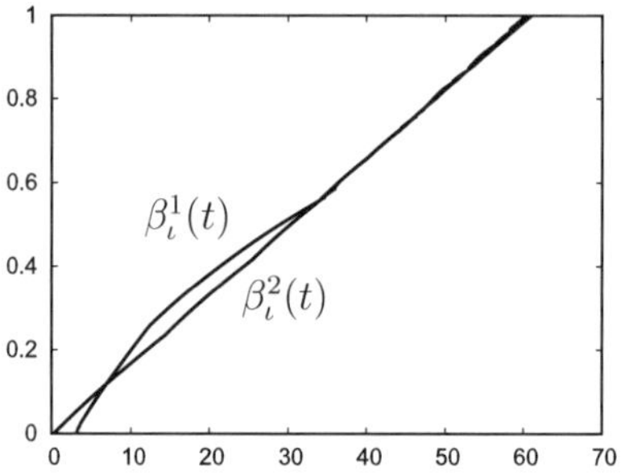

(c) Ice fractions in the extracellular space and in the cell versus time

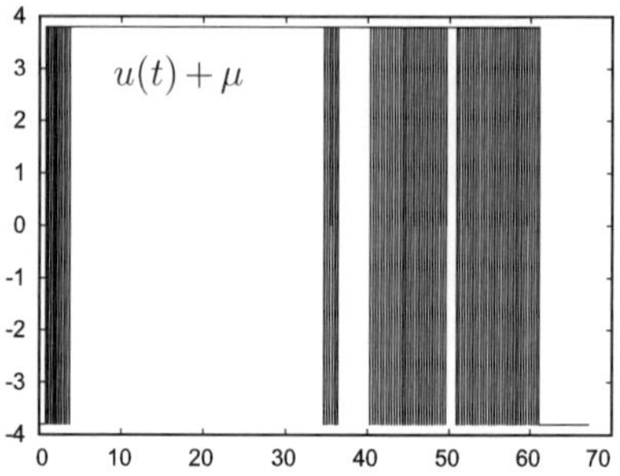

(d) Realization of the control

FIGURE 4. The case of infinite cooling rate, $\theta_{1s} - \theta_{2s} = -13°\text{C}$.

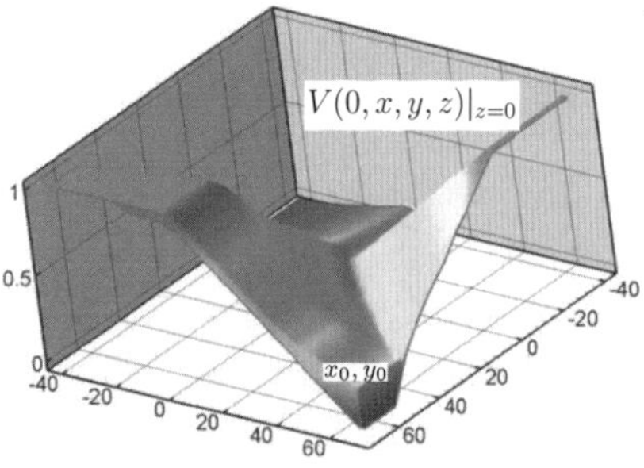

(a) Graph of the value function at $t=0$, $z=0$

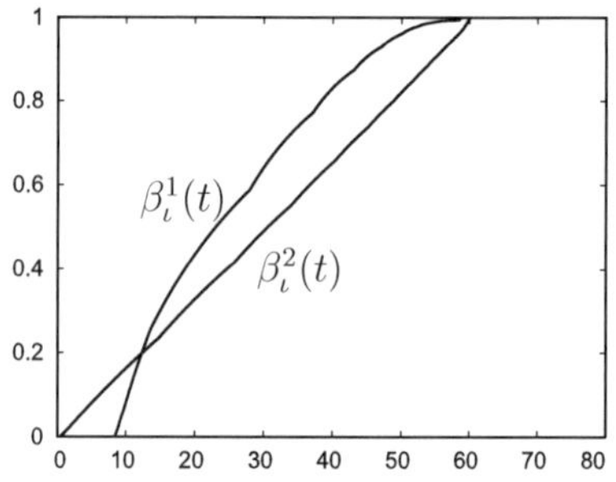

(b) Ice fractions in the extracellular space and in the cell versus time

FIGURE 5. The realistic case of finite cooling rate (three dimensions), $\theta_{1s} - \theta_{2s} = -13°\text{C}$.

7. Mathematical models of dehydration and rehydration of cells

This section is devoted to the modeling of the change of cell shape and volume caused by the dehydration and rehydration processes. Each biological tissue cell is located inside a pore or cannel filled with a saline solution called extracellular fluid. The cell interior is separated from the outer liquid by a cell membrane whose structure ensures a very good permeability of water, which makes possible its easy inflow and outflow due to the osmotic pressure caused by the difference of the salt concentrations inside and outside the cell.

7.1. Dehydration of cells

In the freezing phase, the mechanism of the osmotic effect is the following. Ice formation occurs initially in the extracellular solution. Since ice is practically free of salt, the water-to-ice phase change results in the increase of the salt concentration (c_{out}) in the remaining extracellular liquid. The osmotic pressure forces the outflow of water from the cell to balance the intracellular (c_{in}) and extracellular (c_{out}) salt concentrations. Modeling of the cell shrinkage is based on free boundary problem techniques. The main relation here is the so-called Stefan condition: $\mathcal{V} = \alpha(c_{\text{out}} - c_{\text{in}})$, where $\mathcal{V}$ is the normal velocity of the cell boundary (directed to the cell interior), and the right-hand side represents the osmotic flux that is proportional to the difference of the concentrations. The coefficient α is the product of the Boltzmann constant, the temperature, and the hydraulic conductivity of the membrane (see, e.g., [14]). Note that α is practically a constant in our case. The extracellular salt concentration c_{out} depends on the unfrozen fraction β_ℓ of the extracellular liquid (see the phase field model (1)). Remember that the function $\beta_\ell(\theta)$ is a constitutive material law that, e.g., in frozen-soil science, is measured directly by nuclear magnetic resonance.

The intracellular and extracellular salt concentrations are estimated using the mass conservation law as follows:

$$c_{\text{in}} = c^0_{\text{in}} W^0_c / W_c, \quad c_{\text{out}} = c^0_{\text{out}} W^0 / W,$$

where W^0_c and W_c are the initial and current cell volumes, respectively, W^0 and W are the initial and current volumes of the unfrozen part of the pore. The current volume W at the time t is computed as

$$W(t) = \int_{W^0} \beta_\ell(\theta(t,x))dx,$$

where the distribution of the temperature $\theta(t,x)$ is found from the phase field model (1). A typical form of the function $W(t)$ is shown in Figure 6.

The cell region $\Sigma(t)$ is searched as the level set of a function $\Psi(t,x)$, i.e.,

$$\Sigma(t) = \{x\colon\ \Psi(t,x) \leq 1\}, \qquad x \in R^3 \text{ (or } R^2).$$

Assuming that the cell boundary propagates with the normal velocity $\mathcal{V}$ yields the following Hamilton-Jacobi equation for the function $\Psi(t,x)$:

$$\Psi_t - \alpha(c_{\text{out}} - c_{\text{in}})|\nabla\Psi| = 0, \;\; \Psi(0,x) = \inf\{\lambda > 0\colon\, x \in \lambda \cdot \Sigma(0)\}. \tag{11}$$

Here $|\nabla\Psi|$ denotes the Euclidean norm of the gradient.

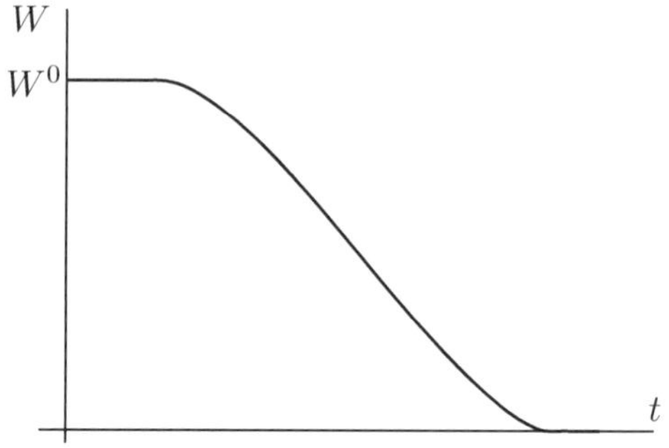

FIGURE 6. Evolution of the volume of the unfrozen extracellular liquid versus time.

7.2. Rehydration of cells

In the thawing phase, the osmotic effect results in the inflow of water into cells and hence in their swelling. We use the following mass conservation law for the salt content:

$$W_c^0\, c_{\text{in}}^0 = W_s\, c_{\text{in}}^0 + W_\ell\, c_{\text{in}}, \quad W_c = W_s + W_\ell, \tag{12}$$

where W_c^0 is the initial volume of the frozen cell, W_c the current volume of the cell, W_s and W_ℓ are volumes of the frozen and unfrozen parts of the cell, respectively, c_{in}^0 and c_{in} the salt concentrations in the frozen and unfrozen parts of the cell, respectively. From (12), one obtains

$$c_{\text{in}} = c_{\text{in}}^0 \left(1 + (W_c^0 - W_c)/W_\ell\right).$$

We assume that W_c is calculated from the current cell shape, and

$$W_\ell(t) \approx \int\limits_{W_c^0} \beta_\ell\,(\theta(t,x))dx,$$

where the function β_ℓ shows now the volume fraction of unfrozen intracellular fluid. The salt concentration c_{out} outside the cell is supposed to be a constant.

Admitting that the propagation velocity $\mathcal{V}$ of the cell boundary is proportional to the difference of the concentrations c_{in} and c_{out}, one arrives at the equation of the form (11).

8. Finite-difference scheme for solving Hamilton-Jacobi equation

Hamilton-Jacobi equation (11) is solved numerically using a finite difference scheme for finding viscosity solutions. The aim of this section is to describe a proper scheme. Let us agree that the symbol "$|\ |$" denotes the absolute value in the case of scalar or the Euclidian norm in the vector case.

Consider a general form of the Hamilton-Jacobi equation (initial value problem) and assume for simplicity that the spatial variable is three dimensional:

$$\Psi_t + H(t, x, \Psi_x) = 0, \qquad \Psi(0, x) = \sigma(x). \tag{13}$$

Here $x = (x_1, x_2, x_3) \in R^3$, $t \in [0, \infty)$, $\Psi : [0, \infty) \times R^3 \to R$, and $\sigma : R^3 \to R$ is some given function. Assume that the Hamiltonian H is defined as

$$H(t, x, p) = \max_{v \in Q} \min_{u \in P} \langle p, f(t, x, u, v) \rangle,$$

where f is the right-hand side of the following conflict controlled system:

$$\dot{x} = f(t, x, u, v), \quad x \in R^3, \quad u \in P \subset R^p, \quad v \in Q \subset R^q, \quad p, q \leq 3. \tag{14}$$

The function f is assumed to be uniformly continuous on $[0, \infty) \times R^3 \times P \times Q$, bounded and Lipschitz-continuous in t, x. The function σ is bounded and Lipschitz-continuous in x.

Let τ, Δ_{x_i}, $i = 1, 2, 3$, are time and space discretization steps. Similar to Section 3, introduce the notation

$$\Psi^n(x_1^i, x_2^j, x_3^k) = \Psi(t_n, i\Delta_{x_1}, j\Delta_{x_2}, k\Delta_{x_3}), \quad t_n = n\tau$$

and consider a difference scheme

$$\Psi^{n+1}(x_1^i, x_2^j, x_3^k) = \Psi^n(x_1^i, x_2^j, x_3^k) - \tau H(t_n, x_1^i, x_2^j, x_3^k, \Psi^n_{x_1}, \Psi^n_{x_2}, \Psi^n_{x_3})$$
$$\Psi^0(x_1^i, x_2^j, x_3^k) = \sigma(x_1^i, x_2^j, x_3^k).$$

Here, the symbols $\Psi^n_{x_1}, \Psi^n_{x_2}, \Psi^n_{x_3}$ denote finite difference approximations (left, right, central, etc.) of the corresponding partial derivatives. Note that the $(n+1)$th function is computed on the base of the nth function, and the Hamiltonian appears with the sign "$-$" in contrast to Section 3 because the initial value problem is studied.

The scheme can be considered as the successive application of an operator Π to the grid functions:

$$\Psi^{n+1} = \Pi(\Psi^n; t_n, \tau, \Delta_{x_1}, \Delta_{x_2}, \Delta_{x_3}).$$

Note that such an operator can be naturally extended to continuum functions. Let us remember the monotonicity and generator properties (comp. Section 3) of the operator Π. The definition of monotonicity is just the same as in Section 3. The generator property is formulated similar to that from Section 3 but with the sign "$+$" in front of the Hamiltonian, i.e.,

$$\left| \frac{\Pi(\phi; t, \tau, a_1\tau, a_2\tau, a_3\tau)(x) - \phi(x)}{\tau} + H(t, x, D\phi(x)) \right| \leq C\big(1 + \|D\phi\| + \|D^2\phi\|\big)\tau$$

for smooth functions ϕ (see Section 3).

Theorem 2 (Convergence, [19]). *Assume that the operator* $\Pi(\cdot; t, \tau, a_1\tau, a_2\tau, a_3\tau)$ *is monotone for any* $t, \tau > 0$ *and satisfies the generator property, then the grid function obtained by the procedure*

$$\Psi^{n+1} = \Pi(\Psi^n; t, \tau, a_1\tau, a_2\tau, a_3\tau), \ n = 0, 1, \ldots, \quad \Psi^0 = \sigma,$$

converges point-wise to a viscosity solution of Hamilton-Jacobi equation (13) *as* $\tau \to 0$, *and the convergence rate is* $\sqrt{\tau}$.

We consider an upwind finite difference scheme similar to that given in Section 4 (distinctions arise because of the initial value problem formulation). Thus,

$$\begin{aligned}&\Pi(\Psi^n; t_n, \tau, \Delta_{x_1}, \Delta_{x_2}, \Delta_{x_3})(x_1^i, x_2^j, x_3^k)\\ &\quad = \Psi^n(x_1^i, x_2^j, x_3^k) - \tau \max_{v\in Q} \min_{u\in P} \sum_{m=1}^{3} (p_m^L \cdot f_m^+ + p_m^R \cdot f_m^-),\end{aligned} \tag{15}$$

where f_1, f_2, f_3 are the right-hand sides of system (14) computed at (t_n, x_1^i, x_2^j, x_3^k, u, v); p_1^R, p_2^R, p_3^R and p_1^L, p_2^L, p_3^L the right and the left divided differences, respectively. The arguments $t_n, x_1^i, x_2^j, x_3^k, u, v$ of f_m^- and f_m^+ are omitted for brevity.

Let us prove that the operator Π meets the requirements of Theorem 2. The following lemmas hold.

Lemma 2. (Monotonicity). *Let* M *be the bound of* $|f|$. *If* $a_1, a_2, a_3 \geq M\sqrt{3}$, *then the operator* $\Pi(\cdot; t, \tau, a\tau, b\tau, c\tau)$ *given by* (15) *is monotone.*

Proof. Suppose $\Psi \leq \Phi$. Let us show that

$$\Pi(\Psi; t, \tau, a_1\tau, a_2\tau, a_3\tau) \leq \Pi(\Phi; t, \tau, a_1\tau, a_2\tau, a_3\tau).$$

Denote

$$\mathfrak{h}_1 = (a_1, 0, 0), \quad \mathfrak{h}_2 = (0, a_2, 0), \quad \mathfrak{h}_3 = (0, 0, a_3).$$

We have

$$\begin{aligned}&\Pi(\Psi; t, \tau, a_1\tau, a_2\tau, a_3\tau)(x) - \Pi(\Phi; t, \tau, a_1\tau, a_2\tau, a_3\tau)(x)\\ &\quad = \Psi(x) - \Phi(x)\\ &\qquad - \tau \max_{v\in Q} \min_{u\in P} \sum_{m=1}^{3} \Big(\frac{\Psi(x) - \Psi(x - \mathfrak{h}_m\tau)}{a_m\tau} f_m^+ + \frac{\Psi(x + \mathfrak{h}_m\tau) - \Psi(x)}{a_m\tau} f_m^- \Big)\\ &\qquad + \tau \max_{v\in Q} \min_{u\in P} \sum_{m=1}^{3} \Big(\frac{\Phi(x) - \Phi(x - \mathfrak{h}_m\tau)}{a_m\tau} f_m^+ + \frac{\Phi(x + \mathfrak{h}_m\tau) - \Phi(x)}{a_m\tau} f_m^- \Big).\end{aligned}$$

By rearranging terms and using the obvious relations $f_m^+ - f_m^- = |f_m|$ and $\max_v \min_u g_1(u,v) - \max_v \min_u g_2(u,v) \le \max_v \max_u \big[g_1(u,v) - g_2(u,v)\big]$, one obtains

$$\begin{aligned}
&\Pi(\Psi; t, \tau, a_1\tau, a_2\tau, a_3\tau)(x) - \Pi(\Phi; t, \tau, a_1, a_2, a_3)(x) \\
&\le \Psi(x) - \Phi(x) \\
&\quad + \tau \max_{v\in Q} \max_{u\in P} \sum_{m=1}^{3} \Big[\Big(\frac{\Psi(x - \mathfrak{h}_m\tau) - \Phi(x - \mathfrak{h}_m\tau)}{a_m\tau} - \frac{\Psi(x) - \Phi(x)}{a_m\tau}\Big) f_m^+ \\
&\quad + \Big(\frac{\Psi(x + \mathfrak{h}_m\tau) - \Phi(x + \mathfrak{h}_m\tau)}{a_m\tau} - \frac{\Psi(x) - \Phi(x)}{a_m\tau}\Big)(-f_m^-)\Big] \\
&\le \Psi(x) - \Phi(x) - \tau \sum_{m=1}^{3} \frac{|f_m|}{a_m\tau}\big(\Psi(x) - \Phi(x)\big) = \Big(1 - \sum_{m=1}^{3} \frac{|f_m|}{a_m}\Big)\big(\Psi(x) - \Phi(x)\big).
\end{aligned}$$

With $\sum_{m=1}^{3} |f_m| \le \sqrt{3}M$ one comes to $1 - \sum_{m=1}^{3} \frac{|f_m|}{a_m} > 0$, which finally implies the required inequality. □

Lemma 3. (Generator property). *The generator property holds for the operator* Π.

Proof. Let $\phi \in C_b^2(R^3)$. Denote $\mathfrak{d}_1 = (\Delta_{x_1}, 0, 0), \mathfrak{d}_2 = (0, \Delta_{x_2}, 0), \mathfrak{d}_3 = (0, 0, \Delta_{x_3})$. We have

$$\begin{aligned}
&\Pi(\phi; t, \tau, \Delta_{x_1}, \Delta_{x_2}, \Delta_{x_3})(x) \\
&\quad = \phi(x) - \tau \max_{v\in Q} \min_{u\in P} \sum_{m=1}^{3} \Big(\frac{\phi(x) - \phi(x - \mathfrak{d}_m)}{\Delta_{x_m}} f_m^+ + \frac{\phi(x + \mathfrak{d}_m) - \phi(x)}{\Delta_{x_m}} f_m^-\Big).
\end{aligned}$$

Estimate

$$\begin{aligned}
&\Big|\frac{\Pi(\phi;t,\tau,\Delta_{x_1},\Delta_{x_2},\Delta_{x_3})(x) - \phi(x)}{\tau} + \max_{v\in Q} \min_{u\in P} \langle D\phi(x), f\rangle\Big| \\
&= \Big| - \max_{v\in Q} \min_{u\in P} \sum_{m=1}^{3} \Big(\frac{\phi(x) - \phi(x - \mathfrak{d}_m)}{\Delta_{x_m}} f_m^+ + \frac{\phi(x + \mathfrak{d}_m) - \phi(x)}{\Delta_{x_m}} f_m^-\Big) \\
&\qquad + \max_{v\in Q} \min_{u\in P} \sum_{m=1}^{3} \frac{\partial\phi}{\partial x_m}(f_m^+ + f_m^-)\Big| \\
&\le \max_{u\in P} \max_{v\in Q} \sum_{m=1}^{3} \Big|\Big(\frac{\partial\phi}{\partial x_m} - \frac{\phi(x) - \phi(x - \mathfrak{d}_m)}{\Delta_{x_m}}\Big) f_m^+ + \Big(\frac{\partial\phi}{\partial x_m} - \frac{\phi(x + \mathfrak{d}_m) - \phi(x)}{\Delta_{x_m}}\Big) f_m^-\Big| \\
&\le M \,\| D^2\phi\| \sum_{m=1}^{3} \Delta_{x_m}.
\end{aligned}$$

Here M is the bound of $|f|$. Choosing $\Delta_{x_m} = a_m\tau$ and letting $C = \sum_{m=1}^{3} a_m$ yields

$$\Big|\frac{\Pi(\phi; t, \tau, \Delta_{x_1}, \Delta_{x_2}, \Delta_{x_3})(x) - \phi(x)}{\tau} + H(t, x, D\phi(x)\Big| \le M\, C\ \| D^2\phi \|\, \tau. \qquad \square$$

9. Three-dimensional simulation of cell shrinkage

The finite difference scheme based on the operator Π is implemented as a parallelized program on a Linux cluster. We compute the evolution of the cell boundary during freezing. The function f in our case is

$$f(t,u,v) = \alpha\big(c_{\text{out}}(t) - c_{\text{in}}(t)\big)^{+}u + \alpha\big(c_{\text{out}}(t) - c_{\text{in}}(t)\big)^{-}v,$$
$$u, v \in R^3, \quad |u| \le 1, \quad |v| \le 1.$$

Here the controls u and v are responsible for the outflow and inflow, respectively. The spatial grid $200 \times 200 \times 200$ for the cubic region $0.3 \times 0.3 \times 0.3$ was utilized, and 1000 time steps of the size $\tau = 0.001$ were done. The running time is about 26 minutes on 30 threads. The initial shape is presented in Figure 7a. Figures 7b and 7c show the computed shape at the time instants $t = 0.8$ and $t = 0.92$.

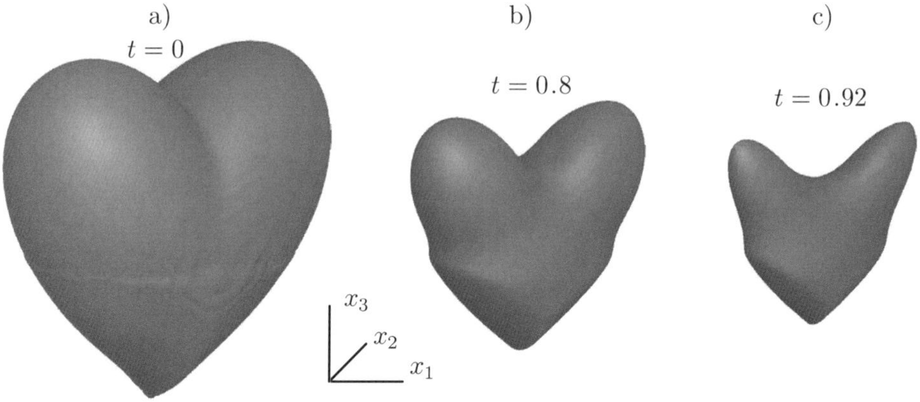

FIGURE 7. Osmotic cell shrinkage during freezing.

10. Accounting for the membrane tension using reachable set approach

In reality, the deformation of the cell membrane depends on the membrane tension which is a function of the curvature. Therefore, a more realistic expression for the normal velocity of the cell boundary would be:

$$\mathcal{V}(t,x) = \alpha(c_{\text{out}}(t) - c_{\text{in}}(t)) + \gamma\sigma(x),$$

where $\sigma(x)$ is the angular curvature at the current point x of the cell boundary, and γ is a constant. The angular curvature is explained in Figure 8. The corresponding Hamiltonian is

$$H(t,x,p) = -\big(\alpha(c_{\text{out}}(t) - c_{\text{in}}(t)) + \gamma\sigma(x)\big)|p|. \tag{16}$$

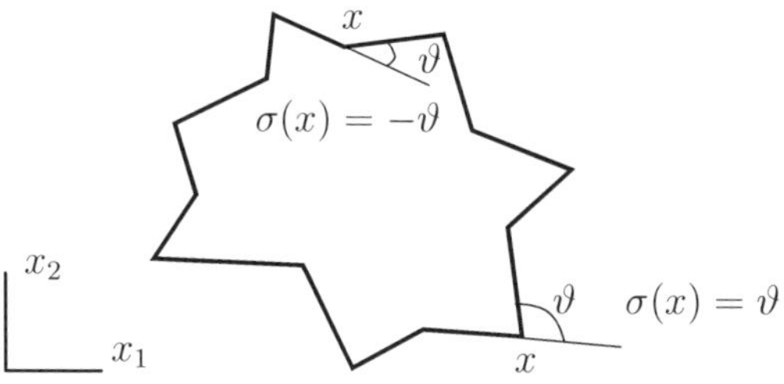

FIGURE 8. Explanation of the angular curvature.

Note that accounting for the curvature can alter the convexity/concavity structure of the Hamiltonian depending on the state x.

Instead of finding viscosity solutions to equation (13), level sets of the value function of an appropriate conflict controlled problem with the Hamiltonian (16) will be computed. We will treat this problem within the framework of [15], [17], [21]. The consideration will be carried out in R^2.

It is easy to see that the conflict controlled system of the form

$$\begin{gathered}\dot{x} = \alpha\big(c_{\text{out}}(t) - c_{\text{in}}(t) + \gamma\sigma(x)\big)^{+} u + \alpha\big(c_{\text{out}}(t) - c_{\text{in}}(t) + \gamma\sigma(x)\big)^{-} v; \\ x, u, v \in R^2, \ |u| \le 1, \ |v| \le 1\end{gathered} \tag{17}$$

has the Hamiltonian (16). The initial cell shape $\Sigma_0 = \Sigma(0)$ is considered as the target set of this differential game where the control u strives to bring the state vector of (17) to Σ_0, and the objective of the control v is opposite.

Since only one of the controls u and v is active at every time t and every point x (see the definition of the operations "+" and "−"), one can reduce the construction of level sets of the value function of system (17) to finding reachable sets of the controlled system

$$\dot{x} = \operatorname{sign}\mathcal{V} \cdot u\,; \quad x, u \in R^2, \quad u \in |\mathcal{V}|\, S, \quad S = \{x \in R^2 : |x| \le 1\} \tag{18}$$

with $\mathcal{V} = \alpha\big(c_{\text{out}}(t) - c_{\text{in}}(t) + \gamma\sigma(x)\big)$ and the initial set Σ_0.

Denote by $x(t; x_0, u(\cdot))$ the state vector of system (18) at time $t \ge 0$ provided that x_0 is the initial point at time $t = 0$ and $u(\cdot)$ is an admissible measurable control acting on the time interval $[0, t]$. The set of points reachable from Σ_0 at time t is

$$G(t, \Sigma_0) = \bigcup_{x_0 \in \Sigma_0} \bigcup_{u(\cdot)} x(t; x_0, u(\cdot)).$$

The algorithm for the numerical construction of the sequence

$$\{\mathcal{G}_i = G(i\Delta t, \mathcal{G}_{i-1}), \quad i = 1, 2, \ldots, \quad \mathcal{G}_0 = \Sigma_0\},$$

is similar to that described in [18]. The difference is that the treatment of the cases of local concavity and local convexity alters depending on the sign of $\mathcal{V}$.

Simulation results showing the time evolution of the cell shape during freezing are presented in Figure 9. The sequence of reachable sets is computed on the time

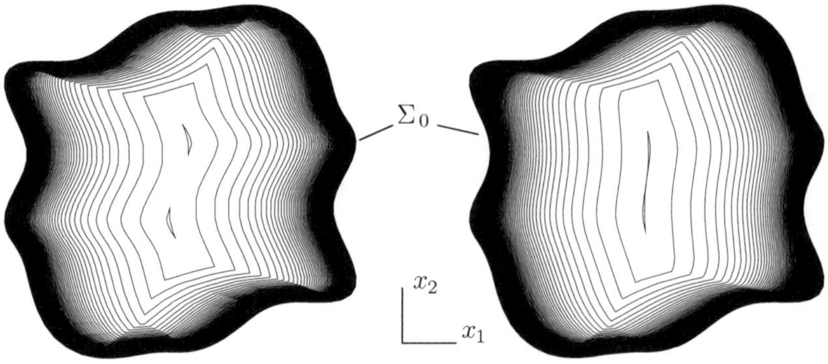

FIGURE 9. Time evolution of the cell shape during freezing. To the left: without accounting for the curvature. To the right: with accounting for the curvature.

interval $[0, 0.645]$ with the time step $\Delta t = 0.001$, every fourth set is drawn. The restriction on the control vector is $u \in |\mathcal{V}|\, P \cdot S$, where $P = \{p_{ij}\}$ is the 2×2-diagonal matrix introducing anisotropy ($p_{11} = 3$, $p_{22} = 1$). For the left picture, $\gamma = 0$, i.e., the curvature independent case is considered. For the right picture, $\gamma = 0.06$. Note that the low propagation velocity at the beginning of the process results in the accumulation of lines (dark regions) near the initial cell boundary.

In Figure 10, an example of the simulation of cell rehydration during thawing is presented. The computation is done on the time interval $[0, 9.7]$ with the time step $\Delta t = 0.002$. Every 100th reachable set is drawn. At the end of the process, the stabilization of the sets is observed, because the equilibrium between the salt concentrations inside and outside the cell is achieved.

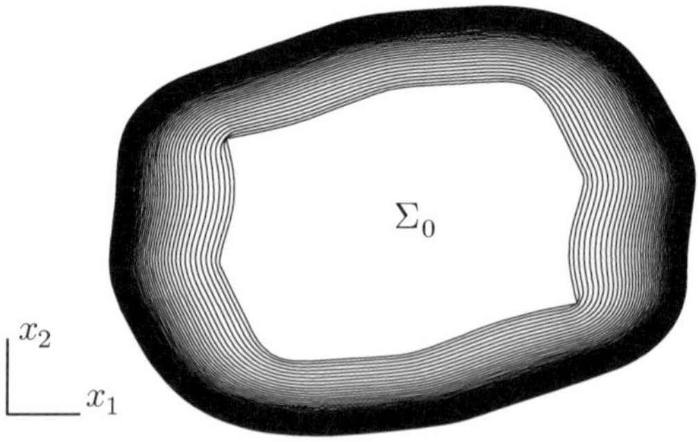

FIGURE 10. Time evolution of the cell during thawing with accounting for the curvature.

11. Conclusion

Problems considered in this paper show that spatially distributed models can be reduced to optimal control problems for ordinary differential equations with control parameters and disturbances using appropriate averaging techniques. Because of tabular form of nonlinear dependencies appearing in the right-hand sides of such equations, the application of Pontryagin's maximum principle is difficult, whereas appropriate grid methods or reachable set techniques do work here. The usage of stable algorithms allows us to obtain convincing results. Optimal cooling protocols designed can be implemented in real cooling devices such as, e.g., IceCube developed by SY-LAB, Geräte GmbH (Austria).

References

[1] Batycky R.P., Hammerstedt R., Edwards D.A., *Osmotically driven intracellular transport phenomena*, Phil. Trans. R. Soc. Lond. A. 1997. Vol. 355. P. 2459–2488.

[2] Botkin N.D., *Approximation schemes for finding the value functions for differential games with nonterminal payoff functional*, Analysis 1994. Vol. 14, no. 2. P. 203–220.

[3] Botkin N.D., Hoffmann K-H., Turova V.L., Stable Solutions of Hamilton-Jacobi Equations. Application to Control of Freezing Processes. German Research Society (DFG), Priority Program 1253: Optimization with Partial Differential Equations. Preprint-Nr. SPP1253-080 (2009).
`http://www.am.uni-erlangen.de/home/spp1253/wiki/images/7/7d/Preprint-SPP1253-080.pdf`

[4] Caginalp G., *An analysis of a phase field model of a free boundary*, Arch. Rat. Mech. Anal. 1986. Vol. 92. P. 205–245.

[5] Chen S.C., Mrksich M., Huang S., Whitesides G.M., Ingber D.E., *Geometric control of cell life and death*, Science. 1997. Vol. 276. P. 1425–1428.

[6] Crandall M.G., Lions P.L., *Viscosity solutions of Hamilton-Jacobi equations*, Trans. Amer. Math. Soc. 1983. Vol. 277. P. 1–47.

[7] Crandall M.G., Lions P.L., *Two approximations of solutions of Hamilton-Jacobi equations*, Math. Comp. 1984. Vol. 43. P. 1–19.

[8] Frémond M., Non-Smooth Thermomechanics. Berlin: Springer-Verlag, 2002. 490 p.

[9] Hoffmann K.-H., Jiang Lishang., *Optimal control of a phase field model for solidification*, Numer. Funct. Anal. Optimiz. 1992. Vol. 13, no. 1,2. P. 11–27.

[10] Hoffmann K.-H., Botkin N.D., *Optimal control in cryopreservation of cells and tissues*, in: Proceedings of Int. Conference on Nonlinear Phenomena with Energy Dissipation. Mathematical Analysis, Modeling and Simulation, Colli P. (ed.) et al., Chiba, Japan, November 26–30, 2007. Gakuto International Series Mathematical Sciences and Applications 29, 2008. P. 177–200.

[11] Hoffmann K.-H., Botkin N.D., Optimal Control in Cryopreservation of Cells and Tissues, Preprint-Nr.: SPP1253-17-03. 2008.

[12] Isaacs R. Differential Games. New York: John Wiley, 1965. 408 p.

[13] Malafeyev O.A., Troeva M.S. *A weak solution of Hamilton-Jacobi equation for a differential two-person zero-sum game*, in: Preprints of the Eight Int. Symp. on Differential Games and Applications, Maastricht, Netherland, July 5–7, 1998, P. 366–369.

[14] Mao L., Udaykumar H.S., Karlsson J.O.M., *Simulation of micro scale interaction between ice and biological cells*, Int. J. of Heat and Mass Transfer. 2003. Vol. 46. P. 5123–5136.

[15] Krasovskii N.N, Subbotin A.I., Positional Differential Games. Moscow: Nauka, 1974. p. (in Russian).

[16] Krasovskii N.N., Control of a Dynamic System. Moscow: Nauka, 1985. 520 p. (in Russian).

[17] Krasovskii N.N., Subbotin A.I., Game-Theoretical Control Problems. New York: Springer, 1988. 518 p.

[18] Patsko V.S., Turova V.L., *From Dubins' car to Reeds and Shepp's mobile robot*, Comput. Visual. Sci. 2009. Vol. 13, no. 7. P. 345–364.

[19] Souganidis P.E., *Approximation schemes for viscosity solutions of Hamilton-Jacobi equations*, J. Differ. Equ. 1985 Vol. 59. P. 1–43.

[20] Subbotin A.I., Chentsov A.G., Optimization of Guaranteed Result in Control Problems. Moscow: Nauka, 1981. 287 p. (in Russian).

[21] Subbotin A.I., Generalized Solutions of First Order PDEs. Boston: Birkhäuser, 1995. 312 p.

Nikolai D. Botkin, Karl-Heinz Hoffmann and Varvara L. Turova
Technische Universität München
Department of Mathematics
Boltzmannstr. 3
D-85748 Garching, Germany
e-mail: botkin@ma.tum.de
hoffmann@ma.tum.de
turova@ma.tum.de

International Series of Numerical Mathematics, Vol. 160, 541–559

Model Reduction, Structure-property Relations and Optimization Techniques for the Production of Nanoscale Particles

Michael Gröschel, Günter Leugering and Wolfgang Peukert

Abstract. The production of nanoscaled particulate products with exactly pre-defined characteristics is of enormous economic relevance. Although there are different particle formation routes they may all be described by one class of equations. Therefore, simulating such processes comprises the solution of nonlinear, hyperbolic integro-partial differential equations. In our project we aim to study this class of equations in order to develop efficient tools for the identification of optimal process conditions to achieve desired product properties. This objective is approached by a joint effort of the mathematics and the engineering faculty. Two model-processes are chosen for this study, namely a precipitation process and an innovative aerosol process allowing for a precise control of residence time and temperature. Since the overall problem is far too complex to be solved directly a hierarchical sequence of simplified problems has been derived which are solved consecutively. In particular, the simulation results are finally subject to comparison with experiments.

Mathematics Subject Classification (2000). Primary 35R09; Secondary 35Q70.

Keywords. Population balance equations, optimal control, model reduction, parameter identification.

1. Motivation

Particulate products cover a wide field of applications ranging from gross products to novel high performance specialty materials. As a special characteristic of disperse compounds, the specific product properties, and consequently the product value, are not governed by the chemical composition alone. The major features are determined to a large extent by the so-called disperse properties, i.e., particle size, primary particle size, free surface area, morphology etc. Therefore, it is of great importance to control and optimize the synthesis process in order to obtain tailor-made products.

In the nanometer regime various quantum mechanical effects are observed exerting an enormous influence on the product properties. An increase in the band gap of a semiconducting material occurs for example, when its size is reduced, resulting in an entirely different optical behavior. For example, a further improvement of the efficiency of thin film solar cells depends strongly on gaining new insight into this field of research. Therefore the possibility of tuning the performance of a particulate product by varying its size distribution will open up promising opportunities for future applications.

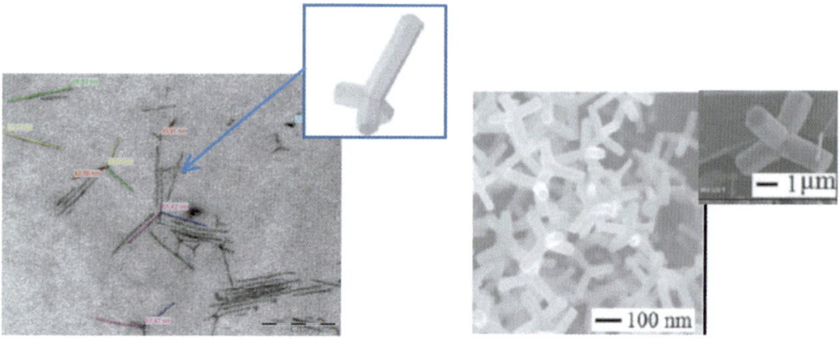

FIGURE 1. Tetrapods made of CdS (left) in liquid phase and ZnO in gas phase (right [28])

2. Population balance equations and model reduction

Particle synthesis comprises in general the following chemical and physical processes: Formation of a supersaturation, nucleation, growth, agglomeration, sintering and ripening. Although the various production routes (precipitation, flame spray pyrolysis, etc.) are quite different, the progress in the reactor may be described in a unified manner. As ripening and sintering are thermodynamically driven slow mechanisms which are predominant in a post reaction step, we are thus focussing on the phenomena of growth and agglomeration.

In this contribution we concentrate on two model-processes, a precipitation reaction representing a widespread process in industry for the production of particles of varying size distributions and/or morphologies and an innovative aerosol processes allowing for a precise control of the residence time and temperature (Fig. 2). Large-scale industrial processes for aerosol synthesis comprise, e.g., the production of carbon black, silica or titania as well as other oxides in flame reactors. Aerosol synthesis processes are typically high-throughput flow reactors operating at very high temperatures with fast kinetics.

If one aims at modeling such processes, one has to choose an appropriate abstraction level. Each individual particle may be characterized by one or more

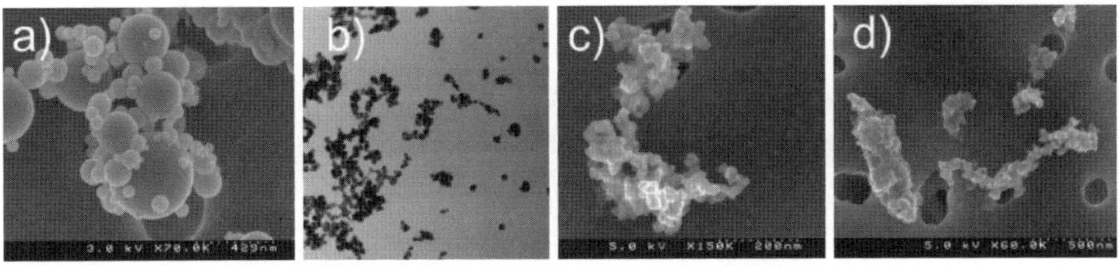

FIGURE 2. Examples of particle synthesis: a) SiO_2 from liquid phase process b) $BaSO_4$ from precipitation c & d) TiO_2 particles formed in a flame process

parameters ranging from the respective particle diameter to complex morphologies. Due to the large number of particles it is reasonable to investigate the evolution of the whole particle ensemble instead of tracking them individually. Therefore, it has to be chosen which quantities will be regarded as distributed or averaged for all particles.

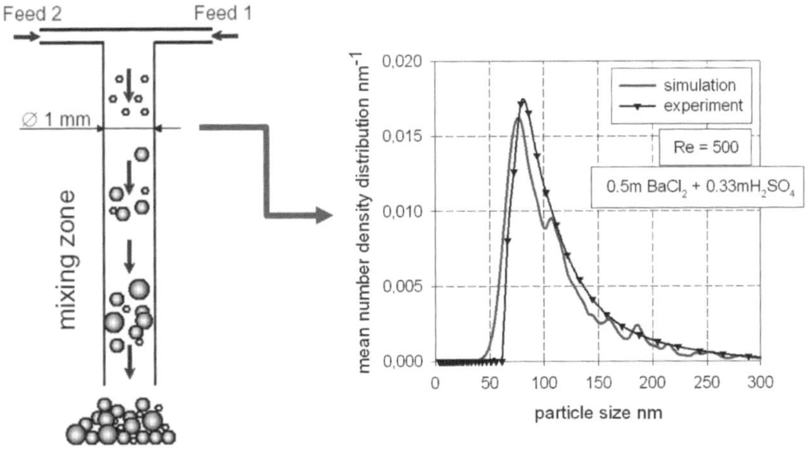

FIGURE 3. Schematic diagram of a precipitation reaction with the corresponding particle size distribution (PSD)

Figure 3 sketches the set up of a precipitation reaction in a T-mixer. The two feed streams mix in the vertical reaction zone, where a local supersaturation is built up. The supersaturation, in turn, constitutes the thermodynamical driving force for the subsequent particle formation and growth processes. More downstream larger particles may collide and form complex agglomerates.

In the simulation of these kinds of reactions the particle ensemble is represented by a particle size distribution (PSD) $y(x_i, r_j, t)$, which is in general depending on several particle properties x_i ($i = 1 \dots I$, usually called 'internal' coordinates), the location r_j (called 'external' coordinates) and the time t. The PSD provides for example information about the number of particles of a specific size at a certain position in the reactor.

Describing the evolution of the whole particle ensemble, the law of conservation for the PSD in an incompressible fluid leads to the general population balance equation:

$$\frac{\partial}{\partial t}y + \frac{\partial}{\partial x_i}\left(\vec{V_x} \cdot y\right) + \frac{\partial}{\partial r_i}\left(\vec{V_r} \cdot y\right) = I + B - D. \tag{1}$$

Here $\vec{V_x}\,(x_i, r_j, t)$ denotes the growth rate in the respective particle coordinates, e.g., the growth of particle diameter due to condensation on the surface of a spherical particle. $\vec{V_r}\,(x_i, r_j, t)$ denotes the fluid velocity causing convection, I represents the source term caused by nucleation (i.e., formation of new particles) and B respectively D denote the so-called birth and death rates accounting for agglomeration (i.e., particles stick to each other after a collision) or breakage. For the agglomeration of two particles the corresponding integral operators are given by:

$$B_{\text{aggl}}(y)(x) := \frac{1}{2}\int_0^x \beta(x-s,s)\,y(x-s)\,y(s)\,ds \tag{2}$$

$$D_{\text{aggl}}(y)(x) := \int_0^\infty \beta(x,s)\,y(x)\,y(s)\,ds. \tag{3}$$

This model accounts for binary particle collisions due to Brownian motion for example. If two colliding particles adhere, they are assumed to form a new one comprising the mass of both collision partners. The collision kernel β in B_{aggl} represents a function describing the probability that two particles of size s and $x - s$ agglomerate adding a new particle of size x to the PSD. The second term D_{aggl} removes afterwards the source particles from the distribution to maintain the mass balance.

Population balance equations represent the state-of-the-art model description of polydisperse particulate processes. The evolution of the particle size distributions (PSD) during the precipitation process as well as the influence of mixing and supersaturation have been studied previous to the project experimentally and numerically by Schwarzer and Peukert [32], using barium sulphate as a test system. Especially in the context of nanoscaled products these techniques have to stay very close to applications in order to provide valuable results. Therefore the acquisition of valid parameter models for complex structures (e.g., Figure 1) is essential for obtaining innovative products. The modeling of the particle parameters often becomes critical due to the lack of theory. In order to specify at least a realistic range, parameter estimation methods have to be taken into account (Section 4).

Numerical techniques for solving 1D-population balances (i.e., one particle property in a well-mixed system) are almost mature nowadays with available solvers based on finite particle size classes (so-called sectional models, see, e.g., [13] [27]), the more advanced cell-average technique [21] or the Finite Element Method (see [42]). Furthermore, there are some attempts to include a second particle property in [1], [4], [20] or dealing with more than one species in the context of biological systems. But the models used to describe these processes raise also the question of a rigorous mathematical analysis subject to existence and uniqueness.

An existence result incorporating the integral terms, which cause the main difficulties, was considered in [29] and consecutively extended in [3] in a very general setting to the case when the distribution function depends additionally on the spatial variable. Since no growth terms have been treated the available results could not be applied to the problems under consideration (Section 3).

A naive discretisation of the integral operators occurring in the models of agglomeration and breakage, leads to full matrices and therefore to quadratic complexity. The most complicated quadratic source term accounting for the agglomeration of particles is therefore subject to ongoing research. For special kernel functions a discretization based on the $\mathcal{H}$-matrix calculus leads to a complexity of $\mathcal{O}(n \log n)$ for large n [19]. Since the population balance equation is of conservation form, its discretization has to be designed in such a way that the mass is preserved [10], [11]. The coupling of the flow field with the population balance in [27], [15] is still restricted to very crude compartment models dividing space in a relatively small number of well-mixed sub-spaces or monodisperse models for the population balance. A first simulation of crystal growth and attrition in a stirred tank in 2d and 3d is considered in [23] using parallel adaptive multigrid methods.

The full problem is far too complex in order to approach it directly. No existence, uniqueness and regularity results for the full system are available, and, consequently, reachability results are not known either. It is therefore necessary to introduce a hierarchy of problems which are themselves tractable. It turns out that each of them focuses on new difficulties in the optimal control of PDEs (Section 5).

A first approach incorporating the adjoint equation to the control of population balance equations is set up in [12]. The initial distribution of different species in a condensational growth process was recovered using various observation modes. In [31], a first step towards the estimation of model parameters was concerned. In the contribution the kernel function is one of the control variables and thus also subject to optimization. All relevant physical quantities are realised being under full control such that the resulting solution might be far away from realistic situations. Therefore adapted solution strategies are required that are capable to handle experimental data (Section 6). The resulting process control should subsequently be benchmarked by its implementation in practise.

3. Mathematical analysis of the considered general PBE

In the modelling of polydisperse particulate processes the common approach is based, as outlined before, on a population balance equation. Including the phenomena of growth and agglomeration the resulting integro differential equation describes the evolution of the particle size distribution $y(t,x) \geq 0$. One specific realisation of the general PBE (1) is the case where the particle size $x \in \mathbb{R}^+$ represents the only internal variable in an ideally mixed system. This approach corresponds to a reduced model for the precipitation reaction shown in Figure 3 under the assumption that the two educts mix instantaneously. For a constant feed rate the residence time t of the particles in the reactor transfers accordingly to a certain position in the mixer.

Therefore, in a first step the homogeneous equations without a spatial transport term are studied. The particles grow in the supersaturated solution due to homogeneous, size independent growth mechanisms (i.e., $V_x = u(t)$). The growth of particles causes a decrease of the supersaturation that affects in reverse again the growth rate. As the formation of new particles is described by the coalescence of smaller ones, we reformulate the general PBE (1) in the spatially homogeneous case to

$$\frac{\partial}{\partial t}y(x,t) + \frac{\partial}{\partial x}(u(t)y(x,t)) = \frac{1}{2}\int_0^x \beta(x-s,s)y((x-s),t)y(s,t)\,ds \qquad (4)$$
$$-y(x,t)\int_0^\infty \beta(x,s)y(s,t)\,ds$$

with the initial data $y(x,0) = y^0(x)$, $\forall x \in \mathbb{R}^+$, accounting for seed particles and an influx boundary condition $y(0,t) = h(t)$ $t \in [0,T]$.

As mentioned, the coalescence kernel $\beta(x_1,x_2)$ corresponds to the collision frequency that result in the coagulation of two particles. For instance the kernel modelling random Brownian motion is given by

$$\beta_B(x_1,x_2) = c(x_1+x_2)\left(\frac{1}{x_1}+\frac{1}{x_2}\right).$$

Considering equation (4) with boundary and initial data we need some additional conditions to guarantee the existence of a weak solution. First discussing the conditions on the coalescence kernel, a natural assumption is the positivity and symmetry of β and the following structure condition

$$\beta(x_1,x_2) \le \beta(x_1,x_1+x_2) + \beta(x_2,x_1+x_2) \qquad \forall x_1,x_2 \in (0,\infty). \qquad (5)$$

To establish the required a priori estimates we introduce a non-negative, non-decreasing and convex function $\phi \in \mathcal{C}^1[0,\infty)$ with $\phi(0) = 0$. We consider the function of time $\varphi(t) := \int_0^\infty \phi(y(s,t))\,ds$ and assume that the following estimate holds for the boundary data

$$\int_0^T y(0,t)\varphi(y(0,t))\,dt + \int_0^T y(0,t)\,dt \le \beta_{bdy}. \qquad (6)$$

Additionally the initial data satisfies

$$\int_0^\infty \phi(y(s,0))\,ds \le \gamma_{in} \qquad (7)$$

and the growth rate $u(\cdot)$ is assumed to be bounded

$$\|u(\cdot)\|_{L^\infty(0,T)} \le \alpha_{\text{vel}}. \qquad (8)$$

Under these assumptions the following estimate can be proven

Theorem 3.1. *Let $\phi \in \mathcal{C}^1([0,\infty))$ be a non-negative and non-decreasing convex function, such that $\phi(0) = 0$ holds and that the estimates (6)–(8) are fulfilled. If $y^0 \in L^1(0,\infty)$ then a solution to (4) will satisfy*

$$\sup_{t\geq 0} \int_0^\infty (\phi(y(s,t)) + y(s,t))\, ds \leq \gamma_{\text{in}} + \alpha_{\text{vel}}\beta_{bdy} < \infty$$

For the precise statement of the existence result a definition of weak solutions (conf. [3]) of the given integro-differential equation (4) is given by

Definition 3.2. *Let $y^0(x)$ be a non-negative function satisfying $y^0 \in L^1((0,\infty),(1+s)ds)$. A weak solution to the stated PBE (4) is given by a non-negative function*

$$y \in \mathcal{C}([0,\infty); L^1(0,\infty)), \qquad y(\cdot,0) = y^0, \tag{9}$$

satisfying the estimate of Theorem 3.1 for a non-negative and non-decreasing convex functions $\phi \in \mathcal{C}^1([0,\infty))$ such that

$$\lim_{u\to\infty} \frac{\phi(u)}{u} = \infty \qquad \text{and} \qquad \phi(0) = 0 \tag{10}$$

and solving (4) in the sense of distributions.

In the proof of the main theorem the convergence of an approximated problem to a unique fixed point is shown using the contraction mapping theorem. Since all the assumptions of Theorem 3.1 are satisfied for the family of approximated problems, the estimate holds uniformly with respect to the regularisation parameter.

Using the structure conditions (5) and the stated a priori result (Thm. 3.1) we can proof by extraction of a subsequence that $y(x,t)$ is a weak solution of (4) in the sense of Definition 3.2:

Theorem 3.3. *Assuming a bounded growth rate, the stated structure conditions on the kernel and $y^0 \in L^1((0,\infty),(1+s)\,ds)$ for the initial data, there exists at least a solution $y \geq 0$ to the population balance equation (4) satisfying*

$$y \in \mathcal{C}([0,T]; L^1(0,\infty)), \qquad Q(y) \in L^1((0,T)\times(0,\infty)) \tag{11}$$

for all $T > 0$.

4. Parameter identification using the notion of flat-systems

One of the substantial tasks of the project was to gain deeper insight into the synthesis process of nanoscaled $BaSO_4$. In-depth studies for the forward modeling have been conducted in order to systematically understand the influence of the leading parameters including the influence of mixing, surface energy, nucleus size and nucleus size distribution as well as their temporal evolution in the process. These evaluations are based on our previous work for the prediction of size distributions and the detailed 3D-reconstructions of all major process variables including supersaturation, nucleation rates, growth rates and particles sizes in the T-mixer [7, 8, 33, 34, 35].

The obtained results give evidence to the fact that, e.g., the critical size in the formation of particles as well as the activity coefficient in the modeling of the supersaturation can be regarded as insignificant. Under the present experimental conditions for the production of $BaSO_4$ aggregation and agglomeration were also found to have negligible influence due to successful electrostatic stabilisation [35]. Therefore, the corresponding terms on the right-hand side vanish in the modeling of the process. In contrary the interfacial energy γ_{PF} exerts a crucial effect on the evolution of the particle size distribution.

For the considered precipitation reaction the Grahame equation yields a validated model for this decisive parameter [14]. In contrast, for the case of more complex molecules or structures no proper model of the interfacial energy is available. This encourages the investigation of techniques to provide at least a reasonable range for this quantity allowing for a deeper insight into the process. The identification of the critical model parameters is fundamental as a first step towards model reduction and subsequently enables suitable optimization techniques.

Under the assumption that new particles are formed having negligible size (i.e., $x = 0$) and that no seed particles are present, the homogeneous nucleation rate can be placed in the boundary term. A reformulation of the general PBE reduces for the case of an ideally mixed batch process to

$$\frac{\partial y(x,t)}{\partial t} = -\frac{\partial (V_x(S,t)y(x,t))}{\partial x} \tag{12}$$

$$y(0,t) = \frac{I_{\mathrm{hom}}(S,\gamma_{\mathrm{PF}},t)}{V_x(S,t)} \tag{13}$$

$$y(x,0) = 0 \tag{14}$$

Here, $y(x,t)$ represents again the number density distribution of particles at time t, I_{hom} the homogeneous nucleation rate and $V_x(x)$ the particle growth rate. The homogeneous nucleation rate I_{hom} for example can be calculated based on the classical nucleation theory [26]:

$$\begin{aligned} I_{\mathrm{hom}}(S,\gamma_{\mathrm{PF}},t) = 1.5 \cdot D\left(\sqrt{K_{\mathrm{SP}}} \cdot S \cdot N_{\mathrm{A}}\right)^{\frac{7}{3}} \sqrt{\frac{\gamma_{\mathrm{PF}}}{k_{\mathrm{B}}T}} \cdot V_{\mathrm{m}} \\ \cdot \exp\left(-\frac{16\pi}{3} \cdot \left(\frac{\gamma_{\mathrm{PF}}}{k_{\mathrm{B}}T}\right)^3 \cdot \frac{V_{\mathrm{m}}^2}{(\nu \ln S)^2}\right). \end{aligned} \tag{15}$$

For a detailed justification of the model derivation of the precipitation experiments under consideration see [34, 39].

Figure 4 shows the influence of the interfacial energy for an educt composition for which the physical model of the interfacial energy, i.e., the *Grahame equation* (Eq. 5.28, taken from Israelachvili [14]), predicts the value $\gamma_{PF} = 0.126\frac{J}{m^2}$. The

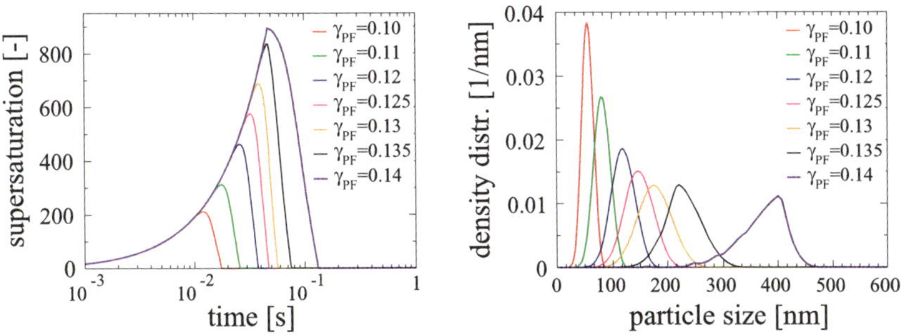

a) supersaturation profile

b) resulting PSD (normalised number density distribution)

FIGURE 4. Influence of varying of interfacial energy

$\pm 20\%$ change of the value of the interfacial energy shifts the mean particle size from 35nm to 220nm and influences also the width of the PSD.

Since there are no proper models for the interfacial energy in the case of more complex molecules or structures as outlined before, parameter estimation methods have to be considered. Therefore we focus on the framework of a specific class of systems, called differentially flat systems [5, 6], to investigate the system dynamics. It is important to point out that many classes of systems commonly used in nonlinear control theory are flat. The system is said to be flat if one can find a set of variables, called the flat outputs, such that the system is (non-differentially) algebraic over the differential field generated by the set of flat outputs. Some applications of this framework stated in [25] are for example the control of nonlinear heavy chain systems and flexible beams or the stabilisation of the water movement in a tank. The method has also been used for example in [41] to control the cooling process of a batch crystallisation process. Basically it is required that the structure of the trajectories of the nonlinear dynamics can be completely characterized, i.e., two distinct trajectories may not intersect.

In order to apply this concept we introduce the method of moments which reduces the description of the model (13)–(14) to the evolution of a number of significant quantities. The kth moment of a number density distribution is defined by

$$\mu_k(t) = \int_0^\infty x^k y(x,t)\,dx.$$

The zeroth moment $\mu_0(t)$ gives per definition the number of particles or the third moment $\mu_3(t)$ respectively is proportional to the total volume of particles per unit suspension volume.

Thus, the presented PBE model (13)–(14) can be approximated by a system of coupled moment equations. In [39] we showed that the system states $\mu_0, \ldots, \mu_3$ together with the control, realised by γ_{PF}, form a finite-dimensional non-linear flat system. In conclusion, the birth and growth rates and therefore also the interfacial energy $\gamma_{PF}(t)$ can be retrieved through the observation of the variable $\mu_3(t)$. Hence an estimated parameter range for the interfacial energy can be recovered using only measurable information on the third moment in the reduced model characterising the evolution of formed particles.

In conclusion the estimation of an optimal value of the interfacial energy becomes feasible using the notion and framework of differentially flat-systems. A forward simulation in PARSIVAL [42] solving the full PBE with predefined values for the educt concentration was conducted, which has been validated against experimental data. The calculated evolution of the third moment acted subsequently as an input for the parameter estimation routine. In Figure 5 the predicted values of the interfacial energy are compared to the physical model. All the values deviate

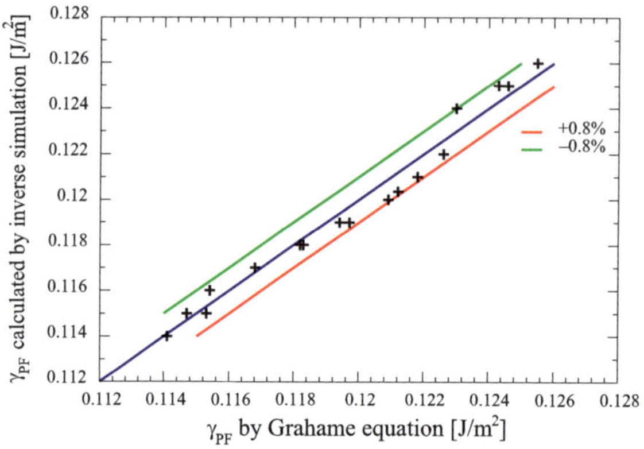

FIGURE 5. Comparison of the interfacial energy by the Grahame equation and the inverse simulation

$\pm 0.8\%$ from the values for the interfacial energy calculated by means of the Grahame equation validating the presented approach. For more complex molecules or structures the stated comparison encourages that this technique may provide at least a reasonable range for the mostly unknown interfacial energy allowing for further investigations.

5. Establishing structure-property relations

Having studied the precipitation of barium sulfate as a reasonable model process, obviously the next step consists in the transfer of the results to issues in applications. Semiconducting ZnO nanoparticles for example are currently of great interest due to their interesting optical and electronic properties. Their scope ranges from sun cream to electronic and photonic devices including solar cells. To study the production chain spanning the entire spectrum from the process conditions to the properties of the final product, ZnO nanoparticles have been prepared by controlled precipitation from zinc acetate and lithium hydroxide in alcoholic solution. The produced particles had sizes between 2 and 8 nm in diameter. The particle size has been obtained from dynamic light scattering (DLS), TEM analysis, UV-Vis spectroscopy (UV-VIS) and Hyper Rayleigh scattering (HRS). The simultaneous measurement of UV-VIS and HRS allow to determine directly the rate of nucleation, growth rate and ripening rates [37].

In the nanometer regime various quantum mechanical effects are observed and exert an enormous influence on the product properties. An increase in the band gap of a semiconducting material occurs when its size is reduced resulting in an entirely different optical behavior. The first approach to a quantitative understanding of these quantum mechanical effects on the band gap as a function of the size was based on the effective mass approximation (EMA). Since EMA has been found to overestimate the size of smaller sized systems Viswanatha et al. elaborated a precise tight-binding model in [40]. The accurate description of the bulk band dispersions using an analysis of the contributions of the partial density of states model and the Crystal Orbital Hamiltonian Population to range the hopping interactions yields a reliable description of the valence and conduction bands.

We studied these relations for semiconducting ZnO nanoparticles establishing the complete chain from the process conditions for particle formation, the dispersed properties (size and shape) to the relevant functional properties of the final particles. The next challenge is now to determine size dependent solubilities and the surface energy of the particles, possibly, also in dependence of the particle size. This is an ideal test case to proof the mathematical concept and to demonstrate the feasibility of the optimization techniques.

6. Optimal control of the flow-rate in an aerosol process

As outlined in the introduction, the possibility of precisely controlling the production of particulate compounds emerges as a fundamental challenge in industry. Since the value of the final product depends mostly on the disperse properties of the particles as outlined before, new techniques are in great demand. Besides precipitation reactions aerosol processes represent a second major field in the production of nanoscaled particles. Furthermore, they are typically modelled by a population balance equation as well describing the evolution of the according particle size distribution along the reactor and time.

The main control elements to influence the process consist in varying the composition of the feed streams, adjusting the temperature of the reactor or regulating the residence time of the particles. In the following we will focus on the last aspect to control the evolution of the particle ensemble. Changing the residence time of the particles in a tubular reactor running along the z-axis is equivalent to the adjustment of the overall feed rate (resp. the flux rate) V_z in time. The corresponding population balance equation reads:

$$\begin{cases} \frac{\partial}{\partial t} y + \frac{\partial}{\partial x}(V_x y) + \frac{\partial}{\partial z}(V_z y) - I(x) = B(y) - D(y) \\ \text{with} \\ B(y) - D(y) = \frac{1}{2}\int_0^x \beta(x-s,s) y(x-s,t) y(s,t)\, ds \\ \qquad -y(x,t)\int_0^\infty \beta(x,s) y(s,t)\, ds. \end{cases} \tag{16}$$

In the investigation of the aerosol process we therefore consider a particle distribution $y(x, z, t)$, where the variable x represents the volume equivalent particle size, z denotes the coordinate running along the reactor and t the time. This formulation includes the prevailing mechanisms of the process, where V_x represents the growth rate through condensation, I the nucleation and B as well as D the influence of agglomeration.

The complexity of the outlined model deserves studying possible reduction techniques to make a control of the process feasible. Referring to the work of A. Kalani and P.D. Christofides [16] we use the fact that the particle size distributions in aerosol processes may be adequately described by the evolution of a log-normal size distribution. Therefore we use the assumption

$$y(x, z, t) = \frac{1}{3\sqrt{2\pi}\ln\sigma} \exp\left(-\frac{\ln^2(x/x_g)}{18\ln^2\sigma}\right)\frac{1}{x},$$

where x_g is the geometric average particle size and σ the standard deviation.

It turns out that the significant parameters of the log-normal distribution may be obtained using only information on the first three moments according to the definition of the kth moment of the size distribution (4). Therefore, it is sufficient to track only these averaged quantities in order to recover the full distribution.

In contrast to the previous precipitation model we cannot eliminate the agglomeration terms since their contribution in an aerosol reaction alters considerably the PSD. Following the demonstration in [16] we thus apply the moment approximation to the given system of PDEs (16) including the source terms to obtain

$$\frac{\partial N_0}{\partial t} = -v_{zl}\frac{\partial N_0}{\partial z} + I' - \xi N_0^2 \tag{17}$$

$$\frac{\partial N_1}{\partial t} = -v_{zl}\frac{\partial N_1}{\partial z} + I'k^* + \eta(S-1)N_0 \tag{18}$$

$$\frac{\partial N_2}{\partial t} = -v_{zl}\frac{\partial N_2}{\partial z} + I'k^{*2} + +2\epsilon(S-1)N_1 + 2\zeta N_1^2 \tag{19}$$

Where N_0, N_1 and N_2 represent the first three moments in an dimensionless formulation describing the evolution of the aerosol concentration, aerosol volume and the second aerosol moment. Completed by a mass and energy balance the equations constitute a reduced model of the process. However, under the assumption of the log-normal distribution the full particle size distribution can be recovered at every point in time of the reaction. In the spatially homogeneous case Kalani and Christofides validated this approach, verifying the simulation results in comparison to a sectional model (see [17]).

For the control of the aerosol process we present a monotonically decreasing algorithm based on the described moment model approximation. Neglecting any fluid dynamics, the resulting PDE constitutes a bilinear optimal control problem with respect to the throughput $u(t) = V_z$ acting as the controlled quantity. Setting $X = [N_0, N_1, N_2]$ as the state variable, we end up with the following bilinear system:

$$\frac{\partial}{\partial t}X + u(t)A\frac{\partial}{\partial z}X = F(X) \tag{20}$$

$$X(0) = X^0, \quad X(t,0) = g(t), \tag{21}$$

including the highly nonlinear right-hand side $F(X)$.

In contrast to the work of Christofides et al. we incorporate the adjoint state Y in order to develop an iterative approach for controlling the particle size distribution. The special properties of the algorithm guarantee that in each iteration step the control is improved to come closer to the required optimal process conditions.

The numerical scheme uses a Strang-splitting approach (see [38]) to separate the solution of the homogeneous advection equation from the incorporation of the right-hand side. For the solution of the advection problem a finite volume approach is used in such a way that the overall scheme is of at least second order.

In the sequel a general cost functional of the following type is assumed:

$$\inf_{u\in\mathcal{U}} J(u,X) := \; g(X(T)) + \int_0^T f(X(t))\,dt + \int_0^T h(u(t))\,dt, \qquad \text{subject to } (20), (21). \tag{22}$$

This cost functional comprises the possibility of tracking the evolution of the particle size distribution along a specific given trajectory or the definition of required properties at the final time T. Additionally the flow-rate may be subjected to penalisation via the function h.

The corresponding adjoint equation is given by

$$\frac{\partial}{\partial t}Y + u(t)A^*\frac{\partial}{\partial z}Y = -DF(X)Y - \nabla f(X) \tag{23}$$

$$Y(T) = \nabla g(X(T)), \quad Y(t,1) = 0. \tag{24}$$

Starting from an initial value u_0 the questions arises how to find an appropriate update rule for the control u. Implementing for example the method of steepest descent, an optimal choice for the step size parameter is indispensable and in most cases quite time-consuming. The choice of a fixed parameter instead can yield poor results.

In the present approach we propose a different first-order optimization algorithm that guarantees, in contrast to the pure gradient method, a monotone decrease of the cost functional even for fixed parameters. In the spirit of the monotone algorithms developed in the context of quantum control (see for example J. Salomon, G. Turinici and Y. Maday in [24, 30]) a solution strategy has been established accounting for the special structure of the underlying model.

To satisfy that $J(u^{k+1}) - J(u^k) < 0$ for all k we introduce an intermediate control $\tilde{u}$ and establish the following iterative approach:

After initialization, first solve the state equation for X^k, u^k

$$\frac{\partial}{\partial t}X^k + u^k A \frac{\partial}{\partial z}X^k = F(X^k)$$

$$X^k(0) = X^0 \quad X^k(t,0) = g(t)$$
(25)

such that the scalar constraint

$$\langle Y^{k-1}(t), (\tilde{u}^{k-1} - u^k)(t)A\frac{\partial}{\partial z}X^k\rangle$$
$$+(h(u^k(t)) - h(\tilde{u}^{k-1}(t))) \leq 0$$
(26)

holds for all t.
Then solve the adjoint equation for Y^k, $\tilde{u}^k$

$$\frac{\partial}{\partial t}Y^k + \tilde{u}^k A^* \frac{\partial}{\partial z}Y^k = -DF(X^k)Y^k - \nabla f(X^k)$$

$$Y^k(T) = \nabla g(X^k(T)), \quad Y^k(t,1) = 0$$
(27)

such that a second scalar constraint holds for all t:

$$\langle Y^k(t), (u^k - \tilde{u}^k)(t)A\frac{\partial}{\partial z}X^k\rangle$$
$$+h(\tilde{u}^k(t)) - h(u^k(t)) \leq 0$$
(28)

Due to the imposed fundamental constraints a monotone decrease of the cost functional is obtained. To match a specified particle size distribution after the process we may choose the set of corresponding values $\bar{X} = [\bar{N}_0, \bar{N}_1, \bar{N}_2, \bar{x}]$ of the log-normal distribution at the final time T. Therefore we set the function g in the cost functional at final time T to

$$g(X(T)) = \frac{1}{2}\|X(T) - X_d\|^2.$$
(29)

In addition, the throughput is penalised via

$$h(u(t)) = \frac{\kappa}{2}\|u\|^2$$

for stability reasons. One can also think of using a different penalisation

$$h(u(t)) = \frac{\kappa}{2}\|u - \bar{u}\|^2 \tag{31}$$

avoiding strong deviations around a mean flow rate $\bar{u}$. We do not impose a specific progress of the state in the aerosol process since tracking the evolution of the distribution is not as relevant as reaching required specifications of the final product.

In the stated iteration procedure we still have to define $\tilde{u}^k$ as well as u^k such that the corresponding constraints are fulfilled. Defining u^k through

$$u^k(t) = (1-\delta)\tilde{u}^{k-1} + \frac{\delta}{\kappa}\int_0^1 Y^{k-1} A \frac{\partial}{\partial z} X^k \, dx. \tag{32}$$

constitutes an update rule that is only based on data in the interval $[0, t]$ whereas the necessary monotonicity condition may also be checked by standard calculations. $\tilde{u}^k$ is then likewise defined by

$$\tilde{u}^k(t) = (1-\eta)u^k + \frac{\eta}{\kappa}\int_0^1 Y^k A \frac{\partial}{\partial z} X^k \, dx. \tag{33}$$

This choice guarantees a monotone decrease of the cost functional for all values of the parameters δ and η in the interval $[0, 2]$.

The given definitions still contain two parameters that may be chosen. In the iteration procedure, the parameters can be fixed for all k without losing the monotonicity. Though using the update rules (32) and (33), a nonlinearity is introduced into the state and adjoint systems (25), (28) which is also affected by the selection of these parameters. This reveals that the choice has to be appropriate to the system dynamics.

Choosing these parameters in each step lead to an improved convergence at the cost of a higher computational effort. For a clear presentation $\eta = 0$ will be set from now on, such that the update rule only affects the calculations for the state equation (25) and not the adjoint system (28), i.e., $\tilde{u}^k = u^k$. Although numerical simulations show that different choices may even lead to better convergence.

Two different approaches have been implemented. Besides an inefficient direct search method that involves many evaluations of the state equation, a different approach is to calculate the optimal value of δ analytically by differentiation of $J(u(\delta))$ using the first-order optimality condition.

$$\frac{\partial}{\partial \delta} J(u^{k+1}(\delta)) = \int_0^T \kappa \left(\frac{u^{k+1} - u^k}{\delta}\right) u^{k+1} \, dt - \int_0^T \frac{u^{k+1} - u^k}{\delta} \left\langle A \frac{\partial}{\partial z} X^{k+1}; W \right\rangle$$
$$- \int_0^T \left\langle \frac{\partial u^{k+1}}{\partial \delta}; \underbrace{-\partial_t W + A^T \frac{\partial}{\partial z} W + \left(\frac{\delta}{\kappa}\left\langle A \frac{\partial}{\partial z} X^{k+1}; W \right\rangle - \delta u^{k+1}\right) A^T \frac{\partial}{\partial z} Y^k}_{(W)} \right\rangle$$

The expression (W) occurring in the calculation denotes a third equation additionally to the state and adjoint system that needs to be solved. However, an approximation using just the known values of the former iteration k

$$\frac{\partial}{\partial t}W - A^T \frac{\partial}{\partial z}W = \left(\frac{\delta}{\kappa}\left\langle A\frac{\partial}{\partial z}X^k; W\right\rangle - \delta u^k\right) A^T \frac{\partial}{\partial z}Y^k \tag{35}$$

$$W(T) = \nabla g(X^k(T)), \quad W(t,1) = 0$$

already yields excellent results without adding to much computational overhead compared to, e.g., the implicit approach or a direct search method. In conclusion, the procedure to calculate an improved δ takes only half the effort which is necessary to solve (25), (28). Thus with the presented algorithm a monotone decrease is obtained converging to the optimal value of the given cost functional.

7. Conclusion and outlook

In this joint project of the mathematics and engineering faculty a comprehensive study of two processes for the preparation of nanoscaled particulate products with exactly pre-defined properties was carried out. Under certain conditions on the initial and boundary data for the considered conservation equation (4) a new existence result has been proven incorporating the growth term. A priori estimates are established in order to show the existence of a global weak solution satisfying (11). In-depth studies for forward modeling have been conducted in order to systematically understand the influence of the leading parameters to the resulting particle size distribution. Since the modeling of some decisive parameters like the interfacial energy is only possible for the simplest molecular structures, an inverse parameter estimation method was established. Providing a valid range for these material properties opens up new perspectives, e.g., for the simulation of macromolecules in pharmaceutical process engineering.

The production of nano-scaled particulate products by aerosol processes provides an economic and innovative approach providing specific pre-required bulk properties. In this contribution a promising algorithm is presented yielding an optimal feed-rate profile to attain the demanded features.

After the detailed study of the semiconducting ZnO nanoparticles the whole chain from process conditions to predefined UV-VIS spectra will be considered. The ZnO nanorods are prepared through oriented attachment, growth and ripening will subsequently be described by a 2D population balance using an anisotropic growth rate. Dealing with complex structures including, for instance, the tetrapods of Figure 1, a generalisation of this approach in higher dimensions is needed posing a challenge for future research.

Acknowledgment

This work was supported by the SPP 1253 "Optimization with Partial Differential Equations" of the German Science Foundation (DFG) through the grant LE 595/23-1.

References

[1] C. Artelt, H.-J. Schmid, W. Peukert, Modelling titania formation at typical industrial process conditions: effect of structure and material properties on relevant growth mechanisms. Chem. Eng. Sci., 61 (2006), 18–32.

[2] P.D. Christofides, Nonlinear and robust control of PDE systems: Methods and applications to transport-reaction processes, Birkhäuser-Verlag 2001, 250p.

[3] M. Escobedo, P. Laurençot, S. Mischler, On a Kinetic Equation for Coalescing Particles, Communications in Mathematical Physics, 246 (2), 2004, 237–267.

[4] T. Fischer, D. Logashenko, M. Kirkilionis and G. Wittum, Fast Numerical Integration for Simulation of Structured Population Equations, Mathematical Models and Methods in Applied Sciences, 16 (12), 2006, 1987–2012.

[5] M. Fliess, J. Levine, P. Martin, P. Rouchon, Sur les systèmes non linéaires differentiellement plats, C.R. Acad. Sci. Paris, 1992, I/315, 619–624.

[6] M. Fliess, J. Levine, P. Martin, P. Rouchon, Flatness and defect of nonlinear systems: Introductory theory and examples, 1995, International Journal of Control, 61 (6), 1327–1361.

[7] J. Gradl, H.-C. Schwarzer, F. Schwertfirm, M. Manhart, W. Peukert, Precipitation of nanoparticles in a T-mixer: Coupling the particle population dynamics with hydrodynamics through direct numerical simulation, Chemical Engineering and Processing, 45 (10), 2006, 908–916.

[8] J. Gradl, W. Peukert, Simultaneous 3D observation of different kinetic subprocesses for precipitation in a T-mixer, Chemical Engineering Science (2009), 64, 709–720.

[9] W. Hackbusch, On the Efficient Evaluation of Coalescence Integrals in Population Balance Models, Computing 78, 2 (Oct. 2006), 145–159.

[10] W. Hackbusch, Fast and exact projected convolution for non-equidistant grids, Computing 80, 2 (Jun. 2007), 137–168.

[11] W. Hackbusch, Approximation of coalescence integrals in population balance models with local mass conservation, Numer. Math. 106, 4 (May, 2007), 627–657.

[12] D.K. Henze, J.H. Seinfeld, W. Liao, A. Sandu, and G.R. Carmichael (2004), Inverse modeling of aerosol dynamics: Condensational growth, J. Geophys. Res., 109, D14201.

[13] M.J. Hounslow, R.L. Ryall, and V.R. Marshall, A discretized population balance for nucleation, growth and aggregation. AIChE Journal, 34 (1988), 1821–1832.

[14] J. Israelachvili, "Intermolecular and Surface Forces", 2nd edition, Academic Press, London, Great Britain.

[15] T. Johannessen, S.E. Pratsinis, and H. Livbjerg, Computational Fluid-particle dynamics for flame synthesis of alumina particles. Chem. Eng. Sci. 55 (2000), 177–191.

[16] A. Kalani, P.D. Christofides, Nonlinear control of spatially inhomogeneous aerosol processes, *CES* **54** (1999), 2669–2678.

[17] A. Kalani, P.D. Christofides, Simulation, estimation and control of size distribution in aerosol processes with simultaneous reaction, nucleation, condensation and coagulation, Com. and Chem. Eng. **26** (2002), 1153–1169.

[18] J. Koch, W. Hackbusch, K. Sundmacher, H-matrix methods for linear and quasi-linear integral operators appearing in population balances, Computers and Chemical Engineering, 31 (7), July 2007, 745–759.

[19] J. Koch, W. Hackbusch, K. Sundmacher, H-matrix methods for quadratic integral operators appearing in population balances, Computers and Chemical Engineering, 32 (8), Aug. 2008, 1789–1809.

[20] J. Kumar, M. Peglow, G. Warnecke, S. Heinrich, E. Tsotsas, and L. Moerl, Numerical solutions of a two-dimensional population balance equation for aggregation, Proceedings of the 5th World Congress on Particle Technology, 2006.

[21] J. Kumar, G. and Warnecke, Convergence analysis of sectional methods for solving breakage population balance equations-II: the cell average technique, Numer. Math. 110, 4 (Sep. 2008), 539–559.

[22] Ph. Laurencot, S. Mischler, The continuous coagulation-fragmentation equations with diffusion, Arch. Rat. Mech. Anal. 162, 2002, 45–99.

[23] D. Logashenko, T. Fischer, S. Motz, E. D. Gilles, and G. Wittum, Simulation of crystal growth and attrition in a stirred tank, Comput. Vis. Sci. 9, 3 (Oct. 2006), 175–183.

[24] Y. Maday, J. Salomon, and G. Turinici. Monotonic time-discretized schemes in quantum control. Numerische Mathematik, 2006.

[25] Ph. Martin, R. Murray, and P. Rouchon, Flat systems, equivalence and trajectory generation, technical report, 2003.

[26] A. Mersmann, K. Bartosch, B. Braun, A. Eble, C. Heyer, "Möglichkeiten einer vorhersagenden Abschätzung der Kristallisationskinetik", 2000, Chemie Ingenieur Technik 71(1-2), 17–30.

[27] H. Mühlenweg, A. Gutsch, A. Schild, and S.E. Pratsinis, Process simulation of gas-to-particle-synthesis via population balances: Investigation of three models, Chem. Eng. Sci., 57 (2002), 2305–2322.

[28] Y. Qiu, S. Yang, ZnO Nanotetrapods: Controlled vapour-phase synthesis and application for humidity sensing, Adv. Functional Materials 2007, 17, 1345–1352.

[29] J.M. Roquejoffre, P. Villedieu, A kinetic model for droplet coalescence in dense sprays, Math. Models Meth. Appl. Sci., 11, 2001, 867–882.

[30] J. Salomon, Contrôle en chimie quantique: conception et analyse de schémas d'optimisation, thesis, 2005.

[31] A. Sandu, W. Liao, G.R. Carmichael, D.K. Henze, J.H. Seinfeld, Inverse modeling of aerosol dynamics using adjoints – theoretical and numerical considerations, Aerosol Science and Technology, 39 (8), 2005,Number 8, 677–694.

[32] H.-C. Schwarzer, W. Peukert, "Combined Experimental/Numerical Study on the Precipitation of Nanoparticles", 2004, AIChE Journal 50 (12), 3234–3247.

[33] H. Schwarzer, W. Peukert, Tailoring particle size through nanoparticle precipitation, Chem. Eng. Comm. 191 (2004), 580–606.

[34] H.-C. Schwarzer, W. Peukert, Combined experimental/numerical study on the precipitation of nanoparticles, AIChE Journal 50 (2004), 3234–3247.

[35] H.-C. Schwarzer, F. Schwertfirm, M. Manhart, H.-J. Schmid, W. Peukert, "Predictive simulation of nanoparticle precipitation based on the population balance equation", 2006, Chemical Engineering Science 61 (1), 167–181.

[36] D. Segets, J. Gradl, R. Klupp Taylor, V. Vassilev, W. Peukert, Analysis of Optical Absorbance Spectra for the Determination of ZnO Nanoparticle Size Distribution, Solubility, and Surface Energy, ACS nano (2009), 3(7), 1703–1710.

[37] D. Segets, L.M. Tomalino, J. Gradl, W. Peukert, Real-Time Monitoring of the Nucleation and Growth of ZnO Nanoparticles Using an Optical Hyper-Rayleigh Scattering Method, J. Phys. Chem. C 2009, 113, 11995–12001.

[38] E.F. Toro, Riemann Solvers and Numerical Methods for Fluid Dynamics, *Springer, Berlin* (2009).

[39] V. Vassilev, M. Gröschel, H.-J. Schmid, W. Peukert, and G. Leugering, Interfacial energy estimation in a precipitation reaction using the flatness based control of the moment trajectories, Chemical Engineering Science (65), 2010, 2183–2189.

[40] R. Viswanatha, S. Sapra, B. Satpati, P.V. Satyam, B. Dev and D.D. Sarma, Understanding the quantum size effects in ZnO nanocrystals, J. Mater. Chem., 14, 2004, 661–668.

[41] U. Vollmer, J. and Raisch, Control of batch cooling crystallization processes on orbital flatness, Int. J. Control 76/16 (2003), 1635–1643.

[42] M. Wulkow, A. Gerstlauer, U. and Nieken, Modeling and simulation of crystallization processes using parsival, Chem. Eng. Sci. 56 (2001), 2575–2588.

M. Gröschel and G. Leugering
Friedrich-Alexander-Universität Erlangen-Nürnberg
Lehrstuhl für Angewandte Mathematik II
Martensstrasse 3
D-91058 Erlangen, Germany
e-mail: groeschel@am.uni-erlangen.de
leugering@am.uni-erlangen.de

Wolfgang Peukert
Friedrich-Alexander-Universität Erlangen-Nürnberg
Lehrstuhl für Feststoff- und Grenzflächenverfahrenstechnik
Cauerstraße 4
D-91058 Erlangen, Germany
e-mail: W.Peukert@lfg.uni-erlangen.de

International Series of Numerical Mathematics, Vol. 160, 561–572

Control of Nanostructures through Electric Fields and Related Free Boundary Problems

Frank Haußer, Sandra Janssen and Axel Voigt

Abstract. Geometric evolution equations, such as mean curvature flow and surface diffusion, play an important role in mathematical modeling in various fields, ranging from materials to life science. Controlling the surface or interface evolution would be desirable for many of these applications. We attack this problem by considering the bulk contribution, which defines a driving force for the geometric evolution equation, as a distributed control. In order to solve the control problem we use a phase-field approximation and demonstrate the applicability of the approach on various examples. In the first example the effect of an electric field on the evolution of nanostructures on crystalline surfaces is considered. The mathematical problem corresponds to surface diffusion or a Cahn-Hilliard model. In the second example we consider mean curvature flow or a Allen-Cahn model.

Mathematics Subject Classification (2000). 35R35; 49M.

Keywords. Geometric evolution, surface diffusion, electromigration, optimal control, phase field approximation.

1. Introduction

The manipulation of nanostructures by macroscopic forces is likely to become a key ingredient in many nanotechnology applications. Understanding and controlling the influence of external fields on the shape evolution of nanoscale surface features is therefore of considerable importance. As a first step in this direction we recently analyzed the effects of an external electric field on single-layer (i.e., atomic height) islands on a crystalline surface [1]. The richness of the computationally discovered behavior in this study leads us to believe that it is possible to control the microscopic island shape evolution through a macroscopic electric field. If applied to real systems, this could have a large technological impact, for example in designing novel electronic devices.

Mathematically the problem leads to the optimal control of a free boundary problem. Here the free boundary is the island edge, which evolves according to edge diffusion (one-dimensional surface diffusion). The electric field as the control parameter enters the system of partial differential equations in form of an additional driving force. The goal is to drive the system utilizing the external electric field (within an allowed range) in such a way that a desired island shape is obtained. The problem will be approximated through a phase-field approach, in which the island edge is represented as a level-set curve of an order parameter. Based on existing theoretical results for the control of phase-field models for solidification, see [2] and references therein, we derive the optimality conditions through utilizing the Lagrangian framework and solve the system of state and adjoint equations by finite elements. A gradient method will be used to update the control.

Even if we concentrate on a specific application, the developed tools are applicable for other control problems for free boundaries or geometric evolution problems. Similar to the electric field any contribution from the adjacent bulk phases can be viewed as a potential control parameter in a driven geometric evolution law. E.g., within solidification the undercooling would play the role of a control parameter within the kinetic Gibbs-Thomson law, which can be viewed as a driven mean curvature flow problem, or within thin film growth the deposition flux would play the role of a control parameter within a driven surface diffusion law, see [3] for further examples. We will demonstrate the applicability of the approach by solving control problems for driven mean curvature flow.

The paper is organized as follows. In Section 2 we focus on the phase-field approximation of the underlying electromigration problem and compare sharp interface and phase-field approximations for various parameter regimes. The numerical results confirm the asymptotic analysis, showing formally the convergence of the phase-field approximation to the sharp interface limit. This motivates to use the phase-field formulation as the underlying model to formulate the control problem. In Section 3 we derive the control problem and show first results for the control of electromigration driven nanoscale islands. In Section 4 we extend the approach to also consider the control of driven mean curvature flow problems. Conclusions are drawn in Section 5.

2. Electromigration – modeling and numerics

The movement and deformation of a single, atomic high island, under the influence of electromigration was first demonstrated in [4, 5]. In [1] resulting instabilities due to the electromigration force are extensively studied using a front tracking parametric finite element algorithm. For simplicity we here consider an isotropic setting for which the fourth-order nonlinear free boundary problem which models electromigration in a nondimensional form reads

$$v = \partial_{ss}(\kappa + E_t) \tag{2.1}$$

with v denoting the normal velocity of an island, κ the mean curvature of the island boundary, E_t the electromigration force (tangential component of a local electric field) and s the arclength. Using matched asymptotic expansion [6] this problem can be approximated by a phase field model. Written as a nonlinear system of two second-order equations it reads

$$\partial_t \phi = \varepsilon^{-1} \nabla \cdot \big(M(\phi) \nabla \omega \big) \tag{2.2}$$

$$g(\phi)\omega = -\varepsilon \Delta \phi + \varepsilon^{-1} G'(\phi) + U_{el} \tag{2.3}$$

for the phase field function ϕ and the chemical potential ω, with initial condition $\phi(x,0) = \phi_0(x)$ and periodic boundary conditions on a rectangular domain Ω in $\mathbb{R}^2$. We have $G(\phi) = 18\phi^2(1-\phi)^2$ a double well potential, $M(\phi) = 36\phi^2(1-\phi)^2$ a degenerate mobility, $g(\phi) = 30\phi^2(1-\phi)^2$ a stabilizing function, $\epsilon > 0$ a small parameter corresponding to the interface thickness, and $U_{el} = -E(t) \cdot x$ the electric potential. For simplicity only the isotropic version is shown (the asymptotics are done in the fully anisotropic setting, with anisotropic stiffness, mobility and kinetic coefficient). The degenerate nonlinear system of equations is discretized in space using linear finite elements and a semi-implicit discretization in time is used, which is based on local variational principles, which ensures that the dissipative properties of the evolution are maintained by the discretization. A general way to formulate such semi-implicit discretizations is described in [7], in which the fully discrete scheme is interpreted as optimality conditions of an optimization problem.

The problem is implemented in the adaptive finite element toolbox AMDiS [8] and validated against the sharp interface model, which is solved by parametric finite elements (also within AMDiS), see Fig. 1.

3. Control problem

Given the agreement of the sharp interface and diffuse interface results we turn to the control problem, which we formulate within the phase-field approximation. Within this formulation controlling the free boundary becomes a standard distributed control problem. The electric field E is our distributed control variable. We use the following cost functional

$$J(\phi, E) = \frac{1}{2} \int_\Omega (\phi(x,T) - \phi_e(x))^2 \, dx + \frac{\nu}{2} \int_0^T E(t) \cdot E(t) \, dt, \tag{3.1}$$

with ϕ_e a phase field function representing the desired shape and location of the island at time $t = T$ and ν a penalty parameter. The goal is to identify an electric field $E = E(t)$ with $E_{\min} \leq E(t) \leq E_{\max}$, such that the corresponding phase field variable ϕ minimizes the cost functional J. The electric field is assumed to be constant in space, since we are looking at conducting materials where the influence of an atomic height island on the macroscopic electric field may be neglected.

Optimal control problem: *Given an initial phase-field $\phi(.,0)$ at time $t = 0$ and a desired phase-field ϕ_e, find a control $E : [0,T] \to \mathbb{R}$, which minimizes the functional J given in* (3.1), *subject to the constraints given by the evolution equations* (2.2) *and* (2.3).

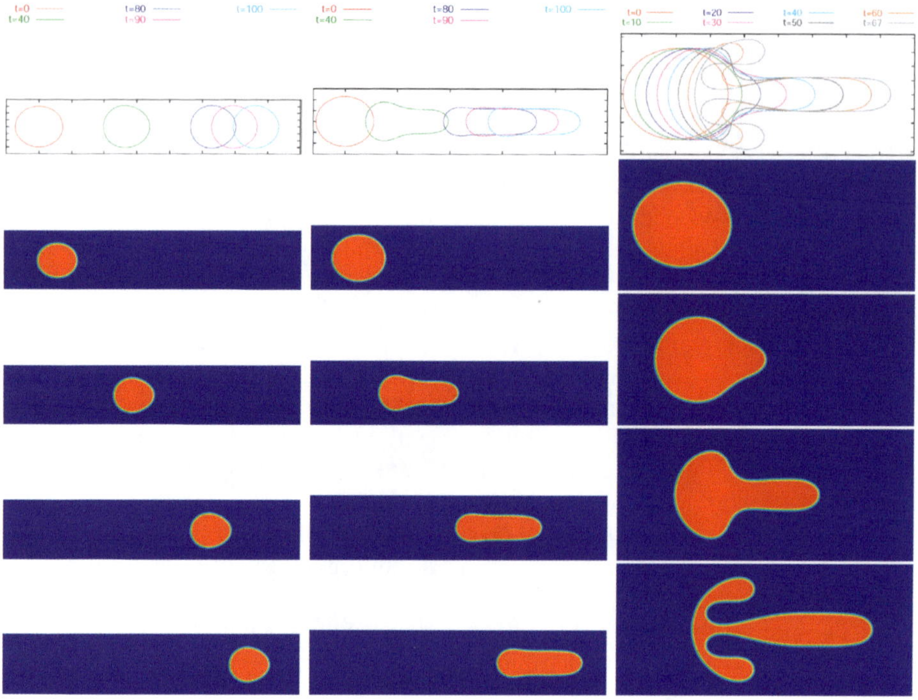

FIGURE 1. Comparison of island evolution under the influence of a constant electric field $E = (1,0)^T$ for radius $r = 3.3$ (left), $r = 5$ (middle) and $r = 7$ (right). With increasing radius the island becomes unstable. Sharp interface results for various time steps (top) and phase field simulation for $t = 0, 40, 80, 100$ (left) and (middle) and $t = 0, 20, 40, 60$ (right). A quantitative comparison of the results shows an excellent agreement of the sharp interface and phase field simulations, see [9] for a detailed description of the results and the numerical methods.

Introducing the Lagrange multipliers q_1 and q_2 (adjoint states) corresponding to the state equations (2.2) and (2.3), respectively, the Lagrange function reads:

$$\begin{aligned} L(\phi, \omega, E, q_1, q_2) = J(\phi, E) &- \int_0^T \int_\Omega \big(\partial_t \phi - \varepsilon^{-1} \nabla \cdot (M(\phi) \nabla \omega)\big)\, q_1 \, dx dt \\ &- \int_0^T \int_\Omega \big(g(\phi)\omega + \varepsilon \Delta \phi - \varepsilon^{-1} G'(\phi) - U_{el}\big)\, q_2 \, dx dt. \end{aligned}$$

From this we formally obtain the necessary first-order optimality conditions for the optimal control $\bar{E}$ with associated optimal states $\bar{\phi}, \bar{\omega}$ as

$$D_\phi L(\bar{\phi}, \bar{\omega}, \bar{E}, q_1, q_2)\phi = 0, \quad \forall \phi : \phi(x, 0) = 0 \tag{3.2}$$

$$D_\omega L(\bar{\phi}, \bar{\omega}, \bar{E}, q_1, q_2)\omega = 0, \quad \forall \omega \tag{3.3}$$

$$D_E L(\bar{\phi}, \bar{\omega}, \bar{E}, q_1, q_2)(E - \bar{E}) \geq 0, \quad \forall E \in E_{ad}, \tag{3.4}$$

where D denotes the functional derivative. The evaluation of (3.2) and (3.3) gives the system of adjoint equations

$$-\partial_t q_1 = -\varepsilon\Delta q_2 - \varepsilon^{-1} M'(\phi)\nabla\omega \cdot \nabla q_1 - g'(\phi)\omega\, q_2 + \varepsilon^{-1} G''(\phi)\, q_2 \tag{3.5}$$

$$g(\phi) q_2 = \varepsilon^{-1} \nabla \cdot \big(M(\phi)\nabla q_1\big) \tag{3.6}$$

with end condition $q_1(T) = \phi(T) - \phi_e$ and again periodic boundary conditions. Evaluating (3.4) yields the following variational inequality for the optimal control $\bar{E}$:

$$\int_0^T \Big[\big(\nu\bar{E} - \int_\Omega x q_2 dx\big) \cdot \big(E - \bar{E}\big)\Big] dt \geq 0 \quad \forall E \in E_{ad}. \tag{3.7}$$

As a first approach we use a gradient method to update the electric field

$$E(t)^{k+1} = E(t)^k + \alpha(-\nu E(t)^k + \int_\Omega q_2(t) x\, dx) \tag{3.8}$$

with k the number of iterations of forward and backward solutions and α the step size of the gradient step. We will follow the common approach to solve the first-order necessary conditions (and therefore the control problem, provided the conditions have also been sufficient) by utilizing the following optimization loop:

(i) Solve the state equations (2.2) and (2.3) forward in time using a given electric field
(ii) Solve the adjoint equations (3.5) and (3.6) backward in time using the computed state variables obtained in (i)
(iii) Update the electric field using (3.8) and proceed with step (i).

The adjoint system is again solved within AMDiS using linear finite elements for discretization in space and an implicit discretization in time. The resulting linear system is solved with a direct solver(umfpack). We further average over various timesteps to update $E(t)^{k+1}$. See [9] for details.

We use a uniform mesh for the state and adjoint problem and run the code for simple test situations, in which an electric field has to be found to move a stable island configuration (radius $r = 1.0$ and $r = 3.0$), see Fig. 2. In both examples the algorithm needs about 10 iterations to reach the desired island position. As the computed state variables for all timesteps are kept in the main memory, the considered examples are restricted in size. The unstable configuration with $r = 7$ requires a much larger time frame and so far cannot be solved without efficient I/O functionality.

We are aware of the fact that due to the simple gradient method the convergence to the optimal solution is slow, but sufficient for giving a proof of concept that islands can be moved with electric fields.

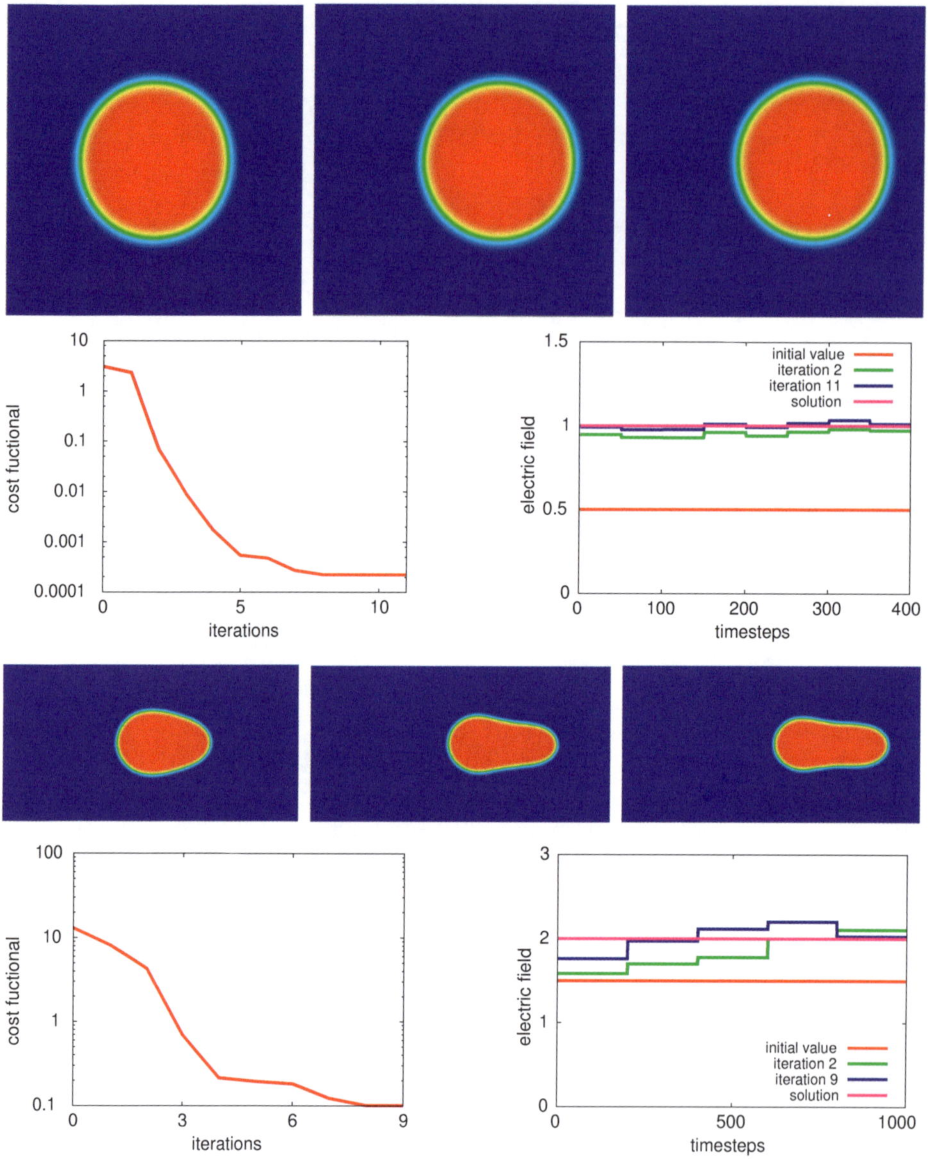

FIGURE 2. Computed optimal island position at $t = T$ (initial, iteration 2 and final iteration), reduction of cost functional and computed electric field (initial, iteration 2 and final iteration) for radii $r = 1.0$ (top) and $r = 3.0$ (bottom).

4. Related control problems for geometric evolution laws

We now consider a less specific application and concentrate on the driven mean curvature flow equation

$$v = \kappa + \eta \tag{4.1}$$

with v the normal velocity, κ the mean curvature of the interface and η the bulk contribution serving here as a control variable, see [10] for details. Using matched asymptotic expansion [11] this problem can again be approximated by a phase field model which is a driven Allen-Cahn equation and reads

$$\partial_t \phi = \varepsilon \Delta \phi - \varepsilon^{-1} G'(\phi) + \eta \tag{4.2}$$

for the phase field function ϕ, with initial condition $\phi(x,0) = \phi_0(x)$ and zero flux boundary conditions on a quadratic domain Ω in $\mathbb{R}^2$. We again have $G(\phi) = 18\phi^2(1-\phi)^2$ as a double well potential and η now denoting an extended control variable. The nonlinear equation is discretized by linear finite elements in space and a semi-implicit discretization in time is used in which $G'(\phi)$ is linearized over the last time-step. The resulting linear system is again solved with a direct solver(umfpack).

We specify the initial and final shape and position of the interface and want to find the corresponding control η such that the initial shape evolves in a given time as close as possible to the final shape. We consider two different type of controls. For simple geometries we take η to be constant in space but changing in time, $\eta = \eta(t)$, as in the above case. For more complex structures, a constant control can not achieve the required changes in the structure, so we use a time and space depending control $\eta = \eta(x,t)$. To formulate this as an optimal control problem, we define the following cost functionals, which have to be minimized:

$$J_1(\phi, \eta) = \frac{1}{2} \int_\Omega (\phi(x,T) - \phi_e(x))^2 \, dx + \frac{\nu}{2} \int_0^T \eta(t)^2 dt \tag{4.3}$$

for a time-dependent control, and

$$J_2(\phi, \eta) = \frac{1}{2} \int_\Omega (\phi(x,T) - \phi_e(x))^2 \, dx + \frac{\nu}{2} \int_0^T \int_\Omega \eta(x,t)^2 dx dt \tag{4.4}$$

for a time and space-dependent control. Here, ϕ_e is again a phase field function representing the desired shape and position of the interface at time $t = T$ and $\nu > 0$ is again a penalty parameter. We thus obtain the following

Optimal control problem: *Given an initial phase-field $\phi(.,0)$ at time $t = 0$ and a desired phase-field ϕ_e, find a control $\eta : [0,T] \to \mathbb{R}$ or $\eta : \Omega \times [0,T] \to \mathbb{R}$, which minimizes the functional J_1 or J_2 given in* (4.3) *and* (4.4), *subject to the constraints given by the evolution equation* (4.2).

Introducing the Lagrange multiplier p the Lagrange function reads:

$$L(\phi, \eta, p) = J_i(\phi, \eta) - \int_0^T \int_\Omega \left(\varepsilon \partial_t \phi - \varepsilon \Delta \phi + \frac{1}{\epsilon} G'(\phi) - \eta \right) p \, dx dt$$

respectively, with $i = 1, 2$. From this we formally obtain the necessary first-order optimality conditions for the optimal control $\bar{\eta}$ with associated optimal states $\bar{\phi}$ as

$$D_\phi L(\bar{\phi}, p, \bar{\eta})\phi = 0, \quad \forall \phi : \phi(x, 0) = 0 \tag{4.5}$$

$$D_\eta L(\bar{\phi}, p, \bar{\eta})(\eta - \bar{\eta}) \geq 0, \quad \forall \eta, \tag{4.6}$$

where D denotes the functional derivative. The evaluation of Eq. (4.5) gives the adjoint equation to (4.2)

$$-\varepsilon \partial_t p = \varepsilon \Delta p - \frac{1}{\epsilon} G''(\phi) p. \tag{4.7}$$

The equation needs to be solved backwards in time with zero flux boundary conditions and end condition $p(T) = \phi(T) - \phi_e$. Evaluating (4.6) gives the following variational inequalities for the optimal control $\bar{\eta}$. For (4.6) and $\eta = \eta(t)$ we get

$$\int_0^T \left[\left(\nu \bar{\eta} + \int_\Omega p dx \right) \cdot (\eta - \bar{\eta}) \right] dt \geq 0 \quad \forall \eta, \tag{4.8}$$

and for $\eta = \eta(x, t)$ we get

$$\int_0^T \int_\Omega \left[(\nu \bar{\eta} + p) \cdot (\eta - \bar{\eta}) \right] dx dt \geq 0 \quad \forall \eta. \tag{4.9}$$

Using a gradient method to update the control, we get

$$\eta^{k+1} = \eta^k - \alpha \left(\nu \eta^k + \int_\Omega p dx \right) \tag{4.10}$$

for $\eta = \eta(t)$ and

$$\eta^{k+1} = \eta^k - \alpha(\nu \eta^k + p) \tag{4.11}$$

for $\eta = \eta(x, t)$. We will again follow the common approach to solve the first-order necessary conditions (and therefore the control problem, provided the conditions have also been sufficient) by utilizing the following optimization loop:

(i) Solve the state equation (4.2) forward in time using a given control η
(ii) Solve the adjoint equation (4.7) backward in time using the computed state variables obtained in (i)
(iii) Update the control by utilizing (4.10) or (4.11) and proceed with step (i).

See [12] for details and more numerical examples.

A typical benchmark for mean curvature flow and thus also for the Allen-Cahn equation is the shrinkage of a spherical droplet. We will use the control η to suppress this shrinkage and instead let the droplet grow. The initial phase field function thereby represents a circle with a radius of $r = 0.5$, whereas the desired radius is $r = 0.7$. We start with a constant control $\eta = 2$ that maintains the size and shape of the droplet. The obtained phase field for $t = T$ after various iterations of the optimization loop is shown in Fig. 3, together with the evolution of the cost functional and the control. In this example the gradient method is sufficient, as we need only 15 iterations of the optimization loop.

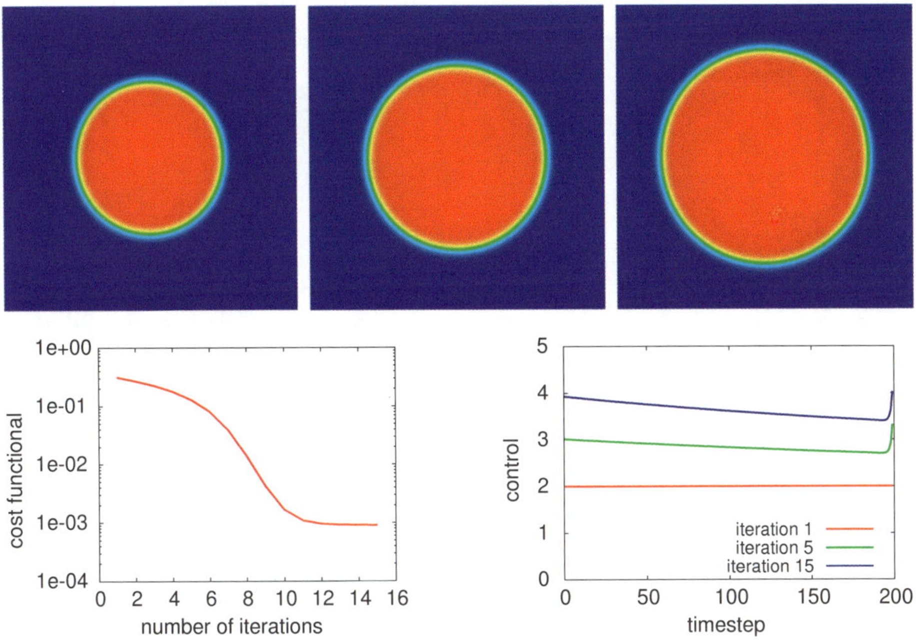

FIGURE 3. Final shape after iteration 1, 5 and the final iteration 15, and cost functional and control over number of iterations and time, respectively.

Another typical test case considers the smoothing of a perturbed circle. Instead of demonstrating smoothing properties we start with a circular domain and ask for a control to obtain a perturbed circle. Again we start with a circle of $r = 0.5$. The desired shape has $r = 0.6$ and is perturbed using a sin-function with amplitude 0.1 and wavenumber 6. We start with a constant control $\eta = 2$ in time and space. The obtained phase field for $t = T$ after various iterations of the optimization loop is shown in Fig. 4 together with snapshots of the final control at different timesteps. The decrease of the cost functional and the minimal and maximal values for the now space-dependent control over time is shown in Fig. 5. Here we need 365 iterations of the optimization loop, which shows the limits of this approach.

Both examples demonstrate the applicability of the approach to control the shape of the evolving interface. The performance of the optimization loop can certainly be improved by using higher-order approaches.

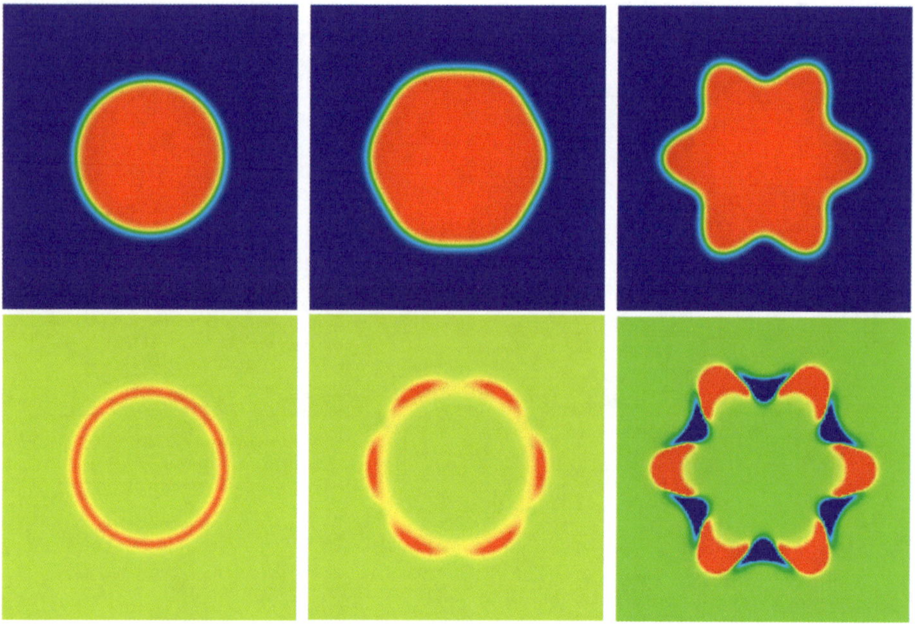

FIGURE 4. Final shape after iteration 1, 8 and the final iteration 365 (top row), final control at different time steps (bottom row).

5. Conclusion

Controlling the interface evolution in free boundary problems is a challenging task. Instead of directly formulating the control problem for the associated geometric evolution equations we first use a phase-field approximation to reformulate the free boundary problem into a more trackable system of partial differential equations and formulate the control problem for this system. This allows to use standard tools to solve the control problem. We demonstrate this by solving a first-order optimality system for the Allen-Cahn and Cahn-Hilliard equations with a distributed control, which plays the role of the bulk contribution and acts as a driving force for the evolution equation. The associated geometric evolution equations are driven mean curvature flow and driven surface diffusion. An open question which remains is how the optimality system formulated for the phase field approximation is related to an optimality system, which is formulated directly for the geometric evolution equations. The approach so far is only based on the results of an asymptotic analysis for the state equations. The next step to improve the approach will be to include adaptive mesh refinement and coarsening to reduce computational time and memory costs. Therefore the state and adjoint equations will be solved on different meshes, see [13].

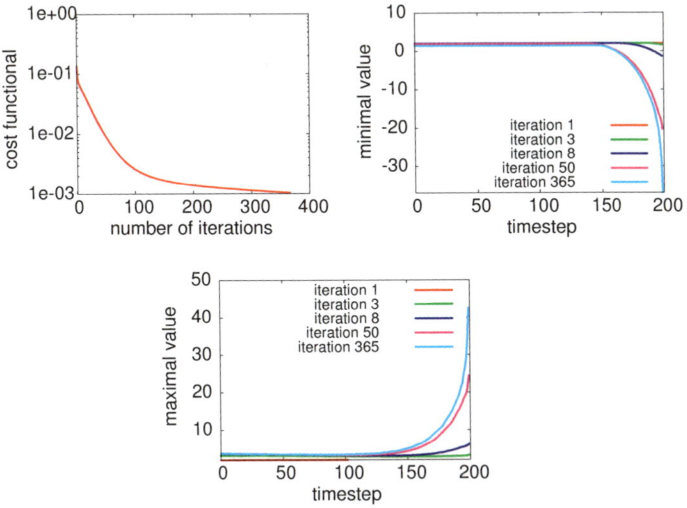

FIGURE 5. Cost functional and minimal and maximal values for the control over number of iterations and time.

Acknowledgment

This work was supported by the SPP 1253 "Optimization with Partial Differential Equations" of the German Science Foundation (DFG) through the grant VO 899/5-1.

References

[1] P. Kuhn, J. Krug, F. Haußer, A. Voigt, *Complex shape evolution of electromigration-driven single-layer islands* Phys. Rev. Lett. **94** (2005), 166105.

[2] F. Tröltzsch *Optimale Steuerung partieller Differentialgleichungen*, Vieweg, Wiesbaden, 2005.

[3] B. Li, J. Lowengrub, A. Rätz, A. Voigt, *Geometric Evolution Laws for Thin Crystalline Films: Modeling and Numerics* Commun. Comput. Phys. **6** (2009), 433–482.

[4] H. Mehl, O. Biham, O. Millo, M. Karimi, *Electromigration-induced flow of islands and voids on the Cu(001) surface* Phys. Rev. B **61** (2000), 4975–4982.

[5] O. Pierre-Louis, T.L. Einstein, *Electromigration of single-layer clusters* Phys. Rev. B **62** (2000), 13697–13706.

[6] A. Rätz, A. Ribalta, A. Voigt, *Surface evolution of elastically stressed films under deposition by a diffuse interface model* J. Comput. Phys. **214** (2006), 187–208.

[7] M. Burger, C. Stöcker, A. Voigt, *Finite Element-Based Level Set Methods for Higher Order Flows* J. Sci. Comput. **35** (2008), 77–98.

[8] A. Voigt, S. Vey, *AMDiS – adaptive multidimensional simulations* Comput. Visual. Sci. **10** (2007), 57–67.

[9] F. Haußer, S. Rasche, A. Voigt, *The influence of electric fields on nanostructures – Simulation and control* Math. Com. Sim. **80** (2010), 1449–1457.

[10] E. Fried, M.E. Gurtin, *A unified treatment of evolving interfaces accounting for small deformations and atomic transport with emphasis on grain-boundaries and epitaxy* Adv. Appl. Mech. **40** (2004), 1–177.

[11] L.C. Evans, H.M. Soner, P.E. Souganidis, *Phase transitions and generalized motion by mean curvature* Comm. Pure Appl. Math. **45** (1992), 1097–1123.

[12] S. Rasche, A. Voigt, *Optimal control of geometric evolution laws – a phase field approach*, submitted.

[13] A. Voigt, T. Witkowski, *A multi-mesh finite element method for Lagrange elements of arbitrary degree*, submitted.

F. Haußer
Beuth Hochschule Berlin
University of Applied Science
Fachbereich II Mathematik – Physik – Chemie
D-13353 Berlin, Germany
e-mail: hausser@beuth-hochschule.de

Sandra Janssen and Axel Voigt
TU Dresden
Fachrichtung Mathemtik
D-01062 Dresden, Germany
e-mail: sandra.janssen@tu-dresden.de
axel.voigt@tu-dresden.de

International Series of Numerical Mathematics, Vol. 160, 573–596

Optimization of Simulated Moving Bed Processes

Achim Küpper and Sebastian Engell

Abstract. In this contribution, the optimization of periodic chromatographic simulated moving bed SMB processes is discussed. The rigorous optimization is based on a nonlinear pde model which incorporates rigorous models of the chromatographic columns and the discrete shifts of the inlet and outlet ports. The potential of the optimization is demonstrated for a separation problem with nonlinear isotherm of the Langmuir type for an SMB process and the ModiCon process. Here, an efficient numerical approach based on multiple shooting is employed. An overview of established optimization approaches for SMB processes is given.

Mathematics Subject Classification (2000). 90C30.

Keywords. Simulated Moving Bed, ModiCon, multiple shooting, non-linear optimization.

1. Introduction

Chromatographic separation is based on the different adsorption affinities of the molecules in the liquid to an adsorbent which is packed in a chromatographic solid bed. Most reported applications of chromatographic separation technology are performed in batch mode. In the batch mode, a pulse of the diluted mixture is injected into the column followed by a stream of pure solvent which desorbs the components from the bed. The components are separated from each other due to their different affinities to the bed while passing through the column, see Figure 1. The purified components are collected during different time intervals at the outlet of the column. Batch chromatography is a simple and flexible separation technology. However, it causes high separation costs due to a high dilution of the products

SPP1253 The financial support of the Deutsche Forschungsgemeinschaft (DFG, German Research Council) within the priority program SPP1253 under grant En152-35/2 and in the context of the research cluster "Optimization-based control of chemical processes" is very gratefully acknowledged.

since most of the time pure solvent is fed and because the adsorption capacity of the solid bed is not fully used. A further disadvantage is the discontinuous mode of operation. The Simulated Moving Bed (SMB) process is an efficient continuous

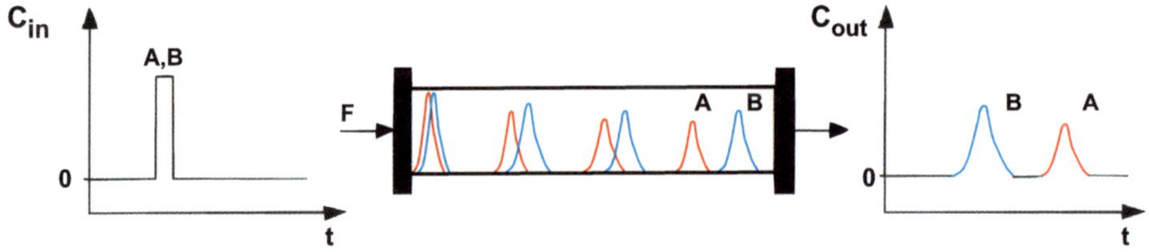

FIGURE 1. Principle of batch chromatography.

chromatographic separation technology that is increasingly applied in the food, fine chemicals, and pharmaceutical industries. Industrial applications have been reported especially for the separation of temperature sensitive components or for the separation of species with similar thermodynamic properties (Juza et al., 2000; Nicoud, 1998). The SMB process was invented (Broughton and Gerhold, 1961) for the separation of aromatic compounds. An effective counter-current movement between the solid bed and the liquid is generated by switching the inlet and the outlet ports in a circle of chromatographic columns periodically by one column in the direction of the liquid flow, as illustrated by Figure 2. Thereby, the columns are moved relative to the ports establishing an approximate counter-current flow of the liquid phase and the solid phase. The movement of the more strongly adsorbed component (depicted by the black circles in Figure 2) is in the same direction as the solid flow, and it is withdrawn at the extract port, while the less strongly adsorbed component (represented by the white circles) moves in the direction of the liquid phase and is withdrawn at the raffinate port. For given equipment (number and dimensions of the columns, number of columns in each section, choice of the adsorbent and of the solvent), the cost of a separation by a SMB process is determined by the throughput and by the solvent consumption. Thus, the goal of process operation is to minimize this cost (in the simplest case maximizing the throughput or minimizing the solvent consumption) while respecting the desired product purities. This contribution deals with the rigorous optimization of the SMB process and the ModiCon variant. It is organized as follows: First the Simulated Moving Bed process and the ModiCon process are presented and the rigorous process model is stated in the following section. In Section 3, the formulation of the optimization problem with an overview of established optimization approaches is given. Section 4 presents simulation results for an enantiomer separation system. Section 5 provides a summary of the work and gives an outlook to open research issues.

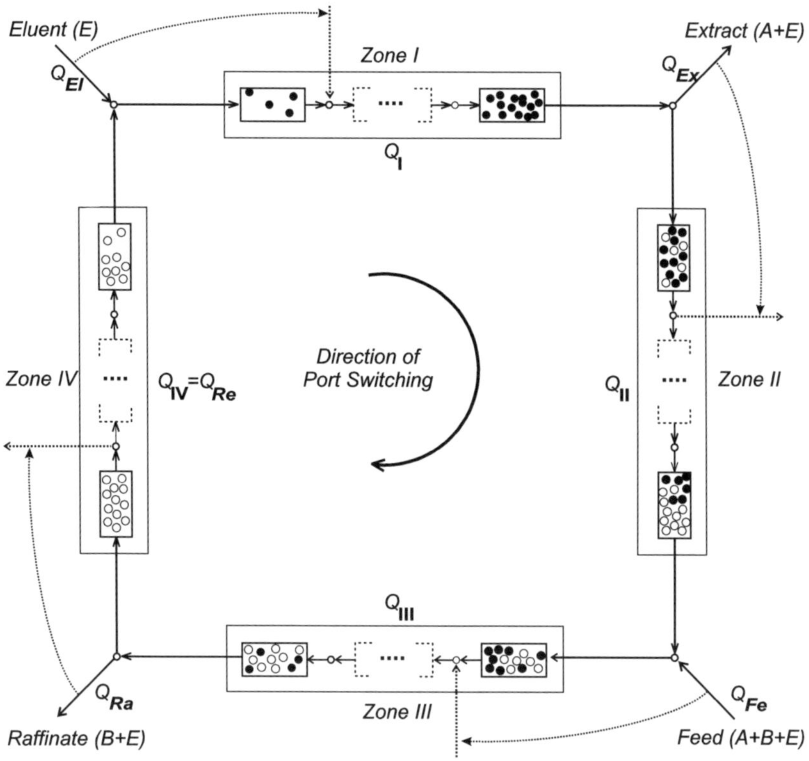

FIGURE 2. Principle of the Simulated Moving Bed process.

2. Simulated Moving Bed

In SMB processes, a counter-current movement of the adsorbent is implemented by switching the ports in the direction of the liquid flow periodically, as illustrated by Figure 2. After a start-up phase, a periodic steady state (PSS) is reached where the outlet concentration profiles change dynamically within a period, but are identical from period to period, as illustrated by the axial concentration profiles and the measurements (extract, raffinate, and internal measurement) in Figures 3 and 4. The different zones of the SMB process are classified according to their function and their position relative to the ports:

- Zone I: regeneration of the solid bed by desorption of the more strongly retained component (A)
- Zone II: desorption of the less retained component (B), purification of the extract product
- Zone III: adsorption of the more strongly retained component (A) purification of the raffinate product

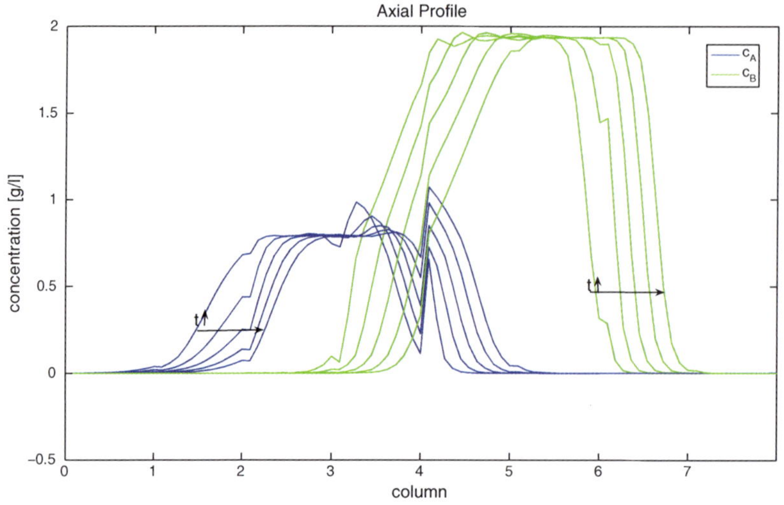

FIGURE 3. Axial concentration profiles of the SMB process during one period.

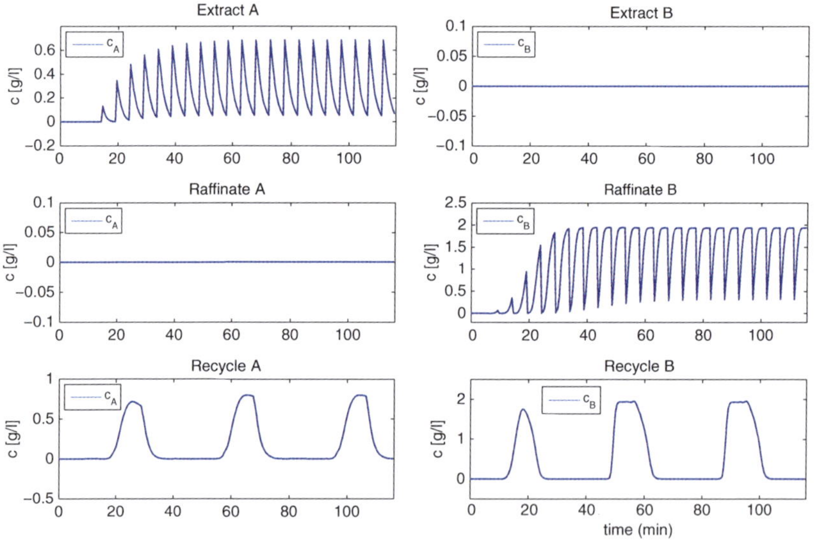

FIGURE 4. Concentrations at the extract port, at the raffinate port, and in the recycle stream during start-up.

- Zone IV: regeneration of the solvent before it is recycled to zone I by adsorbing the less retained component (B).

In a regular SMB operation, it is required that no propagation of components across zone I and zone IV (via the solid flow and the liquid flow) takes place, which is called recycle breakthrough. Established design approaches, e.g., the triangle theory (Mazzotti et al., 1997), provide conditions for the flow rates and the switching period such that such a regular operation is established and pure products are obtained. SMB processes with moderate purity requirements can also be operated in the presence of a recycle breakthrough and may then have a better economic performance than obtained for regular SMB operation.

2.1. ModiCon

The ModiCon process is a variant of the SMB process in which the feed concentration is varied over the switching period. Variable feed concentrations can be achieved by mixing a highly concentrated feed stock with solvent. The ModiCon process enables to issue most of the feed during a small subperiod of the switching time which moves the concentration fronts at the extract and at the raffinate port in the direction of the liquid flow or of the solid flow. This additional degree of freedom leads to an improved economical performance compared to SMB operation, as demonstrated in this paper and reported also, e.g., in (Schramm et al., 2003a,b).

2.2. Other SMB variants

In the original SMB process, all ports are switched synchronously and at a fixed rate, and the external flow rates of the process as well as the concentration of the feed are kept constant. In recent years, further variants of SMB processes have been proposed in which even more flexibility is provided. By switching the ports at different times within one cycle, the VariCol process (Ludemann-Hombourger et al., 2000; Adam et al., 2002; Toumi et al., 2002, 2003) results. Varying the flow rates during the periods leads to the power feed process (Kearney and Hieb, 1992; Kloppenburg and Gilles, 1999; Zhang et al., 2003; Kawajiri and Biegler, 2006a,b), and by varying the zone lengths over a full cycle of operation, the Full-Cycle with Variable Lengths FCVL process (Kawajiri and Biegler, 2006a) is obtained. These variants can be combined to maximize the productivity (Toumi et al., 2007). This paper focuses on the optimization of the ModiCon process as a complex SMB optimization problem as this variant can be easily realized at a real plant.

2.3. Modeling of Simulated Moving Bed processes

Rigorous dynamic models of Simulated Moving Bed processes consist of a description of the port movements that establish the counter-current movement of the solid phase, of mass balances at the inlet and at the outlet ports, and of models of the chromatographic columns in the form of partial differential equations. The concentrations of the components in the liquid phase are coupled with the concentrations in the solid bed via the adsorption isotherm.

2.3.1. Chromatographic column model. A radially homogenous chromatographic column can be modelled very accurately by the *General Rate Model* that accounts for the following effects in the columns:

- convection
- axial dispersion
- pore diffusion
- mass transfer between the liquid phase and the solid phase
- multi-component adsorption.

It is assumed that the particles of the solid phase are uniform, spherical, porous (with a constant particle porosity ϵ_{p}), and that the mass transfer between the particle and the surrounding layer of the bulk is in a local equilibrium. The concentration of component i is denoted by c_i in the liquid phase and by q_i in the solid phase. D_{ax} is the axial dispersion coefficient, u the interstitial velocity, ϵ_{b} the column void fraction, $k_{\mathrm{l},i}$ the film mass transfer resistance, and D_{p} the diffusion coefficient within the particle pores. u_z is the interstitial velocity in zone z of the SMB process (the flow rates differ between the zones). The concentrations within the pores are denoted by $c_{\mathrm{p},i}$. The following partial differential equations of a column can be derived from a mass balance around an infinitely small cross sectional area of the column assuming a constant radial distribution of the interstitial velocity u_z and the concentrations c_i:

$$\frac{\partial c_i}{\partial t} + \frac{(1-\epsilon_{\mathrm{b}})3k_{\mathrm{l},i}}{\epsilon_{\mathrm{b}} r_{\mathrm{p}}}(c_i - c_{\mathrm{p},i}|_{r=r_{\mathrm{p}}}) = D_{\mathrm{ax}}\frac{\partial^2 c_i}{\partial x^2} - u_z\frac{\partial c_i}{\partial x} \tag{1}$$

$$(1-\epsilon_{\mathrm{p}})\frac{\partial q_i}{\partial t} + \epsilon_{\mathrm{p}}\frac{\partial c_{\mathrm{p},i}}{\partial t} - \epsilon_{\mathrm{p}} D_{\mathrm{p}}\left[\frac{1}{r^2}\frac{\partial}{\partial r}\left(r^2\frac{\partial c_{\mathrm{p},i}}{\partial r}\right)\right] = 0, \tag{2}$$

with initial and boundary conditions

$$c_i(t=0,x) = c_{i,0}(x), \quad c_{\mathrm{p},i}(t=0,x,r) = c_{\mathrm{p},i,0}(x,r), \tag{3}$$

$$\left.\frac{\partial c_i}{\partial x}\right|_{x=0} = \frac{u_z}{D_{\mathrm{ax}}}(c_i(0,t) - c_{\mathrm{in},i}(t)), \quad \left.\frac{\partial c_i}{\partial x}\right|_{x=L} = 0, \tag{4}$$

$$\left.\frac{\partial c_{\mathrm{p},i}}{\partial r}\right|_{r=0} = 0, \quad \left.\frac{\partial c_{\mathrm{p},i}}{\partial r}\right|_{r=r_{\mathrm{p}}} = \frac{k_{\mathrm{l},i}}{\epsilon_{\mathrm{p}} D_{\mathrm{p},i}}(c_i - c_{\mathrm{p},i}|_{r=r_{\mathrm{p}}}). \tag{5}$$

It is assumed that the concentrations q_i in the particles are in thermodynamic equilibrium with the liquid concentrations in the pores of the particles and that the equilibrium can be described by an adsorption isotherm. In the application considered here, the adsorption isotherm is of extended Langmuir type

$$q_i = H_i^1 c_{\mathrm{p},i} + \frac{H_i^2 c_{\mathrm{p},i}}{1 + k_{\mathrm{A}} c_{\mathrm{p,A}} + k_{\mathrm{B}} c_{\mathrm{p,B}}} \qquad i = \mathrm{A}, \mathrm{B}, \tag{6}$$

where H_i is the Henry coefficient and k_A, k_B are the saturation coefficients. The denominator terms describe the displacement effect (Schmidt-Traub, 2005) where one component desorbs the other while being adsorbed. The displacement effect is the main reason for the coupling of the partial differential equations. The interstitial velocity u_z in the zones depends on the zone flow rate Q_z, the bulk void fraction ϵ_{b}, and the column diameter D_{col} and is calculated according to

$$u_z = \frac{4Q_z}{\pi D_{\mathrm{col}}^2 \epsilon_{\mathrm{b}}}, \quad z = \mathrm{I}, \mathrm{II}, \mathrm{III}, \mathrm{IV}. \tag{7}$$

The axial dispersion coefficient can be calculated according to the empirical correlation between the Peclet number and the particle Reynolds number given by (Chung and Wen, 1968):

$$D_{\mathrm{ax}} = \frac{2u_z L_{\mathrm{col}}}{Pe}, \tag{8}$$

with

$$Pe = \frac{0.2}{\epsilon_{\mathrm{b}}} + \frac{0.011}{\epsilon_{\mathrm{b}}} \left[Re_{\mathrm{p}}\epsilon_{\mathrm{b}}\right]^{0.48}, \quad Re_{\mathrm{p}} = \frac{2u_z r_{\mathrm{p}} \rho}{\nu}. \tag{9}$$

It should be mentioned that the particle diffusion parameter can only be estimated roughly, but it has a minor influence on the prediction of the model. The *Linear Driving Force Model* (Ruthven, 1984) that assumes a uniform void fraction ϵ (summarizing bulk void fraction and particle porosity) and a uniform solid phase concentration is an alternative. However, the *General Rate Model* describes the process behavior more accurately and can be solved efficiently as described below. For further details regarding the numerical solution of the column model, the reader is referred to (Dünnebier, 2000).

2.3.2. Port mass balances. From mass balances of the components around the inlet and the outlet ports, the internal flow rates and the inlet concentrations can be calculated according to:

$$\text{Eluent node:} \quad Q_{\mathrm{IV}} + Q_{\mathrm{El}} = Q_{\mathrm{I}} \tag{10}$$

$$c_{i,\mathrm{out,IV}} Q_{\mathrm{IV}} = c_{i,\mathrm{in,I}} Q_{\mathrm{I}}, \quad i = \mathrm{A}, \mathrm{B} \tag{11}$$

$$\text{Extract node:} \quad Q_{\mathrm{I}} - Q_{\mathrm{Ex}} = Q_{\mathrm{II}} \tag{12}$$

$$c_{i,\mathrm{out,I}} = c_{i,\mathrm{in,II}}, \quad i = \mathrm{A}, \mathrm{B} \tag{13}$$

$$\text{Feed node:} \quad Q_{\mathrm{II}} + Q_{\mathrm{Fe}} = Q_{\mathrm{III}} \tag{14}$$

$$c_{i,\mathrm{out,II}} Q_{\mathrm{II}} + C_{i,\mathrm{Fe}} Q_{\mathrm{Fe}} = c_{i,\mathrm{in,III}} Q_{\mathrm{III}}, \quad i = \mathrm{A}, \mathrm{B} \tag{15}$$

$$\text{Raffinate node:} \quad Q_{\mathrm{Ra}} + Q_{\mathrm{IV}} = Q_{\mathrm{III}}, \tag{16}$$

$$c_{i,\mathrm{out,III}} = c_{i,\mathrm{in,IV}}, \quad i = \mathrm{A}, \mathrm{B} \tag{17}$$

where $Q_{\mathrm{I-IV}}$ are the flow rates in the corresponding zones, Q_{El}, Q_{Ex}, Q_{Fe}, and Q_{Ra} denote the external flow rates and $c_{i,\mathrm{in}}$ and $c_{i,\mathrm{out}}$ denote the concentrations of the component i in the streams leaving and entering the zones.

2.3.3. Modeling of the counter-current flow. The modeling of the counter-current flow is achieved by periodic shifts of the concentration profiles by one column in opposite direction to the liquid flow after simulating the columns over the switching period:

$$\mathbf{x}_{\text{smb}}^{m+1} = \Phi(\mathbf{x}_{\text{smb}}^{m}) : \begin{cases} \hat{\mathbf{x}}_{\text{smb}}^{m} & = & \mathbf{x}_{\text{smb}}^{m} + \int\limits_{t=0}^{\mathcal{T}} \mathbf{f}(\mathbf{x}(t), \mathbf{u}, \mathbf{p})\, dt, \\ \mathbf{x}_{\text{smb}}^{m+1} & = & \mathbf{P}\hat{\mathbf{x}}_{\text{smb}}^{m} \end{cases} \tag{18}$$

with differential states $\mathbf{x}(t) \in \mathbb{R}^{n_x}$ and parameters $\mathbf{p} \in \mathbb{R}^{n_p}$. Φ is the simulation operator at period m comprising the simulation of the concentrations over the period and the switching of the states at the end of the period by the permutation matrix $\mathbf{P}$ that accounts for the simulated movement of the solid bed. From the operation Φ, the state $\mathbf{x}_{\text{smb}}^{m+1}$ at the beginning of the following period $m+1$ is obtained. The state at the end of period m is referred to as $\hat{\mathbf{x}}_{\text{smb}}^{m}$. The permutation matrix $\mathbf{P}$ switches all states by one column for SMB operation, see the Appendix. $\mathbf{u} \in \mathbb{R}^{n_u}$ is the vector of the inputs to the process including the flow rates at the ports and the recycle flow rate. $\mathbf{f} \in \mathbb{R}^{n_x}$ represents the right-hand side of the set of ordinary differential equations that is obtained after discretising the partial differential equations of the column models.

As can be seen from (1), the SMB model is nonlinear in the flow rates which are degrees of freedom in the optimization and control of SMB processes. In the absence of a chemical reaction, the nature of the SMB model with regard to the states depends on the nature of the adsorption isotherm. For linear isotherms, the model is linear in the states. If the adsorption isotherm is not of multi-component type (the adsorption of one component does not influence the adsorption of the other), the partial differential equations are decoupled. The application considered in this paper is described by a nonlinear coupled adsorption isotherm. This results in partial differential equations that are coupled and nonlinear.

The resulting system of partial differential equations can be efficiently solved by the numerical approach proposed in (Gu, 1995) where a Galerkin finite element discretization of the bulk phase is combined with an orthogonal collocation of the solid phase. This numerical method was first applied to SMB processes in (Dünnebier and Klatt, 2000). The bulk phase is divided into n_{fe} finite elements and the solid phase is discretized by n_{c} internal collocation points. As a result, the set of initial values, boundary values, and partial differential equations (PDE) is transformed into a set of initial values and a system of ordinary differential equations (ODE)

$$\dot{\mathbf{x}} = f(\mathbf{x}, \mathbf{u}, \mathbf{p}), \tag{19}$$

where the flows Q are summarized in the input vector $\mathbf{u}(t) \in \mathbb{R}^{n_u}$. The system output is defined as

$$y = h(\mathbf{x}(\mathbf{u}, \mathbf{p})), \tag{20}$$

with $y \in \mathbb{R}^{n_y}$. For $n_{\text{fe}} = 8$, $n_{\text{c}} = 1$, number of components $n_{\text{sp}} = 2$, and number of columns $n_{\text{col}} = 8$ a typical system order of the SMB process of

$$n_x = n_{\text{col}} * n_{\text{sp}} * (n_{\text{c}} + 1) * (2 * n_{\text{fe}} + 1) = 544 \tag{21}$$

results. The ODE-system is stiff due to large differences in the time scales of the interacting dynamics. In contrast to the *General Rate Model*, the *Ideal Model* that neglects kinetic effects and diffusion effects can be solved analytically via wavefront analysis. However for nonlinear adsorption isotherms, it is not sufficiently precise for optimization or control purposes although it can be used as initialization for the *General Rate Mode* (Dünnebier, 2000).

3. Formulation of the optimization problem

The goal of the optimization is to achieve an optimal economical operation. The objective of the optimization is to minimize the solvent consumption or to maximize the feed throughput, or a combination of both aims (22) while satisfying the product specifications which are stated here as purity constraints (24)–(25). The optimization is performed subject to the process dynamics (18) which is required to be at periodic steady state (23). The degrees of freedom are given by the external flow rates $Q_{\text{El}}, Q_{\text{Fe}}, Q_{\text{Ra}}$, the internal flow rate Q_{Re}, the switching period τ and the initial state $\mathbf{x}_{\text{smb}}^{\text{pss}}$ at periodic steady state. The external flow rate at the extract port results from the total mass balance around the SMB process, see Figure 2. Pump limitations are considered by a constraint on the maximum flow rate which occurs in zone I. The mathematical formulation of the optimization problem can be stated as follows:

$$\min_{Q_{\text{El}},Q_{\text{Fe}},Q_{\text{Ra}},Q_{\text{Re}},\tau,\mathbf{x}_{\text{smb}}^{\text{pss}}} \quad J(Q_{\text{El}}, Q_{\text{Fe}}) \tag{22}$$

$$\text{s.t.} \quad \Phi(\mathbf{x}_{\text{smb}}^{\text{pss}}) - \mathbf{x}_{\text{smb}}^{\text{pss}} = 0 \tag{23}$$

$$\text{Pur}_{\text{Ex}} - \text{Pur}_{\text{Ex,min}} \geq 0 \tag{24}$$

$$\text{Pur}_{\text{Ra}} - \text{Pur}_{\text{Ra,min}} \geq 0 \tag{25}$$

$$Q_{\max} - Q_{\text{I}} \geq 0. \tag{26}$$

The purities Pur_{Ex}, Pur_{Ra} at the extract and the raffinate outlets are defined as follows:

$$\text{Pur}_{\text{Ex}} = \frac{\mathbf{m}_{\text{Ex,A}}}{\mathbf{m}_{\text{Ex,A}} + \mathbf{m}_{\text{Ex,B}}} \tag{27}$$

$$\text{Pur}_{\text{Ra}} = \frac{\mathbf{m}_{\text{Ra,B}}}{\mathbf{m}_{\text{Ra,A}} + \mathbf{m}_{\text{Ra,B}}} \tag{28}$$

where $\mathbf{m}_{j,i}$ are the integrated mass flows over the switching time τ of port j and component i.

$$\mathbf{m}_{j,i} = \int_0^\tau Q_j c_{j,i} dt, \quad j = \text{Ex}, \text{Ra}, \quad i = \text{A}, \text{B},. \tag{29}$$

The main difficulty of the solution of the optimization problem is the computation of the periodic steady state. Three approaches to calculate the periodic steady state have been proposed in the literature.

The first approach is the sequential approach (Dünnebier, 2000) where the simulation is not performed simultaneously with the optimization. The optimization is performed on an upper layer that calls a simulator on a lower layer which simulates the process forward until the periodic steady state is reached with a tolerance ϵ_{pss}

$$\|\mathbf{x}_{\text{smb}}^{\text{pss}} - \Phi(\mathbf{x}_{\text{smb}}^{\text{pss}})\|^2 \leq \epsilon_{\text{pss}}, \tag{30}$$

replacing the constraint (23). $\| \ \|^2$ denotes the Euclidean norm. This approach is also referred to as the Picard approach and is very robust and reliable for inherently stable processes. The SMB process converges to the periodic steady state with a settling time of around 20 switching period initializing the process as a plant filled by pure solvent. For the sequential approach, the initial state $\mathbf{x}_{\text{smb}}^{\text{pss}}$ is not an explicit degree of freedom since it results from simulating the process at the chosen flow rates and the chosen switching time until the periodic steady state is reached. The lower level simulation can be performed using a standard ODE solver, e.g., DVODE (Hindmarsh, 1983), while the optimization problem can be solved by any nonlinear constrained optimization technique, e.g., SQP solvers.

The second approach that is applied in this contribution performs the simulation and optimization simultaneously over one switching period. The initial state $\mathbf{x}_{\text{smb}}^{\text{pss}}$ is explicitly stated as degree of freedom of the optimization problem. The simultaneous approach can be further divided into two types: The first type of the simultaneous approach is the full discretization approach where the dynamic equations are discretized in time transforming the PDE dynamic optimization problem into a large-scale optimization problem with algebraic constraints. In (Kloppenburg and Gilles, 1999), the full discretization was applied to the SMB process using the LANCELOT solver. Poor convergence and insufficient accuracy are reported but the outlook is given that the method might be suitable when better optimization methods are available. Recently, (Kawajiri and Biegler, 2006b) applied an interior point optimization method combined with Radau collocation on finite elements for the temporal discretization to solve the SMB optimization problem. Sufficient accuracy and CPU times of around 10 min are reported. The second type of the simultaneous approach is the multiple shooting method which is implemented, e.g., in the software MUSCOD-II (Leineweber, 1999). Multiple shooting divides the switching period into several subperiods, solves initial value problems on the respective intervals, and enforces the continuity of the solutions upon convergence.

The third approach for solving periodic optimization problems for the SMB process was introduced in (Toumi, 2005; Toumi et al., 2007). The idea is to formulate the optimization problem such that the periodic steady state is directly implemented as part of the simulation of the columns. The columns are simulated one after another in an order opposite to the direction of the liquid flow. Thereby, the axial profile of a column at the end of the simulation period can be taken as the initial profile for the column which lies one column in front in the SMB cycle. The individual columns are successively simulated in this fashion. The periodic steady state is automatically fulfilled except between the first and the last column in the cycle for which it must be imposed as an equality constraint. In this approach, the physical columns are treated decoupled from each other reducing the computational burden of the periodic optimization problem. However, the inlet concentrations of the column are not known and must be assumed as the column one column in front in the cycle is simulated after the considered column. The inlet of the column has to match the respective column outlet. Both are approximated by Legendre polynomials and their equality is imposed by further constraints in the optimization problem. In (Toumi et al., 2007), a faster convergence rate is reported than for the sequential approach and the simultaneous approach. However, this approach was applied for a constant switching period only and is not applied in this paper for this reason.

In the following, the simultaneous approach is applied based on multiple shooting. The periodic optimization problem of the SMB process (22)-(26) is solved by Sequential Quadratic Programming (SQP), see, e.g., (Nocedal and Wright, 1999). The SQP method requires the calculation of the Jacobian and the Hessian matrices. The computational burden grows considerably for the simultaneous approach since the initial state is considered explicitly in the optimization problem leading to large Jacobian and Hessian matrices. The Newton-Picard Inexact SQP Method method uses inexact approximations of the Jacobian of the constraints. For the approximation of the constraint Jacobian, the periodicity operator $\Phi(\mathbf{x}_{\text{smb}}^{\text{pss}})$ is only differentiated with respect to the dominant components of $\mathbf{x}_{\text{smb}}^{\text{pss}}$ dividing the search region of $\mathbf{x}_{\text{smb}}^{\text{pss}}$ into two parts. It is sufficient to update the derivatives only with respect to the slow modes of the constraint Jacobian. More details are given in (Potschka et al., 2008, 2010) where results that show an improved convergence behavior and reduced computational requirements are presented. The approach is still subject to research. In the near future we expect it to be superior.

The natural degrees of freedom of an SMB process are the flow rates $Q_{\text{El}}(t)$, $Q_{\text{Fe}}(t)$, $Q_{\text{Ra}}(t), Q_{\text{Re}}(t)$ (the extract flow results from the global mass balance) and the switching period τ. The numerical tractability of the optimization problem is improved by transforming the natural degrees of freedom of the process into the so-called β-factors via a nonlinear transformation that was first introduced by Hashimoto (Hashimoto et al., 1983). These factors are the ratios, also referred to as m-factors, between the flow rates Q_z in each zone z and the effective solid flow rate Q_{s}, scaled by the Henry coefficients of the adsorption isotherm:

$$\beta_{\text{I}} = \frac{1}{H_{\text{A}}}\left(\frac{Q_{\text{I}}}{Q_{\text{s}}} - \frac{1-\epsilon_{\text{b}}}{\epsilon_{\text{b}}}\right) = \frac{m_1}{H_{\text{A}}} \tag{31}$$

$$\beta_{\text{II}} = \frac{1}{H_{\text{B}}}\left(\frac{Q_{\text{II}}}{Q_{\text{s}}} - \frac{1-\epsilon_{\text{b}}}{\epsilon_{\text{b}}}\right) = \frac{m_2}{H_{\text{B}}} \tag{32}$$

$$\frac{1}{\beta_{\mathrm{III}}} = \frac{1}{H_{\mathrm{A}}}\left(\frac{Q_{\mathrm{III}}}{Q_{\mathrm{s}}} - \frac{1-\epsilon_{\mathrm{b}}}{\epsilon_{\mathrm{b}}}\right) = \frac{m_3}{H_{\mathrm{A}}} \tag{33}$$

$$\frac{1}{\beta_{\mathrm{IV}}} = \frac{1}{H_{\mathrm{B}}}\left(\frac{Q_{\mathrm{IV}}}{Q_{\mathrm{s}}} - \frac{1-\epsilon_{\mathrm{b}}}{\epsilon_{\mathrm{b}}}\right) = \frac{m_4}{H_{\mathrm{B}}} \tag{34}$$

$$Q_{\mathrm{s}} = \frac{(1-\epsilon_{\mathrm{b}})V_{\mathrm{col}}}{\tau}\,. \tag{35}$$

The m-factors are employed in the triangle theory (Mazzotti et al., 1997), a short-cut method for the design of the operating point of SMB processes. In contrast to the original definition of Hashimoto, the beta factors β_{III} and β_{IV} are inverted (Klatt et al., 2000, 2002; Wang et al., 2003), see (33) and (34). The reason for introducing the β-factors is that the front positions (desorption front A in zone I, desorption front B in zone II, adsorption front A in zone III, adsorption front B in zone IV) are then mainly influenced by the corresponding beta factor of the respective zone with rather small sensitivity to the other beta factors (Hanisch, 2002).

4. Example: optimization of a ModiCon process

In this section, a ModiCon process is optimized in order to demonstrate the potential of mathematical optimization based upon rigorous mathematical pde models. The performance of the ModiCon process is evaluated and benchmarked with the SMB process by simulation and optimization studies for the separation of the enantiomers EMD-53986. Enantiomers are chemical molecules that are mirror images of each other, much as one's left and right hands. EMD-53986 is a chemical abbreviation for the enantiomer (5-(1,2,3,4-tetrahydroqinoline-6-yl)-6-methyl-3,6-dihydro-1,3,4-thiadiazine-2-on). It is a precursor for a calcium sensitising agent. The R-enantiomer, here denoted as component B, has the desired pharmaceutical activity in contrast to the S-enantiomer, denoted as component A, which has no positive effect and is removed from the mixture for this reason. The separation of the enantiomers of EMD-53986 was studied experimentally in a joint project by *Merck* (Germany) and Universität Dortmund in 2001. In this work, an accurate simulation model was developed. The parameters of the SMB model are taken from (Jupke, 2004) and are listed in the Appendix. The adsorption of the EMD-53986 components is characterized by a strong displacement effect that is modelled by an extended Langmuir adsorption isotherm of the form (6).

4.1. Simulation study

Before the optimization of the SMB and the ModiCon processes is discussed, the influence of the feed modulation is demonstrated by a simulation study. The EMD-53986 separation system is considered for an operating point of the SMB process that yields purities of [99% 99%] for the extract and raffinate products. A plant configuration with $n_{\mathrm{col}} = 8$ columns and a column distribution of 2/2/2/2 (two columns per zone) are considered. Each chromatographic column is discretized by

8 finite elements with one inner collocation point in the particles resulting in a process description with 544 differential variables. The flow rates, the switching time, and the average feed concentration of the ModiCon process are chosen identical to the optimal operating parameters of the SMB mode in Table 6.

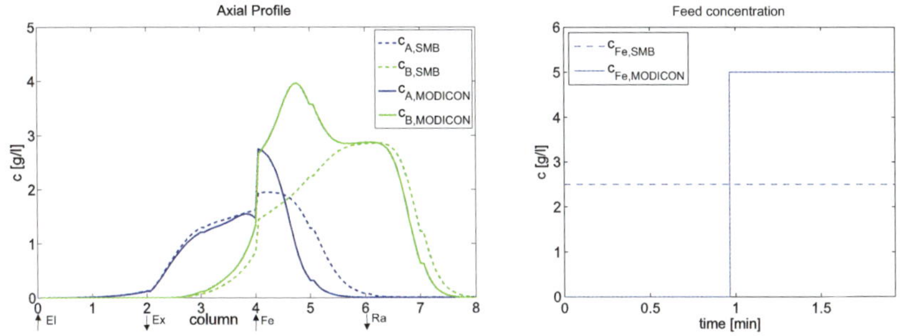

(a) Axial profiles at the end of the period (b) ModiCon feed concentration

FIGURE 5. Case A: Doubled feed concentration in the second half of the period, zero feed concentration in the first half.

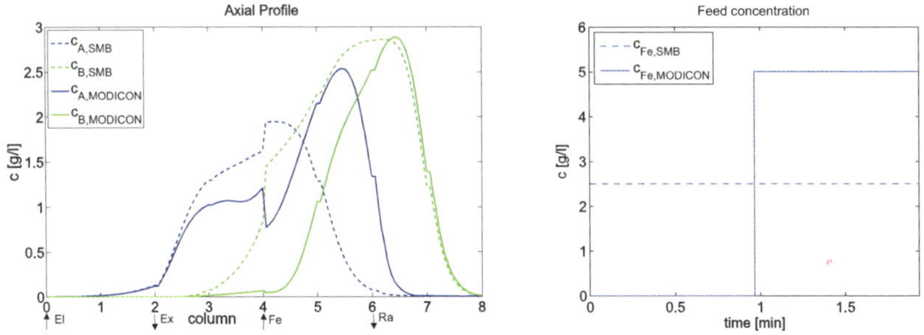

(a) Axial profiles at the end of the period (b) ModiCon feed concentration

FIGURE 6. Case B: Doubled feed concentration in the first half of the period, zero feed concentration in the second half.

For the ModiCon simulation, the feed concentration is set to zero in the first half of the period and doubled in the second half of the period for case A, while it is doubled in the first half and set to zero in the second half for case B, as can be seen in figures 5 and 6. For this separation system and this operating point, the ModiCon operation that issues the concentrated feed in the second half of the period (case A) moves both concentration fronts of the contaminating components (the desorption front of component B in zone II and the adsorption front of component A in zone III) in the direction of the solid flow. The raffinate purity is increased while the extract purity is only slightly reduced compared to the SMB operation, see Table 1, because the adsorption front of component A is moved further than the desorption front of component B. Case B where the concentrated feed is issued in the first half of the period leads to an analogous movement of the contaminating fronts in the liquid flow direction. An improvement of the purity at the extract port and a drastic deterioration of the raffinate purity are obtained because the adsorption front of component A moves further in the direction of the liquid than the desorption front of component B. The much lower purity of the raffinate in case B indicates that issuing the concentrated feed in the first half is not advantageous for this separation task while the only slightly deteriorated extract purity in case A indicates that issuing the concentrated feed in the second half might be a good feeding strategy.

	SMB	ModiCon CASE A	ModiCon CASE B
$\mathrm{Pur}_{\mathrm{Ex}}$ [%]	99.00	98.35	99.66
$\mathrm{Pur}_{\mathrm{Ra}}$ [%]	99.00	99.78	88.29

TABLE 1. Purities of the SMB process and cases A and B of the ModiCon process.

4.2. Optimization study

For the optimization of the MociCon process the mathematical formulation of the optimization problem (22)–(26) is modified as follows:

$$\min_{\beta_{\mathrm{I}},\beta_{\mathrm{II}},\beta_{\mathrm{III}},\beta_{\mathrm{IV}},\mathbf{x}^{\mathrm{pss}}_{\mathrm{smb}},c_{\mathrm{Fe}}(\cdot)} \quad Q_{\mathrm{El}} \tag{36}$$

$$\text{s.t.} \quad \int_0^{\tau} c_{\mathrm{Fe}}(t) Q_{\mathrm{Fe}} dt = c_{\mathrm{Fe,SMB}} Q_{\mathrm{Fe}} \tau \tag{37}$$

$$(23)\text{–}(26).$$

The degrees of freedom are the beta-factors, the initial state $\mathbf{x}^{\mathrm{pss}}_{\mathrm{smb}}$, and the feed concentrations in the subperiods. The average feed concentration of the ModiCon process is required to be the same as the SMB feed concentration (37). The objective is to minimize the eluent consumption (36) for a constant feed flow rate of 30 ml/min. The period is divided into 10 subintervals with individual feed concentrations. The multiple shooting optimization approach is applied. The accuracy of the NLP solution and the integration tolerance in MUSCOD-II are chosen as 10^{-5} and 10^{-7}, respectively. The ModiCon optimization runs were initialized at the corresponding SMB operating points. Conducting further optimization runs with different initializations increased the likelihood of finding a global optimum. In total, 100 optimization runs were performed.

4.2.1. ModiCon optimization results – EMD-53986. For the EMD-53986 system, the ModiCon operation achieves a reduction of the solvent consumption Q_{El} by more than 25% for the case of a maximum feed concentration of two times the average concentration and by more than 43% for the case of a ten times higher

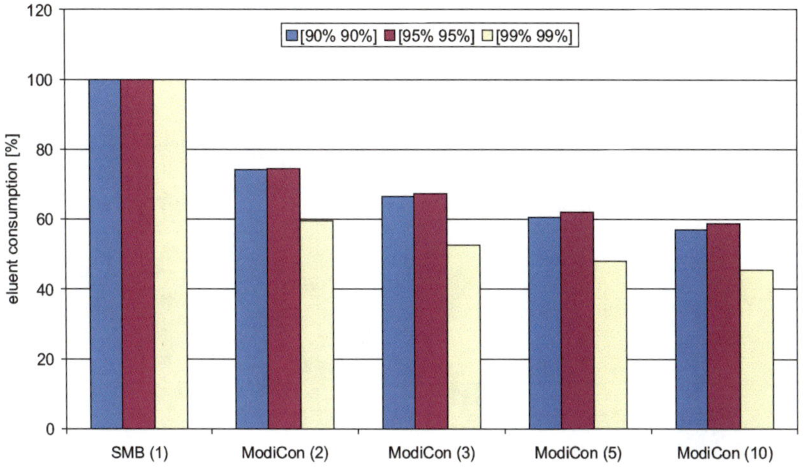

FIGURE 7. Performance of the ModiCon process for EMD-53986; maximum feed concentration as a multiple of the SMB feed concentration in parenthesis; the eluent consumption is scaled to the SMB process (100%) for the respective purity requirement.

maximum feed concentration (compared to the SMB concentration) for the purity requirements of 90% and 95%, as illustrated by Figure 7. The saving in solvent consumption is more pronounced for very strict purity requirements. For extract and raffinate purity requirements of 99%, a 40% reduction for a doubled feed concentration and a 54% reduction for a feed concentrated by a factor of ten is obtained.

The optimum operation is to concentrate the feed at the end of the switching period as indicated by the previous simulation study, see Figure 8. For the investigated enantiomer system, the critical contaminating front is the one close to the raffinate port, which is delayed by issuing the feed at the end of the period, as can be seen from the axial concentration profiles of the SMB process and the ModiCon process over a switching period at [99% 99%] purity in Figures 9 and 10. The feed concentration is zero in the first half of the period and doubled in the second half leading to considerably higher axial concentrations than in SMB operation. The solubility of twice the amount of components in the eluent and the validity of the adsorption isotherm can be expected, since liquid concentrations of 3.5 g/l have been reported in (Jupke, 2004). The CPU requirement[1] for the ModiCon optimization was between 3 h and 6 h due to the explicit consideration of the initial state in the SQP problem resulting in the calculation of large Jacobian and Hessian matrices. The detailed results of the ModiCon optimization of the EMD-53986 system can be found in the Appendix.

[1]Intel Xenon CPU 2.8 GHz, 4.0 GB RAM

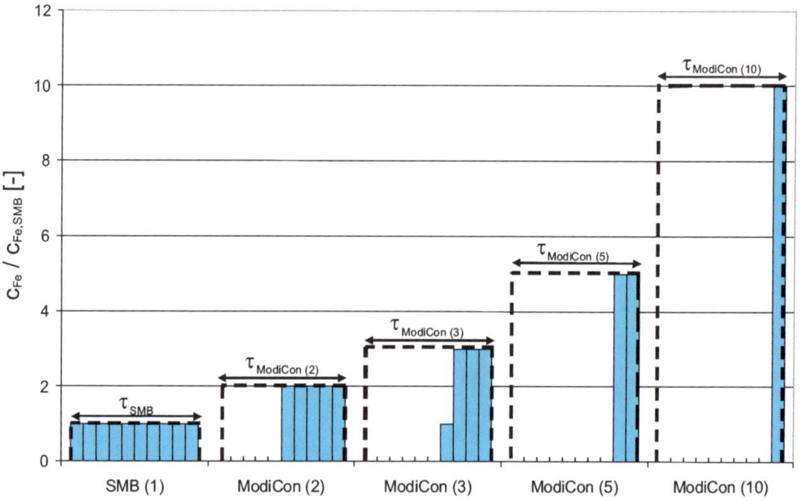

FIGURE 8. Optimal feed strategy of the ModiCon process for purities of [90% 90%], [95% 95%], and [99% 99%], EMD-53986; maximum feed concentration as a multiple of the SMB feed concentration in parenthesis.

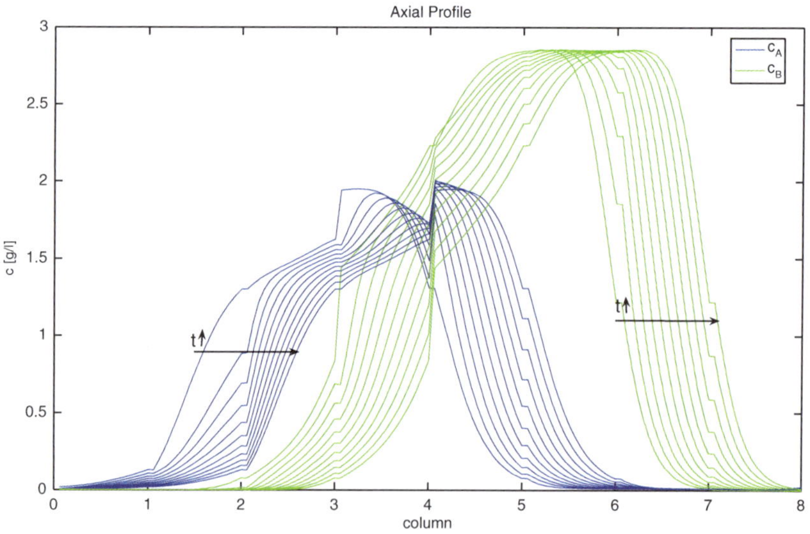

FIGURE 9. Axial concentrations of the SMB process during one switching period for purities of [99% 99%], $c_{Fe} = 2.5$ g/l, EMD-53986.

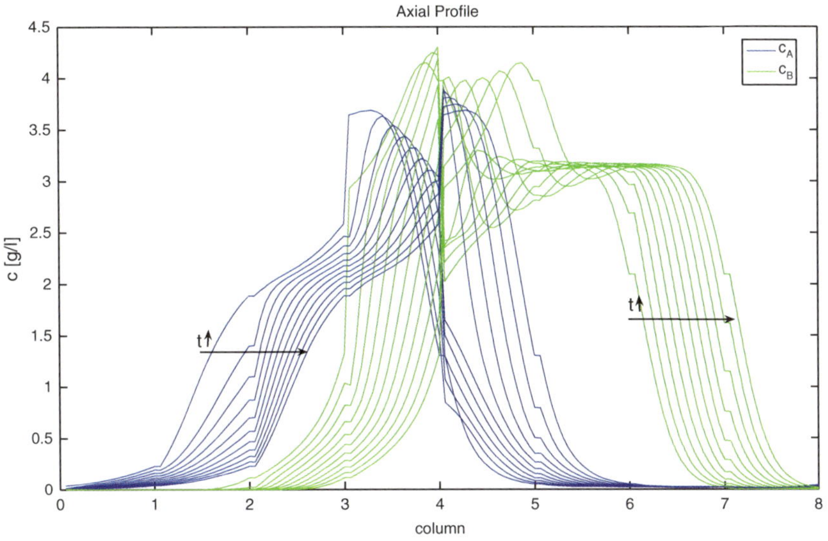

FIGURE 10. Axial concentrations of the ModiCon process during one switching period for purities of [99% 99%], $c_{\mathrm{Fe,max}} = 5$ g/l, EMD-53986.

5. Conclusion

In this paper, optimization approaches to SMB processes were discussed. The SMB process is a challenging chromatographic separation process due to the spatial distribution and due to the hybrid nature of the process. The potential of optimization was demonstrated by the ModiCon process where the operating parameters of the process were optimized such that economic savings in terms of a reduced solvent consumption were achieved. The optimization was performed by multiple shooting after discretization of the pde models of the columns. The multiple shooting approach requires an initialization of the axial profiles with SMB-like profiles in order to calculate sufficiently accurate gradients that are then updated over the iterations. The large CPU times result from the sizes of the Jacobian matrix and of the Hessian matrix of the optimization problem due to the explicit consideration of the initial state $\mathbf{x}_{\mathrm{smb}}^{\mathrm{pss}}$ as degree of freedom in the SQP problem. The application of the inexact Newton-Picard iterations that is still subject to ongoing research promises a significant reduction of the CPU requirements.

The ModiCon process can yield about 40% to 50% savings in solvent consumption for a feed strategy that uses only to doubled feed concentrations. This demonstrates the great potential of the mathematical optimization of SMB processes based on rigorous modeling. Similar results can not be obtained by short-cut methods as the triangle theory for SMB processes because these methods only approximate the hybrid dynamics of the SMB process. Since the solvent consumption is reduced, the optimization results are on the edge of infeasible solutions where the product purity constraints are not satisfied any more. Thus, the correct identification of the model parameters is crucial. Even for small plant-model mismatch, disturbances will lead to a violation of the product purity specifications. Therefore, online control of SMB processes is essential as presented in (Toumi and Engell, 2004; Küpper and Engell, 2007, 2009; Amanullah et al., 2008; Grossmann et al., 2009).

6. Appendix

6.1. State permutation matrix P

$$\mathbf{P} = \begin{pmatrix} \mathbf{0} & \mathbf{I} & \mathbf{0} & \cdots & \mathbf{0} \\ \mathbf{0} & \ddots & \mathbf{I} & \ddots & \vdots \\ \vdots & \ddots & \ddots & \ddots & \mathbf{0} \\ \mathbf{0} & \ddots & \ddots & \ddots & \mathbf{I} \\ \mathbf{I} & \cdots & \mathbf{0} & \mathbf{0} & \mathbf{0} \end{pmatrix} \tag{38}$$

I is a unity matrix with the dimension of a column.

model parameters			
column length L_{col}	18.0 [cm]	column diameter D	2.5 [cm]
particle porosity ϵ_{p}	0.567 [-]	bed porosity ϵ_{b}	0.353 [-]
particle diameter d_{p}	20.0 [μm]	particle diffusion D_{p}	$1.0{\cdot}10^{-3}$ [cm^2/s]
mass transfer coeff. A. $k_{l,A}$	$1.5{\cdot}10^{-4}$ [1/s]	mass transfer coeff. B $k_{l,B}$	$2.0{\cdot}10^{-4}$ [1/s]
Henry coeff. H_{A}^1	2.054 [-]	Henry coeff. H_{B}^1	2.054 [-]
Henry coeff. H_{A}^2	19.902 [-]	Henry coeff. H_{B}^2	5.847 [-]
isotherm param. k_{A}	0.472 [l/g]	isotherm param. k_{B}	0.129 [l/g]
density methanol ρ	0.799 [g/cm^3]	viscosity methanol ν	$1.2{\cdot}10^{-2}$ [g/(cm s)]
solid bed:	Chiralpak AD	solvent:	ethanol
numerical parameters			
collocation points n_{c}	1	finite elements n_{fe}	8
$\mathrm{KKT}_{\mathrm{tol}}$	$1.0{\cdot}10^{-5}$	integration tolerance	$1.0{\cdot}10^{-7}$

TABLE 3. Parameters of the EMD-53986 separation.

$c_{\mathrm{Fe,max}}$	SMB 2.5 [g/l]	ModiCon 5.0 [g/l]	ModiCon 7.5 [g/l]
$\mathrm{Pur}_{\mathrm{Ex}}$ [%]	90.00	90.00	90.00
$\mathrm{Pur}_{\mathrm{Ra}}$ [%]	90.00	90.00	90.00
Q_{El} [ml/min]	45.98	34.12	30.55
Q_{Fe} [ml/min]	30.00	30.00	30.00
Q_{Ra} [ml/min]	20.49	20.16	20.16
Q_{Re} [ml/min]	46.39	36.45	33.30
τ [min]	2.21	2.80	3.07
β [-]	[0.33, 0.39, 4.31, 1.93]	[0.32, 0.35, 3.97, 1.93]	[0.32, 0.34, 3.83, 1.93]
c_{Fe} [g/l]	**[2.5, 2.5, 2.5, 2.5, 2.5, 2.5, 2.5, 2.5, 2.5, 2.5]**	**[0.0, 0.0, 0.0, 0.0, 0.0, 5.0, 5.0, 5.0, 5.0, 5.0]**	**[0.0, 0.0, 0.0, 0.0, 0.0, 0.0, 2.5, 7.5, 7.5, 7.5]**
CPU* [h]	0.55	2.38	3.22
rel. Q_{El} [%]	**100.00**	**74.21**	**66.44**

$c_{\mathrm{Fe,max}}$	SMB 2.5 [g/l]	ModiCon 12.5 [g/l]	ModiCon 25.0 [g/l]
$\mathrm{Pur}_{\mathrm{Ex}}$ [%]	90.00	90.00	90.00
$\mathrm{Pur}_{\mathrm{Ra}}$ [%]	90.00	90.00	90.00
Q_{El} [ml/min]	45.98	27.89	26.19
Q_{Fe} [ml/min]	30.00	30.00	30.00
Q_{Ra} [ml/min]	20.49	20.19	20.29
Q_{Re} [ml/min]	46.39	30.94	29.24
τ [min]	2.21	3.31	3.50
β [-]	[0.33, 0.39, 4.31, 1.93]	[0.32, 0.33, 3.70, 1.92]	[0.31, 0.32, 3.60, 1.92]
c_{Fe} [g/l]	**[2.5, 2.5, 2.5, 2.5, 2.5, 2.5, 2.5, 2.5, 2.5, 2.5]**	**[0.0, 0.0, 0.0, 0.0, 0.0, 0.0, 0.0, 0.0, 12.5, 12.5]**	**[0.0, 0.0, 0.0, 0.0, 0.0, 0.0, 0.0, 0.0, 0.0, 25.0]**
CPU* [h]	0.55	3.84	4.72
rel. Q_{El} [%]	**100.00**	**60.66**	**56.96**

TABLE 4. Optimization of the SMB and the ModiCon processes for EMD-53986, SMB initialized at SMB optimization solution with [65% 65%] purities, ModiCon initialized at SMB optimization solution with [90% 90%] purities.

$c_{Fe,max}$	SMB 2.5 [g/l]	ModiCon 5.0 [g/l]	ModiCon 7.5 [g/l]
Pur_{Ex} [%]	95.00	95.00	95.00
Pur_{Ra} [%]	95.00	95.00	95.00
Q_{El} [ml/min]	67.57	50.30	45.51
Q_{Fe} [ml/min]	30.00	30.00	30.00
Q_{Ra} [ml/min]	22.23	21.17	20.94
Q_{Re} [ml/min]	48.10	39.14	36.21
τ [min]	2.18	2.67	2.89
β [-]	[0.42, 0.43, 4.13, 1.87]	[0.39, 0.39, 3.91, 1.88]	[0.39, 0.38, 3.81, 1.88]
c_{Fe} [g/l]	**[2.5, 2.5, 2.5, 2.5, 2.5, 2.5, 2.5, 2.5, 2.5, 2.5]**	**[0.0, 0.0, 0.0, 0.0, 0.0, 5.0, 5.0, 5.0, 5.0, 5.0]**	**[0.0, 0.0, 0.0, 0.0, 0.0, 0.0, 2.5, 7.5, 7.5, 7.5]**
CPU* [h]	3.94	2.82	3.11
rel. Q_{El} [%]	**100.00**	**74.44**	**67.35**

$c_{Fe,max}$	SMB 2.5 [g/l]	ModiCon 12.5 [g/l]	ModiCon 25.0 [g/l]
Pur_{Ex} [%]	95.00	95.00	95.00
Pur_{Ra} [%]	95.00	95.00	95.00
Q_{El} [ml/min]	67.57	41.96	39.69
Q_{Fe} [ml/min]	30.00	30.00	30.00
Q_{Ra} [ml/min]	22.23	20.80	20.73
Q_{Re} [ml/min]	48.10	33.94	32.41
τ [min]	2.18	3.08	3.22
β [-]	[0.42, 0.43, 4.13, 1.87]	[0.38, 0.36, 3.72, 1.88]	[0.38, 0.35, 3.65, 1.88]
c_{Fe} [g/l]	**[2.5, 2.5, 2.5, 2.5, 2.5, 2.5, 2.5, 2.5, 2.5, 2.5]**	**[0.0, 0.0, 0.0, 0.0, 0.0, 0.0, 0.0, 0.0, 12.5, 12.5]**	**[0.0, 0.0, 0.0, 0.0, 0.0, 0.0, 0.0, 0.0, 0.0, 25.0]**
CPU* [h]	3.94	4.23	3.94
rel. Q_{El} [%]	**100.00**	**62.10**	**58.74**

TABLE 5. Optimization of the SMB and the ModiCon processes for EMD-53986, SMB initialized at SMB optimization solution with [90% 90%] purities, ModiCon initialized at SMB optimization solution with [95% 95%] purities.

$c_{\mathrm{Fe,max}}$	SMB 2.5 [g/l]	ModiCon 5.0 [g/l]	ModiCon 7.5 [g/l]
$\mathrm{Pur_{Ex}}$ [%]	99.00	99.00	99.00
$\mathrm{Pur_{Ra}}$ [%]	99.00	99.00	99.00
Q_{El} [ml/min]	167.14	99.33	87.97
Q_{Fe} [ml/min]	30.00	30.00	30.00
Q_{Ra} [ml/min]	28.79	24.48	23.61
Q_{Re} [ml/min]	53.00	44.08	41.26
τ [min]	1.93	2.41	2.58
β [-]	[0.72, 0.52, 3.94, 1.88]	[0.58, 0.46, 3.81, 1.85]	[0.56, 0.44, 3.75, 1.84]
c_{Fe} [g/l]	**[2.5, 2.5, 2.5, 2.5, 2.5, 2.5, 2.5, 2.5, 2.5, 2.5]**	**[0.0, 0.0, 0.0, 0.0, 0.0, 5.0, 5.0, 5.0, 5.0, 5.0]**	**[0.0, 0.0, 0.0, 0.0, 0.0, 0.0, 2.5, 7.5, 7.5, 7.5]**
CPU* [h]	2.05	4.04	1.66
rel. Q_{El} [%]	**100.00**	**59.43**	**52.63**

$c_{\mathrm{Fe,max}}$	SMB 2.5 [g/l]	ModiCon 12.5 [g/l]	ModiCon 25.0 [g/l]
$\mathrm{Pur_{Ex}}$ [%]	99.00	99.00	99.00
$\mathrm{Pur_{Ra}}$ [%]	99.00	99.00	99.00
Q_{El} [ml/min]	167.14	80.46	76.06
Q_{Fe} [ml/min]	30.00	30.00	30.00
Q_{Ra} [ml/min]	28.79	23.07	22.73
Q_{Re} [ml/min]	53.00	38.98	37.67
τ [min]	1.93	2.73	2.82
β [-]	[0.72, 0.52, 3.94, 1.88]	[0.55, 0.43, 3.70, 1.84]	[0.54, 0.42, 3.67, 1.84]
c_{Fe} [g/l]	**[2.5, 2.5, 2.5, 2.5, 2.5, 2.5, 2.5, 2.5, 2.5, 2.5]**	**[0.0, 0.0, 0.0, 0.0, 0.0, 0.0, 0.0, 0.0, 12.5, 12.5]**	**[0.0, 0.0, 0.0, 0.0, 0.0, 0.0, 0.0, 0.0, 0.0, 25.0]**
CPU* [h]	2.05	3.80	6.31
rel. Q_{El} [%]	**100.00**	**48.14**	**45.51**

TABLE 6. Optimization of the SMB and the ModiCon processes for EMD-53986, SMB initialized at SMB optimization solution with [95% 95%] purities, ModiCon initialized at SMB optimization solution with [99% 99%] purities.

References

Adam, P., Nicoud, R., Bailly, M., Ludemann-Hombourger, O., 2002. Process and device for separation with variable-length chromatographic columns. US Patent 6.413.419.

Amanullah, M., Grossmann, C., Erdem, G., Mazzotti, M., Morbidelli, M., Morari, M., 2008. Cycle to cycle optimizing control of simulated moving beds. AIChE Journal 54 (1), 194–208.

Broughton, D., Gerhold, C., 1961. Continuous sorption process employing a fixed bed of sorbent and moving inlets and outlets. US Patent 2.985.589.

Chung, S., Wen, C., 1968. Longitudinal diffusion of liquid flowing through fixed and fluidized beds. AIChE Journal 14, 857–866.

Dünnebier, G., 2000. Effektive Simulation und mathematische Optimierung chromatographischer Trennprozesse, Dr.-Ing. Dissertation. Fachbereich Bio- und Chemieingenieurwesen, Universität Dortmund. Shaker-Verlag, Aachen.

Dünnebier, G., Klatt, K.-U., 2000. Modelling and simulation of nonlinear chromatographic separation processes: A comparison of different modelling aspects. Chemical Engineering Science 55 (2), 373–380.

Grossmann, C., Langel, C., Mazzotti, M., Morari, M., Morbidelli, M., 2009. Experimental implementation of automatic cycle to cycle control to a nonlinear chiral simulated moving bed separation. J. Chromatogr., A 1217 (13), 2013–2021.

Gu, T., 1995. Mathematical Modelling and Scale-up of Liquid Chromatography. Springer Verlag, New York.

Hanisch, F., 2002. Prozessführung präparativer Chromatographieverfahren, Dr.-Ing. Dissertation. Fachbereich Bio- und Chemieingenieurwesen, Universität Dortmund. Shaker-Verlag, Aachen.

Hashimoto, K., Adachi, S., Noujima, H., Marujama, H., 1983. Models for the separation of glucose/fructose mixture using a simulated moving bed adsorber. Journal of chemical engineering of Japan 16 (15), 400–406.

Hindmarsh, A., 1983. ODEPACK, a systematized collection of ode solvers. In: Stepleman, R. (Ed.), Scientific Computing. North-Holland, Amsterdam, pp. 55–64.

Jupke, A., 2004. Experimentelle Modellvalidierung und modellbasierte Auslegung von Simulated Moving Bed (SMB) Chromatographieverfahren. Dr.-Ing. Dissertation, Fachbereich Bio- und Chemieingenieurwesen, Universität Dortmund. VDI Reihe 3, Nr. 807.

Juza, M., Mazzotti, M., Morbidelli, M., 2000. Simulated moving-bed chromatography and its application to chirotechnology. Trends in Biotechnology 18 (3), 108–118.

Kawajiri, Y., Biegler, L. T., 2006a. Large scale nonlinear optimization for asymmetric operation and design of simulated moving beds. Journal of Chromatography A 1133 (1-2), 226–240.

Kawajiri, Y., Biegler, L. T., 2006b. Optimization strategies for simulated moving bed and powerfeed processes. AIChE Journal 52 (4), 1343–1350.

Kearney, M., Hieb, K., 1992. Time variable simulated moving bed process. US Patent 5.102.553.

Klatt, K.-U., Hanisch, F., Dünnebier, G., 2002. Model-based control of a simulated moving bed chromatographic process for the separation of fructose and glucose. Journal of Process Control 12 (2), 203–219.

Klatt, K.-U., Hanisch, F., Dünnebier, G., Engell, S., 2000. Model-based optimization and control of chromatographic processes. Computers and Chemical Engineering 24, 1119–1126.

Kloppenburg, E., Gilles, E., 1999. A new concept for operating simulated-moving-bed processes. Chemical Engineering and Technology 22 (10), 813–817.

Küpper, A., Engell, S., 2007. Non-linear model predictive control of the Hashimoto simulated moving bed process. In: R. Findeisen, L. B., Allgöwer, F. (Eds.), Assessment and Future Directions of Nonlinear Model Predictive Control. Springer-Verlag, Berlin, pp. 473–483.

Küpper, A., Engell, S., 2009. Optimierungsbasierte Regelung des Hashimoto-SMB-Prozesses. at-Automatisierungstechnik 57 (7), 360–370.

Leineweber, D., 1999. Efficient reduced SQP methods for the optimization of chemical processes described by large sparse DAE models. Vol. 613 of Fortschritt-Berichte VDI Reihe 3, Verfahrenstechnik. VDI Verlag, Düsseldorf.

Ludemann-Hombourger, O., Nicoud, R., Bailly, M., 2000. The Varicol process: a new multicolumn continuous chromatographic process. Separation Science and Technology 35 (16), 1829–1862.

Mazzotti, M., Storti, G., Morbidelli, M., 1997. Optimal operation of simulated moving bed units for nonlinear chromatographic separations. Journal of Chromatography A 769, 3–24.

Nicoud, R., 1998. Simulated Moving Bed (SMB): Some possible applications for biotechnology. In: Subramanian, G. (Ed.), Bioseparation and Bioprocessing. Wiley-VCH, Weinheim-New York.

Nocedal, J., Wright, S., 1999. Numerical optimization. Springer-Verlag, New-York.

Potschka, A., Bock, H., Engell, S., Küpper, A., Schlöder, J., 2008. Optimization of periodic adsorption processes: The Newton-Picard inexact SQP method. Deutsche Forschungsgemeinschaft Priorty Program 1253.

Potschka, A., Küpper, A., Schlöder, J., Bock, H., Engell, S., 2010. Optimal control of periodic adsorption processes: The Newton-Picard inexact SQP method. In: Recent Advances in Optimization and its Applications in Engineering (edited volume of 14th Belgian-French-German Conference on Optimization Leuven, 2009. Springer Verlag, in press.

Ruthven, D. M., 1984. Principles of Adsorption and Adsorption processes. Wiley, New York.

Schmidt-Traub, H., 2005. Preparative chromatography of fine chemicals and pharmaceuticals agents. Wiley-Verlag, Berlin.

Schramm, H., Kaspereit, M., Kienle, A., Seidel-Morgenstein, A., 2003a. Simulated moving bed process with cyclic modulation of the feed concentration. Journal of Chromatography A 1006 (1–2), 77–86.

Schramm, H., Kienle, A., Kaspereit, M., Seidel-Morgenstein, A., 2003b. Improved operation of simulated moving bed processes through cyclic modulation of feed flow and feed concentration. Chemical Engineering Science 58 (23–24), 5217–5227.

Toumi, A., 2005. Optimaler Betrieb und Regelung von Simulated-Moving-Bed-Prozessen, Dr.-Ing. Dissertation. Fachbereich Bio- und Chemieingenieurwesen, Universität Dortmund. Shaker-Verlag, Aachen.

Toumi, A., Engell, S., 2004. Optimization-based control of a reactive simulated moving bed process for glucose isomerization. Chemical Engineering Science 59 (18), 3777–3792.

Toumi, A., Engell, S., Diehl, M., Bock, H., Schlöder, J., 2007. Efficient optimization of simulated moving bed processes. Chemical Engineering and Processing 46 (11), 1067–1084.

Toumi, A., Engell, S., Ludemann-Hombourger, O., Nicoud, R. M., Bailly, M., 2003. Optimization of simulated moving bed and Varicol processes. Journal of Chromatography A 1-2, 15–31.

Toumi, A., Hanisch, F., Engell, S., 2002. Optimal operation of continuous chromatographic processes: Mathematical optimization of the VARICOL process. Ind. Eng. Chem. Res. 41 (17), 4328–4337.

Wang, C., Klatt, K.-U., Dünnebier, G., Engell, S., Hanisch, F., 2003. Neural network-based identification of smb chromatographic processes. Control Engineering Practice 11 (8), 949–959.

Zhang, Z., Mazzotti, M., Morbidelli, M., 2003. PowerFeed operation of simulated moving bed units: changing flow-rates during the switching interval. Journal of Chromatography A 1006 (1-2), 87–99.

Achim Küpper and Sebastian Engell
Process Dynamics and Operations Group
Technische Universität Dortmund
Emil-Figge-Str. 70
D-44227 Dortmund, Germany
e-mail: achim.kuepper@bci.tu-dortmund.de
sebastian.engell@bci.tu-dortmund.de

International Series of Numerical Mathematics, Vol. 160, 597–609

Optimization and Inverse Problems in Radiative Heat Transfer

René Pinnau and Norbert Siedow

Abstract. We discuss the derivation and investigation of efficient mathematical methods for the solution of optimization and identification problems for radiation dominant processes, which are described by a nonlinear integro-differential system or diffusive type approximations. These processes are for example relevant in glass production or in the layout of gas turbine combustion chambers. The main focus is on the investigation of optimization algorithms based on the adjoint variables, which are applied to the full radiative heat transfer system as well as to diffusive type approximations. In addition to the optimization we also study new approaches to the reconstruction of the initial temperature from boundary measurements, since its precise knowledge is mandatory for any satisfactory simulation. In particular, we develop a fast, derivative-free method for the solution of the inverse problem, such that we can use many different models for the simulation of the radiative process.

Mathematics Subject Classification (2000). 35K55, 49J20, 49K20, 80A20.

Keywords. Radiative heat transfer, SP_N-approximation, optimal boundary control, inverse problem, optimality conditions, analysis, numerics, adjoints, integro-differential equations, reduced order modeling.

1. Introduction: Radiative Heat Transfer

The overall goal of each simulation in engineering is the improvement of products or production processes. Therefore, one needs, e.g., in glass production or in the production of gas turbine combustion chambers, accurate simulation algorithms which allow for the simulation of radiation dominant processes in acceptable time [1, 26]. Due to the high numerical complexity of the underlying radiative heat transfer (RHT) equations, the main focus of research lay during the last decades on the development of fast solvers and the investigation of approximative models, which allowed to shorten the design cycle significantly [12, 11, 2, 24]. The interplay of the better performance of the algorithms and the increasing computing power allows now for the construction and investigation of computer-based optimization platforms.

At low temperatures heat flux can be modeled as sole heat conduction, which can be solved with commercial software packages. But radiation plays an important role for the energy balance in many processes at high temperatures, in which materials or devices are heated or cooled. At very high temperatures, like in glass production or combustion, radiation is even dominant, such that one needs more sophisticated models coupling both energy transport mechanisms. In contrast to diffusive heat conduction this yields a more complex model, which can only be simulated with standard software in limit cases, as pure surface radiation or radiation in optically thick media. In semitransparent media, like glass or gas, one cannot disregard volume radiation of heat from the interior. The basic model for its simulation are the radiative heat transfer (RHT) equations [12]

$$\Omega \cdot \nabla I(x,\Omega,t,\nu) = \kappa(\nu)[B(T(x,t),\nu) - I(x,\Omega,t,\nu)] \tag{1.1a}$$
$$+ \sigma(\nu)\left[\frac{1}{4\pi}\int_S \varphi(x,\Omega,\Omega')I(x,\Omega',t,\nu)\,d\Omega' - I(x,\Omega,t,\nu)\right],$$

coupled with an equation for the temperature

$$c_m\rho_m\partial_t T(x,t) = \nabla\cdot(k_h\nabla T(x,t)) - \int_0^\infty \kappa(\nu)\int_S [B(T(x,t),\nu) - I(x,\Omega,t,\nu)]d\Omega d\nu, \tag{1.1b}$$

where we denote by x the space variable, by $\Omega \in S$ a direction vector, by ν the frequency. Here, S is the unit sphere and t the time variable. Further, $\kappa(\nu)$ denotes the frequency-dependent absorption rate, $\sigma(\nu)$ is the scattering coefficient and $\varphi(x,\Omega,\Omega')$ the scattering function. These are assumed to be piecewise constant (frequency bands!). Finally, $I = I(x,\Omega,t,\nu)$ is the radiative intensity and $T(x,t)$ the temperature. Planck's function $B(T,\nu)$ describes the intensity of black body radiation. The RHT system (1.1) is in general seven dimensional and nonlinear, where the coupling with the heat equation is given by its temperature dependence via Planck's function. An analytical investigation of this system can be found, e.g., in [7, 8].

Besides of the full RHT system we also focus on diffusive-type approximations, where the simplest one is given by Rosseland's approximation [12], which is a nonlinear diffusion equation for the temperature. Improved approximations are the so-called *Simplified* P_N (SP_N) approximations, which are derived in [11]. These are given by coupled, nonlinear parabolic/elliptic systems. They can be extended to the case of multiple frequency bands, but one has to encounter that there will arise additional elliptic equations for each frequency band. Further, there exist frequency-averaged version of these approximations [10].

The report is organized as follows. In Section 2 we study optimization problems for radiation problems modeled by the diffusive SP_N approximations. In particular, we provide a thorough analysis of an optimal boundary control problem for the SP_1 system. Numerical results based on Newton's method are presented in

Section 2.1. Due to multiple frequency bands these systems are still a challenging numerical problem. Thus, we discuss in Section 2.2 the applicability of reduced order modeling based on Proper Orthogonal Decomposition. This yields a new a posteriori method for frequency averaging. The numerical solution of an industrial problem, i.e., the minimization of thermal stresses in glass cooling, is shown in Section 2.3. The identification problem for the initial temperature is presented in Section 3. In particular, we discuss a new derivative-free algorithm based on fixed point iterations, which allows for the fast solution of such kind of problems. Section 4 is devoted to the study of the adjoint calculus for the full radiative transfer equation. The link between the optimization problems for the full problem and those for the SP_N systems is discussed in Section 4.1, where tools from asymptotic analysis are used to derive the corresponding first-order optimality conditions.

2. Optimal control of radiation dominant processes based on approximative models

In this section we discuss the optimization of radiation dominant cooling processes modeled by the SP_N-approximations [11]. Starting from the optimality conditions of first and second order we study different numerical methods for the solution of the optimal control problems. Further, we also use model reduction techniques, which allow to reduce drastically the memory requirements and computational times for these high-dimensional problems. Finally, the minimization of thermal stresses in glass cooling is discussed.

The SP_1-approximation to the radiative heat transfer equations is given by the system

$$\partial_t T = k\Delta T + \frac{1}{3\kappa}\Delta\rho, \tag{2.1a}$$

$$0 = -\varepsilon^2 \frac{1}{3\kappa}\Delta\rho + \kappa\rho - \kappa\, 4\pi a \left|T\right|^3 T, \tag{2.1b}$$

with boundary conditions

$$n\cdot\nabla T = \frac{h}{\varepsilon k}(u - T), \tag{2.1c}$$

$$n\cdot\nabla\rho = \frac{3\kappa}{2\varepsilon}\left(4\pi a \left|u\right|^3 u - \rho\right), \tag{2.1d}$$

and supplemented with an initial condition $T(0, x) = T_0(x)$ for the temperature. Here, ρ is the radiative flux, and the prescribed temperature at the boundary is denoted by u. Reasonable regularity assumptions on the data ensure the existence of a unique solution to system (2.1). The proof is based on a fixed point argument in combination with Stampacchia's truncation method for the derivation of the uniform bounds (for details we refer to [18]).

Remark 2.1. We focus here on the SP_1 model, which yields a reasonable approximation of the full RHT problem, as long as the temperature differences are not too large. If one considers a very fast cooling, where strong boundary layers occur in the temperature distribution, then one should better use a higher-order model, e.g., the SP_3 model.

For notational convenience we introduce the following notations and spaces:

$$Q \stackrel{\text{def}}{=} (0,1)\times\Omega, \quad \Sigma \stackrel{\text{def}}{=} (0,1)\times\partial\Omega,$$
$$V \stackrel{\text{def}}{=} L^2(0,1;H^1(\Omega)), \quad W \stackrel{\text{def}}{=} \{\phi\in V : \phi_t \in V^*\}.$$

Based on these we set $Y \stackrel{\text{def}}{=} W \times V$ and as the space of controls we choose $U \stackrel{\text{def}}{=} H^1(0,1;\mathbb{R})$. Further, we define $Z \stackrel{\text{def}}{=} V\times V\times L^2(\Omega)$ and $Y_\infty \stackrel{\text{def}}{=} Y\cap[L^\infty(Q)]^2$ as the space of states $y \stackrel{\text{def}}{=} (T,\rho)$. Finally, we set $\alpha = \frac{h}{\varepsilon k}$ and $\gamma = \frac{3\kappa}{2\varepsilon}$.

We define the state/control pair $(y,u)\in Y_\infty\times U$ and the nonlinear operator $e \stackrel{\text{def}}{=} (e_1,e_2,e_3) : Y_\infty\times U \to Z^*$ via

$$\begin{aligned}\langle e_1(y,u),\phi\rangle_{V^*,V} \stackrel{\text{def}}{=} {}& \langle \partial_t T,\phi\rangle_{V^*,V} + k\,(\nabla T,\nabla\phi)_{L^2(Q)} + \frac{1}{3\kappa}\,(\nabla\rho,\nabla\phi)_{L^2(Q)} \\ & + k\alpha(T-u,\phi)_{L^2(\Sigma)} + \frac{1}{3\kappa}\gamma(\rho-4\pi a u^4,\phi)_{L^2(\Sigma)} \end{aligned} \tag{2.2a}$$

and

$$\begin{aligned}\langle e_2(y,u),\phi\rangle_{V^*,V} \stackrel{\text{def}}{=} {}& \frac{\varepsilon^2}{3\kappa}\,(\nabla\rho,\nabla\phi)_{L^2(Q)} \\ & + \kappa(\rho-4\pi\kappa a T^4,\phi)_{L^2(Q)} + \frac{\varepsilon^2}{3\kappa}\gamma(\rho-4\pi a u^4,\phi)_{L^2(\Sigma)} \end{aligned} \tag{2.2b}$$

for all $\phi\in V$. Further, we define $e_3(y,u) \stackrel{\text{def}}{=} T(0)-T_0$.

Theorem 2.2. *Assume that the domain Ω is sufficiently regular and let $u\in U$ and $T_0\in L^\infty(\Omega)$ be given. Then, the SP_1 system $e(y,u)=0$, where e is defined by (2.2) has a unique solution $(T,\rho)\in Y_\infty$ and there exists a constant $c>0$ such that the following energy estimate holds*

$$\|T\|_W + \|\rho\|_V \le c\left\{\|T_0\|^4_{L^\infty(\Omega)} + \|u\|_U^4\right\}. \tag{2.3}$$

Further, the solution is uniformly bounded, i.e., $(T,\rho)\in[L^\infty(Q)]^2$ and we have

$$\underline{T}\le T\le \overline{T}, \quad \underline{\rho}\le\rho\le\overline{\rho}, \tag{2.4}$$

where

$$\underline{T} = \min\left(\inf_{t\in(0,1)} u(t), \inf_{x\in\Omega} T_0(x)\right),$$

$$\overline{T} = \max\left(\sup_{t\in(0,1)} u(t), \sup_{x\in\Omega} T_0(x)\right),$$

as well as $\underline{\rho} = 4\pi a\,|\underline{T}|^3\,\underline{T}$ and $\overline{\rho} = 4\pi a\,|\overline{T}|^3\,\overline{T}$.

We intend to minimize cost functionals of tracking-type having the form

$$J(T,u) = \frac{1}{2}\left\|T - T_d\right\|^2_{L^2(0,1;L^2(\Omega))} + \frac{\delta}{2}\left\|u - u_d\right\|^2_{H^1(0,1;\mathbb{R})}, \tag{2.5}$$

Here, $T_d = T_d(t,x)$ is a specified temperature profile, which is typically given by engineers. In glass manufacturing processes, T_d is used to control chemical reactions in the glass, in particular their activation energy and the reaction time. The control variable u, which is considered to be space-independent, enters the cost functional as regularizing term, where additionally a known cooling curve u_d can be prescribed. The parameter δ allows to adjust the effective heating costs of the cooling process. The main subject is now the study of the following boundary control problem [18, 6]

$$\begin{aligned} &\min\ J(T,\rho,u) \text{ w.r.t.}(T,\rho,u), \\ &\text{subject to the } SP_1\text{-system (2.1).} \end{aligned} \tag{2.6}$$

Based on standard arguments one can show the existence of a minimizer [18].

Theorem 2.3. *There exists a minimizer $(y^*, u^*) \in Y_\infty \times U$ of the constrained minimization problem* (2.6).

Owing to the fact that the system (2.1) is uniquely solvable, we may reformulate the minimization problem (2.6) introducing the *reduced cost functional* $\hat{J}$ as

$$\begin{aligned} &\text{minimize} \quad \hat{J}(u) \stackrel{\text{def}}{=} J(y(u),u) \quad \text{over} \quad u \in U \\ &\text{where} \quad y(u) \in Y \quad \text{satisfies} \quad e(y(u),u) = 0. \end{aligned} \tag{2.7}$$

The first numerical investigations for the solution of (2.7) can be found in [23, 27, 9]. They were based on the implementation of a descent algorithm using adjoint information to provide the gradient of the reduced cost functional. In fact, one can show that this approach is mathematically sound [18, 25]. Later, also second-order algorithms, like Newton's method, were provided for the solution of (2.7) using the reduced Hessian [22, 25].

2.1. Numerical temperature tracking

Now we want to discuss the numerical results which can achieved using the adjoint approach (for a more detailed discussion see also [6, 22, 23, 25]). The spatial discretization of the PDEs is based on linear finite elements. We use a non-uniform grid with an increasing point density toward the boundary of the medium, consisting of 109 points. The temporal discretization uses a uniform grid consisting of 180 points for the temperature-tracking problem. We employ the implicit backward Euler method to compute the state (T,ρ). For a given (time-dependent) temperature profile T_d we compute an optimal u such that the temperature of the glass follows the desired profile T_d as good as possible. Such profiles are of great importance in glass manufacturing in order to control at which time, at which place and for how long certain chemical reactions take place, which is essential for the quality of the glass.

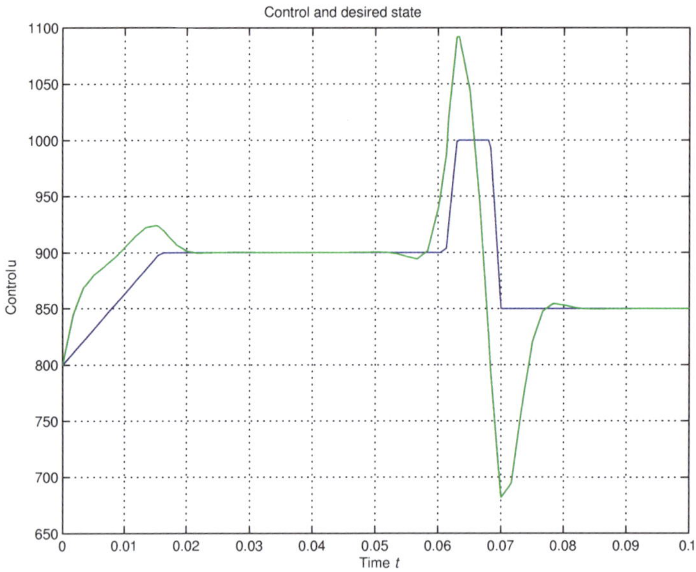

FIGURE 1. Unoptimized (dark) and optimized (light) cooling profile.

The dark line in Figure 1 describes the desired temperature profile $T_d(t)$ which shall be attained homogeneously in space. From the engineering point of view it is an educated guess to use the same profile for the boundary control. Clearly, this leads to deviations which can be seen in the left graphic of Figure 2. Our optimal control approach results now in the light line in Figure 1, which yields in turn the improved temperature differences on the right in Figure 2. One realizes a significant improvement although we have still a large peak. But note that we

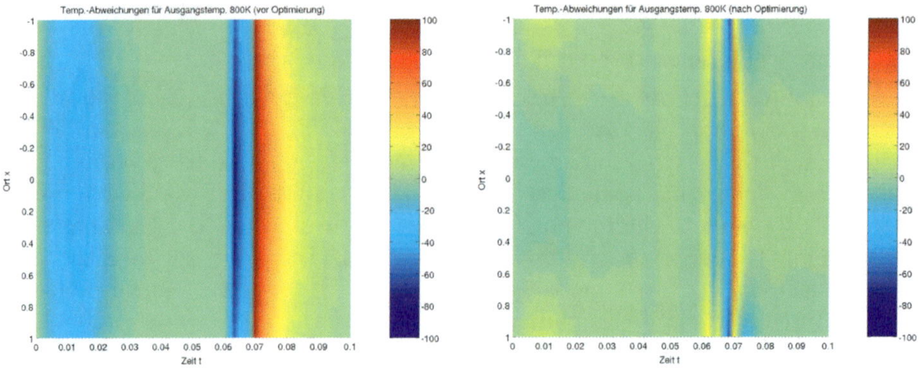

FIGURE 2. Temperature differences for the uncontrolled (left) and controlled (right) state.

want to resolve a very sharp jump in the temperature. Due to diffusive part of the equations it is almost impossible to resolve such a fast change in the cooling.

2.2. Reduced order modeling

Now, we discuss how reduced order modeling can be used to build surrogate models which fasten the optimization algorithms. In particular, we employ the method of *Proper Orthogonal Decomposition* (POD), which is well known in the context of computational fluid dynamics and corresponding optimization problems (see [19, 20] and the references therein).

In [21, 20] a POD based frequency averaging method is discussed for the simulation of temperature and radiation in high temperature processes. Using this method, significantly better results can be obtained with similar or less numerical effort. The main advantage is that POD does not require special engineering knowledge. Hence, POD can be used as a fully automatic black box algorithm for model reduction, requiring no user interaction at all. Even more interesting, POD was also able to outperform frequency averaged equations [10], which have a much stronger theoretical background.

2.3. Reducing thermal stresses in glass cooling

The study of stress reduction in glass cooling requires an extension of the model equations. In [25] the SP_1 system is coupled with the nonstandard viscoelastic model of Narayanaswamy [13]. The Lagrange formalism is formally applied to this model to derive the first-order optimality system. Using the optimization code discussed in detail in [25], it was for the first time possible to calculate a qualitatively, physically reasonable, optimal cooling curve. The corresponding numerical results (for the particular setup and the model parameters, we refer to [25]) are presented in Figure 3. Compared to the linear cooling profile, one gets a stress reduction of 50%. This is due to the fact that the ambient temperature is kept constant near the so-called glass temperature T_g, such that the inherent stresses have time to relax.

3. Identification of the initial temperature for radiation dominant processes using derivative-free methods

This section is devoted to the problem of fast reconstruction of the initial temperature from boundary measurements. For the solution of this inverse problem a derivative-free method based on an appropriate decoupling of the equations is developed and analyzed in [14, 15, 16, 17]. This question is relevant for adequate simulations of many industrial processes, since a direct measurement of the temperature in the interior is in general not possible at these high temperatures. Further, the precise knowledge of the initial temperature distribution is essential for a correct simulation and optimization of the process. The main goal is to reconstruct the spatial distribution of the initial temperature from given data, which are in practice often temperature measurements at the boundary. The difficulties which

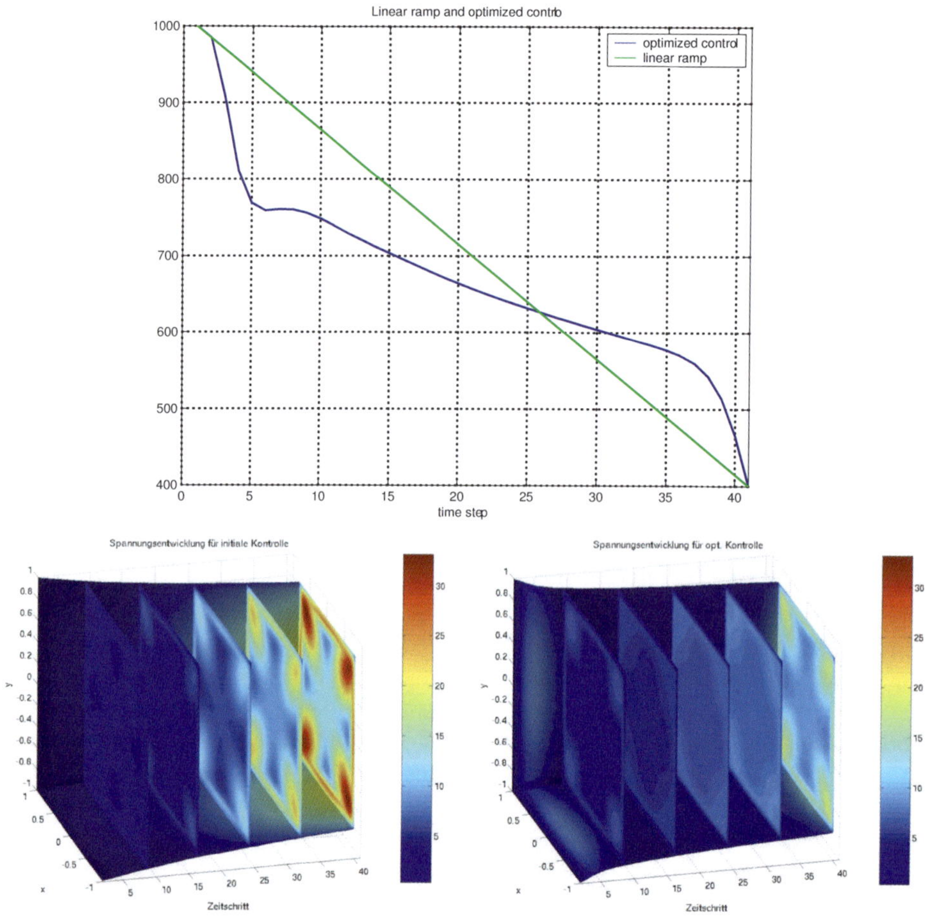

FIGURE 3. Optimal cooling curve (up), unoptimized stress (bottom left) and optimized stress (bottom right).

arise already for linear inverse heat transfer problems are well known and there exist suitable algorithms for their solution [3]. Unfortunately, it is not possible to transfer those techniques to our problem, since we have a different data set (temperatures at the boundary) and we need to incorporate additional radiative effects. Further, the new algorithm ensures that no direct solve of the full nonlinear inverse problem is necessary.

The main advantage of our approach is that one can use fast existing simulation tools for the solution of the forward system. Since also this algorithm will require several solves of the nonlinear forward system, a fast simulation tool for three-dimensional RHT equations is indispensable.

The nonlinear heat transfer problem with nonlinearity $N(T)$ reads [14]:

$$\frac{\partial T}{\partial t}(z,t) = \Delta T(z,t) - N(T), \quad z \in \Omega,\ 0 < t < t_f$$
$$\partial_n T(z_b, t) = 0, \quad \text{for} \quad z_b \in \partial\Omega,$$
$$T(z,0) = u(z), \quad z \in \Omega.$$

The right-hand side depends on the initial condition $u(z)$. For given boundary measurements $D(t)$, $t \in [0, t_f]$ (e.g., at some fixed part of $\partial\Omega$) we can then write the identification problem for $u(z)$ as a nonlinear operator equation

$$L(u, N(u)) = D, \quad L : \mathcal{L}^2(\Omega) \to \mathcal{L}^2(0, t_f).$$

The detailed analytical investigation of the inverse problem and the study of convergence for this algorithm can be found in [14, 16]. There, a derivative-free, iterative method is introduced for the solution of nonlinear ill-posed problems

$$Fx = y,$$

where instead of y noisy data y_δ with $\|y - y_\delta\| \leq \delta$ are given and $F : X \to Y$ is a nonlinear operator between Hilbert spaces X and Y. This method is defined by splitting the operator F into a linear part A and a nonlinear part G, such that $F = A + G$. Then the iterations are organized as

$$Au_{k+1} = y_\delta - Gu_k.$$

In the context of ill-posed problems the situation when A does not have a bounded inverse is considered, thus each iteration needs to be regularized. Under some conditions on the operators A and G the behavior of the iteration error is studied. Its stability with respect to the iteration number k is obtained, as well as the optimal convergence rate with respect to the noise level δ, provided that the solution satisfies a generalized source condition. In [14, 17] the inverse problem of initial temperature reconstruction for a nonlinear heat equation is numerically solved, where the nonlinearity appears due to radiation effects. The iteration error in the numerical results has the theoretically expected behavior.

4. Adjoint methods for radiative transport

Due to the high dimensionality of the phase space of the discretized radiative transfer equation, most optimization approaches are based on approximate models coupled with a heat equation [6, 18]. These approaches allow to compute optimal controls at reasonable numerical costs. Nevertheless, there are applications in which one needs the detailed knowledge of the radiative intensity, like strongly scattering or optically thin media [12]. Then, the mentioned approximate models are no longer appropriate and one has to go back to the full radiative transfer equation (RTE). In particular, in [4] optimal control problems for the multidimensional RTE including scattering are discussed, analytically investigated and numerically solved.

There, a study for optimal control problems of tracking-type in radiative transfer is developed, where the cost functionals have the form

$$\mathcal{F}(R,Q) := \frac{\alpha_1}{2}\int_D \left(R-\bar{R}\right)^2 dx + \frac{\alpha_2}{2}\int_D \left(Q-\bar{Q}\right)^2 dx.$$

Here, $R(x) := \int_{S^2} I(\omega.x)d\omega$ is the mean intensity and the radiative intensity I is given as the solution of the RTE. The control is the source term Q, which can be interpreted as a gain or loss of radiant energy deployed by external forces. Using the adjoint calculus the first-order optimality condition is derived, which proves to be necessary and sufficient. Further, the existence of an optimal control is shown and the first-order derivative information is used in a gradient method for the numerical optimization. There a source Q is computed, which yields the desired distribution of the mean intensity R depicted in Figure 4, where also the optimal state is shown. For details on the numerical setting we refer to [4].

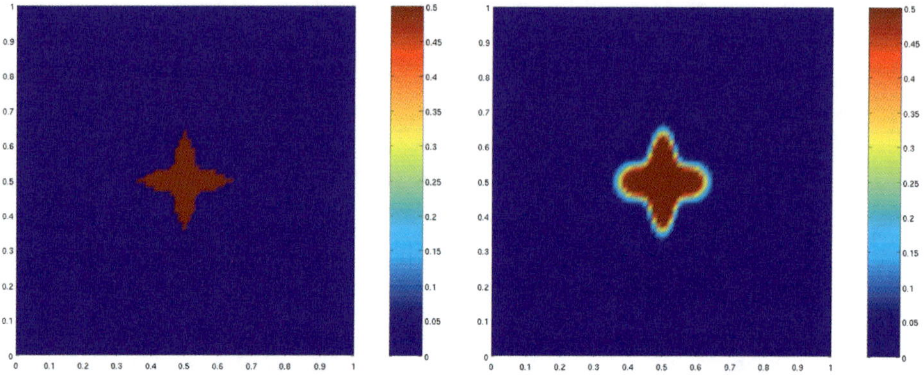

FIGURE 4. Desired shape of R (left) and the optimal state (right).

4.1. Asymptotic and discrete concepts

The link between the optimization problems for the full problem and those for the SP_N systems is discussed in [5], where optimal control problems for the radiative transfer equation and for approximate models are considered. Following the approach *first discretize, then optimize*, the discrete SP_N approximations are for the first time derived exactly and used for the study of optimal control based on reduced order models. Moreover, combining asymptotic analysis and the adjoint calculus yields diffusion-type approximations for the adjoint radiative transport equation in the spirit of the approach *first optimize, then discretize*.

The adjoint calculus is applied to the radiative transfer equation and the continuous SP_N approximation procedure is used to derive the SP_N system for the adjoint problem. Further, the problem of a thorough derivation of the SP_N approximations is discussed by starting from a discretization of the radiative transfer equations and using von Neumann's series to invert the arising matrices of the discrete system. This procedure parallels the derivation of the SP_N approximations in the continuous case and leads to discretisations of the corresponding diffusion equations. The analysis shows that both approaches yield up the discretization error the same discrete KKT systems. The detailed interrelation of the different optimization models on the continuous and discrete level is depicted in Figure 5.

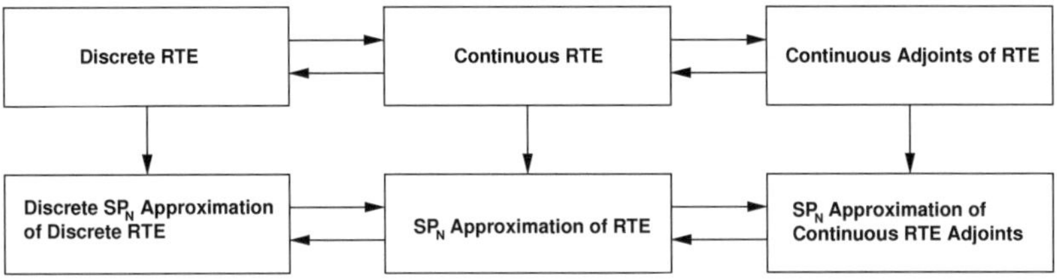

FIGURE 5. Interrelation of the optimization models.

Acknowledgment

The authors would like to thank the PhD students Sergiy Pereverzyev and Alexander Schulze for their essential contribution to the success of this project. Their mathematical ideas and numerical skills had great impact on the analytical and numerical results. Further, we are grateful to our colleagues M. Frank, M. Herty, A. Klar, M. Schäfer, M. Seaid and G. Thömmes for many fruitful discussions influencing this research project. This work was financially supported by the Deutsche Forschungsgemeinschaft (DFG) in the context of the Priority Programme 1253 as project PI 408/4-1, as well as by the DFG project PI 408/3-1.

References

[1] M.K. Choudhary and N.T. Huff. Mathematical modeling in the glass industry: An overview of status and needs. *Glastech. Ber. Glass Sci. Technol.*, 70:363–370, 1997.

[2] B. Dubroca, A. Klar. Half Moment closure for radiative transfer equations. *J. Comp. Phys.*, 180, 584–596, 2002.

[3] H.W. Engl, M. Hanke, and A. Neubauer. Regularization of inverse problems. Kluwer Academic Publishers, Dordrecht, 1996.

[4] M. Herty, R. Pinnau and M. Seaid. On Optimal Control Problems in Radiative Transfer. *OMS* 22(6):917–936, 2007.

[5] M. Herty, R. Pinnau and G. Thömmes. Asymptotic and Discrete Concepts for Optimal Control in Radiative Transfer. ZAMM 87(5), 333–347, 2007.

[6] M. Hinze, R. Pinnau, M. Ulbrich and S. Ulbrich. Optimization with Partial Differential Equations. *Mathematical Modelling: Theory and Applications, Volume* 23, *Springer*, 2009.

[7] C.T. Kelley. Existence and uniqueness of solutions of nonlinear systems of conductive-radiative heat transfer equations. *Transp. Theory Stat. Phys.* 25, No.2, 249–260, 1996

[8] M.T. Laitinen and T. Tiihonen. Integro-differential equation modelling heat transfer in conducting, radiating and semitransparent materials. *Math. Meth. Appl. Sc.*, 21:375–392, 1998.

[9] J. Lang. Adaptive Computation for Boundary Control of Radiative Heat Transfer in Glass, JCAM 183:312–326, 2005.

[10] E.W. Larsen, G. Thömmes and K. Klar. New frequency-averaged approximations to the equations of radiative heat transfer. *SIAM J. Appl. Math.* 64, No. 2, 565–582, 2003.

[11] E.W. Larsen, G. Thömmes, M. Seaid, Th. Götz, and A. Klar. Simplified P_N Approximations to the Equations of Radiative Heat Transfer and applications to glass manufacturing. *J. Comp. Phys*, 183(2):652–675, 2002.

[12] M.F. Modest, *Radiative Heat Transfer.* McGraw-Hill, 1993.

[13] O.S. Narayanaswamy. A Model of Structural Relaxation in Glass. *J. Amer. Cer. Soc.* 54, 491–498, 1981.

[14] S. Pereverzyev. Method of Regularized Fixed-Point and its Application. PhD Thesis, TU Kaiserslautern, 2006.

[15] S. Pereverzyev, R. Pinnau and N. Siedow. Initial Temperature Reconstruction for a Nonlinear Heat Equation: Application to Radiative Heat Transfer. *D. Lesnic (ed.), Proceedings of the* 5th *International Conference on Inverse Problems in Engineering: Theory and Practice*, Cambridge. Vol. III, ch. P02, pp. 1–8, 2005.

[16] S. Pereverzyev, R. Pinnau and N. Siedow. Regularized Fixed Point Iterations for Nonlinear Inverse Problems. *Inverse Problems* 22:1–22, 2006.

[17] S. Pereverzyev, R. Pinnau and N. Siedow. Initial temperature reconstruction for nonlinear heat equation: Application to a coupled radiative-conductive heat transfer problem. *IPSE* 15:55–67, 2008.

[18] R. Pinnau. Analysis of Optimal Boundary Control for Radiative Heat Transfer Modelled by the SP_1-System. Commun. Math. Sci. Volume 5, Issue 4 (2007), 951–969.

[19] R. Pinnau. Model Reduction via Proper Orthogonal Decomposition. In W.H.A. Schilder, H. van der Vorst: Model Order Reduction: Theory, Research Aspects and Applications, pp. 96–109, Springer, 2008.

[20] R. Pinnau and A. Schulze. Radiation, Frequency Averaging and Proper Orthogonal Decomposition. *Proc. Appl. Math. Mech.* 6(1):791–794, 2006.

[21] R. Pinnau and A. Schulze. Model Reduction Techniques for Frequency Averaging in Radiative Heat Transfer. *J. Comp. Phys.* 226(1):712–731, 2007.

[22] R. Pinnau and A. Schulze. Newton's Method for Optimal Temperature-Tracking of Glass Cooling Processes. IPSE 15(4), 303–323, 2007.

[23] R. Pinnau and G. Thömmes. Optimal boundary control of glass cooling processes. *Math. Meth. Appl. Sc.* 27(11):1261–1281, 2004.

[24] M. Schäfer, M. Frank and R. Pinnau. A Hierarchy of Approximations to the Radiative Heat Transfer Equations: Modelling, Analysis and Simulation. *M3AS* 15:643–665, 2005.

[25] A. Schulze. Minimizing thermal stresses in glass cooling processes. *PhD thesis*, TU Kaiserslautern, 2006.

[26] M. Seaid, A. Klar and R. Pinnau. Numerical Solvers for Radiation and Conduction in High Temperature Gas Flows. *Flow, Turbulence and Combustion* 75, 173–190, 2005.

[27] G. Thömmes, *Radiative Heat Transfer Equations for Glass Cooling Problems: Analysis and Numerics*, PhD Thesis, TU Darmstadt, 2002.

René Pinnau
TU Kaiserslautern
Department of Mathematics
Postfach 3049
D-67653 Kaiserslautern, Germany
e-mail: pinnau@mathematik.uni-kl.de

Norbert Siedow
Fraunhofer Institut für Techno- und Wirtschaftsmathematik
Fraunhofer Platz 1
D-67663 Kaiserslautern, Germany
e-mail: siedow@itwm.fraunhofer.de

International Series of Numerical Mathematics, Vol. 160, 611–622

On the Influence of Constitutive Models on Shape Optimization for Artificial Blood Pumps

Markus Probst, Michael Lülfesmann, Mike Nicolai,
H. Martin Bücker, Marek Behr and Christian H. Bischof

Abstract. We report on a shape optimization framework that couples a highly-parallel finite element solver with a geometric kernel and different optimization algorithms. The entire optimization framework is transformed with automatic differentiation techniques, and the derivative code is employed to compute derivatives of the optimal shapes with respect to viscosity. This methodology provides a powerful tool to investigate the necessity of intricate constitutive models by taking derivatives with respect to model parameters.

Mathematics Subject Classification (2000). Primary 76D55; Secondary 90C31.

Keywords. Shape optimization, sensitivity analysis, automatic differentiation, constitutive models.

1. Introduction

The World Health Organization reports cardiovascular diseases as number one cause of deaths worldwide. In 2004, for instance, 12.9 million people died from heart disease or strokes which accounts for an estimated 22% of all deaths. Later stages of heart disease often leave heart transplantation as the only cure. As a response to the dramatic shortage of donor organs, a temporary cure is the implantation of ventricular assist devices (VADs) that support the ailing heart in maintaining sufficient blood flow. Over the last decade, computer simulations of fluid flow have steadily gained acceptance as an effective tool for the design of medical devices. By coupling simulation tools with optimization algorithms, the design of VADs can be automated and an optimal shape can be computed directly. In this context, the focus of our project supported by the DFG priority program 1253 is on two different aspects: (1) the development of a framework for shape optimization of complex geometries in transient, large-scale applications; and (2) the investigation of the influence of constitutive models on the optimization outcome.

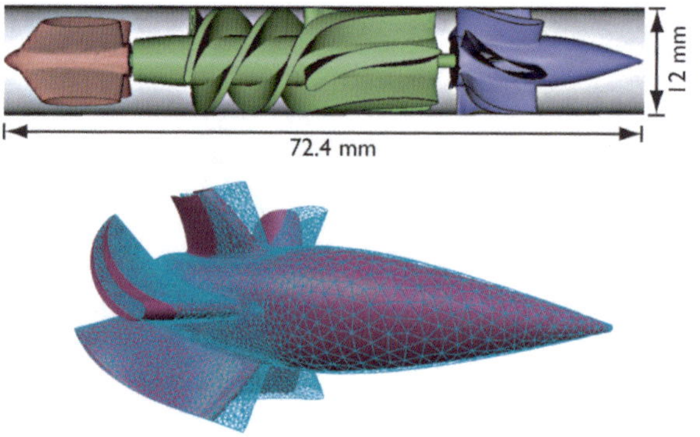

FIGURE 1. Top: Schematic view of the MicroMed DeBakey blood pump with straightener (red), impeller (green) and diffuser (blue) taken from [16]. Bottom: Initial (blue, transparent) and modified (purple, solid) shape of the diffuser component.

The long-term goal of the research carried out in this project is to optimize the design of the MicroMed DeBakey VAD schematically depicted in Figure 1 (top). Analyses based on immersive 3D visualization tools made the diffuser of the pump a primary candidate for shape optimization [16]. A design modification of the diffuser was evaluated over a full revolution of the pump's impeller in terms of hematologic (measured by shear rate) and hydraulic (measured by the generated pressure head) performance. The evaluation required transient simulations of blood flow through the pump for 10 full revolutions of the impeller, which were carried out on 4096 cores of the Blue Gene/P system operated by the Forschungszentrum Jülich. Pulling the diffuser blades backwards, as shown in Figure 1 (bottom), lead to both a reduction of shear rate and an increase in pressure head; this indicates the potential for improving the pump's performance by optimizing the diffuser design. First results from the application of the optimization framework to find an optimal design of the diffuser suggest a minimum at the lower feasibility bound of the deformation parameter. In an ongoing work, different design modifications are investigated.

Previous studies on shape optimization of artificial bypass grafts indicated the need for specific constitutive models to account for the non-Newtonian nature of blood [1, 2]. In some cases, optimal shapes were dependent on the fluid model, which gave rise to the question of whether or not this sensitivity can be predicted beforehand. By successfully transforming an in-house finite element flow solver via automatic differentiation (AD) [12, 22], the sensitivities of the flow field and the squared shear rate with respect to viscosity were computed. Figure 2 shows the spatial distribution of the latter sensitivity. In [20], we showed that elevated sensitivities in the optimal shapes cannot be predicted by the sensitivity of the flow field and directly related quantities. Instead, it is necessary to transform the

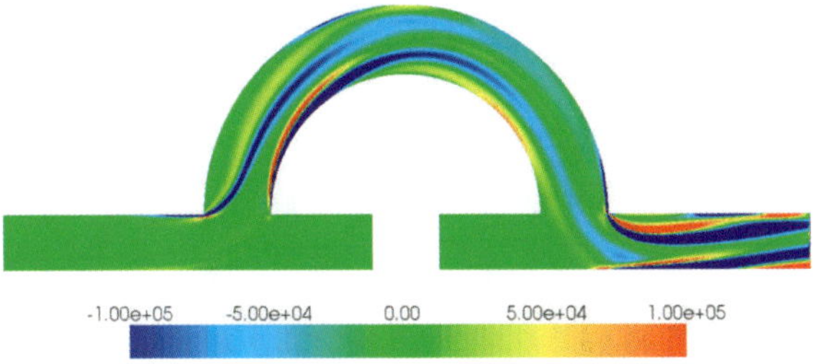

FIGURE 2. Sensitivity of squared shear rate with respect to viscosity in $\mathrm{cm}^{-2}\,\mathrm{s}^{-1}$ for an artificial bypass graft at Reynolds number 300 (adapted from [20]).

entire optimization tool chain via AD to allow for reliable predictions about when it is necessary to use shear-thinning or even more complex models for blood.

In this article, we briefly summarize the main results in the two areas mentioned above. A more detailed description is given in [20] and [21]. After introducing a PDE-constrained shape optimization problem in Section 2 and the sensitivity of the optimal solution in Section 3, we introduce the new shape optimization framework in Section 4. It includes a geometric kernel that offers a flexible way to parameterize and deform complex geometries. In particular, the parameterization is compatible with common CAD software packages and allows for deformations that respect geometric constraints. The entire optimization framework is transformed via AD to compute the sensitivity of the shape optimization outcome with respect to viscosity. Finally, in Section 5, we present results from the application of the shape optimization framework to an artificial bypass graft. We also show that the AD-transformed version of the framework accurately predicts the influence of a viscosity variation on the optimal bypass shape.

2. PDE-constrained shape optimization

Formulation in a general setting. In shape optimization, the goal is to minimize an objective function over a domain $\Omega = \Omega(\boldsymbol{\alpha})$ that depends on a vector of shape or design parameters $\boldsymbol{\alpha} \in \mathbb{R}^n$. The objective function

$$J : X \times \mathbb{R}^n \to \mathbb{R}, \qquad (\mathbf{v}, \boldsymbol{\alpha}) \mapsto J(\mathbf{v}, \boldsymbol{\alpha}), \tag{2.1}$$

typically depends on a state variable $\mathbf{v}$ for some Banach space $X = X(\Omega)$ and, thus, it depends on the design variables as well. The state variable and the design parameters are coupled by a partial differential equation

$$\mathbf{c}(\mathbf{v}, \boldsymbol{\alpha}) = \mathbf{0} \tag{2.2}$$

that is solved on the domain $\Omega(\boldsymbol{\alpha})$. This so-called state equation forms the constraint of the minimization problem

$$\begin{aligned} &\textit{minimize} \quad J(\mathbf{v},\boldsymbol{\alpha}), \\ &\textit{subject to} \quad \mathbf{c}(\mathbf{v},\boldsymbol{\alpha}) = \mathbf{0}, \end{aligned} \tag{2.3}$$

whose solution exists under conditions reported in [15]. By eliminating the PDE-constraint, the problem is reformulated as

$$\textit{minimize} \quad \bar{J}(\boldsymbol{\alpha}) = J(\mathbf{v}(\boldsymbol{\alpha}),\boldsymbol{\alpha}), \tag{2.4}$$

where $\mathbf{v}(\boldsymbol{\alpha})$ is the solution of (2.2). This way, the solution of the state equation is decoupled from the optimization problem.

Formulation for incompressible Navier-Stokes equations. The state variables $\mathbf{v} = (\mathbf{u}, p)$ are velocity and pressure, respectively. On a fixed domain Ω, the stress-divergence form reads

$$\rho\left(\frac{\partial \mathbf{u}}{\partial t} + \mathbf{u}\cdot\boldsymbol{\nabla}\mathbf{u} - \mathbf{f}\right) - \boldsymbol{\nabla}\cdot\boldsymbol{\sigma}(\mathbf{u},p) = \mathbf{0} \quad \text{in } \Omega, \tag{2.5}$$

$$\boldsymbol{\nabla}\cdot\mathbf{u} = 0 \quad \text{in } \Omega, \tag{2.6}$$

where $\boldsymbol{\sigma}$ is the stress tensor, ρ is the density, $\mathbf{f}$ denotes external forces, and suitable boundary conditions are imposed. The stress tensor is composed of a pressure component and a viscous component. For Newtonian fluids, i.e., fluids with constant viscosity, this tensor can be written as

$$\boldsymbol{\sigma}(\mathbf{u},p) = -p\mathbf{I} + 2\mu\boldsymbol{\varepsilon}(\mathbf{u}), \qquad \boldsymbol{\varepsilon}(\mathbf{u}) = \frac{1}{2}(\boldsymbol{\nabla}\mathbf{u} + \boldsymbol{\nabla}\mathbf{u}^T), \tag{2.7}$$

where μ is the dynamic viscosity, the rate of strain tensor $\boldsymbol{\varepsilon}(\mathbf{u})$ is the symmetric part of the velocity gradient, and $\mathbf{I}$ is the identity tensor.

Objective function. We aim to minimize the integral of the squared shear rate $\dot{\gamma}$, given by

$$\bar{J}(\boldsymbol{\alpha}) = J(\mathbf{v}(\boldsymbol{\alpha}),\boldsymbol{\alpha}) = \frac{1}{2}\int_{\Omega(\boldsymbol{\alpha})} \dot{\gamma}^2\, d\mathbf{x} = \int_{\Omega(\boldsymbol{\alpha})} \boldsymbol{\varepsilon}(\mathbf{u}) : \boldsymbol{\varepsilon}(\mathbf{u})\, d\mathbf{x}. \tag{2.8}$$

Ideally, hemolysis would be directly incorporated into the objective function in order to find geometries with minimal blood damage. However, prevalent methods evaluate hemolysis along pathlines [10, 13], among them an enhanced deformation-based model that predicts hemolysis more accurately than purely stress-based measures [3]. Due to the extreme sensitivity of the pathlines on the flow field, pathline-based measures are not easily handled when analytic derivatives are needed. Either a volumetric representation of hemolysis has to be derived [8], or related quantities available over the whole domain can be used instead. Another option is to use pathline integrals in a gradient-free optimization algorithm. These options are not considered here.

Constitutive model. Blood is a highly complex fluid – a mixture composed of red blood cells, white blood cells and platelets that are suspended in plasma. The plasma, with water as its main constituent, is a Newtonian fluid; however, the suspension of plasma and cells exhibits non-Newtonian dynamics [6, 25]. Previous studies show that non-Newtonian effects can influence both global and local flow properties such as the wall dynamics of a blood vessel [9, 17, 18]. Particularly relevant for the aim of our project, Abraham et al. [1] report on the influence of the constitutive model in shape optimization. In generalized Newtonian constitutive models, the viscosity depends on shear rate; these models offer a good compromise between entirely neglecting non-Newtonian properties of blood and solving fully-viscoelastic equations at a high computational expense.

The non-Newtonian behavior of blood motivates the effort to compute sensitivities with respect to viscosity. As a first step, we investigate the influence of variations in viscosity on the outcome of the design optimization for a Newtonian constitutive model. That is, for the sensitivity analysis, we do not consider non-Newtonian models. Rather we treat the viscosity as constant throughout the remainder of this article.

3. Sensitivity of optimal solution with respect to flow parameters

The objective function of the shape optimization, J, given in (2.4) depends on the flow parameters of the Navier-Stokes equations (2.5)–(2.6). For instance, the dynamic viscosity μ influences the state variable $\mathbf{v}$ via (2.7). By introducing a scaling with the constant density ρ, one can consider the kinematic viscosity $\nu = \mu/\rho$ rather than the dynamic viscosity. Assume that the optimization problem (2.4) admits a unique solution $\boldsymbol{\alpha}_*$ and let $\mathbf{v} = \mathbf{v}(\boldsymbol{\alpha}, \nu)$ denote the solution of the state equations for a constant kinematic viscosity. We are now interested in computing the sensitivity of the optimal solution with respect to kinematic viscosity

$$\frac{\mathrm{d}\boldsymbol{\alpha}_*}{\mathrm{d}\nu} \in \mathbb{R}^n. \tag{3.1}$$

We take this sensitivity as an illustrative example for sensitivities of the optimal solution with respect to different flow parameters. Differentiating the necessary optimality condition for a local optimum with respect to viscosity leads to

$$\frac{\mathrm{d}^2 J(\mathbf{v}(\boldsymbol{\alpha}_*, \nu), \boldsymbol{\alpha}_*)}{\mathrm{d}\boldsymbol{\alpha}^2} \frac{\mathrm{d}\boldsymbol{\alpha}_*}{\mathrm{d}\nu} = -\frac{\mathrm{d}^2 J(\mathbf{v}(\boldsymbol{\alpha}_*, \nu), \boldsymbol{\alpha}_*)}{\mathrm{d}\boldsymbol{\alpha}\mathrm{d}\nu}. \tag{3.2}$$

That is, the desired quantity, $\mathrm{d}\boldsymbol{\alpha}_*/\mathrm{d}\nu$, is obtained from the solution of a system of linear equations whose coefficient matrix is given by the Hessian $\mathrm{d}^2 J/\mathrm{d}\boldsymbol{\alpha}^2$ and whose right-hand side is given by the negative mixed second-order derivative $\mathrm{d}^2 J/(\mathrm{d}\boldsymbol{\alpha}\mathrm{d}\nu)$, both evaluated at $\boldsymbol{\alpha} = \boldsymbol{\alpha}_*$.

Rather than following the analytic approach based on (3.2) that would exploit the structure of the problem (2.4) [7, 11], we compute the derivative $\mathrm{d}\boldsymbol{\alpha}_*/\mathrm{d}\nu$ in an alternative way. To this end, we transform all software components of the shape optimization framework by automatic differentiation (AD). The structure of the framework is outlined in the following section. The advantage of this AD-based approach is that one can avoid the explicit computation of the second-order derivative information. The drawback, however, is that one does not take advantage of the particular problem structure.

4. Modular shape optimization framework

To simplify the use of an existing flow solver, the implementation of a framework for the solution of shape optimization problems is based on the formulation (2.4). The overall structure of this framework is illustrated in Figure 3. A geometry kernel updates the computational mesh representing Ω for given design parameters $\boldsymbol{\alpha}$. The value of the objective function J is passed on to an optimization driver that returns an update for the design parameters. On top of the optimization framework, we also study the sensitivity of the optimal solution with respect to kinematic viscosity, $\mathrm{d}\boldsymbol{\alpha}_*/\mathrm{d}\nu$.

Parameterization and mesh update. The current framework has been developed to study realistic applications with complex geometries. In such applications, the following three properties of the mesh deformation method are desirable: it must be possible to impose geometric constraints; resulting optimal shapes must be suitable for machining; and, to ensure the integration of the framework into the production process, the geometry representation used in the optimization should be compatible with CAD software. To meet these requirements, we parameterize the mesh surface by non-uniform rational B-splines (NURBS) [19]. The control points and weights of the NURBS govern the surface deformation; details on the parameterization and the update of interior mesh nodes are found in [21]. Note that the above-mentioned properties are hard to realize in an approach where, for instance, the nodes of the computational mesh form the design parameters.

Flow solver and objective function. For incompressible flows considered here, the state equation $\mathbf{c}(\mathbf{v}, \boldsymbol{\alpha}) = \mathbf{0}$ is given by Navier-Stokes equations (2.5)–(2.6). We employ an in-house flow solver XNS that provides velocity and pressure, $\mathbf{v} = (\mathbf{u}, p)$, on the computational domain $\mathbf{x} = \mathbf{x}(\boldsymbol{\alpha})$. XNS approximates the flow solution by applying Galerkin/Least-Squares stabilized finite-element discretization using conforming piecewise-linear finite elements for velocity and pressure [4]. The functionality of XNS was extended by a routine to evaluate a set of different objective functions including the shear rate integral (2.8).

Optimization driver. As indicated in Figure 3, the framework mainly relies on gradient-based optimization. The results shown in Section 5 were obtained using DONLP2 [23, 24]. The current version of the flow solver does not yet provide an exact gradient of the objective function $\nabla_{\boldsymbol{\alpha}} J := \mathrm{d}J/\mathrm{d}\boldsymbol{\alpha}$. The optimization driver approximates the gradient using finite differences. This numerical approximation is affordable in terms of computing time as long as a small number of optimization

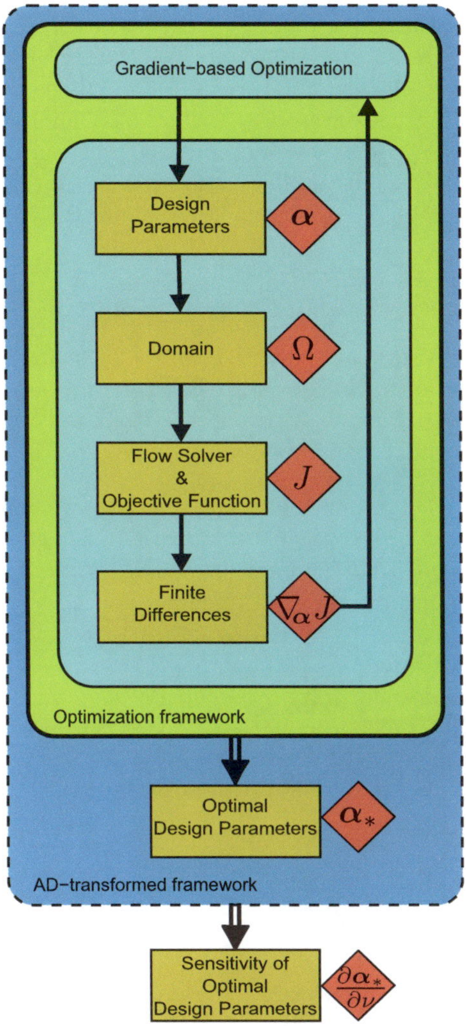

FIGURE 3. Structure of the shape optimization framework to compute $\boldsymbol{\alpha}_*$ and its AD-transformed version to compute $\mathrm{d}\boldsymbol{\alpha}_*/\mathrm{d}\nu$ (taken from [21]).

parameters is considered. Since this technique is not computationally feasible for a large number of parameters, a discrete adjoint approach by hand is currently being developed and will be integrated into the framework. In principle, the discrete adjoint could also be generated via reverse mode of AD. However, this would complicate, in practice, the process of applying AD to the entire optimization framework.

AD transformation of the framework. The geometry kernel, the flow solver XNS, and the optimizer DONLP2 are separately transformed by AD to compute the derivative $\mathrm{d}\boldsymbol{\alpha}_*/\mathrm{d}\nu$. The following mixture of AD tools is used. As described in [20], the flow solver XNS is transformed by Adifor2 [5], and so is the optimization software DONLP2. The geometry kernel contains language constructs beyond the Fortran 77 standard so that it cannot be transformed by this AD tool. Therefore, we transform the geometry kernel by the AD tool TAPENADE [14].

5. Illustrating numerical experiments using the framework

The optimization framework is used to determine the shape of an idealized bypass whose geometry is parameterized by NURBS. In addition, the sensitivity of the optimal shape with respect to viscosity is computed and validated with the AD-transformed version of that framework.

Graft optimization. We optimize the bypass geometry of an idealized artificial graft aiming at minimizing the shear rate integral (2.8). The computational mesh has 1774 nodes forming 3137 triangular elements. We prescribe a constant viscosity, and the inflow conditions are adjusted to generate a flow with Reynolds number 300. The two boundary curves of the bypass are parameterized using NURBS curves of degree two with four control points per curve. Due to restrictions on the control points that are detailed in [21], the bypass shape is governed by four design parameters. The initial and optimal geometry are shown in Figure 4. Compared to the initial shape, the optimal bypass shape is flattened and the artery-to-graft angle is lower. The shear rate integral is reduced by 17%.

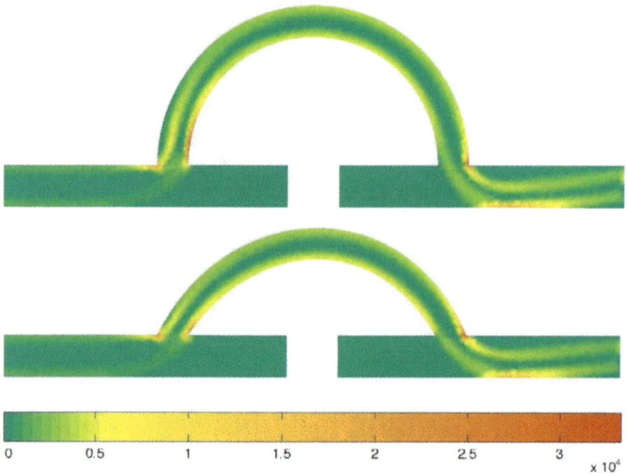

FIGURE 4. Shear rate distribution in s^{-1} for initial shape (top) and optimal shape (bottom) (adapted from [21]).

Sensitivity of the optimal shape. We use the same objective function, computational mesh, and bypass parameterization as before. For simplicity, the movement of the free control points is restricted to be uniform and in the vertical direction only. This leads to a one-parameter optimization problem. The AD-transformed version of the framework is employed to compute the optimal shape represented by $\alpha_* \in \mathbb{R}$ as well as the sensitivity of the optimal shape with respect to kinematic viscosity, $\mathrm{d}\alpha_*/\mathrm{d}\nu \in \mathbb{R}$.

Following [1, 20], the optimal shape is computed for two different bypass-to-artery diameter ratios (0.75 and 1.25), which will be referred to as narrow and wide graft. To provide a reasonable base for the validation of our approach, we compute α_* for different Reynolds numbers ranging from 50 to 300 in uniform steps of 25. The results of the optimization and the sensitivity analysis are depicted in Figure 5. Here, the computed sensitivities $\mathrm{d}\alpha_*/\mathrm{d}\nu$ are indicated by straight lines with the corresponding slope attached to the markers of α_*. These derivatives of the optimal design parameter with respect to viscosity predict the progression of both curves very accurately. The AD-generated sensitivity of the optimal shape can hence indicate the influence of a variation in viscosity on the optimization outcome.

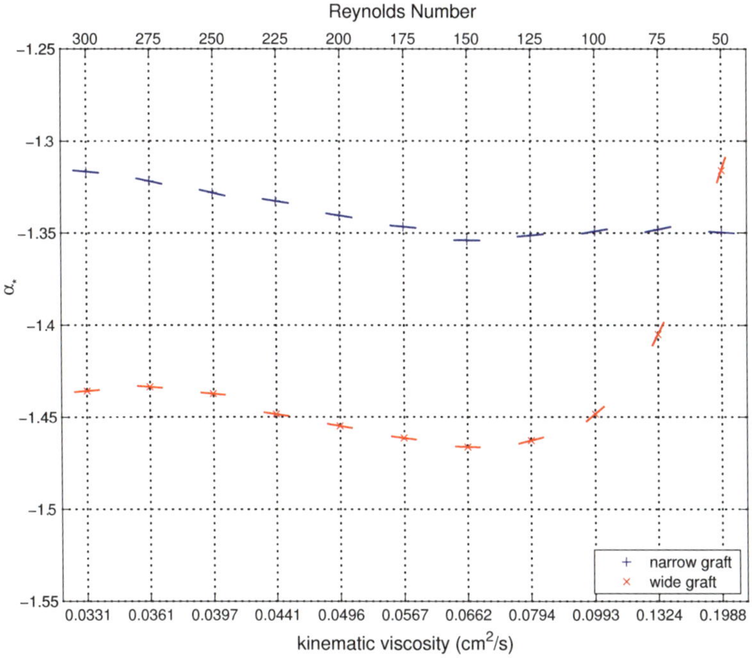

FIGURE 5. Optimal design parameter α_* and its sensitivity $\mathrm{d}\alpha_*/\mathrm{d}\nu$ for narrow (blue markers) and wide (red markers) bypass geometry for different Reynolds numbers (adapted from [21]).

6. Concluding remarks

By summarizing the results of our project supported by the DFG priority program 1253, we show significant progress toward the computational design of medical devices. In particular, a new framework for the solution of PDE-constrained shape optimization problems is implemented. It is validated using carefully-selected simplified problems, including the determination of a bypass geometry of an idealized artificial graft in two space dimensions. The framework has also been applied to more complex, three-dimensional applications: among them a 3D bypass geometry, and flow through a pipe around the diffuser component of the DeBakey blood pump. Furthermore, the entire framework is successfully transformed by automatic differentiation. We demonstrate that this transformed version is capable of reliably quantifying the sensitivity of the optimal bypass shape with respect to kinematic viscosity. The same concept can be used to predict the sensitivity of the optimal shape with respect to parameters of generalized Newtonian constitutive models. This offers previously unavailable means to investigate the role of the non-Newtonian nature of blood in shape optimization. The overall approach paves the way for the future study and analysis of shape optimization problems involving three-dimensional, realistic ventricular assist devices such as the MicroMed DeBakey blood pump.

Acknowledgment

The Aachen Institute for Advanced Study in Computational Engineering Science (AICES) provides a stimulating research environment for our work. This work was supported by the German Research Foundation (DFG) under GSC 111, EXC 128, and especially SPP 1253 programs. Computing resources were provided by the Center for Computing and Communication, RWTH Aachen University, and by the John von Neumann Institute for Computing, Forschungszentrum Jülich.

References

[1] F. Abraham, M. Behr, and M. Heinkenschloss. Shape optimization in steady blood flow: A numerical study of non-Newtonian effects. *Computer Methods in Biomechanics and Biomedical Engineering*, 8(2):127–137, 2005.

[2] F. Abraham, M. Behr, and M. Heinkenschloss. Shape optimization in unsteady blood flow: A numerical study of non-Newtonian effects. *Computer Methods in Biomechanics and Biomedical Engineering*, 8(3):201–212, 2005.

[3] D. Arora, M. Behr, and M. Pasquali. A tensor-based measure for estimating blood damage. *Artificial Organs*, 28:1002–1015, 2004.

[4] M. Behr and T.E. Tezduyar. Finite element solution strategies for large-scale flow simulations. *Computer Methods in Applied Mechanics and Engineering*, 112:3–24, 1994.

[5] C.H. Bischof, A. Carle, P. Khademi, and A. Mauer. ADIFOR 2.0: Automatic differentiation of Fortran 77 programs. *IEEE Computational Science & Engineering*, 3(3):18–32, 1996.

[6] S. Chien, R. King, R. Skalak, S. Usami, and A. Copley. Viscoelastic properties of human blood and red cell suspensions. *Biorheology*, 12(6):341–346, 1975.

[7] A.V. Fiacco. *Introduction to Sensitivity and Stability Analysis*. Number 165 in Mathematics in Science and Engineering. Academic Press, New York, 1983.

[8] A. Garon and M.-I. Farinas. Fast 3D numerical hemolysis approximation. *Artificial Organs*, 28(11):1016–1025, 2004.

[9] F. Gijsen, F. van de Vosse, and J. Janssen. The influence of the non-Newtonian properties of blood on the flow in large arteries: Steady flow in a carotid bifurcation model. *Journal of Biomechanics*, 32:601–608, 1999.

[10] L. Goubergrits and K. Affeld. Numerical estimation of blood damage in artificial organs. *Artificial Organs*, 28(5):499–507, 2004.

[11] R. Griesse and A. Walther. Parametric sensitivities for optimal control problems using automatic differentiation. *Optimal Control Applications and Methods*, 24(6):297–314, 2003.

[12] A. Griewank and A. Walther. *Evaluating Derivatives: Principles and Techniques of Algorithmic Differentiation*. Number 105 in Other Titles in Applied Mathematics. SIAM, Philadelphia, PA, 2nd edition, 2008.

[13] L. Gu and W. Smith. Evaluation of computational models for hemolysis estimation. *ASAIO Journal*, 51:202–207, 2005.

[14] L. Hascoët and V. Pascual. Tapenade 2.1 user's guide. Technical Report 0300, INRIA, 2004.

[15] J. Haslinger and R. Mäkinen. *Introduction to Shape Optimization: Theory, Approximation, and Computation*. Advances in Design and Control. SIAM, Philadelphia, PA, 2003.

[16] B. Hentschel, I. Tedjo, M. Probst, M. Wolter, M. Behr, C. Bischof, and T. Kuhlen. Interactive Blood Damage Analysis for Ventricular Assist Devices. *IEEE Transactions on Visualization and Computer Graphics*, 14:1515–1522, 2008.

[17] B. Johnston, P. Johnston, S. Corney, and D. Kilpatrick. Non-Newtonian blood flow in human right coronary arteries: Steady state simulations. *Journal of Biomechanics*, 37:709–720, 2004.

[18] M. Lukáčová-Medviďová and A. Zaušková. Numerical modelling of shear-thinning non-Newtonian flows in compliant vessels. *International Journal for Numerical Methods in Fluids*, 56(8):1409–1415, 2008.

[19] L. Piegl and W. Tiller. *The NURBS Book*. Monographs in visual communication. Springer-Verlag, 2nd edition, 1997.

[20] M. Probst, M. Lülfesmann, H.M. Bücker, M. Behr, and C.H. Bischof. Sensitivity analysis for artificial grafts using automatic differentiation. *International Journal for Numerical Methods in Fluids*, 62(9):1047–1062, 2010.

[21] M. Probst, M. Lülfesmann, M. Nicolai, H.M. Bücker, M. Behr, and C.H. Bischof. Sensitivity of optimal shapes of artificial grafts with respect to flow parameters. *Computer Methods in Applied Mechanics and Engineering*, 199:997–1005, 2010.

[22] L.B. Rall. *Automatic Differentiation: Techniques and Applications*, volume 120 of *Lecture Notes in Computer Science*. Springer-Verlag, Berlin, 1981.

[23] P. Spellucci. A new technique for inconsistent QP problems in the SQP method. *Mathematical Methods of Operations Research*, 47:355–400, 1998.

[24] P. Spellucci. An SQP method for general nonlinear programs using only equality constrained subproblems. *Mathematical Programming*, 82:413–448, 1998.

[25] G. Thurston. Viscoelasticity of human blood. *Biophysical Journal*, 12:1205–1217, 1972.

Markus Probst, Mike Nicolai and Marek Behr
Chair for Computational Analysis of Technical Systems (CATS)
Center for Computational Engineering Science (CCES)
RWTH Aachen University
D-52056 Aachen, Germany
e-mail: probst@cats.rwth-aachen.de
nicolai@cats.rwth-aachen.de
behr@cats.rwth-aachen.de

Michael Lülfesmann, H. Martin Bücker and Christian H. Bischof
Institute for Scientific Computing (SC)
Center for Computational Engineering Science (CCES)
RWTH Aachen University
D-52056 Aachen, Germany
e-mail: luelfesmann@sc.rwth-aachen.de
buecker@sc.rwth-aachen.de
bischof@sc.rwth-aachen.de